INTRODUCTION TO CONTROL SYSTEMS DESIGN

mcgraw-hill electrical and electronic engineering series

ANGELAKOS AND EVERHART · *Microwave Communications*
ANGELO · *Electronic Circuits*
ANGELO · *Electronics: BJTs, FETs, and Microcircuits*
ASELTINE · *Transform Method in Linear System Analysis*
ATWATER · *Introduction to Microwave Theory*
BENNETT · *Introduction to Signal Transmission*
BERANEK · *Acoustics*
BRACEWELL · *The Fourier Transform and Its Application*
BRENNER AND JAVID · *Analysis of Electric Circuits*
CARLSON · *Communication Systems: An Introduction to Signals and Noise in Electrical Communication*
CHEN · *The Analysis of Linear Systems*
CHEN · *Linear Network Design and Synthesis*
CHIRLIAN · *Analysis and Design of Electronic Circuits*
CHIRLIAN · *Basic Network Theory*
CHIRLIAN · *Electronic Circuits: Physical Principles, Analysis, and Design*
CHIRLIAN AND ZEMANIAN · *Electronics*
CLEMENT AND JOHNSON · *Electrical Engineering Science*
CUNNINGHAM · *Introduction to Nonlinear Analysis*
D'AZZO AND HOUPIS · *Feedback Control System Analysis and Synthesis*
ELGERD · *Control Systems Theory*
ELGERD · *Electric Energy Systems Theory: An Introduction*
EVELEIGH · *Adaptive Control and Optimization Techniques*
EVELEIGH · *Introduction to Control Systems Design*
FEINSTEIN · *Foundations of Information Theory*
FITZGERALD, HIGGINBOTHAM, AND GRABEL · *Basic Electrical Engineering*
FITZGERALD, KINGSLEY, AND KUSKO · *Electric Machinery*
FRANK · *Electrical Measurement Analysis*
FRIEDLAND, WING, AND ASH · *Principles of Linear Networks*
GEHMLICH AND HAMMOND · *Electromechanical Systems*
GHAUSI · *Principles and Design of Linear Active Circuits*

GHOSE · *Microwave Circuit Theory and Analysis*
GLASFORD · *Fundamentals of Television Engineering*
GREINER · *Semiconductor Devices and Applications*
HAMMOND AND GEHMLICH · *Electrical Engineering*
HANCOCK · *An Introduction to the Principles of Communication Theory*
HARMAN · *Fundamentals of Electronic Motion*
HARMAN · *Principles of the Statistical Theory of Communication*
HARMAN AND LYTLE · *Electrical and Mechanical Networks*
HAYASHI · *Nonlinear Oscillations in Physical Systems*
HAYT · *Engineering Electromagnetics*
HAYT AND KEMMERLY · *Engineering Circuit Analysis*
HILL · *Electronics in Engineering*
JAVID AND BROWN · *Field Analysis and Electromagnetics*
JOHNSON · *Transmission Lines and Networks*
KOENIG, TOKAD, AND KESAVAN · *Analysis of Discrete Physical Systems*
KRAUS · *Antennas*
KRAUS · *Electromagnetics*
KUH AND PEDERSON · *Principles of Circuit Synthesis*
KUO · *Linear Networks and Systems*
LEDLEY · *Digital Computer and Control Engineering*
LEPAGE · *Complex Variables and the Laplace Transform for Engineering*
LEPAGE AND SEELY · *General Network Analysis*
LEVI AND PANZER · *Electromechanical Power Conversion*
LEY, LUTZ, AND REHBERG · *Linear Circuit Analysis*
LINVILL AND GIBBONS · *Transistors and Active Circuits*
LITTAUER · *Pulse Electronics*
LYNCH AND TRUXAL · *Introductory System Analysis*
LYNCH AND TRUXAL · *Principles of Electronic Instrumentation*
LYNCH AND TRUXAL · *Signals and Systems in Electrical Engineering*
MCCLUSKEY · *Introduction to the Theory of Switching Circuits*
MANNING · *Electrical Circuits*
MEISEL · *Principles of Electromechanical-energy Conversion*
MILLMAN AND HALKIAS · *Electronic Devices and Circuits*
MILLMAN AND TAUB · *Pulse, Digital, and Switching Waveforms*
MINORSKY · *Theory of Nonlinear Control Systems*
MISHKIN AND BRAUN · *Adaptive Control Systems*
MOORE · *Traveling-wave Engineering*
MURDOCH · *Network Theory*
NANAVATI · *An Introduction to Semiconductor Electronics*
OBERMAN · *Disciplines in Combinational and Sequential Circuit Design*
PETTIT AND MCWHORTER · *Electronic Switching, Timing, and Pulse Circuits*
PETTIT AND MCWHORTER · *Electronic Amplifier Circuits*
PFEIFFER · *Concepts of Probability Theory*

REZA · *An Introduction to Information Theory*
REZA AND SEELY · *Modern Network Analysis*
RUSTON AND BORDOGNA · *Electric Networks: Functions, Filters, Analysis*
RYDER · *Engineering Electronics*
SCHILLING AND BELOVE · *Electronic Circuits: Discrete and Integrated*
SCHWARTZ · *Information Transmission, Modulation, and Noise*
SCHWARZ AND FRIEDLAND · *Linear Systems*
SEELY · *Electromechanical Energy Conversion*
SEIFERT AND STEEG · *Control Systems Engineering*
SHOOMAN · *Probabilistic Reliability: An Engineering Approach*
SISKIND · *Direct-current Machinery*
SKILLING · *Electric Transmission Lines*
STEVENSON · *Elements of Power System Analysis*
STEWART · *Fundamentals of Signal Theory*
STRAUSS · *Wave Generation and Shaping*
SU · *Active Network Synthesis*
TAUB AND SCHILLING · *Principles of Communication Systems*
TERMAN · *Electronic and Radio Engineering*
TERMAN AND PETTIT · *Electronic Measurements*
TOU · *Digital and Sampled-data Control Systems*
TOU · *Modern Control Theory*
TRUXAL · *Automatic Feedback Control System Synthesis*
TUTTLE · *Electric Networks: Analysis and Synthesis*
VALDES · *The Physical Theory of Transistors*
WEEKS · *Antenna Engineering*
WEINBERG · *Network Analysis and Synthesis*

INTRODUCTION TO CONTROL SYSTEMS DESIGN

VIRGIL W. EVELEIGH

Associate Professor of Electrical Engineering
Syracuse University, Syracuse, New York

McGRAW HILL BOOK COMPANY

New York St. Louis San Francisco Düsseldorf Johannesburg
Kuala Lumpur London Mexico Montreal New Delhi
Panama Rio de Janeiro Singapore Sydney Toronto

To
Eleanor and the girls,
who have made this effort worthwhile

INTRODUCTION TO CONTROL SYSTEMS DESIGN

Library of Congress Catalog Card Number 73–160706

07–019773–3

234567890 MAMM 798765432

This book was set in Press Roman by Scripta-Technica, Inc., and printed and bound by The Maple Press Company. The designer was Scripta-Technica, Inc.; the drawings were done by John Cordes, J. & R. Technical Services, Inc. The editor was Charles R. Wade. John A. Sabella supervised production.

CONTENTS

PREFACE

The notes from which this book evolved were developed by the author with the assistance of several associates over a period of several years while teaching basic control theory and practice to senior and first year graduate students at Syracuse University. Many of our students in the large graduate control systems program at Syracuse are full or part time employees of IBM, GE, or one of several other smaller firms in upper New York. These students come to us with a wide diversity of backgrounds, but virtually all of them are looking for a fundamental course which will prove helpful to them in solving their engineering problems. This book was written to help serve this purpose, and the primary objective is to present the material from a straightforward practical point of view. Emphasis is placed upon analysis and synthesis methods which, in the author's opinion, should prove most helpful in solving a majority of control problems.

Most of the material presented herein is similar to that available in several basic control system texts published in the recent past. Although a few novel methods and interpretations are introduced where that seems appropriate, as for example the use of logarithmic Nyquist plots and angular locus transparancies for developing root locus plots, the emphasis upon -1 slope span in the Bode diagram design, and the development of state transition function digital simulation algorithms and a gradient polynomial factoring algorithm, these characteristics of the presentation are of peripheral significance. The fundamental contribution is derived from the choice of topics covered, the manner in which they are interrelated, the establishment of a control system philosophy, and the method of presenting material to the reader.

The Bode diagram method of design is emphasized in this book. Students seem to learn the basic design methods using the Bode diagram faster and more thoroughly than when other techniques such as the root locus procedure are used as the fundamental tool. Once design on the Bode diagram is well understood, all other standard design methods are easily presented. The Root locus plot, the Nyquist plot and a rough visualization of its general shape, and frequency and time response patterns, are suggested primarily as supporting tools. Bode design requires only a straightedge and simple sliderule calculations. The effects of all parameter changes upon system performance are clearly illustrated. Finally, the Bode design methods extend readily to minor loop and other multiple loop systems, which are relatively awkward to treat in other ways. Concentration upon Bode diagram analysis and design methods supported by the other tools available with emphasis placed upon the specification of phase margin, velocity error constant, low frequency steady state errors, and similar system characteristics, essentially summarizes the author's approach to basic control system theory and practice.

A detailed set of problems covering all aspects of the material is included, and a solutions manual is available. The problems run the gamut from routine exercises through complex system designs and analyses. In some problems an analog or digital simulation is the only reasonable way to obtain the answers, whereas in others simulation study is optional, but should prove valuable. Analog and digital simulation techniques are discussed in Appendices C and E and it is suggested that simulation methods be introduced as early as possible whenever computers are readily available. Simulation study adds a dimension to the design problem which cannot be provided by any other means short of extensive hardware experimentation. It also leads readily to straightforward trial and error procedures for imposing even mathematically intractable performance specifications such as unit step response peak overshoot or settling time.

The material in this book can be broken down into several major areas. The first four chapters are devoted to the usual introductory comments and the mathematical background required for developing and solving differential equation and block diagram system descriptions. The material in Chaps. 5 through 8 is related to system stability with emphasis upon closed loop configurations. Bode diagrams and the Routh-Hurwitz and Nyquist criteria are introduced in this section. The design of series equalized systems on the Bode diagram and the effects of performance specifications upon the design are presented in Chaps. 9 and 10. Extension of the standard Bode methods to minor loop design is discussed in Chap. 11. This completes the basic controller design section, and a typical introductory course in control should terminate at about this point.

A number of selected topics of interest to control specialists are included in Chaps. 12 through 15. The standard design procedures are extended to AC controllers in Chap. 12. An introduction to the state space methods of modern control is presented in Chap. 13. The general characteristic of nonlinear systems and a variety of methods for their treatment are provided in Chap. 14. Chap. 15 is devoted to methods, primarily classical, for the analysis and synthesis of sampled data systems. Digital controllers using both discrete and continuous equalizers are considered.

Several appendices are included to support the more basic material in the body of the book. Appendix A provides a table of Laplace transforms. Various methods for factoring polynomials are discussed in Appendix B. Analog and digital simulation techniques are presented in Appendices C and E. Basic matrix operations and relationships are discussed in Appendix D. The comprehensive problem set makes up Appendix F.

Considerable flexibility exists in the use of this book as a text. The first eleven chapters plus selected topics from Chap. 13 and the appendices are suggested as the basis for an introductory course in control at the senior or first year graduate level. Emphasis upon Chaps. 2–4 may be adjusted according to student background, and the additional topics selected from Chap. 13 and the appendices can be used to modify the general tenor of the course between classical and modern, or analytical and computational emphasis. At Syracuse we offer two such courses, one with four semester hours credit for seniors, including a laboratory, and a three credit hour course (without a lab) at a slightly higher level for first year graduate students. Our student backgrounds are so diverse that Chaps. 2–4 must be covered rather thoroughly, and there is little time in either of these courses for consideration of extra topics from the material beyond Chap. 11. An additional 3 credit hour graduate course covering Chaps. 12 and 15 thoroughly and a few topics selected from Chaps. 13 and 14 to round out the presentation is also offered at Syracuse. Although the book is primarily intended as a basic text in classical control theory, it is also applicable as a text or reference for a second course in this area.

It is impossible to prepare a manuscript like this without becoming indebted to many people. Although I cannot possibly mention them all by name, I thank each and every one who has helped me to develop the text and problem presentations. Special thanks are due Dr. M. Krieger, of the University of Ottawa, who wanted to be a co-author of this book had he not become deeply involved in a number of other projects. Dr. Krieger made his extremely thorough and carefully written notes and his problem sets related to the material covered in several of the chapters available to the author as an aid in preparing the final manuscript. The analog simulation appendix was taken with only minor modifications from a manual originally prepared by Dr. Krieger. He also critically reviewed virtually all of the manuscript and offered many valuable suggestions for its improvement. Several other associates, namely Doctors Brulé, Bickart, Huang, and McFee of the Syracuse faculty have critically reviewed parts of the manuscript and suggested a number of modifications.

An enthusiastic vote of thanks is owed to Dr. R. Rohrer of the McGraw-Hill editorial staff who provided the author with an unusually thorough critical review of the manuscript. Virtually all of Dr. Rohrer's suggestions have been taken into consideration in the final draft, and several significant improvements have resulted from his constructive evaluation.

The manuscript was carefully typed by Mrs. Bruce Fairbanks, Mrs. Louise Capra, and Mrs. Vernon Stith. Most of the drafting was capably done by Miss Jeanne Bouteiller. The notes from which the final manuscript was prepared have been used in courses at Syracuse University for several years, and were reproduced by Mr. A. Smith and his staff with the financial support of the Electrical Engineering Department here, Dr. W. LePage, Chairman. I am most appreciative of this support from the SU administration and staff.

Finally, thanks are due my devoted wife Eleanor and our three girls, Laurel Jean, Cheryl Ann, and Amy Jo for their selfless cooperation during this project. Perhaps I can now make up to them for all the attention they have often had to do without as this manuscript was being prepared.

Virgil W. Eveleigh

1 INTRODUCTION

The thing that has allowed man to rise above the other animals with which he was placed in competition, originally in a near desperate battle for food and existence, and more recently primarily as a hunter and sportsman, is man's ability to *control* his environment and its effect upon his life. In even the most primitive of the world's current societies, the effects of control are obvious. Natives in the arctic and the tropics use shelter to protect their families from the bitter cold and blazing sun. Fires and fans are used for heating and cooling. Considerable organization is shown by all civilizations in their approaches to hunting for, and growing, food. Modern societies have developed efficient methods for heating and cooling to provide year-round comfort for all wherever we go. Mass production methods are now used almost exclusively, and they depend heavily upon automation, which uses, in turn, sophisticated control techniques. Control systems abound everywhere we look and the use of increasingly complex controllers is still expanding at a rapid pace. Thus the importance of control system technology is also expanding. We are indeed living in an amazing era, an age in which man has proven his ability to investigate the universe first-hand, landing explorers on the moon and instrument packages on our neighboring planets while there are still those among us who were equally amazed near the start of the twentieth century when the first automobiles and aircraft were introduced.

The development and application of control theory has helped to make such technological advancement possible, and many surprises are in store for the future.

With the rapid development of control theory and its applications over the last half century or so, significantly accelerated in the period since World War II by the expansion of analog computer use and the introduction of general purpose digital computers, a number of changes have taken place in the methods used for teaching control fundamentals and their application to standard design problems. The use of an analog or digital computer to obtain detailed data related to the performance of complicated systems, which would have been prohibitively difficult to investigate in detail only a few years ago, is now a routine task for control system engineers. Time-shared digital computer system consoles are now readily available to virtually all engineers and students, and it should be expected that most extensive numerical calculations will be carried out on a digital computer without undue difficulty. This provides considerable flexibility in the types of problems which are now reasonable to investigate. It is certainly easier, not to mention lots more fun, to leave the tedium of numerical analysis to the computer, learn from the data it develops, study a much larger variety of situations than would otherwise be possible because of the computational speed available, and not have to worry about errors, either! The general purpose digital computer is as much a tool for the engineer and student of today as is his slide rule, and only a trifle less convenient to use.

Just as the digital computer expands the realm of problems considered reasonable for investigation, it also influences the material covered in many of our courses and the emphasis placed upon various topics. It seems appropriate, for example, to consider nonlinear systems, using the computer as a tool to illustrate performance, rather early in the study of system theory, but emphasis upon graphical methods for approximating nonlinear system response is hardly justified. Only a few years ago this was an area of considerable research interest. We must certainly consider sampled data systems, since digital computers process data in sampled form, and the mathematical methods used for the analysis and synthesis of sampled data systems have changed rapidly in the last two decades. Design procedures based upon time domain specifications have generally been avoided whenever possible in the past, since the time response was so difficult to obtain by hand analysis, but performance characteristics such as the unit step response rise time or peak overshoot are readily determined and displayed now, so design methods using the digital (analog) computer as an integral part of the process are not at all unreasonable.

Advancing technology may revolutionize the procedures used for system analysis and design, but the fundamental theory of system performance is not thereby changed. It is true that we design many complex systems using sophisticated computer control concepts for process control, space vehicle guidance, and similar applications, but such situations are rather specialized and the vast majority of control applications involves more mundane problems like temperature control, speed control, voltage regulation, instrument servos, and the like. The material presented in the first 12 chapters of this book is intended to provide a sound theoretical foundation upon which this majority of linear control problem solutions is readily structured. Once this background is well in hand, a number of advanced topics are considered, including ac (carrier) systems, nonlinear systems, and sampled data systems. Our first concern is with the development and interpretation of system models. We then consider several design procedures applicable to a variety of progressively more complex systems. Emphasis is placed upon the development of tools which prove particularly useful in the majority of problems encountered in engineering practice, and with which the control engineer must be familiar to successfully ply his trade. Problems and examples are used throughout to illustrate the analytical and computational aspects of the material considered.

1.1 SOME EXAMPLES AND DEFINITIONS

A control system, or controller, may be defined as follows: "A *control system* is a device in which an output variable, called the *response*, is adjusted as required by a *reference input*, or *control input*." A controller may be either *open loop* or *closed loop*. A *closed loop controller* compares one or more *feedback signals* to the *control input* to develop an *error signal*, which is used to drive the *plant*, or *actuator*, and an *equalizer*, or *processing filter*, in such a way as to (hopefully) reduce the *response error*. The block diagram of a general *unity feedback single loop controller* is shown in Fig. 1.1. This is the configuration most often encountered, and our interest is focused almost exclusively upon this problem until *minor loop design*, a procedure for developing *multiple loop controllers*, is considered in Chap. 11. The major requirements of our study program are readily apparent from Fig. 1.1. We must learn how to:

1. Develop mathematical system descriptions and reduce them to block diagram form.
2. Manipulate and solve the resulting system equations.
3. Design systems (choose the plant, when such freedom is available, and the equalizer topology and parameters) to satisfy general performance specifications.
4. Evaluate our results by analytical and simulation studies.

It is thus important that we become familiar with a variety of electrical and mechanical configurations, including *sensors* or *transducers* which are used for observation and measurement of system variables, and then learn how these devices may be used to efficiently satisfy our needs.

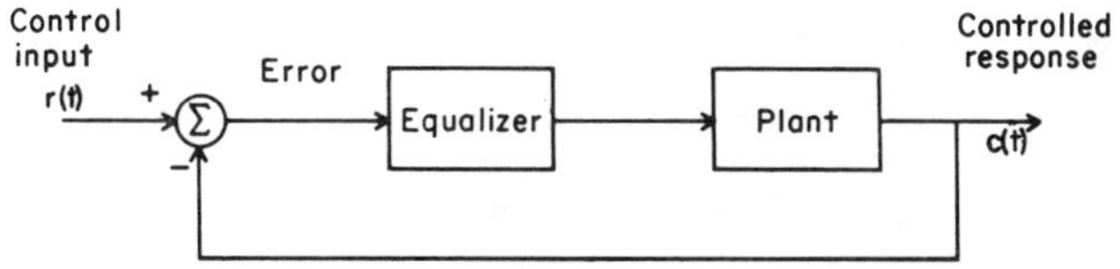

FIG. 1.1. A standard unity feedback controller.

Examples of control systems are extremely common in our immediate environment. The thermostatic temperature control in our homes and public buildings is an example with which we are all familiar. Assume a forced air gas furnace system for further consideration. An electronic *thermostat*, or *temperature sensor*, is placed in a central location, usually on an *inside* wall about five feet from the floor. When the room temperature drops below a preset *reference level*, a relay is actuated to turn on the furnace fire. When the temperature in the furnace air duct system reaches a *reference level*, a blower fan is activated by another relay to force the warm air throughout the building. When the room temperature reaches a *maximum limit*, usually two or three degrees above the *lower limit*, the furnace is turned off. The blower runs until the residual heat in the system has been dissipated and then also turns off automatically. The system cycles on and off as needed, depending upon the outside weather, the opening of doors or windows, etc. Most furnaces are also provided with overheat protection by a temperature actuated relay which turns off the fire should the heat energy not be dissipated rapidly enough, as, for example, if the blower fails to operate. *Air conditioning* systems operate similarly. The gas furnace system described is an *on-off* controller, and is commonly called a *bang-bang controller* for obvious reasons. This approach is opposed to a *proportional controller,* in which the intention is to exert an amount of control effort *proportional* to the difference between the observed and desired responses. The bang-bang controller is only one of many types of *nonlinear* systems, or systems in which superposition does not apply. The *actuators* in our gas heater system are the *furnace fire* and the *blower*. The *relays* might also be called actuators, but they are fundamentally *control devices.* The *thermostat* and the *temperature devices* for blower and overheat control are the *sensors* or *transducers.* The thermostat setting is the control input. The system cycles back and forth between the on and off states, and this mode of operation is called a *limit cycle.* It is a *stable limit cycle* because the range of temperature variation does not grow with time in normal operation. Limit cycles are common in nonlinear systems and must surely occur in *all* bang-bang systems.

Consider the control problems associated with learning to drive a car. The beginning driver must first learn the function of various means provided for controlling the vehicle and the way in which the car reacts under all conditions to each of these inputs. We soon learn, for example, that vigorous application of the brakes or the accelerator on slippery roads is to be avoided, since it results only in the wheels sliding and our losing control of the vehicle's performance. The primary control objective in most driving experiences is to maintain the car in a reasonable state of operation between the highway guidelines, avoiding other vehicles, pedestrians, animals, or other objects which may be in the roadway, while proceeding toward the ultimate destination at a safe velocity. Control inputs are provided by the accelerator, steering wheel, and brakes. In this case the controller is no doubt the most sophisticated known to man, the driver himself. Inputs are obtained through the driver's senses, but primarily from his eyes, ears, and sense of balance or orientation. The controller consists of these sensors, the driver's brain, the appendages used to adjust the controls, and the mechanism by which these changes are made to influence the vehicle's motion. The fundamental actuators are the engine, the brakes, and the steering mechanism.

The ability to easily control a high powered automobile under a variety of environmental conditions is automatically assumed of virtually everyone in our society, but it is far from a trivial problem. The overall control system, including the driver, can readily become *unstable,* where this term is defined loosely at the moment as meaning that the system may yield a response contrary to the control inputs under appropriate environmental conditions. As an example, a skidding performance pattern resulting in total loss of control almost always results from excessive speeds on icy roads, particularly when there is a strong, gusty cross wind. Control also is often lost when the driver's reactions are impaired, as when he is intoxicated, for example. Trouble generally results from late observation of important data, as when the driver turns around to look in the back seat for some purpose and turns back to find that the car he was following closely has slowed down significantly. It is apparent that the ability to control a system depends upon its performance characteristics or *parameters,* the *speed* with which relevant data are provided to the controller, the presence or absence of *disturbing influences,* or *noise,* and the capability or sophistication of the controller, as illustrated in our example by the driver's experience.

A variety of subsidiary control systems are used on most modern automobiles. *Power steering* is a form of *position control* which moves the front wheel orientation as commanded by changing steering wheel position. Power steering controllers are called *fail-safe,* because the driver may mechanically *override* the system to maintain operation should the power support system's operation be interrupted for some reason, as when the fan belt breaks. The engine block water temperature is maintained at about 180°F during normal operation by controlling the rate of water flow through the radiator cooling system. A similar arrangement is used to control the temperature of the intake manifold, and thereby that of the fuel-air mixture reaching the cylinders, by adjusting the ventilating air flow. The battery is automatically kept charged to the desired voltage level by an *alternator* (or *generator*), a *rectifier*, and a *voltage regulator.* When the battery voltage is low the charging rate is maintained high, and vice versa. The fuel-air mixture provided by the carburetor is adjusted according to engine temperature and load, more vaporized gasoline per unit volume of air being provided when the engine is cold and/or when the load is increased, by the *automatic choke* mechanism. The *spark timing* is automatically advanced by the *distributor* as the engine speed increases. A variety of additional electronic equipment is also common. The radio is provided with *automatic gain control* (AGC) to prevent surging and fading of sound volume as the car passes through areas of strong and weak signal reception. AGC is a form of *adaptive control*, because a system parameter, the amplifier gain, is adjusted in an effort to improve overall performance as environment changes. A similar claim may be made for the spark timing and automatic choke controllers.

The control examples presented thus far are simple situations with which we are all familiar. Applications can also be called to attention that stretch the imagination of even the most sophisticated control specialist. Consider the myriad of challenging control problems which must be solved whenever we send an exploration party to the moon or one of the planets. The space vehicle attitude must be controlled, highly critical orbits achieved, a careful landing on the surface of the target accomplished, life support and communications systems provided, the return takeoff and trajectory established and, perhaps most critical of all, the earth's atmosphere must be re-entered at high velocity and a landing reached near the region within which recovery vehicles are deployed. Control systems play a vital role in all of these mission phases. The systems and equations are highly complex and nonlinear, large scale digital and analog computers are integrated into the system, both on board the vehicle and, through communication links, on the ground as well. Data are processed in both continuous and sampled form, and the equipment which must perform satisfactorily to assure success of even the simplest space mission is of fantastic proportions.

It is apparent from these examples that we are surrounded by control system applications. All large scale chemical and industrial processes and automated production facilities depend heavily upon control technology. The expansion of control theory and its application to improve our production capabilities has had tremendous local and world-wide social and economic impact. We are able to maintain our relatively high average standard of living because we can sell a variety of manufactured products at competitive prices. Only through automation is this possible, since our labor costs are very high. If our average production rate per dollar cost for manpower and equipment were to drop below that of other countries in many areas, we would be unable to sell our products on the world market, so a highly automated manufacturing capability is necessary to maintain and improve our living standard. Obviously, automation causes other social problems, such as the need to realign various segments of our labor forces, but these problems would be aggravated, not eliminated, by refusing to make technological improvements.

1.2 A BRIEF HISTORY OF CONTROL DEVELOPMENT

Feedback control systems have been in use for many years, although no significant theoretical foundations were developed prior to the early part of this century. The Babylonian civilization had a float controlled irrigation system[1] in use as early as 2000 BC. The early Egyptians had similar equipment. Andrew Meikle[2] of Holland in 1750 devised a fan-tail gear arrangement to turn large windmills into the wind for maximum operating efficiency. The flyball governor for steam engine speed control was developed by Watt[3] in 1788. The use of large steam engines as power units for ocean-going ships led to the ship steering problem during the 19th century, as well as continued interest in speed control. The large rudders used on ocean ships require some type of powerful controller, and this problem was considered by Maxwell,[4] Routh,[5] and Minorsky,[6] all of whom also worked on the stability problem.

Little control theory existed prior to 1930, but growth since that time has been extremely rapid. Nyquist,[7] in 1932, published his study of stability theory providing a graphical method for determining system stability which is the foundation for many of our design procedures. Although Nyquist was primarily concerned with feedback amplifier design, his results apply equally well to all types of linear systems and have even been expanded under appropriate conditions through *describing function analysis* to special types of nonlinear systems as well. Hazen[8] soon expanded upon Nyquist's work and originated the term servomechanism from servant (slave) mechanism. Developments were particularly rapid during and immediately following World War II, prompted by government interest in such projects as radar, automatically controlled antiaircraft weapons, aircraft autopilot development, automatic bomb release equipment, missile development, and the like. Unfortunately security restrictions kept much of this new knowledge under wraps for a time, and expansion of the field was somewhat retarded for several years.

Developments in control theory and applications were spurred in the 1950's by several factors. Perhaps most significant was the rapid development of both analog and digital computer technology, which promoted the design of increasingly complex controllers, and interest in sampled-data analysis and synthesis methods. Ragazzini and Franklin,[9] Jury,[10] and Tou[11] were instrumental in promoting widespread interest in, and understanding of, sampled systems. Interest in nonlinear systems also developed rapidly during the 1950's, and Kalman[12] helped to emphasize what the Russians had accomplished in this area starting in the 19th century with Liapunov's[13] work. Kalman[14] also promoted interest in state variable analysis with his introduction of the controllability and observability ideas. State representation of system equations is a natural method for working with nonlinear systems, and it is also convenient when a computer simulation is to be developed, so use of the state equations has become relatively standard in recent years.

Interest in adaptive control systems was initiated by Draper and Li[15] in 1951 and has grown rapidly since then. Methods for solving nonlinear boundary value problems using iterative computational methods were developed in the late 1950's and early 1960's by Breakwell,[16] Kelley,[17] and Bryson.[18] The adaptive control area and the boundary value area, or dynamic optimization problem, are closely related, and their development has been somewhat parallel in recent years. Another closely related topic in which considerable interest has arisen during the same period is the *maximum principle* introduced by Pontryagin[19] and his coworkers, which provides necessary and sufficient conditions for optimal solutions to boundary value problems. Bellman[20] introduced *dynamic programming* and the *principle of optimality* in the mid-1950's, and they have also been used to develop solutions to boundary value problems, although dynamic programming applies most directly to discrete systems. Many other interesting areas of study have been initiated more recently, and one of our primary purposes in this book is to establish a sound fundamental background from which all of these problem areas can be pursued.

1.3 AN OUTLINE OF THE MATERIAL TO BE COVERED

The book is divided into four major sections. The first is the study of fundamental mathematical concepts used throughout the later presentation, and comprises Chapters 2 through 7. Procedures for modeling systems in the form of differential equation descriptions are considered in this section. Methods for manipulating and solving differential equations and the equivalent transfer functions using the Laplace transform are also presented, as well as the basic relationships between system characteristics, such as stability and transient response patterns, and model parameters. Design methods for standard linear continuous control applications are considered in the second section, which includes Chapters 8 through 11. All control engineers must develop a sound understanding of all of the concepts discussed in these chapters if they are to efficiently solve application problems. Several advanced topics considered by the author to be of particular importance are included in Chapters 12 through 15. These include carrier, nonlinear, and sampled systems, in addition to the state representation of systems and design methods based upon state variable feedback. The final section is the collection of appendices, which is intended to support the remainder of the book with topics of peripheral interest, tables of useful data, and a comprehensive set of problems, ranging from trivial to difficult, involving everything from a few simple algebraic steps to suggested use of a general purpose digital computer.

The material presented herein is intended to provide a reasonably thorough introduction to the basic area of control system analysis and synthesis appropriate for presentation at the senior or first year graduate level. A minimum amount of background is assumed. The emphasis throughout is placed upon the development of efficient methods for designing the standard types of feedback controllers. Tools which most often prove useful in engineering practice are stressed. This is not a book on modern or optimal control theory, since these topics are logically undertaken only after the foundation provided here has been established. The apprentice control engineer must first learn how to use those tools he will *most often* find useful. A patient pursuit of basic design concepts is well rewarded, however, since the study of advanced topics which most graduate students will subsequently wish to undertake becomes much easier after a proper introduction to the field has been obtained.

REFERENCES

1. Gadd, C. J., Babylonian Law, Encyclopedia Britannica, 14th ed., vol. 2, p. 863, 1929.
2. Wolf, A., "A History of Science, Technology and Philosophy in the XVIIIth Century," The MacMillan Co., New York, 1939.
3. loc. cit.

4. Maxwell, J. C., On Governors, *Proceedings of the Royal Society (London)*, vol. 16, pp. 270–283, 1868.
5. Routh, E. J., "Stability of a Given State of Motion," Adams Prize Essay, MacMillan, London, 1877.
6. Minorsky, N., Directional Stability of Automatically Steered Bodies, *Journal of the American Society of Naval Engineers*, vol. 34, p. 280, 1922.
7. Nyquist, H., Regeneration Theory, *Bell System Technical Journal*, vol. 11, pp. 126–147, 1932.
8. Hazen, H. L., Theory of Servomechanisms, *Journal of Franklin Institute*, vol. 218, pp. 543–580, 1934.
9. Ragazzini, J. R. and G. F. Franklin, "Sampled Data Control Systems," McGraw-Hill, New York, 1958.
10. Jury, E. I., "Sampled Data Control Systems," John Wiley & Sons, Inc., New York, 1958.
11. Tou, J. T., "Digital and Sampled Data Control Systems," McGraw-Hill, New York, 1959.
12. Kalman, R. E. and J. E. Bertram, Control System Analysis and Design via the Second Method of Liapunov, Parts I and II, *Transactions ASME*, vol. 82, series D, pp. 371–400, 1960.
13. Liapunov, A. M., "On the General Problem of Stability of Motion," Ph.D. Thesis, Kharkov, 1892.
14. Kalman, R. E., On the General Theory of Control Systems, *IFAC Proceedings of lst International Congress on Automatic Control*, Moscow, 1960, pp. 481–493, Butterworth and Co., Ltd., London, 1961.
15. Draper, C. S. and Y. T. Li, Principles of Optimalizing Control Systems and an Application to the Internal Combustion Engine, *ASME*, New York, 1951.
16. Breakwell, J. V., The Optimization of Trajectories, *Journal of the Society of Industrial and Applied Mathematics*, vol. 7, pp. 215–247, 1959.
17. Kelley, H. J., Gradient Theory of Optimal Flight Paths, *ARS Journal*, vol. 30, pp. 947–953, 1960.
18. Bryson, A. E. and W. F. Denham, A Steepest Ascent Method for Solving Optimum Programming Problems, *Journal of Applied Mechanics*, June, 1962.
19. Pontryagin, L. S., et.al., "The Mathematical Theory of Optimal Processes," Interscience Publishers, Inc., New York, 1962.
20. Bellman, R. E., "Dynamic Programming," Princeton University Press, Princeton, N.J., 1957.

2 THE DIFFERENTIAL EQUATIONS DESCRIBING VARIOUS PHYSICAL PHENOMENA

The study of control systems is the study of a wide variety of feedback configurations described by one or more differential equations. Feedback is used to compare the system response to the input reference and an error signal is thereby developed, processed by an appropriate filter, and used to drive the actuator, which changes the output variable. Detailed analysis of the variety of situations encountered in practice requires a thorough understanding of many physical phenomena so that a mathematical model of the system under study may be developed. The mathematical model is used in whatever way seems appropriate to develop an acceptable controller design. In this chapter we will define the notation subsequently used in system representation, and review the differential equation descriptions of several phenomena common to control system practice. References where further detail may be found are listed in many cases. General procedures useful in developing mathematical descriptions of components and systems are presented, including Lagrange's equation which is based upon the conservation of energy principle and greatly simplifies the mathematical description of complex systems. It is assumed that the reader is familiar with the standard classical methods for solving differential equations. Solution of linear time invariant differential equations using the Laplace transform is discussed in Chap. 3. Emphasis in this chapter is placed upon the development of a mathematical system description.

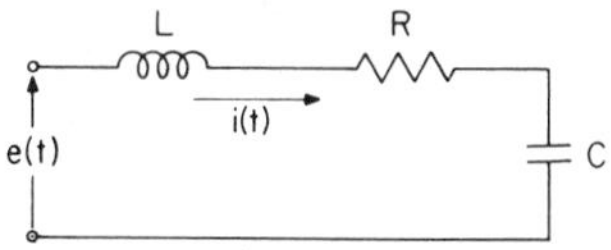

FIG. 2.1. A series RLC circuit with voltage excitation.

2.1 DEFINITION OF NOTATION

Standard control system circuits and components such as amplifiers, motors, and position sensors are described mathematically by differential equations. It is standard practice to use the differential element notation $dx/d\tau, d^2x/d\tau^2, \ldots, d^nx/d\tau^n$ to denote the successive derivatives of $x(\tau)$ with respect to the independent variable τ. In most of our applications the independent variable is time, denoted by t. The use of n dots as a shorthand denoting the nth derivative with respect to time is convenient and is often used in our presentation. Thus

$$\frac{dx}{dt} = \dot{x} \qquad \frac{d^2x}{dt^2} = \ddot{x} \ \ldots \tag{2.1}$$

As a simple example illustrating this notation consider the series resistance-inductance-capacitance (RLC) circuit shown in Fig. 2.1. The integro-differential equation describing this system in terms of $i(t)$ and $e(t)$ is

$$e = L\frac{di}{dt} + Ri + \frac{1}{C}\int_{-\infty}^{t} i\,dt \tag{2.2}$$

Time functions are represented by lower case letters, following the standard practice. It is generally desirable to reduce integro-differential equations of the form given in Eq. (2.2) to equivalent differential equations by differentiating both sides with respect to t, which yields

$$\frac{de}{dt} = L\frac{d^2i}{dt^2} + R\frac{di}{dt} + \frac{i}{C} \tag{2.3}$$

In this form de/dt is an equivalent forcing function obtained by differentiating the original input e with respect to time. We may alternately write Eq. (2.2) in terms of q, the instantaneous charge on C, as

$$e = L\frac{d^2q}{dt^2} + R\frac{dq}{dt} + \frac{q}{C} \tag{2.4}$$

This latter form is generally preferable to Eq. (2.3) because it involves e rather than de/dt. Determination of the general solution for $i(t)$ from Eq. (2.3) may prove awkward when $e(t)$ has one or more discontinuities. All of the three forms presented are equivalent, however, and they yield the same answer. The dot notation equivalent of Eq. (2.4) is

$$e = L\ddot{q} + R\dot{q} + \frac{q}{C} \tag{2.5}$$

and the relative simplicity of this form is immediately apparent.

We assume, unless specifically noted to the contrary, that we are dealing with linear time invariant systems. A system is linear if, and only if, superposition applies. Suppose the response of a given system to an arbitrary input $e_1(t)$ is $f_1(t)$. The system is linear only if its response to input $ae_1(t)$ is $af_1(t)$, where a is a constant. It is easy to tell by inspection of the describing equation if a system is linear. For example, Eq. (2.5) is linear if and only if the coefficients R, L, and C are independent of i and e. Time invariance of Eq. (2.5) requires that R, L and C be independent of time as well. Although no real system is ever truly linear or time invariant, many can be approximated with reasonable accuracy for the conditions normally encountered by linear time invariant models. The general procedures discussed subsequently for manipulating and solving differential equations apply, in general, only to linear time invariant systems. Solutions of nonlinear and/or time varying equations are much more difficult to obtain, and consideration of such problems is postponed until Chaps. 13 and 14.

2.2 DIFFERENTIAL EQUATIONS OF ELECTRICAL NETWORKS

Consider the multiloop network shown in Fig. 2.2. The loop method for writing the describing equations is initiated by assigning a current in each network loop and applying Kirchhoff's voltage law to each loop in turn. Applying this procedure to the system shown in Fig. 2.2 yields

$$v_s(t) = R_1 i_1 + L\frac{di_1}{dt} - R_1 i_2 - L\frac{di_3}{dt} \tag{2.6}$$

$$0 = -R_1 i_1 + (R_1 + R_2) i_2 + \frac{1}{C}\int i_2\,dt - \frac{1}{C}\int i_3\,dt \tag{2.7}$$

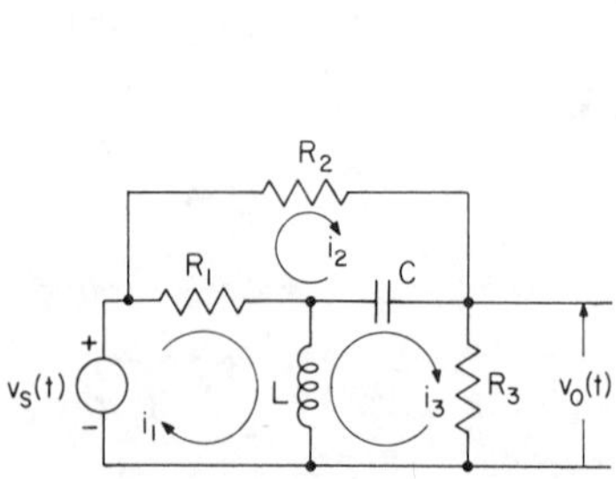

FIG. 2.2. A three-loop electrical network with voltage excitation.

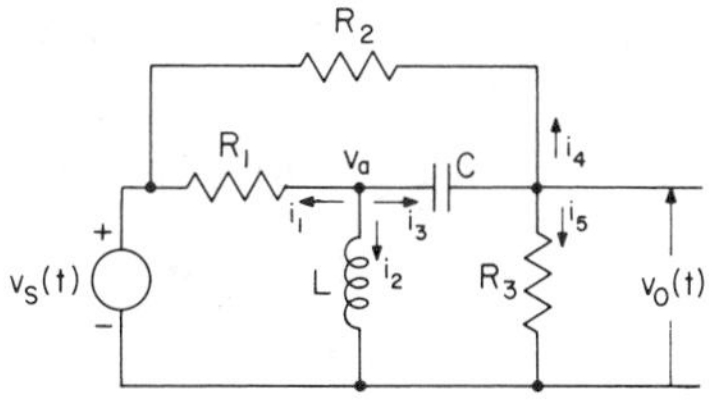

FIG. 2.3. A three-loop network with variables defined for a nodal analysis.

$$0 = -L\frac{di_1}{dt} - \frac{1}{C}\int i_2 dt + R_3 i_3 + L\frac{di_3}{dt} + \frac{1}{C}\int i_3 dt \tag{2.8}$$

$$v_0(t) = R_3 i_3 \tag{2.9}$$

There are thus four equations in four unknowns, the three currents and $v_0(t)$. These equations can be solved easily to find all of the currents and $v_0(t)$ if $v_s(t)$, the initial charge on C, and the initial current through L are known. Application of the Laplace transform to solve problems similar to this is considered in Chap. 3.

A node analysis of the circuit shown in Fig. 2.2 can be carried out by defining nodes and currents as shown in Fig. 2.3. Independently summing the currents into nodes v_a and v_0 to zero provides

$$\frac{v_a - v_s}{R_1} + \frac{1}{L}\int v_a dt + C\left(\frac{dv_a}{dt} - \frac{dv_0}{dt}\right) = 0 \tag{2.10}$$

$$C\left(\frac{dv_0}{dt} - \frac{dv_a}{dt}\right) + \frac{v_0}{R_3} + \frac{v_0 - v_s}{R_2} = 0 \tag{2.11}$$

Although we have only two equations in two unknowns here, Eqs. (2.10) and (2.11) are equivalent to Eqs. (2.6), (2.7), (2.8), and (2.9) and solution of either set of equations provides an identical system response.

These two illustrations of loop and node analysis should provide an adequate review of mathematical model development for simple electrical networks. The basic loop and node procedures apply to all configurations, although the manipulations become progressively more involved as complexity increases.

2.3 THE DIFFERENTIAL EQUATIONS DESCRIBING MECHANICAL TRANSLATION

The term mechanical translation is used to describe motion with a single degree of freedom, or motion in a straight line. Motion in a plane or in three-space is more difficult to define, although the same basic principles apply. Virtually all control applications involve a single degree of freedom or can be broken down into two or three subproblems, each with a single degree of freedom, so our attention is restricted to mechanical translation.

The basis for all translational motion analysis is Newton's second law of motion, which states that the net force, F acting on a body is related to its mass, M and the acceleration a by

$$F = Ma = M\dot{v} = M\ddot{x} \tag{2.12}$$

where v and x denote velocity and position, respectively. This is, in general, a vector or directional relationship, so we must always be careful to assign the proper sign to all terms when writing translation equations. Since we are only concerned with motion in a straight line, it is assumed that all forces and accelerations act along that line. In some cases this requires resolution of F into its vector elements and the development of equations based upon the appropriate components.

The three basic elements used in linear mechanical translation systems are masses, springs, and dashpots, or viscous friction units. The graphical and symbolic notations for all three are shown in Fig. 2.4. A mass M is indicated schematically by a square box and is described mathematically by Eq. (2.12). A linear spring, denoted by K, its elastance or stiffness constant, is drawn like a resistor and exerts a force proportional to its displacement, or

$$F = K(x - x_0) \tag{2.13}$$

where x_0 is its equilibrium position. A dashpot, denoted by B, its viscous friction constant, is depicted schematically as shown in Fig. 2.5c, and exerts a force proportional to the rate of change of the relative displacement of its two terminals, or

$$F = B\dot{x} = Bv \tag{2.14}$$

where x and v are the position and velocity, respectively, of one terminal relative to the other. An automobile shock absorber is a dashpot. It consists of a piston in a cylinder filled with fluid. The fit between piston and cylinder is purposely made poor, usually by putting one or more holes in the piston, to provide fluid leakage at a rate proportional to pressure differential.

Suppose we wish to obtain an equation defining the motion of the mass-spring arrangement shown in Fig. 2.5. Assume the mass can move only straight up and down. Several conditions are possible, and a general answer, valid in all

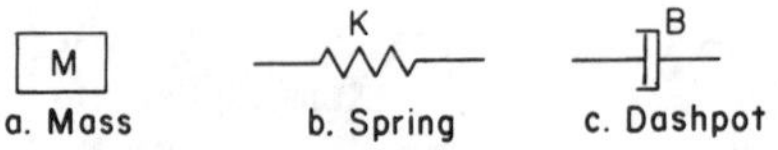

FIG. 2.4. The standard elements used in mechanical translation systems.

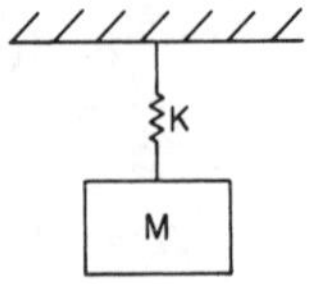

FIG. 2.5. A simple mass-spring arrangement.

cases, is desired. We might, for example, displace M from its equilibrium point and release it. What is the equation describing its motion? If we apply a force $f(t)$ to M, how would it respond? If we displace it from its equilibrium point *and* apply a force, what happens?

To simplify writing the equations describing translational systems, the following procedure is suggested:

1. Assume the system is in equilibrium. This eliminates the effect of gravity with no loss of generality in the result and considerably simplifies the resulting equations.
2. If there is no disturbing or driving force present, assume the system is given an arbitrary displacement from its equilibrium point.
3. Draw a free body diagram for each mass in the system, showing all forces acting on that mass.
4. Apply Newton's second law to each mass (each free body diagram) using the convention that all forces tending to cause movement in the assumed displacement direction are positive.

It is suggested that meticulous attention to detail be given in following this procedure until a considerable amount of experience is gained. Then the free body diagrams can be visualized without drawing them and the equations can be written by inspection.

Let us return to the spring-mass arrangement shown in Fig. 2.5. The equilibrium point is that point where the force due to spring displacement exactly equals the weight (Mg, where g is the gravitational acceleration constant) of M. We will measure all displacements from this point and disregard weight and the counterbalancing spring tension or reaction force. The same arrangement is shown in more detail in Fig. 2.6. We may write the equations of motion directly from the free body diagram of Fig. 2.6d by noting that the total force on M must be zero when the displacement x from the equilibrium position is as shown. The spring exerts an upward force Kx on M. The inertial force of the mass, M, is upward whenever $\ddot{x}$ is positive. In other words the mass opposes any increase in its downward (positive) velocity with a force $M\ddot{x}$. Thus

$$M\ddot{x} + Kx = 0 \tag{2.15}$$

defines the motion of M about its equilibrium point due to an arbitrary displacement, x.

Let us now consider the arrangement shown in Fig. 2.7. M is assumed to slide without friction with one degree of freedom on the baseplate in response to the external force $f(t)$. The positive direction of x, the displacement from the equilibrium position, is assumed as shown. The corresponding free body diagram is shown in Fig. 2.8. The external force $f(t)$ is opposed by the three force components Kx, $B\dot{x}$, and $M\ddot{x}$. Setting $f(t)$ equal to the reaction forces provides

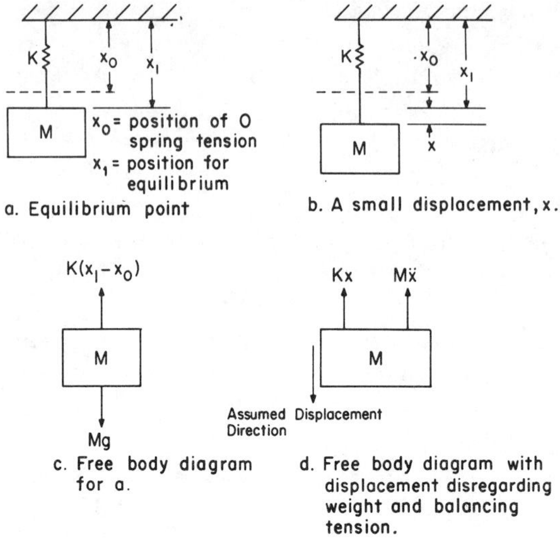

FIG. 2.6. Force relationships for the system of Fig. 2.5.

$$f(t) = M\ddot{x} + B\dot{x} + Kx \tag{2.16}$$

as the describing equation.

Finally, consider the situation shown in Fig. 2.9. It is left as a student exercise to draw the corresponding free body diagrams. Summing the forces on M_1 and M_2 to zero provides

$$f(t) = K_1x_1 + B_1\dot{x}_1 + M_1\ddot{x}_1 + K_2(x_1 - x_2) \tag{2.17}$$

and

$$0 = K_2(x_2 - x_1) + B_2\dot{x}_2 + M_2\ddot{x}_2 \tag{2.18}$$

as the equations defining system response to the force $f(t)$.

Some discussion of units is necessary. Both English and meter-kilogram-second (MKS) units are in common use, so it is important to be able to use either with facility. The standard units in both systems are summarized in Table 2.1 along

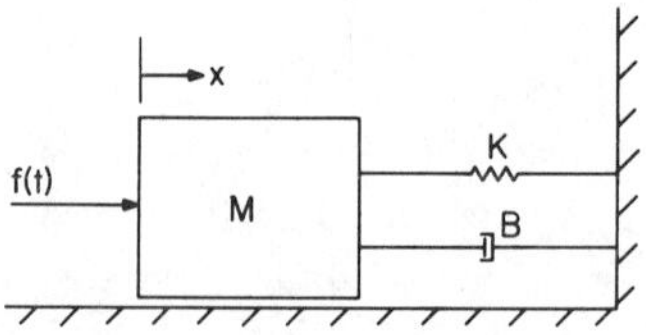

FIG. 2.7. A mass-spring-dashpot sytem.

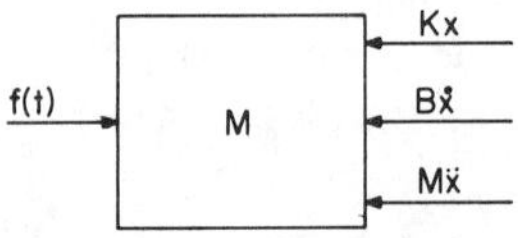

FIG. 2.8. Free body diagram for the system of Fig. 2.7.

Table 2.1. Components, symbols, units, and conversion factors for mechanical translation systems.

Symbol	*Name*	*English*	*MKS*	*Conversion*
M	mass	slugs = lb-sec^2/ft or pounds (32.174 lbs = 1 slug)	kilograms	1 kg = 2.205 lbs = 0.0686 slugs
K	stiffness coefficient	lb/ft	Newtons/m	$\frac{1 \text{ Newton}}{\text{m}} = \frac{0.0686 \text{ lbs}}{\text{ft}}$
B	damping or viscous friction coefficient	lb/ft-sec	Newtons/(m/sec)	$\frac{1 \text{ Newton}}{\text{m/sec}} = \frac{0.0686 \text{ lb}}{\text{ft/sec}}$
f	force	lb	Newtons	1 Newton = 0.2248 lbs
x	distance	ft	m	1 foot = 0.3048 ms
v	velocity	ft/sec	m/sec	1 m/sec = 3.281 ft/sec
a	acceleration	ft/sec^2	m/sec^2	1 m/sec^2 = 3.281 ft/sec^2

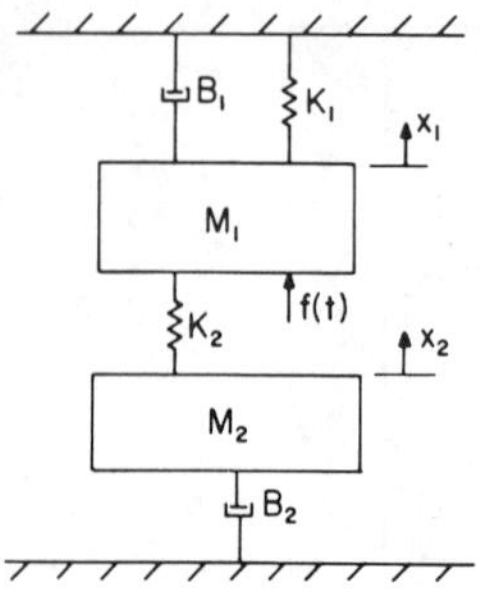

FIG. 2.9. A complex mechanical translation system.

with the factors required to convert from one system to the other. In all problems and examples it is assumed, unless explicitly indicated to the contrary, that a consistent set of units is used.

It is interesting to note that the differential equations describing electrical and mechanical systems (and in fact many other systems as well) are of precisely the same form. Although it is equally easy to write the equations for either form of system, and thus there is no need to consider equivalent networks, or analogs, to simplify the analysis, there are significant advantages to the use of electrical equivalents of mechanical systems, nonetheless. For example it is not particularly convenient to set up a mechanical spring-mass-dashpot system and test its response in the laboratory because such components are not available in a wide variety of sizes and are inconvenient to work with in any event. Since electrical components are readily available, as are current and voltage signals in a variety of forms for test inputs, and since currents and voltages are accurately measured with ease, it is often convenient to study the response characteristics of an electrical network equivalent to the mechanical system of interest, adjusting component values as required to provide the desired results. The equivalent mechanical design is then specified. In such studies a relatively simple form of electrical network is being used as a crude, special purpose, analog computer. One possible assignment of electrical and mechanical equivalences is summarized in Table 2.2.

Table 2.2. Mechanical translation and electrical analogs.

Mechanical translation element		*Electrical element*	
Symbol	*Name*	*Symbol*	*Name*
f	force	i	current
$\dot{x} = v$	velocity	e or v	voltage
M	mass	C	capacitance
K	stiffness	$1/L$	reciprocal inductance
B	damping	$G = 1/R$	conductance

To illustrate the use of analogs, we will develop an electrical circuit equivalent to the mechanical system shown in Fig. 2.7. We see from Eq. (2.16) and Table 2.2 that

$$i = C\dot{v} + Gv + \frac{1}{L}\int v\,dt \tag{2.19}$$

is the equivalent equation in terms of electrical components and the corresponding network is shown in Fig. 2.10. It is apparent that we could also use the dual of this network as an analog, and this leads us to the development of an alternate set of equivalence relationships in which force and voltage are analogous. The corresponding electrical circuit analog is a series RLC network. It is left as an exercise for the reader to develop the detailed analog assignments.

2.4 THE DIFFERENTIAL EQUATIONS DESCRIBING ROTATING MECHANICAL SYSTEMS

The equations describing rotating systems are similar to those of translational systems except that torque replaces force as the driving term and displacement, velocity, and acceleration are expressed in angular units. Equations are written summing the torques on each mass in the system to zero. The three basic elements of rotational systems are the spring or linear torsion element, K, the viscous friction dashpot, B, and the moment of inertia, J. These elements are represented schematically as shown in Fig. 2.11. A torque T is applied to the shaft, which has connected to it a mass configuration with moment of inertia J. The angle of shaft rotation, denoted by θ_1, is measured positive in the direction shown. Torque is applied through the torsion spring, K, to a shaft extension (perhaps the shaft itself is relatively flexible and twists, exerting torque proportional to the relative angular displacement of its two ends) which is connected to the damping element, B. The angular position of the damping element is denoted by θ_2. We will develop a mathematical description of this system as soon as the characteristics of its elements have been discussed.

When torque is applied between the two ends of a spring, a proportional relative angular displacement is obtained, or

$$T_K = K(\theta_2 - \theta_1) \tag{2.20}$$

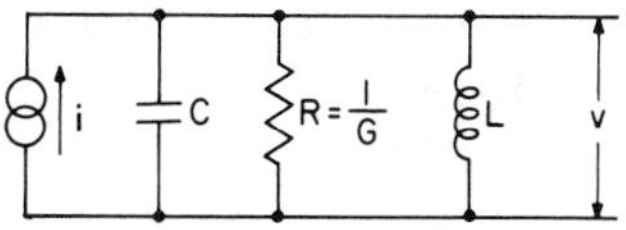

FIG. 2.10. Electrical equivalent of the mechanical network shown in Fig. 2.7.

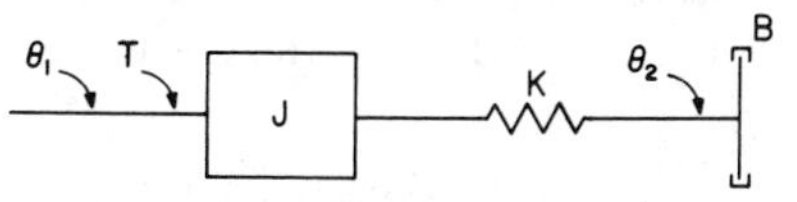

FIG. 2.11. A typical rotational system illustrating the notation to be used.

where θ_2 and θ_1 denote the angular positions of the two ends. The spring transmits any torque applied at one terminal to the second terminal, and it is assumed throughout our discussion that this torque transmission occurs instantaneously, which is equivalent to considering an inertialess spring. It is standard procedure to lump the spring inertia, J_s, with other inertia elements at the ends of the spring.

Rotational viscous friction occurs as a result of bearing drag, or the torque required to rotate a shaft with elements on it through a fluid. The fluid may be air (it takes power to turn a fan), although the effect is more pronounced when gears turn in an oil bath, for example. The torque required to counteract viscous damping is given by

$$T_B = B\omega = B\dot{\theta} \tag{2.21}$$

where ω and θ are shaft speed and shaft position, respectively, relative to the reference frame.

When torque is applied to a mass with moment of inertia J, the angular position of the mass is described by

$$T_J = J\ddot{\theta} = J\dot{\omega} = J\alpha \tag{2.22}$$

where α denotes angular acceleration.

We are now prepared to mathematically describe the system shown in Fig. 2.11. J has torque $T(t)$ applied to it at the input and a reaction torque $K(\theta_1 - \theta_2)$ applied in the opposite direction due to the spring. We can thus sum the torques on J to zero to obtain

$$T(t) = K(\theta_1 - \theta_2) + J\ddot{\theta}_1 \tag{2.23}$$

The spring also applies torque $K(\theta_1 - \theta_2)$ to the damper in the positive θ_2 direction, so

$$K(\theta_1 - \theta_2) = B\dot{\theta}_2 \tag{2.24}$$

Eqs. (2.23) and (2.24) provide a complete mathematical description of the system. They may be reduced to many alternate forms.

Consider the more complex situation shown in Fig. 2.12. This might be interpreted as the representation of an automobile engine coupled through a flywheel, J_1, a torsion tube, K_1, a fluid transmission, J_2, J_3, B_2, B_3 and B_4, and a second torsion tube to the wheels, which are assumed blocked in this case. Summing torques on the inertial elements provides

$$T(t) = J_1\ddot{\theta}_1 + B_1\dot{\theta}_1 + K_1(\theta_1 - \theta_2) \tag{2.25}$$

$$K_1(\theta_1 - \theta_2) = J_2\ddot{\theta}_2 + B_2\dot{\theta}_2 + B_4(\dot{\theta}_2 - \dot{\theta}_3) \tag{2.26}$$

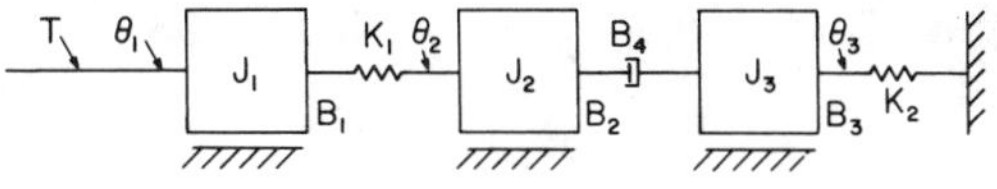

FIG. 2.12. A more complex rotational system.

and

$$B_4(\dot{\theta}_2 - \dot{\theta}_3) = J_3\ddot{\theta}_3 + B_3\dot{\theta}_3 + K_2\theta_3 \tag{2.27}$$

as the mathematical description.

In many applications the load is coupled to the drive unit through a gear train. In specifying the motor characteristics and choosing the gear ratios desirable for driving a given load it is important to know the effective load (inertia, damping, and/or load torque) at the motor as it relates to the J's, B's, and load torques on the various shafts and the gear ratios between shafts. The following assumptions are made: since the shafts between gears are generally short, they are assumed infinitely stiff ($K = \infty$); and the number of gear teeth is directly proportional to the gear radius. A simple gear train is shown schematically in Fig. 2.13. The following notation will be used:

N_i = number of teeth on ith gear.
θ_i = angular position of ith gear.
$\omega_i = \dot{\theta}_i$ = angular velocity of ith gear.
T_i = torque at the ith gear.
F_i = force at the ith gear.
r_i = radius of the ith gear.

At the point of contact between any two gears there are two equal and opposite force components. These result in two torque components which are generally different, since the radii of meshing gears are seldom equal. For example at the point of contact between gears N_1 and N_2 in Fig. 2.13, equal and opposite forces f_1 and f_2 are acting upon gears N_1 and N_2 to produce torques T_1 and T_2, respectively. T_1 is the retarding effect of the load on shaft 1 and T_2 is the torque relayed from shaft 1 to shaft 2 through the gear set N_1N_2. The ratio $T_1/T_2 = r_1/r_2$. Since the number of teeth is proportional to the radius, $T_1/T_2 = \omega_2/\omega_1$. The equations describing the gear train shown in Fig. 2.13 are obtained by

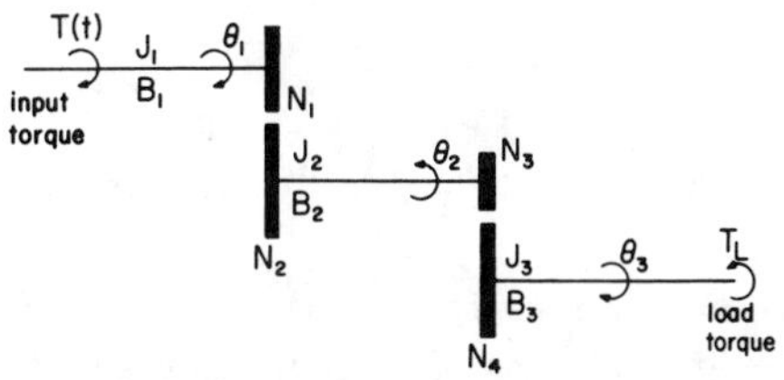

FIG. 2.13. Schematic representation of a gear train.

summing the torques on each shaft to zero as

$$T(t) = J_1\ddot{\theta}_1 + B_1\dot{\theta}_1 + T_1 \tag{2.28}$$

$$T_2 = J_2\ddot{\theta}_2 + B_2\dot{\theta}_2 + T_3 \tag{2.29}$$

$$T_2 = \frac{N_2}{N_1} T_1 \tag{2.30}$$

$$\theta_2 = \frac{N_1}{N_2} \theta_1 \tag{2.31}$$

$$T_4 = J_3\ddot{\theta}_3 + B_3\dot{\theta}_3 + T_L \tag{2.32}$$

$$T_4 = \frac{N_4}{N_3} T_3 \tag{2.33}$$

$$\theta_3 = \frac{N_3}{N_4} \theta_2 = \frac{N_1}{N_2} \frac{N_3}{N_4} \theta_1 \tag{2.34}$$

Using Eq. (2.33) for T_4 in Eq. (2.32), solving for T_3, substituting into Eq. (2.29), substituting T_2 as given by Eq. (2.30) into the result, solving for T_1, and substituting the result into Eq. (2.28) provides

$$T(t) = J_1\ddot{\theta}_1 + B_1\dot{\theta}_1 + \frac{N_1}{N_2}\left[J_2\ddot{\theta}_2 + B_2\dot{\theta}_2 + \frac{N_3}{N_4}(J_3\ddot{\theta}_3 + B_3\dot{\theta}_3 + T_L)\right] \tag{2.35}$$

Using Eqs. (2.31) and (2.34) we can express Eq. (2.35) entirely in terms of θ_1 as

$$T(t) = \left[J_1 + \left(\frac{N_1}{N_2}\right)^2 J_2 + \left(\frac{N_1}{N_2}\frac{N_3}{N_4}\right)^2 J_3\right]\ddot{\theta}_1 + \left[B_1 + \left(\frac{N_1}{N_2}\right)^2 B_2 + \left(\frac{N_1}{N_2}\frac{N_3}{N_4}\right)^2 B_3\right]\dot{\theta}_1 + \frac{N_1}{N_2}\frac{N_3}{N_4} T_L \tag{2.36}$$

It is clear from Eq. (2.36) that the load on several shafts coupled by gears may be treated as a single equivalent load at the drive shaft if each inertia and damping term is multiplied by the gear ratio (to the shaft on which it actually appears) squared and if all load torques are multiplied by the appropriate gear ratio. If the load must be turned faster than the drive shaft, the load inertia is increased by the speed ratio squared (the gear ratio) and vice versa, which is intuitively satisfying from an energy point of view. Load torques on shafts turning faster than the drive shaft are increased by the speed ratio. In this particular case we see from Eq. (2.36) that

$$J_{eq} = J_1 + \left(\frac{N_1}{N_2}\right)^2 J_2 + \left(\frac{N_1}{N_2}\frac{N_3}{N_4}\right)^2 J_3 \tag{2.37}$$

$$B_{eq} = B_1 + \left(\frac{N_1}{N_2}\right)^2 B_2 + \left(\frac{N_1}{N_2}\frac{N_3}{N_4}\right)^2 B_3 \tag{2.38}$$

and

$$T_{eq} = \frac{N_1}{N_2}\frac{N_3}{N_4} T_L \tag{2.39}$$

The equivalent load on shaft 2 or shaft 3 may be obtained by the same procedure. Extension of these results to any number of shafts and gear combinations is straightforward.

English and mks units are commonly used for mechanical rotational systems. The proper units in both systems are summarized in Table 2.3, which also provides the appropriate conversion factors in each case.

Design of systems involving gear trains is an interesting problem. A prime mover capable of providing the power to drive the load must be chosen, but this still leaves much freedom of choice. Should we choose a fast motor and "gear it down" to the load speed or a slow motor and drive the load directly, or "gear it up" to the load? Is the desired load speed comparable to motors available in the proper power range for direct drive use? Would it be more economical to choose a faster or slower motor? How complex (and expensive) is the required gear train, and how do its inertia and damping alter the power requirements? If a specific gear ratio is chosen, say 20:1, should this speed ratio be obtained using one gear mesh or two? How much effect will backlash (space between gear teeth) have on load response? These and other interesting aspects of this problem are discussed in detail by Gibson and Tuteur.[1]

In subsequent discussions on system design we will talk about equalizers, or filter networks, placed in the feedback control loop to provide desired stability and response characteristics. We tacitly assume the signals involved are electrical in nature and electrical networks are specified for equalization purposes. As we have seen, an exact analogy exists between electrical and mechanical translational

Table 2.3. Components, symbols, units and conversion factors for mechanical rotational systems.

Symbol	*Name*	*English units*	*MKS units*	*Conversion factor*
J	moment of inertia	slug-ft^2 (occasionally given as lb-ft-sec^2)*	kg-m^2	1 slug ft = 1.27 kg-m
K	stiffness coefficient	lb-ft/rad	Newton-m/rad	1 lb-ft/rad = 0.738 Newton-m/rad
B	damping coefficient	lb-ft/(rad/sec)	Newton-m/(rad/sec)	1 lb-ft/rad = 0.738 Newton-m/(rad/sec)
T	torque	lb-ft	Newton-m	1 lb-ft = 0.738 Newton-m
θ	angle	rad	rad	rad
ω	angular velocity	rad/sec	rad/sec	rad/sec
α	angular acceleration	rad/sec^2	rad/sec^2	rad/sec^2

*J has the dimensions of mass-$distance^2$, or slug-$feet^2$ in English units. Occasionally J is specified in lb-ft^2. In such cases it must be converted to slug-ft^2 using the relationship 1 slug = 32.174 lbs. The English unit set given here is consistent only if J is specified in slug-ft^2.

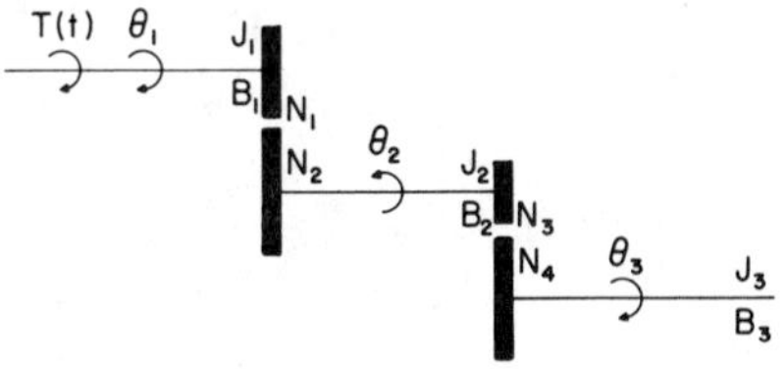

FIG. 2.14. The system considered in Ex. 2.1.

systems and the same is true for electrical and mechanical rotational systems. Mechanical equalization is thereby feasible, and is sometimes used. A table of mechanical networks and their transfer characteristics is available in Gibson and Tuteur.[2]

Example 2.1

Consider the rotational system shown in Fig. 2.14. A 20:1 speed reduction is desired between shafts 1 and 3. It is impossible to provide more than 10:1 speed reduction with a single gear pair. Thus $N_2/N_1 \leq 10$ and $N_4/N_3 \leq 10$ is required. (It becomes increasingly difficult to design precision gear pairs as the tooth ratio increases.) Assuming that J_i and B_i do not depend upon our choice of gear ratios, what values should be used for N_2/N_1 and N_4/N_3? (Note that assuming the J_i and B_i independent of the ratios chosen is valid only if the gear friction and inertia is a negligible part of the total friction and inertia on each shaft. This is generally true whenever large loads are being driven, but would not normally be true in the design of an instrument servo to drive a potentiometer, for example.)

The equivalent inertia and friction values on shaft 1 are given by

$$J_e = J_1 + \left(\frac{N_1}{N_2}\right)^2 J_2 + \frac{1}{400} J_3 \tag{2.40}$$

and

$$B_e = B_1 + \left(\frac{N_1}{N_2}\right)^2 B_2 + \frac{1}{400} B_3 \tag{2.41}$$

Obviously we should make N_1/N_2 as small as possible to minimize the equivalent load upon shaft 1, or set $N_1/N_2 = 1/10$ and $N_3/N_4 = 1/2$.

Although at a casual glance it may appear that we are getting something for nothing here, this is not the case. Note that we have simply minimized the power delivered to shaft 2 (the speed of shaft 2) within the restrictions imposed. The conditions of this problem might be encountered, for example, if the motor and load are reasonably far apart and it is desired to find optimum values for the gear ratios at the motor and loads ends. In such an application, J_2 and B_2 would be influenced little by the ratios used and our assumptions are valid. Although the percentage of total power saved is small, the total amount would be significant over years of use.

2.5 THE DIFFERENTIAL EQUATIONS DESCRIBING THERMAL SYSTEMS

It is generally impossible to describe thermal systems by linear differential equations since this requires that all bodies in the system have uniform temperature and this condition is seldom satisfied. If, however, the system of interest primarily consists of relatively small bodies in a circulating medium such as air or water, a linear approximation is reasonably accurate. In such cases only small temperature differences exist in the bodies and the medium unless conditions are changing rapidly.

The equilibrium equations for thermal systems are written by observing that the total heat input must equal that lost through the insulation plus that carried away by mass leaving the system plus that stored within the system in the form of a temperature increase. The terms and units used in thermal systems are shown in Table 2.4. English and MKS units and the required conversion factors are given, since both sets of units are common.

A linear approximation to the rate of heat flow from side 1 to side 2 through the uniform body (insulation strip) shown in Fig. 2.15 in terms of the two

Table 2.4. Units used in thermal systems.

Symbol	*Name*	*English units*	*MKS units*	*Conversion factor*
M	mass	lb	kg	1 kg = 2.205 lbs
S	specific heat	(Btu/lb)/°F	(Kcal/kg)/°C	none
C	thermal capacitance $C = MS$	Btu/°F	Kcal/°C	1 Btu/°F = 0.4536 Kcal/°C
R	thermal resistance	°F/(Btu/min)	°C/(Kcal/min)	1°F/(Btu/min) = 2.22°C/(Kcal/min)
θ	temperature	°F	°C	1°F = 5/9 °C
h	heat energy	Btu	Kcal	1 Btu = 0.252 Kcal
q	rate of heat flow	Btu/min	Kcal/min	1 Btu/min = 0.252 Kcal/min

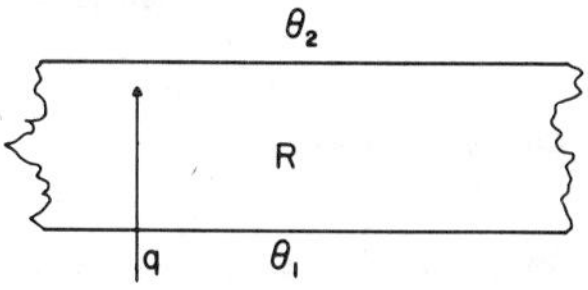

FIG. 2.15. Heat flow through a thermal insulation strip.

boundary temperatures $\theta_1$and θ_2 is given by

$$q = \frac{\theta_1 - \theta_2}{R} \tag{2.42}$$

The value of R obviously depends upon the strip thickness, the strip material, and the surface area over which temperatures θ_1 and θ_2 are effective.

The relationship between the temperature θ (assumed uniform throughout) of a body with mass M and the heat input to that body q is given by

$$q = MS\dot{\theta} = C\dot{\theta} \tag{2.43}$$

where S is the specific heat of the body material, and $C = MS$ is the thermal capacitance of m. Thus, for a constant rate of temperature change, a heat input proportional to the product of change rate, body mass, and specific heat is required.

A thin-walled glass medical thermometer filled with mercury is stabilized at room temperature of 75°F. At $t = 0$ it is placed in a stirred vat of chemicals with temperature θ_c. Assume for convenience that the entire outer thermometer surface is exposed to temperature θ_c, the thin glass shell has thermal resistance R and negligible thermal inertia relative to that of the mercury, (the heat input required to raise the temperature of the glass is negligible) and the mercury with mass M_m and specific heat S_m is all at temperature θ_m. This situation is schematically depicted in Fig. 2.16. The equation for heat flow through the glass wall is

$$q = \frac{\theta_c - \theta_m}{R} \tag{2.44}$$

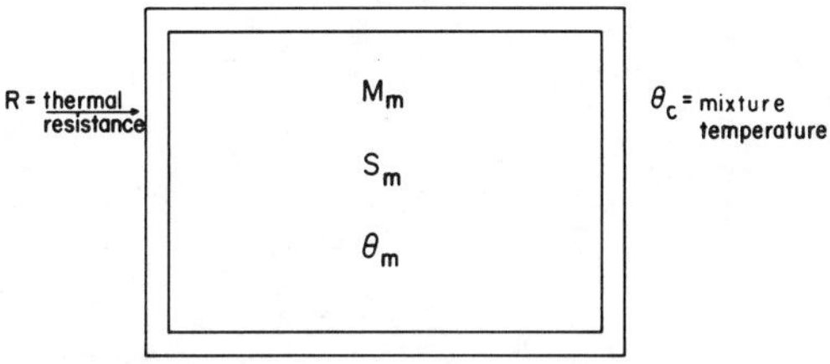

FIG. 2.16. Schematic representation of the thermometer problem.

This heat increases the temperature θ_m according to the relationship

$$q = M_m S_m \dot{\theta}_m = C_m \dot{\theta}_m \tag{2.45}$$

Combining Eqs. (2.44) and (2.45) allows us to write

$$RC_m \dot{\theta}_m + \theta_m = \theta_c \tag{2.46}$$

as the equation defining $\theta_m(t)$ in terms of θ_c. The initial value of θ_m is known to be 75°F, so the solution is completely defined.

We could make a higher-order linear approximation of this nonlinear problem by assuming the heat flow, q, passes through the glass walls with resistance R_g, acts upon the mass, M_g, of the glass wall to raise its temperature, θ_g, which in turn acts through a thermal resistance, R_m, to raise the mercury temperature, θ_m. The result is a second-order differential equation, as readily justified by the reader.

Consider the heat balance equation associated with the heater shown schematically in Fig. 2.17, which is used to provide a catalyst to a chemical process at a desired temperature. Assume that the catalyst volume is constant with uniform temperature θ throughout the tank. The heater is turned on and off automatically by a thermostat to control θ. The catalyst mixture is withdrawn from the tank as needed to control the chemical process. We assume no heat storage in the insulation. These assumptions are reasonable under the conditions of normal use. We define the following notation in addition to that shown in Fig. 2.17.

C = thermal capacitance of the tank full of catalyst mixture.
R = thermal resistance of the insulated tank walls.
S = specific heat of the catalyst mixture.

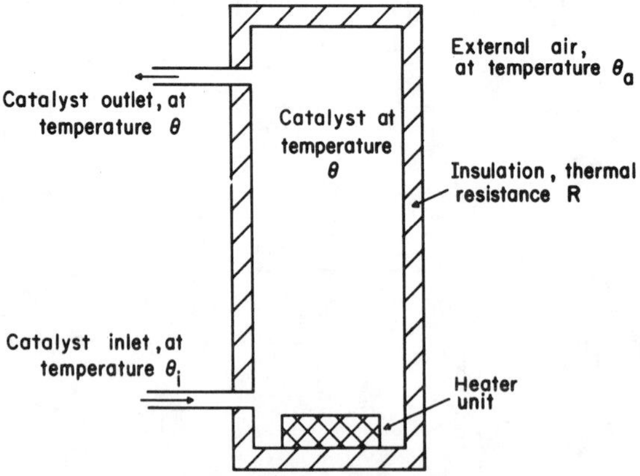

FIG. 2.17. Heat balance schematic for a catalyst heater.

q = heating element heat flow rate.
q_0 = heat flow due to catalyst leaving the tank.
q_i = heat flow due to catalyst entering the tank.
q_e = heat loss rate through the tank insulation.
q_t = rate at which heat is used to increase θ.
f_c = catalyst input and output flow rate.

The heat balance equation is

$$q = q_0 - q_i + q_e + q_t \tag{2.47}$$

where

$$q_0 = f_c S\theta \tag{2.48}$$

$$q_i = f_c S\theta_i \tag{2.49}$$

$$q_e = \frac{\theta - \theta_a}{R} \tag{2.50}$$

$$q_t = C\dot{\theta} \tag{2.51}$$

We may combine these equations to obtain

$$q = C\dot{\theta} + \theta\left(f_c S + \frac{1}{R}\right) - f_c S\theta_i - \frac{\theta_a}{R} \tag{2.52}$$

In general θ_i, θ_a, C, S, and R are known. Thus we may solve Eq. (2.52) for $\theta(t)$ if the initial value of θ is given and if f_c and q are known. In fact a solution can also be obtained if a control law is given defining q as a function of θ.

The procedure for solving thermal problems may be summarized as follows: (1) Set up a schematic of the system, defining the parameters and unknowns to be used, (2) Make such approximations as seem appropriate, (3) Set up the heat balance equation, and (4) Reduce the results to whatever form is desired. Although the examples considered here are rather simple, the same approach is used in more complex situations.

2.6 SIMPLIFIED DIFFERENTIAL EQUATIONS DESCRIBING HYDRAULIC SYSTEMS

The study of hydraulic actuators is a rather specialized, but nonetheless important, area of control system theory. Mathematical complexity arises because typical hydraulic components are much more nonlinear than the analogous electrical units. In some cases, however, a linear approximation is valid, and we will consider only such simple situations here. The reader is referred to Gibson and Tuteur[3], in which the last four chapters are devoted to pneumatic systems, for further detail. Hydraulic actuators are often used when large forces must be applied to move things relatively small distances as, for example, in missile thrust vector control, raising and lowering aircraft landing gear, and many similar applications. The power input required to move the control valve is a small part of the power available at the end of the moving piston in a typical hydraulic actuator, so unusually large power amplification is possible with relatively light and simple equipment. These advantages easily justify the somewhat awkward analysis procedures required for nonlinear systems.

Consider the schematic representation of a hydraulic actuator shown in Fig. 2.18. The three-way input valve controls inlet and outlet orifices such that the fluid flow is into the left side and away from the right side of the output cylinder when $x(t)$ is positive and the opposite when $x(t)$ is negative. In most applications the differential pressure, $P_h - P_s$, is sufficiently large that large loads, limited by $(P_h - P_s)A_p$, where A_p denotes piston area, can be driven by the output piston. As $x(t)$ increases, fluid from the storage reservoir at P_h enters the left side of the output piston, increasing P_1. Simultaneously fluid from the right side of the piston flows out to the sump at P_s through the outlet orifice, decreasing P_2. The induced force on the piston must equal the load force f_L, or

$$(P_1 - P_2)A_p = f_L \tag{2.53}$$

The flow of fluid through the inlet orifice is given by[4]

$$q = cA_0\sqrt{2g\frac{\Delta p}{w}} \tag{2.54}$$

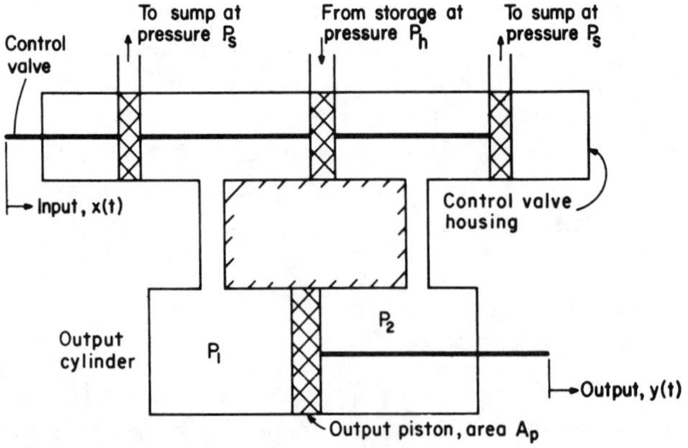

FIG. 2.18. A simple hydraulic actuator.

where q is the fluid flow rate, c is the orifice coefficient, a number related to orifice geometry, A_0 is orifice area, g is the gravitational acceleration, Δp is the orifice pressure drop, and w is the fluid specific weight. As a zeroth-order approximation we can assume that c is a constant independent of valve position (note that orifice geometry changes significantly with changes in x for any valve realization we might think of), Δp is independent of valve position (the load is not sufficiently large to cause P_1 to change appreciably, and P_b is a constant independent of load), and A_0 can be expressed as a linear function of valve position x. Under these rather restrictive conditions we may rewrite Eq. (2.54) as

$$q = K_0 x \tag{2.55}$$

where K_0 is the equivalent valve constant. If we neglect fluid leakage around all pistons (often reasonable) and disregard fluid compressibility (also reasonable for most applications where the fluid used is a liquid as opposed to a gas) the velocity of the main piston is directly proportional to q, or

$$q = A_p \dot{y} \tag{2.56}$$

Combining Eqs. (2.55) and (2.56) provides

$$A_p \dot{y} = K_0 x \tag{2.57}$$

Thus, for the assumptions we have made, the actuator shown in Fig. 2.18 is a hydraulic integrator.

The hydraulic integrator is only one basic type of hydraulic actuator. It is beyond the scope of our present discussion to consider other devices of similar form or to provide a more detailed analysis of the situation shown in Fig. 2.18. The basic procedure used in an exact analysis is identical to that given here, but more detail is required. The flow balance equations are set up, including leakage, compressibility, and load-induced pressure change effects. These equations can be linearized to provide a first-order mathematical system description.

2.7 ROTATING POWER AMPLIFIERS

A standard dc or ac generator is often used as a power amplifier. A control signal is applied to the field circuit, the generator is driven at constant speed, and the output signal is taken from the armature. Large power amplification is available. The voltage, e_g, generated in the armature circuit is given by (we assume a dc generator. Analysis is identical in the ac case, except that the output is an ac signal at frequency ω_1.)

$$e_g = K_1 \phi \omega \tag{2.58}$$

where K_1 is a constant depending upon the generator circuit, ϕ is the field induced flux level, and ω is the armature angular velocity. Over the range of normal operation it is reasonable to assume a linear relationship between field current i_f and ϕ, or

$$\phi = K_2 i_f \tag{2.59}$$

Combining Eqs. (2.58) and (2.59) provides

$$e_g = K_1 K_2 \omega i_f \tag{2.60}$$

If, as in most applications, $\omega = \omega_0$ is a constant, we may rewrite Eq. (2.60) as

$$e_g = K_1 K_2 \omega_0 i_f = K_g i_f \tag{2.61}$$

which provides a linear relationship between the input i_f (or e_f, the voltage applied to the field, which is linearly related to i_f through the field resistance and inductance), and the output e_g.

The schematic representation of a dc generator is shown in Fig. 2.19. The constants L_f, R_f, L_a, and R_a represent the field inductance and resistance and the armature inductance and resistance, in that order. K_g is the constant defined in Eq. (2.61) relating e_g and i_f. The describing equations are

$$e_f(t) = R_f i_f + L_f \frac{di_f}{dt} \tag{2.62}$$

$$e_g(t) = K_g i_f \tag{2.63}$$

$$e_g(t) = R_g i_a + L_g \frac{di_a}{dt} + e_0(t) \tag{2.64}$$

These equations may be manipulated as desired. If, for example, $\mathbf{E}_0(j\omega)$ is the voltage developed across an arbitrary impedance expressed in the usual phasor notation, denoted by $\mathbf{Z}_0(j\omega)$, we observe that

$$\mathbf{E}_0(j\omega) = \mathbf{Z}_0(j\omega)\,\mathbf{I}_a(j\omega) \tag{2.65}$$

Solving Eq. (2.62) through Eq. (2.65) for E_0 in terms of E_f and the system

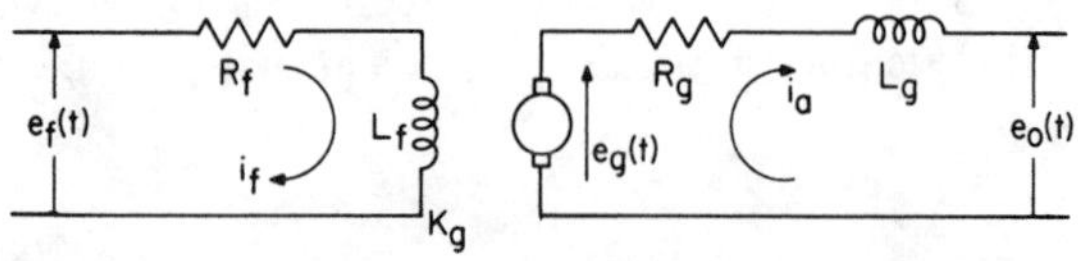

FIG. 2.19. Schematic representation of a DC generator used as a power amplifier.

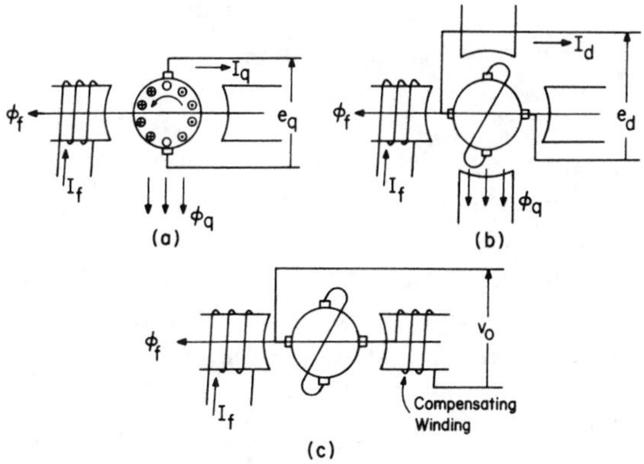

FIG. 2.20. Basic principles of amplidyne operation. See text for explanation.

parameters, including Z_0, provides

$$E_0(j\omega) = \frac{K_g Z_0(j\omega)}{(R_f + j\omega L_f)[R_g + j\omega L_g + Z_0(j\omega)]} E_f(j\omega) \tag{2.66}$$

Many alternate situations are readily treated.

Two stage rotating power amplifiers are available which provide power gains in the range of 10,000 to 100,000. Such devices can be built as a single unit; they are available commercially from several sources; and they are known by a variety of names, the most common of which are Amplidyne, Rototrol, and Regulex.[5] We choose the amplidyne as a representative example of these devices. Emphasis is placed upon the general principles of operation and reduction of the equations to transfer function form. The amplidyne uses one armature with one set of windings upon it, but has two sets of brushes to tap off the output, thereby providing two stages of power gain. The armature is wound as for a two-pole machine and is rotated at constant speed through the fields developed by two pairs of poles. This situation is shown in Fig. 2.20. A simple dc two-pole generator with control field current I_f is shown in Fig. 2.20a. A flux ϕ_f is produced by I_f and, for rotation as shown, the voltage e_g induced in the armature conductors is as shown by the conventional dots and crosses. If a current is taken from the brushes, an armature reaction flux ϕ_q at right angles to ϕ_f results. In standard generators every effort is made to suppress armature reaction flux, but in the amplidyne a magnetic path is provided to encourage it. The brushes passing the current I_q are short circuited to maximize I_q, therefore providing maximum reaction flux ϕ_q for a given control and current input. A voltage is induced in the armature conductors due to ϕ_q as well as due to ϕ_f. This voltage is in the direction shown by the dots and crosses in Fig. 2.20a. The second set of brushes is located on the armature to deliver this induced voltage e_d called the direct axis voltage, to the load. Since the armature circuit resistance is small, the armature is short circuited, and a low reluctance magnetic path is provided for ϕ_q, which is much larger than ϕ_f. Thus e_d is much larger than e_q, the quadrature axis voltage acting within the shorted armature circuit. If a load is connected to the output brushes shown in Fig. 2.20b, any load current I_d would result in a reaction flux tending to counteract ϕ_f, thereby radically decreasing the overall gain. A compensating winding designed to counteract this effect is added as shown in Fig. 2.20c, and this completes the design. Without this compensating winding the machine would act as a constant current generator, the load current being controlled at that level necessary to produce a reaction flux just able to counteract the flux ϕ_f induced by the control input.

An approximate equivalent circuit of the amplidyne is shown in Fig. 2.21. The conversion from e_f to e_q is treated as the first stage of gain and that from e_q to e_d as the second stage. The describing equations are

$$e_f = R_f i_f + L_f \frac{di_f}{dt} \tag{2.67}$$

$$e_q = K_q i_f = R_q i_q + L_q \frac{di_q}{dt} \tag{2.68}$$

$$e_d = K_d i_q = R_d i_d + L_d \frac{di_q}{dt} + e_0 \tag{2.69}$$

These equations may be manipulated as desired, depending upon the application. For example, it is convenient to represent e_d as a general function of e_f, thereby providing an equivalent voltage source for the load (output) circuit. Combining Eqs. (2.67), (2.68) and (2.69) provides

$$\mathrm{E}_d(j\omega) = \frac{K_q K_d}{(R_f + j\omega L_f)(R_q + j\omega L_q)} \mathrm{E}_f(j\omega) \tag{2.70}$$

This often proves useful in working with systems where an amplidyne is used as an amplifier.

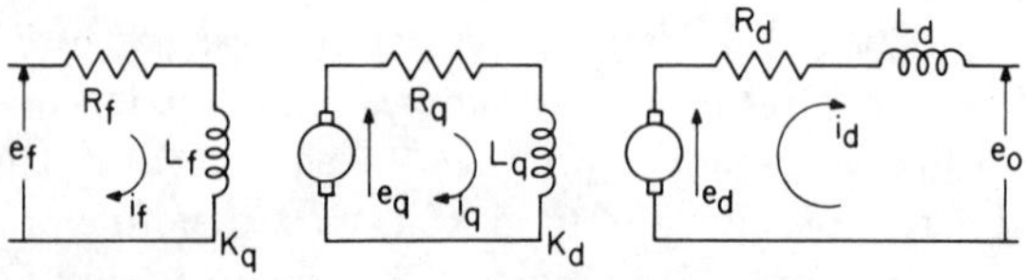

FIG. 2.21. Schematic representation of the amplidyne as a two-stage amplifier.

2.8 THE DC SERVOMOTOR

As a first-order linear approximation it is standard procedure to express the torque T produced by a dc motor as

$$T = K_3 \phi i_m \tag{2.71}$$

where K_3 depends upon the motor characteristics, ϕ is the field flux, and i_m is the armature current. Two common modes of operation will be considered. In the first, i_f and ϕ are held constant and control is achieved by changing i_m. In the second, i_m is held constant and i_f is the control input.

Constant i_f may be obtained by exciting the field circuit from a regulated dc source. Since ϕ is proportional to i_f, we may rewrite Eq. (2.71) in this case as

$$T = K_{Ta} i_m \tag{2.72}$$

The relationship expressed by Eq. (2.64) forms the basis for constant field excitation dc servomotor analysis. When the motor armature rotates, a back emf voltage proportional to ϕ and angular velocity ω [see Eq. (2.58)] is generated. Mathematically

$$e_m = K_1 \phi \omega_m = K_e \omega_m = K_e \dot{\theta}_m \tag{2.73}$$

Motor shaft position is controlled by applying an input voltage e_a to the armature circuit, as shown in Fig. 2.22. The equations relating θ_m to e_a are

$$e_a = R_m + L_m \frac{di_m}{dt} + e_m \tag{2.74}$$

$$e_m = K_e \frac{d\theta_m}{dt} \tag{2.75}$$

$$T = K_{Ta} i_m \tag{2.76}$$

where R_m and L_m are the motor armature resistance and inductance, respectively. Once the motor load is specified, these equations may be solved for the desired input-output relationships. For example, if the load consists of an inertia J and

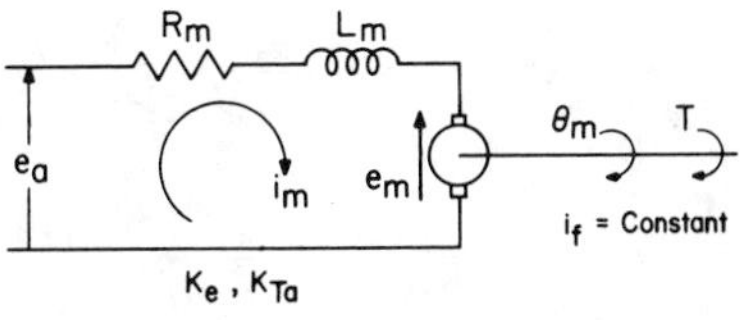

FIG. 2.22. Schematic representation of an armature driven dc servomotor.

damping B, the load torque is related to θ_m by

$$T = J\ddot{\theta}_m + B\dot{\theta}_m \tag{2.77}$$

and the relationship between e_a and θ_m is readily obtained by solving Eqs. (2.74) through (2.77) as

$$\frac{L_m J}{K_T}\dddot{\theta} + \frac{R_m J + L_m B}{K_T}\ddot{\theta}_m + \frac{R_m B + K_e K_T}{K_T}\dot{\theta}_m = e_a \tag{2.78}$$

In most cases L_m is negligible and Eq. (2.78) reduces to a second-order equation. We may also write Eq. (2.78) in terms of ω_m by substituting $\dot{\theta}_m = \omega_m$.

A field-controlled dc motor with constant armature current and load consisting of inertia J and damping B is shown schematically in Fig. 2.23. The mathematical description of this situation is readily written down as

$$e_f = R_f i_f + L_f \frac{di_f}{dt} \tag{2.79}$$

$$T = K_{Tf} i_f = J\ddot{\theta}_m + B\dot{\theta}_m \tag{2.80}$$

Solving Eqs. (2.79) and (2.80) for $\boldsymbol{\theta}_m(j\omega)$ in terms of $\mathbf{E}_f(j\omega)$ provides

$$\boldsymbol{\theta}_m(j\omega) = \frac{KT_f}{(R_f + j\omega L_f)(-\omega^2 J + j\omega B)} \mathbf{E}_f(j\omega) \tag{2.81}$$

as the form in which the results are generally desired. Thus we once again have a third-order equation relating θ_m and the input voltage, but in this case reduction to a second-order equation is not possible because the field time constant is generally significant.

Field control requires far less power from the control circuit than does armature control, and is often used for that reason. The armature time constant is generally negligible, whereas the field time constant can seldom be disregarded. Both approaches are often encountered.

The detailed analysis of ac servomotors is much more complex than that for comparable dc units.[6] The standard ac servomotor is a two-phase induction unit with two stator field coils placed 90 electrical degrees apart. An ac voltage of constant amplitude is applied to one coil (the reference coil) and the voltage

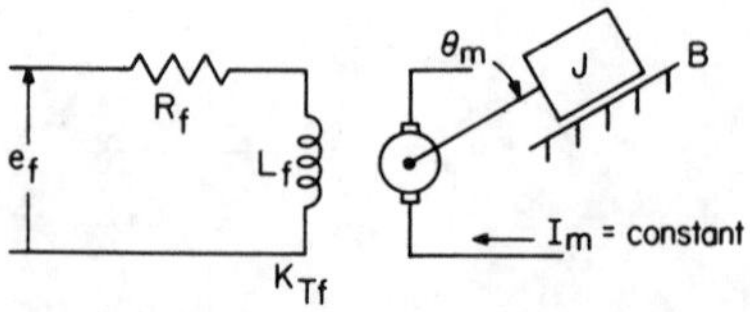

FIG. 2.23. A field driven dc servomotor.

amplitude applied to the second coil (the control coil) is used to control motor speed and torque. To a first-order approximation the developed torque is linearly related to current in the control coil and steady-state speed is linearly related to control coil applied voltage. Thus the same model we have used for the field-controlled dc motor is applicable as a first-order approximation to the ac motor as well, if we keep in mind that the voltages and currents referred to are effective ac rather than dc values. The linear approximation of ac servomotors is generally less accurate over the range of operation than that of equivalent dc units, but they are often used to take advantage of readily available ac power and the relatively simple amplifiers (they need not pass dc) that can be used to provide the necessary control input. Components used in ac servo systems are discussed further in Chap. 12.

2.9 LAGRANGE'S EQUATION AND THE ENERGY BALANCE METHOD FOR WRITING DIFFERENTIAL EQUATIONS

In many problems it is easier to write the equations describing system performance if energy balance relationships are used. In physics, for example, the velocity of a metal sphere after rolling down a frictionless inclined plane is much more difficult to obtain directly than if we equate the total values of initial and final energy. Energy balance relationships simplify the development of a mathematical system description considerably when differential equation relationships are involved and the system contains both electrical and mechanical components. Fortunately, Lagrange developed a general equation which we may use as a systematic approach to the solution of a wide range of physical systems. Development of a mathematical description using Lagrange's equation is generally straightforward irrespective of problem complexity.[7]

Lagrange's energy balance equation is

$$\frac{d}{dt}\left(\frac{\partial T}{\partial \dot{q}_n}\right) - \frac{\partial T}{\partial q_n} + \frac{\partial D}{\partial \dot{q}_n} + \frac{\partial V}{\partial q_n} = Q_n \tag{2.82}$$

where

T = total system kinetic energy
D = total system dissipation factor
V = total system potential energy
n = 1, 2, . . . defines the independent coordinates or degrees of freedom in the system
Q_n = generalized forcing function relative to coordinate n
q_n = generalized coordinate
$\dot{q}_n$ = generalized velocity.

The total kinetic energy T includes energy due to masses in motion in mechanical systems and energy due to current flow through coils in electrical systems. D is defined as one half the rate at which energy is dissipated as heat in the system. Heat is generated by friction in mechanical systems and by resistance in electrical systems. Potential energy V in mechanical systems is due to masses located above the reference plane and by spring tension or compression. In electrical systems, V is the total energy due to charge stored on capacitors. The forcing functions Q_n are forces and torques in mechanical systems and charge in electrical networks. Thus the $\dot{q}_n$ represent velocities in mechanical systems and currents in electrical systems. The various electrical and mechanical components and the corresponding energy terms are summarized in Table 2.5 as an aid in applying Lagrange's equation.

As an example illustrating the general procedure used in applying Lagrange's equation, consider the electromechanical configuration of the capacitor microphone shown in Fig. 2.24. The capacitor plate a is rigidly fastened to the frame. Sound waves pass through the mouthpiece exerting a force on plate b, which is connected to the frame by a spring-damping combination with parameters chosen to give desirable response characteristics over the audible frequency range. We know that $q = C_0 E$ defines the steady-state charge that would appear on the capacitor plates if the plates were rigid, where C_0 is the capacitance between the plates. Sound vibrations cause plate b to move, and q tends toward the corresponding new steady state value. Thus current flows back and forth through the electrical circuit as plate b moves. These current changes can be coupled inductively from L or observed across R to develop electrical signals linearly related to the audio sound waves. These electrical signals are generally manipulated in many ways before they are converted back to sound in a radio or television receiver, for example. Accurate mathematical description of this situation would seem to be rather complex, but use of Lagrange's equation greatly facilitates development of the describing differential equations.

The voltage E appears between the capacitor plates in the equilibrium state, causing a charge q_0 to appear upon them. Let C_0 denote the equilibrium value of C. The charge q_0 causes a force of attraction to appear between the two plates and stretches the spring an amount denoted by x_1. Let x_0 denote the equilibrium distance between the plates. Motion from the equilibrium point is denoted by x and is defined as positive when the plates move closer together. The change in charge on the plates is denoted by q, so the instantaneous charge is $q_0 + q$. The

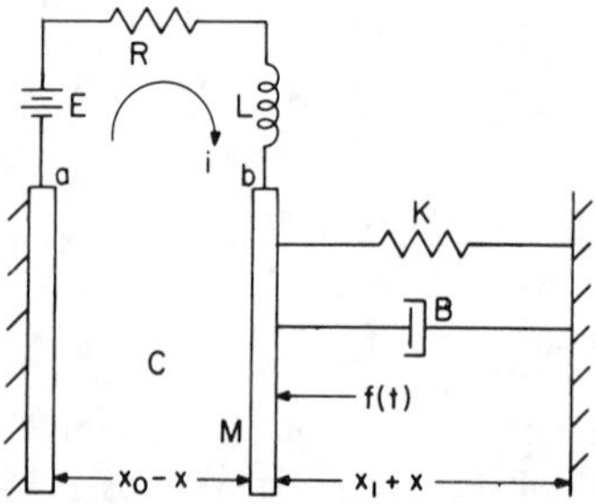

FIG. 2.24. The electromechanical system of the capacitor microphone.

Table 2.5. Summary of energy storage and dissipation factors for electrical and mechanical systems.

Electrical components (Loop analysis) q = charge $i = dq/dt = \dot{q}$ = current e = voltage = forcing function	Inductance, L	$T = \frac{1}{2} Li^2 = \frac{1}{2} L\dot{q}^2$
	Capacitance, C	$V = q^2/2C$
	Resistance, R	$D = \frac{1}{2} i^2 R = \frac{1}{2} \dot{q}^2 R$
	Mutual inductance, L_M	$T = \pm L_M i_1 i_2 = \pm L_M \dot{q}_1 \dot{q}_2$ *
*Use plus if both currents enter (or leave) the coils through the dotted terminals, otherwise use minus.		
Electrical Components (Node analysis) v = voltage drop $\int v\,dt = \phi$ = position coordinate $v = d\phi/dt = \dot{\phi}$ = velocity coordinate i = current = forcing function	Inductance, L	$V = \frac{1}{2L} \phi^2$
	Capacitance, C	$T = \frac{1}{2} Cv^2 = \frac{1}{2} C\dot{\phi}^2$
	Conductance, $G = \frac{1}{R}$	$D = \frac{1}{2} Gv^2 = \frac{1}{2} G\dot{\phi}^2$
Translational mechanical components x = displacement $v = dx/dt = \dot{x}$ = velocity f = force = forcing function	Mass, M	$T = \frac{1}{2} Mv^2 = \frac{1}{2} M\dot{x}^2$
	Elastance, K	$V = \frac{1}{2} K(x_1 - x_2)^2$ where $x_1 - x_2$ = spring displacement
	Damping, B	$D = \frac{1}{2} B(v_1 - v_2)^2$ where $v_1 - v_2$ = velocity of dashpot terminals
Rotational mechanical components θ = angular displacement $\omega = d\theta/dt = \dot{\theta}$ = angular velocity τ = torque = forcing function	Inertia, J	$T = \frac{1}{2} J\omega^2 = \frac{1}{2} J\dot{\theta}^2$
	Elastance, K	$V = \frac{1}{2} K(\theta_1 - \theta_2)^2$ where $\theta_1 - \theta_2$ = torsional displacement
	Damping, B	$D = \frac{1}{2} B(\omega_1 - \omega_2)^2$ where $\omega_1 - \omega_2$ = relative velocity of dashpot terminals.

instantaneous distance between plates is $x_0 - x$ and the instantaneous spring displacement is $x_1 + x$, as indicated in Fig. 2.24.

Assume the capacitor plates are sufficiently close together that the parallel plate approximation is valid, or

$$C = \frac{\epsilon A}{x_0 - x} \qquad C_0 = \frac{\epsilon A}{x_0} \tag{2.83}$$

where A is the plate area and ϵ is the dielectric constant for air. Also,

$$q_0 = C_0 E \tag{2.84}$$

Referring to Eq. (2.82), the energy terms needed for this system are

$$T = \tfrac{1}{2} L\dot{q}^2 + \tfrac{1}{2} M\dot{x}^2 \tag{2.85}$$

$$D = \tfrac{1}{2} R\dot{q}^2 + \tfrac{1}{2} B\dot{x}^2 \tag{2.86}$$

and, using Eq. (2.83) for C,

$$V = \frac{1}{2\epsilon A}(x_0 - x)(q_0 + q)^2 + \frac{1}{2} K(x_1 + x)^2 \tag{2.87}$$

The two degrees of freedom in this example are the motion x of capacitor plate b and current $i = \dot{q}$. Coupling between the electrical and mechanical circuits appears in the potential energy term associated with capacitor, C. Plate motion causes current to flow and changes in charge modify the force on plate b. Applying Eq. (2.82) for the electrical and mechanical coordinates provides

$$L\ddot{q} + R\dot{q} + \frac{1}{\epsilon A}(x_0 - x)(q_0 + q) = E \tag{2.88}$$

and

$$M\ddot{x} + B\dot{x} - \frac{1}{2\epsilon A}(q_0 + q)^2 + K(x_1 + x) = f(t) \tag{2.89}$$

as the two nonlinear differential equations describing the system. We may linearize these results, since we observe that x and q are small relative to the equilibrium values x_0 and q_0, and disregarding the x^2, q^2 and xq terms obtained in expanding Eqs. (2.88) and (2.89) results in little loss of accuracy. Thus we will use the approximations

$$(x_0 - x)(q_0 + q) \simeq x_0 q_0 + x_0 q - q_0 x \tag{2.90}$$

and

$$(q_0 + q)^2 \simeq q_0^2 + 2q_0 q \tag{2.91}$$

Substituting Eqs. (2.90) and (2.91) into Eqs. (2.88) and (2.89) provides

$$L\ddot{q} + R\dot{q} + \frac{1}{\epsilon A}(x_0 q_0 + x_0 q - q_0 x) = E \tag{2.92}$$

and

$$M\ddot{x} + B\dot{x} - \frac{1}{2\epsilon A}(q_0^2 + 2q_0 q) + K(x_1 + x) = f(t) \tag{2.93}$$

We may further simplify Eq. (2.92) using Eqs. (2.83) and (2.84) to eliminate the equilibrium conditions, thereby obtaining

$$L\ddot{q} + R\dot{q} + \frac{q}{C_0} - \frac{q_0 x}{\epsilon A} = 0 \tag{2.94}$$

Noting that

$$Kx_1 = \frac{q_0^2}{2\epsilon A} \tag{2.95}$$

is the equilibrium force balance between the capacitor plates and the spring K allows us to similarly reduce Eq. (2.93) to

$$M\ddot{x} + B\dot{x} + Kx - \frac{q_0 q}{\epsilon A} = f(t) \tag{2.96}$$

Equations (2.94) and (2.96) are the approximate linear equations describing how small changes in the capacitive microphone system are interrelated. It is clear from these results that $q_0/\epsilon A$ is the coefficient defining the linearized intercoupling between the electrical and mechanical parts of the microphone.

The general application of Lagrange's equation is illustrated by the procedure used in attacking the capacitor microphone problem. Linearization of the results is also accomplished in the standard way by making a first-order linear approximation of the changes taking place around an equilibrium condition. The principle advantage of Lagrange's equation is that it provides an orderly general procedure for developing the equations of systems in which energy appears in several forms.

A second form of electromechanical coupling often encountered is illustrated in Fig. 2.25. This could be used, for example, as an electro-magnetic microphone. In this case current flow through the coil produces a force on the mechanical system and motion of the mechanical system, in turn, induces an electromotive

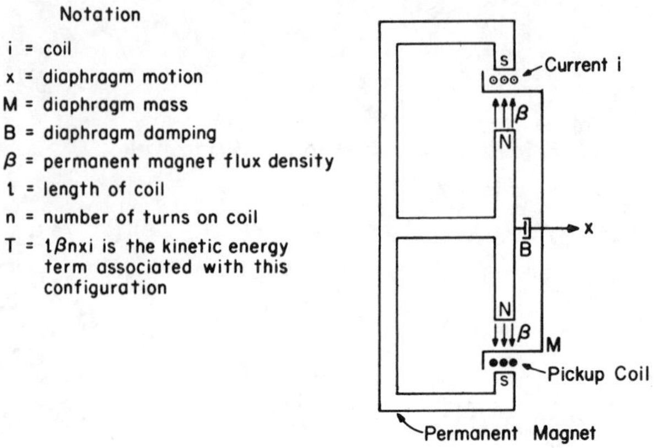

FIG. 2.25. The electromechanical configuration of a microphone using magnetic coupling.

force (EMF) in the coil. A kinetic energy term

$$T = l\beta n x i \tag{2.97}$$

is associated with this configuration and must be used in applying Lagrange's equation, where

l = coil length.
β = permanent magnet flux density linking the coil.
n = number of turns on the coil.
x = coil displacement.
i = coil current.

It is left as a student exercise (see Prob. 2.23) to develop the equations of a system of the form shown in Fig. 2.25.

2.10 SUMMARY AND CONCLUSIONS

Control system analysis and synthesis is always based upon mathematical models of the variety of phenomena encountered. The first step in any analysis is the development of one or more differential equations describing the performance of each basic system element. Subsequent steps often involve solution of these equations or their reduction to transfer function form using operator notation, the choice of one or more equalization or compensation networks to provide stable closed loop operation with the specified response characteristics, and hand analysis or simulation study to check such things as the unit step response pattern. We have considered the first in this chain of closely related steps in this chapter. The detailed approach to each of the subsequent steps is considered in the chapters which follow.

The basic concepts involved in writing the differential equations describing various types of electromechanical systems are reviewed here. The relatively standard differential, dot and $j\omega$ operator notations are used throughout our presentation. The detail presented should prove adequate for our present purposes, but is by no means exhaustive, and references are provided in several cases where additional detail relative to rather special phenomena can be obtained. Lagrange's equation is presented and often proves useful in providing an orderly procedure for writing differential equations describing complex electromechanical systems. We should be able, with the basic concepts reviewed here firmly in hand, to describe most systems encountered in practice by an adequate mathematical model.

Nothing has been said thus far about solving the equations we have developed. A knowledge of the standard classical solution methods is assumed. Application of the Laplace transform to the solution of linear time invariant differential equations is considered in Chap. 3.

REFERENCES

1. Gibson, J. E., and F. B. Tuteur, "Control System Components," McGraw-Hill Book Company, Inc., New York, pp. 315–333, 1958.
2. Loc. cit., pp. 309–311.
3. Loc. cit., pp. 363–480.
4. "Flow Meters: Their Theory and Application," ASME, New York, 1937.
5. Gibson, J. E., and F. B. Tuteur, "Control System Components," McGraw-Hill Book Company, Inc., New York, pp. 192–207, 1958.
6. Loc. cit., pp. 276–303.
7. Ogar, G. W., and J. J. D'Azzo, A Unified Procedure for Deriving the Differential Equations of Electrical and Mechanical Systems, *IRE Transactions on Education*, pp. 18–26, March, 1962.

3 LAPLACE TRANSFORM METHODS

Up to this point we have emphasized the development of mathematical models in the form of one or more integro-differential equations describing system performance. It is assumed that the readers can solve such equations using the classical methods discussed in basic differential equation and network analysis courses. Many of you are also no doubt familiar with the Laplace transform and its use in solving linear time invariant integro-differential equations. Since these methods provide the fundamental background for virtually all of the subsequent material presented, it is appropriate that the subject be discussed here. Because the material is so mathematically oriented, we revert in this chapter to the rather concise method of presentation which is standard in texts on applied mathematics.

Solution of differential equations by classical procedures rapidly increases in difficulty with equation order and with forcing-function complexity. When the Laplace transform is used, the integro-differential equations are transformed into algebraic equations including all boundary conditions. Simple algebraic manipulations of these equations provide the transformed solution including both the transient and the steady state components. The Laplace transform also allows us to define the transfer function or the unit step response of a system, and thus to consider its basic characteristics independent of the input.

3.1 THE LAPLACE TRANSFORM AND ITS CHARACTERISTICS

The *Laplace transform* of a general function $f(t), t > 0$, is denoted by $\mathcal{L}\{f(t)\}$ or by $F(s)$ and is defined as

$$\mathcal{L}\{f(t)\} = F(s) = \int_0^\infty f(t)\epsilon^{-st}dt \tag{3.1}$$

where s is a complex number of the form $s = \sigma + j\omega$. Not all functions $f(t)$ are transformable, since it may not be possible to obtain $F(s)$ from $f(t)$ using Eq. (3.1). If the integral given in Eq. (3.1) converges to a definite functional value, the Laplace transform of $f(t)$ exists. Sufficiency conditions which guarantee the existence of $F(s)$ are:

1. $f(t)$ is *piecewise continuous* over every finite interval $0 \leq t_1 \leq t \leq t_2$
2. $f(t)$ is of *exponential order.*

A function is *piecewise continuous* in a finite interval if the interval can be divided into a finite number of subintervals over each of which the function is continuous and bounded. A function $f(t)$ is said to be of *exponential order* α if there exists a real constant α such that $\lim_{t \to \infty} |\epsilon^{-\alpha t}f(t)| = 0$. In other words $f(t)$ must not approach ∞ faster than exponentially as $t \to \infty$. These conditions are satisfied by virtually all signals and functions of engineering interest.

In control engineering $f(t)$ and $F(s)$ are said to be defined in the time and frequency domains respectively, because t nearly always denotes time and the corresponding s parameter can be interpreted as a complex frequency variable. More generally, the t in Eq. (3.1) can be any variable and it is appropriate to refer to $f(t)$ and $F(s)$ as being defined in the t-domain and the s-domain, respectively. It should also be noted that the Laplace transform as defined by Eq. (3.1) depends only upon $f(t)$ for $t > 0$, and because of this it is referred to as the *one-sided Laplace transform.* Functions $f(t)$ specified for both positive and negative t can be treated using the *two-sided Laplace transform,* which is similar to the Fourier transform, but we have no immediate use for this more general tool, and it is not considered here.

Starting from the Laplace transform definition, we will develop some of its main characteristics using a sequence of theorems.

Theorem 3.1 Uniqueness

The Laplace transform of a transformable function $f(t)$ is unique.

Proof:

We assume the existence of the integral in Eq. (3.1) by stating that $f(t)$ is a transformable function. Thus the uniqueness property is satisfied by definition.

In the following theorems all functions introduced are assumed to be transformable. We shall denote the Laplace transform of a given function using capital letters. That is, $F(s)$, $G(s)$ and $H(s)$ denote the Laplace transforms of $f(t)$, $g(t)$ and $h(t)$, respectively.

Theorem 3.2 Linearity

The Laplace transform is a linear transformation and the Laplace operator $\mathfrak{L}$ is a linear operator. Thus if a and b are constants

$$\mathfrak{L}\{af(t) \pm bg(t)\} = a\mathfrak{L}\{f(t)\} + b\mathfrak{L}\{g(t)\} = aF(s) + bG(s) \tag{3.2}$$

The proof follows directly from the definition.

Corollary 3.1

The Laplace transform of a piecewise continuous function $f(t)$ with finite discontinuities is independent of the values of the function at the points of discontinuity.

Proof:

Let the points of discontinuity of $f(t)$ be at $t_1, t_2 \ldots$ with the corresponding specified values $f(t_1), f(t_2) \ldots$. This function can be written as the sum of two functions $g(t)$ and $h(t)$, where $g(t)$ has the same values as $f(t)$ wherever $f(t)$ is continuous and is zero at the points of discontinuity, and $h(t)$ has the values of $f(t)$ at the points of discontinuity and is zero elsewhere. From Theorem 3.1, the Laplace transform of $f(t)$ is given by

$$F(s) = \mathfrak{L}\{f(t)\} = \mathfrak{L}\{g(t) + h(t)\} = G(s) + H(s) \tag{3.3}$$

But $h(t)$ is zero except at a countable set of points, where it takes on finite values. Thus,

$$H(s) = \mathfrak{L}\{h(t)\} = \int_0^\infty h(t)\epsilon^{-st}dt = 0 \tag{3.4}$$

and $F(s) = G(s)$, which completes the proof.

From this corollary it can be seen that in deriving the Laplace transform of a function with finite discontinuities we need not consider its values at the discontinuities. Note that this does not apply when the function is not finite at the point of discontinuity.

Example 3.1 Laplace transform of the impulse function

The unit impulse function, also called the Dirac delta function, is defined by

$$\delta(t) = 0 \qquad t \neq 0 \qquad \int_{-\infty}^{\infty} \delta(t)\,dt = 1 \tag{3.5}$$

and has the property that

$$\int_{-\infty}^{\infty} \delta(t) \quad f(t) \quad dt = f(0) \tag{3.6}$$

Application of Eq. (3.1) provides

$$\mathcal{L}\{\delta(t)\} = \int_{0}^{\infty} \delta(t)\,\epsilon^{-st}\,dt = \epsilon^{-st}\Big|_{t=0} = 1 \tag{3.7}$$

Example 3.2 Laplace transform of the unit step function
The unit step function is defined as

$$u(t) = \begin{cases} 0\,, & t < 0 \\ 1\,, & t \geq 0 \end{cases} \tag{3.8}$$

Its Laplace transform as defined by Eq. (3.1) is

$$\mathcal{L}\{u(t)\} = \int_{0}^{\infty} \epsilon^{-st}\,dt = -\frac{\epsilon^{-st}}{s}\Big|_{0}^{\infty} = \frac{1}{s} \tag{3.9}$$

The convergence requirement of the integral is $\sigma > 0$. Note that any finite value can be assigned to $u(t)$ at $t = 0$ without changing its Laplace transform.

Example 3.3 Laplace transform of $\epsilon^{-\alpha t}$
The Laplace transform of $\epsilon^{-\alpha t}$ is

$$\mathcal{L}\{\epsilon^{-\alpha t}\} = \int_{0}^{\infty} \epsilon^{-\alpha t}\epsilon^{-st}\,dt = -\frac{\epsilon^{-(s+\alpha)t}}{s+\alpha}\Big|_{0}^{\infty} = \frac{1}{s+\alpha} \tag{3.10}$$

The convergence requirement of the integral is $\sigma > -\text{Re}\{\alpha\}$.

Example 3.4 Laplace transform of $\cos \omega t$

We can write $\cos \omega t$ in exponential form as

$$f(t) = \cos \omega t = \frac{\epsilon^{j\omega t} + \epsilon^{-j\omega t}}{2} \tag{3.11}$$

so,

$$F(s) = \mathcal{L}[\cos \omega t] = \int_0^\infty \frac{\epsilon^{j\omega t} + \epsilon^{-j\omega t}}{2} \epsilon^{-st}\, dt \tag{3.12}$$

which is easily evaluated term by term as in Ex. 3.3 to give

$$F(s) = \frac{1}{2}\left[\frac{\epsilon^{(j\omega - s)t}}{j\omega - s} - \frac{\epsilon^{-(j\omega + s)t}}{j\omega + s}\right]\Bigg|_0^\infty = \frac{1}{2}\left(\frac{-1}{j\omega - s} + \frac{1}{j\omega + s}\right)$$

$$= \frac{s}{s^2 + \omega^2} \tag{3.13}$$

The convergence requirement is $\sigma > 0$.

These examples illustrate the direct application of Eq. (3.1) to obtain Laplace transforms for functions of engineering interest. A short list of Laplace transform pairs for the most common functions is provided in Table 3.1. This list is readily extended using Eq. (3.1) and any of the theorems on Laplace transform characteristics. A much more complete list of Laplace transform pairs is provided in Appendix A.

Theorem 3.3 Time Displacement (Shifting Theorem)

Assume a transformable $f(t)$ with Laplace transform $F(s)$. Then for any positive real number a,

$$\mathcal{L}\{f(t - a)\, u(t - a)\} = \epsilon^{-as} F(s) \tag{3.14}$$

Proof:

$$\mathcal{L}\{f(t - a)\, u(t - a)\} = \int_0^\infty f(t - a)\, u(t - a)\, \epsilon^{-st}\, dt$$

Table 3.1. A short list of Laplace Transform pairs.

$f(t)$	$F(s)$
$\delta(t)$	1
$u(t)$	$\frac{1}{s}$
ϵ^{-at}	$\frac{1}{s+a}$
t^n	$\frac{n!}{s^{n+1}}$
$\cos \omega t$	$\frac{s}{s^2+\omega^2}$
$\sin \omega t$	$\frac{\omega}{s^2+\omega^2}$
$\cosh at$	$\frac{s}{s^2-a^2}$
$\sinh at$	$\frac{a}{s^2-a^2}$

$$= \int_a^\infty f(t-a)\,\epsilon^{-st}\,dt$$

$$= \int_0^\infty f(\tau)\,\epsilon^{-s(\tau+a)}\,d\tau \qquad (t-a=\tau)$$

$$= \epsilon^{-as}\int_0^\infty f(\tau)\,\epsilon^{-s\tau}\,d\tau$$

$$= \epsilon^{-as}F(s) \tag{3.15}$$

Thus multiplication by ϵ^{-as} in the s-domain corresponds to a delay a in the t-domain. This theorem often proves useful because time delay due to communication links and similar phenomena is common in control applications. The theorem is also useful in obtaining the Laplace transforms of composite functions made up of standard functions shifted in time.

Example 3.5

Find the Laplace transform of the function $f(t)$ shown in Fig. 3.1. This

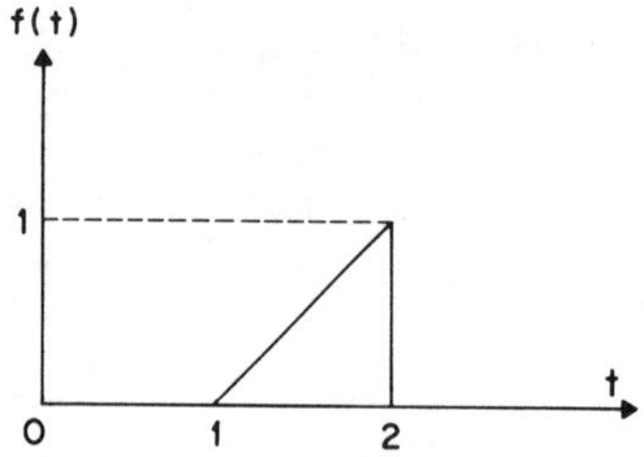

FIG. 3.1. Sawtooth signal used in Ex. 3.5.

function may be expressed as

$$\begin{aligned} f(t) &= (t-1)\,u(t-1) - (t-1)\,u(t-2) \\ &= (t-1)\,u(t-1) - [(t-2)+1]\,u(t-2) \end{aligned} \tag{3.16}$$

Application of Theorem 3.3 shows that

$$\begin{aligned} \mathcal{L}\{f(t)\} &= \frac{1}{s^2}\epsilon^{-s} - \left[\frac{1}{s^2} + \frac{1}{s}\right]\epsilon^{-2s} \\ &= \frac{1}{s^2}[\epsilon^{-s} - (s+1)\epsilon^{-2s}] = F(s) \end{aligned} \tag{3.17}$$

A direct result of the Time Displacement Theorem is the following corollary defining the Laplace transform of a periodic function.

Corollary 3.2 Laplace Transform of Periodic Functions

Let $f(t)$ be a periodic function with period T and Laplace transform $F(s)$. Then $F(s)$ equals $1/(1-\epsilon^{-Ts})$ times the Laplace transform of the first cycle of $f(t)$, or

$$F(s) = \mathcal{L}\{f(t)\} = \frac{\int_0^T f(t)\,\epsilon^{-st}\,dt}{1-\epsilon^{-Ts}} \tag{3.18}$$

The proof is left as an exercise. Hint: Evaluate $1/(1-\epsilon^{-Ts})$ by long division.

Theorem 3.4 *s*-Domain Displacement

Given an $f(t)$ with Laplace transform $F(s)$, for any constant α, real or complex,

$$\mathcal{L}\{\epsilon^{-\alpha t} f(t)\} = F(s+\alpha) \tag{3.19}$$

Proof:

$$\mathcal{L}\{\epsilon^{-\alpha t} f(t)\} = \int_0^\infty \epsilon^{-\alpha t} f(t)\,\epsilon^{-st}\,dt$$

$$= \int_0^\infty f(t)\,\epsilon^{-(s+\alpha)t}\,dt$$

$$= F(s+\alpha) \tag{3.20}$$

Theorem 3.5 Scaling

Given $f(t)$ with Laplace transform $F(s)$, for any constant a

$$\mathcal{L}\{f(at)\} = \frac{1}{a} F\left(\frac{s}{a}\right) \tag{3.21}$$

Proof:

$$\mathcal{L}\{f(at)\} = \int_0^\infty f(at)\,\epsilon^{-st}\,dt$$

$$= \int_0^\infty f(\tau)\,\epsilon^{-(s/a)\tau}\,\frac{1}{a}\,d\tau \qquad (at = \tau)$$

$$= \frac{1}{a} F\left(\frac{s}{a}\right) \tag{3.22}$$

Theorem 3.6 Parameter Differentiation

Given $f(a, t)$ with Laplace transform $F(a, s)$, then

$$\mathcal{L}\left\{\frac{d}{da} f(a,t)\right\} = \frac{d}{da} F(a,s) \tag{3.23}$$

The proof of Theorem 3.6 follows immediately upon interchanging the order of integration and differentiation. The conditions required for Laplace transformability are sufficient to justify interchanging the order of operations.

Example 3.6

Find the Laplace transform of $f(t) = t \sin at$. Note that

$$t \sin at = -\frac{d}{da}(\cos at) \tag{3.24}$$

From Table 3.1

$$\mathcal{L}\{\cos at\} = \frac{s}{s^2 + a^2} \tag{3.25}$$

Applying Theorem 3.6 yields

$$\mathcal{L}\{t \sin at\} = \mathcal{L}\left\{-\frac{d}{da}(\cos at)\right\}$$

$$= -\frac{d}{da}\left(\frac{s}{s^2 + a^2}\right)$$

$$= \frac{2as}{(s^2 + a^2)^2} = F(s) \tag{3.26}$$

With these theorems we can extend the list of Laplace transform pairs given in Table 3.1 to include all functions of interest. The process is more tedious than difficult. An extended table of Laplace transform pairs is included in Appendix A. Since our main purpose in using Laplace transforms is to solve integro-differential equations, we shall now consider the transforms of the derivative and the integral.

Theorem 3.7 Real Differentiation

Given $f(t)$, a continuous Laplace transformable function with a piecewise continuous derivative $\dot{f}(t)$ over every finite interval $0 \leq t_1 \leq t \leq t_2$, then

$$\mathcal{L}\{\dot{f}(t)\} = s\mathcal{L}\{f(t)\} - f(0^+)$$

$$= sF(s) - f(0^+) \tag{3.27}$$

where $f(0^+)$ represents the value of $f(t)$ as $t \to 0$ from the positive side. This allows a finite discontinuity in $f(t)$ at $t = 0$.

Proof:

$$\mathcal{L}\{\dot{f}(t)\} = \int_0^\infty \dot{f}(t)\,\epsilon^{-st}\,dt \tag{3.28}$$

Integrating Eq. (3.28) by parts with $u = \epsilon^{-st}$ and $dv = \dot{f}(t)dt$ provides

$$\mathcal{L}\{\dot{f}(t)\} = \epsilon^{-st} f(t)\Big|_0^\infty - \int_0^\infty f(t)(-s\,\epsilon^{-st})dt \tag{3.29}$$

Since $f(t)$ is of exponential order, $\epsilon^{-st}f(t)|_{t\to\infty} = 0$ for appropriately-chosen σ, so Eqs. (3.27) and (3.29) are equivalent.

Theorem 3.7 can be generalized to provide the Laplace transform of the nth derivative $f^{(n)}(t)$ assuming the $f^{(i)}(t)$, $0 \leq i \leq n$ are continuous Laplace transformable functions. The result is

$$\mathcal{L}[f^{(n)}(t)] = s^n F(s) - s^{n-1}f(0^+) - s^{n-2}\dot{f}(0^+) - \cdots - f^{(n-1)}(0^+) \tag{3.30}$$

where $f^{(i)}(0^+)$ denotes the ith derivative of $f(t)$ relative to t evaluated as $t \to 0$ from the positive side.

Theorem 3.8 Real Integration

Given $f(t)$ with Laplace transform $F(s)$,

$$\mathcal{L}\left\{\int_0^t f(t)dt\right\} = \frac{F(s)}{s} \tag{3.31}$$

Proof:

As in the proof of Theorem 3.7, we start from the definition of the Laplace transform, integrate by parts, and evaluate the limits, or

$$\begin{aligned}\mathcal{L}\left\{\int_0^t f(t)dt\right\} &= \int_0^\infty \left[\int_0^t f(t)dt\right]\epsilon^{-st}dt \\ &= \frac{\epsilon^{-st}}{s}\left[\int_0^t f(t)dt\right]\Bigg|_0^\infty + \int_0^\infty \frac{\epsilon^{-st}}{s}f(t)dt \\ &= -0 + 0 + \frac{1}{s}\int_0^\infty f(t)\epsilon^{-st}dt = \frac{1}{s}F(s)\end{aligned} \tag{3.32}$$

Theorem 3.8 can be generalized to include repeated integration, and

$$\underbrace{\int_0^t \cdots \int_0^t}_{n} f(t)dt = \frac{F(s)}{s^n} \tag{3.33}$$

Also, since the indefinite integral of a function can be written as

$$\int f(t)dt = \int_0^t f(t)dt + f^{(-1)}(0^+) \tag{3.34}$$

where $f^{(-1)}(0^+)$ is the value of the integral as $t \to 0$ from the positive side, by Theorems 3.2 and 3.8

$$\mathcal{L}\left[\int f(t)dt\right] = \mathcal{L}\left\{\int_0^t f(t)dt\right\} + \mathcal{L}\{f^{(-1)}(0^+)\}$$

$$= \frac{F(s)}{s} + \frac{f^{(-1)}(0^+)}{s} \tag{3.35}$$

The theorems on real differentiation and real integration are of significant general importance. They show that differentiation in the t-domain corresponds to multiplication by s in the s-domain, and integration in the t-domain corresponds to multiplication by $1/s$ in the s-domain. These results provide physical interpretations of the terms s^i or s^{-i} which are encountered in using the Laplace transform and they assure us that integro-differential equations in the t-domain transform into polynomial equations in the s-domain.

Theorem 3.9 Multiplication by t (Complex Differentiation)
Given $f(t)$ with Laplace transform $F(s)$

$$\mathcal{L}\{t f(t)\} = -\frac{d}{ds}[F(s)] \tag{3.36}$$

The proof involves only differentiation of the integral defining $F(s)$.

Thus multiplication by t in the t-domain is equivalent to differentiation with respect to s in the s-domain.

Theorem 3.10 Division by t (Complex Integration)
Given $f(t)$ with Laplace transform $F(s)$. If $f(t)/t$ possesses a limit as $t \to 0^+$,

$$\mathcal{L}\left\{\frac{f(t)}{t}\right\} = \int_s^\infty F(s)ds \tag{3.37}$$

Thus division by t in the t-domain corresponds to integration with respect tos in the s-domain.

The theorems relating to basic operations are summarized for easy reference in Table 3.2.

Table 3.2. Laplace transforms of the basic operations.

$g(t)$	$G(s)$
$f(t-a)\,u(t-a)$	$\epsilon^{-as}F(s)$
$\epsilon^{-\alpha t}f(t)$	$F(s+\alpha)$
$f(at)$	$\frac{1}{a}F\left(\frac{s}{a}\right)$
$\frac{d}{da}f(a,t)$	$\frac{d}{da}F(a,s)$
$\frac{d}{dt}f(t)$	$sF(s)-f(0^+)$
$\int_0^t f(t)dt$	$\frac{F(s)}{s}$
$\int f(t)dt$	$\frac{F(s)}{s}+\frac{f^{(-1)}(0^+)}{s}$
$tf(t)$	$-\frac{d}{ds}F(s)$
$\frac{1}{t}f(t)$	$\int_s^\infty F(s)ds$

3.2 THE INVERSE LAPLACE TRANSFORM AND ITS EVALUATION

The inverse Laplace transform of $F(s)$, denoted by $\mathcal{L}^{-1}\{F(s)\} = f(t)$ is defined as

$$\mathcal{L}^{-1}[F(s)] = f(t) = \frac{1}{2\pi j}\int_{\delta - j\infty}^{\delta + j\infty} F(s)\epsilon^{st}ds \tag{3.38}$$

The integration is to be executed along a line $\sigma = \delta$ in the complex plane parallel to the imaginary axis and is to be so displaced that all singularities of $F(s)$ lie to the left of this line. δ must be chosen such that $\delta > \alpha$ where α is the exponential order of $f(t)$. By using the term inverse Laplace transform, we imply that $f(t)$ as obtained by the above integral is a unique function corresponding to $F(s)$. This is true for all Laplace transformable functions. The value obtained for $f(t)$ at finite discontinuities using Eq. (3.38) is $\frac{1}{2}[f(t+) - f(t-)]$, where $t+$ and $t-$ indicate the right and left hand limits. Thus we can treat the Laplace transform and the inverse Laplace transform as a pair of mutually inverse operators. We will speak in general about the Laplace transform pair $f(t)$ and $F(s)$.

As seen from Eq. (3.38), the inverse Laplace transform is defined as the integral of a complex variable over a complex set of limits. Evaluation of this integral directly requires a background in complex function theory, and the desired result is obtained by applying residue theory. We will not use the direct approach here, but will develop a method of inverse transformation based upon the uniqueness of Laplace transform pairs. Our objective is to decompose all functions of s into a summation of simpler functions, each with a known inverse, after which these components are added to obtain the desired function of t. This procedure is based upon inverse Laplace transform linearity.

Theorem 3.11 Linearity

The inverse Laplace transform is a linear transformation and the inverse operator $\mathcal{L}^{-1}$ is a linear operator. Equivalently, if a and b are constants

$$\mathcal{L}^{-1}\{aF(s) \pm bG(s)\} = a\mathcal{L}^{-1}\{F(s)\} \pm b\mathcal{L}^{-1}\{G(s)\} = af(t) \pm bg(t) \tag{3.39}$$

The proof follows directly from the definition of the inverse Laplace transform.

The inversion method presented here assumes functions of s of the form

$$F(s) = \frac{P(s)}{Q(s)} = \frac{a_n s^n + a_{n-1}s^{n-1} + \cdots + a_1 s + a_0}{s^n + b_{n-1}s^{n-1} + \cdots + b_1 s + b_0} \tag{3.40}$$

where the a_i and b_j are real constants. The order of the numerator polynomial must not exceed that of the denominator. The structure of $F(s)$ as given by Eq. (3.40) is sufficiently general to include all problems of interest. Thus we need only to develop a general method for inverting ratios of polynomials in s of the form given in Eq. (3.40).

Inversion of $F(s)$ is accomplished by expanding Eq. (3.40) into partial fractions. Thus, as a first step we must find the n roots of $Q(s)$ and rewrite $F(s)$ as

$$F(s) = \frac{P(s)}{Q(s)} = \frac{P(s)}{(s - s_1)(s - s_2) \cdots (s - s_{n-1})(s - s_n)} \tag{3.41}$$

The roots s_i may be real or complex and the points defined by $s = s_i$ are called the poles of $F(s)$. Since the b_j of $Q(s)$ are real constants, all complex roots must occur in conjugate pairs. We assume hereafter that the roots of $Q(s)$ are known. If this is not the case, they can be obtained as outlined in Appendix B. The n roots of $Q(s)$ may not all be distinct and the general procedure used when there are repeated roots is sufficiently different that we will consider two cases.

Case 3.1 All roots of $Q(s)$ are distinct

When no two of the roots $s_1, s_2, \ldots, s_{n-1}, s_n$ are equal, the rational function $F(s)$ can be expanded as follows

$$F(s) = \frac{P(s)}{Q(s)} = A_0 + \frac{A_1}{s - s_1} + \cdots + \frac{A_k}{s - s_k} + \cdots + \frac{A_n}{s - s_n} \tag{3.42}$$

where the A_i are real or complex numbers, called the residues of $F(s)$, according to whether s_i is real or complex. Since each term on the right side of Eq. (3.42) has known inverse Laplace transform, determination of $\mathcal{L}^{-1}\{F(s)\}$ reduces to evaluating the A_i.

To determine $A_k (k \neq 0)$, multiply both sides of Eq. (3.42) by $(s - s_k)$ and take the limit of each side as $s \to s_k$, which yields

$$\lim_{s \to s_k} \left[\frac{(s - s_k) P(s)}{Q(s)}\right] = \lim_{s \to s_k} \left[A_0(s - s_k) + \cdots + A_k + \cdots + \frac{A_n(s - s_k)}{s - s_n}\right] \tag{3.43}$$

Since $s_i \neq s_j$ for $i \neq j$, all terms on the right side of Eq. (3.43) except A_k

approach 0 as $s \to s_k$. Thus

$$A_k = \lim_{s \to s_k} \left[\frac{(s - s_k) P(s)}{Q(s)} \right] = \left[\frac{(s - s_k) P(s)}{Q(s)} \right] \Bigg|_{s = s_k} \qquad k = 1, 2, \ldots, n \tag{3.44}$$

Note that no limiting process need be carried out to obtain A_k, since the $(s - s_k)$ term is cancelled by the same factor in $Q(s)$. Observe that by taking the limit of both sides of Eq. (3.42) as $s \to \infty$, all terms on the right side except A_0 approach 0. Thus

$$A_0 = \lim_{s \to \infty} \left[\frac{P(s)}{Q(s)} \right] = a_n \tag{3.45}$$

Note that the term A_0 is only necessary when $P(s)$ and $Q(s)$ are of the same order. The inverse Laplace transform of Eq. (3.42) is thus

$$\mathcal{L}^{-1}\{F(s)\} = a_n \delta(t) + \sum_{k=1}^{n} \left[\frac{(s - s_k) P(s)}{Q(s)} \right] \Bigg|_{s = s_k} \epsilon^{s_k t} \tag{3.46}$$

Case 3.2 $Q(s)$ has repeated roots

Assume that $Q(s)$ has an α-th order root at $s = s_1$, with $n - \alpha$ additional simple roots. In this case $F(s)$ can be written as

$$F(s) = \frac{P(s)}{Q(s)} = A_0 + \frac{A_{1\alpha}}{(s - s_1)^{\alpha}} + \frac{A_{1(\alpha - 1)}}{(s - s_1)^{\alpha - 1}} + \cdots + \frac{A_{11}}{s - s_1} + \frac{A_2}{s - s_2} + \cdots + \frac{A_{n-\alpha}}{s - s_{n-\alpha}} \tag{3.47}$$

The constants $A_0, A_2, \ldots, A_{n-\alpha}$ are obtained as in Case 3.1 using Eqs. (3.44) and (3.45). To obtain the other constants, we first multiply both sides of Eq. (3.47) by $(s - s_1)^{\alpha}$ to obtain

$$(s - s_1)^{\alpha} \frac{P(s)}{Q(s)} = A_0(s - s_1)^{\alpha} + A_{1\alpha} + A_{1(\alpha-1)}(s - s_1) + \cdots + A_{11}(s - s_1)^{\alpha - 1} + \frac{A_2(s - s_1)^{\alpha}}{s - s_2} + \cdots + \frac{A_{n-\alpha}(s - s_1)^{\alpha}}{s - s_{n-\alpha}} \tag{3.48}$$

Letting $s \to s_1$ we see that

$$A_{1\alpha} = \lim_{s \to s_1} \frac{(s - s_1)^\alpha P(s)}{Q(s)} = \left[\frac{(s - s_1)^\alpha P(s)}{Q(s)}\right]\Bigg|_{s=s_1} \tag{3.49}$$

Differentiating both sides of Eq. (3.49) with respect to s gives

$$\begin{aligned}\frac{d}{ds}\left[(s - s)^\alpha \frac{P(s)}{Q(s)}\right] &= \alpha A_0(s - s_1)^{\alpha-1} + A_{1(\alpha-1)} + 2A_{1(\alpha-2)}(s - s_1) + \\ &\quad \cdots + (\alpha - 1) A_{11}(s - s_1)^{\alpha-2} \\ &\quad + \frac{d}{ds}\left[\frac{A_2(s - s_1)^\alpha}{s - s_2} + \cdots + \frac{A_{n-\alpha}(s - s_1)^\alpha}{s - s_{n-\alpha}}\right]\end{aligned} \tag{3.50}$$

Taking the limit of both sides of Eq. (3.50) as $s \to s_1$ yields

$$A_{1(\alpha-1)} = \lim_{s \to s_1}\left[\frac{d}{ds}(s - s_1)^\alpha \frac{P(s)}{Q(s)}\right] \tag{3.51}$$

The differentiation and limiting processes may be repeated $\alpha - 1$ times to obtain $A_{1(\alpha-2)}, A_{1(\alpha-3)}, \ldots, A_{11}$. The resulting general formula is

$$A_{1(\alpha-k)} = \lim_{s \to s_1}\left\{\frac{1}{k!}\frac{d^k}{ds^k}\left[(s - s_1)^\alpha \frac{P(s)}{Q(s)}\right]\right\} \qquad k = 1, 2, \ldots, \alpha - 1 \tag{3.52}$$

The inverse Laplace transform of elements involving repeated roots is obtained by using Theorem 3.4 and the third Laplace transform pair in Table 3.1, or

$$\mathcal{L}^{-1}\left[\frac{A_{li}}{(s - s_l)^i}\right] = \epsilon^{s_i t}\mathcal{L}^{-1}\left(\frac{A_{li}}{s^i}\right) = \frac{A_{li}}{(i - 1)!}\, t^{i-1}\epsilon^{s_i t} \tag{3.53}$$

When more than one multiple root exists, the corresponding constants A_{ij} are obtained by direct application of Eqs. (3.49) and (3.52). It should be noted that if $Q(s)$ has a complex multiple root it must also contain the complex conjugate root of the same multiplicity.

Example 3.7

Find the inverse Laplace transform of

$$F(s) = \frac{P(s)}{Q(s)} = \frac{20(s+1)(s+3)}{(s+1+j)(s+1-j)(s+2)(s+4)} \tag{3.54}$$

In this example the order of $Q(s)$ exceeds that of $P(s)$ thus $A_0 = 0$ and $F(s)$ can be decomposed into

$$F(s) = \frac{A_1}{(s+1+j)} + \frac{A_2}{(s+1-j)} + \frac{A_3}{(s+2)} + \frac{A_4}{(s+4)} \tag{3.55}$$

where

$$\begin{aligned} A_1 &= \left[\frac{(s+1+j)P(s)}{Q(s)}\right]\Bigg|_{s=-1-j} = \frac{20(-j)(2-j)}{(-2j)(1-j)(3-j)} \\ &= \frac{20(2-j)}{2(2-4j)} = 5\,\frac{(2-j)(1+2j)}{(1-2j)(1+2j)} = 4+3j \end{aligned} \tag{3.56}$$

$$A_2 = \left[\frac{(s+1-j)P(s)}{Q(s)}\right]\Bigg|_{s=-1+j} = \frac{20(j)(2+j)}{(2j)(1+j)(3+j)} = 4-3j \tag{3.57}$$

$$A_3 = \left[\frac{(s+2)P(s)}{Q(s)}\right]\Bigg|_{s=-2} = \frac{20(-1)(1)}{(-1+j)(-1-j)(2)} = -5 \tag{3.58}$$

$$A_4 = \left[\frac{(s+4)P(s)}{Q(s)}\right]\Bigg|_{s=-4} = \frac{20(-3)(-1)}{(-3+j)(-3-j)(-2)} = -3 \tag{3.59}$$

The corresponding time function is thus

$$\begin{aligned} f(t) &= (4+3j)\epsilon^{(-1-j)t} + (4-3j)\epsilon^{(-1+j)t} - 5\epsilon^{-2t} - 3\epsilon^{-4t} \\ &= \epsilon^{-t}[4(\epsilon^{-jt} + \epsilon^{jt}) + 3j(\epsilon^{-jt} - \epsilon^{jt})] - 5\epsilon^{-2t} - 3\epsilon^{-4t} \\ &= \epsilon^{-t}[8\cos t + 6\sin t] - 5\epsilon^{-2t} - 3\epsilon^{-4t} \end{aligned} \tag{3.60}$$

Note that the complex conjugate root residues in Ex. 3.7 are also complex conjugate numbers. This is always true when the coefficients of $P(s)$ and $Q(s)$ are real numbers, as we have assumed. Thus if $Q(s)$ has a pair of complex conjugate roots we only need to evaluate the residue corresponding to one root and the other residue is its conjugate. It is also clear from Ex. 3.7 that, even though $f(t)$ is a real function of t, it includes complex parts if $Q(s)$ has complex roots. Obviously the existence of complex pole pairs adds algebraic difficulty to the inverse transformation problem. Since the residues associated with conjugate pole pairs are conjugate, such a pair can be replaced by a single element in the partial fraction expansion, leading to considerable simplification. Such a term is of the form

$$\frac{\gamma + j\delta}{s + \alpha + j\beta} + \frac{\gamma - j\delta}{s + \alpha - j\beta} = \frac{(\gamma + j\delta)(s + \alpha - j\beta) + (\gamma - j\delta)(s + \alpha + j\beta)}{(s + \alpha)^2 - (j\beta)^2} = \frac{2\gamma(s + \alpha) + 2\delta\beta}{(s + \alpha)^2 + \beta^2} \tag{3.61}$$

With the inverse Laplace transform

$$\mathcal{L}^{-1}\left[\frac{2\gamma(s + \alpha) + 2\delta\beta}{(s + \alpha)^2 + \beta^2}\right] = \epsilon^{-\alpha t}\mathcal{L}^{-1}\left[\frac{2\gamma s + 2\delta\beta}{s^2 + \beta^2}\right] = 2\epsilon^{-\alpha t}[\gamma \cos\beta t + \delta \sin\beta t] \tag{3.62}$$

In carrying out the partial fraction expansion we can assume such a general second order term and solve for its constants. The following example illustrates the procedure.

Example 3.8

Find the inverse Laplace transform of the function used in Ex. 3.7 by combining the complex conjugate terms.

Using the values of the real residues as previously computed, $F(s)$ can be written as

$$F(s) = \frac{20(s + 1)(s + 3)}{[(s + 1)^2 + 1](s + 2)(s + 4)} = \frac{B_1(s + 1) + B_2}{(s + 1)^2 + 1} + \frac{-5}{s + 2} + \frac{-3}{s + 4} \tag{3.63}$$

That is

$$\frac{B_1(s + 1) + B_2}{(s + 1)^2 + 1} = \frac{20(s + 1)(s + 3)}{[(s + 1)^2 + 1](s + 2)(s + 4)} - \left[\frac{-5}{s + 2} + \frac{-3}{s + 4}\right]$$

[*Eq. (3.64) continued*]

$$= \frac{20(s^2 + 4s + 3)}{[(s+1)^2 + 1](s+2)(s+4)} + \frac{8s + 26}{(s+2)(s+4)}$$

$$= \frac{8s^3 + 62s^2 + 148s + 112}{[(s+1)^2 + 1](s+2)(s+4)}$$

$$= \frac{8s + 14}{(s+1)^2 + 1} = \frac{8(s+1) + 6}{(s+1)^2 + 1} \tag{3.64}$$

The inverse Laplace transform of this function now can be obtained in one step, and it checks our previous result.

Example 3.9

Find the inverse Laplace transform of

$$F(s) = \frac{P(s)}{Q(s)} = \frac{8(s^2 + 4s + 2)}{(s+1)^3(s+3)} \tag{3.65}$$

From Eq. (3.47) we may write

$$F(s) = \frac{P(s)}{Q(s)} = \frac{A_{13}}{(s+1)^3} + \frac{A_{12}}{(s+1)^2} + \frac{A_{11}}{s+1} + \frac{A_2}{s+3} \tag{3.66}$$

where

$$A_2 = \left[\frac{(s - s_2)P(s)}{Q(s)}\right]_{s=s_2} = \left[\frac{8(s^2 + 4s + 2)}{s+1}\right]_{s=-3} = 1 \tag{3.67}$$

$$A_{13} = \left[\frac{(s - s_1)^3 P(s)}{Q(s)}\right]_{s=s_1} = \left[\frac{8(s^2 + 4s + 2)}{s+3}\right]_{s=-1} = -4 \tag{3.68}$$

$$A_{12} = \frac{1}{1!}\left\{\frac{d}{ds}\left[\frac{(s - s_1)^3 P(s)}{Q(s)}\right]\right\}_{s=s_1} = \frac{d}{ds}\left[\frac{8(s^2 + 4s + 2)}{s+3}\right]\Bigg|_{s=-1}$$

$$= 8\,\frac{s^2 + 6s + 10}{s+3}\Bigg|_{s=-1} = 10 \tag{3.69}$$

$$A_{11} = \frac{1}{2!}\left\{\frac{d^2}{ds^2}\left[\frac{(s - s_1)^3 P(s)}{Q(s)}\right]\right\}\Bigg|_{s=s_1}$$

$$= \frac{1}{2}\frac{d^2}{ds^2}\left[\frac{8(s^2 + 6s + 10)}{s+3}\right]\Bigg|_{s=-1} = \left[4\,\frac{-2}{(s+3)^3}\right]_{s=-1} = -1 \tag{3.70}$$

The corresponding time function is thus

$$f(t) = \frac{-4}{(3-1)!} t^{(3-1)} \epsilon^{-t} + \frac{10}{(2-1)!} t^{(2-1)} \epsilon^{-t} - \epsilon^{-t} + \epsilon^{-3t}$$

$$= \epsilon^{-t}(-2t^2 + 10t - 1) + \epsilon^{-3t} \tag{3.71}$$

Sometimes it is more convenient to compute the A_{ij} without using differentiation as required in Eq. (3.52). An alternative method is to compute first only those constants that do not require differentiation and subtract the corresponding terms from $F(s)$. For the case under consideration this yields

$$\frac{A_{11}}{s+1} + \frac{A_{12}}{(s+1)^2} = \frac{8(s^2+4s+2)}{(s+1)^3(s+3)} - \frac{-4}{(s+1)^3} - \frac{1}{s+3}$$

$$= \frac{-s^3 + 5s^2 + 33s + 27}{(s+1)^3(s+3)}$$

$$= \frac{-s+9}{(s+1)^2} = \frac{-1}{s+1} + \frac{10}{(s+1)^2} \tag{3.72}$$

These examples illustrate the procedure for obtaining the inverse Laplace transforms of general rational functions. The process is more tedious than difficult, and a table of Laplace transform pairs often encountered is included in App. A, so it is often not necessary to carry out the detailed steps of the solution.

Theorem 3.12 The Initial Value Theorem

Assume $f(t)$ has Laplace transform $F(s)$, and that $\dot{f}(t)$ is Laplace transformable. The initial value of $f(t)$ is

$$f(0^+) = \lim_{t \to 0^+} f(t) = \lim_{s \to \infty} sF(s) \tag{3.73}$$

Proof:

From Theorem 3.7 we have

$$\mathfrak{L}[\dot{f}(t)] = \int_0^\infty \dot{f}(t)\,\epsilon^{-st} dt = sF(s) - f(0^+) \tag{3.74}$$

Taking the limit of both sides of Eq. (3.74) as $s \to \infty$ provides

$$\lim_{s \to \infty} \int_0^\infty \dot{f}(t)\,\epsilon^{-st} dt = \lim_{s \to \infty} [sF(s) - f(0^+)] \tag{3.75}$$

Since $\dot{f}(t)$ is Laplace transformable, the integration and limiting processes may be interchanged to provide

$$\int_0^\infty \lim_{s\to\infty} [\dot{f}(t)\epsilon^{-st}]dt = 0 = \lim_{s\to\infty} [sF(s) - f(0^+)] \tag{3.76}$$

which is what we set out to show.

Theorem 3.13 The Final Value Theorem

Let $f(t)$ have Laplace transform $F(s)$, and assume $\dot{f}(t)$ is Laplace transformable. If $F(s)$ has only poles with negative real parts, with the possible exception of a single pole at $s = 0$, the final value of $f(t)$ is given by

$$\lim_{t\to\infty} f(t) = \lim_{s\to 0} sF(s) \tag{3.77}$$

Proof:

From Theorem 3.7 we have

$$\mathcal{L}[\dot{f}(t)] = \int_0^\infty \dot{f}(t)\epsilon^{-st}\,dt = sF(s) - f(0^+) \tag{3.78}$$

Taking the limit of the integral in Eq. (3.78) as $s \to 0$ and interchanging the order of operations provides

$$\lim_{s\to 0} \int_0^\infty \dot{f}(t)\epsilon^{-st}\,dt = \int_0^\infty \dot{f}(t)\,dt = \lim_{t\to\infty} \int_0^t \dot{f}(t)\,dt = \lim_{t\to\infty} f(t) - f(0^+) \tag{3.79}$$

We must assume here that the $\lim_{t\to\infty} f(t)$ exists, otherwise these steps have no meaning. This is why the condition on the poles of $F(s)$ is included. Applying the same limit to the other side of Eq. (3.78) yields

$$\lim_{t\to\infty} f(t) - f(0^+) = \lim_{s\to\infty} sF(s) - f(0^+) \tag{3.80}$$

which completes the proof.

Example 3.10

Find the initial and final values of the $f(t)$ corresponding to

$$F(s) = \frac{2}{s(s^2 + 2s + 2)} \tag{3.81}$$

Application of Eqs. (3.73) and (3.77) yields

$$\lim_{t \to 0^+} f(t) = \lim_{s \to \infty} sF(s) = \lim_{s \to \infty} \frac{2}{s^2 + 2s + 2} = 0 \tag{3.82}$$

$$\lim_{t \to \infty} f(t) = \lim_{s \to 0} sF(s) = \lim_{s \to 0} \frac{2}{s^2 + 2s + 2} = 1 \tag{3.83}$$

3.3 USE OF THE LAPLACE TRANSFORM TO SOLVE GENERAL INTEGRO-DIFFERENTIAL EQUATIONS

Solution of general integro-differential equations using the Laplace transform is straightforward. Each term of the original equation is replaced by its equivalent term in the s-domain, the results are manipulated algebraically to solve for the desired response signal, and the inverse transform is obtained either by direct evaluation or through reference to tabulated results. It is important to recall from Sec. 3.1 that

$$\mathcal{L}[f^n(t)] = s^n F(s) - s^{n-1} f(0^+) - \cdots - f^{n-1}(0^+) \tag{3.84}$$

and

$$\mathcal{L}\left[\int_0^t f(t)\,dt\right] = \frac{F(s)}{s} + \frac{f^{-1}(0^+)}{s} \tag{3.85}$$

Although higher-order integrals are occasionally encountered, it is usually best to differentiate the equations in which they appear before carrying out the Laplace transform operation. The results presented in Eqs. (3.84) and (3.85) will be adequate for our purposes here.

Consider the system described by

$$\ddot{x} + 2\dot{x} + 5x = u(t) \tag{3.86}$$

Laplace transforming Eq. (3.86) term by term using Eq. (3.84) and inserting the initial conditions as they arise provides

$$s^2 X(s) - sx(0^+) - \dot{x}(0^+) + 2[sX(s) - x(0^+)] + 5X(s) = \frac{1}{s} \tag{3.87}$$

Solving Eq. (3.87) for $X(s)$ yields

$$X(s) = \frac{s^2 x(0^+) + s[2x(0^+) + \dot{x}(0^+)] + 1}{s(s^2 + 2s + 5)} \tag{3.88}$$

Let $x(0^+) = -1$ and $\dot{x}(0^+) = 1$. Substituting these values into Eq. (3.88) and rearranging provides

$$X(s) = -\frac{s^2 + s - 1}{s[(s+1)^2 + 2^2]} \tag{3.89}$$

which corresponds to entry 28 in App. A. The resulting time response is

$$x(t) = 0.2 - 1.205\epsilon^{-t} \sin\left(2t + \frac{85.2^\circ \pi}{180^\circ}\right) \tag{3.90}$$

This answer may also be obtained by breaking Eq. (3.89) up into partial fraction components and evaluating the result term by term, which involves a considerable amount of algebra.

We see from the simple problem given in Eq. (3.86) that the solution of general differential equations is straightforward. In fact two or more simultaneous differential equations are solved similarly by reducing them to algebraic equations through application of the Laplace transform, manipulation of the equations to solve for the desired variables, and inverse Laplace transformation of the results to obtain the corresponding time responses.

Example 3.11

Consider the field controlled motor shown in Fig. 3.2, which is described by

$$v(t) = Ri + L\frac{di}{dt} \tag{3.91}$$

$$T = K_T i = B\frac{d\theta}{dt} + J\frac{d^2\theta}{dt^2} \tag{3.92}$$

Find $\theta(s)$ in terms of $V(s)$, assuming all initial conditions are zero.

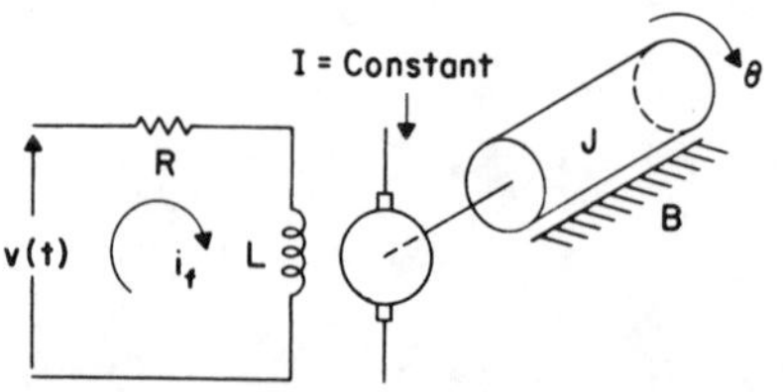

FIG. 3.2. A field-controlled motor.

Laplace transforming both sides of Eqs. (3.91) and (3.92) yields

$$V(s) = RI(s) + LsI(s) \tag{3.93}$$

$$K_T I(s) = Bs\theta(s) + Js^2\theta(s) \tag{3.94}$$

Solving for $I(s)$ provides

$$I(s) = \frac{V(s)}{R + Ls} \tag{3.95}$$

Substituting this into Eq. (3.94) and simplifying gives

$$\theta(s) = \frac{K_T V(s)}{s(B + Js)(R + Ls)} \qquad \frac{\theta(s)}{V(s)} = \frac{K_T}{s(B + Js)(R + Ls)}$$

as the desired result.

If $V(s)$ is specified, $\theta(t)$ is obtained by inverse transforming $\theta(s)$ as given by Eq. (3.95). The general function $\theta(s)/V(s)$ is called the s-domain transfer function relating response $\theta(s)$ to input $V(s)$. The transfer function of a system or network is of particular significance as an analysis and synthesis tool and this concept is considered further after several fundamental principles which will prove useful in our discussion are presented.

3.4 THE IMPULSE FUNCTION, THE CONVOLUTION INTEGRAL, AND TRANSFER FUNCTION ANALYSIS

The unit impulse function occurring at $t = \tau$, denoted by $\delta(t - \tau)$, is defined by

$$\delta(t - \tau) = 0 \quad t \neq \tau \qquad \int_{-\infty}^{\infty} \delta(t - \tau)dt = 1 \tag{3.96}$$

Thus $\delta(t - \tau)$ exists only at $t = \tau$ and the area under $\delta(t - \tau)$ is one. Obviously $\delta(t - \tau)$ takes on infinite value at $t = \tau$. One other useful property of $\delta(t - \tau)$ is

$$\int_{-\infty}^{\infty} \delta(t - \tau)f(t)dt = f(\tau) \tag{3.97}$$

assuming $f(t)$ is single-valued and continuous in an ϵ neighborhood of the point $t = \tau$.

Although impulses do not actually occur in a physical sense, many natural phenomena are near-impulsive in nature. Collisions between space vehicles and micrometerorites provide an interesting example in which the duration of impact, although finite, is negligibly short. The real value of the impulse concept, however, is as a tool for use in system analysis and synthesis. It is often convenient to represent an arbitrary signal by a weighted sequence of near equivalent impulses in order to evaluate system response to the given signal. This explains our interest in the convolution or superposition integral and the use of the impulse response function as a fundamental system characteristic.

We may approximate $\delta(t)$ by the narrow rectangular pulse of unit area shown in Fig. 3.3, which is described by

$$f(t) = \frac{u(t) - u(t - a)}{a} \tag{3.98}$$

The corresponding Laplace transform is

$$F(s) = \frac{1 - \epsilon^{-as}}{as} \tag{3.99}$$

The Laplace transform of $\delta(t)$ is obtained from Eq. (3.98) by taking $\lim_{a \to 0} F(s)$, or

$$\mathcal{L}[\delta(t)] = \lim_{a \to 0} \frac{1 - \epsilon^{-as}}{as} = \lim_{a \to 0} \frac{s\epsilon^{-as}}{s} = 1 \tag{3.100}$$

where L'Hospital's rule is applied to the original indeterminate form. The importance of this result cannot be overemphasized. It is the basis for general transfer function analysis. Suppose we find, in a particular problem, that

$$\frac{C(s)}{R(s)} = \frac{s + 1}{(s + 2)(s + 3)} \qquad C(s) = R(s)\frac{s + 1}{(s + 2)(s + 3)} \tag{3.101}$$

If $r(t) = \delta(t)$, $R(s) = 1$, and we see that the transfer function $C(s)/R(s)$ is also the unit impulse response of the system expressed in the s-domain. This helps to justify our interest in the impulse response function as a standard test characteristic.

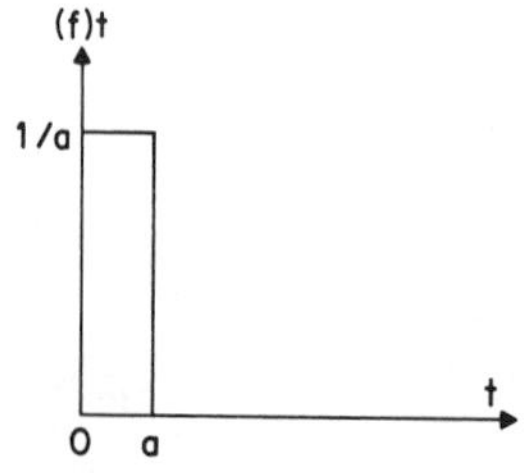

FIG. 3.3. Rectangular pulse approximation of the unit impulse.

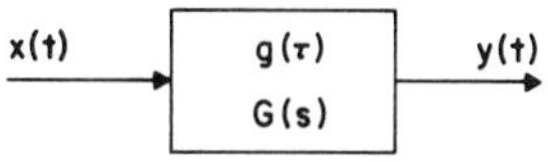

FIG. 3.4. A general system.

Consider the system shown in Fig. 3.4. Let $g(\tau)$ and $G(s)$ denote the impulse response characteristic of this system and the Laplace transform of $g(\tau)$, respectively. Thus $g(\tau)$ signifies the response $y(\tau)$ which occurs when $x(\tau) = \delta(\tau)$ with all initial conditions assumed zero. We are interested in the relationships between system input, system output, and $g(\tau)$. We may approximate the input by a sequence of rectangular segments $\Delta\sigma$ units in width and $x(t - n\Delta\sigma)$ units high for all time prior to $\sigma = t$, as shown in Fig. 3.5, determine an equivalent impulse area for that segment as $x(t - n\Delta\sigma)\Delta\sigma$, weight the result by $g(n\Delta\sigma)$, and sum these response components over all past values of the input to obtain

$$y(t) = \sum_{n=1}^{N} x(t - n\Delta\sigma)\Delta\sigma \quad g(n\Delta\sigma) \tag{3.102}$$

Although there are obvious sources of error in Eq. (3.102) due to the approximations required, in the limit as $\Delta\sigma \to 0$ and $n \to \infty$, $y(t)$ is given by the exact form

$$y(t) = \int_0^\infty x(t - \tau)\, g(\tau) d\tau = \int_\infty^t x(\tau)\, g(t - \tau) d\tau \tag{3.103}$$

where the second integral is obtained from the first by changing variables. These are the convolution or superposition integrals and they are of great significance in linear system analysis. Note the analogy between $g(\tau)$ and a memory device. $g(\tau)$ is a direct measure of system response to a unit impulse input τ units in the past, or a memory of that input and its current effect, and this characteristic of $g(\tau)$ is put to direct use in the convolution integrals.

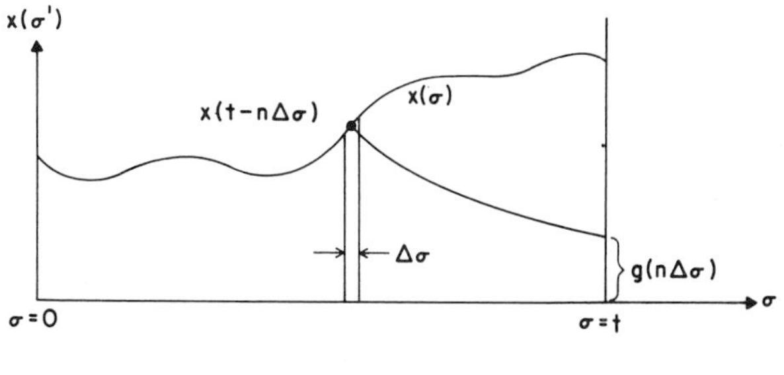

FIG. 3.5. Development of the superposition integral.

Theorem 3.14 Laplace Transform of the Convolution Integral

If $f(t)$ and $g(t)$ have Laplace transforms $F(s)$ and $G(s)$, respectively,

$$\mathcal{L}\left[\int_0^t f(t-\lambda)\,g(\lambda)\,d\lambda\right] = F(s)\,G(s) \tag{3.104}$$

Thus the Laplace transform of the output signal from a system with transfer function (Laplace transform of its unit impulse response) $G(s)$ in response to an input signal with Laplace transform $F(s)$ is the product $F(s)\,G(s)$. Herein lies the most significant mathematical advantage of transform methods, which is the substitution of multiplication in the s-domain for convolution in the time-domain. This result is so important in our subsequent work that the proof, although somewhat involved, is presented.

Proof:

By definition

$$H(s) = \mathcal{L}\left[\int_0^t f(t-\lambda)\,g(\lambda)\,d\lambda\right] = \int_0^\infty \int_0^t f(t-\lambda)\,g(\lambda)\,d\lambda\,\epsilon^{-st}\,dt \tag{3.105}$$

Note that

$$f(t-\lambda)\,g(\lambda)\,u(t-\lambda) = \begin{cases} f(t-\lambda)\,g(\lambda) & \lambda < t \\ 0 & \lambda > t \end{cases} \tag{3.106}$$

Thus we can extend the upper limit of the inner integration in Eq. (3.105) from t to ∞ if we multiply the integrand by $u(t-\lambda)$, or

$$H(s) = \int_0^\infty \int_0^\infty f(t-\lambda)\,g(\lambda)\,u(t-\lambda)\,d\lambda\,\epsilon^{-st}\,dt \tag{3.107}$$

Interchanging the order of integration in Eq. (3.107) yields

$$H(s) = \int_0^\infty g(\lambda)\,d\lambda \int_0^\infty f(t-\lambda)\,u(t-\lambda)\,\epsilon^{-st}\,dt \tag{3.108}$$

Since $u(t - \lambda) = 0$ for all $t < \lambda$, the lower limit on the inner integral in Eq. (3.108) can be changed to λ and the $u(t - \lambda)$ set equal to 1, or

$$H(s) = \int_0^\infty g(\lambda)\, d\lambda \int_\lambda^\infty f(t - \lambda)\, \epsilon^{-st} dt \tag{3.109}$$

In the integral with respect to time in Eq. (3.109) let

$$t - \lambda = \tau \qquad dt = d\tau \tag{3.110}$$

Then

$$H(s) = \int_0^\infty g(\lambda)\, d\lambda \int_0^\infty f(\tau)\, \epsilon^{-s(\tau+\lambda)} d\tau$$

$$= \int_0^\infty g(\lambda)\, \epsilon^{-s\lambda} d\lambda \int_0^\infty f(\tau)\, \epsilon^{-s\tau} d\tau = G(s)\, H(s) \tag{3.111}$$

as desired.

Consider the situation shown in Fig. 3.6. It is clear from Theorem 3.14 that

$$Y(s) = X(s)\, G_1(s) \tag{3.112}$$

and

$$Z(s) = Y(s)\, G_2(s) = X(s)\, G_1(s)\, G_2(s) \tag{3.113}$$

Thus $G(s)$, the transfer function relating $Z(s)$ to $X(s)$ is

$$G(s) = \frac{Z(s)}{X(s)} = G_1(s)\, G_2(s) \tag{3.114}$$

and the transfer function of elements placed in cascade is given by the products of their individual transfer functions. This principle, based upon superposition, the impulse response function, and Laplace transform theory, is the fundamental reason why we are inclined to work in the s-domain. It is much easier to multiply several functions of s than to convolve the corresponding time functions.

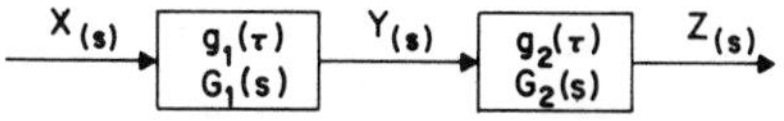

FIG. 3.6. A cascade of two transfer functions.

3.5 IMPULSE AND STEP RESPONSE CHARACTERISTICS OF SECOND-ORDER SYSTEMS

The response characteristics of second-order systems are often used as a reference in designing controllers, even when the system of interest is of third-order or higher. Although the systems encountered in practice are seldom second-order, it is often possible to approximate systems of third-order or higher by "equivalent" second-order systems with reasonable accuracy. The general response patterns for second-order systems with unit step and unit impulse inputs are developed here to illustrate the application of Laplace transform methods to a problem of particular interest.

Consider a typical second-order system with impulse response expressed in the s-domain as

$$G(s) = \frac{\omega_0^2}{s^2 + 2\zeta\omega_0 s + \omega_0^2} \tag{3.115}$$

Dividing by s provides the unit step response as

$$F(s) = \frac{G(s)}{s} = \frac{\omega_0^2}{s(s^2 + 2\zeta\omega_0 s + \omega_0^2)} \tag{3.116}$$

or

$$F(s) = \frac{\omega_0^2}{s[(s + \zeta\omega_0)^2 + \omega_0^2(1 - \zeta^2)]} \tag{3.117}$$

and the corresponding time function is obtained from App. A as

$$f(t) = 1 + \frac{1}{(1 - \zeta^2)^{1/2}} \epsilon^{-\zeta\omega_0 t} \sin[\omega_0(1 - \zeta^2)^{1/2} t - \psi] \tag{3.118}$$

$$\psi = \tan^{-1}\left[\frac{\omega_0(1 - \zeta^2)^{1/2}}{-\zeta\omega_0}\right] \tag{3.119}$$

Several typical second-order responses are shown on a normalized time scale in Fig. 3.7. The peak overshoot approaches 100 percent as $\zeta \to 0$. Peak overshoots of 20 to 50 percent are reasonably common in practice.

We may obtain $g(t)$, the impulse response function, by differentiating $f(t)$ with respect to t or by inverting $G(s)$ as given in Eq. (3.115), either of which yields

$$g(t) = \frac{1}{\omega_0(1 - \zeta^2)^{1/2}} \epsilon^{-\zeta\omega_0 t} \sin \omega_0(1 - \zeta^2)^{1/2} t \tag{3.120}$$

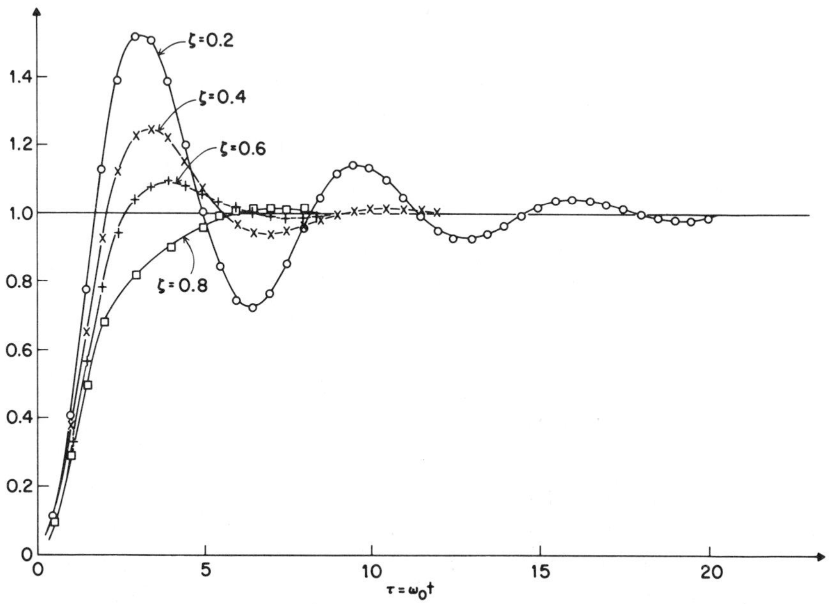

FIG. 3.7. Typical second-order system responses on a normalized time scale versus damping factor ζ.

The unit step response peak overshoot point occurs the first time $g(t) = 0$, and thus at

$$t_{p0} = \frac{\pi}{\omega_0(1 - \zeta^2)^{1/2}} \tag{3.121}$$

Substituting t_{p0} into Eq. (3.118), simplifying, and converting the result to a percentage provides

$$\text{percent overshoot} = 100 \exp\left[\frac{-\pi\zeta}{\sqrt{1-\zeta^2}}\right] \tag{3.122}$$

Note that the percent overshoot for a second-order system depends only upon ζ. We are not able to draw such general conclusions about the time response characteristics of higher order systems. A plot of unit step response percent overshoot versus ζ for this second-order case is provided in Fig. 3.8.

3.6 SUMMARY AND CONCLUSIONS

In this chapter we have presented the Laplace transform method for solving linear time invariant differential equations. We have found that the solution of an nth order differential equation involves n constants which must be chosen to

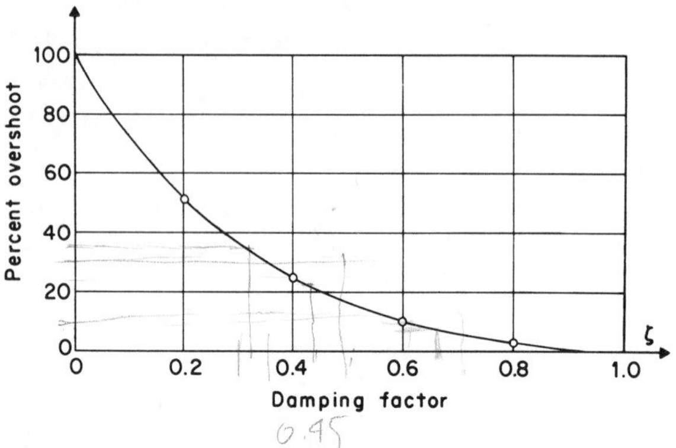

FIG. 3.8. Unit step response percent overshoot versus damping factor ζ for a second-order system.

satisfy the problem boundary conditions. When a complete set of boundary conditions is not specified, one or more of these constants may be carried along as solution parameters. Classical solution methods are based upon a known solution form. Solution by Laplace transforms, on the other hand, is an algebraic process based upon known relationships between functions in the time domain and the corresponding transformed functions in the s-domain. Classical methods are often superior when treating systems of first or second-order, whereas transform methods are generally more easily applied to systems of third order or higher. Many special characteristics of transform solutions may be applied to simplify the algebra.

This completes our general mathematical review. Additional concepts are introduced later as the need arises. The application of Laplace transform theory to block diagram analysis is considered in Chap. 4. The relationships between Laplace transforms and steady-state sinusoidal analysis are presented in Chap. 6. Application of the Laplace transform in solving vector differential equations is considered in Chap. 13.

REFERENCES

1. Wylie, C. R., Jr., "Advanced Engineering Mathematics," McGraw-Hill Book Company, Inc., New York, p. 150, 1951.
2. *Ibid.*, Chap. 6.

4 TRANSFER FUNCTIONS, BLOCK DIAGRAMS, AND FLOW GRAPHS

Classical control theory is based upon a block diagram or transfer function representation of the system of interest. Methods are derived for studying and manipulating this mathematical model to determine response characteristics, margin of stability, and compensation elements that can be used to improve overall performance. Control systems are often complex, consisting of electrical and mechanical components interacting in many ways. In most applications the system can be broken down into its basic elements, each of which can be represented by a transfer function relating its input and output in a compact mathematical form. A shaft position controller, for example, consists of an amplifier, a motor driving the load, and a position error sensor, each of which is readily represented by a transfer function. The resulting mathematical system model may then be studied in detail, if desired, with only limited further consideration of the hardware involved. In this chapter we are concerned primarily with the processes by which mathematical models of systems in transfer function, block diagram, or flow graph form, all of which are equivalent as will be shown, are developed. The rules for manipulating these models are also derived.

4.1 DEVELOPMENT OF TRANSFER FUNCTIONS

The transfer function of a system or subsystem is a mathematical description of its input to output performance. The transfer function may be given in the time domain as the impulse response function, generally denoted by $g(\tau)$, or in the s-domain as the Laplace transform of $g(\tau)$, denoted by $G(s)$. Suppose, for example, that we are interested in the resistance-capacitance (RC) network shown in Fig. 4.1. Assuming the input signal comes from a zero impedance source and no load is connected across the output terminals, the currents through R and C are identical. The equations describing this situation are

$$e_i = \frac{1}{C}\int i\,dt + Ri \qquad e_0 = Ri \tag{4.1}$$

Laplace transforming Eq. (4.1) provides

$$E_i(s) = \frac{1}{C}\left[\frac{I(s)}{s} + \frac{q_0}{s}\right] + RI(s) \qquad E_0(s) = RI(s) \tag{4.2}$$

where q_0 is the initial charge on C. Solving Eq. (4.2) for $E_0(s)$ in terms of $E_i(s)$ yields

$$E_0(s) = \frac{RCs}{RCs + 1}E_i(s) - \frac{q_0 R}{RCs + 1} \tag{4.3}$$

Since $q_0 = C[e_i(0) - e_0(0)]$, we may alternately write Eq. (4.3) as

$$E_0(s) = \frac{RCs}{RCs + 1}E_i(s) - RC\,\frac{e_i(0) - e_0(0)}{RCs + 1} \tag{4.4}$$

The first term on the right side of Eq. (4.4) multiplying $E_i(s)$ is the so-called transfer function of the network shown in Fig. 4.1. The second term illustrates the dependence of the output upon initial capacitor charge and is zero whenever

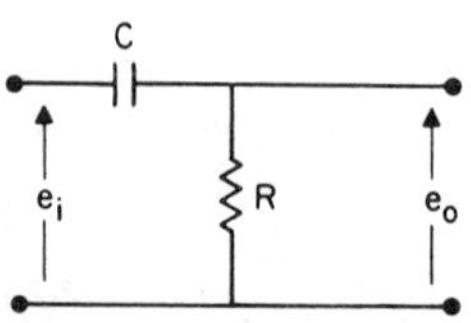

FIG. 4.1. A simple RC network.

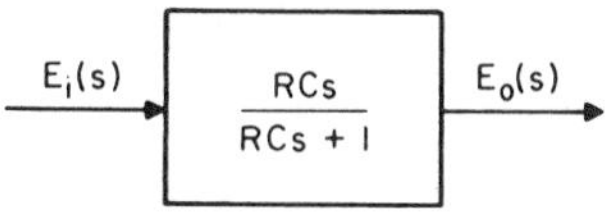

FIG. 4.2. Block diagram of the RC network shown in Fig. 4.1.

$q_0 = 0$. When $q_0 = 0$, Eq. (4.4) reduces to

$$\frac{E_0(s)}{E_i(s)} = \frac{RCs}{RCs + 1} \tag{4.5}$$

which is a form often used. It is clear that $E_0(s)$, and therefore $e_0(t)$ as well, depends upon both $e_i(t)$ *and* q_0. The *s*-domain transfer function is defined as the ratio $E_0(s)/E_i(s)$ assuming all initial conditions are zero. It is the impulse response in the *s*-domain.

We may represent the relationship given in Eq. (4.5) schematically as shown in Fig. 4.2. The arrows indicate the direction of signal flow. The figure is called a *block diagram* of the system, and the output of any box in such a schematic representation is the product of the input to the box times the transfer function shown in the box. It is always assumed, unless specified to the contrary, that all initial conditions are zero. It is always possible, when faced with a specific situation where one or more non-zero initial conditions must be included in the solution, to go back to the original differential equation and retransform it, including the initial condition terms in the results.

Note that a zero impedance source and no loading were assumed in deriving Eq. (4.5). If the input to this network is obtained from one amplifier and the output coupled to another, the ratio $E_0(s)/E_i(s)$ is not given by Eq. (4.5) if source impedance and load are not negligible. Thus the transfer functions used in our analyses must be obtained under the actual source and load conditions pertaining to the system in which they appear. In many cases both source impedance and load are negligible and they are readily included in the original analysis even when this is not so.

The steady-state sinusoidal analysis transfer function of the network shown in Fig. 4.1 is obtained from Eq. (4.5) by substituting $s = j\omega$ and interpreting $\mathbf{E}_0(j\omega)$ and $\mathbf{E}_i(j\omega)$ as phasor quantities (complex numbers) representing the sinusoidal output and input signals, respectively. This provides

$$\mathbf{G}(j\omega) = \frac{\mathbf{E}_0(j\omega)}{\mathbf{E}_i(j\omega)} = \frac{j\omega RC}{1 + j\omega RC} \tag{4.6}$$

where bold faced letters are used for phasor quantities. The complex number $\mathbf{G}(j\omega)$ directly relates the steady-state sinusoidal input and output of the RC network in relative magnitude and phase. We reserve detailed consideration of this aspect of the problem for a later chapter.

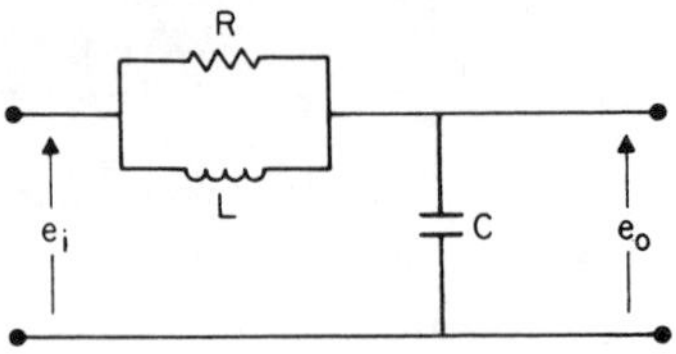

FIG. 4.3. The RLC network considered in Ex. 4.1.

Example 4.1

Find the transfer function of the RLC network shown in Fig. 4.3.

In solving this problem we will bypass much of the rather lengthy process applied in the previous paragraphs to the system shown in Fig. 4.1. Let

$$Z_1(s) = \frac{RLs}{R + Ls} \qquad Z_2(s) = \frac{1}{Cs} \tag{4.7}$$

Note that the same current passes through $Z_1(s)$ and $Z_2(s)$, so

$$\frac{E_0(s)}{E_i(s)} = \frac{Z_2(s)}{Z_1(s) + Z_2(s)} = \frac{1 + Ls/R}{LCs^2 + Ls/R + 1} \tag{4.8}$$

is the desired transfer function.

Example 4.2

The transfer function of a control system is given as

$$\frac{C(s)}{R(s)} = \frac{2}{s^2 + 3s + 2} \tag{4.9}$$

Find $c(t)$ for $r(t) = u(t)$ if $c(0) = -1$ and $\dot{c}(0) = 0$.

We must go back to the original differential equation to insert the condition on $c(0)$. The equation from which Eq. (4.9) was derived by assuming all initial conditions zero is

$$\ddot{c} + 3\dot{c} + 2c = 2r \tag{4.10}$$

as obtained by cross multiplying and treating s as the differential operator. Laplace transforming term by term including the given value of $c(0)$ provides

$$s^2C(s) + s + 3[sC(s) + 1] + 2C(s) = 2R(s) = \frac{2}{s} \tag{4.11}$$

Solving for $C(s)$ yields

$$C(s) = \frac{2 - 3s - s^2}{s(s^2 + 3s + 2)} = \frac{2 - 3s - s^2}{s(s + 1)(s + 2)} \tag{4.12}$$

Inverse transforming gives $c(t)$ as

$$c(t) = 1 - 4\epsilon^{-t} + 2\epsilon^{-2t} \tag{4.13}$$

which satisfies the conditions $c(0) = -1$ and $\dot{c}(0) = 0$, as readily checked.

In many applications it is convenient to break down the transfer characteristics of the overall system into two or more cascade components, as illustrated in Fig. 4.4. Sometimes the describing equations are such that several transfer functions are called for, while in other cases the hardware itself may lead to a logical step by step reduction. Referring to Fig. 4.4, we see that

$$E_3(s) = E_2(s)G_2(s) = E_1(s)G_1(s)G_2(s) \tag{4.14}$$

which makes it clear that the transfer function of two elements in cascade is the product of the individual transfer functions, where $G_1(s)$ and $G_2(s)$ must be obtained assuming appropriate loading conditions.

Suppose we consider the field-controlled dc motor driven inertial load shown in Fig. 4.5. The describing equations are

$$E_1(s) = I_f(s)[R_f + L_f s] \tag{4.15}$$

and

$$T(s) = K_T I_f(s) = Js^2\theta + Bs\theta \tag{4.16}$$

A logical block diagram representation of this system is shown in Fig. 4.6. The individual transfer functions are apparent directly from Eqs. (4.15) and (4.16). Although this is a particularly simple example, it illustrates how the block diagram is established directly from the system and the equations describing it. *The*

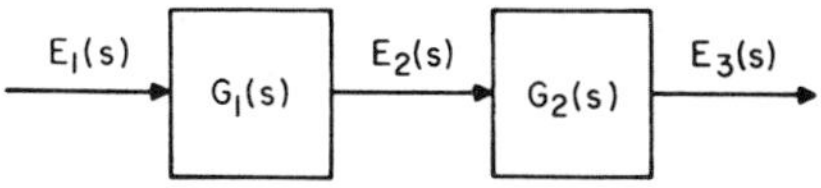

FIG. 4.4. Two cascaded transfer functions.

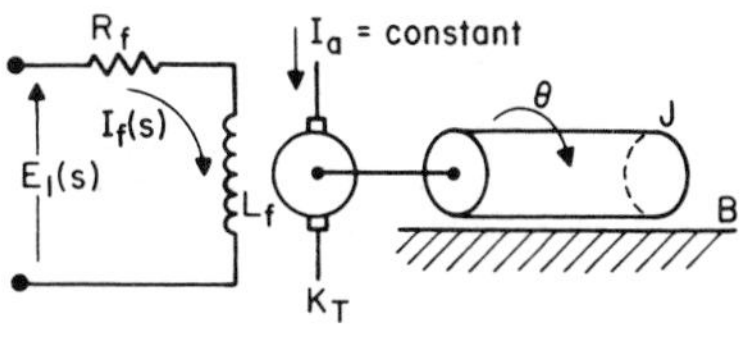

FIG. 4.5. Inertial load driven by a field-controlled dc motor.

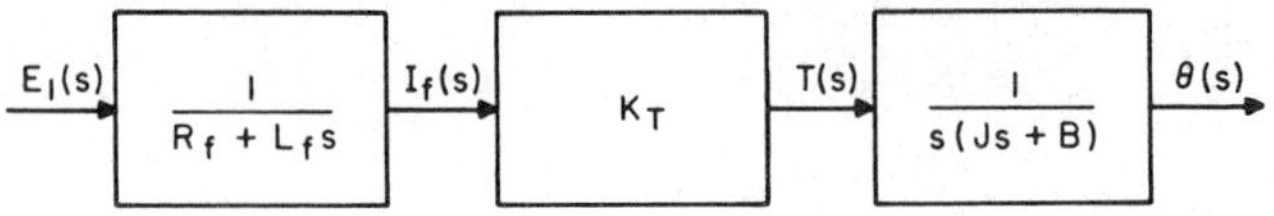

FIG. 4.6. One possible block diagram of the system shown in Fig. 4.5.

block diagram is an alternate form of those equations, indicating how signals are processed in passing from one point in the system to another.

Thus far we have considered only the simplest possible forms of block diagram arrangements with all elements appearing in cascade. A more general situation is illustrated in Fig. 4.7. The circle with the plus and minus inputs denotes addition with the indicated sign, so this system is described by

$$E(s) = R(s) - C(s)H(s) \tag{4.17}$$

and

$$C(s) = E(s)G(s) \tag{4.18}$$

Substituting $C(s)$ from Eq. (4.18) into Eq. (4.17) and solving for $E(s)/R(s)$ yields

$$\frac{E(s)}{R(s)} = \frac{1}{1 + G(s)H(s)} \tag{4.19}$$

Similarly substituting $E(s)$ from Eq. (4.17) into Eq. (4.18) and solving for $C(s)/R(s)$ provides

$$\frac{C(s)}{R(s)} = \frac{G(s)}{1 + G(s)H(s)} \tag{4.20}$$

The configuration shown in Fig. 4.7 is encountered often in control systems. We call $G(s)$ the *forward transfer function,* and $H(s)$ is called the *feedback transfer function.* In many cases $H(s) = 1$, in which case $E(s)$ is called the *error signal,* because it provides a direct indication of the difference between input and output. The *closed loop transfer function,* or *control ratio,* is defined

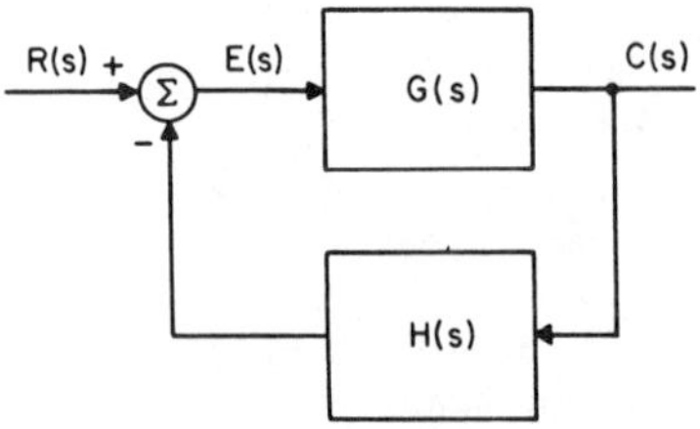

FIG. 4.7. Block diagram of a system with feedback.

as the ratio of the controlled variable to the input variable. Thus $C(s)/R(s)$, as given by Eq. (4.20), is the control ratio for the system shown in Fig. 4.7. The *open loop transfer function* of a closed loop controller is defined as the negative of the transmittance obtained by breaking the loop at an arbitrary point and traversing the entire loop back to the same point. Thus the open loop transfer function of the system shown in Fig. 4.7 is $G(s)H(s)$. When $H(s) = 1$, the open loop and forward transfer functions are identical.

Example 4.3

Find the $C(s)/R(s)$ transfer function for the system shown in Fig. 4.8.

We start by observing that the equivalent forward transfer function of the internal feedback loop consisting of the transfer functions $G_2(s)$ and $H_1(s)$ is

$$\frac{C(s)}{X(s)} = \frac{G_2(s)}{1 - G_2(s)H_1(s)} \tag{4.21}$$

Thus the equivalent forward transfer function is

$$G_{eq}(s) = \frac{G_1(s)G_2(s)}{1 - G_2(s)H_1(s)} \tag{4.22}$$

and

$$\frac{C(s)}{R(s)} = \frac{\dfrac{G_1(s)G_2(s)}{1 - G_2(s)H_1(s)}}{1 + \dfrac{G_1(s)G_2(s)}{1 - G_2(s)H_1(s)}H_2(s)}$$

$$= \frac{G_1(s)G_2(s)}{1 - G_2(s)H_1(s) + G_1(s)G_2(s)H_2(s)} \tag{4.23}$$

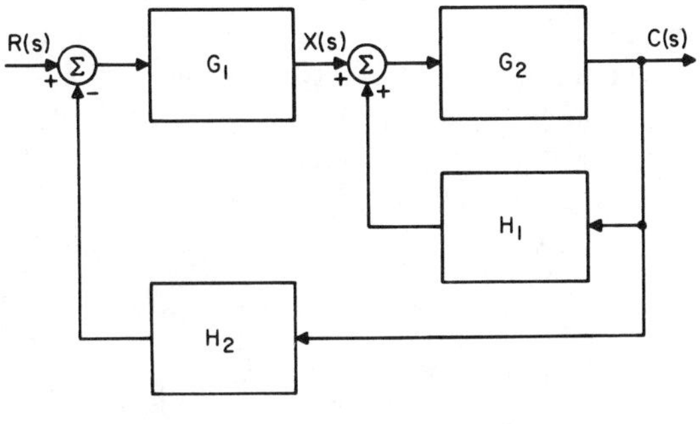

FIG. 4.8. A system with two feedback loops.

4.2 SIGNAL FLOW GRAPHS

The block diagram is an extremely useful tool in control systems analysis and is used more often than any other mathematical representation in the development of classical design procedures. It has two disadvantages however, which occasionally lead us to seek an alternate mathematical model:

1. In a complex system with many interacting effects it is often difficult to determine which group of elements should be gathered together into a single transfer function.
2. The blocks used usually relate signals at points in the system separated by considerable amounts of hardware, and the effects of parameter changes and noise inputs upon signals other than those chosen for reference in the original representation are lost in the subsequent analyses that are carried out.

Signal flow graphs, as developed by Mason,[1] provide an alternate graphic representation of signal flow within the system which shows considerably more detail than a typical block diagram. The initial flow graph construction breaks the system into virtually all of its basic components. The effects of input and/or parameter changes upon signals at all points in the system are thereby clearly illustrated. Flow graphs may be reduced using standard procedures to obtain transfer characteristics between arbitrary points in the system. The flow graph is easily constructed from the system equations and even directly from the electrical or mechanical schematic diagram once the basic concepts are understood. The flow graph is equivalent to a highly detailed block diagram.

A sample flow graph is shown in Fig. 4.9. The signals are represented by *nodes,* which are shown schematically by small circles on the diagram. The nodes are connected by *branches,* schematically represented by arrows. Signals flow only in the direction of the arrow on each branch. Each branch has the *transmittance* or *gain* indicated next to its arrow, and all signals passing along that branch are multiplied by the branch transmittance. The signal at each node is the sum of all signals entering the node. Each branch leaving a node carries the signal represented by that node. *Source nodes* are nodes with outputs and no inputs, which represent independent variables. *Sink nodes* are nodes with inputs and no outputs and are always dependent variables. In Fig. 4.9, e_1 and e_4 are source and sink nodes, respectively. A *mixed node* has both inputs and outputs.

These rules make it readily apparent that the flow graph of Fig. 4.9 is equivalent to the block diagram shown in Fig. 4.10. The equations of the corresponding system are written down immediately from the flow graph as

$$e_2 = ae_1 + ce_3 \qquad e_3 = be_2 \qquad e_4 = de_3 \tag{4.24}$$

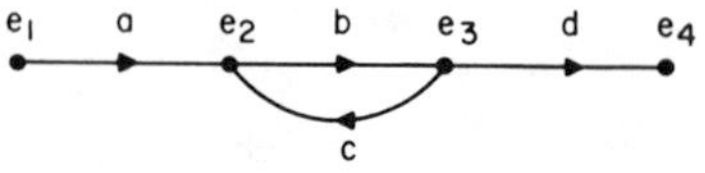

FIG. 4.9. A sample flow graph.

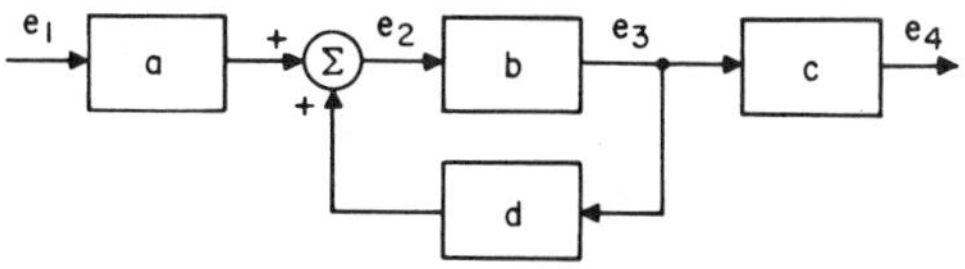

FIG. 4.10. Block diagram equivalent to the flow graph of Fig. 4.9.

Any manipulations we carry out on the flow graph representation are equivalent to steps we might take in solving Eq. (4.24) for one or more of the variables e_i in terms of the others.

The procedure for constructing the flow graph of a system is straightforward. Assign a node on the flow graph for each variable of interest. In electrical systems these variables are voltages and currents. In mechanical systems they are positions, forces, velocities, etc. The nodes are then intercoupled as required by the specific system configuration. Once all of the node inputs and outputs are properly connected, the flow graph is complete. Examples are used to illustrate the procedure.

Example 4.4

Construct a flow graph for the RC network shown in Fig. 4.11.

We initiate the solution procedure by assigning five nodes from left to right to denote $E_1(s)$, $I_1(s)$, $I_2(s)$, $I_3(s)$ and $E_2(s)$, respectively, as shown in Fig. 4.12a. We next observe that I_1 and I_2 are given by

$$I_1(s) = [E_1(s) - E_2(s)]\,Cs \tag{4.25}$$

and

$$I_2(s) = \frac{E_1(s) - E_2(s)}{R} \tag{4.26}$$

It is thus convenient to establish another node equal to $E_1(s) - E_2(s)$ and solve for $I_1(s)$ and $I_2(s)$, as shown in Fig. 4.12b. Since

$$I_3(s) = I_1(s) + I_2(s) \tag{4.27}$$

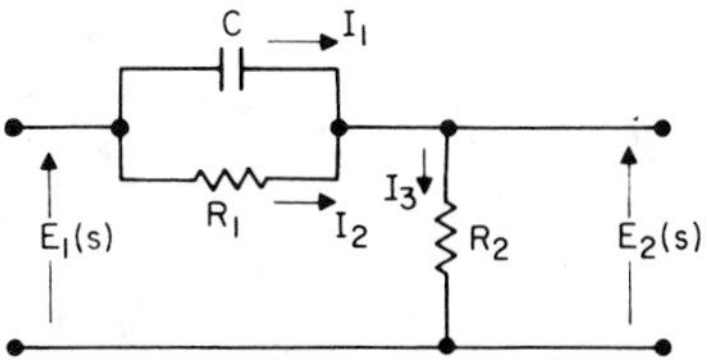

FIG. 4.11. The RC network used in Ex. 4.4.

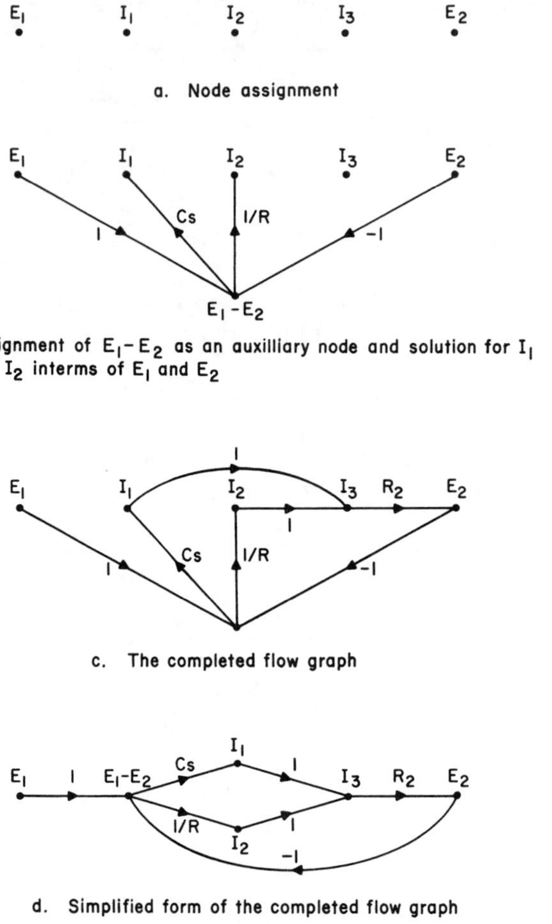

FIG. 4.12. Step by step development of a flow graph for the system shown in Fig. 4.11.

and

$$E_2(s) = R_2 I_3(s) \tag{4.28}$$

the flow graph may be completed as shown in Fig. 4.12c, and as shown in even simpler form in Fig. 4.12d.

Example 4.5

Construct a flow graph for the mechanical system shown in Fig. 4.13.

In this problem there are only three node assignments, which are $F(s)$, $X_1(s)$, and $X_2(s)$, as shown in Fig. 4.14a. In this case it is convenient to assign four auxiliary nodes. The total force F_1 on M_1 is given by

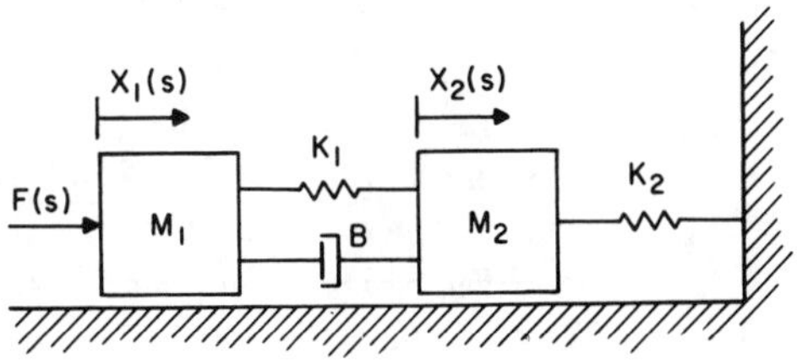

FIG. 4.13. The mechanical network used in Ex. 4.5.

$$F_1 = F - (K_1 + Bs)(X_1 - X_2) \tag{4.29}$$

Similarly the total force F_2 on M_2 is

$$F_2 = (K_1 + Bs)(X_1 - X_2) - K_2 X_2 \tag{4.30}$$

Also,

$$X_1(s) = \frac{F_1}{M_1 s^2} \tag{4.31}$$

$$X_2(s) = \frac{F_2}{M_2 s^2} \tag{4.32}$$

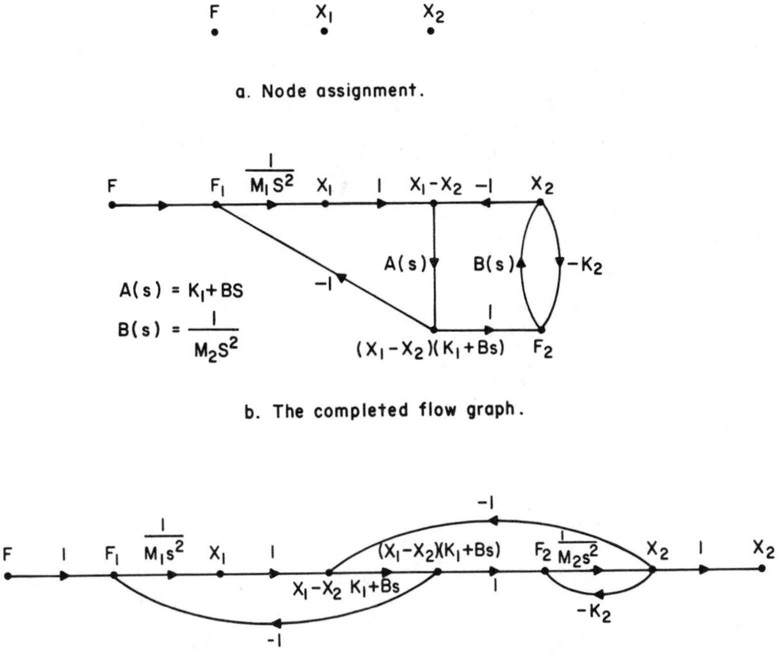

a. Node assignment.

b. The completed flow graph.

c. Simplified form.

FIG. 4.14. Step by step flow graph development for the system shown in Fig. 4.13.

The flow graph is drawn directly from these equations as shown in Fig. 4.14b, and, in simplified form, in Fig. 4.14c. It is suggested that the reader return to this system after we have considered the general reduction procedure and develop an overall transfer function relating $F(s)$ and $X_2(s)$.

It is apparent from Exs. 4.4 and 4.5 that the flow graphs of electrical, mechanical, and electromechanical systems are easily constructed from the describing equations. With a little practice the flow graphs for most situations encountered can be developed without writing any equations at all, since the basic relationships satisfied by electrical and mechanical components are well known. The information flow must, of course, be set up such that the dependent variables are obtained as functions of the independent ones. We must never connect an input branch to an independent (signal input) node, since this changes the signal at that node. It is also standard procedure to make the output a sink node even if this requires isolation from the rest of the diagram by a unit transmittance branch, as illustrated by output node $X_2(s)$ in Fig. 4.14c.

The reduction rules for flow graphs are straightforward and will be presented one at a time:

1. Two parallel paths may be combined by adding the transmittances, as shown in Fig. 4.15.

2. Two cascade transmittances are equivalent to the single transmittance obtained by forming the product of the two original transmittances, as shown in Fig. 4.16. Note that this reduction eliminates the node e_2, and is equivalent to reducing a set of simultaneous equations by substituting for one dependent variable in terms of another.

3. The termination of a branch at a node may be eliminated if equivalent transmittances are added to take care of all signal paths out of that node to the next node encountered on each path. This situation is illustrated in Fig. 4.17. Suppose it is desired to eliminate the branch between e_5 and e_2 through transmittance t_{52}. The signal into e_2 through this branch passes through t_{23} to e_3 and

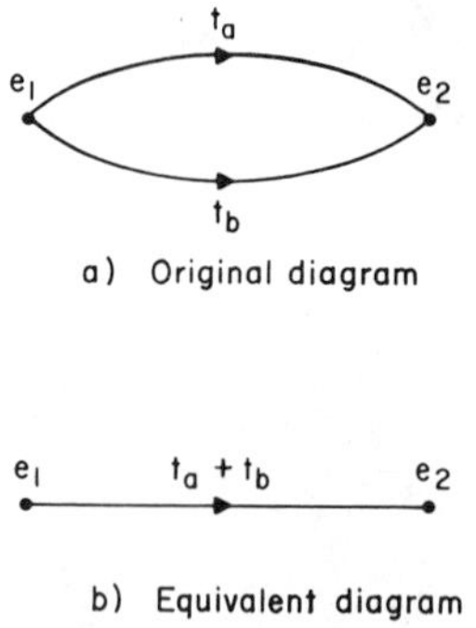

FIG. 4.15. Combination of two parrellel paths.

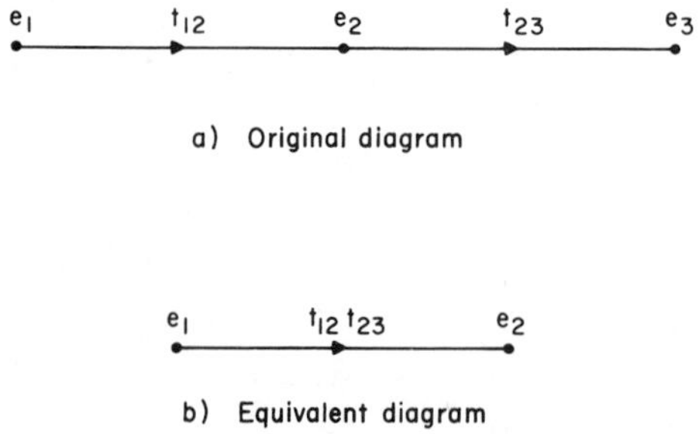

FIG. 4.16. Combination of two cascade transmittances.

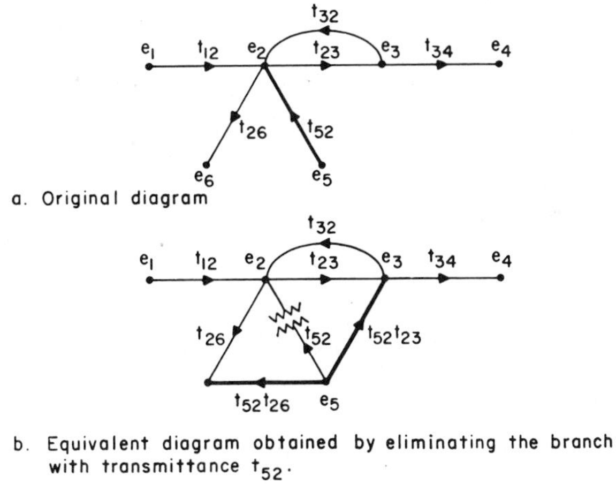

FIG. 4.17. Procedure for eliminating the termination of a transmittance branch at a node.

t_{26} to e_6. Each of these signal paths must be maintained, although it is not required that the signals pass through e_2, assuming, of course, that e_2 is not a required output. This is accomplished by adding the two new transmittances $t_{52}t_{26}$ to $t_{52}t_{23}$, as shown in Fig. 4.17b, and removing t_{52} from the diagram. The signals at all nodes other than e_2 remain unchanged.

4. The starting end of a branch may be moved one node back from the node on which it originates, along each signal input path to that node, by adding a new transmittance to the diagram for each input to the original node. This procedure is illustrated in Fig. 4.18, assuming the starting end of the branch with transmittance t_{24} to be moved back from node e_2 along each input path.

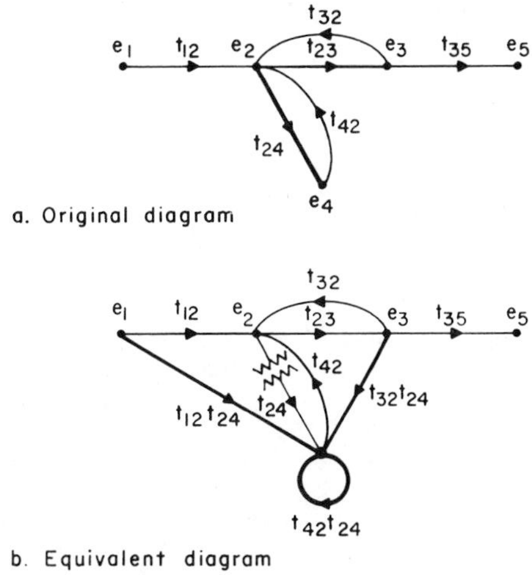

FIG. 4.18. Changes required to move the starting point of transmittance t_{24} back one node along each input to node e_2.

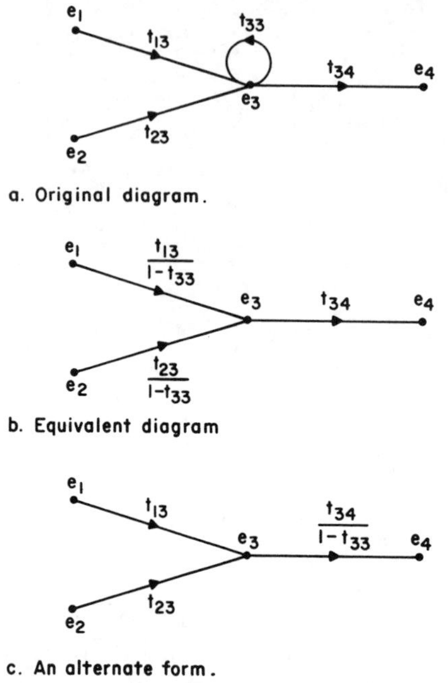

FIG. 4.19. Elimination of a self loop.

There are three inputs to e_2 coming from e_1, e_3, and e_4, respectively. Thus three new transmittances must be added to take these three paths into account, as shown in Fig. 4.18b. In this particular rearrangement none of the variables are changed. We have also generated a so-called *self loop* around node e_4. A self loop is characterized by a node which has as one of its inputs the node variable multiplied by a transmittance in a direct feedback path.

5. Elimination of a self loop is illustrated in Fig. 4.19. The equations describing this situation are

$$e_3 = t_{13}e_1 + t_{23}e_2 + t_{33}e_3 \tag{4.33}$$

and

$$e_4 = t_{34}e_3 \tag{4.34}$$

Solving for e_4 in terms of e_1 and e_2 provides

$$e_4 = t_{34}\left(\frac{t_{13}}{1 - t_{33}}e_1 + \frac{t_{23}}{1 - t_{33}}e_2\right) = \frac{t_{34}}{1 - t_{33}}(t_{13}e_1 + t_{23}e_2) \tag{4.35}$$

Thus, as far as the relationship between output e_4 and inputs e_1 and e_2 is concerned, either of the equivalent forms shown in Fig. 4.19 is equivalent to the original system. A self loop of transmittance t_s may be eliminated by multiplying all inputs to the node, or all outputs from the node, by $1/(1 - t_s)$. In the first case all node variables remain unchanged. In the second case the variable at the node around which the self loop existed is changed.

We now have available all of the tools necessary to manipulate flow graphs into alternate equivalent forms relating the source and sink nodes. In most applications our objective is to obtain an equivalent transfer function from the original diagram. Suppose we wish to reduce the flow graph shown in Fig. 4.20a

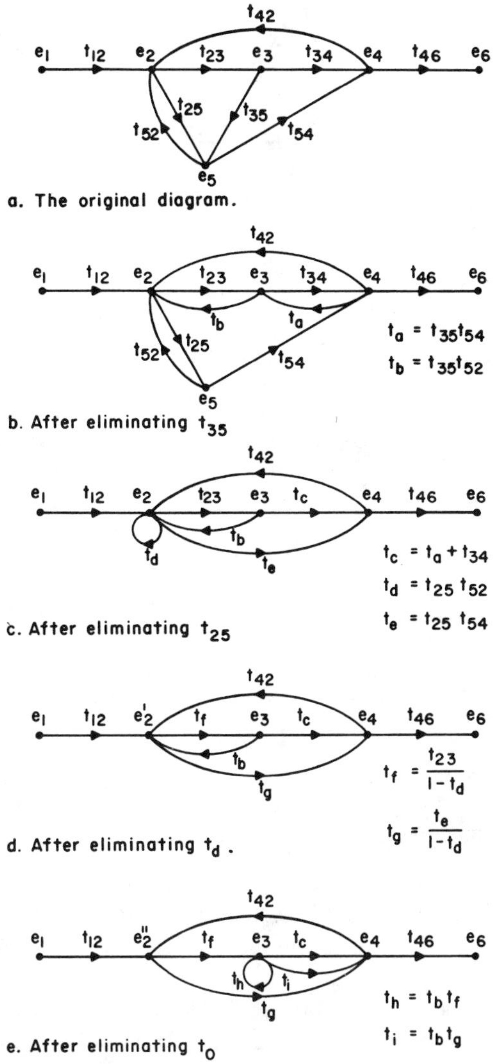

a. The original diagram.

b. After eliminating t_{35}

$$t_a = t_{35}t_{54}$$
$$t_b = t_{35}t_{52}$$

c. After eliminating t_{25}

$$t_c = t_a + t_{34}$$
$$t_d = t_{25}t_{52}$$
$$t_e = t_{25}t_{54}$$

d. After eliminating t_d.

$$t_f = \frac{t_{23}}{1-t_d}$$
$$t_g = \frac{t_e}{1-t_d}$$

e. After eliminating t_0

$$t_h = t_b t_f$$
$$t_i = t_b t_g$$

FIG. 4.20. *(Cont. next page).*

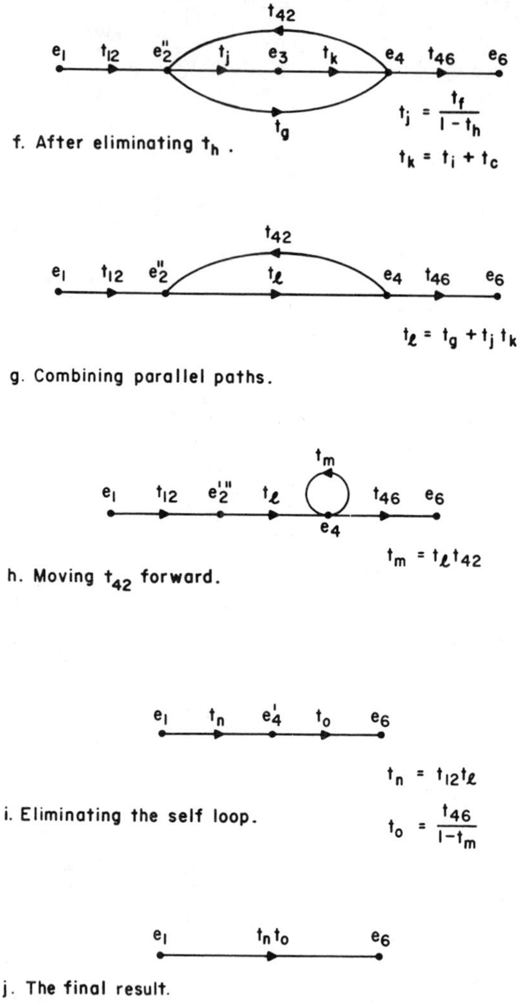

FIG. 4.20. Step by step reduction of a complex flow graph to a transfer function.

to an equivalent transfer function. We arbitrarily choose, as a first step, to eliminate the signal input to e_5 through t_{35}. Since there are two output branches from e_5 this requires the addition of two new branches, as illustrated in Fig. 4.20b. Next combine t_{34} and t_a in parallel and eliminate node e_5 as shown in Fig. 4.20c. Then eliminate t_d by multiplying both outputs from e_2 by $1/(1 - t_d)$ to obtain Fig. 4.20d. The output of t_b is moved forward one node to obtain Fig. 4.20e. We combine t_c and t_i and eliminate t_b to obtain Fig. 4.20f. Combining parallel paths and simplifying provides the final transfer function as given, admittedly in simplified form, in Fig. 4.20j. If the final result is expanded out and simplified it is found that

$$\frac{e_6}{e_1} = \frac{t_{12}t_{46}(t_{23}t_{34} + t_{23}t_{35}t_{54} + t_{25}t_{54})}{1 - t_{25}t_{52} - t_{25}t_{54}t_{42} - t_{23}t_{34}t_{42} - t_{23}t_{35}t_{54}t_{42}} \tag{4.36}$$

which can be expressed in many alternate forms.

The solution provided by Eq. (4.36) is complex, and it is not a trivial matter to obtain it. The problem from which this result was derived is also complex, however, and solution by any method is involved. Several advantages of flow graphs not previously mentioned are brought out here.

1. The flow graph makes it immediately apparent where and how feedback appears in the system.
2. Reduction of the flow graph is an organized and well-defined procedure, whereas it is often not at all apparent where to begin in solving the system equations.
3. Although we did not take full advantage of the fact in the interest of a clear solution procedure, making relatively small changes in the flow graph at each step, flow graphs may be reduced rapidly in large steps by the experienced user since the situations encountered are limited and their implications well understood. In fact, the solution may be written down in one step if one remembers Mason's formula, which is presented in the next paragraph.

S. J. Mason[1] has developed a general formula which allows us to write the overall transmittance of any flow graph almost by inspection. The overall transfer function is given by

$$T = \frac{\Sigma t_n \Delta_n}{\Delta} \tag{4.37}$$

where

1. T_n is the forward transmittance of the nth forward path between the desired source and sink nodes.
2. Δ is the graph determinant found from

$$\Delta = 1 - \Sigma L_{1i} + \Sigma L_{2j} - \Sigma L_{3k} + \ldots \tag{4.38}$$

in which

a. L_{1i} is the loop transmittance of the ith feedback loop, and ΣL_{1i} is the sum of the transmittances of all feedback loops in the system.
b. L_{2j} is the product of the loop transmittances of two nontouching loops. Loops with no common nodes are nontouching. ΣL_{2i} is the sum of transmittance products for all possible nontouching loops taken two at a time.

c. L_{3k} is the product of the loop transmittances of three nontouching loops. ΣL_{3k} is the sum of transmittance products for all possible nontouching loops taken three at a time. The higher-order terms L_{ne} are defined similarly.

3. Δ_n is the cofactor of T_n. It is obtained by evaluating the determinant (Δ) remaining when all loops touching (having nodes in common with) T_n are disregarded. The notation $\Sigma T_n \Delta_n$ indicates summation of the products of all path transmittances and their cofactors.

Application of Mason's formula is straightforward. Consider the flow graph shown in Fig. 4.21. There are two loops with transmittances $t_{23}t_{32}$ and $t_{56}t_{65}$, respectively, so

$$\sum_{i=1}^{2} L_{1i} = t_{23}t_{32} + t_{56}t_{65} \tag{4.39}$$

The two loops are nontouching, so

$$\Sigma L_{2j} = t_{23}t_{32}t_{56}t_{65} \tag{4.40}$$

Since there are only two loops, there are no more nonzero terms in Eq. (4.38) and

$$\Delta = 1 - t_{23}t_{32} - t_{56}t_{65} + t_{23}t_{32}t_{56}t_{65} \tag{4.41}$$

There are two forward paths from e_1 to e_7 with transmittances $t_{12}t_{23}t_{34}t_{45}t_{56}t_{67}$ and $t_{12}t_{24}t_{45}t_{56}t_{67}$. Each path has nodes in common with both loops. Thus

$$\Delta_1 = \Delta_2 = 1 \tag{4.42}$$

The overall transmittance may now be written down as

$$T = \frac{e_7}{e_1} = \frac{(t_{12}t_{45}t_{56}t_{67})(t_{23}t_{34} + t_{24})}{1 - t_{23}t_{32} - t_{56}t_{65} + t_{23}t_{32}t_{56}t_{65}} \tag{4.43}$$

With a little practice transfer functions can be written down by inspection using

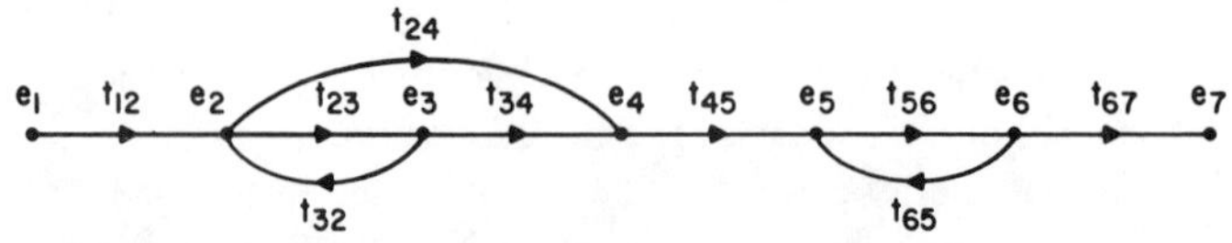

FIG. 4.21. A sample flow graph for application of Mason's formula.

Eq. (4.37), even for relatively complex situations. It is left as a student exercise to check Eq. (4.43) using the more pedestrian methods previously discussed.

Let us return to the situation depicted in Fig. 4.20. There are four feedback loops with transmittances $t_{25}t_{52}$, $t_{25}t_{54}t_{42}$, $t_{23}t_{34}t_{42}$ and $t_{23}t_{35}t_{54}t_{42}$. The loops are all touching. Thus

$$\Delta = 1 - t_{23}t_{52} - t_{25}t_{54}t_{42} - t_{23}t_{34}t_{42} - t_{23}t_{35}t_{54}t_{42} \tag{4.44}$$

There are three forward paths between e_1 and e_6 with transmittances $t_{12}t_{23}t_{34}t_{46}$, $t_{12}t_{23}t_{35}t_{54}t_{46}$ and $t_{12}t_{25}t_{54}t_{46}$. Each forward path has nodes in common with all loops, so $\Delta_1 = \Delta_2 = \Delta_3 = 1$. Substituting these results into Eq. (4.37) provides Eq. (4.36), as easily verified.

Although we could spend considerably more time discussing special characteristics of flow graphs,[2] the material presented here is adequate to serve our needs. We will use flow graphs later when convenient. Subsequent discussions about stability, sensitivity to parameter changes, and disturbance effects are readily interpreted in terms of either a flow graph or a block diagram system representation.

4.3 CONCLUDING COMMENTS

Classical control system analysis and design procedures are almost completely based upon transform mathematics, system representation in the s-domain, transfer functions, and block diagrams or flow graphs relating the signals between key points in the system. We have reviewed briefly the methods for developing and manipulating block diagram models. Flow graph representation has been shown to provide an equivalent, and generally more complete, model. The procedures for developing and reducing flow graphs are identical to the corresponding block diagram development and reduction, but the notation is more compact. More detail is conveniently shown on a flow graph. Use of Mason's formula allows overall transmittance between source and sink nodes to be written down almost by inspection in even the most complex situations, thereby eliminating much algebraic detail when this approach is taken.

Development of a flow graph is a logical first step in system analysis. The rules for constructing flow graphs are easily followed and all system variables are retained in the model at this point. The flow graph may be reduced to a block diagram or an equivalent simplified flow graph as a second step, each transmittance in many cases combining several nodes and branches of the original flow graph. This block diagram may then be used in the subsequent design and analysis. We will often start from the block diagram in later chapters, thereby bypassing these steps, but the block diagram must be developed in this or some equivalent way. Standard computer programs are available for reducing flow graphs and block diagrams.

REFERENCES

1. Mason, S. J., Feedback Theory–Some Properties of Signal Flow Graphs, *Proceedings of IRE*, vol. 41, no. 9, pp. 1144–1156, Sept. 1953.
2. Truxal, J. G., "Automatic Feedback Control System Synthesis," McGraw-Hill Book Co., Inc., New York, pp. 88–160, 1955.

5 SYSTEM STABILITY BY THE DIRECT METHOD

The term *stability* may be defined in many ways, depending upon the interests and inclinations of the user and the particular application involved. The designer of an oscillator, for example, is concerned with both amplitude and frequency stability. It is important in this case to choose the circuit configuration to minimize the effects of loading, voltage changes, and other environmental effects upon output amplitude and frequency. If relatively constant amplitude oscillations are obtained for a wide range of output loads, the circuit is said to be amplitude stable. A similar interpretation applies for frequency stability. However, our interests are specifically focused upon feedback control system stability. Control systems are generally used to position an output variable relative to a reference or control input through some actuator, such as an electric motor, a hydraulic piston, or a voltage source. Control of oscillator output amplitude using feedback methods is an example of this type of problem. We need a general stability definition according to which any controller reasonably well able to follow bounded frequency limited inputs is stable. Conversely, we shall call unstable all controllers which potentially provide unbounded outputs in response to bounded inputs. The main purposes of this chapter are to formally define what we mean by stability and to indicate what is required to assure a stable system.

Our discussion is limited at this point to linear time-invariant systems. Some of the developed ideas apply equally well to linear time varying systems, or even perhaps to nonlinear systems, but s-domain notation is used whenever convenient, which means that the functions introduced must be Laplace transformable and thus the differential equations from which they are derived are implicitly assumed to be linear and time invariant. Our ultimate objective is to provide a simple procedure for determining system stability from the transfer function and block diagram or flow graph representations. A more general stability interpretation is provided in Chap. 14.

5.1 STABILITY IN LINEAR SYSTEMS AS RELATED TO POLE LOCATIONS

We define stability as follows: *a linear system is stable if, and only if, its output in response to every bounded input remains bounded.* This definition must be applied for each input-output pair of interest. It is conceivable that a particular system could be stable relative to some input-output pairs and unstable relative to others, although such conditions are not common. This stability definition is obviously idealized, because it is applied to our mathematical system model. Most control systems are limited in response amplitude by saturation nonlinearities of one form or another, so the actual system output to any bounded input usually remains bounded in any event. If, however, the system goes into an oscillation for a particular parameter setting which tends, on the linear model basis, to grow without bound, we shall call this system unstable. This is generally an undesirable situation in which the system is not performing as the designer intended. Of course there are controllers with outputs that can easily grow without bound for a bounded input. Motor shaft position (in an open loop arrangement) for a unit-step armature-current input with constant field excitation is an example.

A necessary and sufficient condition for stability of a linear system, assuming no right half plane (rhp) or $j\omega$ axis pole-zero cancellations, is that its input to output transfer function has poles only in the left half of the s-plane (lhp), not including the $j\omega$ axis. We will sketch the proof of this statement, which seems intuitively obvious. If a transfer function $G(s)$ has lhp poles only, $g(\tau)$ approaches zero as $\tau \to \infty$ rapidly enough that

$$\int_0^\infty |g(\tau)|d\tau \leq M < \infty \tag{5.1}$$

where M is a positive constant. We may express the output $y(t)$ of this system in terms of $g(\tau)$ and the input $x(t)$ as

$$y(t) = \int_0^\infty g(\tau)\,x(t-\tau)d\tau \tag{5.2}$$

The system is stable if $y(t)$ is bounded for all bounded $x(t)$. We are thus concerned only with $|y(t)|$. Clearly

$$|y(t)| \leq \int_0^\infty |g(\tau)|\,|x(t - \tau)|\,d\tau \tag{5.3}$$

Assume $x(t)$ is bounded and $|x(t)| \leq X$ for all t. Then we may rewrite Eq. (5.3) as

$$|y(t)| \leq X \int_0^\infty |g(\tau)|\,d\tau \leq XM \tag{5.4}$$

Thus the output is bounded. This shows that requiring all poles to lie in the lhp is sufficient to satisfy the stability definition. Necessity of this condition can be shown if we assume the integral of $|g(\tau)|$ is not bounded and find an arbitrary bounded input that violates the stability conditions. Suppose we go back to Eq. (5.2) and choose $x(t - \tau)$ as $+1$ when $g(\tau)$ is positive and -1 when $g(\tau)$ is negative. This input is bounded, and thereby admissible, and Eq. (5.2) becomes

$$y(t) = \int_0^\infty |g(\tau)|\,d\tau \tag{5.5}$$

Since the integral is unbounded, the system is unstable. This completes the proof.

To assure the stability of linear time invariant systems, therefore, we need only to make sure that all system poles lie in the lhp. Alternately we require that $G(s)$ be analytic (have no poles, or singularities) in the rhp and on the $j\omega$ axis. Thus if we can reduce our system to a transfer function in the complex variable s and factor the characteristic equation (denominator) of that transfer function to find the pole locations, we can assuredly determine whether any system is stable or not. In fact the pole locations provide an excellent indication of overall system response characteristics. Unfortunately there are several factors we have over-looked here. As shown in Chap. 3 and App. B, it is difficult to factor polynomials of third order or higher, so finding pole location is easier said than done. Adding to this the fact that systems with transport delay or distributed parameters, as occasionally encountered, do not lead to simple polynomial characteristic equations, it is apparent that we have made little progress. In many applications the exact pole locations are not critical as long as the system is stable. Analytical and graphical techniques for assuring stability and even for measuring relative stability, or stability margin, are available. They are based upon complex variable theory and their use is illustrated in subsequent sections and chapters. Our first concern is to provide a means for determining system stability. We will also consider the problem of stability margin, or a general distance measure of the poles of a system from the $j\omega$ axis in the s-plane, which provides some indication of the transient response to be expected. Ultimately, of course, we must develop procedures for synthesizing systems to satisfy stringent performance specifications.

5.2 THE ROUTH AND HURWITZ STABILITY CRITERIA

The Routh and Hurwitz stability criteria are algebraic procedures which have been developed for determining the number of roots the polynomial

$$s^n + a_{n-1}s^{n-1} + \dots + a_1 s + a_0 = 0 \tag{5.6}$$

has in each half of the s-plane, where a_i indicates real constants with a_n normalized to unity for convenience. Both methods are limited to lumped-parameter networks. Development of these criteria is an involved exercise in complex function theory,[1] so we will state the results without proof and illustrate their use with examples.

Suppose we rewrite Eq. (5.6) in factored form as

$$s^n + a_{n-1}s^{n-1} + \dots + a_1 s + a_0$$
$$= (s - s_1)(s - s_2)\dots(s - s_{n-1})(s - s_n) = 0 \tag{5.7}$$

where the s_i denote the polynomial roots. Assume, for the moment, that all roots are real and located in the lhp, so all s_i are negative real numbers. Expanding the right side of Eq. (5.7) provides

$$(s - s_1)(s - s_2)\dots(s - s_{n-1})(s - s_n)$$
$$= s^n - \left(\sum_{i=1}^{n} s_i\right) s^{n-1} + \dots + (-1)^n \prod_{i=1}^{n} s_i = 0 \tag{5.8}$$

The coefficient of s^{n-1} is $-\sum_{i=1}^{n} s_i$, which must be positive when all roots are in the lhp, since all of the s_i are negative in this case. The coefficient of the s^{n-2} term in Eq. (5.8) consists of the summation of all $s_i s_j$ products which must also be positive for stability. Continuing this procedure we observe that all of the polynomial coefficients must be positive (none can be zero, either) when all roots are in the lhp, since each consists of sums of products of positive real numbers. This is our first stability check, since it is so easily carried out. A similar argument leads to the same conclusion when conjugate pairs of roots exist since they may be combined first to obtain quadratic terms with positive real coefficients whenever the roots have negative real parts. It must be noted that, except for the second-order case, the condition of all positive coefficients is *not* sufficient to assure that all roots lie in the 1hp, although one or more negative coefficients assuredly indicates that all roots *cannot* lie in the 1hp.

The Routh procedure for determining the root locations of the general polynomial

$$b_n s^n + b_{n-1} s^{n=1} + \dots + b_1 s + b_0 = 0 \tag{5.9}$$

is initiated by constructing the coefficient array

$$\begin{array}{c|cccc}
s^n & b_n & b_{n-2} & b_{n-4} \cdots & \\
s^{n-1} & b_{n-1} & b_{n-3} & b_{n-5} \cdots & \\
s^{n-2} & c_1 & c_2 & c_3 & \cdots \\
s^{n-3} & d_1 & d_2 & d_3 & \cdots \\
\cdot & \cdot & \cdot & \cdot & \\
s^1 & g_1 & & & \\
s^0 & h_1 & & &
\end{array}$$

The first and second rows of the array are extended to the right until the b_1 and b_0 terms of the polynomial are reached. Zeros may be added to balance row lengths as required. The constants c_i in the third row are developed from the coefficients in the first two rows using

$$c_1 = \frac{b_{n-1}b_{n-2} - b_n b_{n-3}}{b_{n-1}} \qquad c_2 = \frac{b_{n-1}b_{n-4} - b_n b_{n-5}}{b_{n-1}}$$

$$c_3 = \frac{b_{n-1}b_{n-6} - b_n b_{n-7}}{b_{n-1}} \cdots \tag{5.10}$$

This pattern is continued until all remaining c's are zero. The d_i's are formed in exactly the same way using the coefficients in the s^{n-1} and s^{n-2} rows. The procedure is continued down to the s^0 row. The completed array is roughly triangular, terminating at the s^0 row. The s^1 and s^0 rows always consist of only one nonzero term each. The Routh criterion is stated: *The number of polynomial roots with positive real parts equals the number of coefficient sign changes in the first column of the coefficient array we have developed.* Thus the polynomial has all roots in the 1hp if, and only if, all terms in the first column have the same sign. Application of this result to determine system stability is straightforward.

Example 5.1

Is the system with transfer function

$$G(s) = \frac{P(s)}{Q(s)} = \frac{3s^3 - 12s^2 + 17s - 20}{s^5 + 2s^4 + 14s^3 + 88s^2 + 200s + 800} \tag{5.11}$$

stable? If unstable, how many poles does it have in the rhp?

The Routh array is formed as

s^5	1	14	200	0
s^4	2	88	800	0
s^3	−30	−200	0	
s^2	74.7	800	0	
s^1	121	0		
s^0	800	0		

There are two sign changes in the left column, one from 2 to −30 and the second from -30 to 74.7. The system thus has two rhp poles, and is unstable. Note that, although $P(s)$ has negative coefficients and assuredly has roots in the right half plane, system stability depends only upon the zeros of $Q(s)$. Even if there is an apparent rhp zero-pole cancellation between $P(s)$ and $Q(s)$ the system would still be considered unstable because this cancellation could never be perfect. If $Q(s)$ is factored, using the methods discussed in App. B, its roots are found to be

$$s_1 = -4 \quad s_{2,3} = 2 \pm j4 \quad \text{and} \quad s_{4,5} = -1 \pm j3 \tag{5.12}$$

confirming the fact that there is a rhp complex conjugate root pair. The Routh criterion provides no indication whether roots are real or complex. In many cases we can obtain this information from other sources, such as the root locus procedure discussed in Chap. 6.

In evaluating a Routh array the numbers often become too large or small to work with conveniently. Several types of scaling are possible to simplify the algebra. Any complete row in the array may be multiplied by an arbitrary constant before proceeding, without changing the conclusions as to root locations. Frequency scaling can also be carried out before setting up the array, if desired, by substituting $s = \alpha p$ in the characteristic equation, where α is an arbitrary constant. The p-space roots lie in the same pattern as those in s-space, with the same angles from the origin, but their distances from the origin differ by the factor α. Stability conclusions are unchanged, and frequency scaling often reduces algebraic complexity.

Example 5.2

A particular system has characteristic equation

$$4s^5 + 6s^4 + 8s^3 + 3s^2 + 8s + 6 = 0 \tag{5.13}$$

Is this system stable? Discuss the root locations.

The Routh array is established as

~~s^5~~	~~4~~	~~8~~	~~8~~	~~0~~
s^5	1	2	2	0
~~s^4~~	~~6~~	~~3~~	~~6~~	~~0~~
s^4	2	1	2	0
s^3	3/2	1	0	
s^2	−1/3	2	0	
s^1	10	0		
s^0	2			

Note that the s^5 and s^4 rows were divided by 4 and 3, respectively, considerably simplifying the subsequent algebra. There are two sign changes, so the system is unstable with two rhp poles.

Example 5.3

Consider a system with the characteristic equation

$$s^4 + 3.5 \times 10^3 s^3 + 3.75 \times 10^6 s^2 + 1.75 \times 10^9 s + 0.5 \times 10^{12} = 0 \tag{5.14}$$

Is this system stable?

This problem can be greatly simplified by making a substitution of the form $s = \alpha p$ before setting up the Routh array, choosing α to normalize the polynomial coefficients as nearly as possible. Clearly $\alpha = 10^3$ is appropriate. Setting $s = 10^3 p$ in Eq. (5.14) yields

$$p^4 + 3.5p^3 + 3.75p^2 + 1.75p + 0.5 = 0 \tag{5.15}$$

The Routh array is developed as

p^4	1	3.75	0.5	0
p^3	3.5	1.75	0	
p^2	3.25	0.5	0	
p^1	1.21	0		
p^0	0.5	0		

Since all terms in the left column have the same sign, the system is stable.

Occasionally, a derived term in the left column of the Routh array is zero with other terms in that row nonzero. When this happens the standard procedure for deriving successive terms would involve division by zero, and is no longer applicable. This difficulty can be overcome in either of the following ways:

1. Substitute a small positive number δ for the zero term and evaluate the remaining coefficients as usual. The stability conclusion is made as usual from the resulting sign changes.
2. Substitute $s = 1/p$ in the original polynomial. The resulting polynomial in p has the same numbers of roots located in each half plane as the original polynomial in s.

The following example illustrates both of these procedures.

Example 5.4

Consider the polynomial

$$s^5 + 2s^4 + 3s^3 + 6s^2 + 2s + 1 = 0 \tag{5.16}$$

Find the number of roots in each half plane.

The Routh array is developed as

s^5	1	3	2	0
s^4	2	6	1	0
s^3	$\not{0}\,\delta$	$3/2$	0	
s^2	$\dfrac{6\delta - 3}{\delta}$	1		
s^1	$3/2 - \dfrac{\delta^2}{6\delta - 3}$	0	0	
s^0	1			

Assuming δ approaches zero from the positive side, the left column terms in the s^2 and s^1 rows approach minus infinity and 3/2, respectively, so there are two sign changes and two rhp roots. Thus there are three lhp roots.

Substituting $s = 1/p$ in Eq. (5.16) and simplifying yields

$$p^5 + 2p^4 + 6p^3 + 3p^2 + 2p + 1 = 0 \tag{5.17}$$

The corresponding Routh array is

p^5	1	6	2	0
p^4	2	3	1	0
~~p^3~~	~~9/2~~	~~3/2~~	~~0~~	
p^3	3	1	0	
p^2	7/3	1	0	
p^1	-2/7	0		
p^0	1			

There are two sign changes in the left column, so we reach the same conclusion as before.

The final situation which may arise in developing a Routh array is that all of the coefficients in an odd row ($s^1, s^3, \ldots$) may turn out to be zeros. This condition results because there exists a completely symmetrical array of roots about the complex plane origin such as $s = \pm a$; $s = \pm j\omega$; $s = \pm a \pm j\omega$; etc. Two possible methods for proceeding are as follows:

1. Construct an auxiliary polynomial using the coefficients of the row prior to the zero row encountered. This auxiliary polynomial consists only of even powers of s and has as its highest-order term the power of s indicated in the reference column to the left of its row. The roots of the auxiliary polynomial are also roots of the original polynomial. Thus we may factor the auxiliary polynomial to locate its roots (generally, this is a simple exercise), divide the original polynomial by the auxiliary polynomial, and apply the Routh criterion to the quotient.
2. We may alternately differentiate the auxiliary polynomial with respect to s and substitute the resulting coefficients into the zero row. The Routh procedure is continued using these coefficients, and the results interpreted in the usual way. The auxiliary polynomial must also be factored to locate poles on the imaginary axis, which will always yield a zero row in the array, but do not cause sign changes to occur.

The following example illustrates these procedures.

Example 5.5

A particular system has characteristic equation

$$s^4 + 3s^3 + 6s^2 + 12s + 8 = 0 \tag{5.18}$$

Discuss the location of the roots of this polynomial.

The Routh array is set up as

s^4	1	6	8	0	
s^3	3	12	0		
s^2	2	8	0		
s^1	0	0			stymied!
s^0					

The s_1 row is a zero row. It is important to note that we cannot simply substitute δ for this term and proceed, since all terms in the s_1 row are zero. If we did so we would improperly conclude that the system is stable. The auxiliary equation to be used in completing the Routh array is obtained from the s^2 row as

$$2s^2 + 8 = 0 \quad \text{or} \quad s^2 = -4 \tag{5.19}$$

which has roots at $s = \pm j2$, so the system is unstable. Differentiating the auxiliary polynomial in Eq. (5.19) with respect to s provides the s^1 row as 4, 0, and the s^0 term is derived from this modified array as 8. There are no sign changes in the left column of the final array, so there are no rhp zeros of the polynomial. There are, however, a pair of zeros on the imaginary axis and these might be overlooked unless we factor the auxiliary polynomial. Because of the root pattern symmetry required for a zero row to appear, zero rows always indicate an unstable system with one or more poles in the rhp or on the imaginary axis.

Dividing $s^2 + 4$ out of the polynomial in Eq. (5.18) yields $s^2 + 3s + 2$, which has only lhp roots. In fact those roots are located at $s = -2$ and $s = -1$.

Example 5.6

Discuss the root locations of the polynomial

$$s^7 + 4s^6 + 5s^5 + 2s^4 + 4s^3 + 16s^2 + 20s + 8 = 0 \tag{5.20}$$

The Routh array is set up as

s^7	1	5	4	20	0
~~s^6~~	~~4~~	~~2~~	~~16~~	~~8~~	~~0~~
s^6	2	1	8	4	0
~~s^5~~	~~9/2~~	~~0~~	~~18~~	~~0~~	
s^5	1	0	4	0	
s^4	1	0	4	0	
s^3	0	0	0		

This is as far as we can go without introducing the auxiliary equation derived from the coefficients in the s^4 row, which is

$$s^4 + 4 = 0 \tag{5.21}$$

This equation has roots at $s_{12} = 1 \pm j1$ and $s_{34} = -1 \pm j1$. The s^3 row for the Routh array is obtained by differentiating Eq. (5.21) with respect to s, and the completed array is derived as

s^4	1	0	4	0
s^3	4	0	0	
s^2	~~0~~ δ	4	0	
s^1	$-16/\delta$	0		
s^0	4			

There are two sign changes in the left column and thus two rhp roots known to be located at $s_{12} = 1 \pm j1$. In fact, Eq. (5.20) has, in addition to the four roots already specified, a double root at $s = -1$ and a root at $s = -2$.

The Routh criterion is often used to determine the range of gain, or some other system parameter, for which a closed-loop controller is stable. This information is readily obtained by setting up the Routh array as a general function of the parameter of interest and determining the conditions on that parameter required to force all terms in the left column of the array to have the same sign.

Example 5.7

Consider the controller shown in Fig. 5.1. For what range of K is this system stable? The closed loop transfer function is

$$\frac{C}{R}(s) = \frac{G(s)}{1 + G(s)} = \frac{K}{s^3 + 3s^2 + 2s + K} \tag{5.22}$$

The Routh array, with K as a parameter, is easily developed as

s^3	1	2	0
s^2	3	K	0
s^1	$\left(\dfrac{6 - K}{3}\right)$	0	
s^0	K		

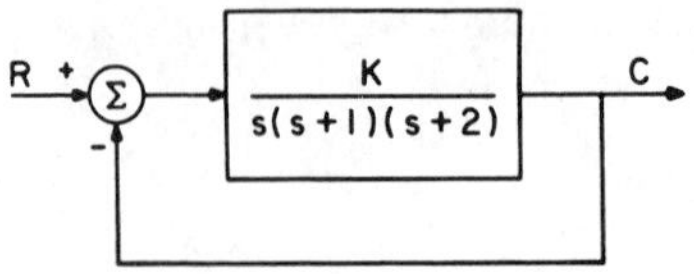

FIG. 5.1. The controller considered in Ex. 5.7.

Since all terms in the left column must be positive, it is apparent that $0 < K < 6$ is required for stability. Actually, $K = 0$ is also allowed, because in this case, although the denominator of Eq. (5.22) then has a root at $s = 0$, the system has no forward transmittance.

The Routh array may be developed with parameters other than gain carried along in general form, or as a function of two or more parameters such as gains or time constants, but complexity builds up so rapidly that this is not particularly rewarding. Graphic methods of stability analysis, discussed later, make it easier to determine the dependence of stability upon several parameters. The Routh procedure is used primarily as a fast stability check when all parameters are fixed and, occasionally, to determine the dependence of stability upon one parameter, such as gain.

The Hurwitz stability criterion provides the same general information regarding stability as the Routh criterion, but the mechanics for obtaining that information are considerably different. From the characteristic equation

$$b_n s^n + b_{n-1} s^{n-1} + \cdots + b_1 s + b_0 = 0 \tag{5.23}$$

we develop the sequence of determinants $D_1, D_2, \ldots, D_{n-1}$, defined as

$$D_1 = b_1 \qquad D_2 = \begin{vmatrix} b_1 & b_0 \\ b_3 & b_2 \end{vmatrix} \qquad D_3 = \begin{vmatrix} b_1 & b_0 & 0 \\ b_3 & b_2 & b_1 \\ b_5 & b_4 & b_3 \end{vmatrix} \cdots \tag{5.24}$$

It is necessary and sufficient to guarantee that all roots of Eq. (5.23) lie in the lhp if; all of the b_i have the same sign, and each of the determinants D_j is positive. This provides an interesting alternate procedure for checking system stability. Since the algebraic complexity of the two approaches is about equal, and since the Routh procedure often provides a considerable amount of supplementary information leading to the precise root locations, the Routh procedure is preferable.

5.3 EXTENSION OF THE ROUTH CRITERION TO THE DETERMINATION OF "RELATIVE" STABILITY

Although direct application of the Routh or Hurwitz criteria in general only indicates how the roots are divided between the two half planes, this information can be extended to pinpoint specific root locations by appropriate changes of variable s to "sweep out" regions of the s-plane, as illustrated in Fig. 5.2. If a succession of variable changes are made and the Routh criterion repeatedly applied, it is possible to determine, as accurately as can be justified by the work required, when the roots pass from one half plane to the other in the new space and thus where they were originally located. The effect of the variable change

$$s_1 = s + \alpha \tag{5.25}$$

is illustrated in Fig. 5.2. The imaginary axis in the s_1-plane is α units to the left of that in the s-plane. Suppose all of the characteristic equation roots are in the left half of the s-plane. Successively larger values of α may be tried until one or more roots pass into the right-half of the s_1-plane, thereby determining their distance from the $j\omega$ axis, and the rate at which transient terms due to the residues at these poles die out. The procedure may be repeated to find more poles if desired. Although the algebra is prohibitively complex in general, the procedure is conceptually straightforward.

Example 5.8

Consider a system with characteristic equation

$$s^4 + 6s^3 + 15s^2 + 18s + 10 = 0 \tag{5.26}$$

Find the root locations by changing variables rather than by direct factoring.

s^4	1	15	10	0
~~s^3~~	~~6~~	~~18~~	~~0~~	
s^3	1	3	0	
~~s^2~~	~~12~~	~~10~~	~~0~~	
s^2	6	5	0	
s^1	13/6	0		
s^0	5			

Thus the system is stable. Suppose we set

$$s = s_1 - \alpha = s_1 - 1 \tag{5.27}$$

as our first trial. The revised polynomial in s_1 is found to be

$$s_1^4 + 2s_1^3 + 3s_1^2 + 2s_1 + 2 = 0 \tag{5.28}$$

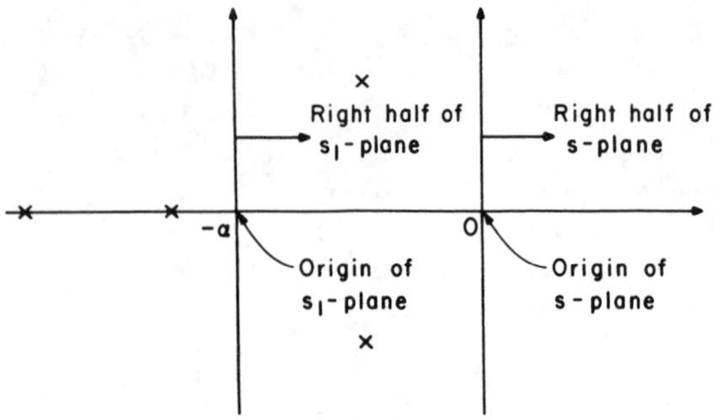

FIG. 5.2. Effect of the translation $s = s_1 - \alpha$.

The Routh array is

s_1^4	1	3	2	0
~~s_1^3~~	~~2~~	~~2~~	~~0~~	
s_1^3	1	1	0	
~~s_1^2~~	~~2~~	~~2~~	~~0~~	
s_1^2	1	1	0	
s_1^1	0	0	0	

We encounter a zero row, so

$$s_1^2 + 1 = 0 \tag{5.29}$$

and $s_1 = \pm j1$ are roots of Eq. (5.28). Thus $s = -1 \pm j1$ and $s = -1 - j1$ are roots of Eq. (5.26). Dividing these roots out of Eq. (5.26) yields

$$(s + 1 + j1)(s + 1 - j1)(s^2 + 4s + 5) = 0 \tag{5.30}$$

and the roots of the quadratic term remaining are readily found to be $s = -2 + j1$ and $s = -2 - j1$.

Obviously we were unusually fortunate in this problem, since our first variable change attempt led to the exact location of a complex conjugate pole pair. Once this pole pair is divided out, the problem is essentially solved. In most applications several variable-change trials will be required before a root, or pair of roots, is accurately located. The procedure is clear, although its application is involved due to the algebraic complexity encountered. It is generally less difficult to factor the original equation, particularly since digital computer programs are readily available for this purpose.

5.4 SUMMARY AND CONCLUSIONS

The word stability is a highly subjective term. We have defined precisely what we mean by the stability of linear control systems and have related stability to the s-plane pole locations for linear time invariant controllers. If all poles are in the lhp, the system is stable. In many control applications we need a quick check on stability, and this is provided by the Routh and Hurwitz criteria, either of which indicate easily whether all poles are in the lhp. Factoring polynomials of fourth order or higher to determine pole locations is a complex and time consuming task. When only a yes-no stability answer is required, the Routh and Hurwitz criteria can save a great deal of work. The Routh criterion is suggested for general use, since it often provides a considerable amount of extra information.

It is painfully obvious at this point that overall control system response characteristics which are normally specified, or at least constrained, by the customer depend upon much more than stability, or even relative stability, however we might define that term. The Routh criterion is by no means an efficient design tool. Neither can we adjust controller parameters by trial and error, repeatedly factoring the resulting characteristic equations, and sketch the corresponding time response patterns in an effort to arrive at a final design. Of course we could resort to an analog or digital computer simulation of the system and use some sort of trial and error process, but this method avoids the issue. What we really want at this point is a relatively simply procedure by which closed-loop controllers satisfying frequency and/or time-domain specifications can be synthesized. It will help greatly in the development of such procedures if we can obtain a better feeling for the manner in which closed loop poles are influenced by gain settings, and open loop pole and zero locations. Our next chapter is devoted to the root locus method, and will help provide this desired background.

REFERENCE

1. Guillemin, E. A., "The Mathematics of Circuit Analysis," J. Wiley & Sons, p. 395, 1949.

6 THE ROOT LOCUS METHOD FOR ANALYSIS AND DESIGN

Establishing the location of closed-loop system poles, or the characteristic equation roots, is of fundamental importance in the design of all controllers. It is generally difficult to obtain the characteristic equation factors for systems of fourth order and higher, so a variety of analysis and design methods have been developed to assure closed-loop stability and reasonable transient response characteristics without actually finding the closed-loop poles. A pole-zero diagram of the final design is valuable as an aid in visualizing the closed-loop time and frequency domain performances. For this reason we are led to consider the *root locus (RL) method* introduced by Evans in 1948.[1] The RL diagram illustrates how the closed-loop poles of the system move about in the complex plane as a function of some parameter, usually open-loop gain. Although in most applications the RL plot is perhaps most appropriately constructed near the end of the design phase, to provide information which supplements that available from the Bode diagram and the Nyquist plot, tools which we shall introduce later, design may be based entirely upon RL diagrams if desired. In situations where the plant is nonminimum phase either because it is open-loop unstable, or has zeros in the rhp, certain standard interpretations of the Bode and Nyquist diagram patterns fail. The RL method is particularly appropriate in such cases.

The RL is considered primarily as an analysis aid. A considerable amount of information is readily available from the RL diagram. The effect of gain upon pole locations is shown directly. The closed-loop frequency response characteristics are easily obtained, either graphically or analytically. The residues at the closed loop poles are also available, requiring only a few simple calculations supported by graphical constructions. The time required for the transient response to decay to a negligible level is directly indicated by the minimum distance from the imaginary axis of a pole (or pole pair). Thus the RL diagram is a useful tool for exhibiting information indicative of closed-loop performance characteristics in both the frequency and time domains.

Several general characteristics of RL diagrams are developed here to provide useful guidelines in the construction process. Although digital computer programs can be used for plotting these diagrams, it is important that we learn to construct them by hand within reasonable accuracy limits. Straightforward graphical methods for sketching the RL are discussed. The spirule, a tool designed as a construction aid by Evans, is often helpful. An alternate construction procedure using transparent sheets with radial lines drawn on them is suggested, and its use is illustrated in Fig. 6.7. This latter procedure is extremely simple, and mechanical errors which may result from using the spirule are eliminated. Its adoption as a general tool is suggested for this reason.

6.1 ILLUSTRATION OF THE ROOT LOCUS USED AS AN ANALYSIS TOOL

Consider the system shown in Fig. 6.1. The closed-loop transfer function for this arrangement is

$$\frac{C(s)}{R(s)} = \frac{G(s)}{1 + G(s)} = \frac{P(s)/Q(s)}{1 + P(s)/Q(s)} = \frac{P(s)}{P(s) + Q(s)} \tag{6.1}$$

We observe from Eq. (6.1) that the zeros of C/R are identical to those of G, and the poles of C/R are located at those values of s for which $P(s)/Q(s) = -1$, or where $P(s) + Q(s) = 0$. It is difficult, in general, to find these closed-loop poles, and this is the reason we are led to develop design procedures based upon the Nyquist and Bode diagrams, as discussed in Chaps. 8 and 9, respectively, thereby avoiding the necessity for obtaining the exact pole locations, and simultaneously assuring that they are located a "comfortable" distance within the lhp. The RL procedure is based upon a variety of graphical methods for determining how the closed-loop pole locations are influenced by open-loop gain and/or open-loop poles and zeros.

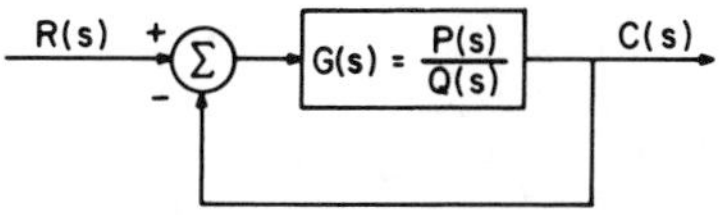

FIG. 6.1. A standard unity feedback controller.

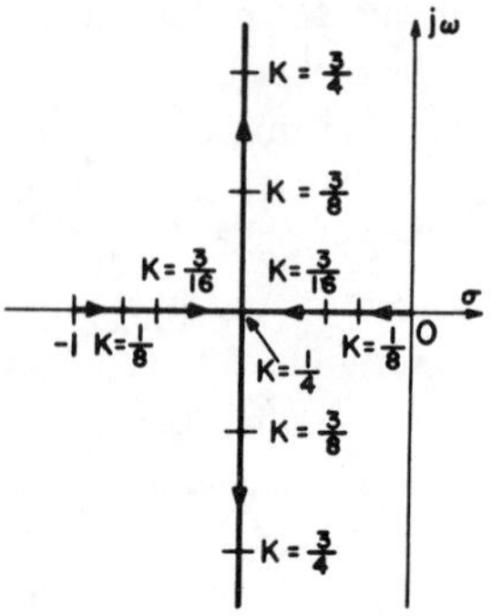

FIG. 6.2. Root locus diagram for the system with $G(s) = K/s(s+1)$.

As a simple example for which the pole movements can be described analytically, consider the system shown in Fig. 6.1 with

$$G(s) = \frac{K}{s(s+1)} \tag{6.2}$$

for which

$$\frac{C(s)}{R(s)} = \frac{K}{s^2 + s + K} \tag{6.3}$$

The closed-loop poles are located at

$$s_1 = -\frac{1}{2} + \sqrt{\frac{1}{4} - K} \qquad s_2 = -\frac{1}{2} - \sqrt{\frac{1}{4} - K} \tag{6.4}$$

The corresponding *root locus plot* obtained from Eq. (6.4) as a function of K is shown in Fig. 6.2. The closed-loop pole locations for a number of specific values of K are shown on the diagram to illustrate how the poles move as K increases from zero toward infinity. It is clear that this system is stable for all positive values of K. A complex conjugate pole pair with any desired damping factor may be obtained by appropriately setting K. The transient response decay rate, which depends upon the real part of the pole located closest to the $j\omega$ axis, is seen to be a constant for all values of $K > 1/4$, and, although the response pattern goes through more cycles per unit time as K increases (the natural frequency increases) the exponential envelope within which this response must fall is invariant for $1/4 < K < \infty$. The transient response decay rate would increase for all large K if the open-loop pole at $s = -1$ were moved further into the lhp.

The second-order example illustrated in Fig. 6.2 is not typical, since the RL is not generally so easy to describe functionally. Fortunately, relatively straightforward graphical construction procedures are available, and many general characteristics of RL diagrams may be established which are very helpful in developing an acceptably accurate sketch.

6.2 GENERAL CHARACTERISTICS OF ROOT LOCUS DIAGRAMS

If we let $A(s) = P(s)/Q(s)$ denote the open-loop transfer function of the system and assume the usual negative feedback arrangement, as shown in Fig. 6.1, the characteristic equation becomes

$$1 + A(s) = 1 + \frac{P(s)}{Q(s)} = 0 \tag{6.5}$$

which is satisfied whenever

$$A(s) = -1 \tag{6.6}$$

The relationship given in Eq. (6.6) involves the complex variable $s = \sigma + j\omega$, and may be interpreted as the two distinct requirements

$$|A| = 1 \qquad \angle A = 180° \pm n360° \tag{6.7}$$

Since gain may be adjusted to satisfy the magnitude requirements for all values of s other than at the poles and zeros of A, we will concentrate upon finding the locus of points for which the angle condition in Eq. (6.7) is satisfied, and this locus will be the desired RL diagram. A typical $A(s)$ is of the form

$$A(s) = \frac{KP(s)}{Q(s)} = \frac{K(s - z_1)(s - z_2)}{(s - p_1)(s - p_2)(s - p_3)} \tag{6.8}$$

Thus $A(s)$ may be interpreted as a ratio of vector products, since the terms $s_1 - z_i$ (or $s_1 - p_i$) are vectors starting from $z_i(p_i)$ and terminating at s_1, as shown in Fig. 6.3. The total phase angle associated with $A(s)$ for a particular complex frequency $s = s_1$ is the sum of the numerator vector angles minus the sum of the denominator vector angles. Points on the RL diagram are obtained by

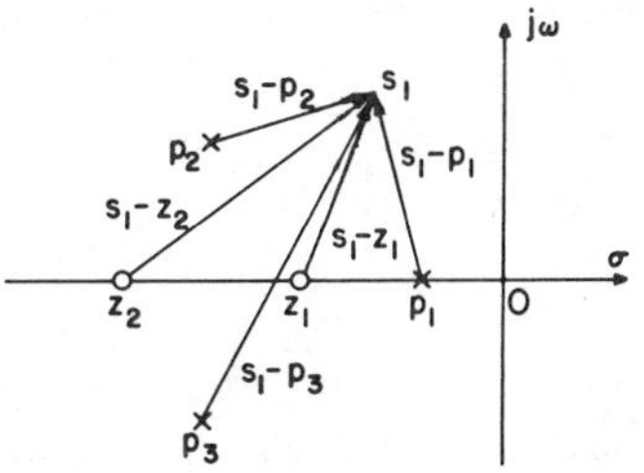

FIG. 6.3. Graphical representation of the vectors making up Eq. (6.8).

finding those values of s for which this algebraic sum equals $180° \pm n360°$, where n may take on positive integer values and zero. It is easy to establish specific points, and even families of points, in the s-plane for which the angle condition is satisfied and, since the RL is always continuous (the roots of a polynomial are a continuous function of its coefficients), the path of the RL between specific points through which it is known to pass can generally be sketched rapidly with acceptable accuracy. The remainder of this section is devoted to the development of specific relationships which will prove helpful in constructing RL diagrams.

(a) Root Locus Characteristics as $K \to 0$: Letting $A(s) = KG(s)$ and observing that $KG(s) = -1$ is required at all points on the locus, it is clear that $|G(s)|$ must approach infinity as $K \to 0$. Thus the RL must start from the open-loop poles as K increases from zero. Since $G(s)$ must have at least as many poles as zeros to be realizable, Eq. (6.5) is the ratio of two nth-order polynomials, where n is the number of open-loop poles of $G(s)$. An nth-order polynomial has n roots, so the closed loop system also has n poles and the RL has n branches, one emanating from each open loop pole.

(b) Root Locus Symmetry: The coefficients of $P(s)$ and $Q(s)$ are real for all cases of interest, so all complex roots of the characteristic equation must occur in complex conjugate pairs. The RL is thus symmetrical relative to the real axis. It is only necessary to construct the upper half plane portion of the RL diagram to obtain a complete system description.

(c) Root Locus Characteristics as $K \to \infty$: As $K \to \infty$, the values of s for which $KG(s) = -1$ become arbitrarily large, since the denominator of $G(s)$ is always of higher order in s than the numerator. We may thus replace the numerator $P(s)$ and the denominator $Q(s)$ by their highest power terms in s to determine the limiting characteristics, or

$$\lim_{K\to\infty}\left[K\frac{P(s)}{Q(s)}\right] = K\frac{s^m}{s^{m+\alpha}} = \frac{K}{s^\alpha} = -1 \tag{6.9}$$

Thus $s^\alpha = -K$ is required in the limit as $K \to \infty$. In general the RL branches approach the open-loop system zeros as $K \to \infty$, including those at infinity, since only near these zeros can Eq. (6.9) be satisfied for large K. The number of RL branches moving toward infinity as $K \to \infty$ equals α, the excess of the number of finite poles over the number of finite zeros in $G(s)$. The RL branch angles for which Eq. (6.9) is satisfied are given by $180°/\alpha, 540°/\alpha, \ldots, [180° + (\alpha - 1)360°]/\alpha$. As $K \to \infty$ the RL branches approach these angles asymptotically. If, for example, $\alpha = 2$, the angles of the asymptotes, denoted by AA, are $\pm 90°$. If $\alpha = 3$, the AA are $\pm 60°$ and $180°$. All cases are easily treated.

The AA do not, in general, emanate from the complex plane origin, but instead emanate from a point on the real axis which may be interpreted as a "center of gravity" of the open-loop pole-zero plot. $A(s)$ may be written as

$$A(s) = K\frac{P(s)}{Q(s)} = K\frac{s^m + a_{m-1}s^{m-1} + \cdots + a_0}{s^{m+\alpha} + b_{m+\alpha-1}s^{m+\alpha-1} + \ldots + b_0} \tag{6.10}$$

Dividing the denominator of Eq. (6.10) by its numerator provides

$$\frac{P(s)}{Q(s)} = \frac{K}{s^{\alpha} + (b_{m+\alpha-1} - a_{m-1})s^{\alpha-1} + \ldots} \tag{6.11}$$

The RL branches are defined by those values of s for which the denominator of Eq. (6.11) equals $-K$, or

$$s^{\alpha} + (b_{m+\alpha-1} - a_{m-1})s^{\alpha-1} + \ldots = -K \tag{6.12}$$

When s is large, the left side of Eq. (6.12) behaves as a polynomial in s of degree α (note that all terms of the form a_i/s^{β} in the expansion become negligible as $|s| \to \infty, \beta \geq 1$), and the sum of its roots is given by $-(b_{m+\alpha-1} - a_{m-1})$. Thus the *origin of the asymptotes*, denoted by OA, is given by

$$OA = -\frac{(b_{m+\alpha-1} - a_{m-1})}{\alpha} \tag{6.13}$$

Since $b_{m+\alpha-1}$ is the negative of the sum of the open-loop poles and a_{m-1} is the negative of the sum of the open-loop zeros,

$$OA = \frac{\Sigma\,\text{poles} - \Sigma\,\text{zeros}}{\text{number finite poles} - \text{number finite zeros}} \tag{6.14}$$

If, for example, $G(s)$ has poles at 0, -1, and -3, and a zero at -2, OA is readily found to be $[(-1 - 3) - (-2)]/2 = -1$.

(d) Root Locus Branches on the Real Axis: It is apparent that the phase contributions from all complex pole and zero pairs cancel out for values of s on the real axis, so the 180° contribution for RL branches on the real axis must come entirely from real poles and zeros. Points on the real axis to the right and left of a critical frequency cause the vector from that singularity to contribute 0° and 180° of phase angle, respectively. Thus the RL includes all points to the left of an odd number of real critical frequencies.

(e) Breakaway Point from the Real Axis: It is often a straightforward algebraic problem to determine the real axis breakaway point for which the condition of 180° phase shift is satisfied. Let ϵ and α denote an arbitrarily small distance from the real axis and the distance from the origin, respectively, as shown in Fig. 6.4. It is necessary that

$$\theta_1 + \theta_2 + \theta_3 = 180° - \tan^{-1}\epsilon/\alpha + \theta_2 + \theta_3 = 180° \tag{6.15}$$

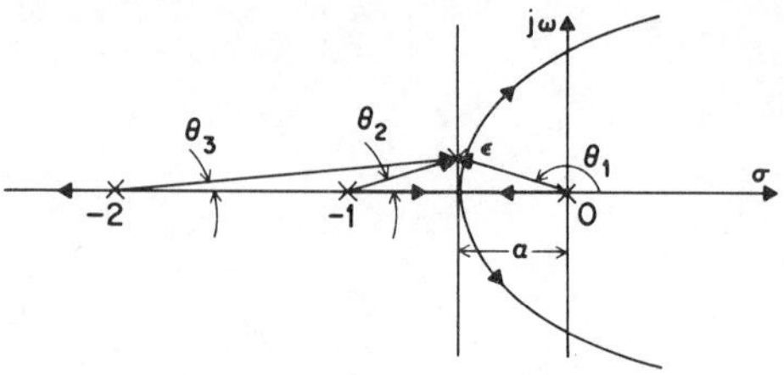

FIG. 6.4. Calculation of the real axis breakaway point.

Since ϵ is small, we may replace the angles by their tangents to obtain

$$\frac{\epsilon}{1-\alpha} + \frac{\epsilon}{2-\alpha} - \frac{\epsilon}{\alpha} = 0 \tag{6.16}$$

which becomes, upon simplification,

$$3\alpha^2 - 6\alpha + 2 = 0 \tag{6.17}$$

Solving Eq. (6.17) for α provides the breakaway point as $\alpha = 1 - 1/\sqrt{3} \simeq 0.423$. The general procedure is similar when there are more critical frequencies, when the critical frequencies include one or more complex conjugate pairs, and when both poles and zeros are involved. Of course the appropriate sign must be associated with each angle, depending upon whether it results from a pole or a zero. Analytical calculation of the breakaway point is impractical when more than four or five poles and zeros are involved. The graphical methods subsequently discussed are straightforward even in complex cases.

(f) Departure Angles from Poles and Arrival Angles at Zeros: The departure angles of the RL branches from the poles and the arrival angles at the zeros are easily calculated using Eq. (6.6). Consider the situation shown in Fig. 6.5. For all s near the pole at $s = -1 + j1$, the angles contributed by all other poles and zeros are essentially given by the phasors drawn from those critical frequencies to the point $-1 + j1$. The total phase shift must be $180° \pm n360°$, and is the sum of the angles contributed by the zeros minus the sum of the angles contributed by the poles, including the contribution from the infinitesimal magnitude phasor due to the pole at $s = -1 + j1$. Thus θ_d, the departure angle,

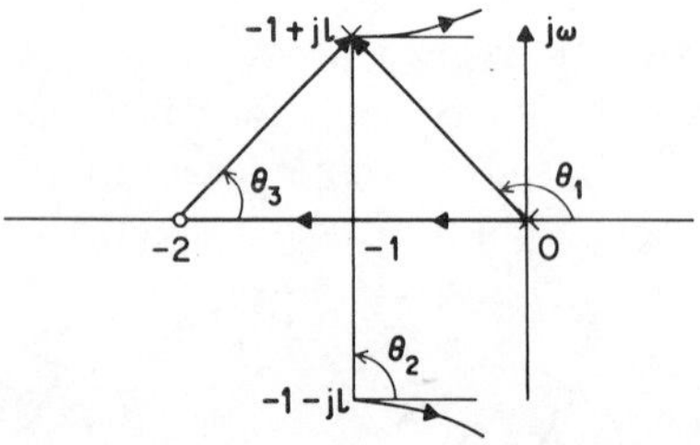

FIG. 6.5. Calculation of the departure angle from a pole.

is obtained by solving

$$\theta_3 - \theta_1 - \theta_2 - \theta_d = 180° \pm n360° \tag{6.18}$$

but $\theta_1 = 135°$, $\theta_2 = 90°$ and $\theta_3 = 45°$, so

$$\theta_d = -360° \pm n360° = 0° \tag{6.19}$$

The same principle is applied to obtain the arrival angle at a zero.

(g) Intersections of the Root Locus with the Imaginary Axis: There are several procedures for analytically obtaining the values of K and ω at the RL intersections with the imaginary axis. The characteristic equation of the closed-loop system is $Q(s) + KP(s) = 0$. If $s = j\omega$ is substituted into this expression and the real and imaginary parts are set to zero, this provides two equations which may be solved for K and ω. An equivalent procedure is to apply the Routh-Hurwitz criterion to the characteristic equation to determine that value of K which provides a pair of imaginary poles, and the corresponding ω. Finally, a term of the form $s^2 + \omega^2$ may be divided out of the characteristic equation and the remainder set to zero to determine the required K and ω values. Of these methods, the first is generally easiest to apply. The same basic procedure may be extended to determine the locus intersections with any line of the form $s = \sigma_1$. Substituting $\hat{s} = -\sigma_1$ or $s = \hat{s} + \sigma_1$ into the characteristic equation translates the line $s = \sigma_1$ to the imaginary axis in the $\hat{s}$ plane, and the solution now may be carried out just as before.

(h) The Sum of the Closed Loop Roots: If $P(s) + KQ(s)$ is of degree n and if the coefficient of s^n is unity, the coefficient of s^{n-1} is the negative of the sum of the zeros of this expression, or the negative of the sum of the closed-loop pole locations. If $Q(s)$ is of degree n and $P(s)$ is of degree $n - \alpha$, with $\alpha \geq 2$, as is often the case in control applications, the s^{n-1} coefficient is independent of K, and the sum of the closed-loop poles is a constant. In this special case, as some branches of the RL move to the right (left), other branches must move to the left (right) to compensate. Use of this characteristic often saves considerable trouble in RL construction.

(i) The Product of the Closed Loop Roots: The constant term in the characteristic equation is the product of the closed-loop poles, assuming the coefficient of s^n is unity. In the special case where the open-loop transfer function $A(s) = KP(s)/Q(s)$ has a pole at the origin (and this situation is common in practice), the product of the roots is directly proportional to K. This also is often helpful in determining how the roots move along the RL branches as $K \to \infty$.

These characteristics of root locus diagrams are summarized in Table 6.1 for convenience.

Example 6.1

To illustrate these RL construction procedures, consider a system with

Table 6.1. Summary of root locus characteristics.

as $K \to 0$	Root locus approaches the open loop poles.
as $K \to \infty$	Root locus approaches the open loop zeros, including those at infinity.
AA	$AA = \frac{180^\circ + n360^\circ}{\alpha} \quad n = 1, 2, \ldots, (\alpha - 1)$
OA	$OA = \frac{\Sigma \text{ poles} - \Sigma \text{ zeros}}{\#\text{poles} - \#\text{zeros}}$
symmetry	The root locus has upper and lower half plane mirror symmetry.
real axis	The root locus for a standard negative feedback system consists of all regions on the real axis to the left of an odd number of real critical frequencies.
real axis breakaway	Feasible to compute analytically for simple systems with no more than three or four real poles and zeros.
departure angle from poles arrival angle at zeros	Easily determined in all cases either analytically or by measurements on a scaled diagram.
imaginary axis intersections	Can be analytically computed for systems of reasonable order in several ways. Graphical methods always applicable.
sum of the closed-loop roots	If the order of $Q(s)$ exceeds that of $P(s)$ by two or more, the sum of the closed loop poles is a constant.
product of the closed-loop roots	If the coefficient of s^n in the Ch. eq. is 1, and if $A(s)$ has a pole at the origin, the product of the closed loop roots is proportional to K.

$$A(s) = \frac{K(1 + s/4)}{s(1 + s/10)(s^2 + 10s + 50)} \tag{6.20}$$

which has a zero at $s = -4$ and poles at $s = 0$, $s = -10$ and $s = -5 \pm j5$. The RL plot for this system is shown in Fig. 6.6. We will proceed step by step to develop the various elements of information which were used in constructing this plot, as suggested in the previous paragraphs. Since there are four poles and one zero, three branches of the RL move off toward infinity as $K \to \infty$ along the asymptotic angles +60°, +180°, and -60°. The origin of the asymptotes is given by

$$OA = \frac{-10 - 5 - j5 - 5 + j5 + 4}{3} = -\frac{16}{3} \tag{6.21}$$

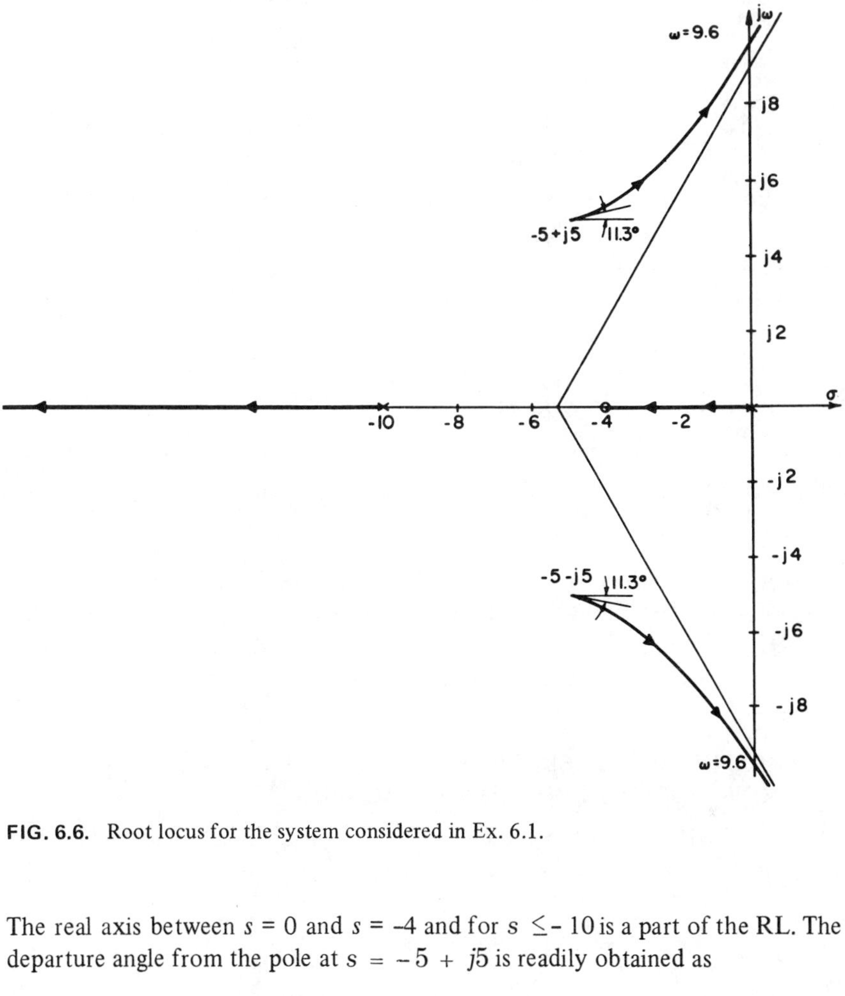

FIG. 6.6. Root locus for the system considered in Ex. 6.1.

The real axis between $s = 0$ and $s = -4$ and for $s \leq -10$ is a part of the RL. The departure angle from the pole at $s = -5 + j5$ is readily obtained as

$$\theta_d = 180° + \underbrace{180° - \tan^{-1}5}_{\text{zero at } -4} - \underbrace{180° + \tan^{-1}1}_{\text{pole at } 0} - \underbrace{90°}_{\text{pole at } -5 - j5} - \underbrace{\tan^{-1}1}_{\text{pole at } -10} \qquad (6.22)$$

where the terms resulting from the poles and the zero are noted individually for added clarity. Solving Eq. (6.22) for θ_d provides $\theta_d = 11.3°$. The next point of interest is the imaginary axis interaction. The characteristic equation is

$$Q(s) + KP(s) = s^4 + 20s^3 + 150s^2 + (500 + 2.5K)s + 10K = 0 \qquad (6.23)$$

Setting $s = j\omega$ and simplifying yields the two requirements

$$\omega^4 - 150\omega^2 + 10K = 0 \tag{6.24}$$

$$\omega(-20\omega^2 + 500 + 2.5K) = 0 \tag{6.25}$$

From Eq. (6.25) we see that $\omega = 0$ or $\omega^2 = (500 + 2.5K)/20$. Obviously the second requirement is the desired one. Substituting this ω^2 into Eq. (6.24) and simplifying provides

$$K^2 - 160K - 200{,}000 = 0 \tag{6.26}$$

Solving Eq. (6.26) for K yields $K = 534$. Thus

$$\omega^2 = \frac{500 + 2.5K}{20} = 91.7 \qquad \omega = \pm 9.6 \tag{6.27}$$

Although the characteristics of the RL diagrams discussed in this section are each obtained analytically without undue effort, they do not completely define the RL paths through the complex plane. Graphical procedures which help in finding points on the RL plot are discussed in the next section.

6.3 GRAPHICAL PROCEDURES FOR CONSTRUCTING ROOT LOCUS DIAGRAMS

A direct trial and error graphical procedure for obtaining points on the root locus diagram may be established from the requirement that all points $\hat{s}$ on the root locus satisfy the condition

$$\sum_{\text{zeros}} \alpha_i - \sum_{\text{poles}} \beta_j = 180° + n360° \tag{6.28}$$

where n may take on any integer value, the α_i are the angles of the phasors drawn from the open-loop zeros to the point $\hat{s}$, and the β_j are the angles of the phasors from the open-loop poles to the point $\hat{s}$. A protractor could be used to check the suitability of test points using Eq. (6.28) to whatever degree of accuracy seems appropriate. The spirule is a transparent special-purpose protractor with a long measuring arm and appropriate scales designed specifically for this purpose by Evans. (Spirules may be obtained from the Spirule Co., 9728 El Venado, Whittier, Calif., 90603.) With a bit of practice, angles may be summed algebraically in rapid succession using a spirule, and trial and error on two or three test points using the method clearly outlined in the instructions enclosed with each spirule converges rapidly to a point for which the angle condition is satisfied. A scale is also provided on the spirule for finding $|P(s)/Q(s)|$, and thus to obtain the required value of K at any point on the RL.

The RL construction procedure may be considerably simplified without requiring a spirule if a bit of trouble is taken to develop appropriate materials. The constant-phase contribution loci for each open-loop pole (zero) are straight lines emanating from the critical frequency location. It is a simple matter to prepare a dozen or so transparent acetate sheets with rays emanating from a point in the middle of the sheet at regular angular intervals of 5°. One of these sheets may be placed over each pole and zero of an appropriately-scaled open-loop pole-zero diagram, with the zero-angle rays parallel to the positive real axis in each case. In most cases of interest there are no more than six or eight open-loop poles and zeros. The angle contribution of each pole and zero at every point in the region of interest is made immediately apparent, and it is a simple matter to determine a number of points on the RL diagram. Accuracy is comparable to that obtained with the spirule; there is no problem with rotor slip; and the procedure is straightforward. The effect of moving a pole or zero is clearly apparent. The RL may be sketched on the uppermost transparency and redrawn on a sheet of graph paper later, if desired, to eliminate the distracting influence of the straight lines showing through the transparent sheets.

The determination of several points on an RL diagram using this procedure is illustrated by the situation shown in Fig. 6.7. A simple case involving only three poles is taken to avoid introducing an excessively confusing grid of constant-angle rays, and our attention is focused upon the RL branch in the second quadrant. Assume that

$$A(s) = \frac{K}{s(s+1)(s+2)} \tag{6.29}$$

We know that $OA = -1$, $AA = 180°$ and $\pm 60°$, and the real axis breakaway point [see Eq. (6.17)] is at -0.423. It is also easily shown that $K = 6$ and $\omega = \pm\sqrt{2}$ are the values at the $j\omega$ axis intersection points. To find a point on the RL above the real axis in the second quadrant, we establish the angles at two or three grid points by adding together all contributions there, and interpolate the results. The values at several grid points are shown on the diagram. For example, we get

$$s = -0.4 + j0.1 \qquad -9° - 166° - 3° = -178°$$
$$s = -0.5 + j0.1 \qquad -10.5° - 169° - 3.5° = -183°$$

so the root locus branch moves up, away from the real axis breakaway point into the second quadrant along a nearly vertical path. Moving further upward,

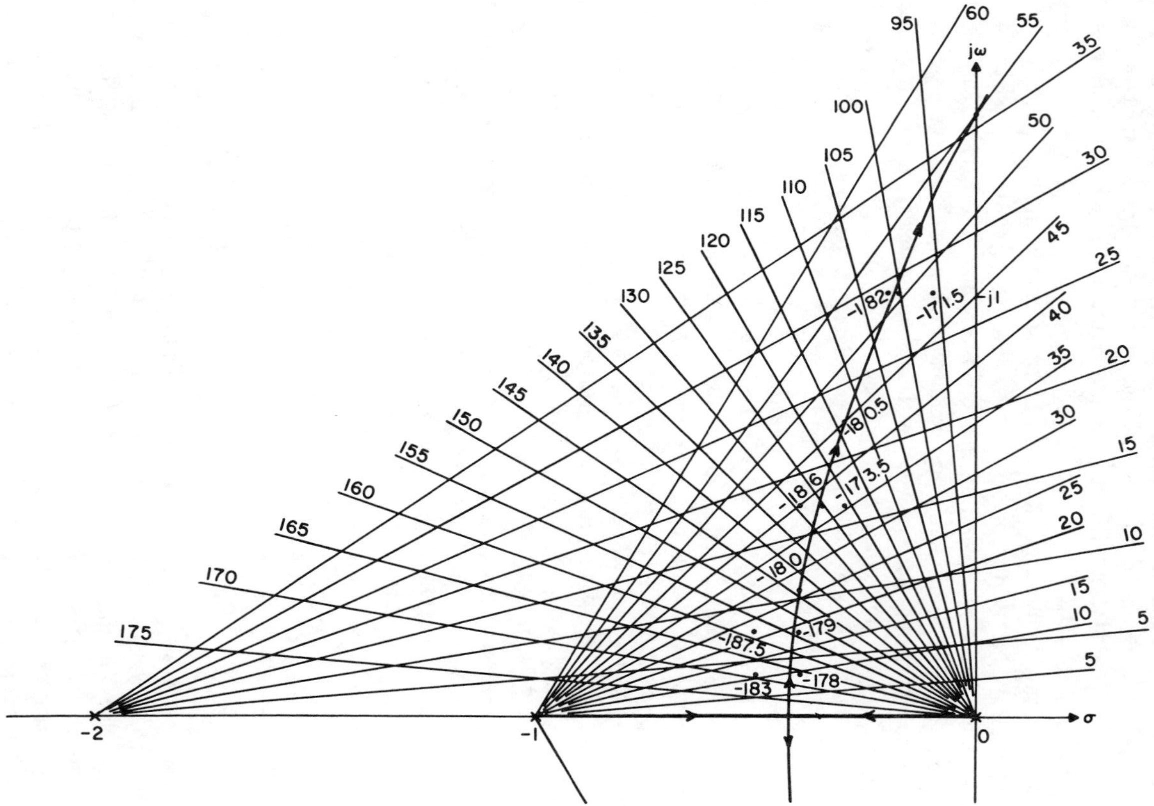

FIG. 6.7. Construction of the root locus from the pole and zero angle contribution loci.

we find:

$s = -0.4 + j0.2$	$-154 - 18 - 7 = -179°$
$s = -0.5 + j0.2$	$-158 - 21.5 - 8 = -187.5°$
$s = -0.4 + j0.3$	$-143 - 11 - 26 = -180°$
$s = -0.4 + j0.5$	$-40 - 128.5 - 17.5 = -186°$
$s = -0.3 + j0.5$	$-16.5 - 121 - 36 = -173.5°$
$s = -0.3 + j0.7$	$-45 - 113 - 22.5 = -180.5°$
$s = -0.2 + j1.0$	$-101 - 52 - 29 = -182°$
$s = -0.1 + j1.0$	$-95.5 - 28 - 48 = -171.5°$

The RL is easily sketched through this region as shown, and all of the calculations required can be carried out in a few minutes. The procedure is not appreciably altered when larger numbers of poles and zeros are involved. The calculations are more difficult when more critical frequencies are introduced only because more numbers must be added and subtracted to find the desired grid point values.

6.4 DESIGN USING ROOT LOCUS METHODS

It is suggested that the RL be used primarily as a tool to aid in establishing final values of gain and equalizer parameters. This is appropriate because of the relative difficulty encountered in accurately constructing RL diagrams as opposed to Bode diagrams or Nyquist plots, as shown later. Equalization requirements are more readily visualized on the Bode diagram but the effects of a particular equalizer upon closed-loop system performance are more thoroughly illustrated by the RL plot. Closed-loop frequency response and performance in the time domain may be obtained easily from the RL plot, as shown in Sec. 6.5. Thus a design procedure which often works well is to "rough out" one or more tentative designs on the Bode diagram, sketch the corresponding RL plots, and adjust the equalizer pole-zero locations and gain level (or choose between several alternatives) using the supplementary frequency and time domain information provided by the RL representation. In many cases this latter step is bypassed in favor of a simulation (or prototype) study, but the net effect is basically the same.

The fundamental theorems developed in Sec. 6.2 are useful in determining the effects of adding or moving equalizer poles and zeros in the s-plane. Each pole added generates a new locus branch, and one of the locus branches must terminate on each zero. The asymptotic performance as gain approaches infinity changes when one or more poles are added, and the asymptotic origin and locus center of gravity shifts with each added pole or zero. A branch of the RL may often be "pulled" out of an undesirable area by careful placement of a zero upon which that branch must terminate. These situations are illustrated by the following example.

Example 6.2

Consider a standard unity feedback configuration with

$$A(s) = \frac{K}{s(s+1)} \tag{6.30}$$

In the absence of compensation, the RL for this system is as shown in Fig. 6.8, with the closed-loop roots proceeding toward infinity along the asymptotes as $K \to \infty$. Suppose we wish to improve the transient performance for a fairly large value of K. This can be done by placing the lead equalizer with transfer function

$$G_e(s) = \frac{1 + s/5}{1 + s/25} \tag{6.31}$$

in series with $A(s)$. The RL is changed to that shown in Fig. 6.9, from which it is obvious that the transient performance will "settle out" much more rapidly for large values of K than in the unequalized system. A reasonable damping level is obtained as long as K is not increased too far, and the system remains stable for all $K > 0$. The effect of the equalizer is to provide conservative transient performance as long as the gain level and closed-loop bandwidth are not pushed up too far. For all positive K within the range where the lead equalization is effective, the rate of transient response decay is increased because the RL is pulled further into the lhp by adding the equalizer pole and zero.

An interesting contrast is provided by considering the lag equalizer described by

$$G_e(s) = \frac{1 + s/0.2}{1 + s/0.04} \tag{6.32}$$

The RL for this case is shown in Fig. 6.10. Whereas it is immediately clear from the RL diagram how lead equalization improves performance, that is not the case when lag equalization is used. A casual comparison of Figs. 6.8 and 6.10 might lead us to believe that the uncompensated system provides superior performance for all values of K. Actually, a closer look shows that lag compensation

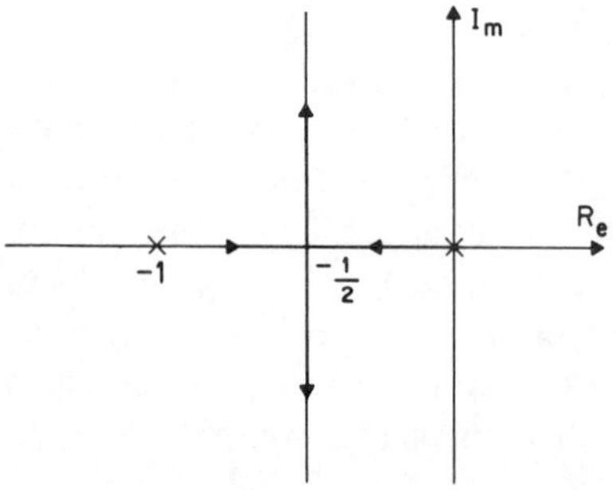

FIG. 6.8. Root locus for the uncompensated system used in Ex. 6.2.

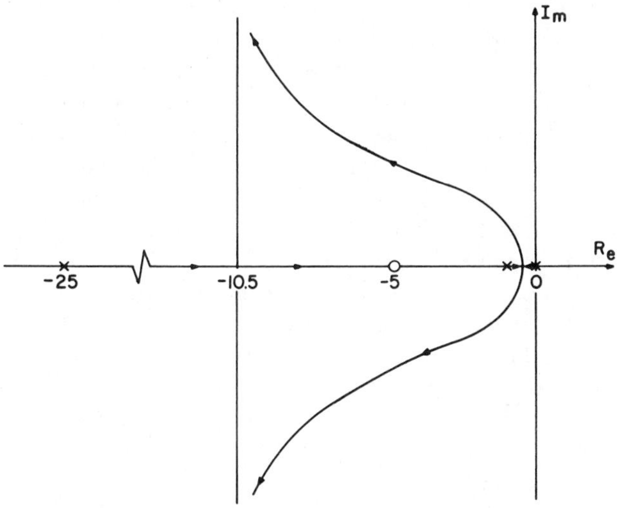

FIG. 6.9. The effect of lead equalization upon the system considered in Ex. 6.2.

reduces the rate at which the RL moves off toward infinity, and provides an improved balance between closed loop bandwidth and damping factor for values of K near the intended setting, but the rationale upon which lag equalization is based is not clearly illustrated by the RL diagrams alone.

Design using the RL diagram is reasonably straightforward, although it is somewhat time-consuming to construct families of RL plots to illustrate the

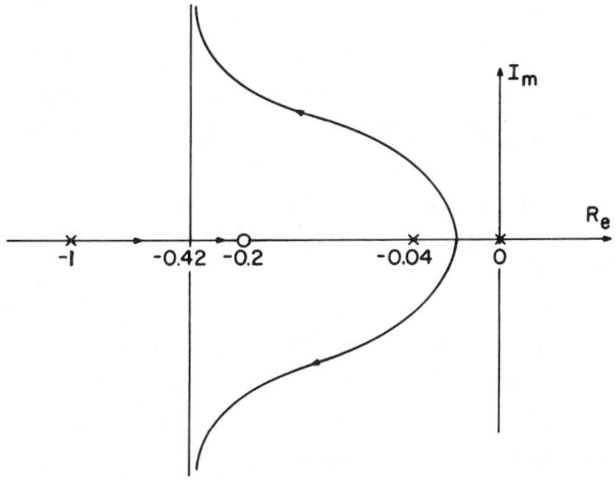

FIG. 6.10. The effect of lag equalization upon the system consider in Ex. 6.2.

effects of adding equalizer poles and zeros in various patterns. As with most design procedures, there is no substitute for experience in helping to visualize what happens to performance as various equalizer choices are made, and it is not generally necessary to worry about RL diagram accuracy until the final configuration has been chosen. Whenever a zero is added to the RL diagram, a branch of the diagram must terminate upon it. Thus the paths of troublesome branches of the RL may often be moved in the desired directions by the judicious placement of zeros. Of course, all equalizers must be realizable, so equalizer poles and zeros must be added in pairs.

It is easy to see how equalizer parameters should be chosen using the Bode diagram methods discussed in Chap. 9, so we will not consider the design problem further at this point. Our primary objective here is to provide a means by which closed-loop controller performance characteristics can be displayed, and that objective has been satisfied. As we use the RL diagram to supplement the analyses considered in subsequent chapters, its utility in the design phase will be further illustrated.

6.5 FREQUENCY AND TRANSIENT RESPONSE FROM THE ROOT LOCUS DIAGRAM

Steady-state, closed-loop magnitude and phase versus frequency (frequency response) plots are easily obtained from the RL diagram. For the usual unity feedback configuration, the zeros of the open-loop transfer function $A(s)$ are also zeros of the closed-loop transfer function

$$\frac{C(s)}{R(s)} = \frac{A(s)}{1 + A(s)} \tag{6.33}$$

These zeros are shown on the RL diagram, as are the closed-loop poles, or the factors of the characteristic equation $1 + A(s) = 0$. A value $s = s_1$ on the $j\omega$ axis provides magnitude and phase for the individual zeros as given by drawing a phasor from the zero of interest to $s = s_1$ and observing its length and angle. Poles are treated similarly. The overall magnitude is obtained by dividing the product of the numerator magnitudes by the product of the denominator magnitudes. The angle is the sum of the numerator angles minus the sum of the denominator angles. This procedure is certainly familiar, and the RL diagram often proves helpful in carrying out the operations graphically.

The RL is also helpful in evaluating the time response of the system, since it may be used, if desired, as an aid in the graphical calculation of the residue at each pole of the closed-loop system. For the usual unity-feedback, single-loop configuration described by Eq. (6.33), the poles of the closed-loop transfer function can be obtained directly from the RL plot for each specific K. The unit-impulse and unit-step responses are readily determined using a partial fraction expansion. The general transfer function form is

$$\frac{C(s)}{R(s)} = \frac{\prod_{i=1}^{M}(s - s_i)}{\prod_{j=1}^{N}(s - s_j)} \qquad N > M \tag{6.34}$$

The value of $C(s)$ for a unit-step input, for example, is thus given by

$$C(s) = \frac{\prod_{i=1}^{M}(s - s_i)}{s\prod_{j=1}^{N}(s - s_j)} = \frac{K_0}{s} + \frac{K_1}{s - s_1} + \ldots + \frac{K_N}{s - s_N} \tag{6.35}$$

where we assume that all poles of $C(s)$ are simple. The constants K_i are obtained as

$$K_i = (s - s_i)\,C(s)\Big|_{s=s_i} \tag{6.36}$$

and they are easily evaluated graphically from the RL plot, if desired, by observing that each factor in the numerator and denominator of $C(s)$ is of the form $(s_i - s_j)$ and is the vector from s_j to s_i. The length and angle of each such term is readily measured with a ruler and protractor (or the spirule) and the ratio of products is obtained directly. The corresponding time response component is given by

$$\mathfrak{L}^{-1}\left[\frac{K_i}{s - s_i}\right] = K_i\epsilon^{s_i t} \tag{6.37}$$

and the total time response is obtained by repeated application of these basic steps and reduction of the results to the standard forms.

6.6 CONCLUDING COMMENTS

The RL provides a valuable means of displaying the closed-loop performance characteristics of typical control systems in the complex plane as a function of the gain level chosen. Because both the time and frequency-response characteristics depend in a known way upon pole and zero locations, the RL diagram shows how performance characteristics in both domains depend upon the open-loop gain. Although design procedures based entirely upon the RL diagram are available, design on the Bode diagram, as discussed in Chap. 9, is generally easier to visualize and carry out. It is thus suggested that the RL be used primarily as a means for illustrating system performance and perhaps as an aid in finalizing the design parameters. Of course, there are cases, such as open-loop unstable systems, in which design on the Bode diagram is not so straightforward, and the RL diagram is very useful as a design tool in these applications. Examples of this nature are considered in later chapters.

The RL method may be extended considerably beyond our presentation. It is possible, for example, to obtain an expression for the slope of the RL at any point as an aid in construction.[2] Since use of the spirule or the alternate phase locus overlay method makes it a simple matter to find points on the RL, the use of RL slope as a construction aid has not been considered. The RL method may be extended to multiple-loop systems, although such applications are cumbersome if gain (or any other parameter for that matter) within an interior loop is to be changed. The RL diagram may be constructed as a function of parameters other than gain, although this is not a common, nor a particularly useful, practice. If, for example, a particular time constant is to be changed, the open-loop transfer function is simply manipulated until that time constant, or its reciprocal, appears as a multiplying factor in the same general way that gain normally does. For each specific value of the time constant, points on the RL satisyfing the angle criterion may be obtained in the usual way. Of these, only one point on each RL branch also satisfies the magnitude requirement, and these are the desired RL points. Each time the time constant is changed, the corresponding pole (or zero) must be moved appropriately, and the process is repeated until the desired RL is obtained to acceptable accuracy. This procedure is too involved to be feasible as a hand analysis tool in a modern computer-oriented society.

The RL construction procedure may be greatly simplified without requiring a spirule if a bit of trouble is taken to develop appropriate materials, and the procedure for doing this is presented here as a suggested alternate approach which the author has found very efficient. Points on the RL are easily found using pole-zero angular locus transparencies, since the angular contribution from each pole and zero at any test point is clearly indicated. Accuracy is comparable to that obtained with the spirule, and there is no problem with rotor slip. The RL is readily sketched with a grease pencil on the uppermost sheet and may be used in that form or copied, whichever is most convenient.

The RL method may be extended to the consideration of systems with time delay, or with other transfer characteristics of a nonrational nature,[2,3] but these are highly specialized situations. The RL method also applies to the more general problem of solving algebraic equations and certain types of transcendental equations, but these problems are of no specific interest to us here.

In summary, we conclude that the RL method provides a valuable tool for the analysis and design of closed-loop control systems. Our interest here is focused upon the analysis problem, since more appropriate design methods are introduced later. Frequency and time-domain information is readily available from the RL plot, and its construction to acceptable engineering accuracy is straightforward.

REFERENCES

1. Evans, W. R., Graphical Analysis of Control Systems, *Transactions AIEE,* vol. 67, pp. 547–551, 1948.
2. Bower, J. L., and P. M. Schultheiss, "Introduction to the Design of Servomechanisms," John Wiley & Sons, Inc., New York, chp. 9.
3. Truxal, J. G., "Automatic Feedback Control System Synthesis," McGraw-Hill Book Co., Inc., New York, chp. 4, 1955.

7 STEADY STATE SINUSOIDAL RESPONSE CHARACTERISTICS

We are interested in the steady-state sinusoidal response characteristics of systems and system elements and their relationships with the transfer function, stability, and transient response patterns for the following reasons:

1. The procedures thus far discussed have failed to provide any straightforward design approach for synthesizing closed-loop systems and we hope to overcome this limitation using frequency-domain methods.
2. It is often necessary or convenient in practice to determine the transfer characteristics of system elements by observing their response to a test input rather than by detailed analysis. A sinusoidal input of variable frequency is readily available as a test signal for this purpose.

Design and analysis in the frequency domain using Bode and Nyquist diagram methods turns out to be particularly straightforward. Our purpose in this chapter is to establish a solid foundation of frequency-domain method understanding.

As a direct consequence of our stability definition, the response of any stable system to a bounded sinusoidal input must be bounded. The steady-state sinusoidal output of a system has both phase and amplitude relationships to the input at each frequency. The patterns these relationships take with frequency are sufficient to define the system transfer function. Furthermore, the amplitude and phase responses of a linear system with all poles and zeros in the lhp are not independent. We shall define the relationship between them. It is this relationship between amplitude and phase responses, perhaps more than any other single factor, that ultimately forms the basis for the frequency-domain (Bode diagram) design procedures introduced in Chap. 9.

In Chap. 9 the development of methods for adequately approximating the amplitude and phase responses of linear systems as a function of frequency are emphasized. It is important to develop a facility for sketching these response patterns directly from the system transfer function. We must also be able to reverse the process, approximating the transfer function from frequency-response data obtained in the laboratory. Since frequency and phase responses are used often in our subsequent discussions, graphical approximations are introduced to reduce the effort involved. The errors resulting from the approximations are evaluated, some correction curves presented, and conditions indicated under which corrections are called for.

7.1 RELATIONSHIPS BETWEEN STEADY-STATE SINUSOIDAL RESPONSE, THE TRANSFER FUNCTION, AND STABILITY

Consider a linear system with input and output terminal pairs. We define the system steady-state sinusoidal response $\mathbf{G}(j\omega)$ as the phasor ratio of its steady-state output $\mathbf{C}(j\omega)$ to the input phasor $\mathbf{R}(j\omega)$, or

$$\mathbf{G}(j\omega) = \frac{\mathbf{C}(j\omega)}{\mathbf{R}(j\omega)} = A + jB = |\mathbf{G}|\epsilon^{j\theta_G} \tag{7.1}$$

where $\mathbf{C}(j\omega)$, $\mathbf{R}(j\omega)$ and $\mathbf{G}(j\omega)$ are complex numbers, or phasors. Since $\mathbf{G}(j\omega)$ is complex, it can only be defined for each ω using two real numbers which are usually chosen as the real and imaginary parts of $\mathbf{G}(j\omega)$, denoted by A and B in Eq. (7.1), respectively, or its magnitude and phase angle, denoted by $|\mathbf{G}|$ and θ_G in Eq. (7.1), respectively, whichever is most convenient. The *frequency response* of the system with transfer function $\mathbf{G}(j\omega)$ is defined by any complete description of $\mathbf{G}(j\omega)$ as ω takes on all values over the range $[0, \infty]$, and is easily obtained in the laboratory using standard test equipment.

Several general comments are in order before proceeding. By *steady-state response*, we mean the response after all transient effects have become negligible. Obviously no steady-state response exists for an unstable system, so a frequency-response test will directly indicate instability. The frequency response of a linear system is measured in the laboratory as follows:

1. A sinusoidal test input of known amplitude and frequency is applied to the system.

2. The system output is allowed to "settle" into a steady-state pattern.
3. The amplitude and relative phase of the sinusoidal output is measured and recorded.
4. This procedure is repeated for values of ω spanning the frequency range of interest.
5. The results are sketched in any of several forms discussed later in this chapter.

It is assumed the reader is familiar with this test procedure and the phasor notation introduced in Eq. (7.1).

We may obtain $\mathbf{G}(j\omega)$ from $G(s)$ by simply substituting $j\omega$ for s. This relationship between the steady-state sinusoidal response and the s-domain transfer function is developed in most introductory circuits courses. It is now apparent that any linear system is completely characterized by $\mathbf{G}(j\omega)$ since $\mathbf{G}(j\omega)$ is equivalent, by setting $s = j\omega$, to $G(s)$, and $G(s)$, or the corresponding impulse response function $g(\tau)$, allows us to obtain the output of the system in response to any known input.

7.2 GRAPHICAL REPRESENTATION OF AMPLITUDE AND PHASE CHARACTERISTICS

Several graphical representations of the phasor $\mathbf{G}(j\omega)$ are available. We could sketch the real and imaginary parts of $\mathbf{G}(j\omega)$ versus ω. This procedure is often applied in network synthesis, but is not generally found useful in controller analysis or design. We can sketch $|\mathbf{G}(j\omega)|$ and θ_G versus ω, or we can construct a polar plot of these data. Both of these methods are common. The polar plot provides the basis for the Nyquist stability criterion,[1] discussed in Chap. 8, from which virtually all classical design procedures have evolved. Magnitude plots, plus a bit of supplementary phase information, are used in the Bode diagram design methods introduced in Chap. 9. We consider $|\mathbf{G}(j\omega)|$ and θ_G versus ω plots in this section. Polar plots are discussed in Chap. 8.

Consider the RC circuit shown in Fig. 7.1. This arrangement is called a lag network because the sinusoidal output lags the input in phase for all positive ω, as shown in Fig. 7.2. The transfer function of this network is

$$\frac{E_0(s)}{E_i(s)} = \frac{1}{1 + RCs} = G(s) \tag{7.2}$$

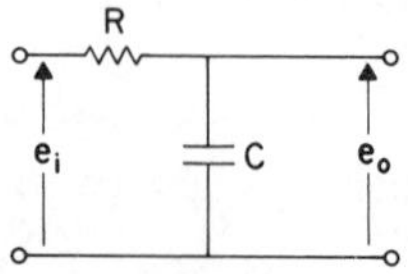

FIG. 7.1. An RC lag network.

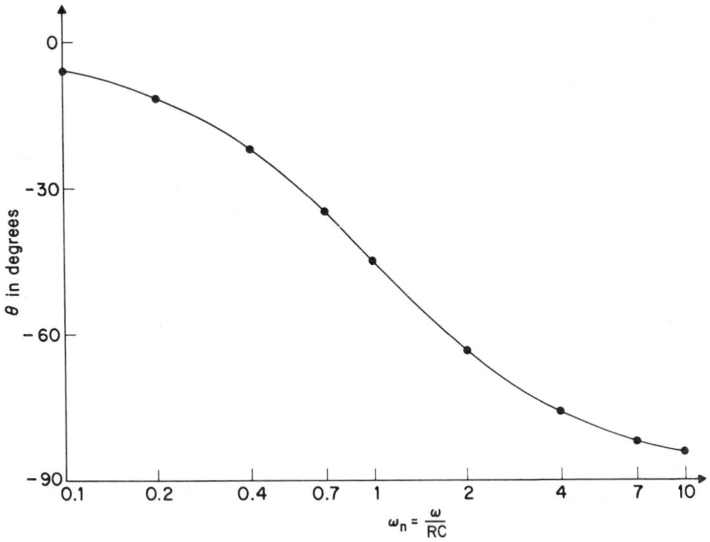

FIG. 7.2. θ versus ω for an RC lag network.

Substituting $s = j\omega$ in Eq. (7.2) provides the harmonic (steady-state) transfer function as

$$\frac{\mathbf{E}_0(j\omega)}{\mathbf{E}_i(j\omega)} = \frac{1}{1 + j\omega RC} = \mathbf{G}(j\omega) \tag{7.3}$$

The magnitude of $\mathbf{G}(j\omega)$ is thus

$$|\mathbf{G}(j\omega)| = \frac{1}{\sqrt{1 + \omega^2 R^2 C^2}} \tag{7.4}$$

which can be approximated as

$$|\mathbf{G}(j\omega)| \simeq \begin{cases} 1 & \omega << 1/RC \\ 1/\omega RC & \omega >> 1/RC \end{cases} \tag{7.5}$$

Only when ω is of the same order of magnitude as $1/RC$ does appreciable error arise in applying one or the other of the approximations given in Eq. (7.5). At $\omega = 1/RC$, $|\mathbf{G}(j\omega)| = \sqrt{2}/2 = 0.707$, and either of the approximations in Eq. (7.5) is in error by roughly 43%. When $\omega \geq 5/RC$ or $\omega \leq 1/5RC$, the approximations in Eq. (7.5) yield results within about 2% of the actual values, and are well within the usual range of desired engineering accuracy. A plot of $|\mathbf{G}(j\omega)|$ versus

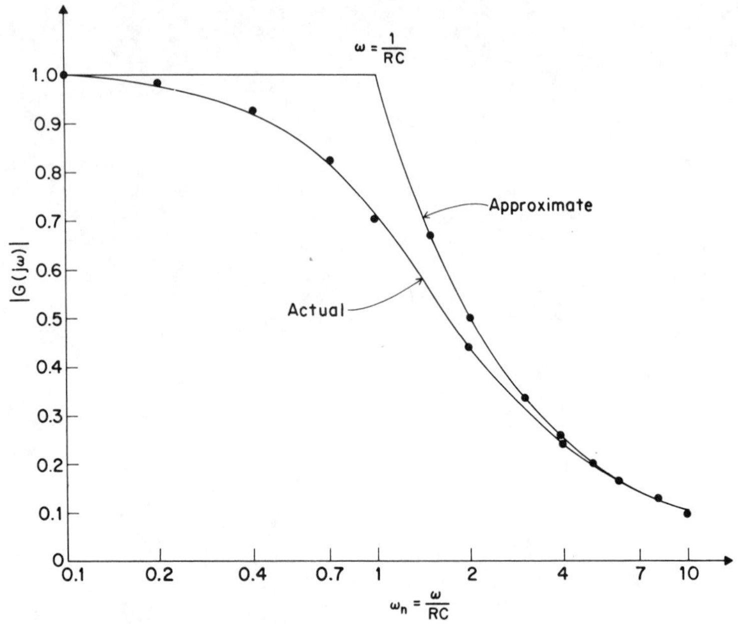

FIG. 7.3. $|G(j\omega)|$ versus ω for an RC lag network.

ω for this case is shown in Fig. 7.3. The approximations in Eq. (7.5) are also shown for comparison, with the transition between the two segments taking place at $\omega = 1/RC$.

The phase of $\mathbf{G}(j\omega)$ as given in Eq. (7.2) is

$$\theta_G = -\tan^{-1}\omega RC \tag{7.6}$$

and a phase plot shown in Fig. 7.2. The phase lag approaches $0°$ as $\omega \to 0, 90°$ as $\omega \to \infty$, and is $45°$ at $\omega = 1/RC$.

Consider the RC network shown in Fig. 7.4. Assume, to simplify the analysis and illustrate several points of interest, that the second RC network does not appreciably load the first. This assumption is exact, for example, if the two networks are separated by an ideal unit-gain-buffer amplifier, and is not unreasonable when the circuit components are arranged as shown, if the impedance level of

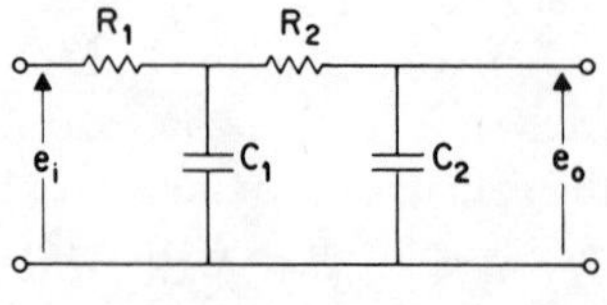

FIG. 7.4. Two RC networks in cascade.

the second stage (R_2 in series with C_2) is very much larger than that of the first stage. The harmonic transfer function, assuming no interstage loading, is

$$\frac{E_0(j\omega)}{E_i(j\omega)} = \frac{1}{1 + j\omega R_1 C_1} \times \frac{1}{1 + j\omega R_2 C_2} = G(j\omega) \tag{7.7}$$

We may evaluate $|G(j\omega)|$ as the product of the magnitudes of the individual terms in Eq. (7.7). Plots of the first and second stage magnitudes, and their product, $|G(j\omega)|$, are shown in Fig. 7.5. It is obviously rather awkward to plot these individual magnitude functions, subsequently evaluating their product, at enough points to obtain the overall harmonic response. When systems of higher order are encountered, and they are common in control applications, this procedure gets completely out of hand. Fortunately this difficulty is easily avoided, as shown in Sec. 7.3.

The total phase shift corresponding to the product of two harmonic response functions is simply the algebraic sum of the phase shifts for the individual terms forming the product. Thus it is easy to determine the phase shift of complex networks. The individual and total phase shift terms corresponding to $G(j\omega)$ as defined in Eq. (7.7) are shown in Fig. 7.6.

The relative ease with which we are able to *add* the phase curves of cascaded harmonic transfer functions to find the overall phase characteristic leads us to seek a similar procedure for treating their magnitudes. Since the logarithm (log) of a product is the sum of the logs of the individual terms making up that product, it seems natural to characterize the magnitudes of product terms by their logs. There are several additional factors in favor of a logarithmic magnitude and frequency representation. The elements of a control system typically involve

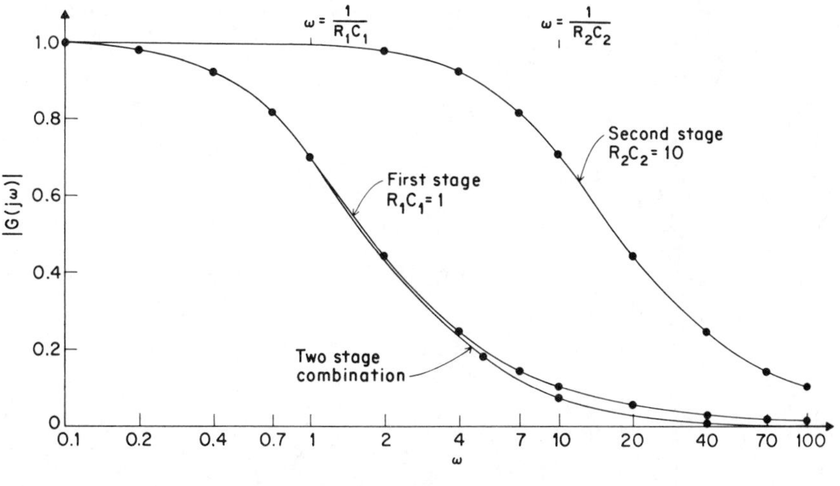

FIG. 7.5. Magnitude plots for the network shown in Fig. 7.4.

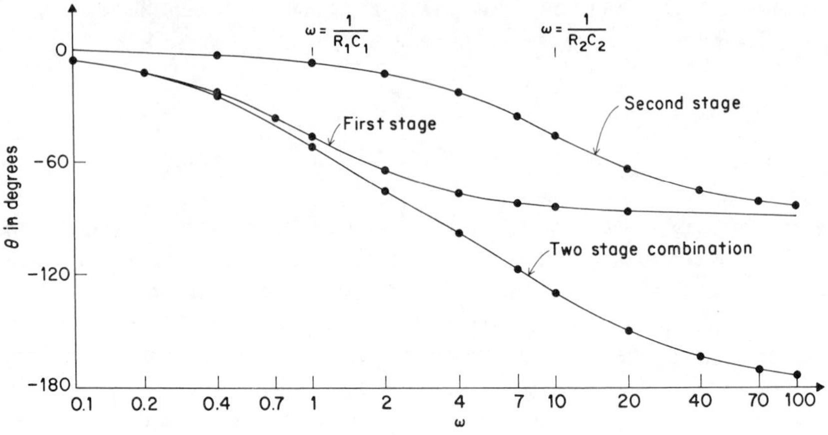

FIG. 7.6. Phase plots for the network shown in Fig. 7.4.

a diversity of frequency-response characteristics. Some have wide bandwidth (BW) while other have narrow BW. The transfer function magnitude often changes by a factor of 1000 or more over the frequency range of interest. It is thus convenient to use logarithmic frequency and magnitude scales which allow us to sketch a greatly expanded range of magnitudes and frequencies accurately on a single plot. The treatment of transfer functions using logarithmic scales is considered in the next section.

7.3 ASYMPTOTIC MAGNITUDE (α) DIAGRAMS, OR BODE DIAGRAMS

It is convenient to consider a specific situation for our introduction to log plots. Extension to the general case follows directly. Consider a system with transfer function

$$\frac{C(s)}{R(s)} = \frac{K(1 + s/\omega_1)}{(1 + s/\omega_2)(1 + s/\omega_3)} = \frac{K(1 + \tau_1 s)}{(1 + \tau_2 s)(1 + \tau_3 s)} \tag{7.8}$$

where the time constants τ_i are related to the frequencies ω_i, $i = 1, 2, 3$, by $\tau_i = 1/\omega_i$. The corresponding harmonic-response function is

$$\frac{C(j\omega)}{R(j\omega)} = \frac{K(1 + j\omega/\omega_1)}{(1 + j\omega/\omega_2)(1 + j\omega/\omega_3)} \tag{7.9}$$

The magnitude of this harmonic-response function is the ratio of the magnitudes of the individual factors, or

$$\left|\frac{C(j\omega)}{R(j\omega)}\right| = \frac{|K|\,|1 + j\omega/\omega_1|}{|1 + j\omega/\omega_2|\,|1 + j\omega/\omega_3|} \tag{7.10}$$

Taking the logarithm of both sides of Eq. (7.10) provides

$$\log\left|\frac{C(j\omega)}{R(j\omega)}\right|$$

$$= \log|K| + \log\left|1 + \frac{j\omega}{\omega_1}\right| - \log\left|1 + \frac{j\omega}{\omega_2}\right| - \log\left|1 + \frac{j\omega}{\omega_3}\right| \tag{7.11}$$

Consider $\log|1 + j\omega/\omega_1|$ as a typical term on the right side of Eq. (7.11) with ω and ω_1 positive. Then

$$\log|1 + j\omega/\omega_1| \simeq \log 1 = 0 \qquad \omega \ll \omega_1 \tag{7.12}$$

and

$$\log|1 + j\omega/\omega_1| \simeq \log\left|\frac{\omega}{\omega_1}\right| = \log\omega - \log\omega_1 \qquad \omega \gg \omega_1 \tag{7.13}$$

Suppose we now make the assumptions, inaccurate as they may seem, that Eq. (7.12) applies for all $\omega < \omega_1$, and Eq. (7.13) applies for all $\omega > \omega_1$. This is called an asymptotic approximation for reasons illustrated in Fig. 7.7, where approximate and actual values of $|1 + j\omega/\omega_1|$ are plotted on a vertical log scale versus ω represented on a horizontal log scale. This is called a log-log plot. Obviously the approximation given by Eq. (7.12) must plot as a straight line and, since the derivative of the right-hand side of Eq. (7.13) with respect to $\log\omega$ is unity, this approximation is represented by a straight line with unit slope on log-log coordinates. The term asymptotic is appropriate, since the approximate and actual magnitudes become arbitrarily close as $\omega \to 0$ and as $\omega \to \infty$. The maximum error between the approximate and actual curves occurs at $\omega = \omega_1$, the so-called *break point* between the asymptotic segments. At $\omega = \omega_1$, $|1 + j(\omega/\omega_1)| = |1 + j1| = \sqrt{2}$, whereas the asymptotic value is 1. Thus the maximum error is approximately 30% and occurs at $\omega = \omega_1$. In our applications of asymptotic diagrams we are seldom interested in accuracy near the break points. Corrections are easily made from the asymptotic representation when necessary. The corrections required over a two decade range are indicated in Fig. 7.7.

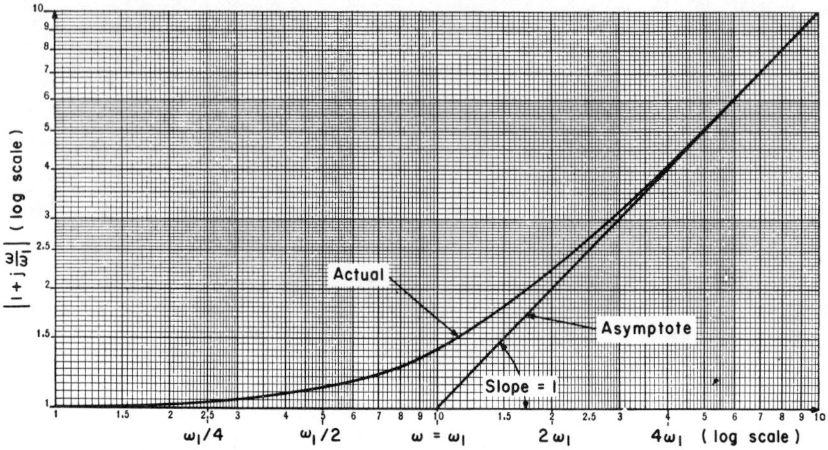

FIG. 7.7. Asymptotic and actual plots of $|1 + j(\omega/\omega_1)|$ on log-log coordinates.

It is now possible to build up an entire asymptotic representation of the function $|\mathbf{C}(j\omega)/\mathbf{R}(j\omega)|$ as given in Eq. (7.10) and (7.11), or any similar function expressed as the product of terms in factored form. Each positive term in Eq. (7.11) comes from the numerator of Eq. (7.10), and its asymptotic representation takes the form shown in Fig. 7.7, with straight line segments having slopes of 0 and +1. The denominator factors in Eq. (7.10) are negative in Eq. (7.11), and they are each represented asymptotically by straight line segments with slopes of 0 and -1. The break point between the two line segments in each case is at that frequency where the real and imaginary parts of the term under consideration are equal in magnitude. It is a simple matter to add these individual terms on logarithmic coordinates to obtain an asymptotic approximation, since we need only add the slopes of the asymptotic representations of all terms in the product function to find the slope of the overall approximation at any frequency. Knowledge of the magnitude at any given frequency provides the necessary reference level. Thus the asymptotic representation of a general product function consists of a series of straight line segments with integer slopes $\pm n$. The break points occur at those frequencies where the real and imaginary parts of the individual terms in the product function have equal magnitudes. The asymptotic and approximate magnitudes of the function introduced in Eq. (7.9) are plotted on log-log coordinates in Fig. 7.8. The actual curve is readily obtained, with reasonable accuracy, from the asymptotic one without detailed calculations by noting the corrections required at the break points and sketching in the remainder, keeping in mind the general characteristics shown in Fig. 7.7. This plot is called a *log magnitude plot*, a *Bode plot*, or an *α plot*, and these terms will be used interchangeably.

Example 7.1

Sketch the asymptotic and actual magnitude curves on log-log coordinates for the transfer function

$$G(s) = \frac{1000(s + 3)}{s(s + 12)(s + 50)} \tag{7.14}$$

It should be noted that Eq. (7.14) is not of the form introduced in Eq. (7.8). It is convenient to first reduce Eq. (7.14) to time-constant form, since this makes the real part in each term unity, thereby normalizing all terms to a common base. The reader should work through the problem directly to gain an appreciation for the relative simplicity of the time-constant form. The time constant form of Eq. (7.14) is

$$G(s) = \frac{5(1 + s/3)}{s(1 + s/12)(1 + s/50)} \tag{7.15}$$

The corresponding harmonic transfer function is

$$G(j\omega) = \frac{5(1 + j\omega/3)}{j\omega(1 + j\omega/12)(1 + j\omega/50)} \tag{7.16}$$

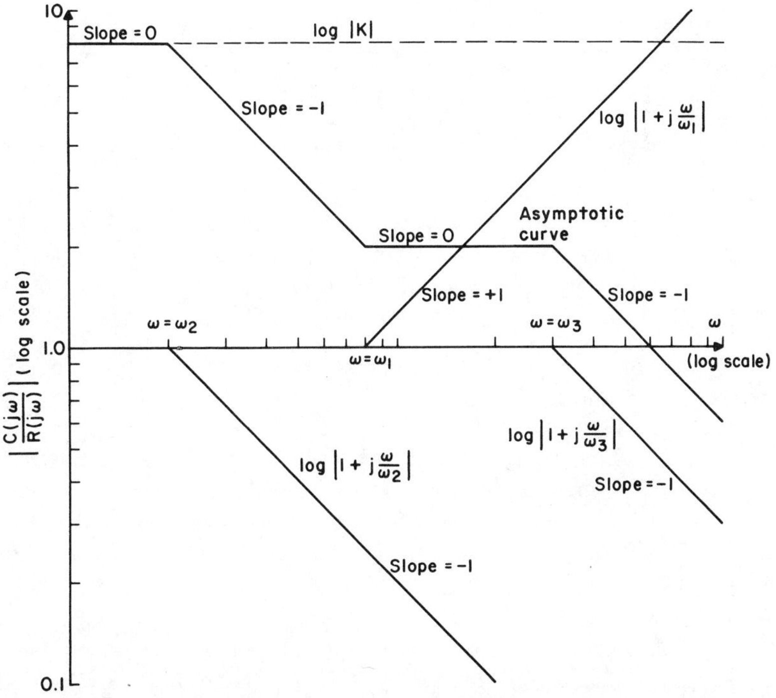

FIG. 7.8. Asymptotic and actual Bode diagram of the function given in Eq. (7.9).

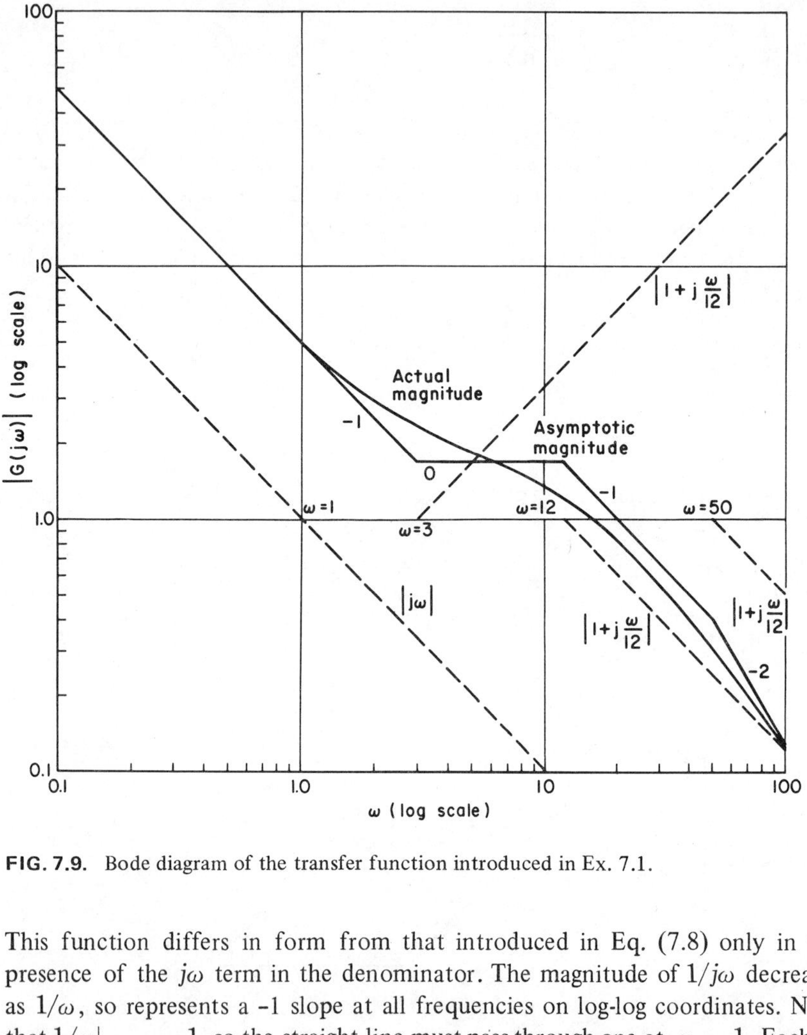

FIG. 7.9. Bode diagram of the transfer function introduced in Ex. 7.1.

This function differs in form from that introduced in Eq. (7.8) only in the presence of the $j\omega$ term in the denominator. The magnitude of $1/j\omega$ decreases as $1/\omega$, so represents a -1 slope at all frequencies on log-log coordinates. Note that $1/\omega|_{\omega=1} = 1$, so the straight line must pass through one at $\omega = 1$. Each of the other terms in Eq. (7.16) is of the form discussed previously. The α plot for this function is shown in Fig. 7.9.

Let us consider further the meaning of the integer slopes encountered on the asymptotic α diagram on log-log coordinates. A +1 slope means the function is increasing linearly with ω, to a first-order approximation, in that frequency range. Thus if we are told that $|\mathbf{G}| = 10$ at $\omega = 1$, and the slope of the Bode plot is +1 between $\omega = 1$ and $\omega = 8$, the Bode plot must pass through $|\mathbf{G}| = 80$ at $\omega = 8$, since $|\mathbf{G}(j8)|/|\mathbf{G}(j1)| = 8/1$ is required. In a region of +2 slope $|\mathbf{G}|$ increases as the *square* of ω. More generally, in a region of slope n the points on the asymptotic α diagram are related by $|\mathbf{G}(j\omega_1)|/|\mathbf{G}(j\omega_2)| = (\omega_1/\omega_2)^n$. Thus if the slope between $\omega_1 = 2$ and $\omega_2 = 5$ is -2, $|G|$ decreases by a factor of $(2.5)^2$, or 6.25, over that frequency range.

In each of the examples considered thus far the transfer function has had only real poles and zeros. Complex conjugate pairs of poles and/or zeros are often encountered, and they must be treated differently. Consider the general quadratic term

$$s^2 + 2\xi\omega_0 s + \omega_0^2 = 0 \tag{7.17}$$

which may alternately be written as

$$\frac{s^2}{\omega_0^2} + \frac{2\xi s}{\omega_0} + 1 = 0 \tag{7.18}$$

The roots of this equation are given by

$$s_{1,2} = -\frac{2\xi\omega_0 \pm \sqrt{4\xi^2\omega_0^2 - 4\omega_0^2}}{2} = -\xi\omega_0 \pm j\omega_0\sqrt{1 - \xi^2} \tag{7.19}$$

assuming $0 \leq \xi \leq 1$. Rewriting Eq. (7.17) in terms of these roots yields

$$\left(1 + \frac{s}{\xi\omega_0 + j\omega_0\sqrt{1 - \xi^2}}\right)\left(1 + \frac{s}{\xi\omega_0 - j\omega_0\sqrt{1 - \xi^2}}\right) = 0 \tag{7.20}$$

Each root has magnitude ω_0 for $0 \leq \xi \leq 1$. The asymptotic representation of each individual term in Eq. (7.20) has unit magnitude for $\omega \leq \omega_0$, changing to a straight line with +1 slope for $\omega > \omega_0$. Thus the composite Bode plot changes from 0 slope to +2 slope at $\omega = \omega_0$.

It is important to note one significant difference between a quadratic term giving rise to a pair of complex conjugate poles with magnitude ω_0 (located ω_0 units from the origin in the complex plane) and a comparable pair of superposed real poles with break frequencies at $\omega = \omega_0$. The asymptotic diagrams are identical, but for the real pole case we have

$$\left|\left(1 + \frac{s}{\omega_0}\right)^2\right|_{\omega=\omega_0} = 2 \tag{7.21}$$

whereas

$$\left|\frac{s^2}{\omega_0^2} + \frac{2\xi s}{\omega_0} + 1\right|_{\omega=\omega_0} = 2\xi \tag{7.22}$$

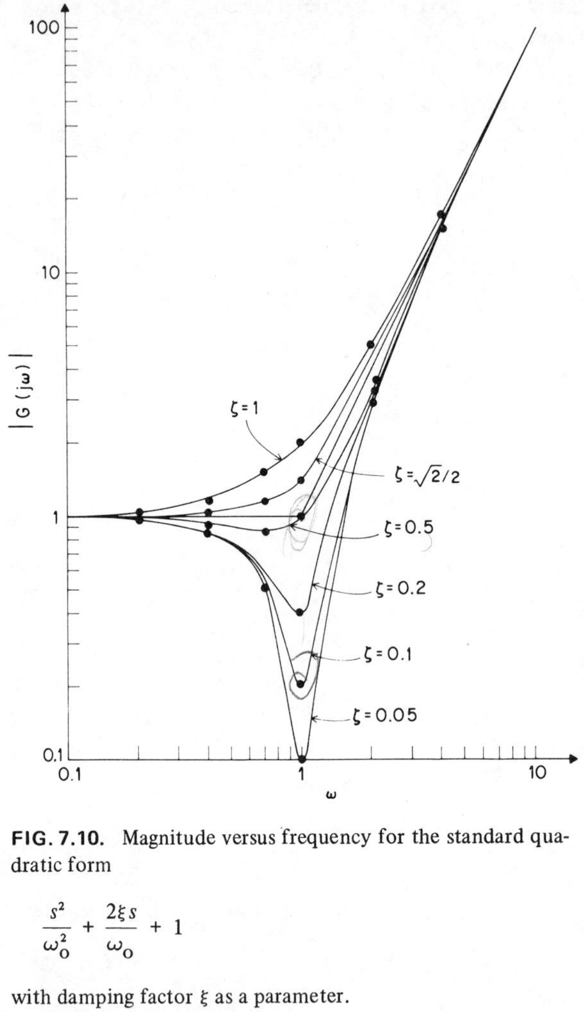

FIG. 7.10. Magnitude versus frequency for the standard quadratic form

$$\frac{s^2}{\omega_o^2} + \frac{2\xi s}{\omega_o} + 1$$

with damping factor ξ as a parameter.

Thus the actual curve cannot be drawn directly from the asymptotic curve without knowledge of ξ in any case where underdamped quadratic terms are involved. This point must be kept in mind whenever we are working with Bode diagrams. A family of curves with ξ as a parameter relating the actual magnitude versus frequency of the quadratic form given in Eq. (7.18) is shown in Fig. 7.10 for $0.05 \leq \xi \leq 1$. Note that significant errors beyond those encountered with real roots can arise at the break point when $\xi \ll 1$. Values of $\xi \geq 1$ give rise to real roots which can be treated in the usual way. Although values of $\xi < 0.05$ are possible, they are not often encountered in control applications. It is clear from Fig. 7.10 that a resonant response dip occurs only if $\xi < \sqrt{2}/2$.

In many cases the quadratic term is in the denominator and the transfer function magnitude is influenced by the reciprocal of the magnitude curves shown in Fig. 7.10. When ξ is small in a denominator quadratic pair, the transfer function tends to exhibit a resonant peak at $\omega = \omega_0$ due to the resonant dip in the denominator term. In some cases other terms in the transfer function may prevent a peak in the composite response pattern despite a rather pronounced peak from a single quadratic pair.

Example 7.2

Consider the transfer function

$$G(s) = \frac{100(s+2)}{s(s^2+5s+100)} \tag{7.23}$$

Sketch the asymptotic and actual magnitude curves.

Converting to time-constant form provides

$$G(s) = \frac{2(1+s/2)}{s(s^2/100+5s/100+1)} \tag{7.24}$$

Observe that $\omega_0 = 10$ and $\xi = 1/4$ in the quadratic term. Thus the Bode plot has -1 slope for $\omega < 2$, 0 slope for $2 \leq \omega \leq 10$, and -2 slope for $\omega > 10$, as shown in Fig. 7.11. The quadratic term magnitude at $\omega_0 = 10$ for $\xi = 1/4$ is seen from Fig. 7.10 to be 1/2, so the resonant peak reaches a maximum of 2. The remainder of the actual curve is also shown in Fig. 7.11.

It is important to emphasize the procedure by which we construct and check asymptotic diagrams on log-log coordinates. A typical term in a harmonic transfer function is factored form is $1 + j\omega/\omega_i$. The magnitude of this term is approximated as 1 for $\omega < \omega_i$, and as ω/ω_i for $\omega > \omega_i$. To check the asymptotic magnitude at any frequency, simply approximate each term this way and evaluate the resulting product. The magnitude of $\omega_i + j\omega$, a term not in time-constant form, is approximated as ω_i for $\omega < \omega_i$ and ω for $\omega > \omega_i$. The quadratic term $s^2/\omega^2 + 2\xi s/\omega_0 + 1$ is approximated as 1 for $\omega < \omega_0$ and as $\omega^2/\omega_0{}^2$ for $\omega > \omega_0$ irrespective of ξ. The following example, involving several forms of terms, illustrates this approximation procedure.

Example 7.3

Determine the asymptotic magnitude of

$$G(s) = \frac{1000(s)(s^2+20s+500)}{(s/0.1+1)(s+10)(s^2/50+s/10+1)} \tag{7.25}$$

at $\omega = 1$ and at $\omega = 10$. Find the actual magnitude at $\omega = 10$. At $\omega = 1$

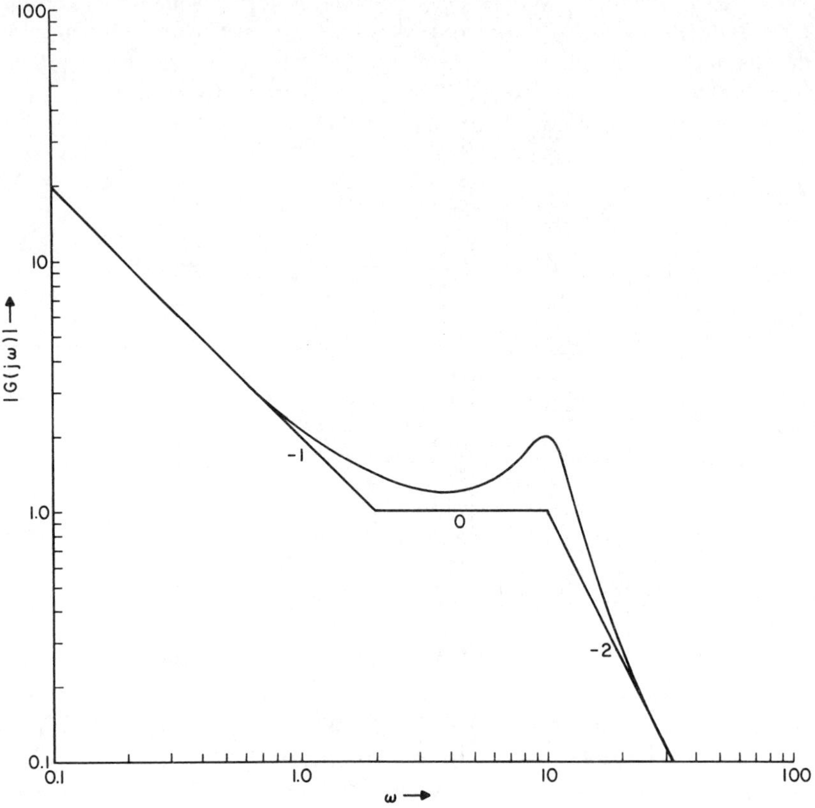

FIG. 7.11. Bode plot of the $G(s)$ introduced in Ex. 7.2.

we have

$$|G(j1)|_{\alpha} = \frac{(1000)(1)(500)}{(10)(10)(1)} = 5000 \tag{7.26}$$

where α denotes the asymptotic approximation. Similarly, at $\omega = 10$,

$$|G(j10)|_{\alpha} = \frac{(1000)(10)(500)}{(100)(10)(100/50)} = 2500 \tag{7.27}$$

The actual magnitude at $\omega = 10$ is

$$|G(j10)| = \frac{1000(10)(400^2 + 200^2)^{1/2}}{(100)(10\sqrt{2})(\sqrt{2})} = 1000\sqrt{5} \simeq 2240 \tag{7.28}$$

Thus there is about 10% error in the asymptotic approximation of Eq. (7.25) at $\omega = 10$.

It is strongly suggested that all α plots be made on log-log paper to facilitate sketching and reading the curves. Semi-log paper may also be used with the magnitude plotted in dB's, and this approach is suggested by some authors. The main advantage of a dB scale is that correction terms are additive in dB's. This advantage is significant when the corrected curves are of primary interest, as is the case in many communications applications involving highly-tuned bandpass amplifiers. In control applications the corrected curves are not often required, however, and log-log coordinates are generally more convenient to use. Several of the design techniques subsequently discussed involve graphical construction of -1, -2, and -3 slope segments on the Bode diagram. These segments are readily drawn using 45°, 26.5°, and 18.4° right triangles, respectively, on log-log coordinates. The triangles required when dB versus log ω scales are used depend upon the dB scale chosen and constructions are a bit more awkward to make in this case. It is also necessary, in checking the asymptotic and actual curves, to convert between magnitude and dB values when a dB scale is used. The reader should try both methods and use that which he finds most convenient; but log-log coordinates are used exclusively in our subsequent discussions.

7.4 PHASE (β) DIAGRAMS

The harmonic response of any system consists of two components, magnitude versus ω and phase versus ω. A plot of phase versus ω is called the β diagram of the system and is generally made on semi-log coordinates, with ω and β plotted on the log and linear scales, respectively. The harmonic magnitude and phase curves of a system over the range $0 \leq \omega \leq \infty$ uniquely define that system. A transfer function is called *minimum phase* if it has no poles or zeros in the rhp. Either the α or β diagram is sufficient to completely define the transfer function of a minimum-phase system, assuming there are no pole-zero cancellations. The following example illustrates the distinctions between minimum-phase and non-minimum-phase transfer functions.

Example 7.4

Sketch the α and β diagrams for the transfer functions

$$G_1(s) = \frac{1 - Ts}{1 + 10Ts} \qquad G_2(s) = \frac{1 + Ts}{1 + 10Ts} \tag{7.29}$$

Compare the results.

The α diagrams of $\mathbf{G}_1(j\omega)$ and $\mathbf{G}_2(j\omega)$ are identical, as shown in Fig. 7.12a, since $|1 - j\omega T| = |1 + j\omega T|$ for all ω. The β diagrams are distinctly different, as shown in Fig. 7.12b, because the numerator of $\mathbf{G}_1(j\omega)$ provides phase lag for

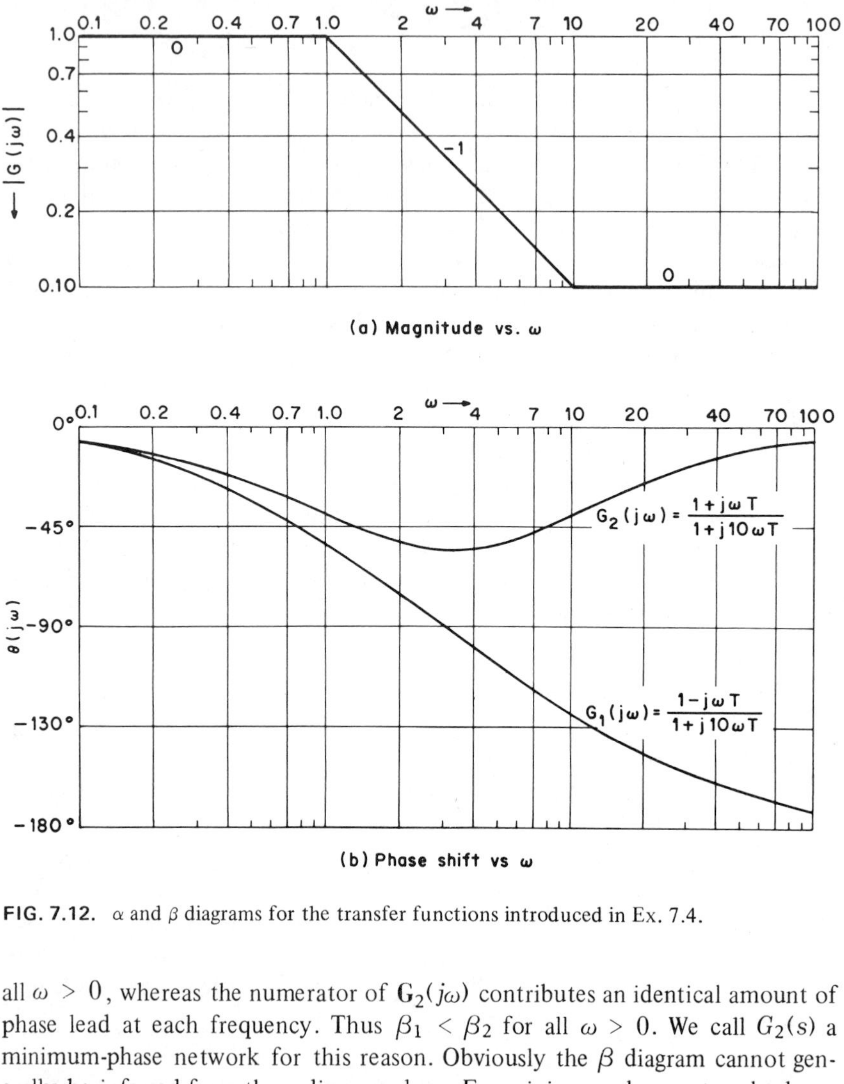

FIG. 7.12. α and β diagrams for the transfer functions introduced in Ex. 7.4.

all $\omega > 0$, whereas the numerator of $G_2(j\omega)$ contributes an identical amount of phase lead at each frequency. Thus $\beta_1 < \beta_2$ for all $\omega > 0$. We call $G_2(s)$ a minimum-phase network for this reason. Obviously the β diagram cannot generally be inferred from the α diagram alone. For minimum-phase networks, however, a unique interrelationship does exist.

It is tempting to try an analysis similar to that used in Ex. 7.4 for the two transfer functions

$$G_1(s) = \frac{1 + Ts}{1 + 10Ts} \qquad G_2(s) = \frac{1 + Ts}{1 - 10Ts} \tag{7.30}$$

to show that poles in the rhp also contribute extra phase shift. Such a test is obviously impossible, however, since $G_2(s)$ is unstable, and a steady-state harmonic response does not exist for this function. We define a system with rhp poles as being nonminimum-phase.

It is a simple matter to extend the results presented for real zeros in Ex. 7.4 to show that all stable minimum-phase transfer functions have less phase shift on the interval $0 < \omega \leq \infty$ than transfer functions having the same α diagrams, but with rhp zeros. We need only show that an rhp complex conjugate pair of zeros in an otherwise minimum-phase function contributes phase lag to the function at all frequencies $0 < \omega \leq \infty$ relative to that of the analogous minimum-phase network having mirror-image lhp zeros. Consider the stable nonminimum-phase transfer function

$$G_1(s) = G(s)(s - a + jb)(s - a - jb) \tag{7.31}$$

where $G(s)$ is minimum-phase. The constants a and b are assumed positive, so the zeros of $G_1(s)$ at $s = a \pm jb$ are in the rhp. Define

$$G_2(s) = G(s)(s + a + jb)(s + a - jb) \tag{7.32}$$

Note that $G_2(s)$ has lhp zeros that are mirror images of the $G_1(s)$ rhp complex zeros. $G_1(s)$ and $G_2(s)$ are related by

$$G_2(s) = G_1(s)\frac{(s + a + jb)(s + a - jb)}{(s - a + jb)(s - a - jb)} = G_1(s)H(s) \tag{7.33}$$

where

$$H(s) = \frac{(s + a + jb)(s + a - jb)}{(s - a + jb)(s - a - jb)} \tag{7.34}$$

Note that $H(s)$ is a mathematical artifice used *only* for relative phase evaluation, and is *not* a realizable network. It is apparent that $|\mathbf{H}(j\omega)| = 1$ for all ω, since the α diagrams of $G_1(s)$ and $G_2(s)$ are identical. If we can show that $\mathbf{H}(j\omega)$ has positive phase shift for all $\omega > 0$, this is sufficient to prove that $G_2(s)$ has less phase lag than $G_1(s)$ for all $\omega > 0$, and thus may be called minimum phase. It is convenient to illustrate the proof with the vector diagram shown in Fig. 7.13. The sketch is made for an arbitrary harmonic frequency $s = j\omega$. The numerator terms of $\mathbf{H}(j\omega)$ are labeled 1 and 2, whereas the denominator terms are labeled 3 and 4. The angles of these terms are denoted by $\theta_1, \theta_2, \theta_3$ and θ_4, respectively. The phase angle of $\mathbf{H}(j\omega)$ is the sum of the angles of the numerator terms minus the sum of the angles of the denominator terms, or

$$\beta_H = \theta_2 - \theta_4 + \theta_1 - \theta_3 = \beta_1 - \beta_2 \tag{7.35}$$

where the terms are grouped together for convenience. It is obvious from Fig. 7.13 that $\beta_1 > \beta_2$ for all $\omega > 0$, so $\mathbf{H}(j\omega)$ contributes phase lead for all $\omega > 0$ and the phase lag of $\mathbf{G}_2(j\omega)$ is less than that of $\mathbf{G}_1(j\omega)$ for all $\omega > 0$. Thus each rhp-zero in a nonminimum-phase transfer function contributes phase lag in

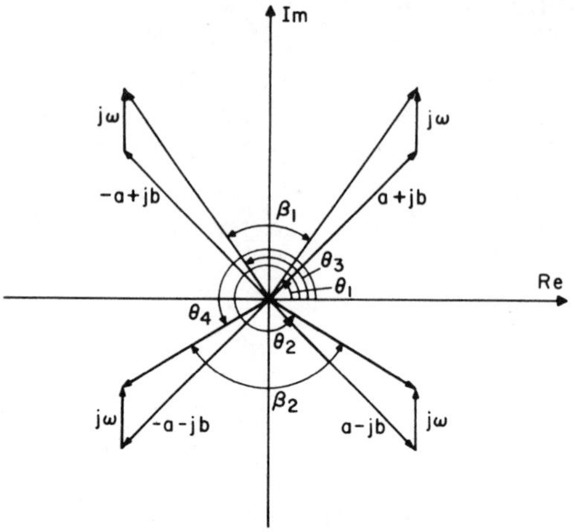

FIG. 7.13. Vector diagram showing the effect of rhp zeros upon phase shift.

excess of that which would apply if the zero were the mirror image in the lhp, as we have now shown for both real and complex zeros. When several rhp-zeros are present the phase lags are additive, so transfer functions with m real and 2n complex rhp zeros, m and n arbitrary non-negative integers, are always nonminimum phase.

It is important to note several factors from Fig. 7.13. $\mathbf{H}(j\omega)$ has a phase shift of 0 at $\omega = 0$ and 2π at $\omega = \infty$. In general, the increase in phase lag between 0 and infinite frequency for a stable nonminimum-phase function relative to that of the corresponding minimum-phase function is $n\pi$, where n is the number of rhp zeros in the function. Observe also that $\beta_1 - \beta_2$ is small for $\omega \ll |a + jb|$, or for frequencies much less than the break frequency at which the nonminimum-phase term becomes effective on the α diagram. Thus the phase shift of a stable nonminimum-phase network is reasonably well approximated by that of the equivalent minimum-phase network for all $\omega \ll \omega_a$, where ω_a is the break frequency of the first nonminimum-phase term to become effective. It is not suggested that this approximation be used for $\omega > \omega_a/10$ without correction, however. The difference at $\omega = \omega_a/10$ is about 12° of extra phase lag, assuming a single rhp-zero with break frequency ω_a. It is fortunate, from a practical viewpoint, that nonminimum-phase zeros have little effect at frequencies significantly below the break frequency, since rhp zeros with large break frequencies arise in many applications. Amplifier transfer characteristics often have rhp-zeros caused by direct interelement capacitive coupling from input to output. These effects generally occur well beyond the frequency range of interest, so the corresponding terms are usually neglected, and this simplification is easily justified.

Most control system components are adequately represented by minimum-phase transfer functions in the frequency range of interest. Cascade combinations of minimum-phase networks are minimum phase. The most common non-minimum-phase arrangements encountered are networks of the lattice, bridged T, or twin T types, and situations resulting in pure time delay caused by sound (or electromagnetic) transmission over significant distances. Nonminimum-phase networks usually provide alternate paths between input and output with significantly different attenuation and phase shift characteristics. One path predominates at low frequencies; the second takes over for high frequencies.

Example 7.5

Consider the lattice network shown in Fig. 7.14. Show that the transfer function $V_0(s)/V_i(s)$ is nonminimum phase.

We see that

$$V_0(s) = V_i(s)\left(\frac{1/Cs}{R + 1/Cs} - \frac{R}{R + 1/Cs}\right) \tag{7.36}$$

which simplifies to

$$\frac{V_0(s)}{V_i(s)} = \frac{1 - RCS}{1 + RCS} \tag{7.37}$$

which has a rhp zero and is nonminimum phase. It is intuitively apparent that $V_0 \simeq V_i$ at very low frequencies where the C's are essentially open circuits, $V_0 \simeq -V_i$ at very high frequencies where the C's are essentially short circuits, and there is a gradual transition between these extremes as ω increases. The critical frequency where the predominant paths change is that where

$$X_c = |\mathbf{Z}_c| = \frac{1}{\omega C} = R, \text{ or } \omega = 1/RC.$$

Unless specifically noted to the contrary, all subsequent discussions pertain to minimum-phase networks. Several mathematical relationships between the α and β curves of minimum phase networks have been derived by Bode,[3] one of

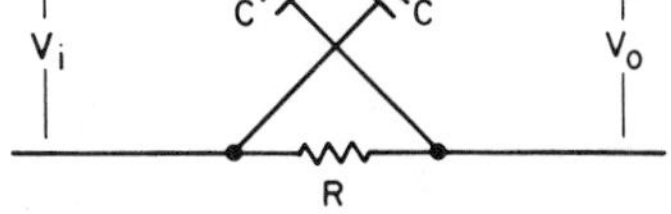

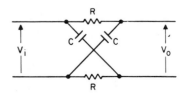

FIG. 7.14. A nonminimum-phase lattice network.

which is

$$\beta(\omega_c) = \frac{1}{\pi}\int_{-\infty}^{\infty} \frac{d\alpha}{du} \ln\left(\coth\left|\frac{u}{2}\right|\right) du \tag{7.38}$$

where

$$u = \ln \frac{\omega}{\omega_c} \tag{7.39}$$

Thus the phase shift at $\omega = \omega_c$ is related to the weighted slope of the magnitude characteristic at all frequencies. The weighting factor is $\ln(\coth|u/2|)$, and Fig. 7.15 shows how this factor depends upon ω. The α curve slope near ω_c is weighted heavily, and its slope far from ω_c has little effect. It is usually possible, by keeping this general idea in mind, to visualize with reasonable accuracy the phase characteristics of minimum-phase functions directly from the α diagram without detailed calculations. Derivation of Eq. (7.38), the Bode phase integral, is rather involved and is not presented since the steps in the derivation add little

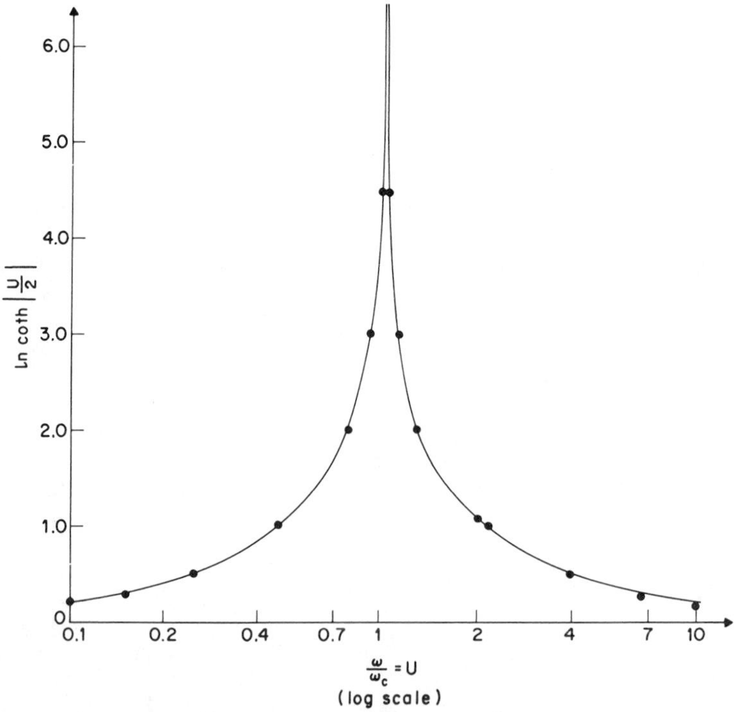

FIG. 7.15. Dependence of the weighting factor ln coth $|u/2|$ upon ω.

to our subsequent understanding of minimum-phase networks and their treatment. Although the concept of weighted dependence of phase shift upon the α curve slope is an important one, it is awkward to apply Eq. (7.38) directly due to the integrand complexity. The β curves of most control system transfer functions are reasonably smooth and readily sketched with acceptable accuracy by computing only a few appropriately chosen points on the curve. A slide rule with a tangent scale is particularly useful in sketching phase curves.

Example 7.6

Sketch the α and β diagrams of

$$G(s) = \frac{100(1 + s/100)}{s(1 + s/10)(1 + s/300)} \tag{7.40}$$

The asymptotic α diagram is shown in Fig. 7.16. The slope is -1 for all $\omega < 10$, -2 for $10 \leq \omega \leq 100$, -1 for $100 < \omega < 300$, and -2 for $\omega \geq 300$. The magnitude reference point is readily obtained by observing that $|G|_\alpha = 100$ at $\omega = 1$. A general expression for β is obtained directly from Eq. (7.40) as

$$\beta(\omega) = -\frac{\pi}{2} - \tan^{-1}\frac{\omega}{10} - \tan^{-1}\frac{\omega}{300} + \tan^{-1}\frac{\omega}{100} \tag{7.41}$$

The phase curve obtained by computing β at a few points spanning the range $1 \leq \omega \leq 1000$, and plotting the results is shown in Fig. 7.16. It is apparent from Eq. (7.41) that $\beta(0) = -\pi/2$ and $\beta(\infty) = -\pi$.

Note that the β curve tends toward $n\pi/2$ when the α curve slope is n. This rough approximation helps in sketching the β curve. The actual phase shift need only be calculated at a few points. Specific points of interest are the break frequencies and the geometric means between adjacent break frequencies which occur on a log scale at the midpoints of the straight line segments on the asymptotic α diagram. The phase is thus computed in Ex. 7.6, for example, at ω values of 1, 3, 10, 30, 100, 200, 300, 1000, and 3000. A smooth interpolation between these points provides an accurate phase curve. In the design procedures introduced later we will be particularly interested in the point where the α diagram passes through unit magnitude and the point where the β diagram passes through -180°. These points are significant in determining closed-loop system stability, as shown subsequently, and we should become familiar with efficiently estimating their location.

Suppose we have

$$G(s) = \frac{100(1 + s/\omega_1)}{s(1 + s/\omega_2)} \qquad \omega_1 \ll \omega_2 \tag{7.42}$$

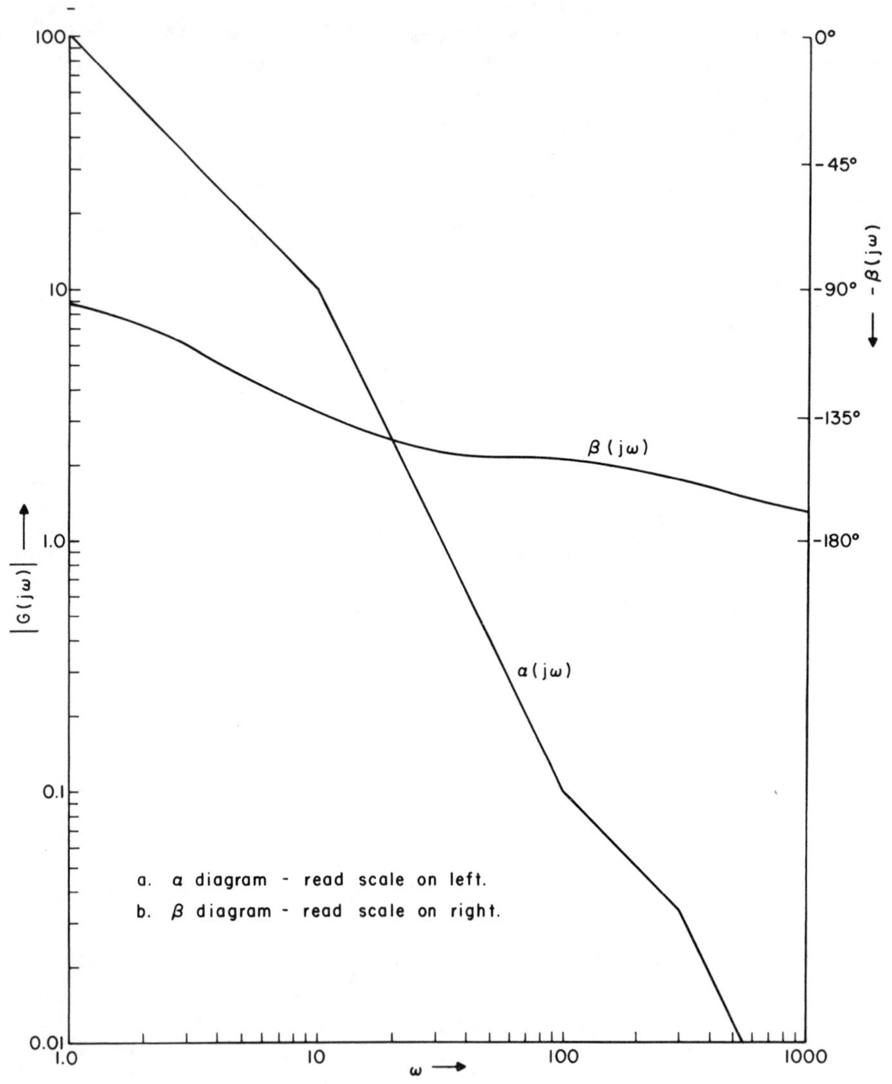

FIG. 7.16. α and β curves for the *G(s)* introduced in Ex. 7.6.

The α diagram has -1 slope for $\omega \leq \omega_1$, 0 slope in the range $\omega_1 < \omega < \omega_2$, and -1 slope for $\omega \geq \omega_2$. The α and β curves are shown in Fig. 7.17. The β curve tends toward -90° as $\omega \to 0$ and as $\omega \to \infty$, and has a maximum greater than -90° and less than 0° in the range $\omega_1 < \omega < \omega_2$. Suppose we wish to find that maximum and the frequency at which it occurs. The general expression for β is

$$\beta(\omega) = -\frac{\pi}{2} + \tan^{-1}\frac{\omega}{\omega_1} - \tan^{-1}\frac{\omega}{\omega_2} \tag{7.43}$$

The maximum β occurs at a point where $d\beta/d\omega = 0$. Note that

$$\frac{d}{d\theta}(\tan^{-1}\phi) = \frac{1}{1+\phi^2}\cdot\frac{d\phi}{d\theta} \tag{7.44}$$

Thus

$$\frac{d\beta}{d\omega} = 0 = \frac{1}{1+(\omega/\omega_1)^2}\cdot\frac{1}{\omega_1} - \frac{1}{1+(\omega/\omega_2)^2}\cdot\frac{1}{\omega_2} \tag{7.45}$$

or

$$\frac{\omega_1}{\omega_1{}^2+\omega^2} = \frac{\omega_2}{\omega_2{}^2+\omega^2} \tag{7.46}$$

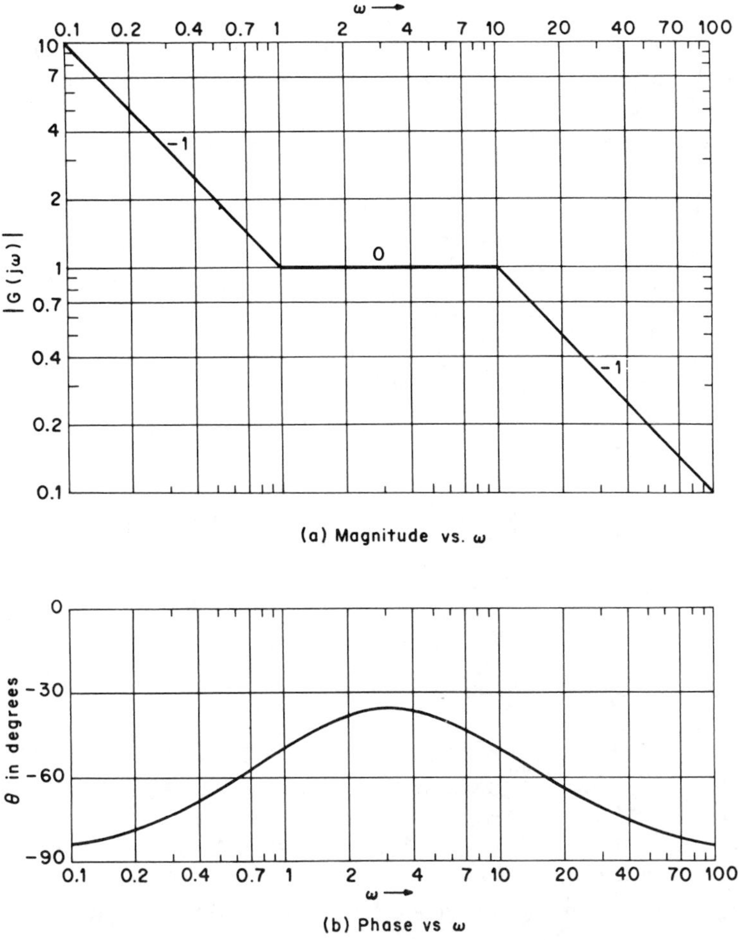

FIG. 7.17. Phase and magnitude curves for $G(s)$ as given in Eq. (7.42).

Cross multiplying the terms in Eq. (7.46) and solving for ω yields

$$\omega = \sqrt{\omega_1 \omega_2} \tag{7.47}$$

Thus the minimum-phase lag occurs at the geometric mean of the 0 slope segment end frequencies, as we should expect from the Bode phase integral and the weighting factor symmetry shown in Fig. 7.15. This result explains our interest in the geometric means between break frequencies as data points in constructing β plots.

7.5 PHASE SHIFT FROM THE α DIAGRAM USING THE ARC TAN APPROXIMATION

In control applications we are often interested in the approximate phase shift of a transfer function at a particular frequency, usually that ω, denoted by ω_c, at which the α curve passes through unit magnitude, and we may have little concern with the overall β curve. Although direct evaluation of β from the transfer function using a slide rule or tables is straightforward, considerable simplification is possible in many cases. The arc tan function can be expressed in product series form as

$$\tan^{-1}\frac{\omega}{\omega_1} = \frac{\omega}{\omega_1} - \frac{1}{3}\left(\frac{\omega}{\omega_1}\right)^3 + \frac{1}{5}\left(\frac{\omega}{\omega_1}\right)^5 - \cdots$$

$$= \sum_{n=1}^{\infty} \frac{(-1)^{n+1}}{2n-1}\left(\frac{\omega}{\omega_1}\right)^{(2n-1)} \quad \text{for } \frac{\omega}{\omega_1} < 1 \tag{7.48}$$

and

$$\tan^{-1}\frac{\omega}{\omega_1} = \frac{\pi}{2} - \left(\frac{\omega}{\omega_1}\right)^{-1} + \frac{1}{3}\left(\frac{\omega}{\omega_1}\right)^{-3} - \frac{1}{5}\left(\frac{\omega}{\omega_1}\right)^{-5} + \cdots$$

$$= \frac{\pi}{2} + \sum_{n=1}^{\infty} \frac{(-1)^n}{2n-1}\left(\frac{\omega}{\omega_1}\right)^{-(2n-1)} \quad \text{for } \frac{\omega}{\omega_1} > 1 \tag{7.49}$$

The error magnitude which results from approximating an absolutely convergent alternating series by n terms is no greater than the magnitude of the first term neglected. Thus if $\omega/\omega_1 = 1/2$, $\tan^{-1} \omega/\omega_1 \simeq \omega/\omega_1 = 1/2$ is accurate at least to within 1/24 radian, or about 2.5 degrees. Approximation of the arc tan function by the first order term in Eqs. (7.48) or (7.49), whichever applies, often allows the β curve to be sketched directly from the transfer function or the α diagram using only frequency ratio calculations. For a right triangle, as shown in

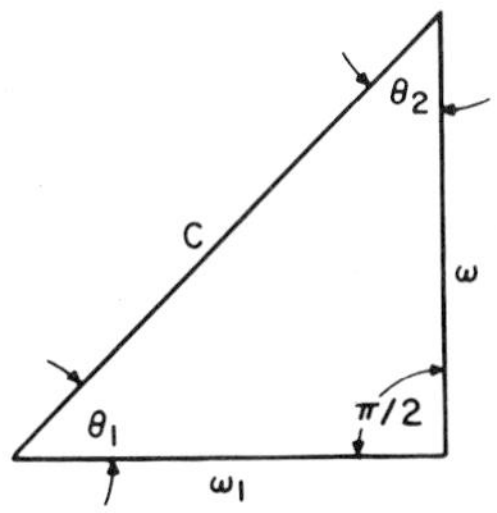

FIG. 7.18. Basic right triangle relationships.

Fig. 7.18, it is necessary that

$$\theta_1 = \tan^{-1}\frac{\omega}{\omega_1} = \frac{\pi}{2} - \tan^{-1}\frac{\omega_1}{\omega} = \frac{\pi}{2} - \theta_2 \tag{7.50}$$

This identity allows us to use either Eq. (7.48) or (7.49) and the arc tan approximation in most cases of interest since we can always choose the ratio which is less than unity. Application of the arc tan approximation is illustrated by the following examples.

Example 7.7

Consider the function

$$G(s) = \frac{K(1 + s/10)}{s(1 + s)(1 + s/100)} \tag{7.51}$$

Determine β at $\omega = 30$.

The asymptotic α diagram of this function is shown in Fig. 7.19. Note that the phase shift at any given frequency is independent of K. We anticipate

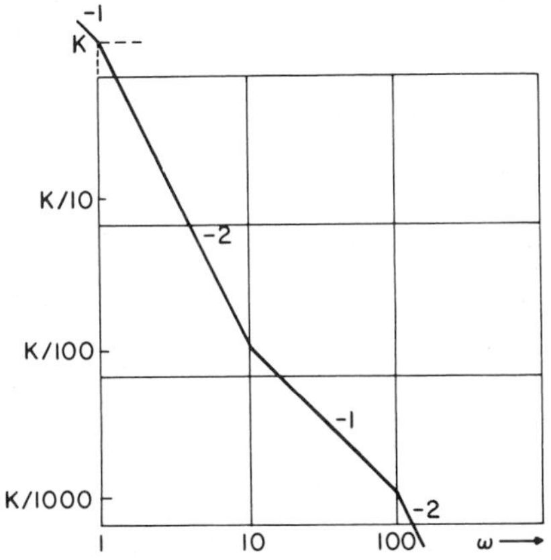

FIG. 7.19. α diagram for $G(s)$ as given by Eq. (7.51).

approximately 90° of phase lag at $\omega = 30$ because of the -1 slope there. The -2 slope regions flanking $\omega = 30$ on each side contribute additional phase lag however, so perhaps $\beta = 120°$ lag is a more appropriate guess. Actually,

$$\beta(30) = \tan^{-1} 3 - \frac{\pi}{2} - \tan^{-1} 30 - \tan^{-1} 0.3 = -123.4° \tag{7.52}$$

Suppose we apply the arc tan approximation here. Then

$$\beta(30) = \frac{\pi}{2} - \frac{1}{3} - \frac{\pi}{2} - \frac{\pi}{2} + \frac{1}{30} - 0.3 = -\frac{\pi}{2} - 0.6 = -2.17 = 124° \tag{7.53}$$

Accuracy of this sort is fairly typical and certainly acceptable.

Efficiency in estimating phase shift will later prove very useful in stability analysis and system design. We observed that $\beta(30)$ should not be radically different than -90°, since the Bode diagram has -1 slope at $\omega = 30$, but corrective adjustments are necessary because there are -2 slopes on either side of $\omega = 30$. We can reduce Eq. (7.52) to a form in which $\beta(30)$ is given as -90° plus 3 corrective terms, one for each break point on the α diagram. Each corrective term can be expressed as the arc tan of a frequency ratio less than 1 in magnitude, so the arc tan approximation is particularly easy to apply. Note that

$$\tan^{-1} 3 = \frac{\pi}{2} - \tan^{-1}\frac{1}{3} \qquad \tan^{-1} 30 = \frac{\pi}{2} - \tan^{-1}\frac{1}{30} \tag{7.54}$$

Substituting these results into Eq. (7.52) and simplifying provides

$$\beta(30) = -\frac{\pi}{2} - \tan^{-1}\frac{1}{3} + \tan^{-1}\frac{1}{30} - \tan^{-1} 0.3 \tag{7.55}$$

Each term can be associated with specific slopes and break points on the α diagram. The $-\pi/2$ term is the 0-th order approximation because the frequency of interest falls in a region of -1 slope. Additional phase lag caused by the change from a -2 slope to a -1 slope at $\omega = 10$ is provided by $-\tan^{-1}(1/3)$. Phase lead correction, necessary because of the change from a -1 slope to a -2 slope at $\omega = 1$, is provided by $\tan^{-1}(1/30)$, and the additional phase lag necessary because the slope changes to -2 at $\omega = 100$ is provided by $-\tan^{-1}(3/10)$. The following example further illustrates phase evaluation using this method.

Example 7.8

Consider the function

$$G(s) = \frac{K(1 + s/30)(1 + s/50)}{s(1 + s/2)^2(1 + s/1000)^2} \tag{7.56}$$

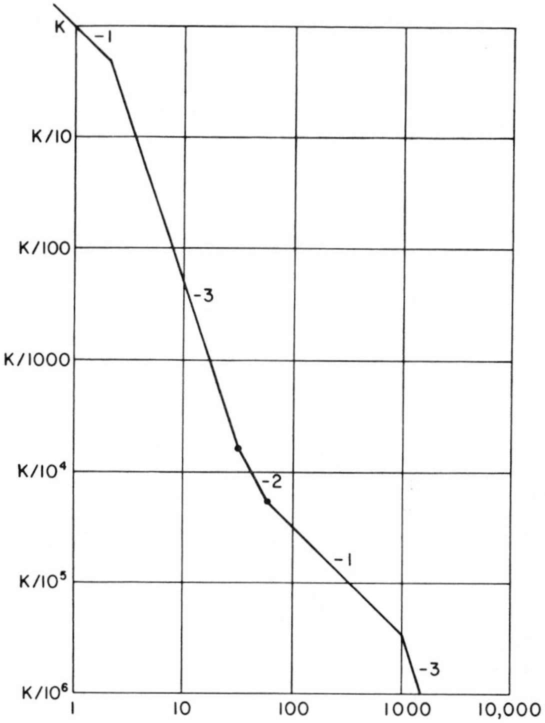

FIG. 7.20. α diagram for $G(s)$ as given by Eq. (7.56).

Express the phase shift at $\omega = 200$ directly from the asymptotic slope at that frequency with corrective terms for each break frequency on the α diagram.

The Bode plot for $G(s)$ is shown in Fig. 7.20. Suppose we attempt to construct an expression for $\beta(200)$ directly from the Bode diagram. Since $\omega = 200$ occurs in a -1 slope region, the 0-th order approximation to $\beta(200)$ is $-\pi/2$ radians. Corrective terms are needed due to the breaks at ω values of 50, 30, 2, and 1000. The breaks at 50 and 30 cause additional lag. That at $\omega = 2$ is a double break and results in less lag, or lead. The corresponding corrective term must be multiplied by 2 (or added in twice). The double break at $\omega = 1000$ also results in additional phase lag. Putting all of this together using the appropriate frequency ratios and the arc tan approximation provides

$$\beta(200) = -\frac{\pi}{2} - \frac{50}{200} - \frac{30}{200} + 2\left(\frac{2}{200}\right) - 2\left(\frac{200}{1000}\right) = -135° \qquad (7.57)$$

The source of each correction term is apparent from Fig. 7.20. We may easily justify the result given in Eq. (7.57) by evaluating the phase directly from Eq. (7.56) as

$$\beta(200) = \tan^{-1}\frac{200}{30} + \tan^{-1}\frac{200}{50} - \frac{\pi}{2} - 2\tan^{-1}\frac{200}{2} - 2\tan^{-1}\frac{200}{1000} \tag{7.58}$$

Expressing each arc tan function in Eq. (7.58) in terms of an argument with magnitude less than unity and applying the arc tan approximation to the result provides the desired comparable form as

$$\beta(200) = \left(\frac{\pi}{2} - \frac{30}{200}\right) + \left(\frac{\pi}{2} - \frac{50}{200}\right) - \frac{\pi}{2} - 2\left(\frac{\pi}{2} - \frac{2}{200}\right) - 2\left(\frac{200}{1000}\right) \tag{7.59}$$

where the individual terms are set off in brackets for emphasis. Comparison of Eqs. (7.57) and (7.59) shows that they are identical. Thus the phase shift at any frequency may be determined directly from the α plot, rather than from the transfer function, assuming $G(s)$ is minimum-phase and has only real poles and zeros. Although the same procedures apply for complex poles and zeros and for nonminimum-phase networks, additional information must be provided.

We are now prepared to formulate a general rule for finding the phase shift directly from the Bode diagram for minimum-phase functions with no underdamped quadratic terms. The result can be expressed as follows. The phase shift $\beta(\omega_0)$ is given by

$$\beta(\omega_0) = n\left(\frac{\pi}{2}\right) - \sum_{i=1}^{A} a_i \tan^{-1}\frac{\omega_i}{\omega_0} + \sum_{j=1}^{B} b_j \tan^{-1}\frac{\omega_0}{\omega_j} \tag{7.60}$$

where

n = the asymptotic slope at ω_0
A = number of break frequencies below ω_0
B = number of break frequencies above ω_0
a_i = change in slope at $\omega = \omega_i$ as ω increases
b_j = change in slope at $\omega = \omega_j$ as ω increases

If ω_0 corresponds to a break frequency, the slope at ω_0 can be taken as the value as $\omega \to \omega_0$ from either the right or the left as long as the corrective term due to the slope change at ω_0 is interpreted properly. In most applications, ω_0 is sufficiently far from all ω_i and ω_j (a factor of two is more than adequate for the accuracies generally required) to justify the arc tan approximation for all terms in Eq. (7.60). We often wish to know the value of phase shift at a critical frequency in terms of the location of one or more break frequencies in the transfer function, so use of the arc tan approximation is often very helpful. Since Eq. (7.60) gives β as a transcendental function, which is difficult to work with analytically, the arc tan approximation reduces the problem to manageable level. This point is emphasized in Chap. 9, where the design problem is introduced.

Underdamped quadratic terms result in a double break on the asymptotic α diagram and require special treatment in determining the actual α curve. It is not surprising, therefore, that they must also be treated carefully in constructing β curves. Determining the phase curve from the transfer function is straightforward, irrespective of quadratic terms with complex roots, but β can be obtained directly from the Bode diagram without additional information only if no such terms exist. The β-curve changes more rapidly near a double break caused by an underdamped quadratic term than for a corresponding pair of real poles or zeros, and this effect increases as the damping factor decreases. Although it is possible to construct asymptotic β curves in much the same way as asymptotic α curves are constructed,[4] they are of less general interest, so the matter is not pursued further. It is sufficient, in most applications, to obtain β at a few key frequencies and connect these data points with a continuous curve, since β curves are generally smooth.

Consider the quadratic factor

$$\frac{s^2}{\omega_0^2} + \frac{2\xi}{\omega_0} s + 1 = G_q(s) \tag{7.61}$$

Substituting $s = j\omega$ provides

$$1 - \frac{\omega^2}{\omega_0^2} + j\frac{2\xi\omega}{\omega_0} = G_q(j\omega) \tag{7.62}$$

The phase as a function of ω is

$$\beta_q(\omega) = \tan^{-1} \frac{2\xi\omega/\omega_0}{1 - \omega^2/\omega_0^2} = \tan^{-1} \frac{2\xi\omega\omega_0}{\omega_0^2 - \omega^2} \tag{7.63}$$

Thus $\beta_q(0) = 0$, $\beta_q(\infty) = \pi$ and $\beta_q(\omega_0) = \pi/2$. Values of β_q at other frequencies depend upon ξ, as shown in Fig. 7.21 in terms of the normalized frequency ω/ω_0. These curves often prove useful in constructing composite β diagrams.

Example 7.9

Consider a system with Bode diagram as shown in Fig. 7.22. Find $\beta(100)$ if the double break at $\omega = 500$ corresponds to a denominator quadratic term with $\xi = 0.5$. There are no other complex terms, and the system is minimum-phase.

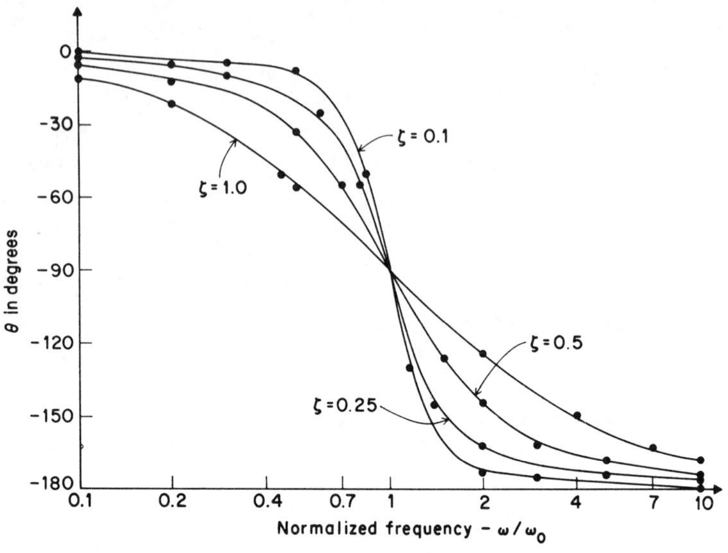

FIG. 7.21. Phase curves for the quadratic term $s^2 + 2\xi\omega_0 s + \omega_0^2$.

We will solve this problem in two steps. First, find $\beta_1(100)$ disregarding the double break at $\omega = 500$. Then add the appropriate correction term, as found from Fig. 7.21, to obtain the desired answer.

Proceeding,

$$\beta_1(100) = -\frac{\pi}{2} - 0.4 + 0.05 = -1.92 \text{ rad} = -110° \tag{7.64}$$

From Fig. 7.21 we see that, for $\omega/\omega_0 = 0.2$ and $\xi = 0.5$, an additional phase lag of 12° must be included. Thus $\beta(100) = -122°$. Note, in comparison, that a double break at $\omega = 500$ due to a pair of superposed real poles would require a phase lag correction of approximately $2(1/5) = 0.4$ radian, or about 22.9° at $\omega = 100$.

Although we have excluded nonminimum-phase networks in our previous discussion, their β diagrams can also be obtained directly from the α diagram, assuming there are $n < \infty$ rhp zeros, if it is known which break points correspond to these zeros. We simply recall that rhp zeros add phase lag rather than phase lead, and change the signs of the appropriate correction terms in the general expression for β.

Example 7.10

Consider a system with α diagram as shown in Fig. 7.23. The zero at $\omega = 200$ is in the rhp. Find $\beta(10)$. Also find $\beta(10)$ for the corresponding minimum-phase system.

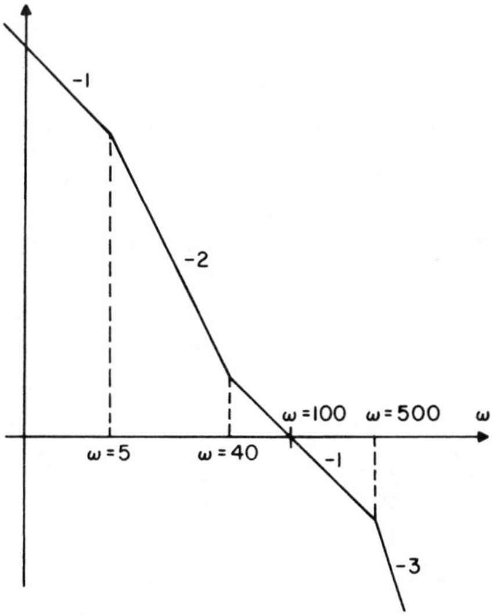

FIG. 7.22. α diagram for the system considered in Ex. 7.9.

The general expression for $\beta(10)$ is

$$\beta(10) = -\frac{\pi}{2} - 0.3 + 0.05 - 0.2 - 0.05 = -118.7° \tag{7.65}$$

The -0.05 rad term corresponds to the zero at $\omega = 200$. If this zero is in the lhp, the sign is plus, and the difference in phase shift between these two cases is 0.1 radian, or approximately 5.7°. This difference in $\beta(\omega)$ rapidly becomes more

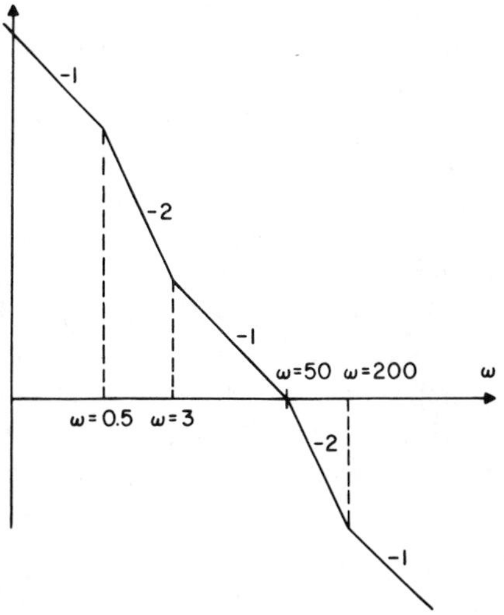

FIG. 7.23. α diagram of the system considered in Ex. 7.10.

significant as $\omega \to 200$. The minimum phase system has $\beta(10) = -113°$.

7.6 SUMMARY AND CONCLUSIONS

In control system applications, we are often interested in the magnitude and phase characteristics of the transfer functions corresponding to basic controller elements over a wide range of frequencies. It is not unusual to desire sketches of magnitude and phase versus ω over two or three decades of frequency, and the magnitude often changes by 10^6 or more over this range. For these reasons it has become standard practice to "compress" the scales on α diagrams by using log-log coordinates. A logarithmic frequency scale is generally used on β diagrams, although the phase shift is plotted on a linear scale. Our major purpose in this chapter is to familiarize the reader with efficient procedures for sketching α and β diagrams. An effort has also been made to develop a sound intuitive feeling for the phase shifts at key frequencies in terms of general Bode diagram shape. This facility will prove useful in later sections on design, since the design of controllers satisfying typical specifications can be reduced to an exercise in shaping the α diagram near-unit magnitude such that the phase lag there is 30° to 50° less than 180°. We now have all the tools available to start considering the design problem.

Closed-loop control systems have a notorious tendency toward instability. Prior to the 1930's, when Nyquist formulated his now-famous stability criterion, feedback arrangements were avoided, if possible, for this reason, and feedback system design was essentially carried out by trial and error. Organized trial and error is still a useful design method often applied on a computer, but the sound theoretical footing upon which current design methods are based greatly improves the designer's efficiency.

The classical design procedures with which this book is fundamentally concerned place major emphasis upon system stability. We have shown that factoring the characteristic equation, or even application of the Routh criterion, to locate closed-loop poles are much too cumbersome methods to find general use as design tools. Since the Routh and Hurwitz criteria are derived using complex function theory, one might intuitively feel that stability, or lack thereof, could be more conveniently illustrated through some alternate application of that theory. The Nyquist criterion is developed from complex function theory and is discussed extensively in the next chapter. It is conceptually straightforward, easily derived, readily applied, and forms the basis for virtually all of the design procedures in common use. It is based upon steady-state magnitude and angle characteristics of system transfer functions, which we are now well prepared to handle. It provides a simple graphical indication of system stability, as well as the margin by which that stability is maintained, and makes the effects of changes in critical parameters upon system performance readily apparent.

REFERENCES

1. Nyquist, H., Regeneration Theory, *Bell System Technical Journal,* vol. 11, pp. 126–147, 1932.
2. Newton, G. C., Jr., L. A. Gould, and J. F. Kaiser, "Analytical Design of Linear Feedback Controls," chap. 1, John Wiley & Sons, Inc., New York, 1957.
3. Bode, H. W., "Network Analysis and Feedback Amplifier Design," D. Van Nostrand Co., 1945.
4. Bower, J. L., and P. M. Schultheiss, "Introduction to the Design of Servomechanisms," John Wiley & Sons, Inc., New York, pp. 88–91, 1958.

8 THE NYQUIST STABILITY CRITERION

Of the procedures thus far considered for assuring closed-loop control system stability, none leads readily to practical design methods. Factoring the characteristic equation to determine all pole locations is time consuming, and impossible to carry out in terms of general parameter values for systems of nominal complexity. The Routh and Hurwitz criteria have similar limitations plus the additional restriction that, although the numbers of poles in each half plane are indicated, the exact locations of these poles can be obtained only through extensive additional work. Each of these methods is a useful analysis tool, but neither leads to a useful synthesis procedure. The ideal synthesis method should be relatively simple to apply, should lead directly to straightforward rules for system synthesis, and should provide a sound intuitive understanding of the changes in closed-loop performance influenced by making changes in those parameters subject to choice in the design phase. The Nyquist stability criterion satisfies these objectives and provides the basis for all standard classical design methods.

The Nyquist criterion is a graphical procedure for illustrating system stability, or lack thereof. The complex plane-polar plots required can usually be sketched quickly without extensive mathematical calculations. System stability and the margin by which that stability is obtained are readily apparent from the sketch.

To the extent that the effects of parameter changes upon the Nyquist plot shape can be visualized, dependence of system stability upon parameters can also be observed. Not only can the numbers of closed-loop poles in each half plane be determined, but some indication is also given of how far they are from the imaginary axis in the complex plane, thereby providing an indirect measure of relative stability and the transient response characteristics to be anticipated.

The Nyquist stability criterion is developed in Sec. 8.1 and its application to a variety of typical situations is considered in the subsequent sections. Examples are included to illustrate the basic concepts introduced. The chapter concludes with a summary of the principles introduced and how they should be used.

8.1 DEVELOPMENT OF THE NYQUIST CRITERION

Consider a closed-loop controller of the general form shown in Fig. 8.1. The input to output transfer function is

$$\frac{C(s)}{R(s)} = \frac{G_e(s)G_p(s)}{1 + G_e(s)G_p(s)H(s)} = \frac{G(s)}{1 + A(s)} \tag{8.1}$$

where

$$G(s) = G_e(s)G_p(s) \qquad A(s) = G_e(s)G_p(s)H(s) = G(s)H(s) \tag{8.2}$$

The notation is chosen to be descriptive and is relatively standard. Most controllers consist of a plant or actuator with transfer function $G_p(s)$, an equalizing or compensating filter denoted by $G_e(s)$ preceding the plant and in series with it, and a feedback transfer function $H(s)$ characterizing the device used to compare the output response $C(s)$ to the input forcing function $R(s)$. In many cases $H(s) = 1$. The plant $G_p(s)$ is usually given as part of the design problem statement. In some applications there may be several actuators that seem appropriate, in which case a tentative design can be established for each and the results compared. The compensating filter $G_e(s)$ is chosen during the design phase to assure closed-loop stability and satisfy the controller specifications (specs). Choosing an appropriate $G_e(s)$ constitutes the design problem. In this chapter we are fundamentally concerned with those factors considered in defining $G_e(s)$ to assure closed-loop stability. Later we will consider in greater

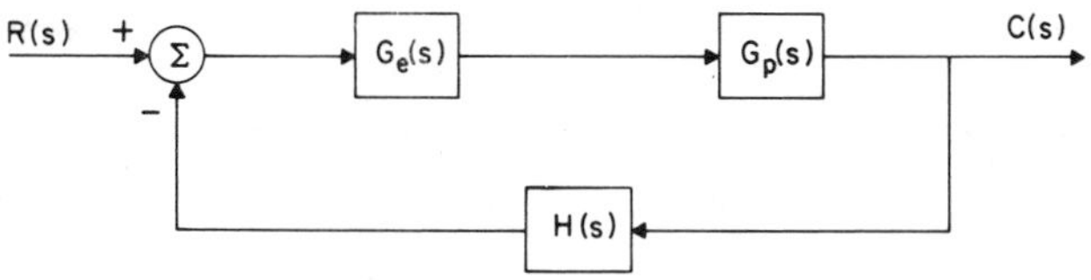

FIG. 8.1. The standard closed-loop controller configuration.

detail how specs upon such closed-loop system characteristics as bandwidth, unit step response peak overshoot, or steady-state error to a unit ramp input influence the final choice of $G_e(s)$.

It is necessary and sufficient to assure stability of $C(s)/R(s)$ as given by Eq. (8.1) that all of its poles lie in the lhp. Poles of $C(s)/R(s)$ arise from two sources: poles of $G(s)$ not cancelled out by poles of $A(s)$ when the function is rationalized; and, zeros of $1 + A(s)$. If we assume that $G(s)$ is open-loop stable, a condition generally satisfied, it follows that the locations of the zeros of $1 + A(s)$ determine the stability of the closed-loop system. It is always necessary for closed-loop stability that all zeros of $1 + A(s)$ lie in the lhp. We thus concentrate for the moment upon developing a procedure by which zeros of $1 + A(s)$ on the $j\omega$ axis and in the rhp can be detected without factoring the characteristic equation. Let

$$A(s) = \frac{P(s)}{Q(s)} \tag{8.3}$$

We assume at this point that $A(s)$ can be represented by a ratio of finite algebraic polynomials in s primarily for convenience. The Nyquist criterion applies to more general situations, as shown subsequently. Using $A(s)$ as given by Eq. (8.3) provides

$$\begin{aligned} 1 + A(s) &= 1 + \frac{P(s)}{Q(s)} = \frac{Q(s) + P(s)}{Q(s)} \\ &= \frac{(s - a_1)(s - a_2)(s - a_3)\ldots}{(s - b_1)(s - b_2)(s - b_3)\ldots} \end{aligned} \tag{8.4}$$

It is apparent from Eq. (8.4) that the poles of $1 + A(s)$ are the factors of $Q(s)$ and are identical to those of $A(s)$. $A(s)$ is the product of the plant, equalizer, and feedback transfer functions, and its pole and zero locations are assumed known since these transfer functions are generally available in factored form. Since $A(s)$ must be realizable, the order of $Q(s)$ always exceeds that of $P(s)$ in a reasonably modeled system. Thus $1 + A(s)$ has an equal number of poles and zeros.

Assume the poles and zeros of $1 + A(s)$ are distributed in the complex plane arbitrarily as shown in Fig. 8.2. Suppose we concentrate for the moment upon determining what happens to the polar plot of $1 + A(s)$ as $s = \sigma + j\omega$ traverses an arbitrary closed-contour Γ in the s-plane. A typical term in the numerator or denominator of $1 + A(s)$ is of the form $s - a_1$, and this vector is also shown in Fig. 8.2. Each pole and zero factor of $1 + A(s)$ contributes a vector of this form, and the complex function $1 + A(s_1)$ may be obtained as the product of the numerator vector magnitudes divided by the product of the denominator vector magnitudes at an angle equal to the sum of the numerator

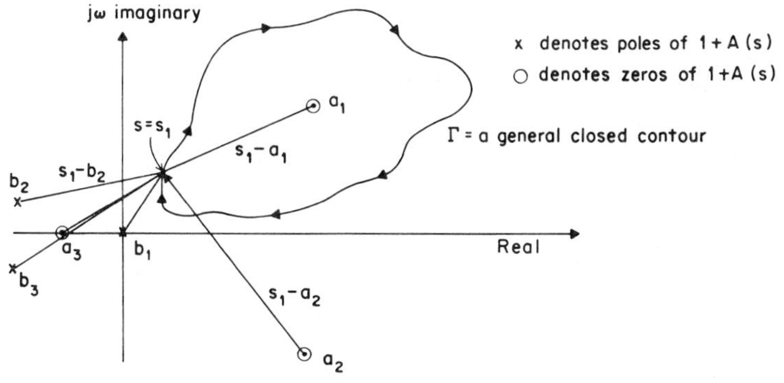

FIG. 8.2. Vector representation of $1 + A(s_1)$.

vector angles minus the sum of the denominator vector angles. Thus if we let

$$(s_1 - a_i) = A_{i1}\epsilon^{j\theta_{i1}} = \mathbf{A}_i \tag{8.5}$$

and

$$(s_1 - b_i) = B_{i1}\epsilon^{j\phi_{i1}} = \mathbf{B}_i \tag{8.6}$$

denote the numerator and denominator vectors, respectively, we have

$$1 + A(s_1) = \prod_{i=1}^{n} \frac{A_{i1}}{B_{i1}} \epsilon^{j\left[\sum_{i=1}^{n} (\theta_{i1} - \phi_{i1})\right]} \tag{8.7}$$

where n is the order of $Q(s)$ and s_1 denotes a specific value of s. Of course, no pole of $1 + A(s)$ [and thus of $A(s)$], may occur on the boundary Γ. Suppose Γ encloses the zero at $s = a_1$, but encloses no other pole or zero of $1 + A(s)$. As s takes on values along Γ, traversing the contour in a clockwise (cw) direction, the angles and magnitudes of the vector components $\mathbf{A}_i$ and $\mathbf{B}_i$ change, and the complex function $1 + A(s)$ traces out a closed path in the complex plane. Note in particular that no net phase shift occurs for those $\mathbf{A}_i$ and $\mathbf{B}_i$ corresponding to poles and zeros not enclosed by Γ. Although the phases of these components change as s takes on progressive values along the Γ contour, phase changes along the "far side" of the contour are counteracted by changes in the opposite direction along the "near side". The origin of the vector $\mathbf{A}_1 = s - a_1$ is contained within Γ, however, so cancellation does not occur in this case and θ_1 changes by -2π as s traverses Γ in a cw direction. Since $\mathbf{A}_1$ corresponds to a zero of $1 + A(s)$, the phase angle of the polar plot of $1 + A(s)$ must also change by -2π as s traverses Γ, thus tracing out a contour Γ' which encircles the origin of the complex plane in a cw direction, as illustrated in Fig. 8.3. More generally, the polar plot of $1 + A(s)$ encircles the origin in the cw direction

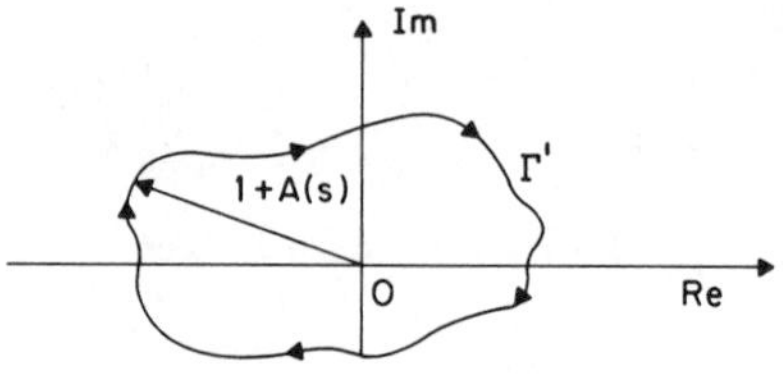

FIG. 8.3. A sketch of the contour Γ′ traced out by $1 + A(s)$ as s traverses Γ.

a number of times equal to $Z - P$, where Z is the number of zeros and P the number of poles of $1 + A(s)$ enclosed within the contour Γ, as s traverses Γ in the cw direction. The Nyquist stability criterion is based upon this simple observation. Thus we may write

$$N = Z - P \tag{8.8}$$

where N denotes the number of cw encirclements of the origin by Γ', the contour traced out by $1 + A(s)$ as s traverses the closed contour Γ in the cw direction.

We may apply Eq. (8.8) to determine directly the number of zeros of $1 + A(s)$ in the rhp. Since critical frequencies in the rhp are of interest, Γ is chosen as the $j\omega$ axis and an infinitely large semicircle enclosing the entire rhp (the Bromwich contour), as illustrated in Fig. 8.4. This infinitely large semicircle causes no difficulty in applications, because physical realizability of $A(s)$ requires that $|A(s)| \to 0$ as $|s| \to \infty$, so the entire infinite semicircle maps into the point $1 + A(s) = 1$ in the complex plane. Development of the polar plot thus reduces to determining $1 + A(s)$ as s takes on values along the $j\omega$ axis. Since $\mathbf{A}(-j\omega) = \mathbf{A}^*(j\omega)$, where * denotes complex conjugation, the polar plot of $1 + A(s)$ for $s = -j\omega$ is the mirror image across the real axis of that obtained for $s = j\omega$, so it is really only necessary to evaluate $1 + \mathbf{A}(j\omega)$ for

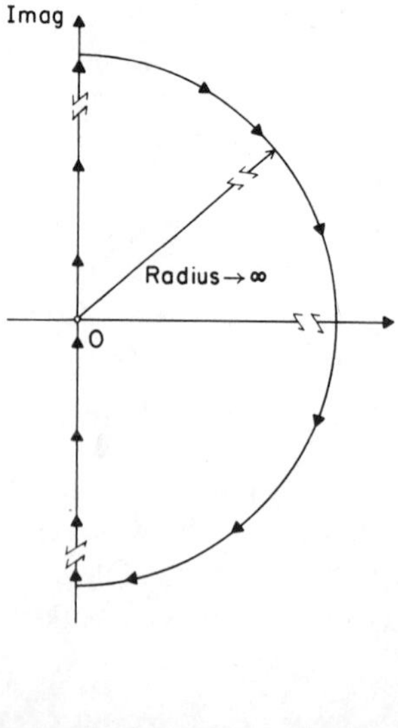

FIG. 8.4. The contour Γ used to enclose the entire rhp in applying the Nyquist criterion.

$0 \leq \omega \leq \infty$. In some cases the polar plot of $1 + A(s)$ may follow a rather complex path, but it is a simple matter to find N by determining how many net cw rotations an observer at the origin would have to undertake to follow a particle as it traverses the closed contour Γ'. P denotes the number of rhp poles of $A(s)$, and is generally known. Thus we may solve for Z as soon as Γ' has been constructed. If $Z \neq 0$, the system is unstable. $Z = 0$ is not sufficient to assure closed-loop stability in certain special cases, as will be shown, although such situations are not often encountered.

It is not convenient to plot $1 + \mathrm{A}(j\omega)$, so it is standard practice to plot $\mathrm{A}(j\omega)$ instead. The two curves are identical in shape, but the $\mathrm{A}(j\omega)$ plot is translated one unit to the left. When $\mathrm{A}(j\omega)$ is plotted, N is the number of cw encirclements of the $-1 + j0$ point (since we refer to this point so often, we will simply call it the -1 point hereafter).

The Nyquist stability criterion may be stated as follows: *If a polar diagram of the complex function $A(s)$ is plotted for values of s traversing the Bromwich contour, the polar diagram encircles the -1 point in the cw direction a number of times N equal to the number of zeros Z minus the number of poles P of the function $1 + A(s)$ enclosed within the rhp. It is necessary for closed-loop stability that all zeros of $1 + A(s)$ lie in the lhp. Thus, for stability, the Nyquist plot must encircle the -1 point in the counterclockwise (ccw) direction a number of times equal to the number of $A(s)$ poles in the rhp.* It should be stressed that these conditions are necessary for closed-loop stability, but not sufficient to guarantee it. The exceptions are considered in detail later.

Most systems encountered in practice are open-loop stable. In this special case the Nyquist criterion can be restated as follows: *An open-circuit, stable, closed-loop controller is stable if, and only if, the polar plot of $A(s)$ as s traverses the Bromwich contour makes no net encirclements of the -1 point, and does not pass through that point.* Contact of the Nyquist plot with the -1 point indicates the presence of one or more zeros of $1 + A(s)$ on the $j\omega$ axis, resulting in closed-loop instability according to our definition. Some reasonable margin of -1 point avoidance is generally desirable, since otherwise small changes in gain or time constant values could result in instability and, even if no changes occurred, the closed-loop transient response would be poor.

In summary, the steps taken in applying the Nyquist criterion are:

1. Obtain the system open loop transfer function, $A(s)$. This is accomplished by breaking the loop at an arbitrary point, preferably where loading is negligible, terminating the open loop with the impedance it normally sees at the point where the break is made, and finding the resulting open-loop transfer function.

2. Construct a polar plot of $A(s)$ for values of s along the Bromwich contour shown in Fig. 8.4. This normally only requires plotting values along the $j\omega$ axis, since the magnitude of realizable $A(s)$ functions approaches zero as $\omega \to \infty$. The values of $A(s)$ obtained for $s = -j\omega$ are the conjugates of those corresponding to $s = j\omega$, so the total curve may be obtained by plotting values of $A(j\omega)$ along the positive $j\omega$ axis and reflecting this piot across the real axis.

3. Count N, the number of cw encirclements of the -1 point by the Nyquist plot, and solve for the number of zeros of $1 + A(s)$ in the rhp using the relationship $N = Z - P$. If $Z \neq 0$, the closed-loop system is unstable. If $Z = 0$, the system is always stable if $A(s)$ is open-loop stable, but may be unstable if $A(s)$ is open-loop unstable and $H(s)$ has an rhp zero which cancels out an rhp pole of $A(s)$. This special case seldom arises and is considered separately in Sec. 8.3.

A minor modification of the Nyquist criterion which sometimes proves useful is the following. Let

$$A(s) = K\frac{P(s)}{Q(s)} = K\hat{A}(s) \tag{8.9}$$

Setting $1 + A(s) = 0$ yields

$$1 + A(s) = 1 + K\hat{A}(s) = 0 \tag{8.10}$$

Thus

$$\hat{A}(s) = -\frac{1}{K} \tag{8.11}$$

and the critical point becomes $(-1/K, j0)$ instead of $(-1, j0)$, and encirclements of this $-1/K$ point by the $\hat{A}(s)$ plot as s traverses the Bromwich contour have the usual interpretation. The advantage of this procedure is that the gain K may be retained as a parameter in the solution and only one point on the diagram needs to be moved when K is changed. This procedure is particularly important when working with nonlinear systems, as shown in Chap. 14.

8.2 APPLICATION OF THE NYQUIST CRITERION TO ELEMENTARY SINGLE LOOP SYSTEMS

In this section we will consider a sequence of progressively more difficult examples illustrating the procedure for applying the Nyquist criterion. Although our primary interest is centered about the -1 point and the number of encirclements of that point by the Nyquist plot, we are also often concerned with the general shape of the plot for magnitudes far greater than unity. We are led to use a logarithmic scale in constructing Bode plots, and it is equally convenient in many cases to use a logarithmic scale in constructing our Nyquist plots. This allows reasonably accurate representation of the magnitude and phase of $A(s)$ over a range of 10^4 or more in magnitude. There is a small problem, however, since zero is infinitely far from unity on a log scale. This causes no real difficulty, because we are not concerned with the detailed structure of the Nyquist plot

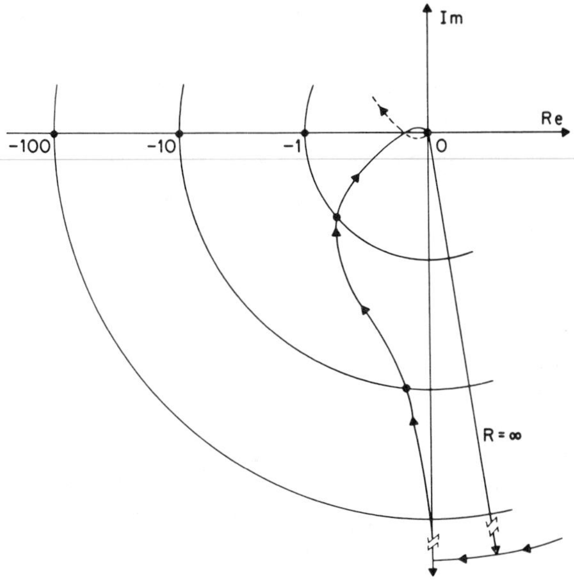

FIG. 8.5. Nyquist plot for the system of Ex. 8.1.

for magnitudes significantly less than unity. We will use a logarithmic scale on all axes in making most of our polar plots, and we will generally let the origin correspond to the "point" denoting all magnitudes of 0.1 or less. Seldom are magnitudes below 0.1 of detailed interest. The polar plot shown in Fig. 8.5 is drawn on the suggested logarithmic coordinates.

The Nyquist plots in most references are drawn on some compressed scale, but the axes are not generally labeled. Use of a log scale allows more accurate plots to be made and more information to be obtained readily from them. Although polar paper is not generally available with a logarithmic scale, it is a simple matter to construct a master with log-scaled coordinates running out from the origin in all directions, and copies are readily run off for general use. The rewards are significant.

Example 8.1

Consider the position controller shown in Fig. 8.6. Analyses and laboratory tests performed upon the system components indicate that the open-loop

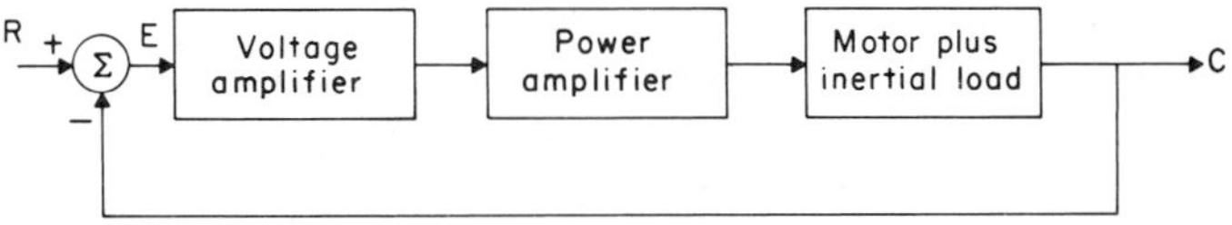

FIG. 8.6. General structure of the controller considered in Ex. 8.1.

transfer function $A(s)$, under normally loaded conditions, is

$$A(s) = \frac{C(s)}{E(s)} = \frac{20}{s(1 + s/20)(1 + s/100)} \tag{8.12}$$

This $A(s)$ is obviously minimum phase, so stability is assured if the Nyquist plot has no net encirclements of the -1 point. We encounter a problem immediately in trying to construct the Nyquist plot due to the pole at the origin, which results in $A(s)$ being undefined for $s = 0$. This problem is resolved by modifying the Bromwich contour slightly, as shown in Fig. 8.7. The radius of the semicircle at the origin is assumed to be infinitesimally small, and the limiting performance of the Nyquist plot as this radius approaches zero is considered. The same general procedure applies for all poles on the $j\omega$ axis.

To construct the Nyquist plot, set $s = \delta\epsilon^{j\theta}$, where δ is assumed very small, and see what happens to $A(s)$ as θ takes on values from 0 to $\pi/2$. The value $\theta = 0$ corresponds to the point labeled 1 in Fig. 8.7, and $s = \delta$ at this point. Thus

$$A(s) = A(\delta) = \frac{20}{\delta(1 + \delta/20)(1 + \delta/100)} \simeq \frac{20}{\delta} \simeq \infty \tag{8.13}$$

Point 2 in Fig. 8.7 corresponds to $\theta = \pi/4$, and $s = (\sqrt{2}/2)\,\delta(1 + j1) = \delta_1 + j\delta_1$ at that point, where $\delta_1 = (\sqrt{2}/2)\,\delta$. Thus

$$\begin{aligned} A(s) &= A(\delta_1 + j\delta_1) \\ &= \frac{20}{(\delta_1 + j\delta_1)[1 + (\delta_1 + j\delta_1)/20][1 + (\delta_1 + j\delta_1)/100]} \\ &\simeq \frac{20}{\delta_1 + j\delta_1} \simeq \infty\epsilon^{-j\pi/4} \end{aligned} \tag{8.14}$$

The Nyquist plot thus proceeds cw at infinite radius as the test point proceeds ccw at infinitesimal radius about the pole at the origin. This is intuitively expected, since $A(s)$ is made up of individual vectors starting from the critical frequencies (in this case there are only poles) and terminating at the test point of interest. As the test point traverses the quarter circle between the real and the imaginary axes, the vectors corresponding to the poles at -20 and -100 are essentially unchanged, whereas the vector corresponding to the pole at $s = 0$ rotates through $\pi/2$ radians. Since the latter vector is in the denominator of $A(s)$, and its magnitude is small, the vector denoting $A(s)$ has large magnitude and rotates $\pi/2$ radians in the negative direction. The Nyquist plot is completed by obtaining several additional points for representative values along the positive $j\omega$ axis and sketching the resulting curve plus its mirror image. Since $|A(s)| \to 0$

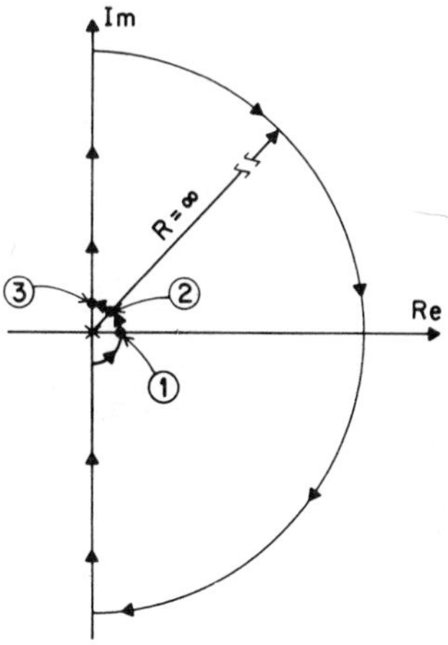

FIG. 8.7. The modified contour used to avoid a pole at the origin.

as $|s| \to \infty$, values of s on the infinite semicircle are of no significance. The resulting Nyquist plot is shown in Fig. 8.5. There are no encirclements of the -1 point, so the system is stable.

Several characteristics illustrated by this example should be stressed. The treatment of poles on the $j\omega$ axis is straightforward. They are simply avoided using a small semicircular indentation into the rhp. The resulting Nyquist plot rotates at infinity on this part of the contour. In most cases, Nyquist plot accuracy is required only near unit magnitude, so sketches can be constructed rapidly. As $\omega \to \infty$ the plot approaches 0 at an angle depending upon the critical frequencies of $A(s)$. In general, for open-loop, stable, minimum-phase systems,

$$\theta(\infty) = -\frac{\pi}{2}(P - \hat{Z}) \tag{8.15}$$

where P and $\hat{Z}$ denote the numbers of poles and zeros, respectively, of $A(s)$. This observation often helps in making approximate sketches. We chose arbitrarily to avoid the pole at the origin by indenting the contour into the rhp. In this case the pole is not included within Γ, and stability requires zero encirclements of the -1 point. This is the standard treatment. We could just as well, however, indent Γ into the lhp and include the pole at the origin within Γ. Then a ccw encirclement of the -1 point would be required for stability. However, since the infinite semicircle in Fig. 8.5 would close around the lhp instead of the rhp in this case, the required encirclement would be obtained and the same answer results either way.

Example 8.2

Consider the system shown in Fig. 8.8. It is desired to choose a compensating filter $G_c(s)$ to assure stability. What general effect must this filter have on the Nyquist plot? Suggest a filter which should prove acceptable.

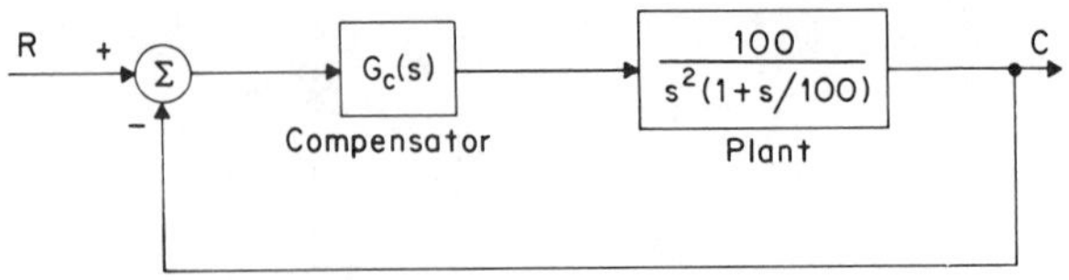

FIG. 8.8. The system considered in Ex. 8.2.

Sketching the Nyquist diagram of the plant alone will indicate what is required of the compensator to stabilize the system. The double pole at the origin is avoided as in Ex. 8.1, and the polar plot is shown in Fig. 8.9. Since there are two poles at the origin, the Nyquist plot rotates through $-\pi$ radians at infinite radius and then, due to the additional pole at $s = -10$, approaches zero along the -270° (imaginary) axis. There are two encirclements of the -1 point, so the uncompensated closed-loop system has two rhp poles. The compensator must provide a phase lead of 30° to 40° near unit magnitude to give some stability margin. Although relatively efficient procedures for choosing compensators are discussed later using Bode diagram design methods, it is not difficult to determine, using trial and error, that

$$G_c(s) = \frac{1 + s/6}{1 + s/60} \tag{8.16}$$

provides stability of the compensated plant, as shown in Fig. 8.10. A $G_c(s)$ of

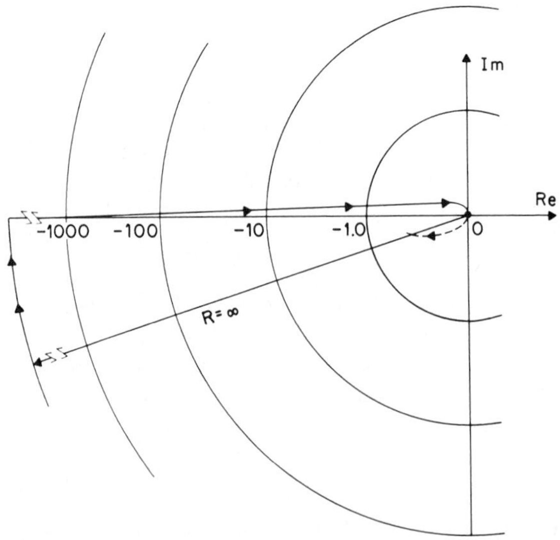

FIG. 8.9. Nyquist plot of the plant introduced in Ex. 8.2.

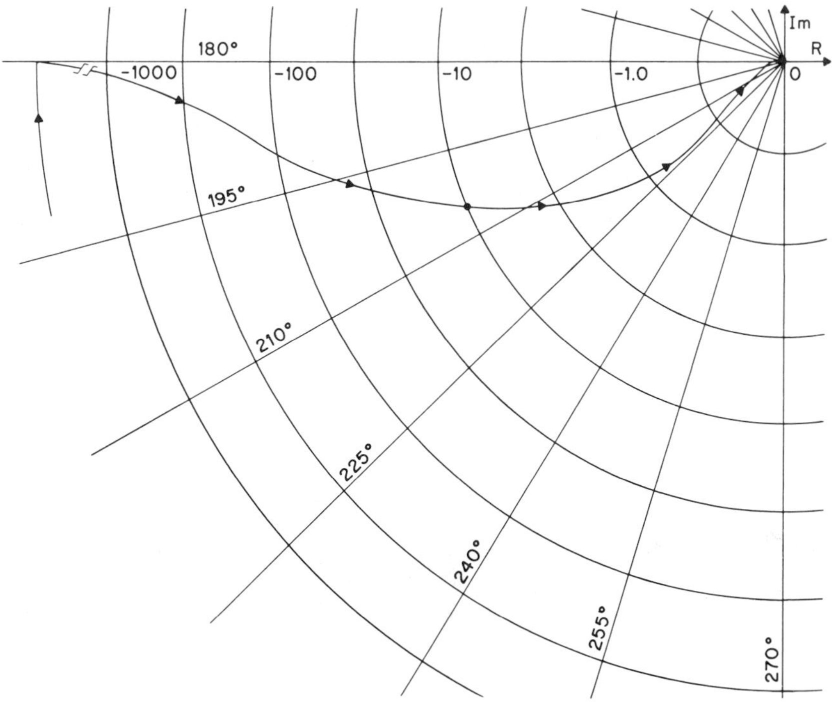

FIG. 8.10. Nyquist plot of the compensated system developed in Ex. 8.2.

the form defined in Eq. (8.16) is called a lead compensator because it provides phase lead over a frequency range centered between the two break frequencies.

Example 8.3

Consider a system with

$$A(s) = \frac{10}{s(1 + s/10)(1 + s/30)(s^2 + 10{,}000)} \tag{8.17}$$

Sketch the Nyquist diagram and determine if the closed-loop system is stable.

This plant is unusual because it contains a pair of poles on the $j\omega$ axis at $\omega = \pm 100$. Although systems with poles on the $j\omega$ axis are not often encountered in practice, many mechanical and electrical configurations are so lightly damped that it is reasonable to approximate their transfer functions in this way. A sketch showing the complex plane pole locations for this $A(s)$ is provided in Fig. 8.11. The modified Bromwich contour chosen to avoid the pole at $s = 0$ and the poles at $s = +j100$ is also shown there. The corresponding Nyquist plot is shown in Fig. 8.12. As $\omega \to 100$, $|A(s)| \to \infty$. The angle of $A(s)$ rotates 180° in a negative (ccw) direction as the semicircle avoiding the pole at $s = j100$ is traversed. The angle at which the plot returns to infinite magnitude is readily

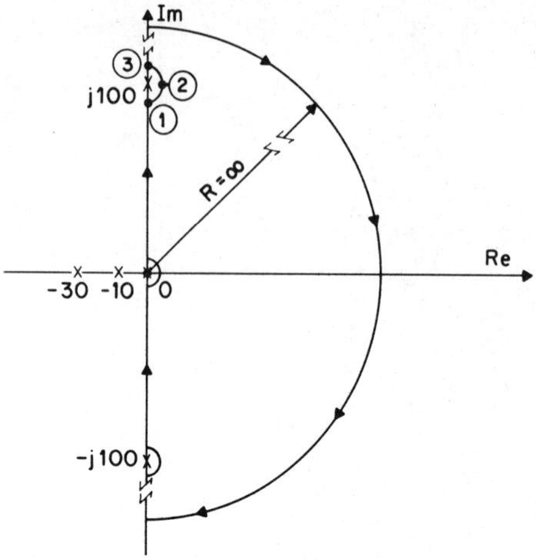

FIG. 8.11. Complex plane plot of the $A(s)$ defined by Eq. (8.17).

obtained by summing the angles of the individual vectors from which $A(s)$ is composed, or by substituting $s = j(100 - \delta)$ into Eq. (8.17), which yields

$$A_1(s) = A_1[j(100 - \delta)]$$

$$= \frac{10}{j(100-\delta)\left(1 + j\,\dfrac{100-\delta}{10}\right)\left(1 + j\,\dfrac{100-\delta}{30}\right)\{[j(100-\delta)]^2 + 10{,}000\}} \tag{8.18}$$

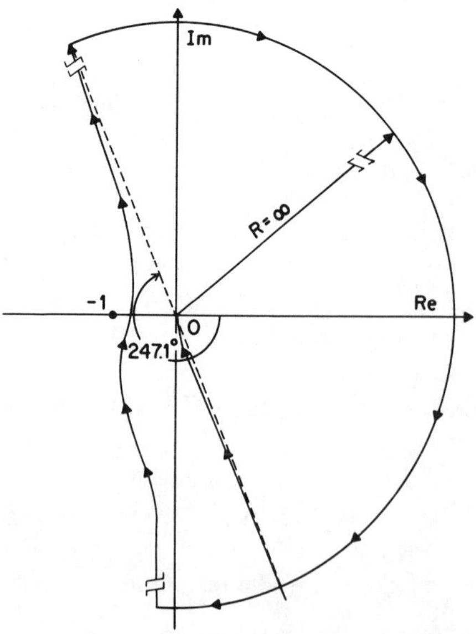

FIG. 8.12. Nyquist plot of $A(s)$ defined by Eq. (8.17).

which simplies, assuming δ very small, to

$$A_1(s) = \frac{10}{j100(1 + j10)[1 + j(10/3)](200\delta)} \tag{8.19}$$

Thus the asymptotic angle, denoted by θ_1, is

$$\theta_1 = -\frac{\pi}{2} - \tan^{-1}10 - \tan^{-1}\frac{10}{3} \simeq -\frac{3\pi}{2} + 0.1 + 0.3 \simeq -247.1^\circ \tag{8.20}$$

The points denoted by two and three on the plot in Fig. 8.11 are 90° and 180° displaced, respectively, from the point denoted by 1. This is readily checked by substituting $s = \delta + j100$ and $s = j(100 + \delta)$ into Eq. (8.17) and determining the corresponding angles of $A(s)$.

The examples presented here illustrate several special conditions encountered in applying the Nyquist criterion. Most systems do not have undamped pole pairs on the $j\omega$ axis, although a pole at the origin is a rather common occurrence. Application of the Nyquist criterion to typical control problems should now be straightforward.

8.3 STABILITY MARGIN AND CONDITIONAL STABILITY

Up to this point we have considered the Nyquist criterion only for obtaining a yes-no type answer to the stability question. Just as the encirclement of the -1 point, assuming an open-loop stable system, indicates instability, the margin by which that encirclement is avoided provides some measure of relative stability. Although it is impossible to define this relative stability concept precisely, an intuitive feeling is readily developed which proves satisfactory in most applications. If the Nyquist plot "narrowly avoids" enclosing the -1 point, the closed-loop system "almost" has poles in the rhp, and thus has one or more poles "near" the complex-plane imaginary axis. Poles near the $j\omega$ axis cause the response of the system to approach steady-state slowly after input disturbances are applied due to the corresponding lightly-damped transient terms, so this condition is generally unacceptable. A reasonable margin of -1 point avoidance is necessary if relatively well behaved closed-loop transient-response characteristics are to be obtained. Two precise measures of -1 point avoidance which have been defined in an effort to relate frequency domain (Nyquist plot) and time domain (transient response, damping) characteristics are *phase margin* and *gain margin*. By imposing restrictions upon one or both, some assurance can be given that the resulting closed loop system transient response characteristics will be "acceptable".

Consider the relatively typical Nyquist plot shown in Fig. 8.13. *Phase margin* θ_m is defined as the cw angle through which the Nyquist plot would have to be

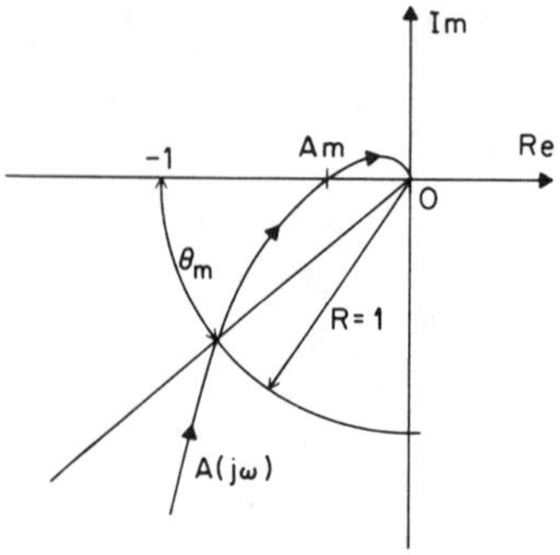

FIG. 8.13. Nyquist plot used to define phase margin and gain margin.

rotated to pass through the -1 point. θ_m is thus a direct angular measure of -1 point avoidance and stability margin. θ_m is readily measured as the angle between a ray from the origin passing through the intersection of the Nyquist plot with the unit circle and the negative real axis. If the Nyquist plot passes through the unit circle two or more times, the third quadrant intersection nearest the negative real axis is used in measuring phase margin. Phase margin is, of course, negative (or undefined if you choose) for unstable systems.

Let A_m denote the magnitude inside the unit circle and nearest the -1 point where the Nyquist plot crosses the negative real axis. *Gain margin* G_m is defined as

$$G_m = \frac{1}{A_m} \tag{8.21}$$

G_m is also a measure of -1 point avoidance, and is the ratio by which open-loop gain must increase to cause instability if all other parameters remain constant. Although neither θ_m nor G_m alone can assure a large stability margin in all cases due to unusual situations which can arise, when both are restricted to reasonable values the closed-loop transient response is generally acceptable. It is typical to require $\theta_m \geq 30°$ and $G_m \geq 3$, although each application must be considered on the basis of detailed performance objectives. Systems are often designed to satisfy a particular phase margin spec. It is somewhat less common to consider G_m directly in choosing a design. This is because a spec on θ_m nearly always results in an acceptable value of G_m, and specifying both is somewhat redundant.

A system is defined as *conditionally stable* if it can become unstable for a *decrease* in gain. This is not a desirable situation, since amplifier and component gains generally tend to decrease with age, reduced supply voltages, or large signal levels (due to saturation). Such systems are difficult to turn on and off, because they tend to be unstable until all power supplies reach a steady-state condition. If the gain decreases for any reason, the closed-loop transient performance is degraded, and instability can eventually result. The Nyquist diagrams of three conditionally stable systems are shown in Fig. 8.14. Single integration systems can be conditionally stable if there are two or more phase lag terms (poles) with break frequencies below unit magnitude crossover (assuming minimum phase conditions) or, equivalently, if the Bode diagram has a -3 slope region of significant duration prior to unit magnitude crossover, which generally occurs in a region of -1 slope. Systems with two integrations are conditionally stable if there are more lag terms than lead terms with break frequencies below crossover. Double integration systems are rare. Systems with three integrations are always conditionally stable, are extremely difficult to compensate, and, fortunately, are seldom encountered outside the classroom.

A reasonably accurate Nyquist diagram can be used to determine the range of gain values over which a system will be stable, and the frequency (or frequencies) at which it will tend to oscillate as it becomes unstable. The oscillatory frequency indicates the point on the $j\omega$ axis where the closed-loop poles enter the rhp, and is the frequency value on the Nyquist diagram at the point of intersection with the -1 point for some appropriately chosen gain. In many cases it is more convenient to find the approximate range of stable gain from the Nyquist plot than by using the Routh criterion. The ratios by which gain must be changed from a nominal value to bring about various intersections of the Nyquist plot with the -1 point are immediately apparent if a reasonably accurate plot is available. In fact, accuracy is only required at points where the Nyquist plot crosses the negative real axis, and such accuracy may be obtained by evaluating the magnitude and phase of $A(s)$ at only a few critical points.

Example 8.4

Consider the detailed Nyquist plot of a conditionally stable system, as shown in Fig. 8.15. The particular $A(s)$ for which this plot applies is not important, but the plot corresponds to a gain value of 100. Find G_m and the range of stable gain. For each marginally unstable value of gain, find the frequency at which the system would tend to oscillate, or the ω coordinate of the corresponding pole pair on the $j\omega$ axis.

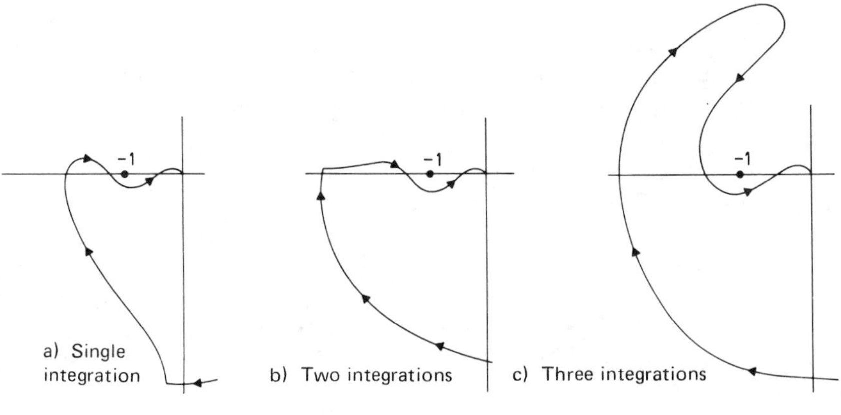

FIG. 8.14. Nyquist diagrams of conditionally stable systems.

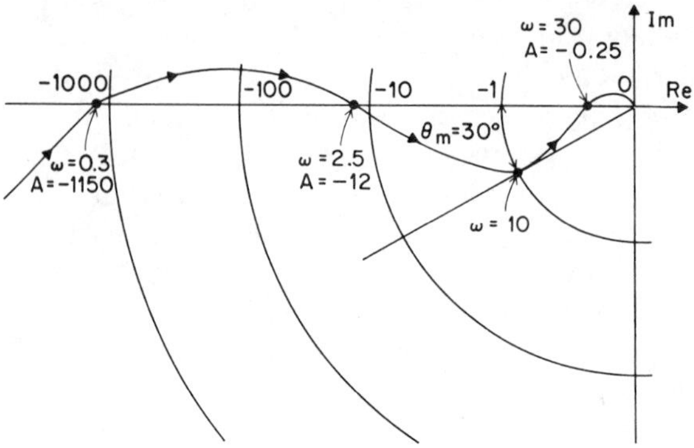

FIG. 8.15. Detailed Nyquist plot used in Ex. 8.4.

First of all,

$$G_m = \frac{1}{A_m} = \frac{1}{0.25} = 4 \tag{8.22}$$

Thus if the gain increases by a factor of 4, or to 400, the system becomes unstable and oscillates at $\omega = 30$. The system is unstable for all $K \geq 400$. Similarly, if the gain decreases by a factor of 12, or to 8.33, the system becomes unstable and oscillates at $\omega = 2.5$. If, however, the gain decreases further to 100/1150, the system once again becomes marginally unstable, this time at $\omega = 0.3$, and the system is stable for all non-negative $K < 100/1150$. Thus the range of stability is given by $0 \leq K < 100/1150$ and $8.33 < K < 400$. The system is unstable for all $K < 0$, for $100/1150 \leq K \leq 8.33$, and for $K \geq 400$.

Note how readily requirements on gain for stability are obtained from the Nyquist plot. Although the Routh criterion will provide the same information, its use requires extensive mathematical manipulations for complex systems. Acceptably accurate answers are obtained rapidly from the Nyquist plot. A bit of practice enables negative real axis crossings of the Nyquist plot (points of 180° phase shift) to be approximated rapidly using a Bode diagram to obtain magnitude and a slide rule to obtain phase. The advantage of this method is most pronounced for systems of high order. Although the range of gain for stability is readily obtained from the Nyquist plot, it is not particularly convenient to use this method for determining, at least in any particularly quantitative way, how other parameters, such as time constants, influence stability. The Routh criterion is generally superior for this purpose.

8.4 CLOSED-LOOP FREQUENCY RESPONSE FROM THE NYQUIST PLOT

If the Nyquist plot is made on *linear* coordinates, the plot may be used directly to obtain the closed-loop frequency response. For the unity feedback case this is particularly simple. In all other cases an auxiliary plot of the forward transfer function $G(s)$ is required, since the closed-loop response function of the standard controller shown in Fig. 8.16 is given by

$$\frac{C(s)}{R(s)} = \frac{G(s)}{1 + G(s)H(s)} = \frac{G(s)}{1 + A(s)} \tag{8.23}$$

and the Nyquist plot is a polar plot of $A(s)$. When $H(s) = 1$, $A(s) = G(s)$, and the closed-loop response is available directly from the linear Nyquist plot. Assume $H(s) = 1$. The procedure presented here need only be modified slightly when $H(s) \neq 1$. Consider a system with linear coordinate Nyquist plot for $G(s) = A(s)$ as shown in Fig. 8.17. Note that

$$\left|\frac{\mathbf{C}(j\omega)}{\mathbf{R}(j\omega)}\right| = M(j\omega) = \left|\frac{\mathbf{G}(j\omega)}{1 + \mathbf{G}(j\omega)}\right| \tag{8.24}$$

The magnitudes of the numerator and denominator terms in Eq. (8.24) are readily obtained from the Nyquist plot as shown in Fig. 8.17. Thus the solution of a rather tedious mathematical problem is reduced to a straightforward graphical procedure once the Nyquist plot has been constructed.

$M(j\omega)$ for a system with $\mathbf{G}(j\omega)$ as shown in Fig. 8.17 is illustrated in Fig. 8.18. As $\omega \to \infty, |\mathbf{G}(j\omega)| \to \infty$, and $M(j\omega) \to 1$. This is true for all unity feedback systems with one or more integrations in $G(s)$. As ω increases, the Nyquist plot eventually approaches the -1 point vicinity. In this region, $|\mathbf{G}(j\omega)|$ decreases as ω increases, but $|1 + \mathbf{G}(j\omega)|$, which appears in the denominator of $M(j\omega)$, decreases even more rapidly, so $M(j\omega)$ has a peak value, denoted by M_p, at some frequency in this vicinity denoted by ω_p. For the situation depicted in Fig. 8.17,

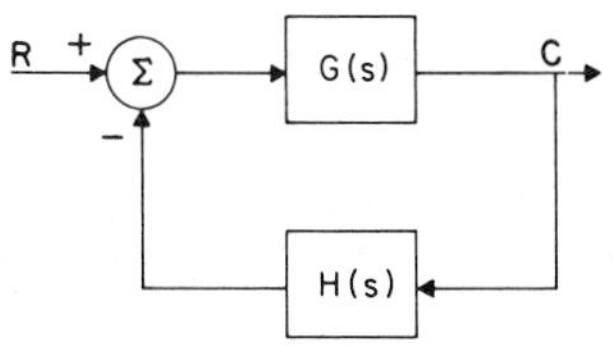

FIG. 8.16. Standard non-unity feedback controller.

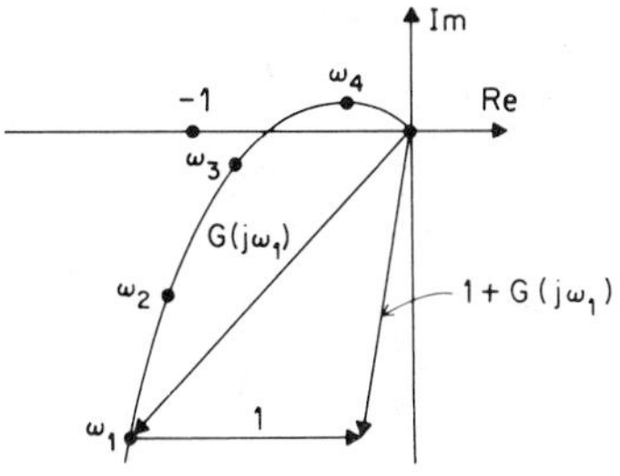

FIG. 8.17. Use of the Nyquist plot to determine frequency response.

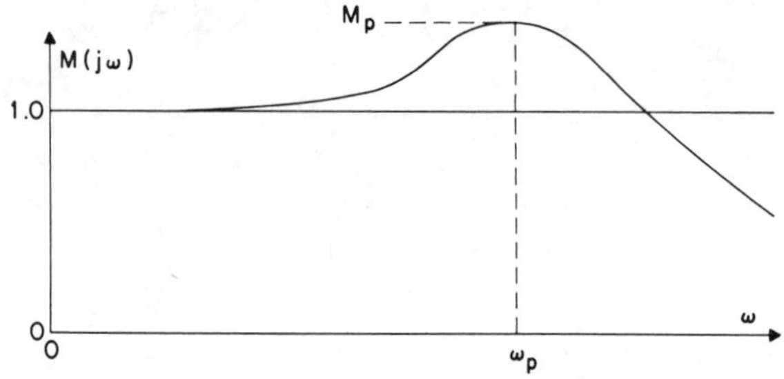

FIG. 8.18. Frequency response of a unity feedback system with Nyquist plot as shown in Fig. 8.17.

ω_p is near the value labeled ω_3 on the diagram and M_p is significantly greater than 1 because the Nyquist plot passes rather close to the -1 point in this region. These conditions are reasonably typical.

Some authors suggest using M_p and ω_p as primary design parameters,[1] and develop detailed design procedures from this point of view. It is often necessary in system design to limit M_p to prevent excessively oscillatory response to inputs containing frequency components near ω_p, but M_p can be controlled indirectly in almost all cases of practical interest by restricting θ_m. Since θ_m is so much easier to determine from the Bode diagram, and since most of our designs are carried out on the Bode diagram, we choose θ_m as our primary design parameter. This is not meant to imply that M_p is unimportant, but only that θ_m is more easily put to use, and specs upon θ_m and M_p are very nearly equivalent. Of course,

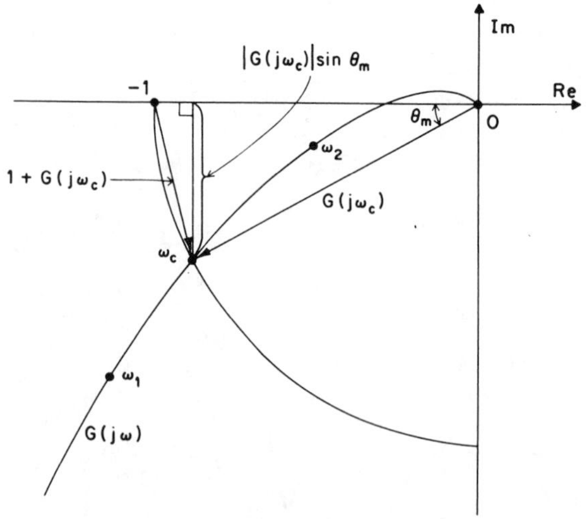

FIG. 8.19. Graphical illustration of the relationship between θ_m and M_p.

the frequency response of most final designs will be checked, as will its unit-step response. Adjustments are readily made when they seem appropriate.

A general rule of thumb relationship between θ_m and M_p which is acceptably accurate in most unity feedback situations encountered is given by

$$M_p \simeq \frac{1}{\sin\theta_m} \tag{8.25}$$

The proof of this follows directly from the construction shown in Fig. 8.19. Assuming θ_m is not greater than 45° or so, the minimum $|1 + \mathbf{G}(j\omega)|$, and thus, to a first-order approximation, the maximum $M(j\omega)$, occurs very near ω_c, the value of ω for which $|\mathbf{G}(j\omega)| = 1$. At ω_c

$$|1 + \mathbf{G}(j\omega)| \simeq |\mathbf{G}(j\omega)| \sin\theta_m \tag{8.26}$$

as shown in Fig. 8.19. Substituting Eq. (8.26) into Eq. (8.24) yields Eq. (8.25). Obviously the approximation error increases nonlinearly with θ_m, but the accuracy is acceptable for the development of a preliminary design which may be adjusted later if that is necessary. A spec on ω_p is equivalent to specifying ω_c, since $\omega_c \simeq \omega_p$ is a close approximation, and it is much easier to restrict ω_c than ω_p. Thus M_p and ω_p specs need not be imposed directly, even when they are given, since near-equivalent specs on θ_m and ω_c, which are easier to use, can always be substituted during the design phase.

There are two situations for which the approximation presented in Eq. (8.25) is poor. They are:

1. G_m is small despite θ_m having a reasonably large value such as 30° or more, as depicted in Fig. 8.20.
2. θ_m is large, such as 60° or more.

Neither of these conditions arises often, and the symptoms are readily apparent from the Bode and Nyquist diagrams when they do. In the first case equalization to yield a more gradual increase in phase lag with frequency following unit

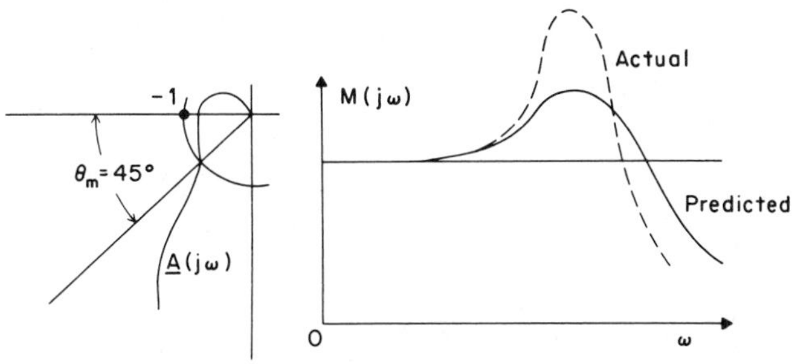

Fig. 8.20. A situation for which the approximation of Eq. (8.25) breaks down.

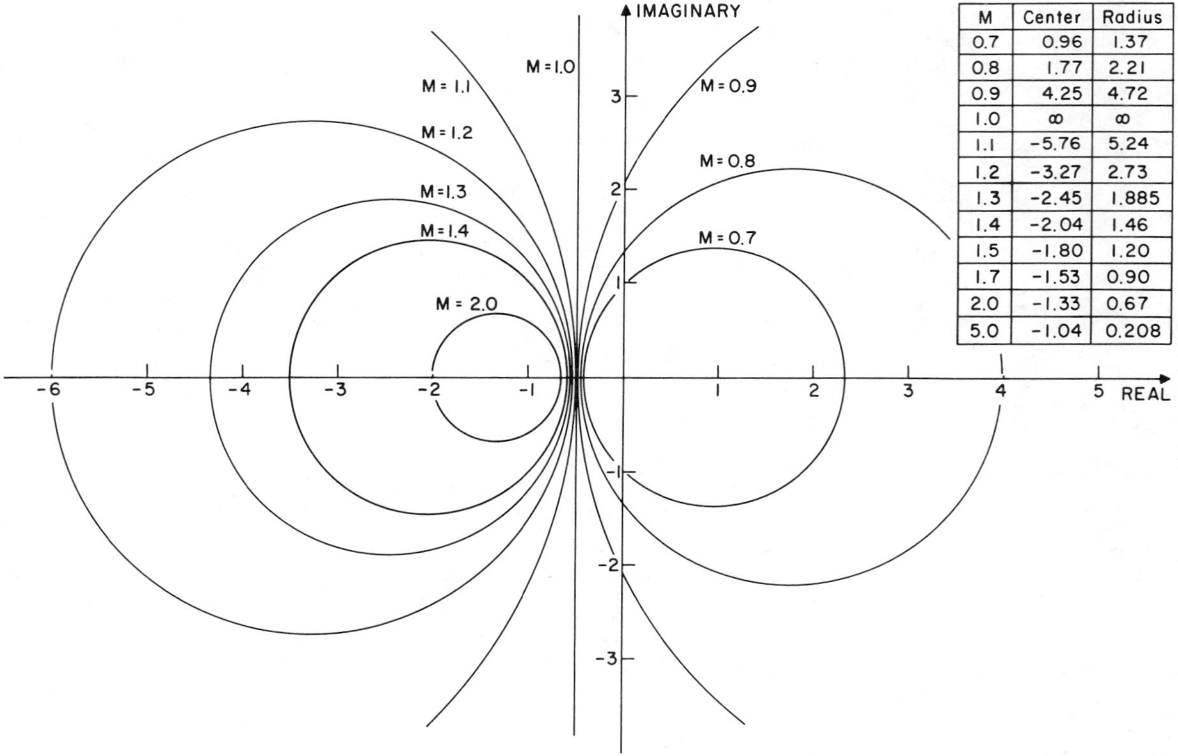

M	Center	Radius
0.7	0.96	1.37
0.8	1.77	2.21
0.9	4.25	4.72
1.0	∞	∞
1.1	-5.76	5.24
1.2	-3.27	2.73
1.3	-2.45	1.885
1.4	-2.04	1.46
1.5	-1.80	1.20
1.7	-1.53	0.90
2.0	-1.33	0.67
5.0	-1.04	0.208

FIG. 8.21. Loci of constant M_p in the complex plane for the unity feedback case.

magnitude crossover is appropriate, and can be easily realized if desired. In the second case M_p is small anyway, and should not generally be a significant consideration in the design.

It is easily shown for the unity feedback case, using the basic concepts of analytical geometry,[2] that the loci of constant M_p in the complex plane on which we make the Nyquist plot of $\mathbf{A}(j\omega) = \mathbf{G}(j\omega)$ are circles (plotted on a linear scale) of radius $M_p/(M_p^2 - 1)$ with centers on the negative real axis at $-M_p^2/(M_p^2 - 1)$. A set of constant M_p loci covering a range of common interest is shown in Fig. 8.21. To design a system with a particular value of M_p, the Nyquist plot is shaped such that it is tangent to that particular circle at one point (there could be more than one such point, but this would be unusual) and is outside the circle everywhere else. Organized trial and error generally yields a solution without difficulty. Establish an initial design on the Bode diagram to provide $\theta_m = \sin^{-1}(1/M_{pd})$, where M_{pd} denotes the desired M_p. Make a Nyquist plot of the result. Construct the M_{pd} locus on the Nyquist plot. If the resulting M_p is sufficiently far from M_{pd}, make appropriate modifications in the design. This involves "shaping" the Bode diagram, and consequently the Nyquist plot, to avoid or approach the M_{pd} locus at those frequencies where adjustment is necessary. Suppose, for example, $M_p = 1.3$ is desired and the situation shown in Fig. 8.22 results after the first trial design. This is easily corrected by adding a small amount of additional phase lead at frequencies above ω_c. If the situation shown in Fig. 8.23 results, additional phase lead below ω_c is necessary. The adjustments required are always readily apparent from the Nyquist plot with the M_{pd} locus superimposed upon it and are easily made using the Bode diagram techniques discussed in Chap. 9. The Nyquist plot must be made on linear coordinates for this purpose, contrary to our usual practice, but this causes no difficulty since only small ranges of magnitudes and frequencies near crossover are involved.

Although the constant M_p loci are not generally circles when $\mathbf{H}(j\omega) \neq 1$, the same basic design procedures apply in this case. Of course, more work is involved

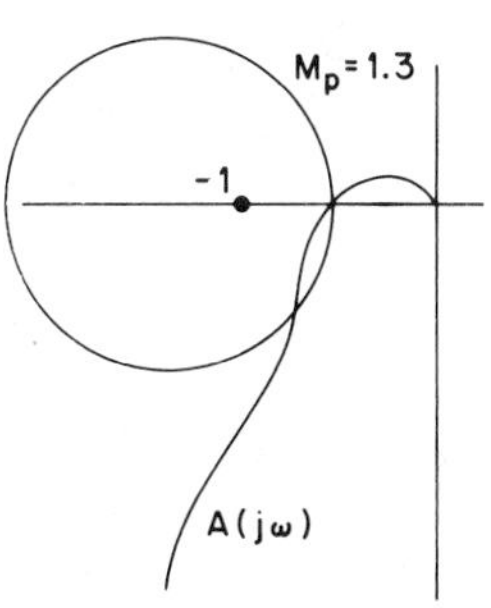

FIG. 8.22. A case where additional phase lead is required at frequencies above ω_c.

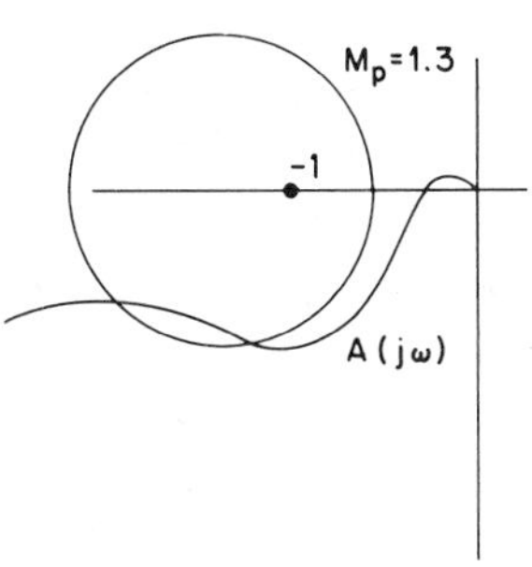

FIG. 8.23. A case where additional phase lead is required at frequencies below ω_c.

in converting from the Nyquist to the $M(j\omega)$ plot to obtain M_p, but this cannot be avoided. Fortunately, unity feedback systems are common, and the more torturous analysis procedures required for the nonunity feedback case are not often necessary. It is straightforward in any event to obtain open and closed-loop frequency response data using a computer and finalize the design on a simulation, if this seems more appropriate.

8.5 THE NYQUIST CRITERION FOR COINCIDENT POLES AND ZEROS

Suppose we have a plant with a pole in the rhp. It is tempting to connect an equalizer with an overlapping rhp zero in series with the plant in an effort to stabilize it, but *this procedure will never work.* The rhp pole location depends upon plant parameters and load effects, and any changes will cause the pole to move. Thus exact cancellation of a pole with a zero is impossible, and this is true no matter where the pole is located. For poles in the lhp, approximate cancellation is adequate since the transient response term corresponding to the residue at that pole decays with time and causes no problem. The response terms corresponding to incompletely canceled rhp poles always grow without bound, however, so cascade cancellation of rhp poles by zeros is not feasible. This does not mean that plants with poles in the rhp cannot be stabilized and used as the nucleus of a stable closed-loop system, but only that cascade stabilization is impossible. Open-loop unstable plants can only be stabilized using a feedback configuration. The Nyquist criterion may yield wrong answers if cascade rhp or $j\omega$ axis pole-zero cancellations are allowed, so this possibility should be checked in all applications. A situation of special interest is illustrated by the following example.

Example 8.5

Consider the system shown in Fig. 8.24. The controller is a standard speed control system, since a constant input level results in a constant rate of change, or velocity, of the output variable, which we shall interpret as shaft position. Note that a pole-zero cancellation occurs at the origin in $A(s) = G(s)H(s)$, so we might anticipate some problem in applying the Nyquist criterion. Is this system stable? Applying the Nyquist criterion directly to

$$A(s) = \frac{10}{(1 + s/10)(1 + s/100)} \tag{8.27}$$

provides the Nyquist plot shown in Fig. 8.25. Since there are no -1 point

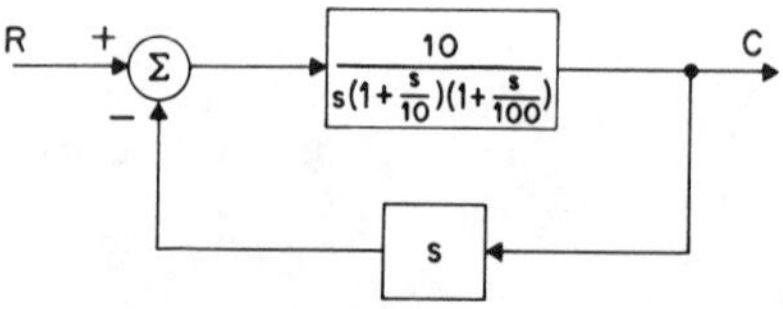

FIG. 8.24. A system with tachometer feedback.

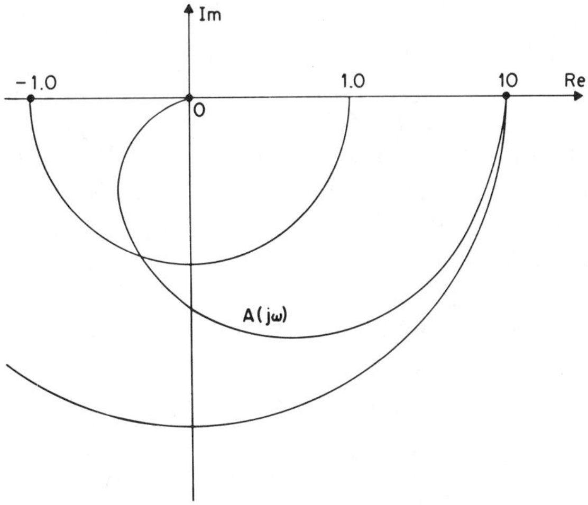

FIG. 8.25. Nyquist plot for the system shown in Fig. 8.24.

encirclements, there are no characteristic equation roots in the rhp. The superficial interpretation that this implies system stability is incorrect, however. This system has a closed-loop pole at the origin that is not a root of $1 + A(s) = 0$, but results from the pole-zero cancellation at the origin in $A(s)$ due to the s (tachometer) term in the feedback path. The closed-loop transfer function is

$$\frac{C(s)}{R(s)} = \frac{G(s)}{1 + A(s)}$$

$$= \frac{\dfrac{10}{s\left(1 + \dfrac{s}{10}\right)\left(1 + \dfrac{s}{100}\right)}}{1 + \dfrac{10}{\left(1 + \dfrac{s}{10}\right)\left(1 + \dfrac{s}{100}\right)}} = \frac{10}{s\left[\left(1 + \dfrac{s}{10}\right)\left(1 + \dfrac{s}{100}\right) + 10\right]} \tag{8.28}$$

Setting $1 + A(s) = 0$ yields the roots of the denominator term, in square brackets in Eq. (8.28), and the Nyquist criterion correctly indicates that both of these roots are in the lhp. The pole at the origin arises because the plant is open-loop unstable, and cancellation of the pole causing that instability is present in the feedback path. We must always be alert to note situations of this type.

Tachometer feedback cancellation of a motor integration is a common phenomenon used in speed control devices. This is the only case where effective cascade cancellation of an unstable open-loop pole is possible, so the situation depicted in Ex. 8.5 is a highly specialized one. The example illustrates nicely the point of interest however. Note that we would consider the same system stable if the output is taken as shaft velocity (the tachometer output) rather than shaft position. This is an interesting situation indeed.

In summary, the Nyquist criterion indicates only if there are rhp roots of the *characteristic equation* $1 + A(s) = 0$. This indication will be wrong if cascade cancellation of rhp plant poles is allowed. We must also check to see if there are any closed-loop poles in the rhp or on the $j\omega$ axis caused by feedback cancellations. Such conditions can only arise if the forward transfer function is open-loop unstable, and the situation most often encountered involves tachometer feedback, as illustrated in Ex. 8.5.

8.6 APPLICATION OF THE NYQUIST CRITERION TO MULTIPLE LOOP SYSTEMS

The Nyquist stability criterion is readily applied to systems containing two or more closed loops to determine not only the stability of the total system, but that of each individual loop as well. Consider the system shown in Fig. 8.26 as an example. The stability of this system can be determined in any one of the following three ways:

1. Find the forward transfer function $C(s)/E(s) = G_1G_2G_3/(1 + G_2H) = A(s)$ and make a Nyquist plot of this $A(s)$ to determine the number of rhp poles.
2. Break both loops together and make one Nyquist plot using $A(s)$ as the sum of the parallel paths.
3. Consider the minor loop alone to determine if it has any rhp poles, and make a Nyquist plot of the entire forward transfer function to determine the number of rhp poles for the overall system.

This latter two-step procedure is preferable, generally requiring less overall effort. The first of the suggested procedures may prove difficult to accomplish. The forward transfer function of the outer loop contains the minor loop characteristic equation, since the transfer function between the points labeled Y and X is

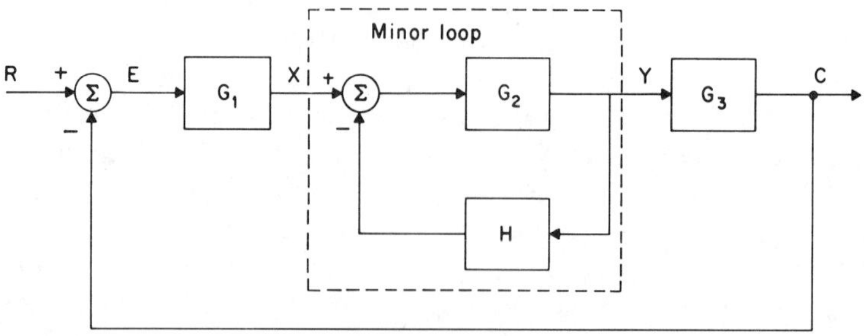

FIG. 8.26. Block diagram of a two loop system.

given by

$$\frac{Y(s)}{X(s)} = \frac{G_2(s)}{1 + G_2(s)H(s)} \tag{8.29}$$

The locations of these transfer functions poles must be determined before the Nyquist criterion can be applied to the outer loop. Application of the Nyquist criterion to the minor loop first is the easiest way to determine in which half plane its poles are located. For this reason it is standard procedure in applying the Nyquist criterion to multiple-loop systems to work outward from the inner loops, defining pole locations for each loop in turn, until the major loop answer is obtained. The second procedure is not always applicable because it may not be possible to break all loops by breaking one point in the system.

Consider the system shown in Fig. 8.26 to illustrate the general treatment of multi-loop configurations one loop at a time. Assume the minor loop open-loop transfer function G_2H has no rhp poles. Assume that the Nyquist plot of G_2H yields one cw encirclement of the -1 point. Thus $Y(s)/X(s)$ for the minor loop has one rhp pole, since $Z - P = Z - 0 = 1$. When we plot

$$A(s) = \frac{G_1G_2G_3}{1 + G_2H} \tag{8.30}$$

we must obtain one ccw encirclement of the -1 point to assure overall stability, assuming G_1G_3 has no rhp poles. If we alternately break both loops at the input to G_2 and plot

$$A_1(s) = G_2(H + G_1G_3) \tag{8.31}$$

we must get no encirclements to yield a consistent answer. Since a plot of G_2H has been assumed to provide one encirclement, the addition of $G_1G_2G_3$ to this plot must "pull" the plot around to avoid the -1 point if stability is to be obtained.

Example 8.6

Consider the system shown in Fig. 8.27. Apply the Nyquist criterion to determine if this system is stable.

We will start with loop 1, the internal feedback loop around the integrator. The Nyquist plot for loop 1 is shown in Fig. 8.28. There are no encirclements of the -1 point, so loop 1 is stable. The Nyquist plot for loop 2 is shown in Fig. 8.29. Since there is one encirclement and no rhp poles of $A_2(s)$, loop 2 is unstable with one closed loop rhp pole. Further support for this conclusion is provided by

$$\frac{Y(s)}{X(s)} = \frac{3/(s+1)}{1 - 3/(s+1)} = \frac{3}{s-2} \tag{8.32}$$

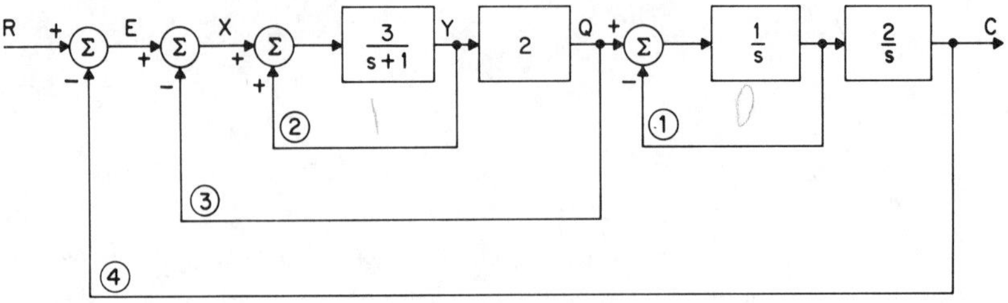

FIG. 8.27. The multiple loop system introduced in Ex. 8.6.

Proceeding to loop 3, a plot of $A_3(s) = 6/(s - 2)$ is shown in Fig. 8.30. There is one ccw encirclement, so $Z - P = -1$. Since $P = 1$, $Z = 0$. Thus loop 3 is stable. In fact

$$\frac{Q(s)}{E(s)} = \frac{6/(s-2)}{1 + 6/(s-2)} = \frac{6}{s+4} \tag{8.33}$$

is the closed-loop transfer function of loop 3. The open-loop transfer function of loop 4 is thus

$$A_4(s) = \frac{12}{s(s+1)(s+4)} \tag{8.34}$$

A Nyquist plot of $A_4(s)$ is shown in Fig. 8.31. The overall system is seen to be stable, but the stability margin is very small.

The situation considered in Ex. 8.6 illustrates the various steps required in solving multiple-loop problems. The example is simple and the answer is almost apparent by inspection, but this eliminates analytical detail and allows us to concentrate instead upon the underlying principles. Typical problems are seldom so simple, and the multiple-loop procedure generally yields significant advantages.

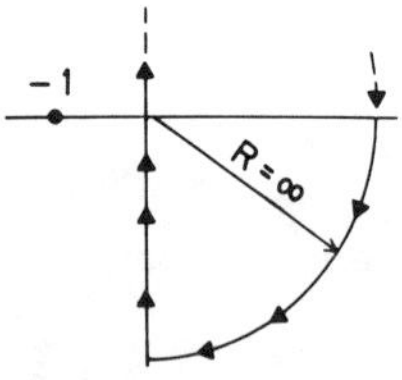

FIG. 8.28. Nyquist plot for loop 1.

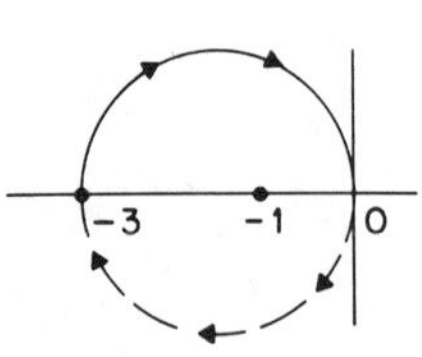

FIG. 8.29. Nyquist plot for loop 2.

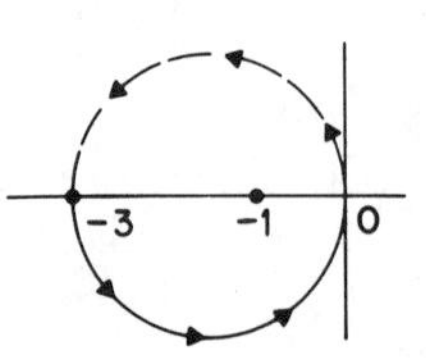

FIG. 8.30. Nyquist plot for loop 3.

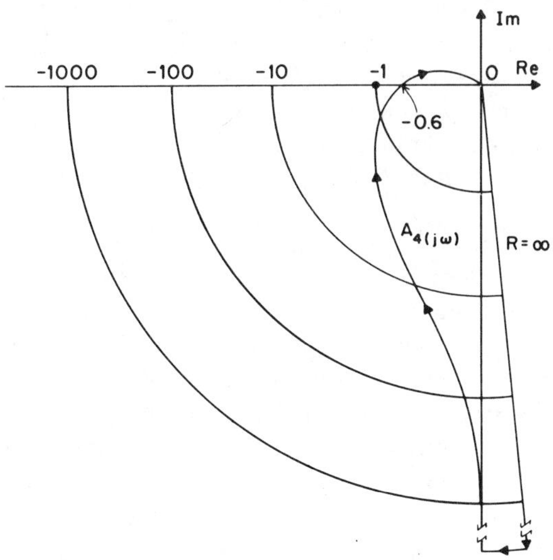

FIG. 8.31. Nyquist plot for loop 4.

8.7 TRANSPORT LAG

All of the systems discussed up to this point have been described by transfer functions expressed as ratios of finite polynomials in s. This is equivalent to assuming that all of our systems can be represented in lumped parameter form. Situations occasionally arise in which a fixed time delay, or transport lag, occurs at some point in the control loop. Although electrical signals travel at approximately the speed of light, and thus so rapidly that time delay due to electrical transmission is usually negligible (consider as an exception a communication link between the earth and a space vehicle to the moon or another planet, in which case the round trip transmission time may become several seconds) information is often carried by other than electrical means in control systems. Consider, for example, control of the catalyst concentration in a chemical process. The catalyst is added to the chemical stream through a valve, the stream flows through a pipe at velocity v_0, and the combination is agitated by fins inside the pipe to disperse the catalyst. An accurate measure of catalyst concentration can only be made after a thorough mixing has taken place, or significantly downstream from the injection point. A time delay of $t_0 = x/v_0$ exists in the control loop, where x is the distance between the injection and measurement points. Time delay of this nature, called transport lag or transport delay, often occurs in control systems.

An element with transport delay t_0 converts a general signal $x(t)$ at its input to $x(t - t_0)$ at its output. By the real translation theorem, $\mathcal{L}[x(t)] = X(s)$ and $\mathcal{L}[x(t - t_0)] = \epsilon^{-t_0 s}X(s)$, so a transport delay of t_0 seconds has transfer function $\epsilon^{-t_0 s}$. The function $\epsilon^{-t_0 s}$ has unit magnitude for all $s = j\omega$, and provides linearly increasing phase lag with frequency. Thus the presence of transport lag in a system adds phase shift to the Nyquist plot, but does not alter the magnitude at any frequency. The Nyquist plots of systems with transport lag spiral in toward the origin with ever increasing phase lag. Transport lag is a nonminimum phase component, since the minimum possible phase lag we could associate with a unity gain transfer function at all frequencies is obviously zero.

Example 8.7

Consider the control of sheet metal thickness at the output of a steel rolling mill, as shown in the diagram of Fig. 8.32. Output thickness is measured by an electronic gauge which measures the average x-ray passage through the rolled material. It is necessary for practical reasons that this gauge be located somewhat downstream from the rollers controlling thickness. The distance downstream is $d = 25$ feet, and the velocity of the output sheet metal is $v = 1500$ ft./min. Thus a transport delay of t_0 seconds exists between the thickness control point (the point or region of roller contact with the sheet metal) and the point at which the output thickness is actually observed by the gauge, and

$$t_0 = \frac{v}{d} = \frac{1500/60}{25} = 1 \text{ sec.} \tag{8.35}$$

One roller is fixed and the position of the second is controlled by a motor and gear arrangement. The thickness controller with individual element transfer functions as obtained from the equipment used is shown in Fig. 8.33. Sketch the Nyquist diagram for this system. Is it stable? If so, what is the phase margin? Can you determine the stability of systems with transport lag by direct application of the Routh criterion?

The Nyquist diagram of the system with and without transport delay included is shown in Fig. 8.34. The system is seen to be stable, with approximately 55° of phase margin. The transport lag effect is clearly evident as the plot spirals in toward zero magnitude. The Routh criterion does not apply to a system with transport lag, since the characteristic equation of such a system cannot be expressed as a finite polynomial in s. Thus we conclude that the Nyquist stability criterion is more general than the Routh criterion.

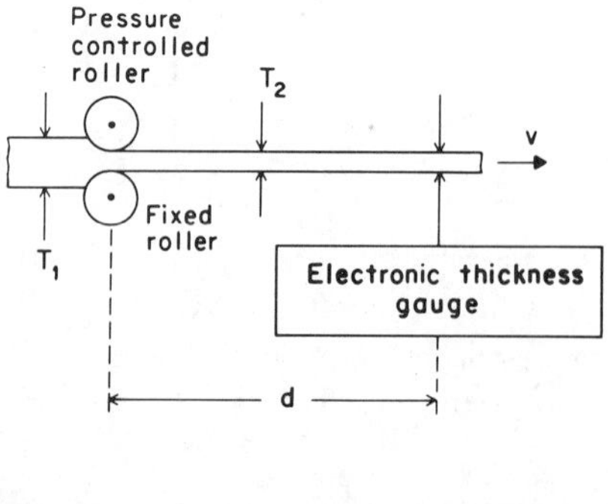

FIG. 8.32. Diagram of the steel rolling process.

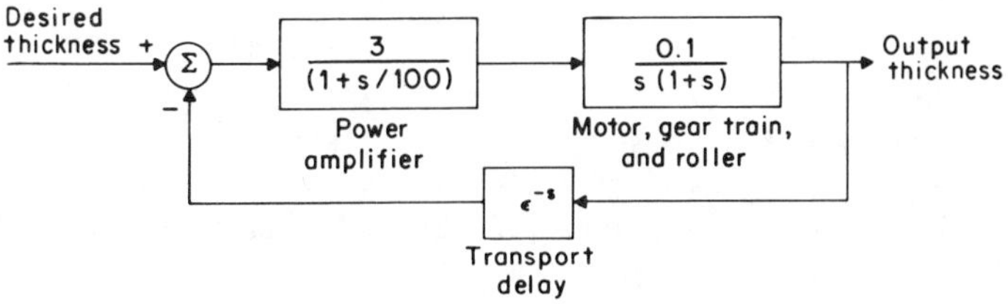

FIG. 8.33. Block diagram of the thickness controller.

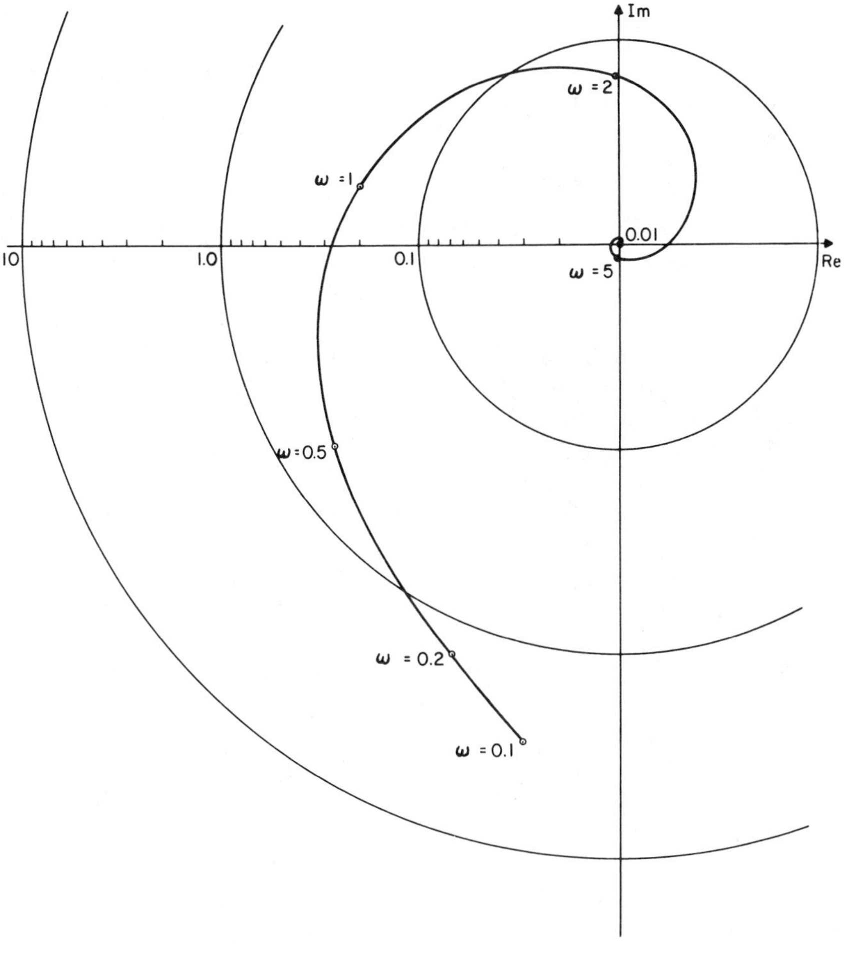

FIG. 8.34. Nyquist diagram of the system with transport delay used in Ex. 8.7.

Transport delay introduces phase lag which increases linearly with frequency, whereas the phase lag due to individual time-constant terms increases almost linearly with frequency until it reaches 30° or so, and then saturates, approaching a final value of 90°. We can thus often overcome a tendency toward instability caused by the phase lag of one or two system time constants by adding phase lead in the appropriate frequency range near unit magnitude of $A(s)$, thereby assuring reasonable closed-loop response characteristics, and increasing closed-loop bandwidth. Phase lag due to transport delay increases so rapidly once it starts to become appreciable that closed-loop bandwidth is severely constrained by its presence. Transport delay should always be minimized when possible, but many situations exist where there is little or no control over it. Seldom are closed-loop systems designed in which the phase lag contribution due to transport delay at the crossover frequency is more than 5° to 10°, since otherwise the relative stability and response characteristics are extremely sensitive to increases in gain. Transport lag is a critical system characteristic with far-reaching implications, and its presence must not be taken lightly.

8.8 CONCLUDING COMMENTS

The Nyquist stability criterion is a powerful analysis tool for determining the stability of feedback controllers. It is derived using only basic complex variable principles, and it is easily applied to all types of systems. Since the primary factor of interest is the number of -1 point encirclements, a crude sketch made directly from the Bode diagram is generally adequate to check system stability. With a bit more care, particularly in the region where $|\mathbf{A}(j\omega)| \simeq 1$, relative stability information can be obtained by finding phase margin and gain margin values. Closed-loop frequency response can be obtained from the Nyquist plot with a minor extension of the results. If care is exercised in defining all negative real axis intersections, the entire range of gain for which the closed-loop system is stable is easily obtained. An accurate Nyquist plot thus provides the same information relative to the stable gain range as the Routh criterion. The effects of other parameter changes upon stability are not so readily apparent from the Nyquist plot, since they generally change both magnitude and phase of $\mathbf{A}(j\omega)$. The Nyquist criterion does not, therefore, entirely supersede the Routh criterion. Transport lag, a condition for which the Routh criterion does not apply, is readily handled using the Nyquist criterion.

Although the Nyquist criterion is an extremely useful analysis tool, its importance in the development of design procedures is even more significant. The general shape of the Nyquist plot is readily apparent from the Bode diagram for all minimum-phase networks and can be visualized without excessive difficulty even for the nonminimum-phase case if the rhp zeros are clearly indicated on the Bode plot. Synthesis of series compensation or equalizer networks which will provide stable closed-loop operation of open-loop stable systems within specs upon such closed-loop characteristics as phase margin can thus be reduced to the graphical problem of "shaping" the Bode diagram in the vicinity of

$|A(j\omega)| = 1$ to avoid encircling the -1 point on the Nyquist plot by whatever margins are required on θ_m and/or G_m. Design of the equalizing filter for an open-loop unstable system is somewhat more involved, but fortunately such systems are not often encountered. Detailed design procedures based upon the Nyquist stability criterion, but carried out upon the Bode diagram using simple graphical methods, are discussed in detail in Chap. 9. These procedures often make possible the rapid specification of equalizing filters for relatively complex systems required to operate within stringent performance limits once the transfer functions of the individual system elements have been determined. The Nyquist criterion is the basis for all standard classical design procedures.

REFERENCES

1. D'Azzo and Houpis, "Feedback Control System Analysis and Synthesis," McGraw-Hill, New York, Chap. 11, 1960.
2. Bower, J. L. and P. M. Schultheiss, "Introduction to the Design of Servomechanisms," John Wiley & Sons, Inc., New York, Chap. 9, 1958.

9 USE OF THE BODE DIAGRAM IN DESIGNING SERIES-EQUALIZED CLOSED-LOOP CONTROLLERS

The general objective in all classical design procedures is to shape the Nyquist plot such that the -1 point is avoided with some reasonable safety margin, thereby providing acceptable closed-loop response characteristics. Direct use of the Nyquist diagram for design is not particularly convenient, since changes in parameters other than gain require extensive plot revisions, so we are led to seek a more efficient synthesis procedure. Fortunately the general characteristics of the Nyquist plot may be visualized with reasonable accuracy in most cases of interest from the Bode diagram shape, and the Bode plot is easily constructed and modified. Our primary objectives in this chapter are to establish more completely the relationship between Bode and Nyquist diagrams, and develop direct procedures by which the general characteristics of series-equalizing filters may be chosen on the Bode diagram to provide the desired closed-loop performance. Series compensation using attenuation, lag, lead, and lag-lead, or composite, networks are each considered in turn. The design procedures are reduced to orderly graphical trial and error processes, which are carried out using the Bode diagram. The only tools required are a pencil, a straightedge, log-log paper, and a few supplementary calculations for which a slide rule is appropriate. We restrict our attention in this chapter to single-loop series-equalized systems. The design of minor-loop (multiple loop or feedback) compensated systems is considered in Chap. 11.

9.1 COMPENSATION TO PROVIDE DESIRED CLOSED-LOOP PERFORMANCE; THE EQUALIZER PROBLEM

The standard closed-loop controller design problem is generally developed in the following form: given a variable to be controlled, shaft position, for example, an actuator or plant capable of providing this control, denoted as $G_p(s)$, and a set of specs upon the final design, such as $\theta_m \geq 40°$, step response peak overshoot less than 20 percent, closed-loop bandwidth at least 100 rad/sec, and other restrictions of a similar nature, determine a controller configuration which will satisfy these objectives. Of course the parameters of all elements used must be specified, so the designer must choose amplifier gain levels, filter network locations and parameters, measurement sensors, and the overall system topology such that all restrictions are met. Several design methods have become standard. The simplest method to develop, and that which is emphasized in this chapter, is the addition of a series equalizer $G_e(s)$ preceding the plant at a point where the voltage and power levels are reasonable. The parameters of $G_e(s)$ are chosen to shape the open-loop Bode and Nyquist diagrams as required to satisfy closed-loop specs. A series equalized unity feedback system is shown in block diagram form in Fig. 9.1. The equalizing (compensating) filter may alternately be placed in the feedback path, as shown in Fig. 9.2, or compensation may be accomplished using the minor-loop configuration shown in Fig. 9.3. Each of the feedback loops in multiple-loop systems may contain one or more equalizing filters. The objective in design using minor-loop compensation is to appropriately shape the open-loop Bode and Nyquist diagrams, just as when series equalizers are used. In this respect the two approaches are virtually identical. The internal, or minor, feedback loop is generally placed around some convenient part of the plant $G_p(s)$, which has been broken up into the component transfer elements G_1, G_2 and G_3 in Fig. 9.3. There are several practical advantages to minor loop designs when disturbances, component costs, parameter drift, and plant nonlinearities are considered, so minor-loop compensation is often used. Minor-loop design requires only a straightforward extension of the series equalizer design concepts developed in this chapter.

The four basic types of series equalization are: 1. gain adjustment, or equal change in gain of the open-loop transfer function $A(s)$ at all frequencies; 2. attenuation of $A(s)$ at high frequencies only using a phase lag compensation

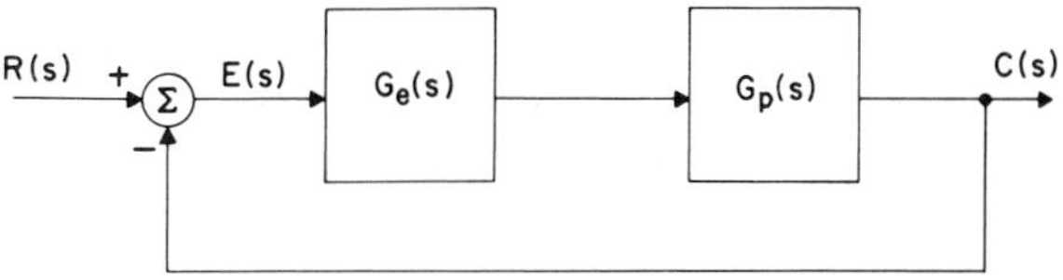

FIG. 9.1. The standard cascade equalized configuration.

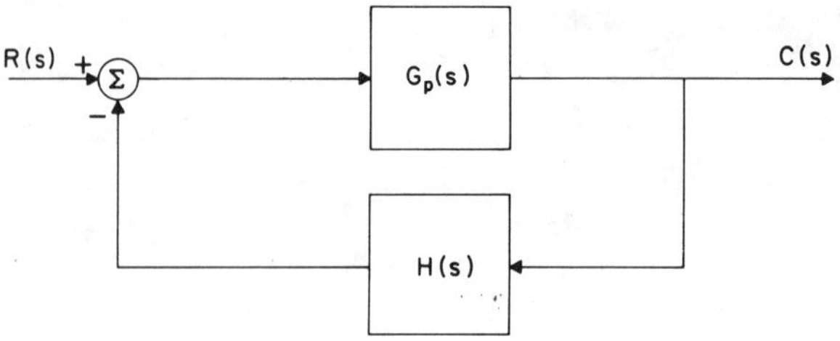

FIG. 9.2. Feedback compensation.

network of the form

$$G_e(s) = \frac{1 + \tau s}{1 + \alpha\tau s} \qquad \alpha > 1 \tag{9.1}$$

3. attentuation of $A(s)$ at low frequencies only using a phase lead network of the form

$$G_e(s) = \frac{1}{\alpha}\left(\frac{1 + \alpha\tau s}{1 + \tau s}\right) \qquad \alpha > 1 \tag{9.2}$$

and 4. composite equalization using a combination of two or more of the first three methods. Each of these approaches is considered in detail in subsequent sections of this chapter. Note that any equalizer added in series with $G_p(s)$ must be isolated from the plant and the error sensor by appropriate coupling networks or buffer amplifiers. We assume hereafter that loading effects are negligible, or have been taken into consideration, and the equalizer and plant transfer functions are not changed when they are interconnected. Care must be exercised in the analysis and equipment development phases not to invalidate this assumption.

System specs are given to the control engineer in many forms, and one of his major problems early in the design procedure is to interpret how each spec influences the desired open-loop Bode diagram. In most cases, as we will show with design examples, specs on a wide variety of performance characteristics are easily

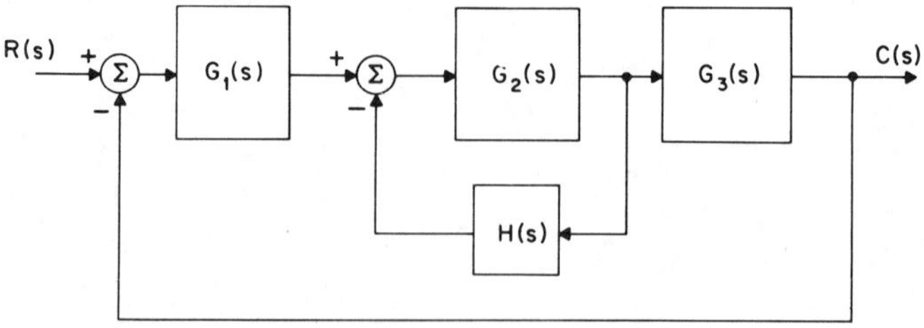

FIG. 9.3. A system with minor-loop equalization.

considered in Bode diagram design, and this is one of the method's major advantages. Performance specs are often imposed upon: 1. phase margin θ_m; 2. gain margin G_m; 3. closed-loop bandwidth BW; 4. dominant closed-loop pole pair damping factor; 5. unit-step response rise time, percent overshoot, and settling time, or response duration; 6. peak sinusoidal steady-state closed-loop transfer function magnitude M_p; 7. noise and steady-state error performance; and 8. velocity error constant K_v. Phase margin, gain margin, and closed-loop BW are easily obtained from the Bode diagram, as is the information necessary to determine steady-state error and noise performance. For values of θ_m in the normal range encountered (30° to 45°), it is reasonably accurate to approximate (see Sec. 9.4) M_p by

$$M_p \simeq \frac{1}{\sin \theta_m} \tag{9.3}$$

so M_p is also easily controlled on the Bode plot. Time domain characteristics such as damping factor and unit-step response percent overshoot are not directly apparent from the Bode plot, but are related in a general way to the margin of Nyquist plot -1 point avoidance, and therefore to Bode diagram "shape". Simulation is the usual means of time domain spec measurement and adjustment, since hand analysis is too complex to be feasible. When time domain specs are to be imposed, a tentative design is generally roughed out using whatever approximations are available, a simulation of the system is set up, and the design parameters are adjusted to satisfy the specs using the simulation model.

The velocity error constant K_v is defined as the reciprocal of the closed-loop steady-state error to a unit ramp input. For a standard unity feedback stable system it is easily shown (see App. F and Sec. 10.5) that $K_v = 0$ if there is no integration in the forward path transfer function $G(s)$, $K_v = \infty$ if there are two or more integrations in $G(s)$, and $K_v = K$ if $G(s)$ contains one integration, where K is the forward transfer function gain constant when all terms are expressed in the usual time constant form. Thus K_v is also readily apparent from the Bode diagram.

The influences of many performance specs upon the design process are considered in detail in Chap. 10. Our purpose here is to introduce the fundamental concepts of design using Bode plots. Rather than cloud the main issue unnecessarily, we will rely heavily in this preliminary work upon the use of θ_m and K_v as our major design specs. Stability is of primary importance. If we assume the open-loop transfer function is stable, as will generally be the case, θ_m provides a direct measure of the margin by which encirclement of the -1 point is avoided, and thus yields a general measure of stability margin for the final design. Although a specific value of θ_m, say 45°, provides no guarantee of acceptable transient performance in a particular problem, conditions are seldom encountered where design for $\theta_m = 45°$ will not provide reasonable values of step response percent overshoot and dominant pole pair damping factor. Therefore our emphasis at this point upon θ_m as a design spec seems justified.

Imposing a design spec upon K_v, such as $K_v = 100$, has two interpretations. Since the unit-ramp response steady-state error equals $1/K_v$, fixing K_v establishes the closed-loop system's ability to follow ramp-like inputs such as triangular or sawtooth waveforms. It is often desirable to control such performance characteristics. Setting K_v also establishes the open-loop gain and the resulting low frequency error performance of the closed-loop system, as we shall show. In general, for stable systems, steady-state errors for sinusoidal inputs decrease as K, the open-loop gain, increases. The values of K_v and/or K are generally limited for these reasons.

It is common practice to design for values of θ_m and/or K_v which engineering judgment dictates should prove satisfactory, check the transient and steady-state error performances of the resulting system, and make corrective adjustments if necessary. In situations where the Nyquist diagram approaches or crosses the negative real axis prior to unit magnitude crossover, or where G_m is small despite a large value of θ_m, as illustrated for three cases in Fig. 9.4, we should anticipate unusual results and plan accordingly. The establishment and satisfaction of realistic design specs based upon the use to which the system under consideration will be put and the requirements of that application is introduced in this chapter and the scope of coverage is expanded significantly in Chap. 10. Although it may seem somewhat fictitious to start out our examples by imposing what may appear to be entirely arbitrary specs on performance characteristics such as θ_m, K_v, and low frequency errors, it is generally rather easy to convert the original problem statement to a set of equivalent or near equivalent limitations of this form.

9.2 THE BODE DIAGRAM AND ITS RELATIONSHIP TO THE NYQUIST PLOT

Unless otherwise specified we always assume that $A(s)$ is minimum phase. The asymptotic and actual $|A(s)|$ curves are acceptably close for our purposes

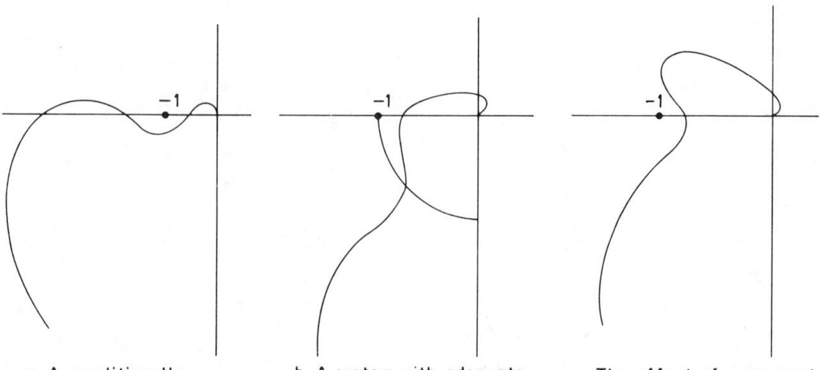

a. A conditionally (Nyquist) stable system.

b. A system with adequate θ_m, but small G_m.

c. The effect of a resonant peak just above $\omega = \omega_c$.

FIG. 9.4. Nyquist diagrams of several potentially troublesome situations.

except near the break points, and even in these regions the actual magnitude is easily obtained if that becomes necessary. Whenever the change in slope is caused by one or more pairs of complex conjugate poles or zeros, the damping factor(s) must be known. Curves providing the actual magnitude and phase contributions as a function of ω/ω_0, the normalized frequency variable, for all types of terms are given in Secs. 7.3 and 7.5 respectively, and should be referred to whenever necessary. With a little practice it is easy to visualize an approximate Nyquist plot directly from the Bode diagram. Considerable graphical simplicity is thus realized by using the asymptotic magnitude diagram without losing the stability margin insight provided by the Nyquist plot.

Consider a system with no complex poles or zeros in $A(s)$. In this case the actual $|A(s)|$ may generally be estimated with acceptable accuracy from the Bode diagram without further calculations. As a gross approximation we know that the phase of $\mathbf{A}(j\omega)$ tends toward -90° in regions of -1 slope, -180° in regions of -2 slope, and in general toward $-n(90°)$ in regions of $-n$ slope, with gradual transitions taking place as the slope changes. We can thus develop a rough picture of the corresponding Nyquist plot with little effort. Since closed-loop stability and transient performance both depend primarily upon the polar plot characteristics near the -1 point, accuracy is needed only near $|\mathbf{A}(j\omega)| = 1$. The ω for which $|\mathbf{A}(j\omega)| = 1$ is denoted by ω_c, and ω_c is nearly always apparent by inspection. Thus, even in marginal cases, only two or three detailed magnitude and phase calculations are required to determine system stability. The Bode diagram of a minimum phase $A(s)$ with no complex poles or zeros (conditions satisfied by most standard design problems) indicates immediately to the practiced designer which system parameters, if any, are critical in providing stability margin. Ease of construction and the simplicity of routine design procedures based upon its use provide two additional advantages of Bode diagram methods.

When $A(s)$ contains one or more complex conjugate pairs of poles or zeros, the exact magnitude and phase curves depend upon the damping factor ζ, as discussed in Chap. 7. Since curves of actual amplitude and phase versus damping factor and normalized frequency are available for use, however, and since appreciable corrections are required for such terms only near the break frequency, this causes little difficulty. The Nyquist plot is still readily visualized from the Bode diagram as soon as all damping factors are known. The range of frequencies where the complex pole (or zero) pair effects are significant is often far from ω_c, the point of primary interest, and in such cases these terms can often be neglected during the design process.

The basic objectives of design on the Bode diagram are simply expressed. The asymptotic $|A(s)|$ plot is generally positioned and shaped by setting the gain and using an equalizing filter to provide a region of -1 slope near ω_c, the unit magnitude crossover. The width or span of the -1 slope region around ω_c is adjusted to control θ_m. As the frequency ratio from ω_c to each break point at which the slope changes to a value of -2 or less increases, θ_m increases, and vice versa. In general it is desirable to avoid final system designs for which the phase lag of $\mathbf{A}(j\omega)$ exceeds 180° for any $\omega < \omega_c$, since such systems can become unstable for an open-loop gain decrease (conditional stability), so an effort is generally made to avoid asymptotic segments of $-n$ slope, $n \geq 3$, for frequencies below ω_c. In some cases this general rule must be violated, but the principle should always be kept in mind. In short, Bode diagram design is the process of constructing an asymptotic diagram shape which assures the satisfaction of all performance specs and then determining the parameters of an equalizer which, when placed in series with the plant, provides that shape.

Example 9.1

The transfer function of a given power amplifier and motor combination which is to be used for shaft position control in the configuration shown in Fig. 9.5 is found to be

$$G_p(s) = \frac{\theta(s)}{V_i(s)} = \frac{250}{s(1 + s/10)} \tag{9.4}$$

The final design must provide 50° of phase margin. It is proposed that this be accomplished in each of the following ways:

1. By adding an appropriate amount of attenuation in series with $G_p(s)$.
2. By adding the lag equalizer

$$G_e(s) = \frac{1 + s/1.25}{1 + s/0.02} \tag{9.5}$$

in series with $G_p(s)$.

3. By adding the lead equalizer

$$G_e(s) = \frac{1 + s/30}{1 + s/225} \tag{9.6}$$

in series with $G_p(s)$.

Construct Bode and Nyquist diagrams for each case, determine the unit-step response of the three systems, and compare the three designs.

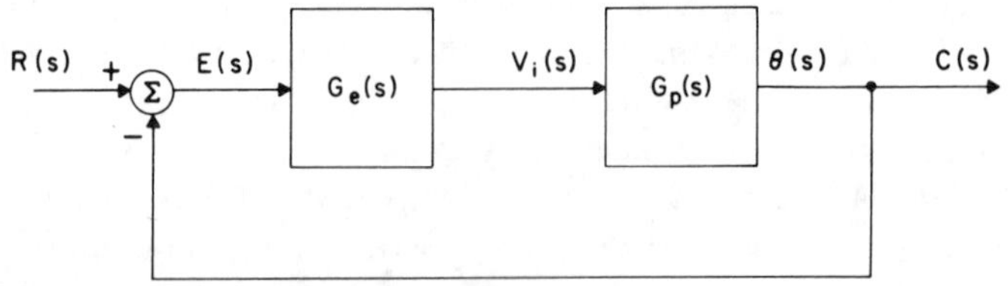

FIG. 9.5. Block diagram of the system considered in Ex. 9.1.

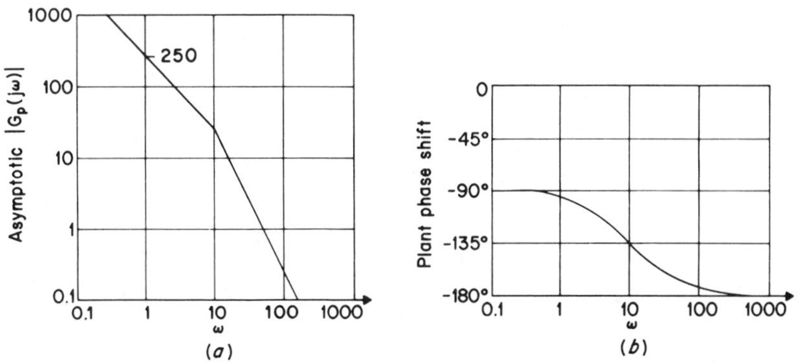

FIG. 9.6. Characteristics of the plant introduced in Ex. 9.1.

Case 9.1 Attenuation

A Bode diagram and a phase versus frequency plot of $G_p(s)$ as obtained directly from Eq. (9.4) are shown in Fig. 9.6. From the phase curve we see that the phase lag is 130° at $\omega = 8.4$, and $|\mathbf{G}_p(j8.4)| \simeq 25$. Thus if an attenuation of 25:1 is introduced in series with $G_p(s)$, the magnitude of the resulting open-loop transfer function $A(s)$ becomes unity where the phase shift is 130° and the desired phase margin of 50° is thereby obtained. This situation is further illustrated by the Nyquist diagram denoted as (a) in Fig. 9.7. Note how the polar plot angle approaches -90° for all small values of ω, and approaches -180° as $\omega \to \infty$, as we should anticipate from Bode diagram slopes of -1 and -2, respectively, in these regions. Note that when attenuation is added in series, the Bode plot moves down an amount directly related to the gain reduction, and the Nyquist plot contracts by the appropriate factor, but the shapes of both diagrams remain unchanged and the phase versus ω curve shown in Fig. 9.6 is not altered by gain changes. The unit-step response of the closed-loop system obtained by a gain reduction of 25:1 is denoted by (a) in Fig. 9.8.

Case 9.2 Lag Equalization

It is clear from the α and β diagrams of $G_p(s)$ shown in Fig. 9.6 and the Nyquist plot labeled (a) in Fig. 9.7 that the phase lag is not excessive at low values of ω, but $|\mathbf{G}_p(j\omega)|$ is too large there to take advantage of this fact. When the lag network defined in Eq. (9.5) is placed in series with the plant, the modified Bode diagram shown in Fig. 9.9 results. Unit magnitude crossover occurs at $\omega_c \simeq 4.0$ in the middle of a -1 slope region which is sufficiently wide to provide the desired phase margin of 50°. The corresponding Nyquist plot is labeled (b) in Fig. 9.7. Note how the phase lag approaches 180° in the region around $\omega = 0.1$, where the asymptotic slope is -2, but then decreases again, assuming a relative minimum value of 130° (50° of phase margin) near the geometric center of the -1 slope region around ω_c before approaching 180° again as $\omega \to \infty$. By choosing the equalizer break frequencies such that ω_c occurs in the middle of

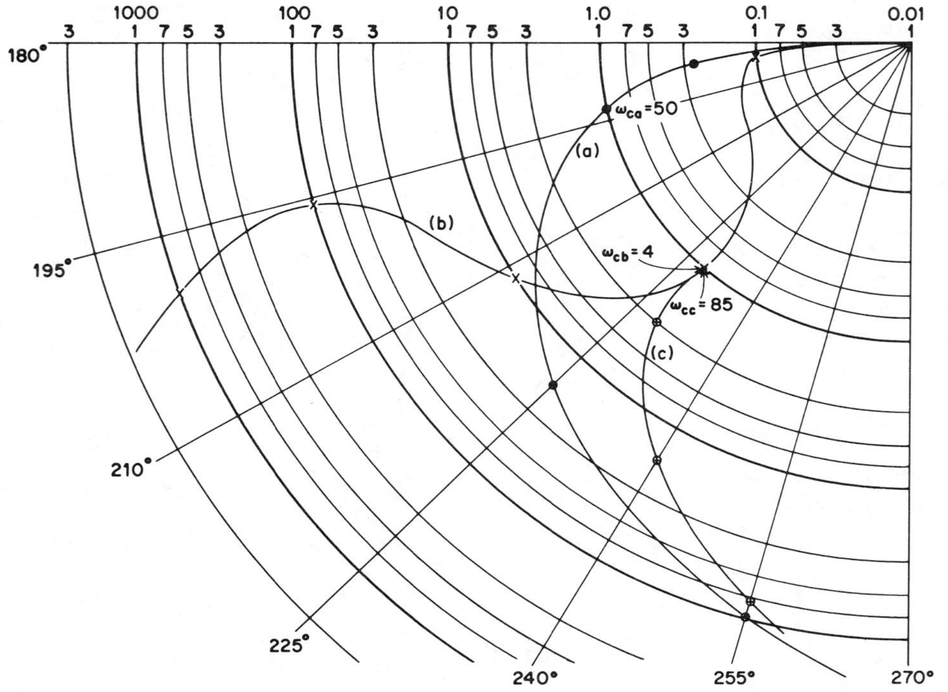

FIG. 9.7. Nyquist plots for Ex. 9.1.

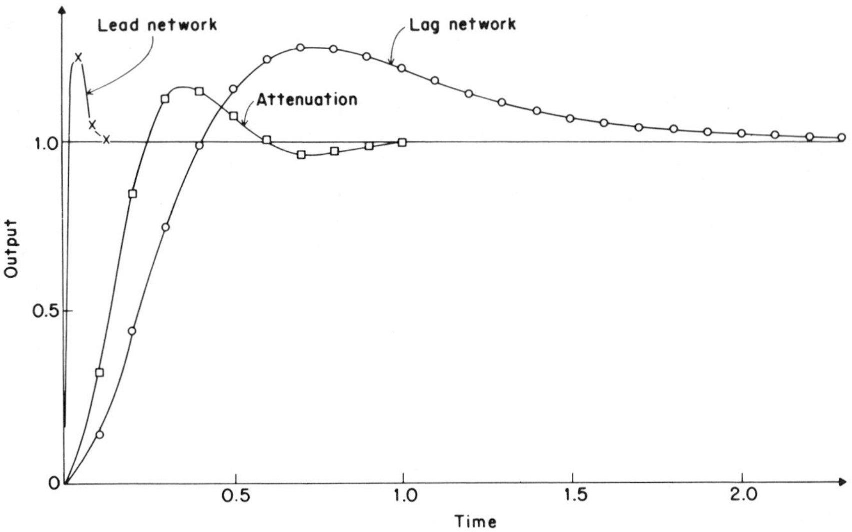

FIG. 9.8. Unit-step responses of the systems considered in Ex. 9.1.

the -1 slope region, θ_m is maximized for a given ratio between the end frequencies of the -1 slope segment. (A given -1 slope *span*.) All of this can be visualized directly from the Bode diagram shape, and it is important that we develop our ability to do so. An asymptotic magnitude diagram of $G_e(s)$ as given by Eq. (9.5) is also shown in Fig. 9.9. The lag compensated system's unit-step response is labeled (b) in Fig. 9.8.

Case 9.3 Lead Equalization

When the lead network defined in Eq. (9.6) is inserted in series with the plant, the Bode diagram near ω_c is converted to -1 slope as shown in Fig. 9.10, the Nyquist plot is modified as shown by the curve labeled (c) in Fig. 9.7, and $\omega_c \simeq 85$, which is over 20 times as great as the value achieved with either the attenuation or lag equalizer approach. This significant difference in closed-loop bandwidth yields the relatively fast unit step response labeled (c) in Fig. 9.8.

We have purposely not attempted to indicate the specific methods by which the lag and lead equalizers introduced in Ex. 9.1 were chosen, since these approaches are discussed in detail in Secs. 9.3 and 9.4, respectively. It is obvious that a graphical trial and error procedure on the Bode diagram should rapidly converge to the desired answer. For the present we are content to help develop an intuitive feeling for the relationships between the Bode diagram, the Nyquist plot and the closed-loop transient performance. We will not generally construct Nyquist plots in our subsequent examples, since the information we need is adequately provided by the Bode diagram alone in most cases.

It should be noted that the lag equalizer used in Ex. 9.1 has a lag span ratio of 62.5:1, which is considerably in excess of the general 10:1 limit for realizability

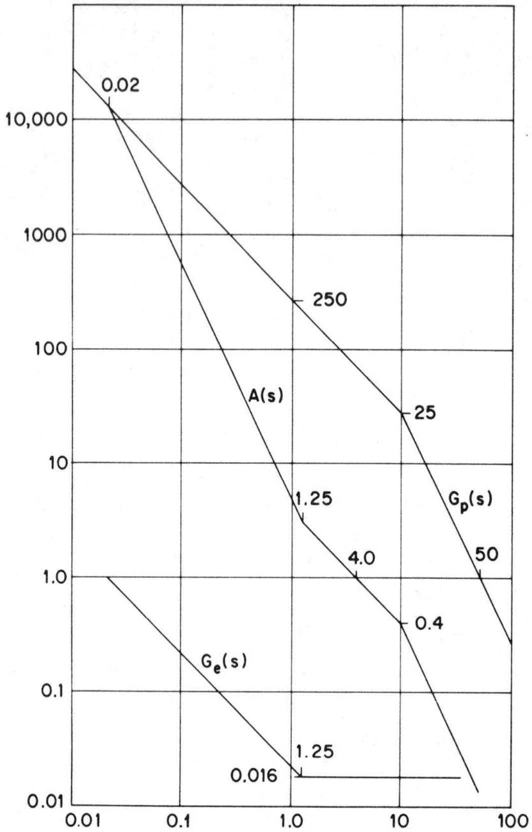

FIG. 9.9. Bode diagrams for the lag equalized system.

as a single RC network stage (see Sec. 9.3). Realization requires either a single stage which will have a very large input to output impedance ratio, or two RC network stages separated by a buffer amplifier of some sort. In either case, careful design should provide the desired transfer function.

The relative effects of attenuation, lag compensation and lead compensation are illustrated clearly by the results of Ex. 9.1. Attenuation reduces the gain at all frequencies and, since

$$\frac{E(s)}{R(s)} = \frac{1}{1 + A(s)} \qquad A(s) = G_p(s)G_e(s) \tag{9.7}$$

for the usual unity feedback configuration, it is clear that reducing $|\mathbf{A}(j\omega)|$ increases the steady-state error components at all frequencies. Thus attenuation results in significant degradiation of error performance, particularly in the low frequency range where $|\mathbf{G}_p(j\omega)| > 1$. Attenuation also reduces the closed-loop BW, thereby leading to a relatively slow unit-step response pattern. Lag compensation provides *selective* attenuation which is not effective at low frequencies, increases linearly (to a first-order approximation; see Fig. 9.9) in the range $\omega_1 \leq \omega \leq \omega_2$, where ω is the denominator (lag) break frequency and ω_2 is the numerator (lead) break frequency, and provides constant attenuation for all $\omega > \omega_2$. The net effect of lag compensation is similar to attenuation in the margin of -1 point avoidance provided near ω_c, but the low frequency gain remains large and the closed-loop BW is somewhat less than that obtained using attenuation. The unit-step responses of the attenuation and lag compensated systems, primarily due to their smaller BW's, are slower than that of the lead compensated system. The differences are roughly the ratios of closed-loop BW's. Although the lag-compensated system's BW is smaller, and its step response slower than that of the attenuated system, lag compensation provides significantly smaller steady-state sinusoidal error at low frequencies (in this case a 25:1 improvement below $\omega = 0.02$), and the lag-compensated system will also respond to a ramp or parabolic input with much smaller errors. The closed-loop BW obtained using lead compensation is significantly larger than in the other two cases, and the error performance is not degraded for any frequency below ω_c. Although a casual consideration of these results makes it appear as though lead compensation should always be used, there are negative factors as well. Increased BW may not be an advantage when disturbance effects are considered. Lead compensation requires additional gain, and is relatively expensive to realize. The increased gain requires a larger dynamic range from the plant elements, and this extra range may not be available. The characteristics of lag and lead compensated systems are considered further in Secs. 9.3 and 9.4.

Although we are not yet prepared to discuss the relative merits of lag and lead compensation in detail, we can conclude at this point that both offer

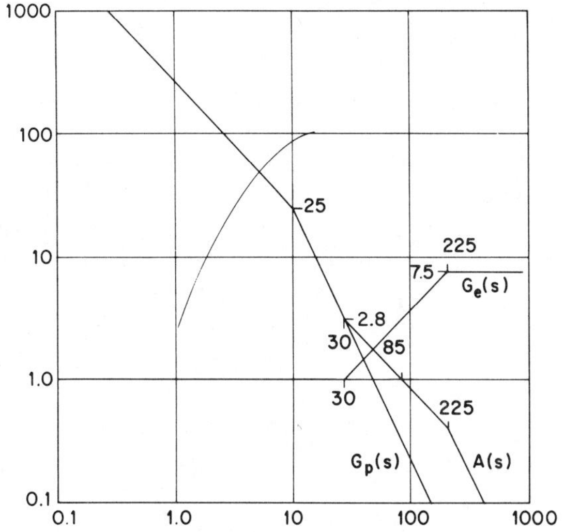

FIG. 9.10. Bode diagrams for the lead-equalized system.

significant advantages over attenuation, which is seldom used as a compensation technique except in applications where the specs provide little design challenge. Attenuation equalizer design is carried out just as illustrated in Ex. 9.1 and we will not consider its use further.

9.3 LAG EQUALIZER COMPENSATION

The simple passive RC network illustrated in Fig. 9.11 has transfer function

$$\frac{E_0(s)}{E_i(s)} = \frac{\tau s + 1}{a\tau s + 1} \tag{9.8}$$

where

$$\tau = R_2 C \qquad a = \frac{R_1 + R_2}{R_2} \geq 1 \tag{9.9}$$

A significant advantage of lag compensation is the relative ease with which lag transfer functions are obtained in passive RC network form. There are some practical limitations, however. As the values of R_1 and R_2 are increased, the output becomes more sensitive to loading. If R_2 is $10K$, for example, it is not particularly difficult to isolate the equalizer output from the plant using an amplifier with input impedance very much greater than $10K$. If R_2 is increased to $1M$, however, this is not so readily accomplished. Thus the maximum value of R_2 is limited by loading effects. The value of C is limited by the sizes available in the voltage range desired. For normal signal levels it is reasonable to specify capacitors in sizes up to a maximum of about $100\mu f$. If $R = 100K$ and $C = 100\mu f$ are taken as practical R and C limits, $R_2 C \leq 10$ is required. It is also reasonable to choose values of R_1 in the general range $R_2 \leq R_1 \leq 10R_2$.

Long term accuracy is sacrificed if $R_1 > 10R_2$ is used, since small changes in R_2 with aging will result in significant lag span (a) changes. Thus the practical limit on lag span is about 10:1. The lag break frequency [denominator term of Eq. (9.8)] is limited to $\omega \simeq 0.01$. All of these numbers are general limits and it is possible to do better with careful design, but it is important to keep in mind the conditions under which network realization may become difficult.

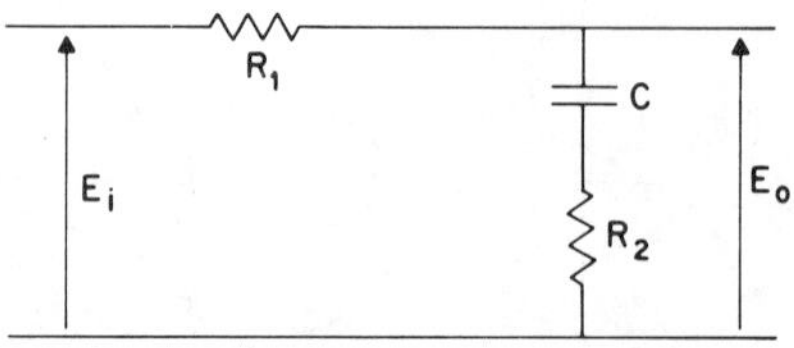

FIG. 9.11. A lag equalizer realized in passive RC network form.

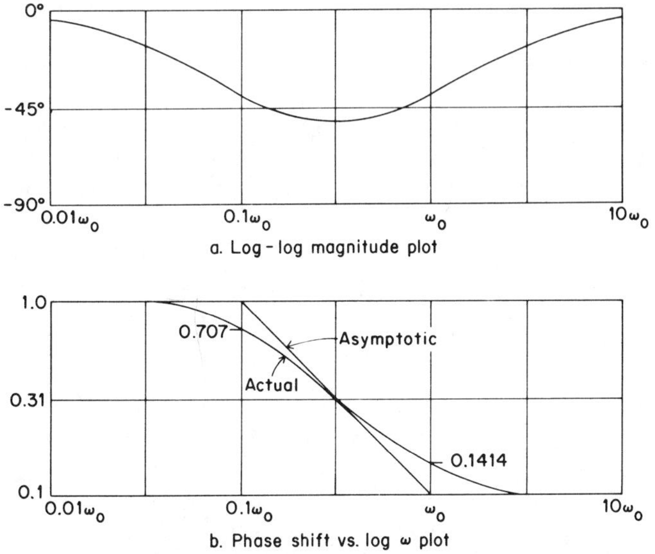

FIG. 9.12. Magnitude and phase diagrams for a 10:1 span lag network.

Asymptotic magnitude and phase plots corresponding to the transfer function presented in Eq. (9.8) are shown in Fig. 9.12 in terms of the normalized frequency variable ω/ω_0, where $\omega_0 = 1/\tau$, for the specific case $a = 10$. It is clear from these plots that the lag network introduces phase lag at all frequencies and selectively adds attenuation for all high frequencies starting at about $\omega = \omega_0/a$. The attenuation factor approaches the constant value a for $\omega > \omega_0$. The phase lag introduced is maximum at the geometric center of the -1 slope region, and this maximum value approaches 90° as $a \to \infty$. Note that lag equalization can only be used to shape the plant Bode diagram in a downward direction. Since unit magnitude crossover in a -1 slope region of the Bode diagram is generally desirable to satisfy a phase margin spec, lag compensation is only applicable if a -1 slope region (or perhaps a 0 slope region in some cases) exists on the uncompensated $G_p(s)$ diagram which can be forced downward using a lag network in the frequency range prior to crossover to provide the desired compensated diagram shape (phase margin) near ω_c. In this respect lag compensation is similar to attenuation. The lag compensation advantage accrues from the fact that only the high frequencies are attenuated, and error performance at the very low frequencies is left unchanged.

Design on the Bode diagram is almost always carried out within a spec on θ_m. Generally there are other specs as well and we will consider their effects later. Our present objective is to minimize problem complexity. The proposed design procedure is as follows:

1. Draw the plant Bode diagram on log-log coordinates.

2. Try to visualize how this diagram shape should be changed to provide a design which satisfies all system specs.
3. Sketch your first estimate of the compensated asymptotic diagram on the same coordinates.
4. Check phase margin and other performance specs for this tentative design and adjust parameters as required to satisfy any specs which have been violated.
5. Specify the transfer function of the equalizer required.

This procedure can be carried out quickly by trial and error in two or three steps to within an acceptable degree of accuracy in almost all cases. We soon learn that a -1 slope span of somewhere between 7:1 and 10:1 with crossover occurring at or near the geometric center of the -1 slope region yields $\theta_m \simeq 45°$, the exact requirement depending upon the desired θ_m and the overall Bode diagram shape. A tentative design providing about an 8:1 span of -1 slope, with ω_c positioned near the middle of the span, often yields results near those desired. An appropriately larger or smaller -1 slope span should be used as the desired θ_m is increased or decreased, respectively. A completely analytical design procedure based upon the arc-tangent approximation is also possible, and both methods are illustrated in the following examples.

Example 9.2

Consider the plant transfer function

$$G_p(s) = \frac{K}{s(1 + s/10)(1 + s/100)} \tag{9.10}$$

Determine the characteristics of the lag network equalizer required to provide $\theta_m = 45°$ with K_v, the velocity error coefficient, set at 250. Minimum degradation of low frequency error performance is also desired.

The spec on K_v requires $K = 250$ if the lag equalizer is to have unity gain at all low frequencies (see Sec. 10.5 for a further discussion of K_v). Minimum low frequency degradation requires that the frequency where the lag network starts to "pull" the compensated diagram downward away from the $G_p(s)$ plot be maximized within the other specs. The θ_m spec requires some minimum -1 slope span near ω_c. Consider initially the general trial and error design procedure. A Bode diagram of $G_p(s)$ with $K = 250$ is shown in Fig. 9.13. We assume arbitrarily that a 10:1 span of -1 slope should be provided near ω_c on the compensated diagram. This assumption is conservative and will provide more than adequate θ_m, but minor adjustments are easily made later, and the 10:1 choice will help us to illustrate the iterative procedure. It is obvious from the plant Bode diagram that the -1 slope region near ω_c must terminate at $\omega_3 \leq 10$, since the G_p plot breaks to a -2 slope there. Thus we choose a -1 slope region from $\omega_{2a} = 1$ to $\omega_3 = 10$. The geometric center of this region is the point where the phase lag is minimum, as proven in Sec. 7.4, so we choose

$$\omega_{ca} = \sqrt{\omega_{2a}\omega_3} = \sqrt{10} = 3.16 \tag{9.11}$$

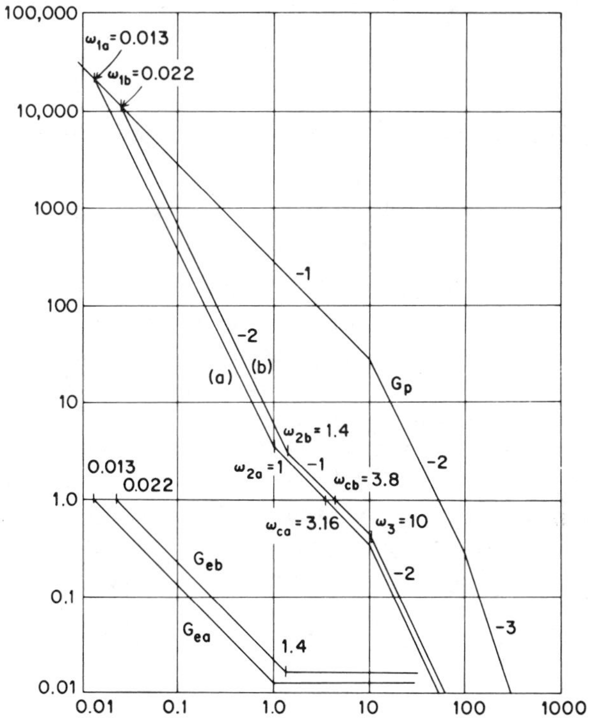

FIG. 9.13. Bode diagram of the system discussed in Ex. 9.2.

Constructing a -1 slope span of 10:1 through the point $\omega_{ca} = 3.16$ and connecting it to the original G_p diagram with a -2 slope segment constructed backward from $\omega_{2a} = 1$ yields the tentative design labeled (a) in Fig. 9.13. The transfer function corresponding to this diagram is

$$A_a(s) = G_{ea}(s)\,G_p(s) = \frac{250(1 + s)}{s(1 + s/0.013)(1 + s/10)(1 + s/100)} \tag{9.12}$$

Assuming actual crossover occurs at $\omega_{ca} = 3.16$, an accurate approximation in this case because of the symmetrical Bode diagram shape around ω_{ca}, θ_m may be computed from Eq. (9.12) as

$$\theta_{ma} = \pi + \tan^{-1}3.16 - \frac{\pi}{2} - \tan^{-1}\frac{3.16}{0.013} - \tan^{-1}0.316 - \tan^{-1}0.0316 \tag{9.13}$$

Observing that

$$\tan^{-1}\frac{x}{y} = \frac{\pi}{2} - \tan^{-1}\frac{y}{x} \tag{9.14}$$

and using the approximation

$$\tan^{-1}\phi \simeq \phi \qquad \phi < 1 \tag{9.15}$$

which is accurate within a few percent for the types of terms we must evaluate

$$\theta_{ma} = \pi + \frac{\pi}{2} - \frac{1}{3.16} - \frac{\pi}{2} - \frac{\pi}{2} + \frac{0.013}{3.16} - 0.316 - 0.0316 \simeq 52^\circ \tag{9.16}$$

We may also compute θ_{ma} directly from the Bode plot by the method introduced in Sec. 7.5. Our first tentative design thus provides considerably more θ_m than is required. Also, by lowering the magnitude of $A(s)$ sooner than necessary, some excess low frequency error degradiation has been introduced. For our next attempt, suppose we try a 7:1 span of -1 slope around ω_c. Since this span must terminate at $\omega_3 = 10$ to yield minimum low frequency error degradation, it starts at $\omega_{2b} = 1.4$. The desired value of ω_{cb} is obtained from Eq. (9.11) as $\omega_{cb} = \sqrt{14} = 3.8$. Placing a 7:1 span of -1 slope through $\omega_{cb} = 3.8$ and connecting it to the original G_p curve with a -2 slope segment provides the result labeled (b) in Fig. 9.13. We find $\theta_m \simeq 45^\circ$ for this case. Bode diagrams of the equalizer networks required for the two designs are labeled (a) and (b), respectively, in Fig. 9.13.

Although it may appear that we were unusually fortunate to obtain the desired answer in only two steps of the general trial and error design process here, the method converges rapidly in most cases, and seldom are more than two or three design iterations required. If -1 and -2 slope templates are constructed out of metal, plastic, or heavy cardboard, the construction of the most common straight line segments on the Bode diagram is expedited, and a little practice makes it possible to rapidly construct and check our tentative design. Note that a long span of lag equalization is required in this example. This results from the large K_v requirement and our decision (perhaps inappropriate) to use a lag equalizer. The lag network will be difficult to realize. We will show in Sec. 9.5 that a composite (combination of lag and lead) equalizer is more appropriate for systems of this type.

Example 9.3

Consider now the design problem introduced in Ex. 9.2 from a more analytical point of view. From the general form of the compensated Bode diagram shown in Fig. 9.13, it is apparent that ω_c near three or four will be required. Observe that the phase contributions at ω_c due to the breaks at $\omega_1 = 0.022$ and $\omega_4 = 100$ are relatively insignificant, and also tend to compensate for each other. Thus it is reasonable to approximate the phase margin at ω_c by

$$\theta_m \simeq \frac{\pi}{2} - \tan^{-1}\frac{\omega_2}{\omega_c} - \tan^{-1}\frac{\omega_c}{\omega_3} \qquad \omega_3 = 10 \tag{9.17}$$

Applying the arc-tan approximation yields

$$\theta_m \simeq \frac{\pi}{2} - \frac{\omega_2}{\omega_c} - \frac{\omega_c}{\omega_3} \tag{9.18}$$

We know that $\omega_c = \sqrt{\omega_2\omega_3}$ maximizes θ_m for the -2, -1, -2 geometry around ω_c used as an approximation here. Since $\omega_3 = 10$ in this case, substituting $\omega_c = \sqrt{\omega_2\omega_3}$ into Eq. (9.18) and setting $\omega_3 = 10$ provides

$$\theta_m \simeq \frac{\pi}{2} - \frac{\omega_2}{\sqrt{10\omega_2}} - \frac{\sqrt{10\omega_2}}{10} \tag{9.19}$$

Solving for the ω_2 required to provide $\theta_m = 45°$ yields $\omega_2 = 1.55$ and $\omega_c = 3.94$, values differing only slightly from those obtained previously. Similar analytical solutions based on the arc-tan approximation are possible in most applications.

A few comments comparing graphical trial and error methods to analytical methods are in order. Graphical solutions can be roughed out quickly with little effort. They are easily revised to provide whatever accuracy is required. When log-log coordinates are used, as strongly recommended, the vertical and horizontal scales are always equal, so -1, -2, and -3 slope templates are readily prepared to facilitate rapid graphical construction. It is suggested that -2 and -3 slope templates be constructed (a 45° plastic triangle provides a convenient -1 slope template) and used to approximate all solutions graphically as a first step. Analysis should be carried out from the approximate graphical solution to finalize required equalizer break point locations in those cases where further adjustments seem appropriate. Approximations which simplify the analysis are much more readily made and justified when the desired answer is nearly known.

When the Bode diagram geometry is -2, -1, -2 as shown in Fig. 9.14a, we have shown that $\omega_c = \sqrt{\omega_1\omega_2}$ is desired for maximum θ_m and this aids in quickly obtaining an approximate solution. A plot of θ_m versus -1 slope span $\gamma = \omega_2/\omega_1$ for the situation depicted in Fig. 9.14a is provided in Fig. 9.15. A

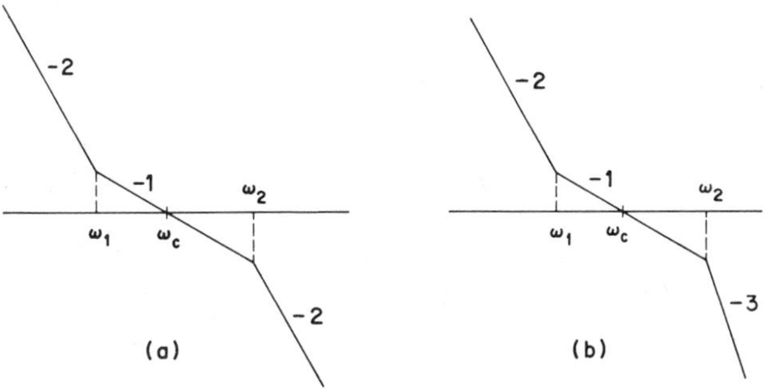

FIG. 9.14. Two common Bode diagram configurations.

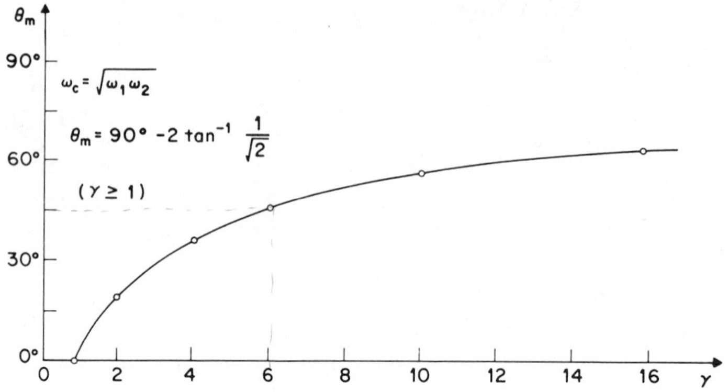

FIG. 9.15. Phase margin versus γ for the –2, –1, –2 slope case.

similar plot for the –2, –1, –3 slope case of Fig. 9.14b (assuming the double break is caused by two real poles) is shown in Fig. 9.16. In this latter case it is easily shown (see Problem 9.8, App. F) that $\omega_c = \sqrt{\omega_1\omega_2/2}$ maximizes θ_m for a given –1 slope span. In practice, of course, there are usually other break points to be considered, but their effects are often negligible (they are far from ω_c) and even if this is not true the –1 slope span and the relative location of ω_c are easily adjusted to appropriately compensate for additional slope changes. It is apparent by comparison of Figs. 9.15 and 9.16 that a given value of θ_m requires a much wider –1 slope span when a –3 slope region exists near ω_c.

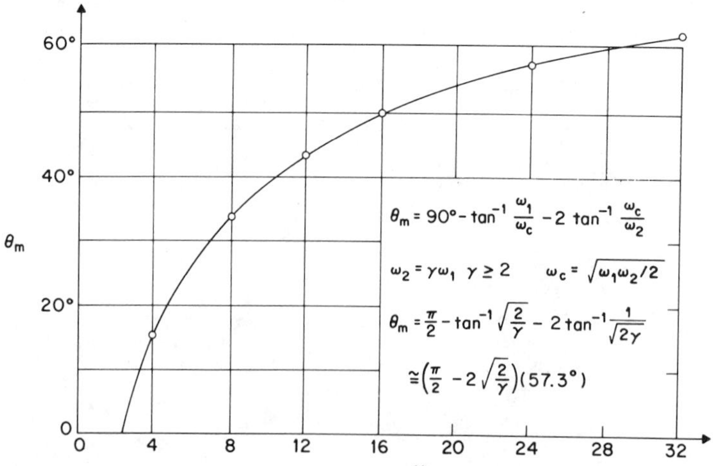

FIG. 9.16. Phase margin versus γ for the –2, –1, –3 slope case.

Example 9.4

Consider a plant with transfer function

$$G_p(s) = \frac{K}{s(1 + s/10)(1 + s/100)} \tag{9.20}$$

Choose the value of K and the parameters of a cascade lag network equalizer to provide $\theta_m \geq 50°$ while maintaining steady-state error in response to sinusoidal inputs for $\omega \leq 0.1$ to 1 percent or less. K should be no larger than necessary.

The low frequency error spec is our first concern. We know that

$$\frac{E(s)}{R(s)} = \frac{1}{1 + A(s)} \tag{9.21}$$

where $A(s) = G_e(s)\,G_p(s)$ is the compensated forward transfer function of a unity feedback system. For steady-state error less than 1 percent at all $\omega \leq 0.1$, $|A(s)| \leq 100$ is required over this frequency range. From the Bode diagram of the plant shown in Fig. 9.17 it is clear that only a small compensation adjustment is necessary near crossover at $\omega_c \simeq 4$, and the Bode diagrams of $A(s)$ and $G_p(s)$ will be identical for all $\omega \leq 0.1$. If $K = 10$ is chosen, the Bode plot of $G_p(s)$ is above the region disallowed by the error spec as denoted by the cross-hatched area in Fig. 9.17. We will attempt to satisfy all specs with $K = 10$,

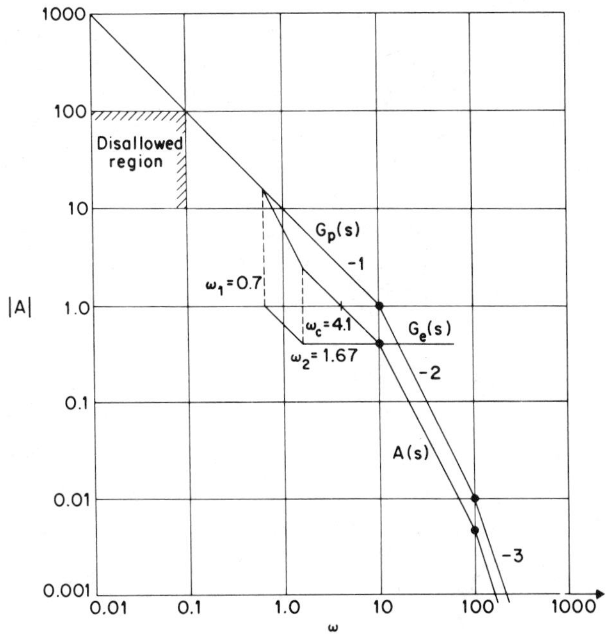

FIG. 9.17. Bode diagram of the system considered in Ex. 9.4.

adjusting this value later if necessary. If $G_p(s)$ is left uncompensated, straightforward calculations show that $\omega_c \simeq 8.0$ and $\theta_m \simeq 47°$. The extra 3° of phase margin may be obtained by adding a small amount of lag compensation. Arbitrarily choosing a 6:1 span of -1 slope symmetrically located around ω_c results in $\omega_2 = 10/6 \simeq 1.67$ and $\omega_c = \sqrt{16.7} \simeq 4.1$. Bode diagrams of the equalized forward transfer function $A(s) = G_e(s)\,G_p(s)$ and of $G_e(s)$, the required equalizer, are also shown in Fig. 9.17. The equalized diagram, as expected, does not violate the error spec. The resulting phase margin is $\theta_m \simeq 53°$, which provides a small margin for error. The -1 slope span around ω_c could be somewhat reduced and still provide acceptable results. Many alternate solutions providing similar performance are possible.

This completes our discussion of lag equalization. The examples presented illustrate the general procedure required to design lag equalizers within specs on performance characteristics such as θ_m, K_v, and low frequency steady-state errors. Lag equalization can *only* be used to shape the Bode diagram downward. Its use is thus limited to systems for which sufficient θ_m can be obtained by forcing ω_c to occur at a frequency lower than that provided by the uncompensated system, and usually within a region of -1 slope on the $G_p(s)$ plot. Lag equalization is particularly appropriate when the phase lag of $G_p(s)$ increases rapidly near ω_c (a condition which makes lead equalization difficult), when response to high frequency disturbances is to be subdued, and whenever design specs require minimum BW. Lag networks are generally easy to realize, although large values of R and C may be required if one or more break frequencies are small. The most significant disadvantage of lag compensation arises from the BW reduction, which increases the transient response duration and the steady-state errors for sinusoidal inputs in that frequency range where $|A(s)| < |G_p(s)|$. Lag equalizers are easily specified based upon a few graphical Bode diagram constructions supported by simple algebraic calculations. Acceptable designs may generally be reached using either trial and error or more sophisticated analytical methods, depending upon personal preference. Extensive calculations to a high degree of accuracy are not often justified, since the parameters of our control system models are not generally known precisely.

9.4 LEAD EQUALIZER COMPENSATION

We have seen that lag compensation uses the existing low frequency -1 slope segment of the uncompensated plant to provide desirable Bode diagram shape near ω_c. This is not always possible for several reasons. Foremost among these is that some plants have no low frequency -1 slope region, or the -1 slope segment occurs so low in frequency that it is impossible to achieve sufficiently rapid transient response and satisfy low frequency error specs if a lag equalizer is used. Lead equalization provides considerably more design flexibility in such cases. The basic function of a lead equalizer is to provide phase lead in the range of frequencies near ω_c. Only if ω_c on the uncompensated diagram occurs in a region of -1 or -2 slope are we inclined to use lead compensation. Although a -3 slope region can be converted to a -1 slope region by using two lead networks, this requires significant gain increases and often leads to noise and dynamic range problems as well. Seldom is lead compensation considered when two or more lead sections would be necessary, or, equivalently when more than 45° or 50° of phase lead is required. Often a compromise solution can be worked out using a *composite equalizer* to provide both lag and lead compensation, as discussed in Sec. 9.5. Lead equalizer design is illustrated by the following example.

Example 9.5

Consider the plant transfer function

$$G_p(s) = \frac{K}{s(1 + s/10)} \tag{9.22}$$

The following specs are to be met: steady-state error to sinusoidal inputs up to 1 rad/sec is not to exceed 1 percent, and $\theta_m \geq 50°$.

The low frequency error spec requires that the Bode diagram pass above $|A| = 100$ at $\omega = 1$. Thus $K \geq 100$ is required. (We assume here that low frequency Bode diagram-shaping to satisfy the error spec is not to be used. Our objective is to satisfy the specs with a single section equalizer if possible.) The Bode diagram of $G_p(s)$ with $K = 100$ is shown in Fig. 9.18. It is apparent that

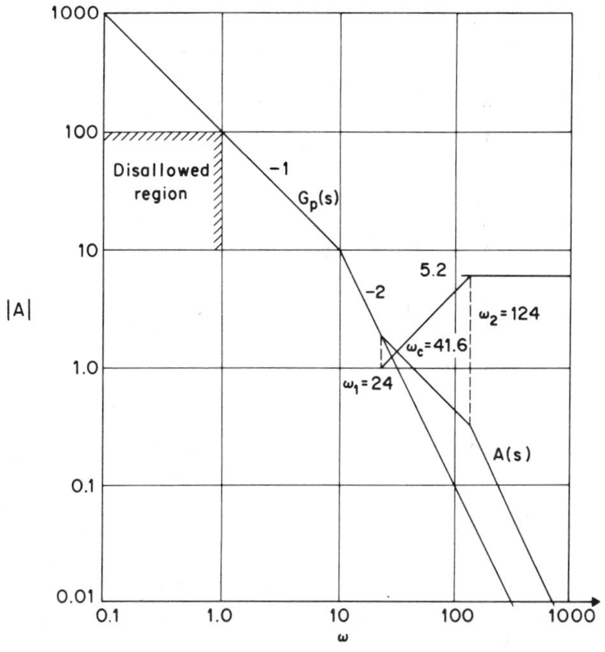

FIG. 9.18. Bode diagrams for the system considered in Ex. 9.5.

the error spec makes lag equalization alone impossible. We will use trial and error to find an acceptable lead equalizer. The lead span must provide about 40° of phase lead, since the compensated ω_c will exceed that of the plant alone. As a first trial it seems appropriate to try a 6:1 lead span. Since the resulting diagram has a short -2 slope segment below ω_c and an infinitely long -2 slope segment above ω_c, it also seems reasonable to use $\omega_c \simeq \sqrt{1/2\omega_1\omega_2}$ where ω_1 and ω_2 are the lower and upper frequency limits, respectively, of the -1 slope segment near ω_c. This places ω_c appropriately closer to the short -2 slope segment with its relatively small phase lag contribution, and is the same ω_c choice we would use with a -2, -1, -3 slope shape. Thus $\omega_2 = 6\omega_1$ and $\omega_c = \sqrt{3}\omega_1$ are desired. Setting $|A| = 1$ at $\omega = \omega_c$ for this configuration (see Fig. 9.18) provides

$$|A(s)| = 1 \simeq \frac{100(\omega_c/\omega_1)}{\omega_c(\omega_c/10)} = \frac{1000\sqrt{3}}{\omega_c^2} \tag{9.23}$$

or $\omega_c = \sqrt{1{,}732} \simeq 41.6$, $\omega_1 = 24$ and $\omega_2 = 124$. The corresponding phase margin is given by

$$\theta_m = \frac{\pi}{2} - \tan^{-1}\frac{24}{41.6} - \tan^{-1}\frac{41.6}{124} + \tan^{-1}\frac{10}{41.6} \tag{9.24}$$

which reduces to $\theta_m \simeq 55°$. The lead span can be reduced somewhat and still satisfy the θ_m spec. Obtaining a more accurate solution is left as an exercise for the reader. It should be obvious that similar answers are easily obtained using graphical trial and error with a minimum of numerical calculations.

The compensation required in Ex. 9.5 is readily synthesized using the RC network and amplifier shown in Fig. 9.19. The network transfer function is

$$\frac{V_1}{V_i}(s) = \left(\frac{R_2}{R_1 + R_2}\right)\left(\frac{1 + R_1Cs}{1 + \dfrac{R_2}{R_1 + R_2}R_1Cs}\right) = \frac{1}{B}\left(\frac{1 + B\tau_1 s}{1 + \tau_1 s}\right) \tag{9.25}$$

where

$$B = \frac{R_1 + R_2}{R_2} \qquad \tau_1 = \left(\frac{R_2}{R_1 + R_2}\right)R_1C \tag{9.26}$$

Since $V_0 = BV_1$, V_0/V_i has the desired unity dc gain lead equalizer form.

Lead equalization works nicely in problems of the type posed in Ex. 9.5, where the plant Bode diagram has -2 slope for a wide range of frequencies (infinite in the example) above ω_c. When there are additional lag breaks near

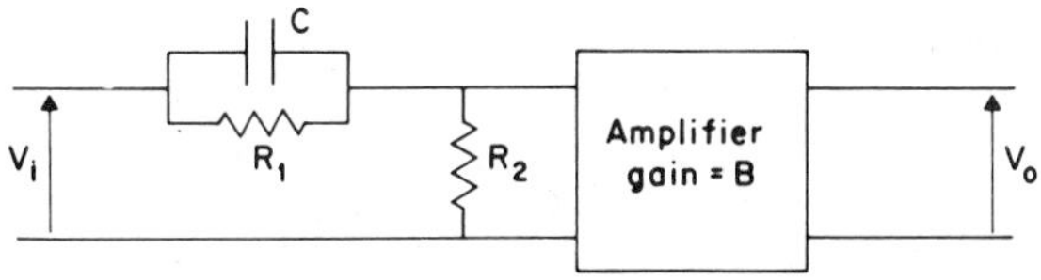

FIG. 9.19. A lead compensator.

ω_c, however, lead compensation requires a complex network and extensive additional gain. In some cases two or more sections of lead compensation may provide the only workable solution. An example will be used to illustrate the design of multiple-section lead equalizers.

Example 9.6

Consider the plant transfer function

$$G_p(s) = \frac{K}{s(1 + s/10)(1 + s/100)} \tag{9.27}$$

The following specs must be satisfied: 1. steady-state error to sinusoidal inputs up to 1 rad/sec is not to exceed 1 percent; 2. $\omega_c \geq 20$, a restriction imposed with the intention of limiting the duration of transient patterns such as the unit-step response; and 3. $\theta_m \geq 50°$.

As in Ex. 9.5, the low frequency error spec is satisfied by setting $K = 100$. A Bode diagram of the resulting $G_p(s)$ is shown in Fig. 9.20. It is obviously going to be more difficult to provide a -1 slope region of sufficient width to satisfy the θ_m spec in this case. The break to a -3 slope at $\omega = 100$ is the major source of trouble. We will use two sections of lead compensation to provide a -1 slope from ω_1 to ω_2 as shown in Fig. 9.20. The first section covers the entire span from ω_1 to ω_2, whereas the second runs from $\omega = 100$ to ω_2. The compensated Bode diagram is thus -1, -2, -1, -3, and the -2 slope span is relatively narrow. We assume $\omega_c = \sqrt{1/2\omega_1\omega_2}$ is desirable since this is known to be optimum for the -2, -1, -3 case, and our first guess is that a 10:1 span of -1 slope is necessary to satisfy the θ_m spec. Thus $\omega_2 = 10\omega_1$ and $\omega_c = \sqrt{5}\omega_1$. Setting the asymptotic $|A| = 1$ at ω_c provides the relationship for ω_1 as

$$|A| = 1 = \frac{100\,(\sqrt{5}\omega_1/\omega_1)}{(\sqrt{5}\omega_1)(\sqrt{5}\omega_1/10)} = \frac{1000}{\sqrt{5}\omega_1{}^2} \tag{9.28}$$

from which we find that $\omega_1 = 21$. This results in $\omega_2 = 210$ and $\omega_c = 47$, as also shown in Fig. 9.20. The corresponding θ_m is given by

$$\theta_m = \frac{\pi}{2} - \tan^{-1}\frac{21}{47} + \tan^{-1}\frac{10}{47} - 2\tan^{-1}\frac{47}{210} \simeq 52.8° \tag{9.29}$$

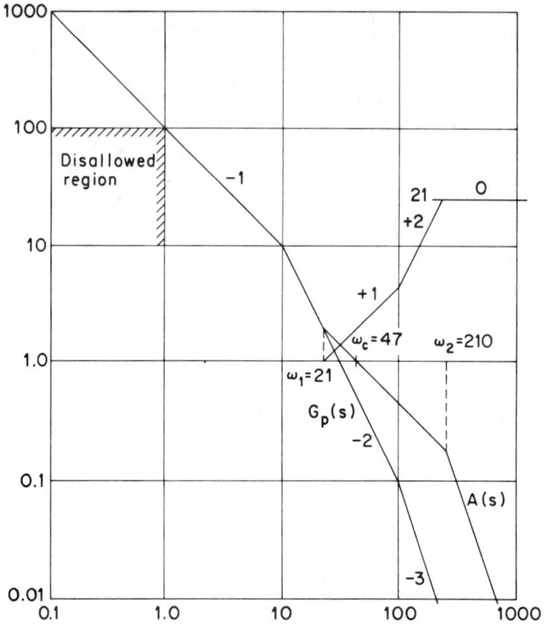

FIG. 9.20. Bode diagrams of the system introduced in Ex. 9.6.

which satisfies the spec with a comfortable margin for error. The asymptotic diagram of the required equalizer is also shown in Fig. 9.20, and its transfer function is

$$G_e(s) = \frac{(1 + s/21)(1 + s/100)}{(1 + s/210)^2} \tag{9.30}$$

This $G_e(s)$ is readily realized using a passive RC network and an amplifier with a gain of 21.

We are now in position to summarize the relative merits of lag and lead compensation. Lag compensation has the following advantages:

1. It is realizable in most applications using passive RC networks.
2. It reduces open-loop BW, thereby decreasing system susceptibility to noise and reducing dynamic range requirements.
3. Steady-state tracking capability at frequencies *below* the lag break is not impaired.

The disadvantages of lag compensation are:

1. Since the closed-loop BW is reduced, responses to inputs such as a unit step are proportionately slower.

2. Steady-state errors to sinusoidal inputs in the frequency range where the compensated Bode plot falls below that of the plant alone are proportionately increased.

3. Lag compensation applies only if a -1 slope (or 0 slope) region is available for use in $G_p(s)$, and therefore cannot generally be used in systems with two or more plant integrations.

4. If the desired equalizer break frequencies are extremely small, it may be difficult to synthesize the network due to prohibitively large values of the RC products (time constants) required.

The general characteristics of lead compensation are complementary to those of lag compensation. Its advantages are:

1. Closed-loop BW is increased, resulting in faster transient response.

2. Low frequency steady-state sinusoidal error characteristics are not degraded.

3. The time constants required are generally easier to realize than those of a lag network to compensate the same system, assuming both forms of equalization are possible.

The disadvantages of lead equalization are:

1. The wider BW often complicates the noise and dynamic range requirements in amplifiers or other circuit components, and saturation effects are more apt to arise during normal operation.

2. Additional gain is always necessary.

3. Since ω_c is increased, phase lag contributions due to high frequency lag breaks become more significant at ω_c and it is generally more difficult to realize a given θ_m.

With regard to disadvantage 3, lead compensation is entirely inappropriate in any system where phase lag is increasing rapidly at and just above the uncompensated ω_c. It is not generally a feasible compensation procedure for nonminimum phase systems such as those resulting when the signal experiences a time delay at some point in the control loop. Nonminimum phase effects are considered further in Sec. 9.6. The fundamental characteristics of lag and lead equalizers are summarized for convenience in Table 9.1.

Control system design generally involves an initial appraisal of all specs and their effects (often approximate) upon the Bode diagram. The required gain is usually set by a spec upon K_v or error. If not, engineering judgment must be used to establish a reasonable value. The relative merits of lag and lead equalization should be considered before either is chosen. In many cases neither seems appropriate. Fortunately a compromise between the two extremes can often be made which provides performance superior to that available using either alone. This leads us to the consideration of lag-lead (composite) equalization, which is by far the most common form of series equalizer.

Table 9.1. Summary of compensation characteristics.

	Lag equalization	*Lead equalization*
Motivation	Reduction of high frequency gain to provide crossover in a region of −1 slope on the plant transfer function.	Introduction of phase lead near the crossover frequency.
Effects	1. Introduces additional phase lag below ω_c. 2. Decreases system BW. 3. Increases low frequency errors.	1. Increases high frequency gain. 2. Increases system BW. 3. Increases dynamic range requirements and noise susceptibility.
Advantages	1. Simple passive structure. 2. Suppresses high frequency noise.	1. Increased BW and faster dynamic response. 2. Superior low frequency error performance.
Disadvantages	1. Decreases BW. 2. Slows down transient performance. 3. Increases low frequency errors. 4. Adds low frequency phase lag. 5. May require large RC products.	1. Requires additional gain. 2. Increases dynamic range requirements. 3. Increases BW and noise susceptibility. 4. It can provide only limited phase lead at a reasonable cost ($\simeq 60^\circ$).
Most Applicable	1. When system phase lag increases rapidly with ω near ω_c, making lead impractical. 2. When reduced BW and slower transient performance is acceptable. 3. When noise and dynamic range considerations are significant. 4. When additional amplification cannot be justified.	1. When system phase lag changes slowly with ω near ω_c. 2. When large BW and fast transient response are required. 3. When noise and dynamic range problems are not of primary significance. 4. When lag equalization is impossible.
Not Applicable	1. When no low frequency region exists in which the phase lag is less than that desired at ω_c. 2. When transient response, BW, or low frequency error specs are excessively stringent.	1. When phase lag increases rapidly with ω near ω_c. 2. When the phase lead requirement is excessive. 3. When noise specs make high frequency gain increases impossible. 4. When dynamic range (saturation) problems are severe.

9.5 CASCADE EQUALIZATION USING COMPOSITE EQUALIZERS

We have seen that a tight spec on steady-state following error makes it more difficult to use a lag equalizer, whereas a BW or noise limitation may make lead compensation impossible. Any combination of the two effects may be used in specific applications, thereby often realizing the most significant advantages of both methods in a single design. The most restrictive disadvantage of lead equalization is the *gain* requirement, whereas the use of lag equalization results only in *downward* shaping of the Bode diagram. It is significant to note that the equalizer can always (at least in theory) be realized as a passive RC network as long as the compensated Bode diagram falls on or below the uncompensated diagram, irrespective of the specific shape of either. We may thus construct an arbitrary compensated Bode diagram shape near ω_c, as long as that shaping is all done *below* the uncompensated diagram, and the equalizer can be realized without requiring additional gain. Thus we do not really need to force ω_c on the compensated diagram down far enough to take advantage of the -1 slope region on the uncompensated diagram, assuming such a region exists, as was found necessary in designing lag compensated networks, since we can construct a network to generate our own -1 slope region around ω_c at a higher frequency if desired. This is a simple combination of lag and lead design techniques. As long as the equalizer lead span is less than or equal to the equalizer lag span, and occurs after the lag span in frequency, the compensating network can be realized in passive form. Equivalently, as long as the compensated diagram is never *above* the uncompensated diagram, a passive equalizer is possible. This gives rise to many interesting possibilities. Maximum dynamic performance capability within a θ_m or relative stability spec using a passive equalizer is generally achieved by shaping the compensated Bode diagram downward from the plant diagram just enough to achieve crossover in a -1 slope region sufficiently wide to satisfy the desired θ_m spec and then bringing the compensated and uncompensated diagrams back together again above ω_c. This maximizes the closed-loop BW available using a passive equalizer within a θ_m spec, and generally provides relatively fast transient response, good settling characteristics, good error performance, and a cheap and simple design. These points are illustrated by the following example.

Example 9.7

Consider the plant

$$G_p(s) = \frac{K}{s(1 + s/10)(1 + s/100)} \tag{9.31}$$

The following specs must be satisfied:

1. Steady-state error to sinusoidal inputs up to 1 rad/sec is not to exceed 1 part in 70.
2. $\theta_m \geq 50°$,

3. 60 Hz input noise components are to be attenuated by at least a factor of 250 at the output.
4. K_v, the velocity error coefficient, is to be 100, and
5. There is to be no section with -3 slope prior to ω_c.

The K_v specification requires that $K = 100$. The Bode diagram of $G_p(s)$ with $K = 100$ is shown in Fig. 9.21. The error spec will allow a compensation lag break no sooner than $\omega = 1.0$. Putting a lag break at $\omega = 1.0$ will not bring the compensated diagram down fast enough to allow us to use the -1 slope region of $G_p(s)$ near ω_c. Spec number 5 disallows the use of two or more lag sections. Thus lag compensation alone will not work. The spec on noise at 60 Hz, or ω = 377 rad/sec, requires that $|A| \leq 0.004$ at $\omega = 377$, since the input to output transfer function is given by

$$\left|\frac{C}{R}(s)\right| = \left|\frac{A(s)}{1 + A(s)}\right| \simeq |A(s)| \qquad |A(s)| \ll 1 \tag{9.32}$$

Thus spec number 3 rules out lead compensation alone, assuming some form of 60 Hz signal trap is not to be used. The restrictions on the compensated Bode diagram are now readily apparent.

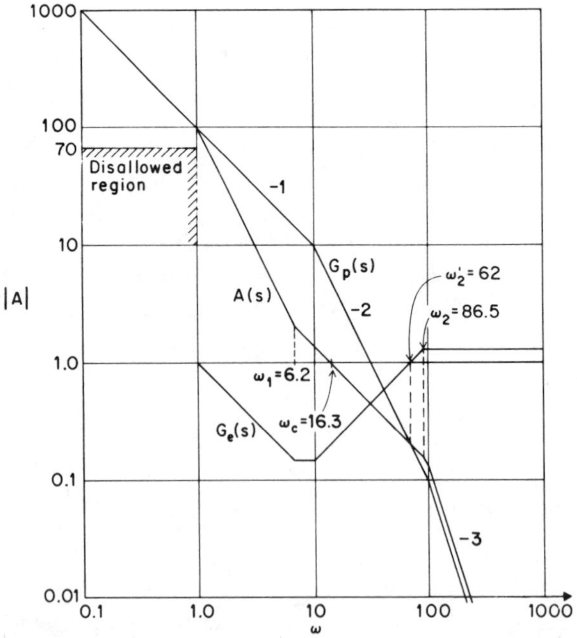

FIG. 9.21. Bode diagrams for the system introduced in Ex. 9.7.

We initiate the design process by placing a lag break at $\omega = 1.0$, which is the lowest frequency for this break within the error spec constraint. Since the final geometry will be very nearly -2, -1, -3 near ω_c and since $\theta_m = 50°$ is required, we will set $\omega_c = \sqrt{\omega_1\omega_2/2}$ and $\omega_2 = 14\omega_1$, which are the approximate values required according to Fig. 9.16 for $\theta_m = 50°$. Thus $\omega_c = \sqrt{7}\omega_1$. Setting $|A| = 1$ at ω_c and simplifying provides $\omega_c = 16.3$, $\omega_1 = 6.17$, and $\omega_2 = 86.5$. A check on phase margin shows that $\theta_m \simeq 53°$ for these values. In fact the $\theta_m = 50°$ spec is more than satisfied if $\omega_2' = 62$ is used, in which case the compensated curve reverts back to the $G_p(s)$ curve for all $\omega > 62$, and the required $G_e(s)$ can be realized as a passive network.

The situation depicted in Fig. 9.21 is typical of composite equalizer applications. Use of a composite equalizer represents a compromise between a lag equalizer, with the associated BW reduction that often may be very significant, and a lead equalizer, which, when used alone, requires additional gain and tends to aggravate dynamic range and high frequency noise problems. Whenever the composite equalizer has no frequency range over which its gain exceeds unity it may be realized as a passive network, so it is an important design objective to see that this is true whenever possible. Note that the composite equalizer design does not generally result in a reduction of ω_c from its uncompensated value by more than a factor of about 3:1. When the -1 slope span near ω_c is chosen to just satisfy the θ_m spec with ω_c properly placed within the -1 slope interval and $\lim_{\omega\to\infty} |G_e(s)| = 1$ is used [which means $A(s)$ and $G_p(s)$ are identical for all ω significantly above ω_c], this corresponds to minimum BW reduction and minimum low frequency error degradation within a θ_m spec and a passive equalizer requirement. Thus composite equalization, although the network required is somewhat harder to synthesize, is superior to lag compensation in low frequency error performance, and preferable to lead equalization because it requires no additional gain.

The steps required in designing composite equalized systems depend to some extent upon the specs to be satisfied. If, for example, phase margin is to be maximized using a passive equalizer with a low frequency error spec, the following steps are taken:

1. The low frequency error spec point is placed on a Bode diagram of the plant.
2. A -2 slope of the $A(s)$ curve is placed as close to the spec point as possible.
3. The desired relationship between ω_c and the end frequencies of the -1 slope segment near ω_c is chosen by considering the Bode diagram shape near ω_c.
4. A -1 slope segment is chosen, either analytically or graphically by trial and error, to connect the -2 slope segment constructed in step 2 with the $G_p(s)$ curve at frequencies ω_1 and ω_2, respectively, such that ω_c has the desired relationships to ω_1 and ω_2.

5. The required $G_e(s)$ diagram is drawn.
6. $G_e(s)$ is specified in transfer function form.
7. Characteristics of the resulting design such as θ_m, BW, unit-step response, etc, are determined as desired.

An alternate design procedure often followed results when a specific θ_m spec is to be satisfied using a passive network with minimum degradation of low frequency error performance. In this case the objective is not to place the low frequency -2 slope segment of $A(s)$ as near the error spec point as possible, since no specific spec point exists, but to place the break to a -2 slope as high as possible in frequency within the θ_m spec. The suggested design procedure in this case is as follows:

1. Appraise the resulting geometry near ω_c and choose the desired relationship between ω_c, ω_1 and ω_2.
2. Set the span, ω_2/ω_1, just wide enough to satisfy the θ_m spec.
3. Determine, either graphically using trial and error, or analytically, the desired values of ω_1, ω_2 and ω_c.
4. Construct the remainder of the $A(s)$ curve by connecting the -1 slope segment at ω_c to the $G_p(s)$ curve with a -2 slope segment intersecting the -1 slope segment at $\omega = \omega_1$.
5. Sketch the $G_e(s)$ curve.
6. Specify $G_e(s)$ in transfer function form.
7. Determine all desired system characteristics and make design modifications if necessary.

Realization of the lag-lead equalizer as an RC network is a simple matter. One possible configuration is shown in Fig. 9.22. The transfer function is readily found to be

$$\frac{V_2}{V_1} = \frac{(1 + R_1C_1s)(1 + R_2C_2s)}{(R_1C_1R_2C_2 + R_3C_1R_2C_2)s^2 + (R_1C_1 + R_2C_1 + R_3C_1 + R_2C_2)s + 1} \tag{9.33}$$

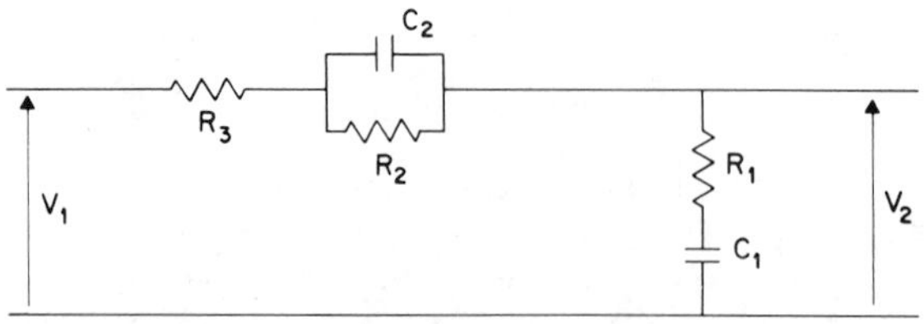

FIG. 9.22. Realization of the lag-lead equalizer as an RC network.

The components are chosen to give the desired break frequencies with reasonable input and output impedance levels. It should be noted that by using a single network we require no additional amplification, whereas the use of separate lag and lead networks would not accomplish this objective.

9.6 CASCADE EQUALIZATION OF NONMINIMUM PHASE NETWORKS

In any problem where $G_p(s)$ is not a minimum phase transfer function, the phase characteristic can not be obtained from the Bode plot of $G_p(s)$ in the usual manner, and the design procedure must be appropriately modified. It is suggested that a phase versus ω plot of $G_p(s)$ also be made in all such cases. This emphasizes the phase contribution an equalizer must provide wherever it might be placed in frequency. The design objective is generally to place ω_c at a point where adequate phase margin can be obtained.

Perhaps the simplest nonminimum phase transfer function encountered regularly is a time delay of T seconds, which has transfer function ϵ^{-Ts}. Note that $|\epsilon^{-Ts}| = 1$ for all frequencies and delay simply provides phase lag which increases linearly with ω. It is suggested that a phase lag versus ω sketch for the time delay term alone be superimposed on the Bode diagram of $G_p(s)$ and the design carried out just as usual, ignoring the time delay except to provide sufficient additional θ_m to take care of its phase lag contribution. A minimum change in the usual design procedure is thereby required. It should be pointed out that, although the effect of a lag term in $G_p(s)$ can be postponed using lead networks to increase closed-loop BW, it is a practical impossibility to increase BW significantly beyond the point where time delay, or any other type of nonminimum phase term, starts to contribute appreciable phase lag. Since phase lag for time delay increases linearly with frequency, whereas lead compensation provides phase lead which increases approximately linearly with span only for spans up to about 6:1, lead compensation can only improve θ_m within reasonable limits. Suppose, for example, the uncompensated ω_c occurs in a -2 slope region and the phase lag there is 170°, 10° of which comes from a pure time delay in the loop. To provide $\theta_m = 50°$ would normally require a lead span of about 10:1, and the compensated ω_c would be roughly three times the uncompensated value. At this new ω_c the time delay phase lag is 30°, and $\theta_m \simeq 30°$ results, instead of the desired 50° value. Increasing the lead span increases ω_c further, and θ_m decreases instead of increasing. Trying to increase BW and maintain reasonable θ_m with time delay in the loop is not feasible beyond the point where the time delay phase lag contribution becomes about 10°. Systems with a significant phase contribution at ω_c from a time delay term generally have small values of gain margin, and relatively oscillatory transient response.

Example 9.8

Consider a system with a plant transfer function

$$G_p(s) = \frac{K\epsilon^{-Ts}}{s(1 + s/10)(1 + s/100)} \tag{9.34}$$

where $K = 100$ and $T = 1/200$. The specs are: 1. $\theta_m = 45°$; 2. the equalizer dc gain is to be unity; and 3. low frequency steady-state sinusoidal error performance is to be degraded no more than necessary. Find the required cascade equalizer.

The first step in the design is to draw a Bode diagram of $G_p(s)$ and a phase shift curve for the ϵ^{-Ts} term as shown in Fig. 9.23. Note that the time delay phase contribution plot is a straight line on log-log coordinates, which is particularly convenient. Whereas the time delay can essentially be ignored for $\omega_c \leq 10$, it rapidly becomes the predominant factor over the decade $10 \leq \omega \leq 100$. It would obviously be difficult to set ω_c much above 10 in this problem. It seems reasonable to use lag-lead compensation and set $\omega_c = 10$ as a first trial, as shown in Fig. 9.23. The double break at $\omega = 100$ provides 11.5° of phase lag at $\omega_c = 10$, the time delay provides 2.9°, the break at $\omega_2 = 6$ contributes 31°,

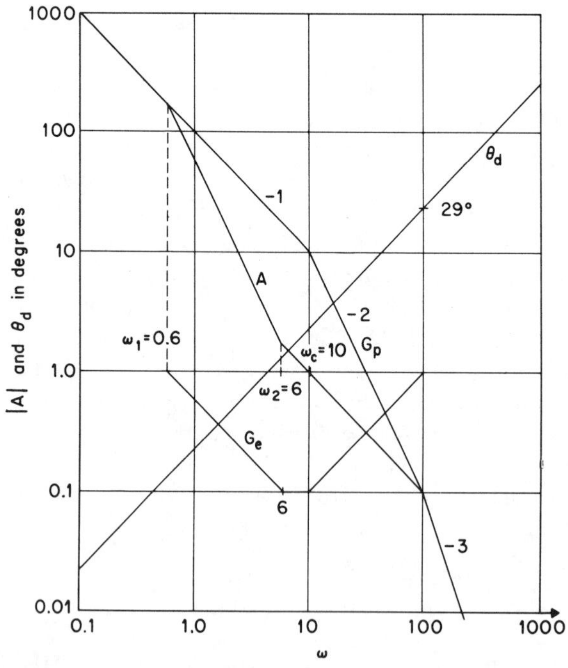

FIG. 9.23. Bode diagram of the system introduced in Ex. 9.8.

and the break at $\omega_1 = 0.6$ reduces the phase lag by 3.4°, so

$$\theta_m = 90° - 11.5° - 2.9° - 31° + 3.4° = 48° \tag{9.35}$$

which satisfies the spec with a bit to spare. Note that ω_c as provided by this design is by no means optimally located in the usual sense relative to the -2, -1, -3 slope configuration, which would normally require $\omega_c = \sqrt{1/2\omega_2\omega_3} = 24.5$, but the time delay terms leads us to set ω_c significantly below this value because θ_m decreases with increasing ω_c just as though the slope above ω_c was much more negative. If, for example, the loop gain were to increase by a factor of 2, ω_c would change to 20 and θ_m would decrease somewhat, but by no means as much as would be the case if ω_c had been placed nearer the -1 slope segment center. We could increase ω_c somewhat, perhaps to $\omega_c \simeq 20$, and still satisfy the spec on θ_m, but the resulting design would be more sensitive to increases in loop gain (small gain margin) and would either require additional gain (lead equalization), or a very long -1 slope segment below ω_c, resulting in poorer low frequency error performance. The design shown represents a workable compromise providing adequate values of G_m, θ_m and low frequency error performance.

The procedure illustrated by Ex. 9.8 for designing systems with a time delay in the loop is readily extended to the design of systems with a rhp zero. If $G_p(s)$ is given by

$$G_p(s) = G_p'(s)(1 - \tau s) \tag{9.36}$$

the effect of the rhp zero is to provide phase *lag* instead of phase *lead* at all frequencies. The nonminimum phase zero should be clearly marked on the Bode diagram to act as a constant reminder that the phase contribution of that break point at ω_c or any other frequency of interest is a *lag* rather than a *lead.* All of the usual procedures apply. The break effect can *not* be cancelled by an equalizer pole. In general a somewhat longer than normal -1 slope span near ω_c is required whenever rhp zeros are present.

9.7 SUMMARY AND CONCLUSIONS

We have seen that, except for quadratic pair breaks, break frequencies due to nonminimum phase zeros, and pure time delay, which does not alter the Bode diagram, the Nyquist plot is easily visualized directly from the Bode plot and design can be carried out entirely on the Bode plot using simple graphical and/or analytical procedures. These same procedures are readily modified to apply when time delay, rhp zeros, or quadratic pair breaks are encountered. Thus, although the Nyquist criterion and the polar plot characteristics form the basis for all of our designs, the mechanics of the design procedure are readily established on the Bode plot without requiring the construction of polar diagrams.

We have thus far considered only cascade equalization. This is the simplest equalization form to work with initially, but there are many practical advantages for using minor-loop equalization, as discussed in Chap. 11. The same basic design objectives can be achieved either way. We have not specifically considered feedback compensation, or the use of other than unity feedback, but virtually all of what has been done applies with only minor modifications to all such systems. Since we have designed closed-loop controllers with reasonable relative stability as our primary objective, it does not really matter from this point of view where the equalizer network is placed in the loop. In most applications the objective is to follow the input, so unity feedback is naturally used. Placing the equalizer in the feedback path changes the input-to-output characteristics, but has no effect on stability. In most applications the equalizer is placed at a point in the loop where the power level is minimum and the signals are easily manipulated. Since power levels generally go up as we move from input to output, it is natural to place the equalizer as early as possible in the forward path, or, when appropriate, in the feedback path. In most applications equalizer transfer functions are realized by using appropriate networks as amplifier interstage coupling units, in feedback arrangements around voltage and power amplifiers, or some combination of the two.

In many design problems it is desired to control steady-state sinusoidal responses and general transient characteristics such as unit-step response percent overshoot or settling time. Although we can obtain frequency response accurately from the Nyquist plot, unit-step response patterns can only be predicted in a general way from the Bode or Nyquist diagrams. It is considered standard procedure in control system design to simulate each proposed design on an analog or digital computer and observe its response to a unit-step input and such other signals as are considered important. Final adjustment of parameters such as gain and time-constant values can be made at this time as desired. The effect on the transfer characteristics of placing some or all of the equalization in the feedback path may also be observed from the simulation with little trouble.

Design relative to many of the standard performance specs has been illustrated by examples. Low frequency error specs tend to push the Bode diagram up, increasing BW and making a given value of θ_m harder to achieve. A spec on K_v sets the gain level. Noise specs force the Bode plot down at high frequencies or may force the use of a special noise filter to eliminate one or more noise components such as those commonly encountered at 60 or 400 Hz. A spec on θ_m essentially fixes the -1 slope span near ω_c. A BW spec fixes ω_c. A spec on M_p may be converted to an approximately equivalent spec on θ_m and a spec on ω_m fixes ω_c. Restrictions on unit-step response peak overshoot, rise time, or settling time are harder to impose on the Bode diagram, and simulation study is suggested. Increasing θ_m tends to decrease peak overshoot and settling time, and increasing BW cuts down rise time and settling time, but the exact relationships are complex. At any rate we now have available general design procedures for approaching all of these problems. The presence of time delay or rhp zeros in $G_p(s)$ complicates things somewhat, but both of these cases are readily handled by drawing an auxiliary phase plot on the Bode diagram and designing for a revised value of θ_m. Although we have not considered the problem in detail, complex conjugate pole or zero pairs are handled similarly. The phase contributions due to all such pole and zero pairs are computed separately and taken into consideration in the usual way. Nonminimum phase terms severely limit the BW which can be achieved with reasonable stability margin. Time delay also considerably complicates transient analysis and simulation (at least analog simulation). The procedures we have discussed here are not applicable to positive feedback systems or to systems with poles of $G_p(s)$ in the rhp. Such cases are seldom encountered. If they are, the Nyquist diagram itself should be used as the basic design tool and compensation chosen to shape it to either provide or avoid -1 point encirclements, as desired. The methods discussed in detail here are applicable to most systems encountered.

10 GENERAL RELATIONSHIPS BETWEEN PERFORMANCE CRITERIA AND SYSTEM DESIGN

The design of complex control systems must usually be carried out within a detailed set of performance specs. The customer imposes a sufficient number of design restrictions to assure satisfactory closed-loop response characteristics for the resulting system. Because controllers can assume many forms, a myriad of performance criteria are available. A close interrelationship exists between several of the criteria in common use. Examples are provided by θ_m and M_p, or ω_p and ω_c. In many cases, however, although one criterion is closely related to others, it is extremely difficult to express those relationships in any formal way. An example is provided by θ_m and unit-step response peak overshoot. Our objective in this chapter is to define the performance criteria in common use, and indicate how they may be taken into consideration during the design phase. Since the Bode diagram is our principle design tool, this requires that we determine how each performance spec restricts the Bode diagram shape and the decisions the designer makes in synthesizing an equalizer.

The values of many performance criteria are not readily apparent from the Bode diagram and Nyquist diagrams. The Nyquist diagram indicates directly the steady-state open-loop frequency response characteristics. As discussed in Sec. 8.4, the closed-loop frequency response is easily obtained from the Nyquist plot. Although the Laplace transform directly relates closed-loop time response and frequency response characteristics, it is not practical to repeatedly go through the inverse transformation process to obtain the values of typical time domain criteria such as unit-step response percent overshoot or rise time for specific parameter settings. General solutions of the transcendental equations relating unit step response peak overshoot to the transfer function coefficients are impossible for systems of third order or higher, so repeated inverse transformation would generally be necessary to exactly satisfy a time domain spec of this nature. Fortunately we may develop reasonably accurate rule of thumb relationships between frequency domain characteristics, which are readily apparent from a Bode diagram, and most other criteria of interest. Although these may provide only an approximate indication, the situation may be appraised more carefully after a preliminary design has been developed, and critical parameters adjusted as necessary to establish the final design. The approximations are sufficiently accurate that an iterative design procedure of this nature converges rapidly to the desired solution in almost all cases.

Performance criteria in the time domain are relatively difficult to evaluate accurately from the Bode diagram. Simulation of the system under study using either an analog (see App. C) or a digital (see App. E) computer is so easily carried out, and computational facilities have become so readily available in recent years, that no non-trivial design should be arrived at without some degree of simulation study. This is not to say that a completely theoretical design is impossible, but only that such an approach is not feasible from an economic point of view. It takes only a few minutes to program an analog or a digital computer to study closed-loop system performance in either the frequency or the time domain. Once the equalization method has been chosen based upon broad general principles and a tentative design proposed, final parameter settings are easily established using organized trial and error on the computer simulation. Time domain performance criteria are readily measured and specific requirements upon them satisfied in this way. A rough Bode diagram design followed by a computer simulation to establish the final equalizer parameters required by performance specs is strongly suggested. Several of the homework exercises are best solved in this way. One of the control engineer's primary functions is to decide where hand analysis should end and computer study begin. The trend in recent years is away from the former and toward extended use of the latter. It will always be important to realize how analysis can be carried out, if necessary, but it is even more important to know when the computer should be used. An engineer is fundamentally concerned with answers, and answers may be obtained in many ways.

10.1 PERFORMANCE SPECS IN THE FREQUENCY DOMAIN

Since design is generally carried out on the Bode diagram, frequency domain performance criteria can be considered directly during the design process. The standard frequency domain criteria are:

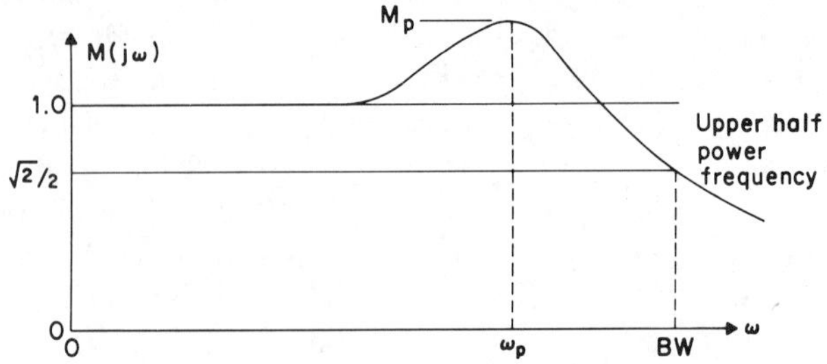

FIG. 10.1. Frequency domain performance criteria derived from the $M(j\omega)$ function.

1. $M(j\omega) = |C(j\omega)/R(\omega)|$, the closed-loop frequency response magnitude function.
2. $E(j\omega)/R(j\omega)$, the closed-loop error response function, and
3. $L(j\omega)$, the closed-loop compliance function.

We have already discussed and applied the first two of these criteria in Chaps. 8 and 9 so it will suffice to review their characteristics briefly here. In some applications the restrictions imposed upon the error response are so stringent that the control configuration must be modified to effect an acceptable solution. This problem is considered in Sec. 10.2. Compliance is analogous to the output impedance of electronic devices. $L(j\omega)$ is a measure of the steady-state error in response to a sinusoidal *load* disturbance. It should ideally be kept as small as possible at all frequencies to assure precise output control relative to the input reference signal. Compliance is considered further in Sec. 10.3.

It is not appropriate in specifying a design to define $M(j\omega)$ for all ω, since this would completely define the final closed-loop system characteristics (assuming a minimum phase realization) in both the frequency and the time domains, leaving essentially no design flexibility. It is common, however, to place limits upon certain characteristics of $M(j\omega)$, such as M_p, its peak value, ω_p, the frequency at which that peak value occurs, and the closed-loop bandwidth (BW) of the system, which for low pass controllers is defined as the frequency at which $M(j\omega)$ becomes $\sqrt{2}/2$ times as large as its value at $\omega = 0$. These criteria are illustrated in Fig. 10.1. It was shown in Sec. 8.4 how readily the $M(j\omega)$ function can be constructed from the Nyquist plot. Since M_p tends to increase as the margin of -1 point avoidance decreases, M_p is readily controlled by restricting θ_m and G_m, both of which are apparent from the Bode and Nyquist plots. In fact we showed in Chap. 8 that

$$M_p \simeq \frac{1}{\sin \theta_m} \tag{10.1}$$

is generally a good approximation in the unity feedback case. Thus design specs on M_p are easily taken into consideration during the design process by imposing suitable restrictions upon θ_m.

The value of ω_p is not often specified. M_p generally occurs near unit magnitude crossover on the Nyquist plot, so $\omega_p = \omega_c$ is the standard approximation. If more accurate values of M_p or ω_p are needed, the loci of constant M_p introduced in Sec. 8.4 may be used on a Nyquist plot to finalize the design.

Specification of closed-loop BW is fairly common. The closed-loop transfer function of the standard unity feedback system is

$$\frac{C(s)}{R(s)} = \frac{G(s)}{1 + G(s)} \tag{10.2}$$

which may be roughly approximated as

$$\frac{C(s)}{R(s)} = \begin{cases} 1 & |G(s)| \geq 1 \\ G(s) & |G(s)| < 1 \end{cases} \tag{10.3}$$

In this approximation $1 + G(s)$ is assumed equal to 1 for $|G(s)| < 1$ and equal to $G(s)$ for $|G(s)| \geq 1$. This approximation, rough as it may be, is relatively standard. Thus BW restrictions are easily imposed during the design phase for the unity feedback case. When $H(s) \neq 1$ the approximate value of BW is not so readily apparent, but direct application of the definition will always provide the answer.

It is standard procedure to impose design specs upon the steady-state sinusoidal error response of control systems. A typical spec might be worded as follows: "The steady state error in response to sinusoidal inputs at $\omega \leq 1$ must not exceed 1 percent of the input signal level." For the standard unity feedback controller illustrated in Fig. 10.2, we know that

$$\frac{\mathbf{E}(j\omega)}{\mathbf{R}(j\omega)} = \frac{1}{1 + \mathbf{G}(j\omega)} \tag{10.4}$$

Error not to exceed 1 percent requires $|\mathbf{E}(j\omega)/\mathbf{R}(j\omega)| \leq 0.01$. This requires, in turn, $|\mathbf{G}(j\omega)| \geq 100$ for all $\omega \leq 1$. This type of restriction is easily imposed directly upon the Bode diagram. The Bode diagram gain level and shape must be chosen such that $|\mathbf{G}(j\omega)|$ is acceptably large for $\omega \leq 1$. In most cases this causes no difficulty, as illustrated by the examples in Chap. 9. In some applications, however, it may seem impossible to satisfy all specs. Error restrictions

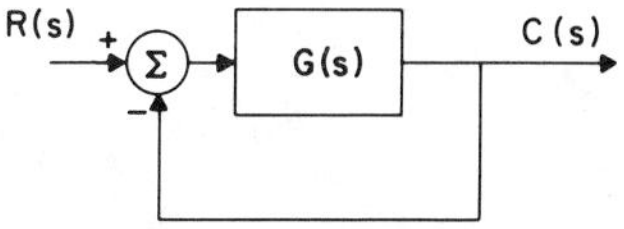

FIG. 10.2. Standard form of single-loop controller.

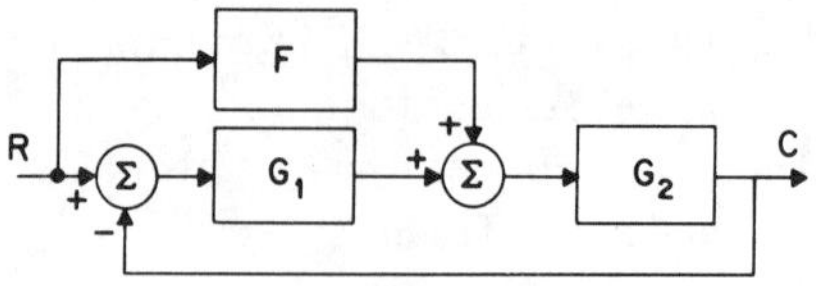

FIG. 10.3. The standard additive correction configuration.

may often be satisfied in such cases by using an alternate controller configuration in which the input signal is applied to the plant through two parallel paths. This method, called *additive correction*, or *feedforward design*, is considered in the next section.

10.2 ADDITIVE CORRECTION OR FEEDFORWARD DESIGN

Additive correction is a method for improving the low frequency error performance of a control system without requiring a gain increase, thereby eliminating certain stability difficulties. Additive correction involves a combination of open and closed-loop control, however, and is particularly sensitive to parameter changes, as are all open-loop controllers. It also requires two parallel signal paths and some circuit redundancy. Because of the unusual parameter drift sensitivity, additive correction is used only as a last resort to satisfy error specs when all other standard closed-loop design procedures have failed.

Consider the system configuration shown in Fig. 10.3. The additive correction element $F(s)$ has been appended to a standard unity feedback controller with forward elements $G_1(s)$ and $G_2(s)$ in series. The closed-loop transfer function, and thus the error characteristics as well, are influenced by $F(s)$ and the input signal it provides to $G_2(s)$. If $F(s)$ is properly chosen, the steady-state tracking capabilities of the system can often be markedly improved. Note in particular that the addition of $F(s)$ to the system introduces a second driving signal into the control loop, but does not alter the open-loop transfer function of the feedback loop involving G_1 and G_2. Thus the stability of the overall system is not affected by additive correction, assuming of course, that $F(s)$ is open-loop stable.

The influence of F upon C/R and E/R is easily determined by applying superposition to the system shown in Fig. 10.3. The output is the sum of two components, or

$$C = \frac{RG_1G_2}{1 + G_1G_2} + \frac{RFG_2}{1 + G_1G_2} \tag{10.5}$$

so

$$\frac{C}{R} = \frac{(G_1 + F)\,G_2}{1 + G_1\,G_2} \tag{10.6}$$

The error is the difference between input and output, or

$$E = R - C = R\left[1 - \frac{G_2(G_1 + F)}{1 + G_1G_2}\right] = R\left(\frac{1 - G_2F}{1 + G_1G_2}\right) \tag{10.7}$$

It is apparent from Eq. (10.7) that dynamic error could be eliminated completely if we set

$$1 - G_2F = 0 \qquad \text{or} \qquad F = \frac{1}{G_2} \tag{10.8}$$

This is obviously impossible, since G_2 is the transfer function of a particular device, and must approach zero as $s \to \infty$. The dynamic error can usually be made small over the frequency range of interest, however, by approximating the equality in Eq. (10.8) for all frequencies within that range. This is the principle upon which error improvement using additive correction is based. Without additive correction the dynamic error can only be reduced [see Eq. (10.7) with $F = 0$] by increasing the magnitude of G_1G_2 over the range of interest. Additive correction provides an alternate procedure for error reduction when it is impossible or undesirable to increase G_1G_2.

Let us consider carefully the practical aspects of additive correction. The additive signal, or the output of $F(s)$, is fed into the forward path at some arbitrary point shown as the junction between G_1 and G_2 in Fig. 10.3. Because it is required that $F = 1/G_2$ over the frequency range for which error improvement is to be obtained, it appears that G_2 should be chosen as simple as possible to facilitate realization of F. Pursuing this point further, we might be inclined to feed the additive signal forward all of the way to the output. In the limit we would be faced with the condition $F(s) = 1$. Although it appears that $F(s) = 1$ is easily realized, this cannot be true. If it were, the original design problem must be trivial and it would not be necessary to resort to additive correction to satisfy the error spec. Synthesis of F is one of the factors which must be considered during design, but it is not generally the most significant one.

As the additive point is moved forward, $G_1(s)$ includes more terms and $G_2(s)$ becomes simpler. Thus $1/G_2(s)$ becomes simpler from a strictly theoretical or transfer function point of view. This does not generally mean that $F(s)$ is more easily realized, however. As the signals move forward in a typical control loop, the power level generally increases rapidly. Consider as an example a radar antenna position controller using a motor rated at 15 horsepower, the voltage and power amplifiers required to drive the motor, and a compensation network to provide the desired response characteristics. It might be reasonable in this case to place the additive point after the equalizer and voltage amplifier, or perhaps even after the power amplifier, although this is much less likely. The additive signal must be provided at the same power level as the signal to which it is added, and this requires amplifier and network redundancy in G_1 and F as the additive point is moved forward. Signal addition is generally carried out at a point of relatively low power level, because it is not economically feasible to duplicate expensive power amplification equipment, and linear addition at high power levels is not easily accomplished.

Error reduction achieved using additive correction is very sensitive to parameter changes, as are all open-loop system designs. This limitation of open-loop system realizations is one of the major reasons for using closed-loop controllers. Suppose, for example, the steady-state error in a particular system is reduced significantly for $\omega \leq 1$ by setting $\mathbf{F}(j\omega) = 1/\mathbf{G}_2(j\omega)$ over the range $0 \leq \omega \leq 1$. If the parameters of either G_2 or F change, the error improvement obtained using additive correction is rapidly degraded. Significant changes in the characteristics of typical actuators over extended periods of use are not at all uncommon, so long-term performance improvement using additive correction is not particularly reliable, and adjustment of F during normal maintenance to compensate for such changes is often necessary to assure continued operation within specs. Since changes in F are critical, it is desirable to synthesize F using only passive elements whenever possible. R, L, and C values are much less apt to change during normal operation than is the gain of an amplifier, or the performance characteristics of other active devices.

The suggested additive correction design procedure involves the following steps:

1. Exhaust all reasonable design possibilities in an effort to satisfy the specs without additive correction before resorting to its use.
2. Establish a conventional feedback design which satisfies all specs other than the one on low frequency error, and which comes as close as possible to satisfying that spec.
3. Choose a point in the forward path where the corrective signal can be added.
4. Determine the desired $\mathbf{F}(j\omega) = 1/\mathbf{G}_2(j\omega)$, and the frequency range over which the equality must be nearly satisfied.
5. Choose an $\mathbf{F}(j\omega)$ which should satisfy the error spec.
6. Check the result using Eq. (10.7).

The following examples illustrate this procedure.

Example 10.1

Consider a position controller with motor transfer function

$$G_2(s) = \frac{5}{s(1 + s/10)} \tag{10.9}$$

Design a controller to satisfy the following specs:

1. $K_v = 100$.

2. The closed-loop BW is not to exceed 40 rad/sec.
3. There must be no segment of $-n$ slope, $n > 2$, for $\omega < \omega_c$.
4. $\theta \geq 45°$.
5. The steady-state sinusoidal error for inputs in the range $\omega \leq 5$ rad/sec is not to exceed 2 percent of the input signal amplitude.
6. Additive correction, if necessary, should be coupled in at the motor input.

The Bode diagram of a compensated system satisfying all specs except that on steady-state error is shown in Fig. 10.4. Without additive correction, the error at $\omega = 5$ is approximately 15 percent. The required system structure is shown in Fig. 10.5. Zero steady-state error requires

$$F^*(s) = \frac{1}{G_2(s)} = \frac{s(s/10 + 1)}{5} \tag{10.10}$$

where the * is used to denote that this transfer function is not realizable. An accurate approximation of $F^*(s)$ for $\omega \leq 5$ will satisfy the requirements of this

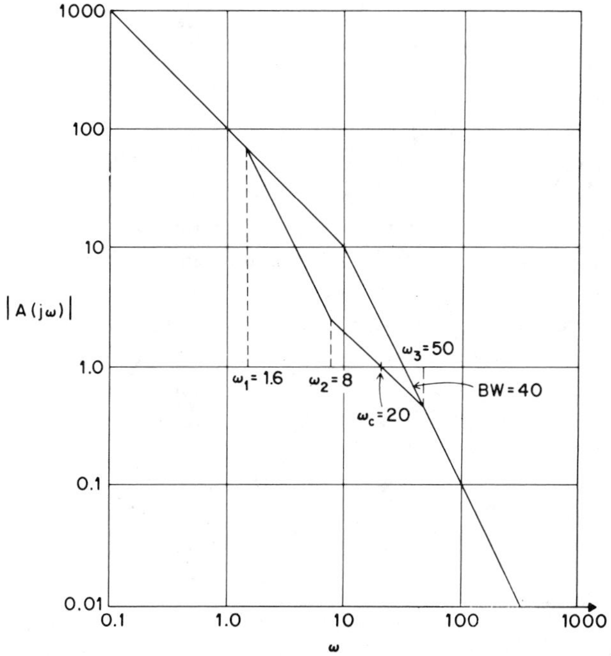

FIG. 10.4. Bode diagram of the compensated system introduced in Ex. 10.1.

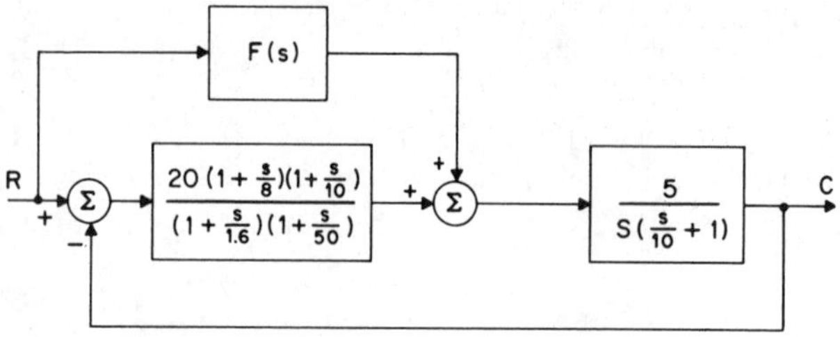

FIG. 10.5. Feedforward structure used in Ex. 10.1.

problem. Suppose we use

$$F(s) = \frac{s(s/10 + 1)}{5(s/100 + 1)^2} \tag{10.11}$$

which approximates $F^*(s)$ reasonably well. Checking the error at $\omega = 5$ using Eq. (10.7) provides

$$|1 + G_1G_2|_{\omega=5} = 7 \quad \text{and} \quad |1 - G_2F|_{\omega=5} = 0.1 \tag{10.12}$$

so the steady-state error at $\omega = 5$ is approximately 1.4 percent, which is well within the spec.

Note that $F(s)$ as given by Eq. (10.11) is a lead network. For steady-state sinusoidal response this network anticipates the error and drives the motor with a leading sinusoidal component such as to reduce that error significantly. Sudden changes in the input, such as a step function, result in relatively large motor drive signals, perhaps causing saturation to occur. This is a characteristic of all lead compensation methods which the designer must consider in defining the dynamic range over which the various components should operate.

The application of feedforward design to limit errors for sinusoidal inputs is seen from Ex. 10.1 to be straightforward. A feedforward signal may also be used to eliminate the steady-state error to a step input for systems with no integration in G_1G_2, to a ramp input for systems with one integration, or to an acceleration input for systems with two integrations. The procedure used is illustrated by the following example.

Example 10.2

Consider the system introduced in Ex. 10.1 with Bode diagram and configuration as shown in Figs. 10.4 and 10.5, respectively. Assume the ramp input $R(s) = v_0/s^2$. Determine the restrictions on $F(s)$ for which the steady-state ramp input error will be zero.

We have from Eq. (10.7) that

$$E(s) = \frac{v_0}{s^2}\left[\frac{1 - F(s)\dfrac{5}{s(1 + s/10)}}{1 + 100\dfrac{(1 + s/8)}{s(1 + s/1.6)(1 + s/50)}}\right] \tag{10.13}$$

and

$$\lim_{t\to\infty} e(t) = \lim_{s\to 0} sE(s) = v_0 \lim_{s\to 0}\left[\frac{1 - F(s)\dfrac{5}{s(1 + s/10)}}{s + 100\dfrac{(1 + s/8)}{(1 + s/1.6)(1 + s/50)}}\right] \tag{10.14}$$

The limit of the denominator term as $s \to 0$ is 100, so the final velocity error approaches zero only if

$$\lim_{s\to 0}\left[1 - F(s)\frac{5}{s(1 + s/10)}\right] = 0 \quad \text{or} \quad \lim_{s\to 0} F(s) = \frac{s}{5} \tag{10.15}$$

But this condition is satisfied by $F(s)$ as given in Eq. (10.11), so the solution obtained in Ex. 10.1 has zero velocity error. If the requirement $F(s) = 1/G_2(s)$ is exactly satisfied at $s = 0$, feedforward compensation reduces the steady-state step input error of a type zero (no integration) system to zero, the steady-state ramp input error of a type one system to zero, or the steady-state acceleration input error of a type two system to zero. This fringe benefit of additive correction is always obtained when the suggested design procedure is followed. It is never possible, no matter how we proceed, to reduce any error which approaches infinity as $t \to \infty$ to zero by using additive correction, since this would require infinite dynamic range.

In summary we emphasize that additive correction is an open-loop design procedure. Its performance is degraded rapidly by parameter changes. Although it reduces steady-state errors for certain types of inputs, it has no effect upon errors caused by load disturbances or noise introduced within the control loop. Its major advantage is error reduction without stability complications. Since the disadvantages far outweigh this advantage, its use is suggested only as a last resort when the error specs can be satisfied in no other way.

10.3 COMPLIANCE, A MEASURE OF DISTURBANCE SENSITIVITY

Thus far we have disregarded the effects of load disturbances or load changes upon controller performance. Since reduced sensitivity to load changes is one of the main reasons for using a feedback configuration, load effects deserve further consideration. The load on the actuator can be treated as an additional input and, since we are concerned with linear systems, the total response is readily obtained by superposition. The general situation is shown in Fig. 10.6, where Q is the load and Z_0 denotes the open-loop transfer characteristics between Q and C. The situation is described by

$$C(s) = A(s)E(s) + Q(s)Z_0(s) \tag{10.16}$$

Focusing our attention for the moment on the effect of Q upon C, denoted by C_q, we set $R = 0$. In this case

$$\frac{C_q(s)}{Q(s)} = \frac{Z_0(s)}{1 + A(s)} \equiv Z_1(s) \tag{10.17}$$

where $Z_1(s)$ denotes the closed-loop transfer characteristic between disturbance $Q(s)$ and the system output. It is obvious from Eq. (10.16) that $Z_0(s)$ may be obtained from

$$Z_0(s) = \left.\frac{C(s)}{Q(s)}\right|_{E(s)=0} \tag{10.18}$$

In other words $Z_0(s)$ is the output response to a disturbance impulse, $Q(s) = 1$, with $R(s) = 0$ and the feedback path broken. It is apparent from Eq. (10.17) that feedback reduces the load effect upon the output by the factor $1 + A(s)$. We have seen that $|1 + A(s)| >> 1$ for most controllers over the frequency range of interest, so load effects are generally reduced significantly by the addition of feedback. An example will be used to illustrate these basic principles.

Example 10.3

Consider the configuration shown in Fig. 10.7, consisting of an armature-controlled motor driven by a field-controlled generator. Find Z_0 for a load torque applied at the output shaft.

Since $Z_0(s)$ is the open-loop transfer function between $Q(s)$ and $C(s)$, which in this case is output shaft position $\theta(s)$, we may solve for $Z_0(s)$ by setting

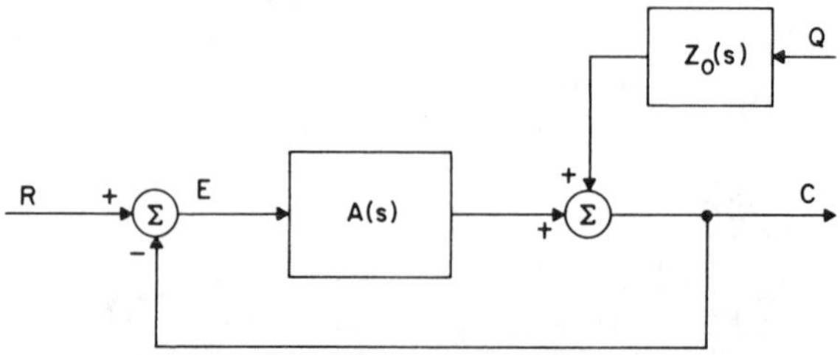

FIG. 10.6. Representation of a load disturbance.

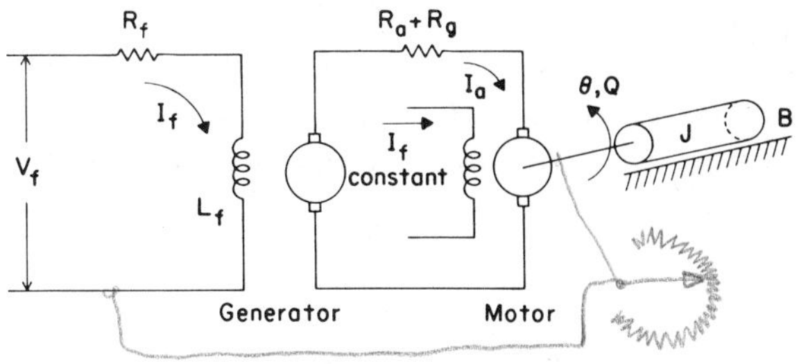

FIG. 10.7. Generator-driven motor used in Ex. 10.3.

$V_f(s) = 0$ and writing the corresponding equations for the system. The torque causes the output shaft to rotate, the motor acts as a generator which develops a voltage proportional to the shaft speed, and the generator armature acts as a load on the motor. We assume that there is no coupling from the generator armature back into its field circuit. Thus we have

$$Q(s) = Js^2\theta_q(s) + Bs\,\theta_q(s) + K_t I_a(s) \tag{10.19}$$

$$K_v s\theta_q(s) = (R_a + R_g)I_a(s) \tag{10.20}$$

Substituting $I_a(s)$ from Eq. (10.20) into Eq. (10.19) and solving for $\theta_q(s)/Q(s)$ provides

$$Z_0(s) = \frac{\theta_q(s)}{Q(s)} = \frac{K}{s(\tau_m s + 1)} \tag{10.21}$$

where

$$K = \frac{R_a + R_g}{K_t K_v + B(R_a + R_g)} \qquad \tau_m = \frac{J(R_a + R_g)}{B(R_a + R_g) + K_t K_v} \tag{10.22}$$

This example illustrates the usual procedure for determining load disturbance effects upon the system. The load is included as a general input in writing the equation describing the output element. The equations are solved for $Z_0(s)$ by setting the feedback signal to zero. A block diagram including $Z_0(s)$ as shown in Fig. 10.6 is set up and the system performance evaluated as desired either in response to $R(s)$ alone, $Q(s)$ alone, or any combination of the two. The time response to a unit disturbance step is readily obtained, the steady-state error caused by a constant load of magnitude Q_0 may be derived by applying the final value theorem and the frequency response for load inputs is also easily evaluated.

The effect of disturbance level Q_0 upon the system output obviously depends upon the type of system under study. A torque load of 1 ft-lb is relatively insignificant on the position controller for a large radar antenna, but would have an overwhelming effect upon a small instrument servo. *Compliance* is a more generally useful performance measure for position controllers obtained by normalizing Z_0 relative to the motor ratings. The normalized output change in response to a step function of torque is designated as the *dynamic compliance* $l(t)$, defined as

$$l(t) = \frac{c(t)/\dot{c}_{\text{max}}}{Q_0/Q_{\text{rated}}} = \frac{c(t)}{Q_0}\,\frac{Q_{\text{rated}}}{\dot{c}_{\text{max}}} \tag{10.23}$$

where $\dot{c}_{\text{max}}$ is the motor rated speed, Q_0 is the magnitude of the torque step, and Q_{rated} is the rated motor torque. The steady-state compliance frequency function is defined as

$$\mathbf{L}(j\omega) = \frac{\mathbf{C}(j\omega)/\dot{c}_{\text{max}}}{|\mathbf{Q}(j\omega)|/Q_{\text{rated}}} \tag{10.24}$$

where $|\mathbf{Q}(j\omega)|$ is the amplitude of the applied sinusoidal torque function. $l(t)$ and $\mathbf{L}(j\omega)$ relate to each other very much as the unit-step response and $\mathbf{L}(j\omega)$ are related for normal inputs. The steady-state compliance for a step input is denoted by $\mathbf{L}_s$, and is given by

$$L_s = \lim_{t\to\infty} l(t) = \lim_{\omega\to 0} \mathbf{L}(j\omega) \tag{10.25}$$

In general we want the compliance to be as small as possible, both dynamically and in steady-state, since this means the controller is relatively insensitive to load disturbances, or "stiff". A high gain feedback controller generally holds compliance well within bounds. Increasing loop gain, retaining a reasonable stability margin, decreases compliance and improves the controller's rejection of load disturbances.

One further point should be emphasized. The usual shaft position controller has an integration in the forward path due to the actuating motor, and thus has zero steady-state error for a unit-step input. In response to a unit-step of load torque, however, the steady-state error is *not zero*. This is easily seen from Fig. 10.6 and Eq. (10.21). A unit step at the input of Z_0 results, as $t \to \infty$, in a ramp as the driving signal into the control loop. The type 1 controller will have a finite nonzero final error for this input. The distinction here is that Z_0 contains the motor integration. Zero signal error for a load torque step input can thus only be achieved if we are dealing with a type 2 system.

Disturbances can occur at any point in a system. All of our previous comments apply equally well if the disturbance of interest appears at a general point in the loop, as shown in Fig. 10.8. The procedure for determining Z_0 and the loop response to disturbance input Q is identical to that previously discussed. The term compliance, however, is used only to characterize loop performance relative

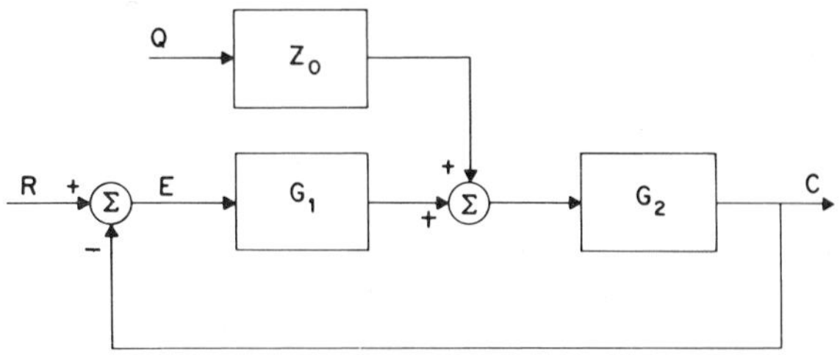

FIG. 10.8. More general disturbance input representation.

to *load* disturbances. The transfer relationship between C and Q for the arrangement shown in Fig. 10.8 is obviously

$$\frac{C_q(s)}{Q(s)} = \frac{Z_0(s)\,G_2(s)}{1 + G_1(s)\,G_2(s)} \tag{10.26}$$

This disturbance effect is also reduced as $|G_1(s)\,G_2(s)|$ is increased, assuming a reasonable stability margin is maintained. The effect of a disturbance spec upon design is thus similar to that of an error spec in limiting $|G_1(s)G_2(s)|$.

10.4 PERFORMANCE SPECS IN THE TIME DOMAIN

A number of performance specs often imposed in the time domain are illustrated by the closed-loop unit step response pattern shown in Fig. 10.9. We wish to relate these time domain characteristics in some general way to the Bode diagram so they may be taken into consideration to establish a tentative design. In most cases the exact relationships are so complex that their use is not feasible, but certain general rules of thumb can be established. A simulation study of the tentative design is suggested as a means for checking time response specs and making final adjustments.

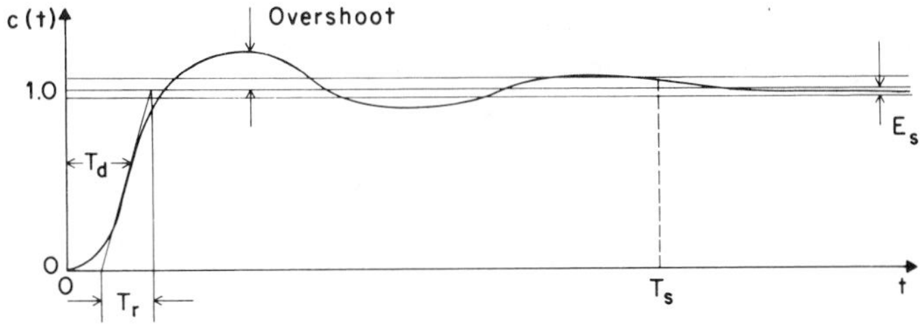

FIG. 10.9. Performance characteristics defined from the unit-step response.

Percent overshoot is defined as

$$\%\text{overshoot} = 100\left(\frac{\text{peak response} - \text{final value}}{\text{final value}}\right) \tag{10.27}$$

It is common design practice to require less than 30 to 50 percent overshoot. Although percent overshoot is readily expressed in terms of system parameters for second-order systems, this is not generally possible for higher-order cases. Percent overshoot is indirectly related to θ_m and G_m and generally decreases as either of these specs increases, although an explicit general relationship is impossible. Organized trial and error using analog or digital computation is the standard design approach.

Rise time T_r is defined as the time axis projection of a line tangent to the step response curve at its steepest point and extending from 0 to the final output level. A rough first-order approximation for T_r is given by

$$T_r \simeq \frac{1}{6} T_n \simeq \frac{2\pi}{6\,\text{BW}} \simeq \frac{\pi}{3\omega_c} \tag{10.28}$$

where BW is the closed-loop bandwidth in rad/sec, and ω_c is the Bode diagram unit magnitude crossover frequency. A spec on T_r is thus easily imposed on the Bode diagram by limiting BW.

Delay time T_d is the time it takes the step response output to rise to half its final value. T_d is seldom used as a design spec. It is typically 0.6 to 0.8 times T_r, although this ratio increases significantly with system order. Simulation study is the only reasonable measurement means.

Settling time T_s is defined as the time necessary for the response to settle and remain within ±5 percent of the final value. T_s is a general measure of the system damping factor as well as the response rate. Simulation is suggested for measurement.

Steady-state error E_s (for a unit step input) is a common constraint and is 0 for a unity feedback type n system, $n \geq 1$. E_s is $1/(K + 1)$, where K is the open-loop gain, for a unity feedback type 0 system. E_s is specified whenever a static correction accuracy is required, and it is equivalent to a direct constraint on K for type 0 systems. Thus E_s is easily imposed on the Bode diagram. The effects of the performance specs introduced up to this point upon the design process are summarized in Table 10.1 for handy reference.

The following example, taken from Eveleigh,[1] illustrates the analytical calculation of several of the performance specs we have discussed here, and serves as an introduction to the basic idea of optimal design. The optimal design concept is considered much more thoroughly in the referenced text.

Example 10.4

Given a standard unity feedback controller with forward transfer function $G(s) = K/s(s + 1)$, find that K which minimizes T_r and T_d while maintaining

Table 10.1. Frequency and time domain performance specs and their effects upon system design.

Spec	*Effect upon the design process*
M_p	$M_p \simeq \dfrac{1}{\sin\theta_m}$ M_p is treated as a near-equivalent spec upon θ_m.
ω_p	$\omega_p \simeq \omega_c$ ω_c is treated as a near-equivalent spec upon ω_c.
BW	Closed-loop BW is approximated by that ω at which the open-loop transfer function magnitude reaches $\sqrt{2}/2$ for the usual unity feedback low pass configuration. BW $\simeq \sqrt{2}\omega_c$.
low frequency error percentage	$\dfrac{E(j\omega)}{R(j\omega)} = \dfrac{1}{1 + G(j\omega)}$ for the unity feedback case. Thus a low frequency error spec is easily converted to a near-equivalent restriction upon $\lvert G(j\omega)\rvert$, or K. In some cases feedforward compensation is required to satisfy particularly stringent error specs.
compliance	$\dfrac{C_q(s)}{Q(s)} = \dfrac{Z_0(s)}{1 + A(s)}$ Frequency domain compliance restrictions are nearly equivalent to restrictions upon $\lvert A(j\omega)\rvert$. Dynamic compliance specs limit the actuators which can be used in terms of their maximum rated capabilities.
percent overshoot	Indirectly related to θ_m, G_m, and the general margin of −1 point avoidance. Trial and error design using a simulation is called for.
T_r	$T_r \simeq \dfrac{\pi}{3\omega_c}$ T_r converts to a near-equivalent spec on ω_c.
T_d	Typically $0.6T_r - 0.8T_r$. Relationship varies significantly with system order. Measure using simulation.
T_s	Given approximately by $3\tau_s$, where τ_s is the largest closed-loop time constant. Obtain τ_s from a root locus plot or measure T_s directly using a simulation.
E_s	$E_s = 1/(1 + K)$ for type 0 systems, and $E_s = 0$ for type n systems, $n \geq 1$. Thus equivalent to a gain spec for type 0 systems.

$\theta_m \geq 30°$. Determine all of the standard performance specs for the resulting design.

In general, T_r and T_d are decreased by increasing BW, or equivalently by increasing K. Thus we wish to find the maximum K providing $\theta_m = 30°$. A straightforward approach is based upon the Bode diagram with an auxiliary phase shift curve superimposed upon it as shown in Fig. 10.10. The asymptotic diagram is drawn for $K = 1$ and is readily adjusted as required to give the proper

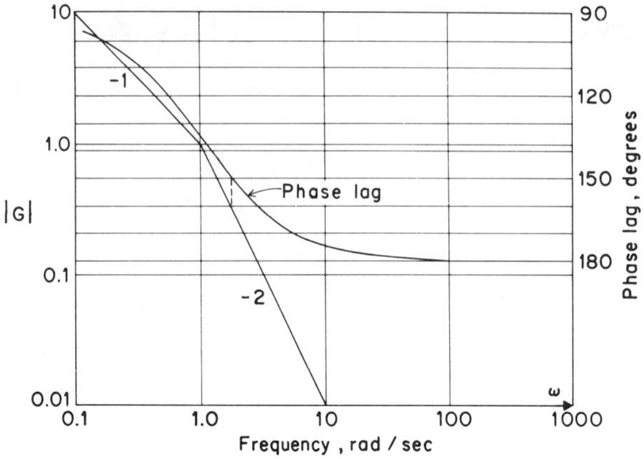

FIG. 10.10. Bode diagram of the system used in Ex. 10.4 with phase lag superimposed.

phase margin. At $\theta_m = 30°$ (phase lag = $150°$) corresponding to $\omega = 1.74$ rad/sec, the asymptotic open-loop transfer function magnitude is 0.32. Actually, $|\mathbf{G}(j1.74)| = 0.286$ if the proper correction is made. Thus $K = 1/0.286 = 3.5$ is the desired gain value. The following calculations are based upon $K = 3.5$.

It is apparent from Fig. 10.10 that G_m is infinite. Of course no system actually has infinite gain margin, and we mean only that stability is anticipated for extremely large gain values. For $K = 3.5$, BW = 2 rad/sec. The approximation for M_p is close in this case, so $M_p \simeq 1/\sin 30° = 2.0$. The peak frequency response occurs near crossover, so $\omega_p \simeq \omega_c = 1.75$ rad/sec. The steady-state step input error is $E_s = 0$, because this is a type 1 system.

It is considerably more difficult to determine the family of time domain specs. One method is to find the closed-loop transfer function, obtain its inverse transform, and determine, either analytically or graphically, the various criterion values. Although all results can be obtained analytically in this case (T_s causes some difficulty) a sketch of the step response is more generally applicable and is the method used here. Obtaining the unit-step time response is an involved procedure, especially in more complex systems. A simulation study is generally more appropriate.

The closed-loop transfer function for this system is

$$\frac{C(s)}{R(s)} = \frac{3.5}{(s + 1/2)^2 + 3.25} \tag{10.29}$$

The corresponding response for a unit step input at $t = 0$ is

$$c(t) = 1 + 1.04\epsilon^{-t/2} \sin\left(1.8t - \frac{74.5\pi}{180}\right) \tag{10.30}$$

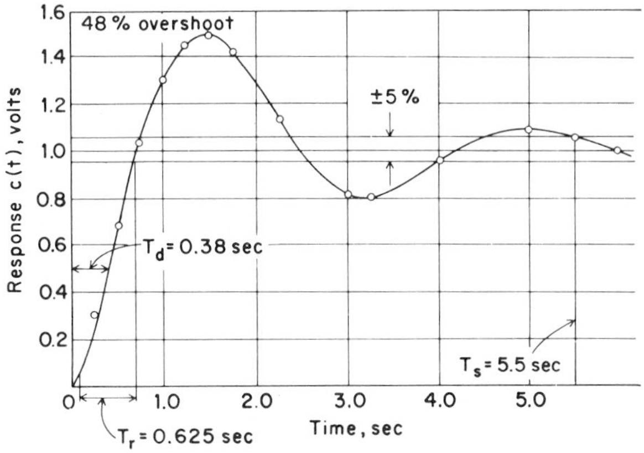

FIG. 10.11. Time domain performance characteristics for the system of Ex. 10.4.

as obtained from tables or by direct inverse transformation using residues. A sketch of $c(t)$ is shown in Fig. 10.11, where the values of T_d, T_r, T_s and percent overshoot are also illustrated. The overshoot is nearly 50 percent, a relatively large value, but this is not unexpected, since $\theta_m = 30^\circ$ is not particularly large. T_s is 5.5 sec., or nearly three times the major closed-loop time constant, as we should normally anticipate. From Eq. (10.28) we predict $T_r = 0.60$, as compared to the actual value of $T_r = 0.625$. $T_d = 0.38$, which is approximately 0.61 T_r, as anticipated.

10.5 STEADY-STATE ERRORS FOR SINGULARITY INPUTS (STEPS, RAMPS, ETC.)

In many applications the steady-state, closed-loop errors for the standard singularity function inputs are of particular significance. It is desirable that position controllers have zero final error for a fixed change in desired position (step input). In some applications the input approximates a ramp or a parabola over an extended period of time, and the steady-state errors for these inputs are also of considerable interest. The entire area of steady-state errors is easily handled using the *final value theorem,* and a brief discussion of the general concepts involved is presented here to indicate how these situations can be treated.

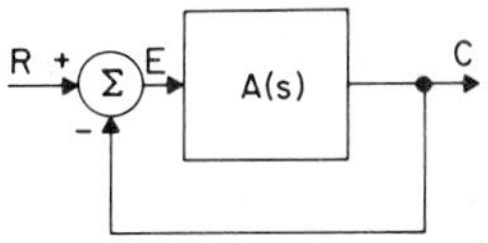

FIG. 10.12. A standard unity feedback controller.

Consider the standard unity feedback configuration shown in Fig. 10.12. The final value theorem provides

$$\lim_{t \to \infty} e(t) = \lim_{s \to 0} sE(s) = \lim_{s \to 0} s \frac{R(s)}{1 + A(s)} \tag{10.31}$$

This provides a basis for determining the steady-state error of arbitrary linear system structures to any transformable input. Consider the special case of a general stable type 0 system with

$$A(s) = \frac{K(1 + \alpha_1 s)(1 + \alpha_2 s) \ldots (1 + \alpha_m s)}{(1 + \beta_1 s)(1 + \beta_2 s) \ldots (1 + \beta_n s)} \tag{10.32}$$

where $n > m$, as required for realizability. Assume a unit-step input, or $R(s) = 1/s$. Then $\lim_{s \to 0} A(s) = K$, so

$$\lim_{t \to \infty} e(t) = 1/(1 + K) \tag{10.33}$$

For a unit ramp, $R(s) = 1/s^2$ and $e(t) \to \infty$ as $t \to \infty$. Similarly $e(t) \to \infty$ as $t \to \infty$ for all higher-order inputs.

Consider a general stable realizable type 1 system with

$$A(s) = \frac{K \prod_{i=1}^{m} (1 + \alpha_1 s)}{s \prod_{j=1}^{m} (1 + \beta_j s)} \tag{10.34}$$

For a unit-step input,

$$\lim_{t \to \infty} e(t) = \lim_{s \to 0} \frac{1}{1 + A(s)}$$

$$= \lim_{s \to 0} \frac{s \prod_{j=1}^{m} (1 + \beta_j s)}{s \prod_{j=1}^{m} (1 + \beta_j s) + K \prod_{i=1}^{m} (1 + \alpha_i s)} = 0 \tag{10.35}$$

Similar calculations show that $\lim_{t \to \infty} e(t) = 1/K$ for a ramp input to the type 1 system and all higher-order inputs result in infinite steady-state error. A summary of steady-state error results for stable unit feedback systems in response to

Table 10.2. Summary of steady-state error data for singularity inputs.

System type (integrations)	*Position error*	*Velocity error*	*Acceleration error*
0	$p_0/(1+K)$	∞	∞
1	0	v_0/K	∞
2	0	0	a_0/K
3	0	0	0

the standard singularity inputs is presented in Table 10.2. It is apparent that steady-state error, whenever finite, is inversely proportional to K and can be decreased in each case by increasing K. Thus steady-state error specs on singularity input responses are easily imposed during design.

In many cases a finite steady-state error to a singularity input can be reduced to zero if a nonunity feedback configuration is allowed. The following example provides a simple illustration.

Example 10.5

Consider the system shown in Fig. 10.13. What are the general conditions on $H(s)$ for zero position error if

$$G(s) = \frac{K}{1 + \tau s} \tag{10.36}$$

Since the error is defined as the input minus the output, it is not the signal applied to $G(s)$ in this case, but is given by

$$E(s) = R(s) - C(s) = R(s)\left[1 - \frac{G(s)}{1 + G(s)H(s)}\right] \tag{10.37}$$

The final error in response to input $R(s) = p_0/s$ is

$$\lim_{t \to \infty} e(t) = \lim_{s \to 0} \left[\frac{1 + G(s)H(s) - G(s)}{1 + G(s)H(s)}\right] \tag{10.38}$$

If the system is stable, all roots of the characteristic equation are in the lhp, so

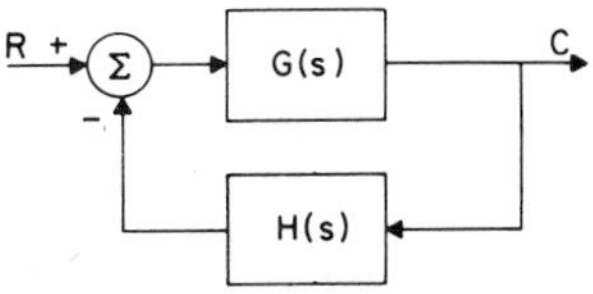

FIG. 10.13. The system discussed in Ex. 10.5.

$1 + G(s)H(s) \neq 0$ for $s = 0$. Zero position error is assured if the numerator of Eq. (10.38) approaches 0 as $s \to 0$, so

$$\lim_{s \to 0} [1 + G(s)H(s) - G(s)] = 0 \quad \text{or} \quad \lim_{s \to 0} H(s) = \lim_{s \to 0} \left[1 - \frac{1}{G(s)}\right] \tag{10.39}$$

For the unity feedback case $H(s) = 1$, so it is apparent from Eq. (10.39) that $\lim_{s \to 0} 1/G(s) = 0$ is required, and $G(s)$ must contain at least one integration. Substituting $G(s)$ as given by Eq. (10.36) into Eq. (10.39) shows that zero position error results if

$$H(0) = 1 - \frac{1}{K} \tag{10.40}$$

and any $H(s)$ with this characteristic at $s = 0$ and which provides closed-loop stability will do. $H(s)$ should be made as simple as possible to limit parameter drift effects if zero position error is to be obtained over a long period of use, so a precision component attenuator is particularly appropriate here.

The physical implications of this result are worth pursuing further. When $G(s)$ has no integrations, a constant input to $G(s)$ inversely proportional to K in level is required to hold the output at a desired level, and with unity feedback there must be some error between input and output to hold the output at a fixed level. If, however, only an appropriate part [see Eq. (10.40)] of the output is fed back, the signal applied to $G(s)$ is exactly that needed to maintain the output at the desired level when the error is zero. An alternate interpretation of Eq. (10.40) is illustrated in Fig. 10.14, where the feedback effects are broken down into a unity feedback loop and a loop with positive feedback $1/K$. The inner loop has transfer function

$$A(s) = \frac{K/(1 + \tau s)}{1 - 1/(1 + \tau s)} = \frac{K}{\tau s} \tag{10.41}$$

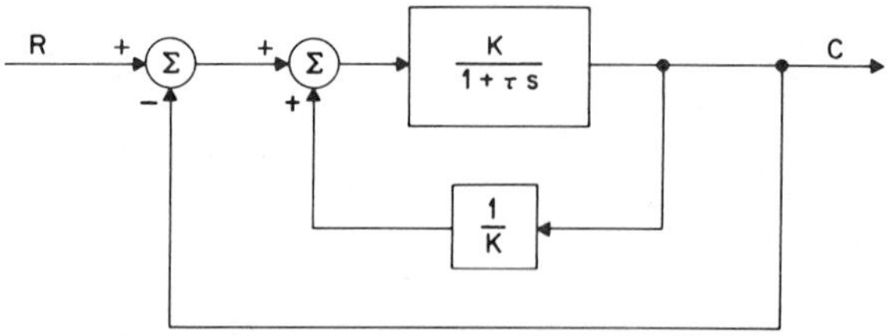

FIG. 10.14. Alternate representation of the nonunity feedback system of Ex. 10.5.

so subtracting $1/K$ from the unity feedback has the effect of replacing the original forward transfer function by a mathematically equivalent one with an integration in it. Of course this effect is only possible if the feedback gain is set precisely. The following example illustrates how this error reduction concept using non-unity feedback extends to systems with one or more integrations.

Example 10.6

Consider the configuration shown in Fig. 10.13 with

$$G(s) = \frac{K}{s(1 + \tau s)} \tag{10.42}$$

What conditions upon $H(s)$ assure zero final error for the velocity input $R(s) = v_0/s^2$?

The final error is

$$\lim_{t \to \infty} e(t) = \lim_{s \to 0} \frac{v_0}{s} \left\{ \frac{1 + \dfrac{K}{s(1 + \tau s)}[H(s) - 1]}{1 + \dfrac{KH(s)}{s(1 + \tau s)}} \right\}$$

$$= v_0 \lim_{s \to 0} \left\{ \frac{1 + \tau s + (K/s)[H(s) - 1)]}{s(1 + \tau s) + KH(s)} \right\} \tag{10.43}$$

Since the system must be stable, the denominator can have no root at $s = 0$, so the numerator must approach zero as $s \to 0$, and

$$\lim_{s \to 0} \left\{ 1 + \tau s + \frac{K}{s}[H(s) - 1] \right\} = 0 \tag{10.44}$$

Simplifying provides

$$\lim_{s \to 0} H(s) = 1 - \frac{s}{K} \tag{10.45}$$

Although $H(s) = 1 - s/K$ is unrealizable, any $H(s)$ with this limiting characteristic is satisfactory. Assume, for the present, that $H(s) = 1 - s/K$ is used. The system may be interpreted as shown in Fig. 10.14 with the inner loop feedback replaced by s/K. The forward transfer function becomes

$$A(s) = \frac{K/s(1 + \tau s)}{1 - 1/(1 + \tau s)} = \frac{K}{\tau s^2} \tag{10.46}$$

and the overall system is obviously unstable. It may be stabilized, however, by adding a lead network in the forward path. Since a lead network is of the form

$$G_L(s) = \frac{1 + Ts}{1 + \alpha Ts} \qquad \alpha < 1 \tag{10.47}$$

$\lim_{s \to 0} G_L(s) = 1$, and the conditions upon $H(s)$ for zero velocity error are not altered when a lead network is used. The required values of α and T in Eq. (10.47) are determined to provide a reasonable θ_m and acceptable Bode diagram crossover region characteristics in the usual way.

It is interesting to note that a simple lag term may be expanded in a power series as

$$\frac{1}{1 + \tau_1 s} = 1 - \tau_1 s + (\tau_1 s)^2 - (\tau_1 s)^3 + \cdots \quad |\tau_1 s| < 1 \tag{10.48}$$

Thus a lag term provides the limiting performance of Eq. (10.45) as $s \to 0$ if $\tau_1 = 1/K$ is chosen, and zero velocity error can also be obtained by adding *lag* in the feedback path. Although this may complicate the stability problem, equalization is still accomplished in the usual way.

These examples illustrate how steady-state errors can be controlled. The final value theorem is the basic tool used in all of our analyses. Any special situations encountered can be handled using the tools introduced here. Two factors should be kept in mind. It is never possible to reduce an infinite final error to a finite value by any of these procedures. Also, since the gain and/or one or more time constants in the feedback path in all applications depends upon the gain in the forward path, the compensation effect upon error is degraded by any changes in forward gain, and such changes will always occur. It is important to consider whether a reduction to zero error is worth the trouble when the error is (usually) already small, the improvement is imperfect, and the sensors used to observe the controlled characteristic (position, temperature, etc.) are also subject to error. Zero error performance is an idealistic objective never truly realized in practice.

It is apparent from the information discussed in this section that the steady-state error of a stable type 0 system to a step input is inversely proportional to K. Similar statements can be made regarding type 1 systems and ramp inputs, type 2 systems and acceleration inputs, etc. It is possible to extend these concepts, using the so called *error series approach,*[2] to develop analytical procedures for approximating the closed-loop response errors to specific inputs by breaking down those inputs into their components of position, velocity, etc. Unfortunately, the procedures are cumbersome, apply only to limited special situations, and indicate little about system performance that cannot be learned more easily. Errors are reduced by designing "tighter" controllers with more integrations, wider BW, higher gain, and faster settling time. When system response to complex special inputs is of interest, the most significant input components should be considered in deciding upon a "type" of system (0, 1, 2, . . .), the desired response rate in deciding upon BW, etc., but general principles usually provide an adequate design guide. A simulation study is suggested in those cases where further information is desired rather than embarking upon extensive analyses which often prove of limited value.

10.6 CONCLUDING COMMENTS

Control system performance specs are given in many forms. Since the fundamental design tool in most of our applications is the Bode diagram, frequency domain specs are easily imposed upon the design. Restrictions on steady-state error, θ_m, BW, M_p, and other similar characteristics can be imposed directly on the Bode plot. Time domain specs often can be roughly satisfied by controlling the general Bode plot characteristics. A preliminary design should be carried out on this basis and the time domain specs checked before final parameter choices are made. A simulation study is generally the most appropriate method for imposing time domain specs upon a system. Simulation on an analog or digital machine is straightforward (see Appendices C and E, respectively), computers are readily available and easily programmed, and the cost of a few minutes of computer running time is far less than an engineer's salary for hours of analysis. We should always consider using a computer when faced with an extensive analysis task. Its use has rapidly increased in recent years as availability has increased and cost decreased. The suggested problems cover the range from situations easily handled analytically to others where computer study is the only feasible way to obtain an answer.

REFERENCES

1. Eveleigh, V. W., "Adaptive Control and Optimization Techniques," McGraw-Hill, New York, pp. 44–45, 1967.
2. Bower, J. L. and P. M. Schultheiss, "Introduction to the Design of Servomechanisms," John Wiley & Sons, Inc., New York, 1958, Chap. 9.

11 MINOR LOOP DESIGN

Thus far we have accomplished controller equalization only by adding a series network with the desired frequency and phase characteristics at an appropriate point in the control loop. Equalization is also possible using one or more feedback loops, called minor loops, within the primary, or major, control loop. The standard design procedures based upon Bode diagram shaping near ω_c are essentially the same whether a series or minor-loop equalizer is to be used. Therefore the material in Chap. 9 provides a solid foundation for our discussion of minor loops and is assumed well understood by the reader. The point of departure between series and minor-loop procedures is reached only when we consider how the desired compensated Bode plot is to be achieved. The minor-loop analysis problem will be considered initially, after which a design procedure is developed by simply reversing the analysis steps.

It is logical to ask why minor loops are used. First of all, multiple-loop systems often arise in practice, and analysis methods must therefore be developed. More important, however, is the fact that several significant advantages are provided by a minor-loop system realization. Minor-loop feedback reduces the performance degradation caused by drift of those parameters enclosed by the feedback. Thus if the gain of an amplifier within the forward path of the minor loop tends to decrease with age, the minor-loop configuration provides a self-compensation effect. Similar advantages are obtained when minor-loop feedback is used around nonlinear elements where the incremental or small signal gain depends upon the operating point and the average gain depends upon excitation amplitude. The effects of noise and load disturbances arising within the minor loop are also reduced significantly. Noise, where this term is used at this point simply to denote any undesired signal component, arises in a variety of ways. It results from power supply ripple, from filament ripple, and shot effects when vacuum tubes are used, from resistor noise when signal levels are very low, and is often introduced as a by-product by actuators, sensors, and transducers. In applications where all signal levels are relatively high, noise is not usually a factor. In many applications, however, its effects must be considered, and minor-loop design offers potential advantages in such cases.

Load disturbances, as discussed in Sec. 10.3 where the concept of compliance is introduced, are a form of noise input to the system. There are many practical illustrations with which we are all familiar. Wind effects upon an antenna which our system is position-controlling will tend to modify its instantaneous position error. A "tight" system design will keep this effect small, and minor-loop designs tend to be tight in this sense. Variations in the current load drawn from a voltage-regulated power source will cause changes in the voltage output. Friction or other load torque variations on the output shaft of a position controller, such as an instrument servo used for remote display purposes or a television volume remote controller, are also load disturbances. A myriad of other examples could be used such as the influence of road surface variations upon power steering performance, the effects of ingot hardness and temperature changes upon the output thickness of a sheet metal rolling process, etc. Load disturbances are present in some form or another in almost all applications, or control would not be a problem, and in many cases the minimization of load disturbance effects upon closed-loop system performance is a significant factor in the design phase. If a design can be established in which minor-loop feedback is placed around the element where the disturbance originates, and this is generally possible, the resulting disturbance effects are often reduced significantly from the level which would result using the near-equivalent series-compensated design. When the practical advantages of minor-loop design are considered, minor-loop compensation is often used, and should no doubt be used even more.

11.1 ANALYSIS OF THE MINOR-LOOP CONFIGURATION

Analysis is always easier than synthesis. All references in this section are to Fig. 11.1 unless explicitly indicated to the contrary. The system shown is a standard position controller, and it will be used as a representative example. G_1 denotes the position transducer gain, G_2 is the transfer function of the power amplifier and generator, G_3 and G_4 are the time constant and integration, respectively, of the loaded position actuator (motor), and H is the transfer function of the minor-loop feedback network. In this minor-loop arrangement, the feedback element H is used for Bode diagram shaping in a manner entirely

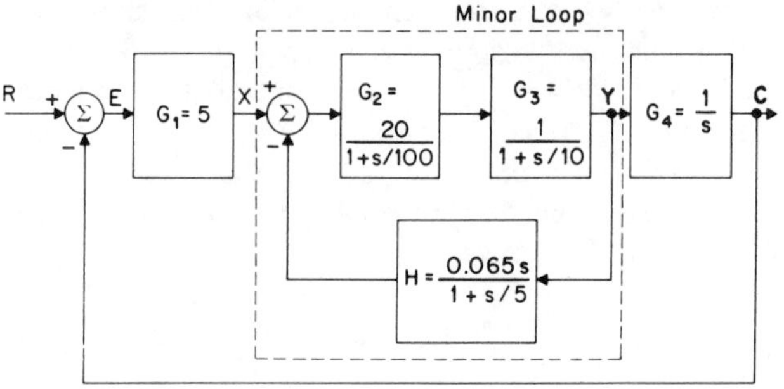

FIG. 11.1. The system used to introduce minor-loop analysis.

analogous to our previous application of series equalizers. The likenesses and differences between the two design methods are pursued thoroughly in this chapter.

Some definitions are in order. The transfer function of the forward path, disregarding H, or assuming the minor-loop feedback path open, is called A major uncorrected, denoted by A_{Mu}, and is given by

$$A_{Mu} = G_1 G_2 G_3 G_4 \tag{11.1}$$

The total transfer function around the minor loop is called A minor, is denoted by A_m, and in this case is given by

$$A_m = G_2 G_3 H \tag{11.2}$$

The forward transfer function between the points labeled X and Y, which is also the closed-loop minor-loop transfer function, is seen to be

$$\frac{Y}{X} = \frac{G_2 G_3}{1 + G_2 G_3 H} = \frac{G_2 G_3}{1 + A_m} \tag{11.3}$$

Thus the open-loop forward transfer function including the effect of minor-loop feedback, called the corrected major-loop transfer function, and denoted by A_{Mc}, is

$$A_{Mc} = \frac{G_1 G_2 G_3 G_4}{1 + G_2 G_3 H} = \frac{A_{Mu}}{1 + A_m} \tag{11.4}$$

The stability and response characteristics of the major loop are indicated by A_{Mc} in the usual way, assuming minor-loop stability, so our objective is to develop a procedure for constructing an asymptotic plot of A_{Mc}. In making this plot, we will use an approximation for A_{Mc} which is consistent with the rules established in Chap. 7 for constructing Bode plots for terms of the form $1 + \tau s$. Assume that

$$1 + A_m \simeq \begin{cases} 1 & |A_m| \leq 1 \\ A_m & |A_m| > 1 \end{cases} \tag{11.5}$$

This approximation is least exact in the region where $|A_m| \simeq 1$, but generally $|A_m| >> 1$ near the crossover point of A_{Mc}, and the use of Eq. (11.5) yields acceptably accurate results in most cases of interest. Since Bode diagram shape near ω_c is the most critical consideration in determining stability and general transient characteristics, use of the approximations in Eq. (11.5) generally provides an adequate indication of closed-loop response characteristics by direct interpretation of the A_{Mc} Bode diagram shape near ω_c. Subsequent examples illustrate the degree of approximation involved.

The Bode diagram analysis procedure for a minor-loop configuration can now be set down in an orderly fashion. We construct asymptotic plots of A_{Mu} and A_m on a common set of (preferably log-log) coordinates. The asymptotic plot of A_{Mc} is derived from these two plots using Eqs. (11.4) and (11.5). When $|A_m| \leq 1$, $A_{Mc} \simeq A_{Mu}$. When $|A_m| > 1$, A_{Mc} is approximated as the ratio A_{Mu}/A_m. The log of this ratio may be obtained by subtracting the log of A_m from that of A_{Mu}. An alternate interpretation is made possible by observing that the asymptotic A_{Mc} diagram slope is given at each point over the range where $|A_m| > 1$ by

$$n_c = n_u - n_m \tag{11.6}$$

where n_c, n_u and n_m denote the slopes of the compensated, uncompensated and minor loop asymptotic diagrams, respectively. We may rapidly construct A_{Mc} from A_{Mu} and A_m by application of Eq. (11.6). A complete set of Bode diagrams for the system shown in Fig. 11.1 is provided in Fig. 11.2. Note in particular that $|A_m| \simeq 4$ at ω_c, thereby apparently justifying our use of Eq. (11.5).

The steps required in constructing and using A_{Mc} are summarized here for emphasis:

1. Construct the asymptotic A_{Mu} diagram.
2. Construct the asymptotic A_m diagram.
3. Construct A_{Mc} from A_{Mu} and A_m by using Eq. (11.6) to determine the slope of A_{Mc} at all points over the interval where $|A_m| > 1$.
4. Interpret A_{Mc} as it relates to closed-loop performance.

Let us now apply these steps to derive the Bode diagrams shown in Fig. 11.2. The asymptotic A_{Mu} diagram is constructed for the function

$$A_{Mu} = G_1G_2G_3G_4 = \frac{100}{s(1 + s/10)(1 + s/100)} \tag{11.7}$$

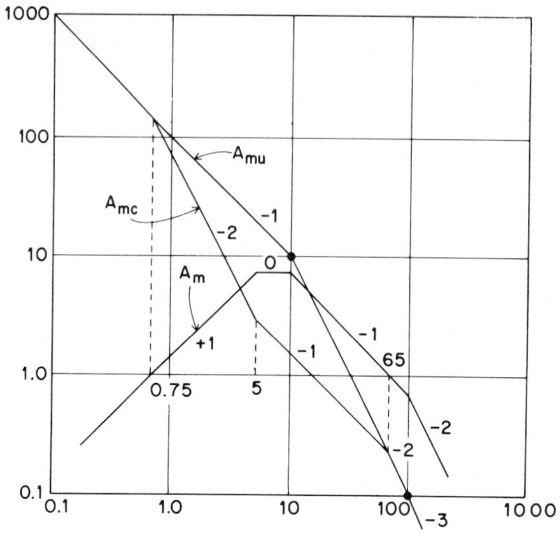

FIG. 11.2. Bode diagrams for the system shown in Fig. 11.1.

Obviously without some form of compensation the closed-loop system would be highly unsatisfactory, since $\theta_m \simeq 0$ for A_{Mu}. The open-loop minor-loop transfer function is

$$A_m = G_2 G_3 H = \frac{1.3s}{(1 + s/5)(1 + s/10)(1 + s/100)} \tag{11.8}$$

A_m passes through unit magnitude in an increasing direction at $\omega = 0.77$ and at that point the A_{Mc} asymptotic diagram breaks away from A_{Mu}, changing from -1 to -2 slope as seen from Eq. (11.6). At $\omega = 5.0$ the slope of A_m becomes 0 and the slope of A_{Mu} remains unchanged at -1, so the slope of A_{Mc} returns to -1. At $\omega = 10$, the slope of A_{Mu} changes to -2 and the slope of A_m changes to -1. These two effects compensate for each other, and the A_{Mc} slope remains at -1. A_m passes through unity in a negative direction at $\omega = 65$, so A_{Mc} coincides to A_{Mu} for $\omega \geq 65$. Thus the construction of A_{Mc} is straightforward for the assumptions made. We simply check for changes of slopes on the A_{Mc} diagram at the points where $|A_m| = 1$ and at all break frequencies of A_{Mu} and A_m on the interval where $|A_m| > 1$. (Although it is conceivable that $|A_m| > 1$ over two or more spans of frequency, this situation is not often encountered.)

We have assumed minor-loop stability in this development. Although analysis situations are occasionally encountered where this is not true, we are fundamentally concerned here with the design problem. In designing minor-loop systems, we will virtually always choose the minor-loop elements to assure stability. The Nyquist stability criterion or some equivalent basic procedure should be used for the analysis of systems with unstable minor loops.

Suppose we check the validity of the assumption given in Eq. (11.5). The actual Y/X transfer function through the minor loop, as given by Eq. (11.3) is

$$\frac{Y}{X} = \frac{20{,}000(s + 5)}{s^3 + 115s^2 + 8050s + 5000} \tag{11.9}$$

where the usual time constant form is abandoned for simplicity in factoring the denominator polynomial. Factoring provides the dominator roots as approximately $s = -0.625$, $s = -57 + j68$, and $s = -57 - j68$. The values used in constructing A_{Mc} based upon Eq. (11.5) were $s = -0.75$, $s = -65$, and $s = -100$. The values $s = -0.625$ and $s = -0.75$ are very close. The complex pole pair at $s = 57 \pm j68$ provides a double break on the actual asymptotic diagram at $\omega \simeq 90 \simeq \sqrt{57^2 + 68^2}$, instead of the individual breaks at $\omega = 65$ and $\omega = 100$ used in the approximation, but this difference is also of little consequence. This situation is typical. The maximum $|A_m| \simeq 6$ and $|A_m| \simeq 4$ at $\omega = \omega_c$. Since the approximate and exact A_{Mc} asymptotic diagrams are so nearly identical in this case, we will generally use the approximation suggested in Eq. (11.5) without testing the results, although it should be kept in mind that minor differences in closed-loop response will exist between the equivalent series and minor-loop-equalized systems. The approximation is usually excellent when the maximum $|A_m|$ is 5 or more, and adequate as a first approximation even for maximum levels down to 2 or 3.

Several characteristics of minor-loop configurations should be noted. The synthesis of an appropriate H can be used to shape A_{Mc} down below A_{Mu}, but A_{Mc} can never be pushed above A_{Mu} using minor-loop feedback alone. In most cases $|A_m|$ approaches zero as $\omega \to 0$ and $|A_m|$ must always approach zero as $\omega \to \infty$ for realizability. Thus the minor loop typically shapes A_{Mu} downward starting from the frequency ω_1 where $|A_m|$ increases through unity, provides the desired A_{Mc} shape near ω_c, and reunites with A_{Mu} at $\omega_2 > \omega_c$ where $|A_m|$ decreases through unity. In our example, $\omega_1 \simeq 0.75$ and $\omega_2 \simeq 65$. Minor-loop equalization is almost directly analogous to a lag-lead cascade equalizer in which the lag and lead spans are equal. Any equalization which requires shaping A_{Mu} upward to provide A_{Mc}, which can be accomplished using a series lead network and additional gain, can *never* be achieved by minor-loop methods alone. Thus lead compensation, if necessary, must be added in the major-loop forward path.

When $|A_m| > 1$ we say the minor loop is "in control," and the transfer function between input and output of the minor loop is governed by the elements in H. Minor-loop forward path time constants in the frequency range where $|A_m| > 1$ do not cause slope changes in A_{Mc}, since the slopes of A_{Mu} and A_m both change at all such points cancelling each other in their effect upon A_{Mc}. Similarly, lag and lead breaks in the minor-loop feedback path appear as lead and lag breaks in A_{Mc}, respectively [see Eq. (11.6)], in the range where $|A_m| > 1$.

11.2 SYNTHESIS OF MINOR-LOOP EQUALIZERS

Equation (11.6) provides a straightforward basis for minor-loop design. In fact the design process is identical to that for series equalizers up to the point where the compensation element is to be realized. The Bode diagram of A_{Mu} is drawn and all performance specs indicated on it. The desired A_{Mc} is then constructed to satisfy all specs. The method by which A_{Mc} is to be realized is not considered up to this point other than to tend toward an A_{Mc} which falls entirely below A_{Mu} and returns to A_{Mu} for all $\omega >> \omega_c$. The forward path elements around which H is to be placed are chosen, and a Bode diagram of the required H determined from A_{Mu} and A_{Mc} by applying Eq. (11.6). Modifications are made in the original design if this appears necessary, as for example if the resulting minor loop has less stability margin than desired.

Example 11.1

Consider an angular position controller with the transfer function elements shown in Fig. 11.3. This is essentially the same system considered in the previous section, but we are concerned with the synthesis problem now, and the specs imposed will require the resulting solution to differ significantly from that used previously. For the present, assume the decision to place H around the two lag terms with break frequencies at $\omega = 10$ and $\omega = 100$ is part of the problem statement. The factors which must be considered in choosing the location of H are discussed later. The following specs must be satisfied:

1. Zero steady-state position error.
2. $K_v = 200 \text{ sec}^{-1}$.
3. $\theta_m \geq 45°$.
4. Minor-loop compensation in the form shown in Fig. 11.3 must be used.
5. Steady-state low frequency error performance is to be degraded no more than necessary.

Each of these restrictions is similar in form and effect to specs considered in the design of series-compensated controllers in Chap. 9. The procedures used to satisfy them in the minor-loop case are essentially identical to those with which

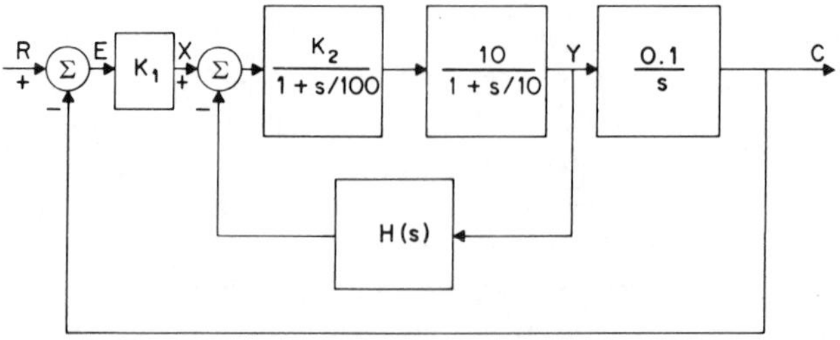

FIG. 11.3. Block diagram of the system used in Ex. 11.1.

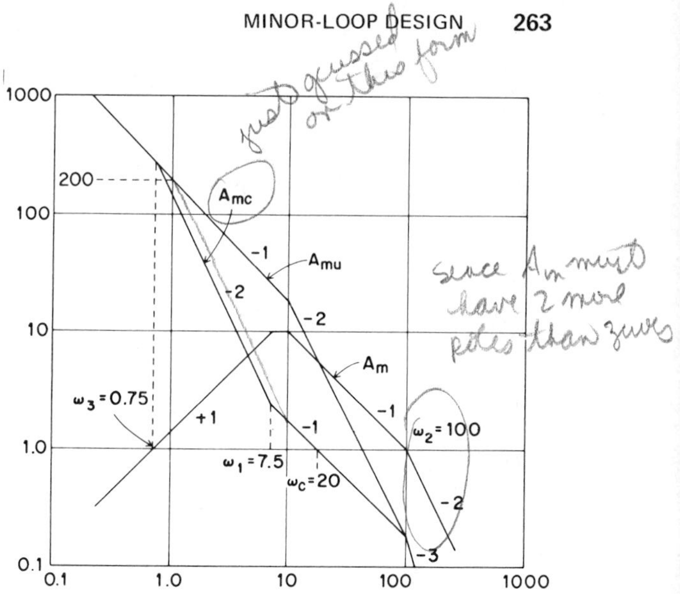

FIG. 11.4. Bode diagrams for the system used in Ex. 11.1.

we are already familiar. Before proceeding with the Bode diagram construction and design, let us interpret the implications of each spec in detail. Zero steady-state position error requires that A_{Mc}, the compensated major-loop transfer function, must have at least 1 integration, and thereby must have -1 slope for all low frequencies. Thus the minor loop must not eliminate the actuator (motor) integration effect by introducing the equivalent of forward path differentiation. We will construct our solution such that A_{Mc} and A_{Mu} are identical for all low frequencies to assure that this condition is satisfied. The condition that $K_v = 200$ is equivalent to demanding retention of a single forward path integration plus requiring $K_1K_2 = 200$, where K_1 and K_2 are the gain components shown in Fig. 11.3. The phase margin spec determines the -1 slope span required near ω_c. The minor-loop compensation restriction rules out any form of series or feed-forward equalization, thereby fixing the controller configuration. The low frequency error spec requires that the A_{Mc} diagram have maximum value at all frequencies. Thus we must choose the optimal shape of A_{Mc} near $\omega = \omega_c$.

The design process is initiated by constructing a Bode plot of

$$A_{Mu} = \frac{K_1K_2}{s(1 + s/10)(1 + s/100)} = \frac{200}{s(1 + s/10)(1 + s/100)} \tag{11.10}$$

as shown in Fig. 11.4, where the required A_{Mc} and A_m are also shown. Even before starting to construct A_{Mc}, we know that it will no doubt consist of a sequence of straight line segments with slopes of -1, -2, -1 and -3 in that order as ω increases from zero toward infinity. There is some question about whether the slope above ω_c changes in one step from -1 to -3, or whether there is a short -2 slope segment between the -1 and -3 slope regions. Even if a -2 slope segment appears in that region, however, it will be short, so our analysis is based upon the assumption of a -2, -1, -3 slope profile near ω_c. The resulting solution justifies this assumption as illustrated by the final A_{Mc} shown in Fig. 11.4. We know from the solution to Prob. 9.5 that

$$\omega_c = \sqrt{\frac{\omega_1\omega_2}{2}} \tag{11.11}$$

is required to obtain maximum θ_m for a given -1 slope span near ω_c for the -2, -1, -3 slope profile, ω_1 and ω_2 (see Fig. 11.4) denoting the frequencies where the slope changes from -2 to -1 and from -1 to -3, respectively. Let γ denote the -1 slope span, or

$$\omega_2 = \gamma\omega_1 \tag{11.12}$$

From Eqs. (11.11) and (11.12) we find that the -1 slope span should be divided up as

$$\omega_c = \sqrt{\frac{\gamma}{2}}\,\omega_1 \qquad \text{and} \qquad \omega_2 = \sqrt{2\gamma}\,\omega_c \tag{11.13}$$

The arc tan approximation for phase shift can be applied to the assumed Bode diagram configuration to solve for γ such that $\theta_m > 45°$. We will disregard the break at ω_3, the frequency where A_{Mc} departs from A_{Mu}, thereby assuming a -2 slope for all $\omega < \omega_1$. This simplification is conservative and will assure satisfying the θ_m spec with a small error margin. Setting $\theta_m = 45°$ provides

$$\frac{\pi}{4} = \frac{\pi}{2} - \tan^{-1}\frac{\omega_1}{\omega_2} - 2\tan^{-1}\frac{\omega_c}{\omega_2} \tag{11.14}$$

Applying Eqs. (11.12), (11.13) and the arc tan approximation to Eq. (11.14) yields

$$\sqrt{\frac{\gamma}{2}} = \frac{8}{\pi} \qquad \text{or} \qquad \gamma \simeq 13 \tag{11.15}$$

Thus we should use [see Eq. (11.13)]

$$\frac{\omega_2}{\omega_c} = \sqrt{26} \simeq 5.1 \qquad \text{and} \qquad \frac{\omega_c}{\omega_1} = \sqrt{6.5} \simeq 2.55 \tag{11.16}$$

It is easy to position the -1 slope segment of A_{Mc} using a symmetrical 45° right triangle such that the ratio of ω_c to ω_2 is 5 to 1. This results in $\omega_c = 20$ and

$\omega_2 = 100$. Since $100/13 \simeq 7.7$, $\omega_1 = 7.5$ will be used. Passing a -2 slope segment through $\omega_1 = 7.5$ results in $\omega_3 = 0.75$ and completes the determination of A_{Mc}. Although analytical methods are used here, a graphical solution by trial and error readily yields essentially equivalent results.

A Bode diagram of the A_m required to realize A_{Mc} can be constructed directly from A_{Mu} and A_{Mc}. A_m must pass through unit magnitude at $\omega_3 = 0.75$ with a +1 slope to cause A_{Mc} to depart from A_{Mu} at that frequency. Since the slopes of A_{Mu} and A_{Mc} remain constant thereafter until $\omega = 7.5$, the slope of A_m must also remain +1 until that frequency is reached. At $\omega = \omega_1 = 7.5$ the slope of A_{Mu} does not change, but the A_{Mc} slope becomes -1. This requires that A_m assume 0 slope until the next break frequency on either A_{Mu} or A_{Mc} is reached. At $\omega = 10$ the A_{Mu} slope changes to -2, so the A_m slope must change to -1 at $\omega = 10$ to compensate for this effect and allow A_{Mc} to retain a -1 slope. At $\omega_2 = 100$, $|A_m| = 1$, and $A_{Mc} = A_{Mu}$ for all $\omega \geq 100$. Although the A_m slope changes to -2 at $\omega_2 = 100$, this does not contribute to the shape of A_{Mc} because $|A_m(100)| = 1$. The shape of A_m for $|A_m| < 1$ is of no consequence in defining the asymptotic A_{Mc}, although it obviously influences minor-loop stability margin, and thereby imposes secondary effects upon the major-loop performance.

The desired A_m transfer function is obtained directly from the Bode diagram in Fig. 11.4 as

$$A_m = \frac{1.3s}{(1 + s/7.5)(1 + s/10)(1 + s/100)} \tag{11.17}$$

We observe that the minor loop is comfortably stable, with a phase margin in excess of 45°. The required H is obtained by equating A_m as given by Eq. (11.17) to A_m as obtained from Fig. 11.3, or

$$\frac{10K_2H}{(1 + s/10)(1 + s/100)} = \frac{1.3s}{(1 + s/7.5)(1 + s/10)(1 + s/100)} \tag{11.18}$$

Solving Eq. (11.18) for H provides

$$H = \frac{1.3s}{10K_2(1 + s/7.5)} \tag{11.19}$$

H should be made passive if possible, for reasons developed later. We know that

$$\frac{E_0}{E_i} = \frac{s/7.5}{1 + s/7.5} \tag{11.20}$$

is the transfer function of the series RC network shown in Fig. 11.5. Assume $k(E_0/E_i)$ is used for H, which requires

$$\frac{1.3}{10K_2} = \frac{k}{7.5} \quad \text{or} \quad K_2 = \frac{0.975}{k} \tag{11.21}$$

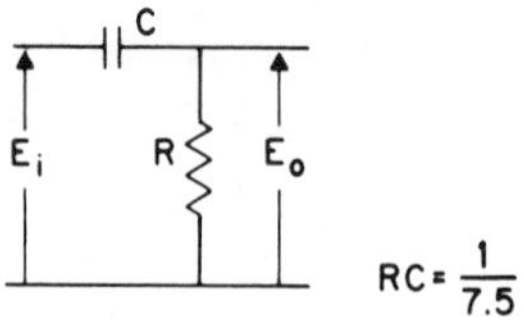

FIG. 11.5. Series RC network used for H in Ex. 11.1.

if $k = 0.975$, a value easily realized by tapping the output off across only $0.975\,R$, $K_2 = 1$ results. Then

$$H = \frac{0.13s}{1 + s/7.5} \tag{11.22}$$

and the $K_v = 200$ spec requires

$$K_1K_2 = K_1 = 200 \tag{11.23}$$

This completes the synthesis procedure, since all of the component characteristics have been defined in an arrangement which should easily satisfy all design specs.

The minor-loop design developed in Ex. 11.1 should be interpreted as a preliminary approximation. If we solve for the actual minor-loop transfer function pole location and plot the corresponding A_{Mc}, the results will differ somewhat from the approximate diagrams shown in Fig. 11.4. Since $|A_m| = 5$ at ω_c and reaches a maximum value of 10 near $\omega = 10$, however, the approximations used are very good in this case. If more precise results are necessary, the minor-loop pole locations can be obtained, and final adjustments can be made upon this more exact representation. A simulation study should be carried out if time domain performance is of interest. Although such further study requires additional time and effort, the methods required are straightforward and similar to those used in series equalizer design.

Several factors must be considered in deciding how the various gain components should be allotted between K_1, K_2 and H. In most applications, the gain level of A_{Mu} is limited by a K_v spec or a low frequency error spec. It is usually necessary, in order to satisfy such specs, to add gain in the forward path above and beyond that provided by the error sensor and actuator transfer functions. This gain should be divided between K_1 and K_2 as required by other considerations. The standard minor-loop design procedure results in a Bode diagram description of the required A_m for $|A_m| > 1$, with freedom retained to shape A_m outside this region to obtain the desired minor-loop stability characteristics. The gain required in A_m may be placed either in K_2 or in H. As the gain allotted to H increases, however, K_2 decreases proportionately, and K_1 increases proportionately. When the point is reached where increasing the gain in H requires attenuation in the A_m forward path, it is obviously undesirable to proceed further. It is not desirable for H to be active (contain an amplifier), since changes of parameters in H are not self-compensating, and active networks are much more subject to parameter drift than passive networks. Although signal and noise levels may require some minimum H gain level, we should not go out of our way to increase the feedback gain. Since it is about equally difficult to realize a gain of 1.25 or 12.5, gain levels should be established such that amplifier stages are efficiently used whenever such freedom exists.

In Ex. 11.1, we chose the feedback gain as $K = 0.975$ rather than unity to avoid the need for forward path signal attenuation. This results in $K_1 = 200$. If $k = 1$ is used $K_1 = 205$ is required. Although this difference is not particularly significant, it does not make sense to amplify signals and then attenuate them again in a supposedly well-designed controller.

Note that the A_{Mc} shown in Fig. 11.4 is also easily realized using a cascade equalizer and the methods discussed in Chap. 9. There are several advantages to a minor-loop realization, however, some of which we are in position to observe at this point. Additional advantages relative to parameter variations, nonlinearities, noise and load disturbances are given detailed consideration in subsequent sections. A lag-lead series equalizer with transfer function

$$G_e(s) = \frac{(1 + s/7.5)(1 + s/10)}{1 + s/0.75)(1 + s/100)} \tag{11.24}$$

would provide the same asymptotic A_{Mc}. This equalizer is more complex than H. $G_e(s)$ also has its lowest break frequency at $\omega = 0.75$, and the large RC product thereby required in $G_e(s)$ is difficult and expensive to realize. Large time constants result in slow filter transient response, and linear operation of such systems can only be obtained during periods of vigorous control activity if all system elements have relatively large dynamic ranges. Saturation tendencies are enhanced by the sluggish filter response. H, on the other hand, has its lowest break frequency at $\omega = \omega_1 = 7.5$, a value more readily achieved with reasonable sizes and impedance levels, and the minor-loop response is enough faster than that of $G_e(s)$ to significantly reduce the dynamic range problem. The unit step responses of the minor loop and equivalent cascade equalized systems for Ex. 11.1 are shown in Fig. 11.6. The comparison is typical, with the minor-loop system providing a somewhat less oscillatory response pattern.

Consider the question of minor-loop stability margin. The minor loop used in Ex. 11.1, with Bode diagram as shown in Fig. 11.4, has $\theta_m \simeq 45°$ and should perform satisfactorily. It is typical for the minor loop, as in Ex. 11.1, to have an ω_c exceeding that of the compensated major loop by a factor of 3 or 4 to 1. Thus minor-loop transients generally decay much more rapidly than those in the major loop. Even when the minor-loop θ_m is small, and its response is lightly damped, the larger minor-loop BW results in relatively fast settling time. A rapid, lightly-damped minor-loop response should not adversely affect major-loop performance, since the major loop, with its smaller BW, is unable to follow the oscillatory response pattern of the minor loop. Thus the minor-loop phase margin need not generally be as large as we strive to attain in the major loop, although some reasonable stability margin is necessary to keep the minor loop from becoming unstable if some parameter, such as a time constant or gain level, changes somewhat. Vigorous minor-loop oscillations in response to transient inputs may cause undue stress upon circuit components, perhaps increasing the failure rate. The minor-loop M_p should not be so large as to cause the actual A_{Mc} curve near the point where A_{Mc} and A_{Mu} reunite (where $|A_m| = 1$; see Fig. 11.4) to peak up to unit magnitude, as that would cause significant major-loop response characteristic differences from those we have anticipated, perhaps even resulting in major-loop instability. With all of these factors in mind, however, a minor-loop phase margin of 20° to 30° is generally adequate.

Suppose we need to improve the minor-loop stability margin by adding an equalizer network. Should the equalizer be placed in the forward path, or in series with H? This question is best answered by considering a specific situation such as that shown in Fig. 11.7. For this system A_m crosses unit magnitude at $\omega = 180$ with a -2 slope, the minor-loop phase margin is about 30°, and we might consider taking some steps to improve it. Assume that a 3:1 lead span is to be added in the minor loop starting at $\omega = 100$. This modifies A_m as indicated by the dashed line extension. Placing the lead network in the feedback path of A_m results in the A_{Mc} denoted by (a) on the diagram, and the shape of A_{Mu} does not change. When the lead is placed in the forward path of A_m, both A_m and A_{Mu} are modified as denoted by (b) on the diagram. Placing the lead in the forward path of A_m generally results in improvements

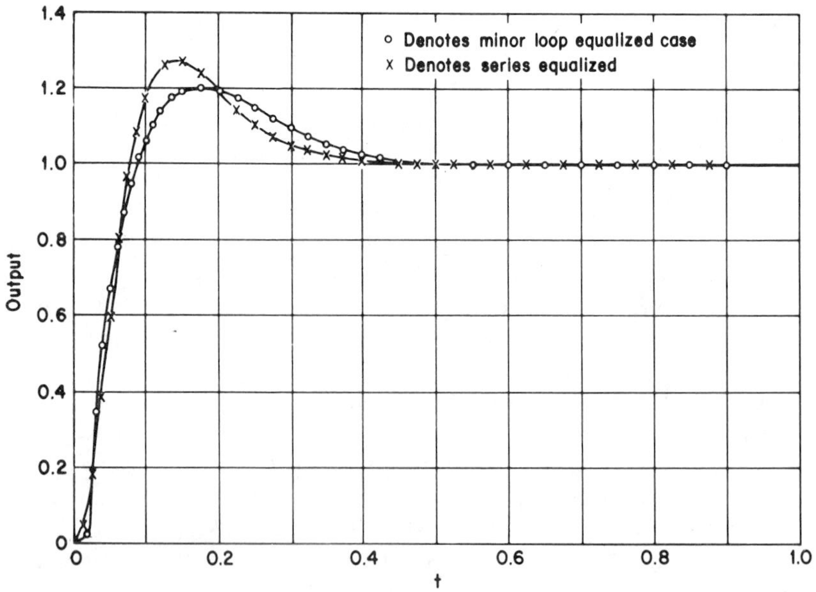

FIG. 11.6. Comparison of unit-step responses of minor-loop and equivalent cascade-equalized systems with Bode diagrams as shown in Fig. 11.4.

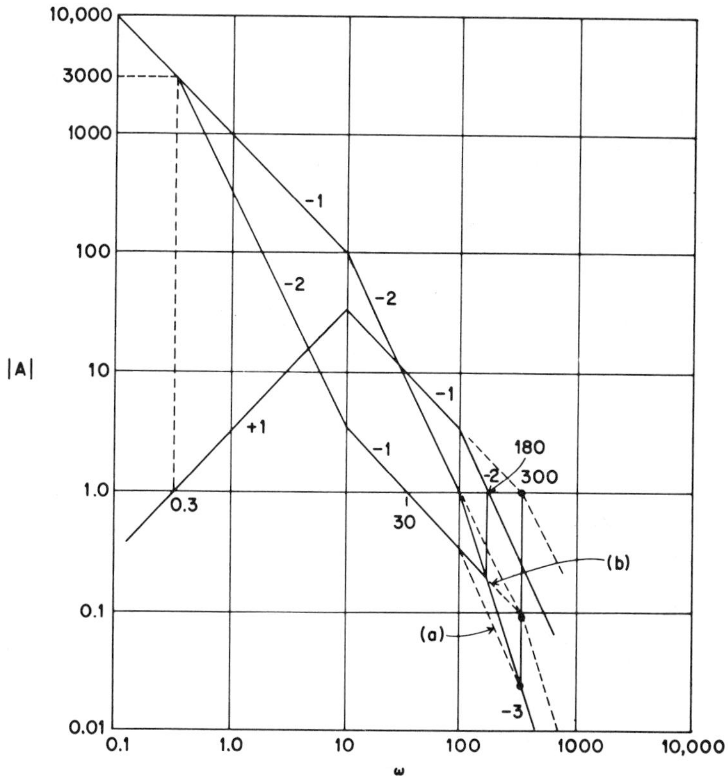

FIG. 11.7. The effect of a lead network in the forward and feedback paths of the minor loop.

of both the major-loop and minor-loop phase margins, whereas major-loop phase margin and performance is slightly degraded when a compensation lead network is placed in the minor-loop feedback path. Even if phase lead is placed in the feedback path, the required gain should be put in the forward path to make the feedback elements passive, and thus as much drift-free as possible. This allows an equivalent gain reduction in that part of the major loop around which H is not connected. In most cases it is advantageous to place the lead in the forward path.

11.3 CHOOSING THE MINOR-LOOP ELEMENTS

Several factors should be considered in choosing those system elements around which the minor loop feedback is to be placed. Since the minor loop is usually introduced for design convenience, it is advisable not to add complexity to the design problem. The minor-loop forward elements should not generally include more than two of the significant major-loop time constants, thereby assuring that minor-loop stability will not be a serious problem. It is not sensible to choose a minor-loop configuration in which extensive attention

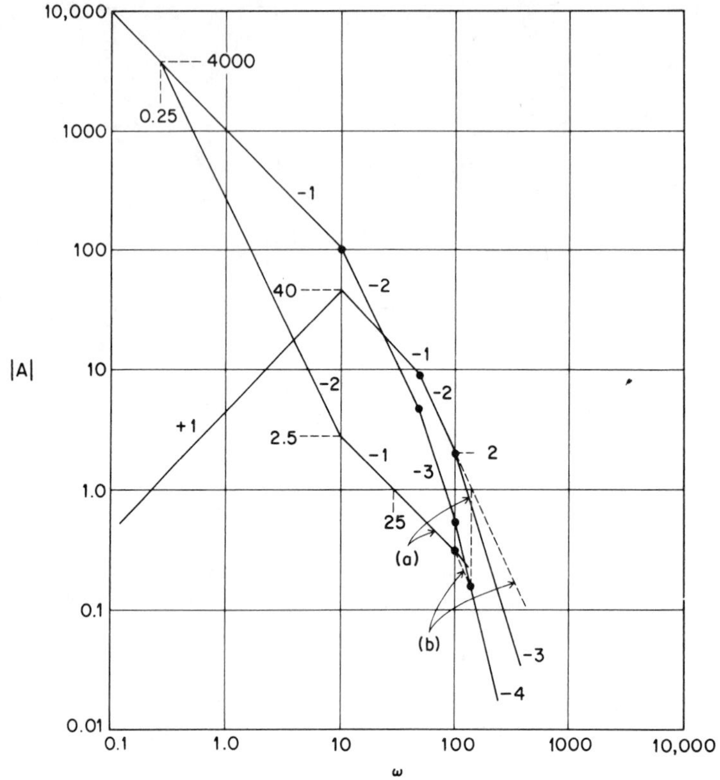

FIG. 11.8. A case where A_m stability margin is significant.

must be devoted to minor-loop equalization. In many cases the two lowest major-loop break frequencies are included in the minor loop, and additional lag terms are avoided, if possible, in an effort to assure a reasonable minor-loop stability margin.

When A_{Mu} has a slope of $-n$, $n \geq 3$, at the point where $|A_{Mu}| = 1$, it may be necessary, if minor-loop stability problems are to be avoided, to settle for A_{Mc} curves that are not of optimum shape. Consider the case shown in Fig. 11.8. If we try to realize the design labeled (a), A_{Mc} is optimally symmetrical near $\omega = \omega_c$, but A_m is unstable as shown and would have to be compensated by adding a pair of lead breaks just above minor-loop crossover if this A_{Mc} is to be realized. The problem arises because *three* major-loop time constants are included in A_m. If the break frequency at $\omega = 100$ is not included in A_m, the resulting curves are labeled (b) in Fig. 11.8. A_{Mc} is changed only slightly and A_m is stable, with approximately 30° of phase margin, so design (b) should be an acceptable compromise. This example illustrates how minor-loop stability problems are indicated in advance by the A_m shape near crossover. When trouble arises, adjustments are easily made.

In some cases the desired minor-loop connections may be difficult to make. Not all signals are readily available, and compromises must sometimes be made using measurable signals that are closely related to those desired. Suppose, for example, the motor in a position controller has transfer function

$$G = \frac{K}{(\tau s + 1)\, s} \tag{11.25}$$

If only the time-constant term is to be included within the minor loop, we must measure or approximate the output derivative. None of the voltages or currents in the motor circuit provide the desired signal directly, assuming we are dealing with the standard armature-driven motor configuration with a load inertia and viscous friction. If load friction is negligible, the output derivative is proportional to the motor back electromotive force (emf), which can perhaps be measured if armature resistance is small, but this situation illustrates how the signals directly related to all possible points on a block diagram are not always readily available. When the desired signals are difficult or impossible to measure, it may be necessary to synthesize them as best we can from available signals. In the motor example we can either measure the armature drive voltage and process that signal using a filter with transfer function $K_f/(\tau s + 1)$ to obtain an output derivative estimate, or place a tachometer on the output shaft to generate a voltage proportional to shaft speed. In most cases a bit of ingenuity will provide an acceptable solution.

11.4 MULTIPLE MINOR LOOPS

Multiple minor loops are often used in complex control systems. Two representative multiple-loop configurations are shown in Figs. 11.9 and 11.10. Both arrangements can be analyzed and designed using straightforward extensions of the standard minor-loop methods. A minor loop within a minor loop, as shown in Fig. 11.9 may be used when minor-loop stability becomes a problem and a stabilizing network is needed within the minor loop. In the case illustrated, A_{m1} is used to stabilize A_{m2} and its characteristics are obtained in the usual way.

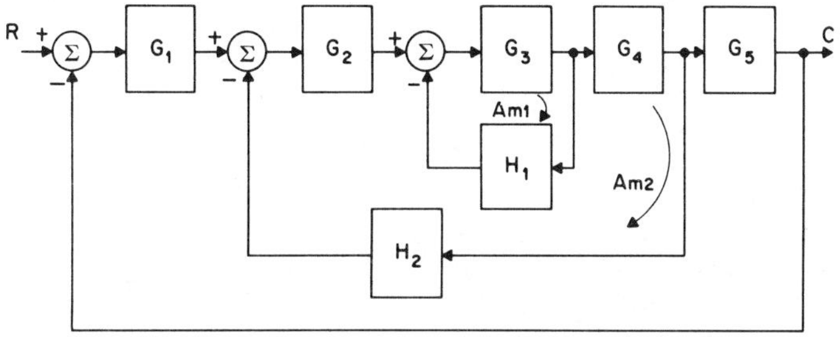

FIG. 11.9. Use of two minor loops, one within the other.

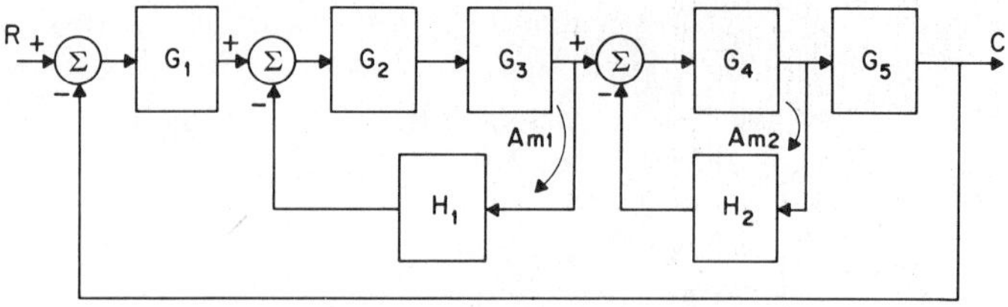

FIG. 11.10. Use of two minor loops in series.

Although theoretically this procedure can be extended to three or four minor loops, significant difficulties arise. Recall that A_{Mc} returns to A_{Mu} at ω_{cm}, the minor-loop crossover frequency. Thus ω_{cm} is significantly greater than ω_{cM}, the major-loop crossover frequency, and each successive minor loop must have substantially greater BW than the previous one. Time constants which could normally be disregarded may become significant, and stability of the inner loop becomes progressively more difficult to achieve. Within these general restrictions, however, multiple loops may be used in the configuration shown in Fig. 11.9.

The use of two or more minor loops in cascade, as shown in Fig. 11.10, avoids the minor-loop stability problem, since the lag terms included in each minor loop are limited. A_{Mc} is obtained from A_{Mu} by subtracting (on a log scale), A_{m1} and A_{m2} from A_{Mu}, so the synthesis procedure remains straightforward. There are limitations, however. Signal power levels generally increase from input to output along the forward path, so it may be difficult to measure and process the signals in the minor loop nearest the output. There is a rapid accumulation of phase lag in the major loop near those frequencies where the minor loop gain levels pass through unity, and this may complicate the design. It may also be difficult to get the magnitude of both minor-loop transfer functions sufficiently large in the region of interest that the usual approximations are acceptably accurate. When multiple minor loops are used the situation should be studied carefully to make sure that all performance objectives have been satisfied.

Neither of the minor-loop configurations shown in Figs. 11.9 and 11.10 is universally best, and both should be considered carefully when a multiple-loop realization seems appropriate. Because of the problems inherent in multiple-loop systems, we might be tempted to abandon the consideration of minor-loop arrangements in complex systems and revert to the standard series-equalization methods. So many advantages can be realized by using minor-loop equalization, however, that a considerable amount of design effort is justified to arrive at an acceptable minor-loop design.

11.5 THE EFFECT OF MINOR-LOOP FEEDBACK UPON PARAMETER CHANGES AND NONLINEAR ELEMENTS

We have assumed heretofore, and in fact naively so, that the system components we work with are either linear, or can be adequately approximated by a linear model. We generally justify this simplification by pointing out that it is prohibitively difficult to develop a mathematical analysis for most systems containing nonlinear (NL) elements. The principle of superposition, one of our most versatile analytical tools, does not apply for the input-output performance of a NL element. Although certain special NL systems can be treated accurately, most analysis procedures involve approximations based upon some type of near-equivalent linear model. We have implicitly assumed that such an equivalent model was available, and all of our analyses have been based upon it. More sophisticated methods for treating NL systems are discussed in Chaps. 13 and 14.

No physical element is truly linear over infinite dynamic range. Many control system components such as motors, generators, amplidynes, and synchos have highly NL performance characteristics. Our linear representation of such elements is a gross approximation which provides only a rough estimate of system performance. Fortunately the use of feedback reduces the effect of NL characteristics upon the closed-loop performance, and this reduction is further enhanced when appropriately-chosen minor-loop feedback is placed around the NL element. We have thus far represented the dc transfer characteristics of a generator by the gain constant K_g, thereby assuming the generated voltage versus field current relationship is a straight line. In actuality, this transfer characteristic typically appears as shown in Fig. 11.11. The equivalent gain relating small changes Δe_g and Δi_g actually is highly dependent upon the reference point. If we are only concerned with performance over a small signal range, a linear model is reasonably accurate. If, however, operation over a wide dynamic range is encountered, a compromise must be made if a linear model is to be used, and the corresponding analysis provides less dependable answers. The effect of changing gain upon system performance in a standard-unity feedback controller is readily apparent from the Bode diagram, which is moved up and down proportionally with the gain changes. When the gain of a device changes significantly over the operating range, the controller must be designed more carefully to provide acceptable operating characteristics under all conditions encountered. A minor-loop realization provides significant advantages in such cases. Note that the NL characteristic shown in Fig. 11.11 essentially results in a control-loop gain which varies with the operating point. A NL component may often be treated as a signal or time-dependent gain.

An example will be used to illustrate the relative effects of a gain change upon minor-loop and cascade-equalized systems. Consider once again the system

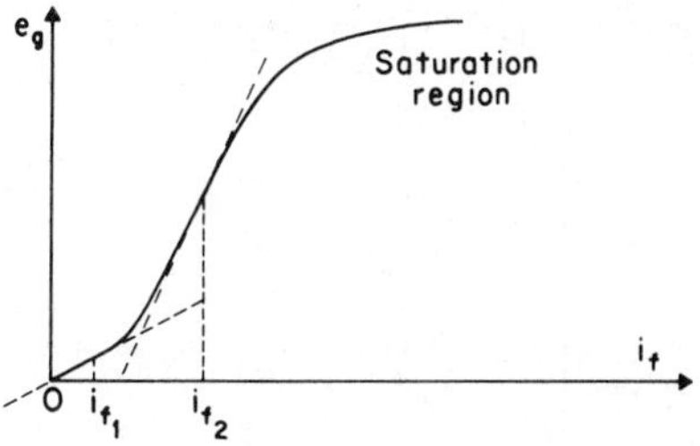

FIG 11.11. Generated voltage versus field current for a typical dc generator.

shown in Fig. 11.1, and the corresponding minor-loop design Bode diagrams shown in Fig. 11.2. If the gain constant in G_2 decreases from 20 to 10, the effect on the minor-loop-compensated system is as shown in Fig. 11.12. The effect of the same change upon a series-equalized system is shown in Fig. 11.13. Note how the Bode diagram symmetry near ω_c is altered significantly in the series-equalized case, but remains essentially unchanged for the minor-loop-compensated system. Low frequency error performance and K_v are degraded by a factor of 2 in both cases. Closed-loop BW and ω_c are reduced by a factor of 2 and θ_m is significantly decreased in the series-equalized system, whereas BW and ω_c are virtually unchanged and θ_m is decreased only slightly in the minor-loop case. The minor-loop advantage is realized because the changes in A_{Mu} and A_m tend to compensate for each other in their net effect upon A_{Mc}. Note in particular that this minor-loop advantage disappears if the parameter change takes place in H, since no self-compensating change in A_{Mu} occurs in this case. Thus, every effort should be made to keep H passive. The comparative results shown in Figs. 11.12 and 11.13 illustrate one of the principle advantages of minor loop design. *Minor-loop-compensated systems are significantly less sensitive to parameter changes in the forward path elements around which the minor-loop feedback is placed.* Although a gain decrease was used as an illustration, similar sensitivity reduction occurs for gain increases (which occur less often) or

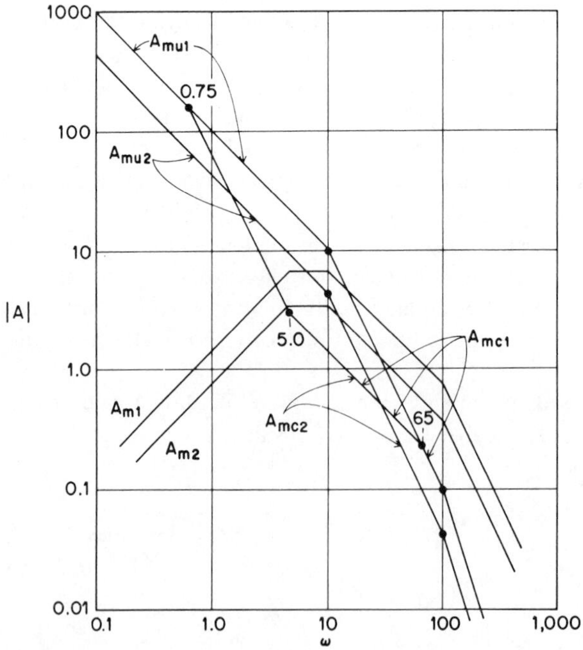

FIG. 11.12. Effect of a 2:1 gain reduction in G_2 for the system shown in Fig. 11.1.

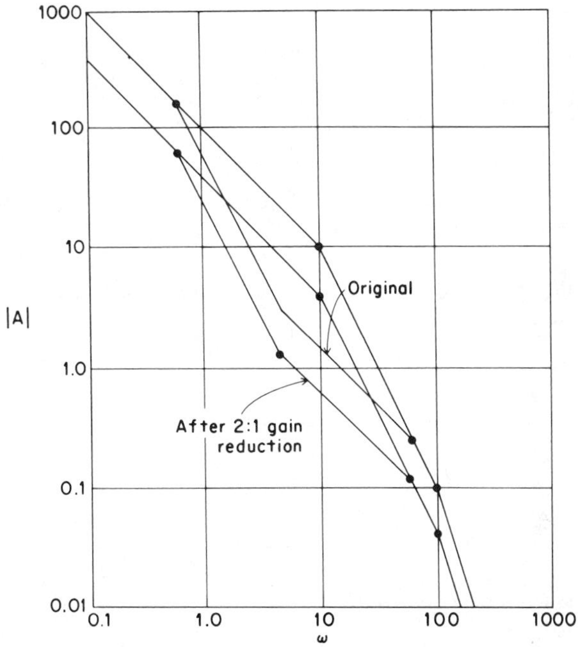

FIG. 11.13. Effect of a 2:1 gain reduction when series equalization is used.

time-constant changes *as long as they occur in the minor-loop forward path.* In fact the minor-loop advantage is more pronounced for gain increases, as the reader should justify with a set of diagrams. Although minor-loop θ_m generally decreases with increasing gain, it should not be difficult to design the minor loop such that θ_m remains acceptable for a gain increase of 2:1 or more from the nominal value. When the NL element primarily results in an operating-point-dependent gain level, and such situations are common, minor-loop feedback placed around the NL element provides a form of linearization or input to output performance-averaging. When gain variations are large, series compensation is not particularly appropriate, and only by using minor-loop equalization can reasonably constant response characteristics be obtained.

The situation shown in Fig. 11.14 illustrates the use of minor-loop feedback to reduce the effects of gain changes. The forward element has a nominal gain level of K which may decrease or increase by a factor of 2 during normal operation. This amounts to a total of 400% from the minimum value, or -50% and +100% from the nominal value. Such gain changes, rather common for control system devices like generators, motors, and amplidynes, make cascade equalization of a closed-loop configuration difficult to achieve. The use of feedback as shown in Fig. 11.14 reduces the net gain change at the expense of

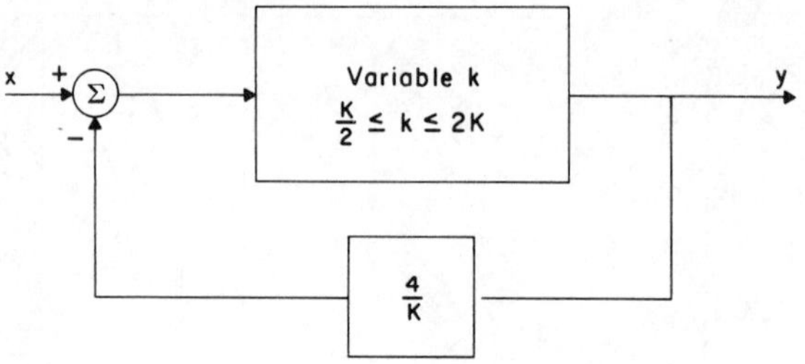

FIG. 11.14. Use of feedback to reduce the effect of a gain change.

reduced forward gain. When $k = K$,

$$\frac{y}{x} = \frac{K}{1+4} = \frac{K}{5} \tag{11.26}$$

when k decreases to $K/2$,

$$\frac{y}{x} = \frac{K/2}{1+2} = \frac{K}{6} \tag{11.27}$$

and when k increases to $2K$

$$\frac{y}{x} = \frac{2K}{1+8} = \frac{2K}{9} \tag{11.28}$$

It is apparent from Eqs. (11.26), (11.27), and (11.28) how feedback reduces the percent change in y/x. Increasing feedback gain reduces the percent variation in y/x, but also reduces its nominal value. In most applications, gain changes of 20% or less are easily tolerated. If the gain variation in the forward path is known, we can solve for the feedback gain required to hold closed-loop y/x gain changes to within any desired limits. The corresponding reduction in forward gain can be made up in the amplifier stages.

Example 11.2

Assume that it is necessary to reduce the net change in y/x as shown in Fig. 11.14 to 20% for the range $K/2 \leq k \leq 2K$. Let the feedback gain be denoted by h/K. Then

$$\left(\frac{y}{x}\right)_{\text{min}} = \frac{K/2}{1 + h/2} \qquad \left(\frac{y}{x}\right)_{\text{max}} = \frac{2K}{1+2h} \tag{11.29}$$

and we require

$$\frac{\left(\frac{y}{x}\right)_{max} - \left(\frac{y}{x}\right)_{min}}{\left(\frac{y}{x}\right)_{min}} = 0.2 = \frac{\dfrac{2K}{1+2h} - \dfrac{K/2}{1+h/2}}{\dfrac{K/2}{1+h/2}} \tag{11.30}$$

Simplifying Eq. (11.30) and solving for h provides $h = 7$. This is a particularly stringent example, because the range of change in k is large, and the limit on y/x change is small. The nominal reduction in y/x we must accept is a factor of 8, and this gain must be made up elsewhere. In most applications the range of k change is smaller and changes in y/x can often be held within bounds using a nominal minor-loop gain of 3 or 4.

We will sometimes face situations in which it is desired to use one or more minor loops to reduce the effects of operating-point-dependent gain changes in a NL element and also to shape the A_{Mc} Bode diagram near ω_c for equalization purposes. This may be accomplished with a single minor loop serving both purposes, by using cascades of minor loops, or by constructing two or more minor loops with each loop enclosing the previous one. If a single-loop design is attempted, several factors should be kept in mind. Equivalent linearization of a NL characteristic requires dc gain in the minor loop, typically of the order of 3 or 4. This limits to some extent our ability to shape A_{Mc} with the minor loop. Also, the approximations made in minor loop analysis are poor unless $|A_m| \geq 5$ or so in the region near major loop ω_c. When a NL element with a fairly large range of gain change is present in the minor loop, minor-loop stability can become a sticky problem, and this limits the amount of phase lag we can allow in A_m. These factors must all be considered in choosing a final configuration. It is common to use one minor loop to reduce gain changes caused by the NL element and a second loop, often enclosing the first, to realize the desired A_{Mc}.

11.6 MINOR-LOOP PERFORMANCE WITH LOAD AND NOISE DISTURBANCES

We have already considered two basic advantages of minor-loop equalizers over series equalizers:

1. The time constants required in the minor-loop feedback element H are more easily realized than those needed in the comparable series element G_c.
2. Minor-loop systems are less sensitive to parameter changes occurring within the minor loop, thereby providing a partial linearization of any enclosed NL elements.

A third major advantage of minor loops is their relative insensitivity to load and

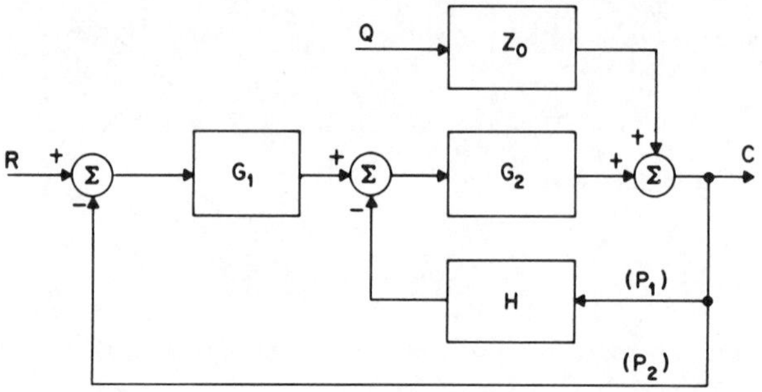

FIG. 11.15. A system with load disturbance enclosed within the minor loop.

noise disturbances. Most controllers are exposed to load disturbances, such as the effects of wind upon a radar antenna positioning unit, and all systems are subjected to the noise generated within their basic components.

Consider the situation shown in Fig. 11.15, where Q denotes an arbitrary load disturbance and Z_0 is the open-loop transfer function (open-loop impedance to this disturbance) between the disturbance and output C. Opening both loops at points (P_1) and (P_2) yields

$$\frac{C_Q}{Q} = Z_0 \tag{11.31}$$

where C_Q denotes the output caused by disturbance Q with input $R = 0$. Closing the minor loop and leaving the major loop open reduces the output impedance by the factor $1 + A_m$, as shown in Sec. 10.3, so

$$\frac{C_Q}{Q} = \frac{Z_0}{1 + A_m} \tag{11.32}$$

Closing the major loop reduces the equivalent impedance to Q further by the factor $1 + A_{Mc}$, so with both loops closed

$$\frac{C_Q}{Q} = \frac{Z_0}{[1 + A_m][1 + A_{Mc}]} \tag{11.33}$$

But

$$A_{Mc} = \frac{A_{Mu}}{1 + A_m} \tag{11.34}$$

so

$$\frac{C_Q}{Q} = \frac{Z_0}{[1 + A_m][1 + A_{Mu}/(1 + A_m)]} \tag{11.35}$$

The usual approximations may be used for the terms in Eq. (11.35). At low frequencies where $|A_{Mc}| > 1$

$$1 + \frac{A_{Mu}}{1 + A_m} \simeq \frac{A_{Mu}}{1 + A_m} \tag{11.36}$$

and Eq. (11.35) becomes

$$\frac{C_Q}{Q} \simeq \frac{Z_0}{A_{Mu}} \tag{11.37}$$

It is apparent from Eq. (11.37) that low frequency disturbance effects are reduced by the factor A_{Mu} rather than by A_{Mc}, as would be the case if the equivalent series equalizer were used. Since A_{Mc}, A_{Mu} and A_m are related as defined in Eq. (11.34), the advantage in reducing Z_0 in the minor-loop system is essentially $|A_m|$ over that frequency range where $|A_m|$ and $|A_{Mc}|$ both exceed unity. In fact this improvement is also obtained when $|A_{Mc}| < 1$ and $|A_m| > 1$, since in this case

$$1 + \frac{A_{Mu}}{1 + A_m} \simeq 1 \tag{11.38}$$

Thus the system with minor-loop feedback around the disturbance effect provides an improvement in the reduction of that disturbance effect approximately equal to $|A_m|$ over the entire frequency range where $|A_m| > 1$. At those frequencies where $|A_m| < 1$, there is no significant reduction of Z_0 when either approach is used.

Load disturbances are seldom sinusoidal or harmonic in nature. More often they are random variables which can only be described by probability density and frequency power spectrum functions. If the disturbances contain significant frequency components over the range where $|A_m| > 1$, the dynamic load disturbance response characteristics of the minor-loop-equalized system are superior to those of the equivalent series-equalized arrangement. Thus a significant advantage is realized in the typical case of a unit step disturbance, for example, which contains frequency components over the entire range $0 \leq \omega \leq \infty$. The minor-loop arrangement is definitely "stiffer," and able to respond more rapidly to counteract the effects of random load changes.

Control systems are typically exposed to noise in addition to load disturbances. Consider the general situation shown in Fig. 11.16. It is easily found, using steps analogous to those applied in treating load disturbances, that

$$\frac{C_N}{N} = \frac{G_3 G_4}{(1 + A_m)(1 + A_{Mc})} \tag{11.39}$$

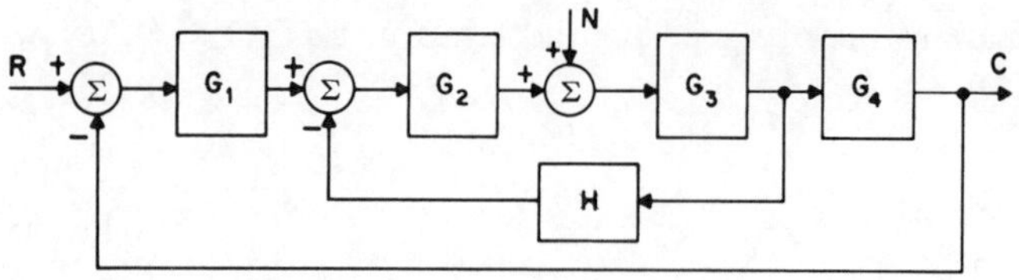

FIG. 11.16. Noise disturbance in a minor-loop-equalized system.

where C_N denotes the response to input N assuming R = 0. Applying the usual approximations to Eq. (11.39) provides

$$\frac{C_N}{N} \simeq \frac{G_3 G_4}{A_{Mu}} = \frac{1}{G_1 G_2} \qquad |A_{Mc}| > 1 \tag{11.40}$$

and

$$\frac{C_N}{N} \simeq \frac{G_3 G_4}{A_m} \qquad |A_{Mc}| < 1 \qquad |A_M| > 1 \tag{11.41}$$

An equivalent series-equalized controller is shown in Fig. 11.17, where G_e, the equalizing filter, precedes the disturbance input. In this case we have

$$\frac{C_N}{N} = \frac{G_3 G_4}{1 + A_{Mc}} \simeq \frac{1}{G_1 G_2 G_e} \qquad |A_{Mc}| > 1 \tag{11.42}$$

in the range where $|A_m| > 1$, $G_e \simeq 1/A_m$ so Eq. (11.42) becomes

$$\frac{C_N}{N} \simeq \frac{A_m}{G_1 G_2} \qquad |A_{Mc}| > 1 \qquad |A_m| > 1 \tag{11.43}$$

Similarly, when $|A_{Mc}| < 1$ the feedback effect in the series-equalized system is essentially negligible, so

$$\frac{C_N}{N} \simeq G_3 G_4 \qquad |A_{Mc}| < 1 \tag{11.44}$$

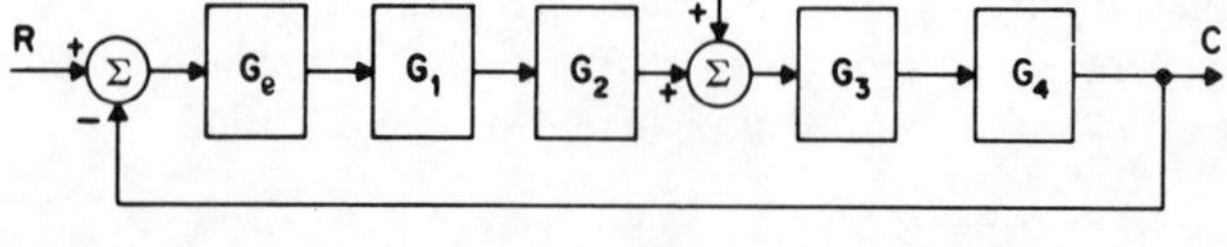

FIG. 11.17. Series-equalized arrangement equivalent to that shown in Fig. 11.16.

Comparing Eq. (11.40) with Eq. (11.43) and Eq. (11.41) with Eq. (11.44) shows that the minor-loop response to a general disturbance N occurring within the minor loop is smaller by the factor $|A_m|$, when $|A_m| > 1$, than that of the equivalent series-equalized system in which G_e precedes the disturbance point. There is no significant difference in disturbance response when $|A_m| < 1$.

Suppose G_e appears between the disturbance and the output. In this case the transfer given in Eqs. (11.42) and (11.44) are each multiplied by $G_e \simeq 1/A_m$, and there is no apparent advantage to the minor-loop arrangement. Power levels typically increase as the signal progresses through the forward path, however, and equalization must generally be accomplished where the power level is small. As the power level increases the impedance level decreases, and the large time constant typically used in equalizing filters require impractically large capacitor values at low impedance levels. In many cases realization of the desired $G_e(s)$ is so difficult even at high impedance levels that we are strongly motivated to use a minor-loop design. Although it would be beneficial to place $G_e(s)$ as far forward as possible to reduce disturbance effects, we are not often able to take advantage of this fact in practice. Only when the disturbance arises in the error sensing process or very soon thereafter in the signal processing sequence is it feasible to place $G_e(s)$ after the disturbance occurs.

11.7 CONCLUDING COMMENTS

We have shown in this chapter that it is usually no more difficult to design a minor-loop-equalized system than to design a series-equalized system with the same A_{Mc}. In fact the two design procedures are essentially identical up to the point where the equalizer, either $G_e(s)$ or $H(s)$, is to be realized. Minor-loop feedback cannot shape A_{Mc} above A_{Mu}, so upward-shaping of the Bode diagram must always be accomplished with a series-lead equalizer. This is the most significant limitation of minor-loop equalization, and it is not as flexible as series equalization. It can, however, be used to perform almost any compensation desired where A_{Mc} falls entirely below A_{Mu}. Series and minor-loop equalizers may be combined if desired, and there are so many advantages to the use of a minor-loop-equalized system that we are often inclined to accomplish as much of the equalization as possible with minor loops even in those cases where an additional series equalizer is required. Only in relatively complex situations where a considerable amount of Bode diagram shaping is required near ω_c are significant problems encountered in the minor-loop design, and these problems are readily overcome in most cases by making simple design adjustments, by using two or more minor loops, or by adding an appropriate equalizer in the minor loop. (Usually in the forward path, resulting in a combination of series and minor-loop equalization.) It is often expedient to make small compromises in A_{Mc} near ω_c to assure minor-loop stability, or occasionally to use a bit of series lead in the minor loop for the same purpose. The signal desired for minor-loop feedback may not be readily available, but an approximation of that signal or an acceptable substitute can usually be synthesized from the available signals.

There are several significant advantages to minor-loop designs. The required H is generally easier to realize than the equivalent G_e. More important, however, the minor-loop configuration is less sensitive to parameter changes in the forward path elements around which compensating feedback is placed, and its response to noise and load disturbances occurring within the minor loop is reduced by the factor $|A_m|$ when $|A_m| > 1$ compared to all series-equalized cases where G_e precedes the disturbance. Because these advantages are so important, we might wonder why series equalization is ever used. Minor-loop design should be used more often than it is, but the specs in many applications are so easily satisfied using series equalization that little or no consideration is given to further system refinement. Since series compensation is learned first, we tend to feel more comfortable in using it despite the fact that the opposite should be true. Unfortunately the advantages of minor-loop design are not fully appreciated, and series-compensated systems are often developed without anyone checking to see if the job could be done better.

The general concept of minor-loop feedback can be extended further to the feedback of all significant signals appearing in the processing and actuating elements of the forward path. The feedback gains may be adjusted to yield the most desirable (optimum) input to output performance characteristics. This relatively new design procedure, called state-variable feedback, sometimes offers significant advantages, and its use is discussed in Chap. 13.

12 CARRIER CONTROL SYSTEMS

In all previous discussions we have assumed the information passing through our systems is carried by frequency components centered around $\omega = 0$. Such systems are generally referred to as *dc* or *baseband* systems. In many situations encountered in practice the information is present in the form of a modulated carrier signal. In such cases stability depends upon what happens to the modulated signal as it traverses the control loop, so the usual analysis and synthesis procedures must be appropriately modified. The same basic principles apply, however, so this task is not as formidable as it might at first appear. Three distinct types of modulation often encountered in control applications are:

1. Suppressed carrier amplitude modulation (AM).
2. Pulse amplitude modulation, called *sampled data* by control specialists.
3. Pulse width modulation.

We are primarily concerned with AM in this chapter. Sampled data systems, including pulse width modulation, a special type of nonlinear sampled data modulation, are considered in Chap. 15. We will consistently refer to suppressed carrier amplitude modulated systems as simply *ac systems* or *carrier systems* throughout this chapter.

Several advantages of ac systems are:

1. It is often easier to make precise position measurements using ac components.
2. It is generally less difficult to amplify a modulated carrier signal than the equivalent baseband signal, since dc amplifier drift is a limiting factor in the latter case.
3. Many useful control system components such as synchros are inherently ac devices.
4. It is often desired to send signals through air, free space, or the sea (sonar), and transmission through such media is only feasible if an appropriate carrier frequency is used.
5. Components such as motors, generators, and amplifiers used in ac servos are often lighter, cheaper, and more efficient than their equivalent dc counterparts.

The use of ac systems is not a panacea, however, and their most significant disadvantages are:

1. Component characteristics of ac systems are often significantly more nonlinear than those of the comparable dc elements, and response predictions based upon a linear model are generally less exact.
2. Equalization of ac systems requires frequency-dependent amplitude and phase adjustments upon the envelope signal. The ac equalizer component values must be precisely set, and equalization is more readily accomplished at baseband (around zero frequency). Carrier frequency shift is critical when ac compensation is used.

When the advantages and disadvantages are considered, ac systems are chosen most often where size and weight are restrictive, as in aircraft, missile or satellite applications; where reasonably accurate 360° angular position control is desired, as in many instrument servo applications; where the control problem is so straightforward that equalization causes little or no difficulty; and when so many systems of a particular type are to be offered for sale in such a highly competitive market that component costs are critical.

Our consideration of ac systems is initiated by discussing briefly the characteristics of several standard ac components. The suppressed carrier amplitude modulation and demodulation processes are introduced. The analysis problem is considered using an equivalent baseband mathematical model and the principles upon which this model is based are presented. Equalization methods are developed and the problems related to equalization at the carrier frequency are discussed. Several ways to overcome these difficulties are suggested. A design example is presented.

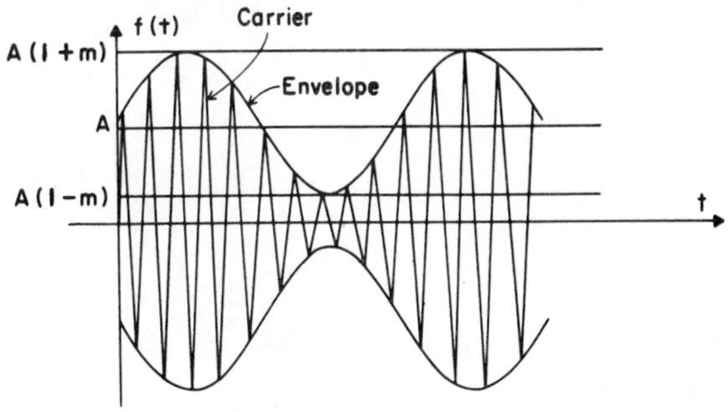

FIG. 12.1. Standard amplitude modulation.

12.1 SUPPRESSED CARRIER AMPLITUDE MODULATION

We are concerned in control application with suppressed carrier amplitude modulation because it arises in standard ac position transducers such as the synchro generator and control transformer pair, and also because regular amplitude modulation is inadequate to provide bidirectional control of an ac motor. The signal $f(t)$ transmitted by your local AM radio station is of the form

$$f(t) = A[1 + g(t)] \sin\omega_0 t \tag{12.1}$$

where $g(t)$ denotes the baseband envelope modulating signal and ω_0 is the carrier frequency. This relationship is often expressed alternately in terms of a constant amplitude modulating tone at frequency ω_m, which yields

$$f(t) = A[1 + m \sin\omega_m t] \sin\omega_0 t \tag{12.2}$$

where m is called the modulation index, and $|m| \leq 1$ is required for linear modulation. A typical $f(t)$ of this form is shown in Fig. 12.1. The amplitude modulation processes in common use are such that $f(t) = 0$ when $1 + m \sin\omega_m t \leq 0$, and this condition, called overmodulation and analogous in a sense to saturation limiting at baseband, is illustrated in Fig. 12.2.

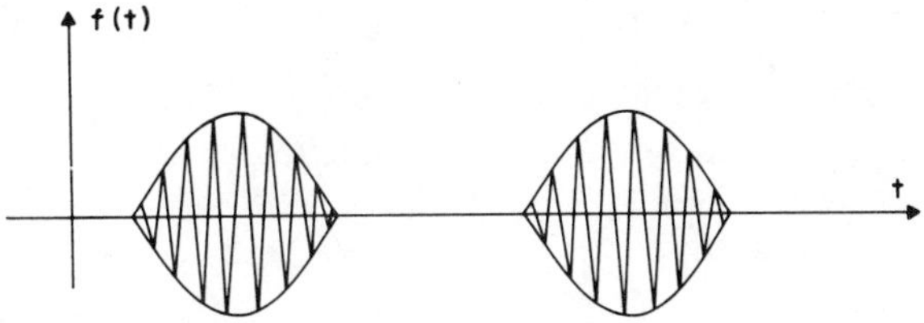

FIG. 12.2. Amplitude modulation with saturation.

The general suppressed carrier amplitude modulated signal is of the form

$$f(t) = Ag(t) \sin \omega_0 t \tag{12.3}$$

When $g(t)$ is a modulating tone of amplitude B and frequency ω_m, this becomes

$$f(t) = AB \sin \omega_m t \sin \omega_0 t \tag{12.4}$$

This $f(t)$ is shown in Fig. 12.3. Note that the envelope and the modulating signal pass through zero together, a phenomenon not present in normal amplitude modulation, and the only indication of modulating signal sign available is the carrier phase, which changes by 180° each time the envelope passes through zero. Whereas the amplitude of regular AM as given by Eq. (12.2) is A when the modulating signal amplitude is zero and envelope phase is indicated by the instantaneous envelope value relative to A, suppressed carrier AM is identically zero when no modulating signal is present and envelope phase can be obtained only by relating carrier phase to that of a reference signal. In many communication applications envelope phase is relatively unimportant. Either choice may be made arbitrarily as long as we stick with the result, but in a typical controller a 180° phase inversion of the information signal would cause instability (positive, instead of negative, feedback). Fortunately the desired reference signal is readily available in most control applications, and a standard phase detector provides both amplitude and phase information.

Breaking $f(t)$, as given by Eq. (12.4), down into its individual frequency components yields

$$f(t) = \frac{AB}{2}[\cos(\omega_0 - \omega_m)t - \cos(\omega_0 + \omega_m)t] \tag{12.5}$$

The effect of the modulation process upon the baseband information signal spectrum is directly apparent from Eq. (12.5). If we interpret the signal power in each incremental frequency as an equivalent sinusoidal signal component, it is clear that the baseband signal spectrum is shifted to the frequencies around ω_0 as shown in Fig. 12.4. The signal components in the modulated form are symmetrically located about the carrier frequency components $\pm\omega_0$. The conditions

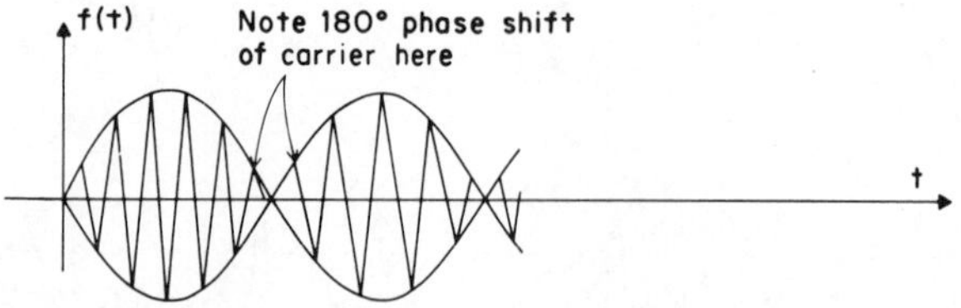

FIG. 12.3. A typical suppressed carrier amplitude modulation signal.

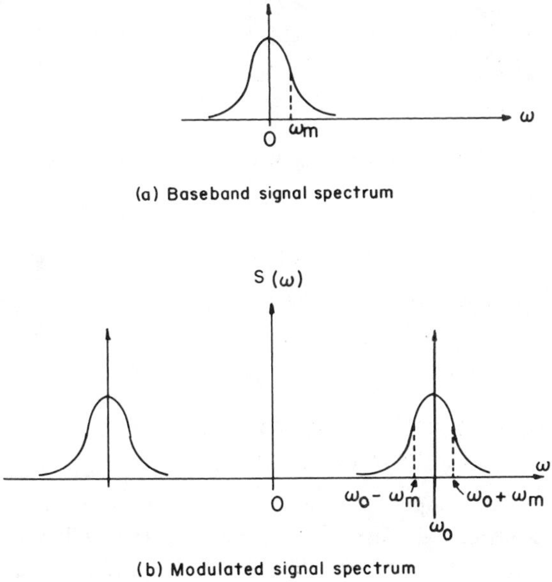

FIG. 12.4. Effect of modulation upon the signal spectrum.

which must be satisfied by networks used to amplify or process the modulated signal can be visualized from Fig. 12.4. An amplifier, if it is not to introduce distortion or phase lag, must amplify all frequencies near ω_0 equally well. Since the BW of typical controllers is small, this is not generally a serious limitation. The advantages gained by being able to use ac coupling between amplifier stages and frequency selective (resonant) load networks, provided the load impedance remains essentially constant over the frequency range of interest near ω_0, are significant.

One difficulty associated with ac system design is clearly illustrated by the spectra shown in Fig. 12.4. At baseband, equalization is realized by shaping the system Bode diagram near ω_c to provide satisfactory θ_m. Bode diagram shaping is simple at baseband, but this is not generally true when attempted at ω_0 in ac systems. Since the information is broken up into two sinusoidal sideband components at $\omega_0 - \omega_m$ and $\omega_0 + \omega_m$ by the modulation process, Bode diagram shaping near ω_0 must treat these components symmetrically to assure the desired compensation effect upon the envelope signal. A typical value of $f_0 = 2\pi\omega_0$ is 400 Hz, and the controller baseband BW might be 10 Hz or less. If equalization of the modulated signal is required, we are faced with the difficult task of shaping the forward path frequency response of the system symmetrically and precisely within a range of 10 Hz or so on each side of 400 Hz. This can perhaps be accomplished by using carefully designed networks, but ac servo system equalization is not as simple as the corresponding design problem at baseband. Carrier drift by more than a small percentage of the information BW

FIG. 12.5. Physical arrangement of a synchro pair.

destroys the equalizer's symmetrical sideband characteristic, and cannot be tolerated when extensive compensation of the modulated signal is used. Design of compensators for ac systems is considered in Sec. 12.6.

12.2 POSITION TRANSDUCERS FOR AC SYSTEMS—SYNCHROS

The modulation process in an ac controller generally takes place in the position transducer. We assume that an angular position controller is under consideration, and a synchro unit is being used to detect angular error. The basic operation of a synchro pair is readily explained using the typical arrangement shown in Fig. 12.5. The standard schematic representation of this combination is shown in Fig. 12.6. The ac signal applied to the generator rotor sets up a varying flux, and ac voltages are induced in the stator windings by the resulting transformer action. These voltages cause currents to flow in the stator windings. The induced currents are all in the same phase, but their magnitudes depend upon the net induced voltages in the stator windings in each closed path. The induced voltages depend, in turn, upon the rotor position. When the rotor head is centered on a stator winding, half of the induced flux flows through the winding in each direction, and the net induced voltage is zero. As the rotor is turned the voltages induced in each coil vary accordingly. The net result is a combination of currents in the stator of the control transformer setting up a flux pattern which, ideally, has the same space orientation as that in the synchro generator. When the rotor of the control transformer is aligned with this flux, maximum output voltage is generated. When it is aligned perpendicular to the induced flux, no net flux linkages occur and the induced voltage is zero. In

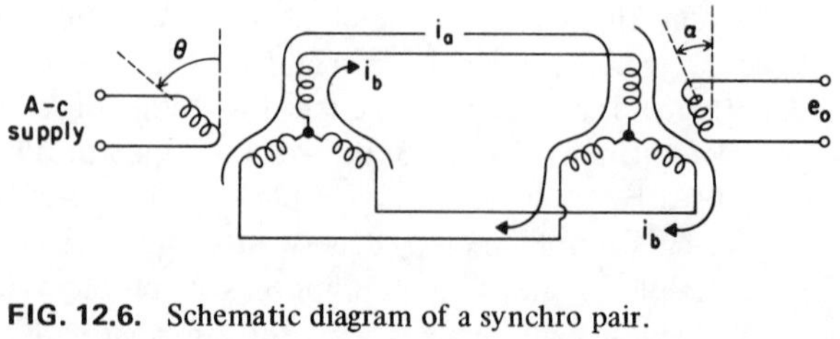

FIG. 12.6. Schematic diagram of a synchro pair.

general, assuming the generator rotor voltage is $E \sin \omega t$, the induced output is

$$e_0 = E_{om} \cos(\theta - \alpha) \sin(\omega t - \beta) \tag{12.6}$$

where θ and α are the angles, measured from a common reference, of the generator and control transformer rotors, respectively. E_{om} denotes the peak output caused by winding resistance, which is kept small relative to reactance in standard synchros, so β is typically only a few degrees and is often neglected. Since β remains constant, it is easily compensated for in synchro applications.

To use a synchro pair in a servo system, the control transformer rotor is connected to the output shaft and the desired input is connected to the generator rotor. The output voltage e_0 is applied to an amplifier feeding the motor which drives the output shaft. The servo system tends to drive the control transformer output to zero, which requires $\cos(\theta - \alpha) = 0$ [see Eq. (12.6)] or $\theta - \alpha = \pi/2$. It is apparent from this that our angular reference on the control transformer is more convenient if it is chosen 90° displaced from that of the generator, since the output null occurs when the two rotors are 90° apart. Thus we let $\delta = \alpha + \pi/2$ and Eq. (12.6) becomes

$$e_0 = E_{om} \sin(\delta - \theta) \sin(\omega t - \beta) \simeq E_{om} (\delta - \theta) \sin(\omega t - \beta) \tag{12.7}$$

where the approximation is valid if $\delta - \theta$ is small. The error output of a synchro pair is thus an ac voltage at the carrier frequency with peak (envelope) value proportional to shaft error for small errors. The carrier phase is shifted from the reference by β radians, and shifts 180° as the error signal passes through zero. Thus a phase-sensitive detector (a correlation detector or discriminator) is necessary for demodulation purposes. A two-phase ac motor is a special form of phase-sensitive detector often used. The operating characteristics of two-phase ac motors and a general-purpose detector are discussed in Secs. 12.3 and 12.4, respectively.

Note the direct relationship between the synchro error voltage as given by Eq. (12.7) and the general suppressed carrier AM signal given in Eq. (12.3). When the system is in normal operation, the error $\delta - \theta$ is a time function, and for a sinusoidal excitation signal at ω_m the error becomes a sinusoid at ω_m and Eq. (12.7) relates directly to Eq. (12.4). Thus the basic concepts of suppressed-carrier AM modulation discussed in Sec. 12.1 will prove useful in the analysis and synthesis of systems operating upon an error signal of the form given in Eq. (12.7).

We have considered only the most basic aspects of synchro performance in an effort to achieve an intuitive understanding of synchro operation. It is apparent from Eq. (12.7) that the envelope output magnitude is proportional to error, and, since information is carried in the output envelope, a gain is often adequate to describe the transfer characteristics of a synchro combination. In some cases it is more appropriate to describe the synchro transfer function by a gain and a time constant rather than by a gain alone, since there is some delay in the synchro's response when the relative shaft positions are changed. Several other minor sources of error should be considered if extreme precision is necessary. For an extensive discussion of ac system components, including synchros, refer to Gibson and Tuteur.[1]

12.3 TWO PHASE AC SERVOMOTORS AND THE AC TACHOMETER

The standard ac servomotor is a two-phase squirrel cage unit. It is common practice to define its transfer function as the relationship between the input envelope amplitude of the carrier, modulated by a tone at frequency ω_m, and the output shaft position. This equivalent baseband transfer function is typically found to be

$$\frac{\theta(s_m)}{E(s_m)} = \frac{K_m}{s_m(\tau_m s_m + 1)(JBs_m + 1)} \tag{12.8}$$

where JB is the time constant due to inertia and viscous friction, and τ_m is the stator inductance time constant, usually much smaller than JB, and often neglected completely. We use s_m rather than s to indicate that this is the *envelope* transfer function. A detailed development of the characteristics of ac servomotors may be found in Chap. 7 of Ref. 1. The ac servomotor thus acts as a demodulator, converting an input suppressed carrier AM signal with envelope signal $E(s_m)$ applied to its control winding into shaft position through the transfer function given by Eq. (12.8). A constant amplitude signal in quadrature with the carrier is applied to the reference winding. This situation is shown schematically in Fig. 12.7. To a first-order approximation the blocked rotor torque varies linearly with the magnitude of the modulated control input.

It is, unfortunately, often found that ac servomotor performance differs considerably from the ideal linear model. A common characteristic is a decided hysteresis tendency in the input voltage–torque characteristics. As the control-phase voltage magnitude is increased the motor often does not respond until perhaps 20 percent of maximum voltage is reached, after which the speed builds

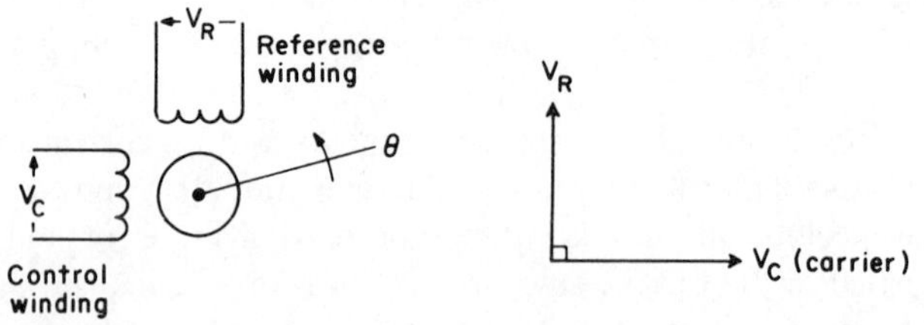

FIG. 12.7. Two phase servomotor and signal phase relationships.

up with increasing voltage until maximum speed is reached. As the control-phase voltage is decreased, the unloaded motor may continue to turn when the control voltage is set to zero. A small amount of breaking torque or negative (carrier phase reversed) voltage is often required to stop this single phase effect, and there may be a "dead space" as the voltage is adjusted further negative before rotation in the opposite direction begins. These hysteresis and single phasing tendencies can be reduced by careful design at some expense in motor efficiency, but ac servomotors are typically rather nonlinear at best. To reduce this effect, and also because it is the most convenient method available, minor-loop feedback around the motor using an ac tachometer is a relatively standard compensation method with ac servos, and ac servomotors often have a built in tachometer for this purpose.

It is a simple matter (see Sec. 7.10 of Ref. 1) to design a tachometer which produces an ac output voltage at the reference frequency with peak level proportional to shaft speed. The output phase angle relative to that of the reference signal depends upon the stator power factor and load but is constant and is easily adjusted to provide an appropriate compensating signal. The tachometer lag time constant is generally negligible, and its transfer function is

$$\frac{V_t(s_m)}{\theta(s_m)} = K_{\text{tach}} s_m \tag{12.9}$$

where K_{tach} denotes the gain in volts per rad/sec. The ac tachometer is used for minor-loop compensation in a manner directly analogous to the equivalent dc application, although the relative simplicity of tachometer feedback makes it more widely used in the ac case, where equalization using lag or lead networks is much less readily realized.

12.4 THE DEMODULATION PROCESS

It is often convenient to demodulate the suppressed-carried AM signal, manipulate the resulting envelope signal to achieve the desired compensation, and then re-modulate before amplifying to drive the ac servomotor. Compensation of the signal in modulated form is relatively difficult under any conditions, and virtually impossible if the carrier is subject to significant frequency shift. All demodulators (see Chap. 6 of Ref. 1) are mathematically equivalent to the situation depicted schematically in Fig. 12.8. The multiplier output is

$$x(t) = C_m \sin \omega_m t \sin^2 \omega_0 t = \frac{C_m}{2} \sin \omega_m t (1 - \cos 2\omega_0 t) \tag{12.10}$$

Since $\omega_0 >> \omega_m$, the low pass filter can be chosen to eliminate the components

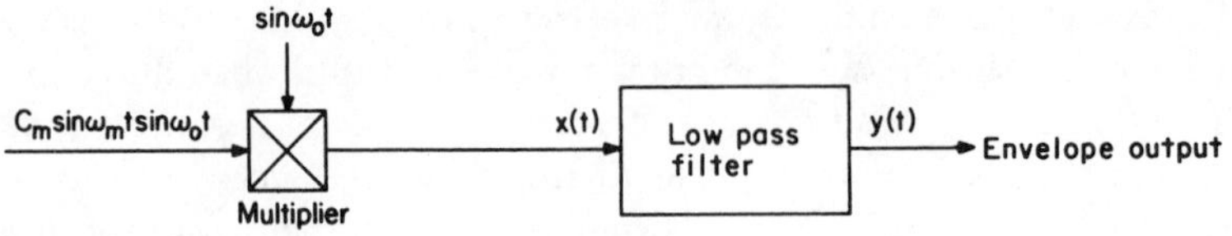

FIG. 12.8. The demodulation process.

in $x(t)$ near $\omega = 2\omega_0$, so

$$y(t) \simeq \frac{K_f C_m}{2} \sin \omega_m t \tag{12.11}$$

where K_f is the filter gain constant. The filter output is thus proportional to the modulation signal. The baseband equalizing filter required to satisfy system specs may be placed at this point and its output used to suppressed-carrier AM modulate a signal at ω_0 if an ac actuator is to be used. Modulation is also a straightforward process. Compensation ease at baseband is obtained at the expense of the demodulation and modulation circuitry required and the noise and errors introduced by these processes.

12.5 BASEBAND EQUIVALENT CIRCUITS OF NETWORKS OPERATING UPON AM SIGNALS

The analysis and synthesis of ac servos is generally carried out at baseband, which is equivalent to focusing our attention upon the envelope of the modulated signal and the manner in which it is modified as it passes through the various devices it encounters. We have already pointed out that a synchro position error detector may be represented by a gain, or a gain and a lag time constant, and the ac servomotor has a baseband equivalent transfer function similar to that of a dc motor. Demodulators are represented by a gain and the lag characteristics of the smoothing filter used. In this section we will develop general procedures by which filters operating upon the AM signal may be represented by baseband equivalents. We will also consider the inverse process, or the methods by which a baseband filter may be converted to an equivalent ac filter for shaping the AM signal envelope. Once these methods have been developed, we may bring all of our standard design procedures to bear upon the problem, working entirely with an equivalent baseband circuit. The desired equalization may then be accomplished at baseband (by first demodulating), or at ω_0, whichever seems most appropriate.

Consider the modulated signal component

$$x(t) = C_m \sin \omega_m t \sin \omega_0 t = \frac{C_m}{2}[\cos(\omega_0 - \omega_m)t - \cos(\omega_0 + \omega_m)t] \tag{12.12}$$

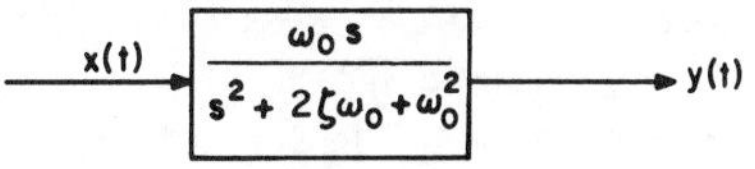

FIG. 12.9. A standard bandpass filter.

Suppose this signal passes through the standard bandpass amplifier shown in Fig. 12.9 with transfer function

$$G(s) = \frac{\omega_0 s}{s^2 + 2\zeta\omega_0 s + \omega_0^2} \tag{12.13}$$

The amplitude and phase characteristics of $G(s)$ near $\omega = \pm\omega_0$ are shown in Fig. 12.10. If the filter BW is much smaller than ω_0, and assuming $\omega_m \ll \omega_0$, it is apparent from Fig. 12.10 that the attenuation factors at the two frequencies $\omega_0 \pm \omega_m$ are essentially equal. Let $\alpha(\omega_m)$ denote $|G[j(\omega_0 \pm \omega_m)]|$. It is also clear that the phase shift at $\omega = \omega_0 - \omega_m$, denoted by $\beta(\omega_m)$, is approximately the negative of that at $\omega = \omega_m$. (More generally, if the phase shift at $\omega_0 \neq 0$, the *changes* in phase for equal ω_m excursions on the two sides of ω_0 are opposite in sign and approximately equal in magnitude.) The filter output $y(t)$ is thus given by

$$y(t) = \frac{C_m \alpha(\omega_m)}{2} \{\cos[(\omega_0 - \omega_m)t + \beta(\omega_m)]$$

$$- \cos[(\omega_0 + \omega_m)t - \beta(\omega_m)]\} \tag{12.14}$$

or

$$y(t) = C_m \alpha(\omega_m) \sin[\omega_m t - \beta(\omega_m)] \sin \omega_0 t \tag{12.15}$$

Thus the envelope signal is amplified (attenuated) by the factor $\alpha(\omega_m)$ and experiences a phase lag of $\beta(\omega_m)$ in passing through G. In the more general case where G has a nonzero phase shift at $\omega = \omega_0$, the carrier signal experiences that phase shift in passing through G. Shift of the carrier phase causes no difficulty, since it is readily compensated for by arranging an equivalent shift in the demodulation reference signal.

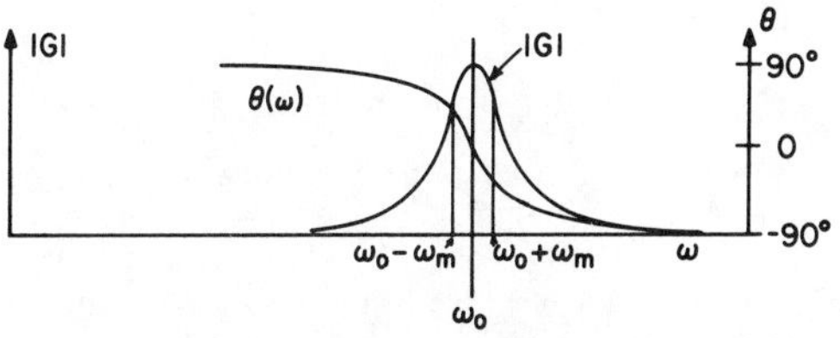

FIG. 12.10. Magnitude and phase shift versus ω for a typical bandpass amplifier.

We desire the expression for an equivalent baseband filter $G_e(s_m)$ representing the envelope signal adjustments caused by $G(s)$. $G_e(s_m)$ must have the same attenuation and phase characteristics around $\omega = 0$ as provided by $G(s)$ around $\omega = \omega_0$. This simple observation is the key to all procedures we will describe for translating back and forth between baseband (dc) and bandpass (ac) filters providing near-equivalent operations upon the information signal. One straightforward method for deriving near-equivalent filter forms is based upon their relative pole-zero patterns. The phase and amplitude response characteristics of any system described by a rational ratio of polynomials is easily determined from the pole-zero plot. Amplitude and phase characteristics in a specific region, say near $\omega = \omega_0$, depend primarily upon the poles and zeros closest to that region. Since $\omega_m << \omega_0$ in most ac servo applications, the range of frequencies around ω_0 within which the critical changes in envelope amplitude and phase take place is small. Thus these changes can be influenced by a few poles and zeros appropriately placed near $\omega = \omega_0$. The fact that other poles and zeros are required at complex plane locations remote from ω_0 for filter realization causes little difficulty, because such poles and zeros contribute essentially constant phase and magnitude effects at all frequencies near $\omega = \omega_0$.

Consider the situation shown in Fig. 12.11. Identical phase and magnitude, characteristics centered around $\omega = \omega_0$ could be obtained by placing a pole at $s = a + j\omega_0$. Unfortunately, this is impossible, since s-plane poles off the real axis introduced by realizable networks must appear as complex-conjugate pairs. Since $\omega_0 >> \omega_m$ and therefore $\omega_0 >> a$, if the pole at $-a$ in the baseband case is of any significance in altering the information signal, the additional pole required at $s = -a - j\omega_0$ in the bandpass case influences the phase and amplitude characteristics near $\omega = \omega_0$ by contributing nearly constant-phase angle and attenuation components. Thus, to a first-order approximation at least, this second pole may be neglected in synthesizing the desired amplitude and phase characteristic *shapes* near $\omega = \omega_0$. The situation is further illustrated in Fig. 12.12. The amplitude and phase curves have the desired shapes, although the phase shift at $\omega = \omega_0$ is nearly -90° instead of 0°, the phase shift of $G_e(s_m)$ at $\omega = 0$.

Returning to the question of realizing an equivalent baseband transfer function for a $G(s)$ of the general form given in Eq. (12.13), we see that

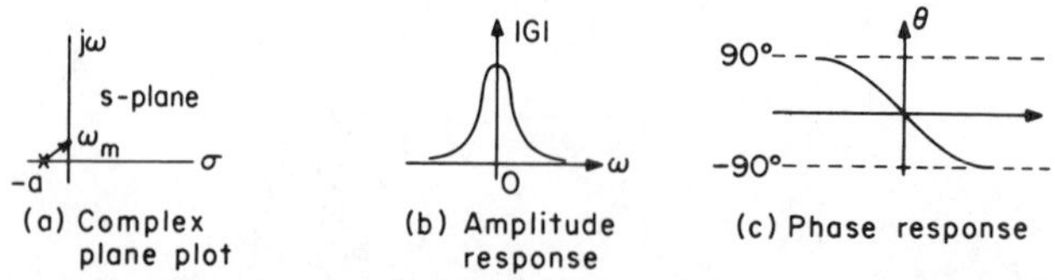

FIG. 12.11. A single pole (lag time constant) and its amplitude and phase response.

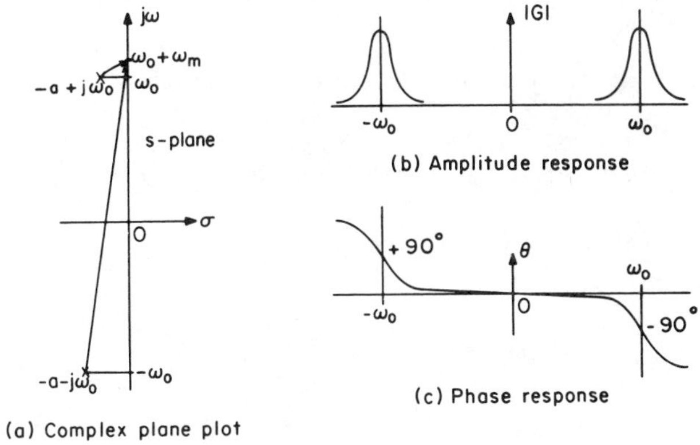

FIG. 12.12. Amplitude and phase response of a complex-conjugate pole pair.

$$G_e(s_m) = \frac{K_e}{1 + s_m/a} \tag{12.16}$$

where a is the negative real part of the complex-conjugate pole pair $G(s)$, K_e is the gain of $G(s)$ at $\omega = \omega_0$, and s_m is used as usual to indicate the equivalent baseband frequency variable. The same basic procedure may be applied to derive equivalent baseband networks to represent an arbitrary $G(s)$ acting upon the AM signal. The steps required are as follows:

1. Determine the pole-zero configuration of $G(s)$.
2. Define the poles and zeros "near" $\omega = \omega_0$ as critical in the process of shaping the amplitude and phase characteristics of the information sidebands.
3. Shift this critical pole-zero pattern to the origin, retaining the same relative pole-zero positions, to obtain an equivalent baseband network.
4. Match the gain of $G_e(s_m)$ at $\omega = 0$ to that of $G(s)$ at $\omega = \omega_0$.

Note that this process may be reversed to convert from a baseband transfer function, often a desired equalizer, to the equivalent bandpass network. In this latter procedure additional poles and/or zeros may be added at points remote from $\omega = \omega_0$ more or less arbitrarily if that makes it more convenient to realize the resulting network.

The intuitive arguments based upon pole-zero patterns for relating ac and dc networks may be developed from a rigorous mathematical point of view. Consider the baseband signals $X_1(s)$ and $X_2(s)$, the baseband filter $G_e(s_m)$, as shown in Fig. 12.13a, the modulated signals $X_1'(s)$ and $X_2'(s)$, and the filter $G(s)$ which acts upon them, as shown in Fig. 12.13b. Our objective is to relate $G(s)$ and $G_e(s_m)$. The input to $G(s)$ is the modulated signal

$$x_1'(t) = x_1(t) \cos \omega_0 t \tag{12.17}$$

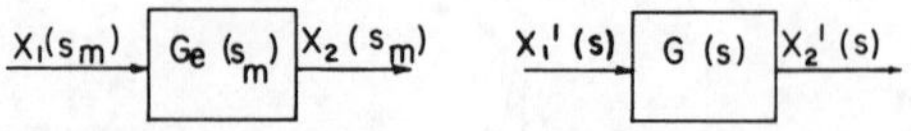

(a) Baseband compensator (b) Bandpass compensator

FIG. 12.13. Notation used in deriving useful relationships between baseband and bandpass filters.

If $G(s)$ acts upon the envelope $x_1(t)$ as $G_e(s_m)$ does, the output must be of the form

$$x_2'(t) = x_2(t) \cos \omega_0 t \tag{12.18}$$

Although a carrier phase shift will in general occur, it causes no concern, and may be set to zero with no loss of generality. Expressing Eqs. (12.17) and (12.18) in equivalent exponential form provides

$$x_1'(t) = x_1(t) \frac{\epsilon^{j\omega_0 t} + \epsilon^{-j\omega_0 t}}{2} \tag{12.19}$$

and

$$x_2'(t) = x_2(t) \frac{\epsilon^{j\omega_0 t} + \epsilon^{-j\omega_0 t}}{2} \tag{12.20}$$

Laplace transforming Eqs. (12.19) and (12.20) yields

$$X_1'(s) = \frac{X_1(s + j\omega_0) + X_1(s - j\omega_0)}{2} \tag{12.21}$$

and

$$X_2'(s) = \frac{X_2(s + j\omega_0) + X_2(s - j\omega_0)}{2} \tag{12.22}$$

Thus

$$G(s) = \frac{X_2'(s)}{X_1'(s)} = \frac{X_2(s + j\omega_0) + X_2(s - j\omega_0)}{X_1(s + j\omega_0) + X_1(s - j\omega_0)} \tag{12.23}$$

If $\pm\omega_m$ denote the frequencies of the baseband signal components, the corresponding frequency components in the ac system are at $\pm\omega_0 \pm\omega_m$. However, $\omega_0 >> \omega_m$, and the detection process eliminates all frequency components other than those near $\omega = \pm\omega_0$, so $G(s)$ as given by Eq. (12.23) may be interpreted as consisting of two parts. For $\omega \simeq \omega_0$, $X_1(s + j\omega_0)$ and $X_2(s + j\omega_0)$ are well outside the range of interest, will be highly attenuated by the system, and may be disregarded. Similarly, when $\omega \simeq -\omega_0$, $X_1(s - j\omega_0)$ and $X_2(s - j\omega_0)$ are unimportant. Thus

$$G(s) = \frac{G_e(s - j\omega_0) + G_e(s + j\omega_0)}{2} \tag{12.24}$$

is a convenient equivalent form of Eq. (12.23). It is apparent from Eq. (12.24) that $G(s)$ may be realized by placing the poles and zeros of $G_e(s_m)$ around the points $s = \pm j\omega_0$ in the same pattern as they make around $s_m = 0$ in the desired baseband filter. This is the same conclusion reached using a geometrical argument based upon the *s*-plane pole-zero diagram.

Example 12.1

Find an equivalent baseband network for the bandpass filter centered at $\omega_0 = 100$ given by

$$G(s) = \frac{100s}{s^2 + 20s + 10{,}000} \tag{12.25}$$

The poles of $G(s)$ are located at $s \simeq -10 \pm j100$. The zero at the origin results in carrier phase shift and attenuation, but need not be considered in obtaining $G_e(s_m)$, since it does not result in any appreciable *variation* in envelope phase shift or attenuation as ω traverses the range near $\omega = \omega_0 = 100$. $G_e(s_m)$ must have a pole at $s_m = -10$. The equivalent baseband gain is $|G(j100)| = 5$. Thus

$$G_e(s_m) = \frac{5}{1 + s_m/10} \tag{12.26}$$

is the equivalent baseband network desired.

Suppose we use Eq. (12.24) to obtain a bandpass network equivalent to Eq. (12.26). This yields

$$G(s) = \frac{5/2}{1 + \dfrac{s - j100}{10}} + \frac{5/2}{1 + \dfrac{s + j100}{10}} \simeq \frac{500 + 50s}{s^2 + 20s + 10{,}000} \tag{12.27}$$

Comparing Eqs. (12.25) and (12.27) we see that, although the desired poles are retained, the gain is changed, and an extra zero at $s = -10$ is introduced. The gain should always be checked and set to the appropriate value. Although application of Eq. (12.24) yields the proper pole-zero pairs near $\omega = \pm\omega_0$, additional remote zeros are often introduced because many $G(s)$ functions will serve the same equalization purpose, and Eq. (12.24) provides one filter with the appropriate response pattern near $\omega = \omega_0$. The remote zeros, if any, can be ignored in realizing $G(s)$.

Example 12.2

In the design of a particular ac servo system, it is desired to realize

$$G_e(s_m) = \frac{10(1 + s_m/10)^2}{(1 + s_m/1)(1 + s_m/100)} \tag{12.28}$$

at a carrier frequency of 10,000 rad/sec. Find the transfer function of the required bandpass filter.

The bandpass filter must have poles at $s = -1 \pm j10{,}000$ and at $s = -100 \pm j10{,}000$, must have double zeros at $s = -10 \pm j10{,}000$, and must have a gain of 10 at $\omega = \omega_0 = 10{,}000$. Thus

$$G(s) \simeq \frac{10(s^2 + 20s + 10^8)^2}{(s^2 + 2s + 10^8)(s^2 + 200s + 10^8)} \tag{12.29}$$

is required.

Although Eq. (12.24) could have been used to obtain $G(s)$, it is often easier to simply write down the desired answer as done in Ex. 12.2. One of the difficulties of equalization at ω_0 is illustrated by Ex. 12.2. Equalizer breaks at low frequencies result in complex-pole and zero pairs near the imaginary axis which are not easily realized. It is virtually impossible to precisely position the desired pole pair at $s = -1 \pm j10{,}000$ in Eq. (12.29). This difficulty can be overcome when it arises by either decreasing ω_0 or by demodulation and baseband compensation, whichever seems most appropriate.

12.6 REALIZATION OF AC COMPENSATION NETWORKS THROUGH A FREQUENCY TRANSFORMATION

Thus far we have not indicated how ac equalizers should be realized. A straightforward procedure can be developed for obtaining the ac equalizer component arrangement and values directly from the equivalent baseband compensating network. Each baseband circuit element must be transformed into an equivalent combination of elements with the same impedance at $\omega = \omega_0 + \omega_m$

as noted for the original element at $\omega = \omega_m$. Thus we seek a frequency translation of the form

$$p(j\omega) = j(\omega - \omega_0) \qquad \omega_0 - \omega_{BW} < \omega < \omega_0 + \omega_{BW} \tag{12.30}$$

where $2\omega_{BW}$ is the frequency range about ω_0 within which significant information is contained in the modulation sidebands. The desired translation cannot be realized in this form, because it does not treat negative frequencies symmetrically, but a near-equivalent realizable transformation is

$$p(j\omega) = j(\omega - \omega_0)\left(\frac{\omega + \omega_0}{2\omega}\right) \tag{12.31}$$

Note that $\omega \simeq \omega_0$ in the range of interest for ac system components and $(\omega + \omega_0)/2\omega \simeq 1$ for all frequencies near ω_0. Thus, we may alternately write Eq. (12.31) as

$$p(j\omega) = j\omega\left(\frac{1}{2} - \frac{\omega_0^2}{2\omega^2}\right) \tag{12.32}$$

from which the effect of this transformation is more readily apparent. An inductance which has impedance $Z_L(j\omega) = j\omega L$ before the transformation becomes an equivalent element combination with impedance

$$Z_L[p(j\omega)] = Z_L'(j\omega) = \frac{j\omega L}{2} + \frac{L\omega_0^2}{2j\omega} \tag{12.33}$$

which is recognized as the impedance of a series LC circuit with

$$L' = \frac{L}{2} \qquad C'' = \frac{2}{\omega_0^2 L} \tag{12.34}$$

Note that this series circuit resonates at ω_0, since $1/\sqrt{L'C''} = \omega_0$. Similarly, a capacitance C is transformed into an equivalent element combination with admittance

$$Y_C[p(j\omega)] = Y_C'(j\omega) = \frac{j\omega C}{2} + \frac{\omega_0^2 C}{2j\omega} \tag{12.35}$$

which is recognized as a parallel LC circuit with

$$C' = \frac{C}{2} \qquad L'' = \frac{2}{\omega_0{}^2 C} \tag{12.36}$$

This circuit is anti-resonant at $\omega = \omega_0$, since $1/\sqrt{L''C'} = \omega_0$. Since resistances are not frequency dependent, they are unchanged by the frequency transformation. A general design procedure is thus:

1. Determine baseband-equivalent transfer functions for all system components exclusive of the equalizer.
2. Synthesize the controller at baseband using standard design concepts to arrive at an equalizer network.
3. Replace each capacitance in the baseband equalizer by a lossless, parallel LC circuit with $C' = C/2$ and inductance chosen to resonate at $\omega = \omega_0$.
4. Replace each inductance in the baseband equalizer by a lossless series LC circuit with $L' = L/2$ and capacitance chosen to resonate with L' at $\omega = \omega_0$.

Although this provides a straightforward theoretical procedure for realizing ac equalizers, there are practical difficulties. At least one inductor is required in all nontrivial applications. For carrier frequencies of 400 Hz or higher, and 400 Hz is relatively standard in aircraft and missile applications, the required accuracy is difficult to obtain with components of standard tolerance. As ω_0 decreases, the L's required become prohibitively large. Fortunately, as shown in Sec. 12.7, equalizer networks not requiring inductors are sometimes possible.

Example 12.3

In a particular 400 Hz ac servo design, analysis at baseband shows that

$$G_e(s_m) = \frac{1 + s_m/100}{2(1 + s_m/10)} \tag{12.37}$$

is required as an equalizer. Find an RLC network which will provide equivalent envelope compensation at 400 Hz using the frequency translation procedure.

The RC network shown in Fig. 12.14 has transfer function $G_e(s_m)$, as readily checked. To convert to the bandpass equivalent, each C is replaced by $C/2$ and

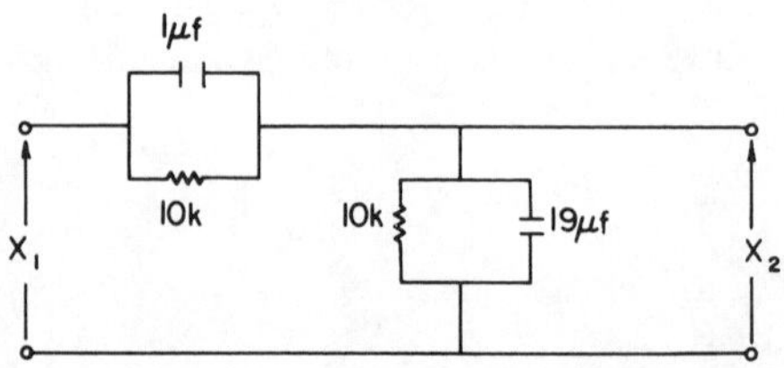

FIG. 12.14. Baseband network used in Ex. 12.3.

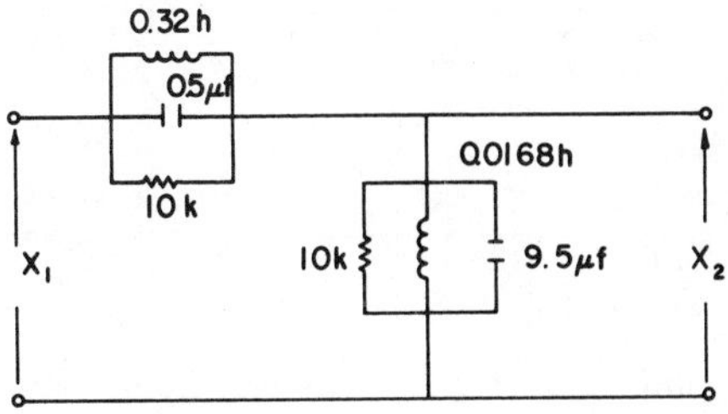

FIG. 12.15. Bandpass equivalent of the network in Fig. 12.14 for $\omega_c = 2500$.

parallel resonated with an L at $\omega = 400(2\pi) \simeq 2500$ rad/sec. The result is as shown in Fig. 12.15. It is important that the L's and C's be adjusted very precisely. The losses in the L's are readily compensated for by making small adjustments (increases) in the parallel R's.

It is now clear that ac servo design is readily carried out at baseband using all of our standard procedures. Baseband-equivalent circuits are obtained for all sensors, transducers, amplifiers and actuators. Baseband (dc) compensation is chosen in the usual way. Compensation is realized either by demodulating the signal, compensating at baseband, and remodulating the result, a procedure which must be used when the carrier frequency tends to shift and when it is not feasible to build the desired ac equalizer; or by applying an equivalent ac equalizer at ω_0. Several procedures have been discussed for accomplishing the latter. All of the methods considered thus far require the use of one or more LC circuits, which may be difficult to realize. In the next section we consider several special RC networks which are sometimes useful as ac equalizers.

12.7 SPECIAL TYPES OF RC NETWORK AC COMPENSATORS

It is generally desirable to realize ac compensators using only R's and C's if possible, since L's are bulky, heavy, often nonlinear, particularly in large sizes, and cannot be set as precisely as R's and C's. We can often approximate the desired ac transfer function using RC networks. Consider, for example, the dc lead equalizer and the corresponding ac compensator shown in Fig. 12.16. The RC network shown in Fig. 12.17 provides an approximately equivalent ac

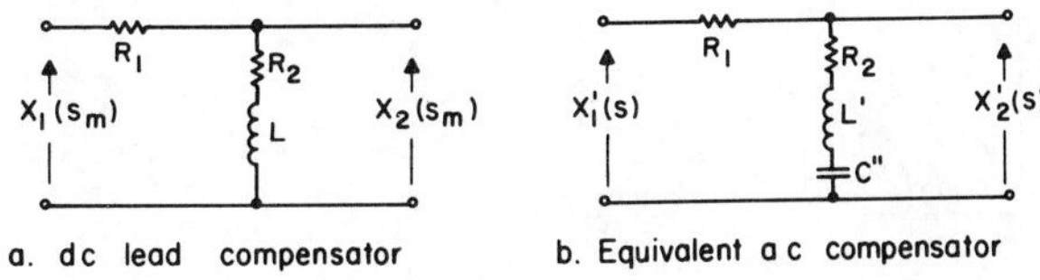

FIG. 12.16. Dc and ac lead networks involving an inductor.

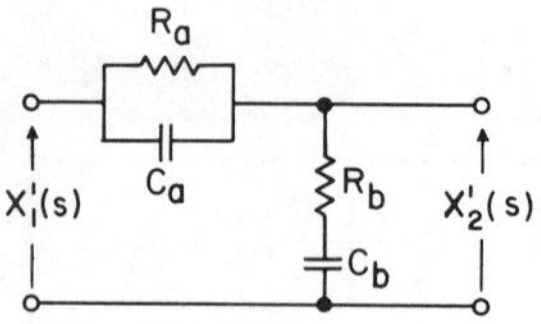

FIG. 12.17. Lead network approximately equivalent to that shown in Fig. 12.16.

compensation effect. We initiate the proof of this claim by observing that

$$G(s) = \frac{X_2'(s)}{X_1'(s)} = \frac{(1 + R_bC_bs)(1 + R_aC_as)}{(1 + R_bC_bs)(1 + R_aC_as) + R_aC_bs} \tag{12.38}$$

is the transfer function of the network shown in Fig. 12.17. Letting $s = j\omega$ in Eq. (12.38) and simplifying provides

$$G(j\omega) = \frac{(1 - \omega^2R_aC_aR_bC_b) + j\omega(R_aC_a + R_bC_b)}{(1 - \omega^2R_aC_aR_bC_b) + j\omega(R_aC_a + R_bC_b + R_aC_b)} \tag{12.39}$$

The polar plot of $G(j\omega)$ is a circle, as shown in Fig. 12.18, with unity transmittance at $\omega = 0$ and $\omega = \infty$. For comparison, the phase characteristic of an ideal ac lead network at $\omega = \omega_0$ is shown in Fig. 12.19. Comparison of Figs. 12.18 and 12.19 shows that the phase shift introduced by $G(j\omega)$ provides lead compensation approximating that of an ideal lead network over a range of frequencies near $\omega = \omega_0$.

We must be able to "tune" $G(j\omega)$ to the desired frequency. $G(j\omega_0)$ is real, and thus the parameters are properly adjusted to realize lead compensation if the real parts of the numerator and denominator in Eq. (12.39) are made zero, or if

$$\omega_0{}^2 = \frac{1}{R_aC_aR_bC_b} \tag{12.40}$$

to simplify the notation, let

$$T_1 = R_aC_a + R_bC_b \qquad T_2 = R_aC_a + R_bC_b + R_aC_b \tag{12.41}$$

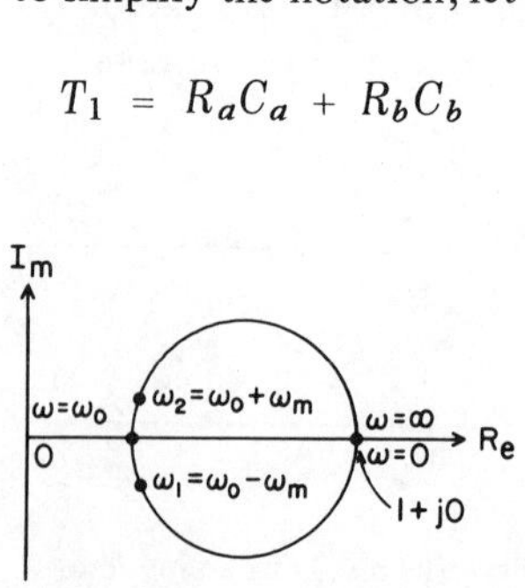

FIG. 12.18. Polar plot of $G(j\omega)$ as given by Eq. (12.39).

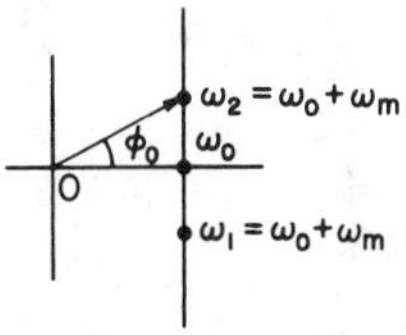

FIG. 12.19. Phase versus frequency near $\omega = \omega_0$ for an ideal ac lead compensator.

Now Eq. (12.39) may be rewritten as

$$G(j\omega) = \frac{\left(1 - \dfrac{\omega^2}{\omega_0^2}\right) + j\omega T_1}{\left(1 - \dfrac{\omega^2}{\omega_0^2}\right) + j\omega T_2} = \frac{T_1}{T_2} \frac{1 + j\dfrac{(\omega^2/\omega_0^2) - 1}{\omega T_1}}{1 + j\dfrac{(\omega^2/\omega_0^2) - 1}{\omega T_2}} \tag{12.42}$$

Note that

$$\frac{\omega^2}{\omega_0^2} - 1 = \frac{\omega^2 - \omega_0^2}{\omega_0^2} = \frac{(\omega - \omega_0)(\omega + \omega_0)}{\omega_0^2} \simeq \frac{2\omega\,\omega_m}{\omega_0^2} \tag{12.43}$$

The approximation used in Eq. (12.43) is valid for $\omega_0 >> \omega_m$, since $\omega + \omega_0 \simeq 2\omega$ and $\omega - \omega_0 = \omega_m$. Using the approximation introduced in Eq. (12.43), Eq. (12.42) may be written in terms of ω_m as

$$G_e(j\omega_m) \simeq \frac{T_1}{T_2} \frac{1 + j\omega_m(2/T_1\omega_0^2)}{1 + j\omega_m(T_1/T_2)(2/T_1\omega_0^2)} \tag{12.44}$$

or

$$G_e(j\omega_m) = \gamma \frac{1 + j\omega_m T}{1 + \gamma j\omega_m T} \tag{12.45}$$

which has the form of a baseband lead compensator. The parameters γ and T are defined by

$$\gamma = \frac{T_1}{T_2} = \frac{R_aC_a + R_bC_b}{R_aC_a + R_bC_b + R_aC_a} < 1 \tag{12.46}$$

and

$$T = \frac{2}{T_1\omega_0^2} = \frac{2R_aC_aR_bC_b}{R_aC_a + R_bC_b} \tag{12.47}$$

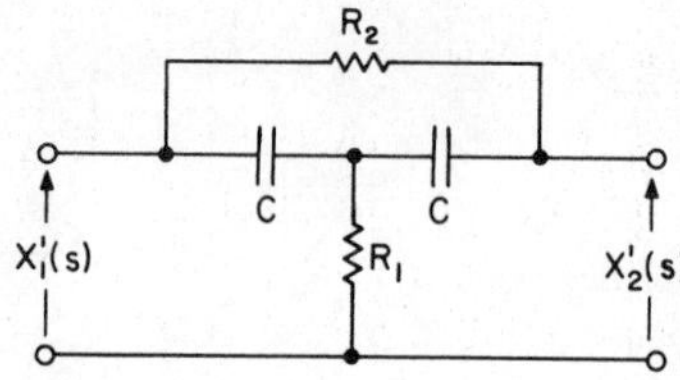

FIG. 12.20. A resistor shunt bridged T network.

Once the γ and T of the desired baseband lead equalizer are determined, parameters of the RC network shown in Fig. 12.17 can be chosen to accomplish equivalent equalization at ω_0 by imposing the conditions given in Eqs. (12.40), (12.46), and (12.47) to solve for R_a, R_b, C_a and C_b. As usual one parameter, and thus the impedance level, may be chosen arbitrarily, since there are four free parameters and only three restrictions to be satisfied.

The RC bridged T network shown in Fig. 12.20 may also be used as an ac compensator. Its transfer function is

$$G(j\omega) = \frac{(\omega^2 R_1 R_2 C^2 - 1) - j2\omega R_1 C}{(\omega^2 R_1 R_2 C^2 - 1) - j\omega C(2R_1 + R_2)} \tag{12.48}$$

This $G(j\omega)$ has the same form as that given in Eq. (12.39), and the polar plot shown in Fig. 12.18 applies here, also. Setting $\omega_0{}^2 = 1/R_1 R_2 C^2$ tunes the network to ω_0, and Eq. (12.48) becomes

$$G(j\omega) = \frac{\dfrac{R_1 R_2 C^2(\omega^2 - \omega_0{}^2)}{\omega} - j2R_1 C}{\dfrac{R_1 R_2 C^2(\omega^2 - \omega_0{}^2)}{\omega} - jC(2R_1 + R_2)} \tag{12.49}$$

Using the same approximation $[(\omega + \omega_0)/\omega \simeq 2]$ as in the previous development yields $G_e(j\omega_m)$ as

$$G_e(j\omega_m) = \gamma \frac{1 + j\omega_m T}{1 + \gamma j\omega_m T} \tag{12.50}$$

where

$$\gamma = \frac{2R_1}{2R_1 + R_2} < 1 \qquad T = R_2 C \tag{12.51}$$

Since Eqs. (12.50) and (12.45) are identical the networks shown in Figs. 12.17 and 12.20 can be used interchangeably to provide phase lead compensation at ω_0.

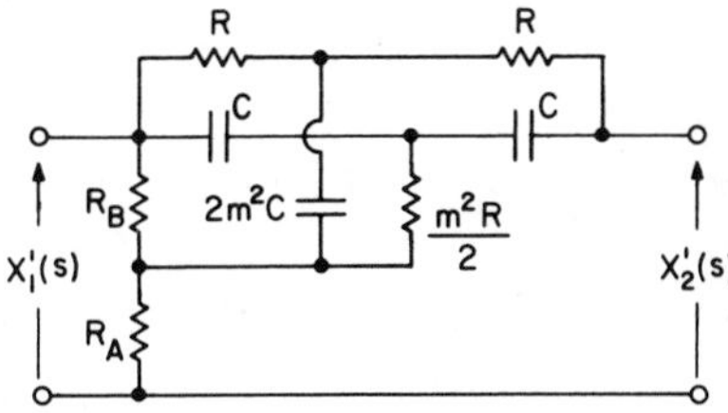

FIG. 12.21. Modified parallel T network for use as an ac lead compensator.

One final lead compensation network is shown in Fig. 12.21. The standard parallel T network shown in Fig. 12.22 has transfer function

$$G(j\omega) = \frac{\omega^2 m^2 R^2 C^2 - 1}{\omega^2 m^2 R^2 C^2 - 1 - j2\omega C(m^2 - 1)} \tag{12.52}$$

which is zero when

$$\omega = \frac{1}{mRC} \tag{12.53}$$

If the network is tuned to ω_0 by setting $\omega_0 = 1/mRC$, Eq. (12.52) becomes

$$G_e(j\omega_m) \simeq \frac{m}{(m^2 - 1)\omega_0}\left[\frac{j\omega_m}{j\dfrac{m\omega_m}{(m^2 - 1)\omega_0} + 1}\right] \tag{(12.54)}$$

Since the parallel T has no transmission at $\omega = \omega_0$, however, it is not generally useful in this form. The modification shown in Fig. 12.21 bypasses some of the input signal around the network at all frequencies. R_a and R_b make up a voltage divider at a low impedance level relative to the network input impedance. The modified network is described by

$$G_e(j\omega_m) \simeq \gamma\frac{j(\beta\omega_m/\gamma) + 1}{j\beta\omega_m + 1} \tag{12.55}$$

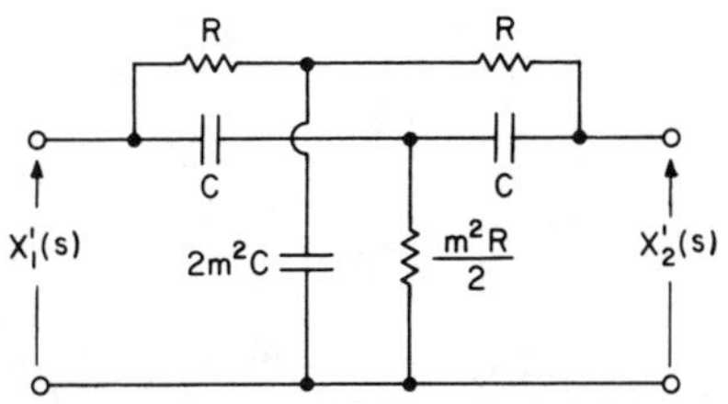

FIG. 12.22. Standard parallel T network.

where

$$\gamma = \frac{R_a}{R_a + R_b} < 1 \qquad \beta = \frac{m}{(m^2 + 1)\omega_0} \tag{12.56}$$

Since $\gamma < 1$, this network also provides equivalent phase lead compensation at ω_0 at the expense of attenuation equal to the lead span realized.

Although phase lead near ω_0 is relatively easily obtained with a variety of RC network arrangements, it is difficult to obtain a lag equivalent with passive RC networks. A phase lag equivalent network must peak up rather sharply at $\omega = \omega_0$, as shown in Fig. 12.23. Since RC networks can have only real poles, this requires that *many* zeros be located below $\omega = \omega_0$. The rapid decrease of gain above $\omega = \omega_0$ may be realized by placing one or more zeros near the $j\omega$ axis just above $\omega = \omega_0$, or by using several real poles with break frequencies near $\omega = \omega_0$. No matter which way we attempt to realize the network it is prohibitively complex, and has a large amount of flat loss (loss or attenuation at all frequencies near $\omega = \omega_0$) which requires that additional gain be added at considerable extra expense.

The RC compensation networks introduced here do not provide nearly the same degree of flexibility in shaping the envelope frequency and phase characteristics as is available using RLC networks. Placement of the network zeros determines how rapidly phase increases with frequency near $\omega = \omega_0$, and thus where the equivalent lead break (or breaks) occurs, but the equivalent lag break location depends to a considerable extent upon ω_0, over which we generally have no control. This condition results because all RC network poles must be on the negative real axis. Thus, although phase lead is possible, the range over which it is obtained is not adjustable with any great precision by controlling poles so far from ω_0. It is often necessary to accept large signal attenuation to realize the desired phase shifts. These points are illustrated by the design example considered in Sec. 12.7.

Since it is not generally possible to solve explicitly for all of the components in ac equalizers, it is convenient to have an organized trial and error design procedure to follow. Several ac networks, their equivalent baseband transfer characteristics, and suggested trial and error design procedures for each case are presented in Table 12.1.

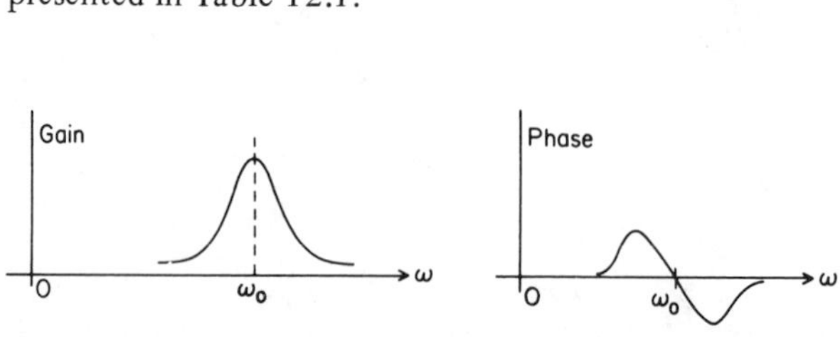

FIG. 12.23. Gain and phase characteristics of a phase lag network near $\omega = \omega_0$.

Table 12.1. A summary of ac compensator networks and design procedures.

Network	Transfer function	Design procedure
1. 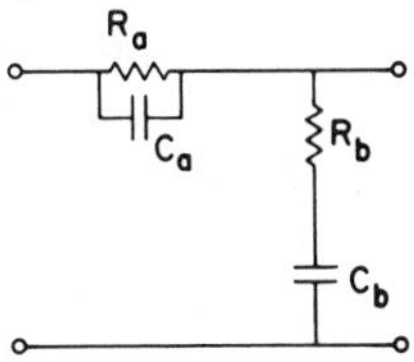	$G_e(s_m) = \gamma \dfrac{1+Ts_m}{1+\gamma Ts_m}$ $T_a = R_aC_a$ $T_b = R_bC_b$ $T_{ab} = R_aC_b$	1. Find T_b from $T = 2T_b/(\omega_0^2 T_b^2 + 1)$ 2. Solve for T_a from $T_a = 1/\omega_0^2 T_b$ 3. Solve for T_{ab} from $\gamma = \dfrac{T_a + T_b}{T_a + T_b + T_{ab}}$ 4. Choose R's and C's for appropriate Z level.
2. 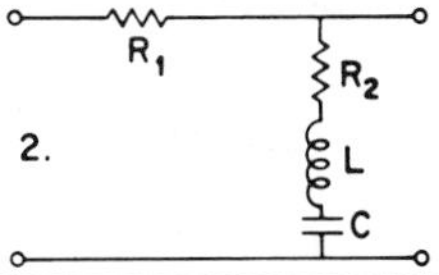	$G_e(s_m) = \gamma \dfrac{1+Ts_m}{1+\gamma Ts_m}$ $0<\gamma<1$	1. Choose a value for L 2. Set $C = 1/\omega_0^2 L$ 3. Set $R_2 = 2L/T$ 4. Set $R_1 = R_2(1-\gamma)/\gamma$
3. 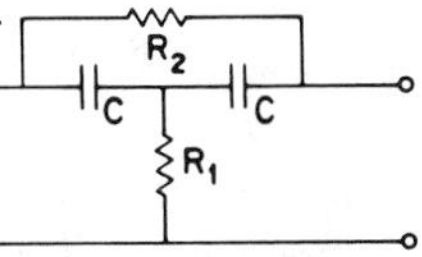	$G_e(s_m) = \gamma \dfrac{1+Ts_m}{1+\gamma Ts_m}$ $0<\gamma = \dfrac{2}{2+\omega_0^2 T^2} < 1$	1. Choose a value for C 2. Set $R_2 = T/C$ 3. Set $R_1 = 1/TC\omega_0^2$ Note that you must take what you get for γ
4. 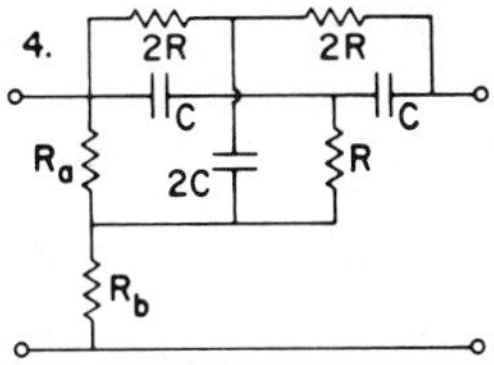	$G_e(s_m) = \gamma \dfrac{1+Ts_m}{1+\gamma Ts_m}$ $0<\gamma = \dfrac{1}{2\omega_0 T} < 1$	1. Choose a value for R_a 2. Set $R_b = R_a/(2\omega_0 T - 1)$ 3. Choose a value for C 4. Set $R = 1/\omega_0 C$ Once again you are not free to choose a value for γ.
5. 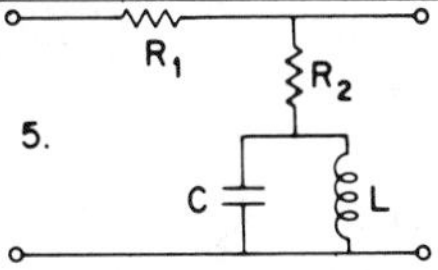	$G_e(s_m) = \dfrac{1+Ts_m}{1+Ts_m/\gamma}$ $0<\gamma<1$	1. Choose a value for L 2. Set $C = 1/\omega_0^2 L$ 3. Set $R_2 = T/2C$ 4. Set $R_1 = (1-\gamma)R_2/\gamma$

12.8 AN AC DESIGN EXAMPLE

The results of our ac system study are summarized here by designing the instrument servo system shown in Fig. 12.24. Analysis of the motor and load shows that

$$G_m(s_m) = \frac{0.01}{s_m(1 + s_m/10)} \tag{12.57}$$

The synchro pair provides a gain of 50 volts/rad and has a lag time constant so small that it can be disregarded. The amplifier time constants are also insignificant.

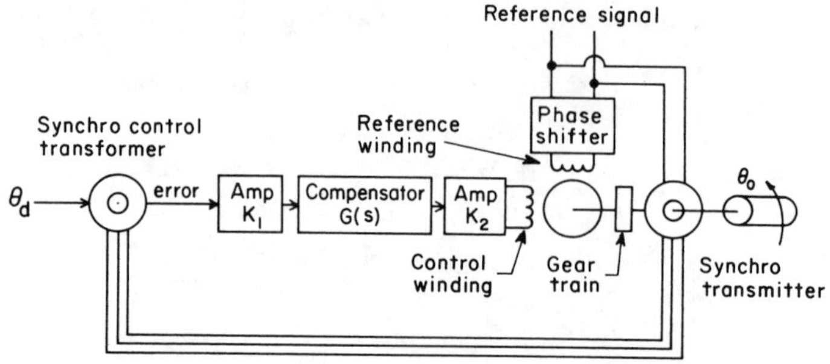

FIG. 12.24. Schematic diagram of a simple ac angular position controller.

The carrier frequency is 400 Hz $\simeq$ 2500 rad/sec. Assume ω_0 does not drift. The following specs must be satisfied:

1. $K_v = 100$
2. $\theta_m \geq 45°$.

We will determine the compensator $G(s)$ and the gain settings required.

This system may be represented by the block diagram shown in Fig. 12.25. The K_v requirement fixes the loop gain at 100, so, assuming the equalizer will have unit dc gain, $K = 200$ is required. The uncompensated baseband-equivalent Bode diagram may now be drawn as shown in Fig. 12.26. It is obvious that the θ_m spec is not satisfied without an equalizer. An acceptable compensated diagram using lead equalization and the corresponding equalizer Bode diagram are also shown in Fig. 12.26. The baseband lead equalizer required is

$$G_e(s_m) = \frac{1 + s_m/20}{1 + s_m/100} \tag{12.58}$$

Suppose we try to realize this transfer characteristic around ω_0 using the bridged T configuration listed as item 3 in Table 12.1. Choose $C = 0.1\mu f$. Solving for R_2 and R_1 provides

$$C = 0.1\mu f \qquad R_2 = 500\,K\Omega \qquad R_1 = 32\Omega \tag{12.59}$$

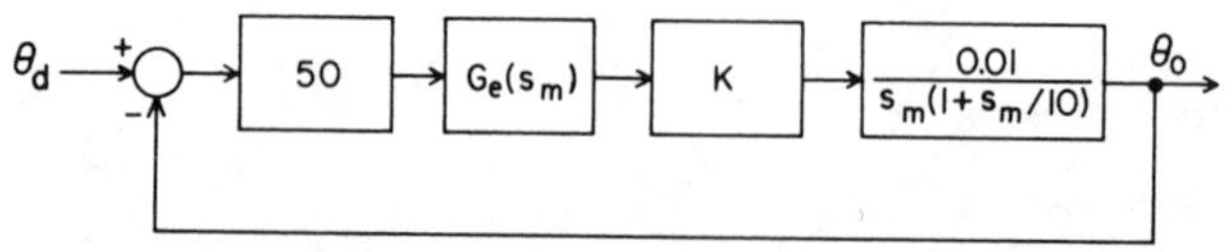

FIG. 12.25. Equivalent baseband block diagram of the system shown in Fig. 12.24.

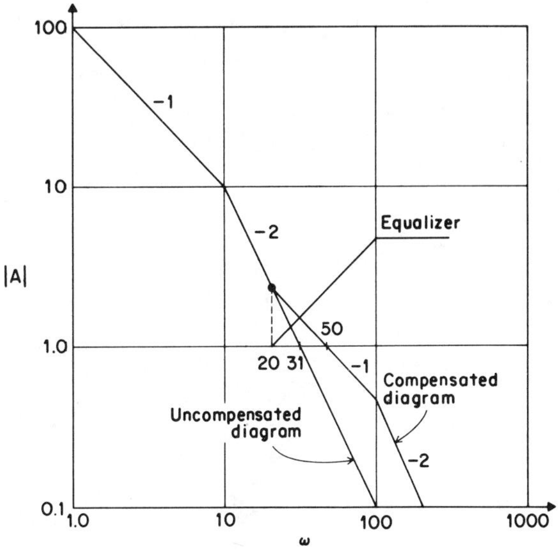

FIG. 12.26. Bode diagrams for the system shown in Fig. 12.25.

This wide divergence of parameter values is common, and they must be set precisely. Furthermore,

$$\gamma = \frac{2}{2 + \omega_0{}^2 T^2} = 1.28 \times 10^{-4} \qquad (12.60)$$

which means we must either revise the design or accept a lead span and an attenuation loss of nearly 10,000 through the network. The excess lead span causes no problem, but the attenuation is another matter. The only way we can decrease this attenuation, and the additional gain required to overcome it, is by decreasing ω_0, over which we are assumed to have no control, or by increasing T, thereby postponing the introduction of lead compensation. At best we can increase T to only about 1/60 without violating the θ_m spec, assuming the resulting -1 slope segment due to the compensation remains extremely long. This assumption is only valid provided there are no lag terms which have been disregarded with break frequencies prior to $\omega_m \simeq 1000$. Suppose we set T = 1/60. The corresponding $\gamma \simeq 10^{-3}$. This seems more reasonable, but an additional gain of even 1000 is something to think about. We thus conclude that, although this RC bridge network will provide an equalization arrangement, significant problems are thereby introduced and compromises must be made. The Bode diagram for $T = 1/60$ and $\gamma \simeq 10^{-3}$ is shown in Fig. 12.27. It is straightforward to show that, although the results are far from what we set out to obtain, the specs are satisfied.

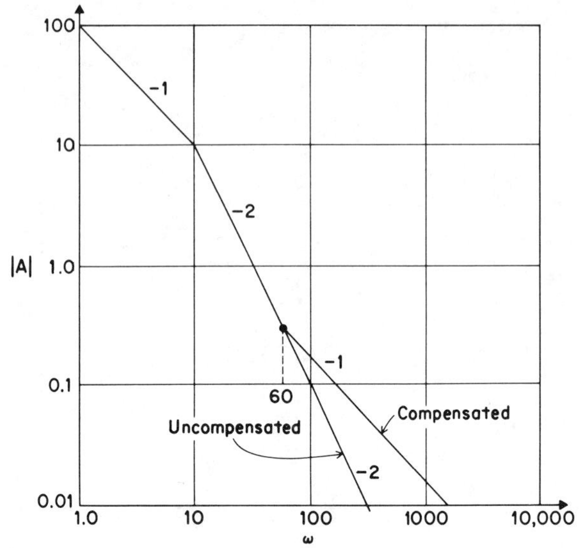

FIG. 12.27. Bode diagrams using $T = 1/60$.

Now let us use the RLC lead network shown as entry 2 in Table 12.1. In this case we can come much closer to achieving the compensated diagram shown in Fig. 12.26, within the limits of component tolerances. Following the suggested design procedure, choose $L = 0.1h$. Then $C = 1/\omega_0{}^2 L = 1.6\mu f$, $R_2 = 2L/T = 4\Omega$, and $R_1 = R_2(1 - \gamma)/\gamma = 16\Omega$. These values are not convenient to work with, since the impedance level is so low, but the only alternatives are to either increase L, which also is not very desirable, or decrease γ, which increases the signal attenuation through the filter. Decreasing γ also increases the -1 slope span near ω_c, but this causes no difficulty. Realization of ac compensation networks is not a trivial network synthesis problem, but the requirements are clearly defined and the compromises apparent.

Suppose we use ac tachometer feedback around the motor in an effort to equalize the system. The tachometer signal can be added at the input to the amplifier shown with gain K_2 in Fig. 12.24. If the tachometer has transfer function $K_t s_m$, the closed-loop transfer function of the resulting minor loop becomes

$$G_m(s_m) = \frac{0.01K_2}{s_m(1 + 0.01K_2K_t + s_m/10)} = \frac{0.01K_2/(1 + 0.01K_2K_t)}{s_m[1 + s_m/10(1 + 0.01K_2K_t)]} \tag{12.61}$$

It is clear from Eq. (12.61) that tachometer feedback decreases the forward gain and increases the lag time constant break frequency. From Fig. 12.26 we see that $\theta_m \simeq 45°$ if the motor break frequency is pushed out to $\omega_m = 100$, which requires

$$0.01K_2K_t = 9 \qquad K_2 = \frac{900}{K_t} \tag{12.62}$$

The value of K_1 must be adjusted to retain $K_v = 100$. The use of tachometer feedback is clearly the simplest design solution.

Once the desired compensation is established, the phase shift through the system is checked and the motor reference input signal phase appropriately set. With this final adjustment, the system should be ready to perform as predicted analytically.

12.9 SUMMARY AND CONCLUSIONS

We have developed methods which allow us to analyze and design ac systems at baseband using familiar procedures. Although a bit more time is required for the analysis, and the accuracy with which performance can be predicted is generally poorer than usual, ac system design involves the same basic steps carried out in designing standard ac systems. Equalizer realization for AM systems is considerably more troublesome, but standard synthesis methods are available and the principles are straightforward. The components used in ac systems are often cheaper, lighter, smaller, and more efficient than comparable dc devices, and ac systems are used where these advantages outweigh the disadvantages associated with equalization. Learning to design ac systems is largely an exercise involving familiarization with the components used and their mathematical descriptions, since no fundamentally new analysis principles are involved.

REFERENCES

1. Gibson and Tuteur, "Control System Components," McGraw-Hill Book Co., New York, 1958, Chap. 5.
2. Airato, A. J., "An error Study of AC System Compensation Methods," M.S.E.E. Thesis, Syracuse University, Syracuse, New York, August, 1969.

13 BASIC STATE-SPACE CONCEPTS

In our previous discussions of plant dynamics we have considered the system almost exclusively as defined by its subsystem and overall transfer functions. All analysis and design procedures are thus initiated by developing transfer functions from the integro-differential equations describing each system component. Thereafter the plant is considered as a "black box" described by these transfer functions. In this chapter we introduce a new point of view in which n first-order differential equations (DE's) are used to describe an nth-order plant. These equations are generally called the *state space representation* and the n variables chosen to describe the system are called *state variables.* The state variables are defined as elements of an $n \times 1$ *state vector,* and system performance is evaluated by manipulating the differential and algebraic equations expressed in matrix form. Our fundamental purpose in this chapter is to develop a basic understanding of the methods available for working with equations of this type.

The state-space representation of a given system is not unique and standard forms can be developed in several ways. We could develop an algorithm for choosing the independent variables used in the original set of integro-differential equations such that all of the resulting equations are of the first order. An algorithm of this form is often used in network analysis, for example, where capacitor voltages and inductor currents are chosen as the standard state variables.

An alternate method is to start from the original set of describing integro-differential equations of arbitrary order and transform each of them into an equivalent set of first-order DE's. The latter method is chosen here, since it allows us to start from the usual system description. We will use simulation diagrams as an aid in state equation development, thereby arriving at several standard forms. Standardization is important because an infinity of state representations is possible and manipulative simplicity is often promoted by structuring the problem carefully. General forms also allow us to make comparisons readily and make it much easier to arrive at general conclusions.

A brief review of the standard matrix operations used throughout this chapter is presented in App. D to help refresh the memories of those who may not use matrices regularly. Development of the state equations in several forms from the original integro-differential equation or transfer function system description is discussed initially. Transformations of the state equations into several canonical forms are developed. The general solution structure for linear time-invariant systems expressed in state equation form is presented. The transition matrix concept is introduced and used in describing the general solution for forced linear time-varying systems. Controllability and observability are defined and the condition which must be satisfied by the state equations to assure a controllable and observable system are presented. The use of state-variable feedback in the design of closed-loop systems, rather than the classical series or minor-loop equalization techniques, is also considered.

A strong unifying influence throughout this chapter is the placement of equal emphasis upon all state elements rather than considering the output response as the only significant system characteristic. The study of systems in state equation form leads to a variety of interesting and powerful design procedures. Although we must generally pay a price in solution complexity for the added generality of state representation, the reward obtained is a fuller representation of system performance, which often leads to an improved design. The state equation representation is a valuable supplement to the standard classical analysis and design procedures discussed in previous chapters. It is important to note, however, that it is not intended as a substitute for them. The relative ease with which we can handle a majority of our control system design problems using classical methods is the reason they are retained as fundamental background. The apprentice in any craft is first taught to handle those tools he uses most often. As time passes our emphasis tends more toward the state representation and away from classical methods, however, because the ready availability of digital computers is rapidly influencing the methods we use to structure and solve our problems. State representation is particularly convenient when a simulation study or a computational solution is intended.

13.1 GENERAL DEVELOPMENT OF THE STATE-SPACE FORMULATION

In this section we discuss the general procedure for reducing DE and transfer function representations of systems to state vector form. If the input to output characteristics of the system are described by an nth-order DE, it is called an nth-order system. Alternately, the denominator polynomial in the s-domain input to output transfer function representation has s^n as its highest-power term in s. The transfer relationship for a multivariable system with q inputs and p outputs is described by a $p \times q$ transfer matrix $\mathbf{G}(s)$. Each matrix element $g_{ij}(s)$ is a transfer function relating the ith output to the jth input and has a denominator polynomial in s of degree l_{ij}. A multivariable system is of nth order if the common denominator of the $g_{ij}(s)$ terms is of degree n. (We tacitly assume the system is controllable and observable. If this is not the case, n, as defined here, is only a lower limit on system order.) The DE description of this nth-order multiple input, multiple output system is given by a set of p simultaneous DE's, each of nth-order or less.

The block diagram of a general nth-order q input, p output plant is shown in Fig. 13.1. If we wish to obtain the output vector $\mathbf{y}(t)$ for this system at some time point t_1, we must have complete knowledge of the input vector $\mathbf{m}(t)$ and its effect upon the system for all time prior to t_1. This is provided, for example, if the system equations are available and $\mathbf{m}(t)$ is known over the closed interval $[-\infty, t_1]$. If $\mathbf{m}(t)$ is known only on $[t_0, t_1]$, additional information must be provided at some time t_a, where $t_0 \leq t_a \leq t_1$. In general t_a is chosen as t_0 and the additional information provided is the system state at t_0, denoted by $\mathbf{x}(t_0)$ or $\mathbf{x}^\circ$. We call $\mathbf{x}^\circ$ the system initial condition vector. The n components of $\mathbf{x}^\circ$ provide the n conditions required to completely define system response. System state is defined as: *the state of a system, denoted by* $\mathbf{x}(t)$, *is an n element vector containing all relevant information concerning the past history of the system required to completely define the response for an arbitrary input starting at time t.* It should be noted that although system dimensionality is unique, system state may be chosen in an infinity of ways.

The state equations describing a general linear system are of the form

$$\dot{\mathbf{x}}(t) = \mathbf{A}(t)\,\mathbf{x}(t) + \mathbf{B}(t)\,\mathbf{m}(t) \qquad \mathbf{x}(t_0) = \mathbf{x}^\circ \tag{13.1}$$

$$\mathbf{y}(t) = \mathbf{C}(t)\,\mathbf{x}(t) + \mathbf{D}(t)\,\mathbf{m}(t) \tag{13.2}$$

where

$\mathbf{x}(t)$ = the n-element state vector
$\mathbf{m}(t)$ = the q-element input vector
$\mathbf{y}(t)$ = the p-element output vector

and $\mathbf{A}(t)$, $\mathbf{B}(t)$, $\mathbf{C}(t)$, and $\mathbf{D}(t)$ are time varying matrices of $n \times n$, $n \times q$, $p \times n$, and $p \times q$ dimensions, respectively. If the coefficients of the linear-ordinary DE's describing the system are constants, the matrices $\mathbf{A}$, $\mathbf{B}$, $\mathbf{C}$ and $\mathbf{D}$ do not depend upon t and the system is called time invariant.

Simulation diagrams are introduced at this point to aid in our development of the state equations. The five basic operational blocks used in simulation

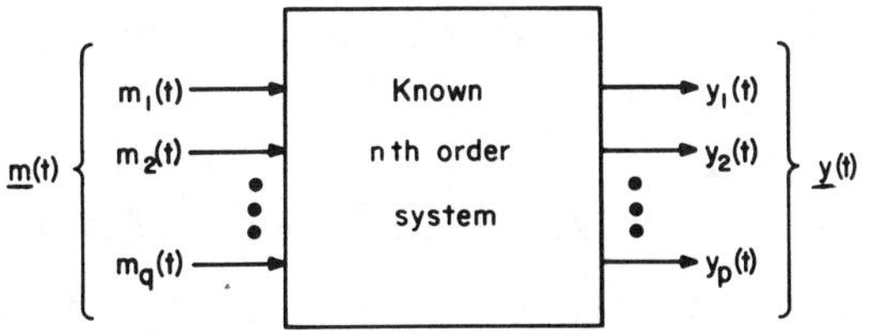

FIG. 13.1. Block diagram of a general multiple-input-multiple-output system.

diagrams are: 1. integrators; 2. amplifiers; 3. adders; 4. multipliers; and 5. function generators. These elements are all shown schematically in Fig. 13.2. They are operationally similar to those used in developing analog computer flow diagrams. The phase inversion present in analog computer operational amplifier units is omitted here to provide a mathematical tool of minimum complexity. Simulation diagrams of DE's are developed by relating the variables in the equations functionally using the standard operational blocks. Only integrators, amplifiers, and adders are needed to describe linear time-invariant ordinary DE's. Many of the operations discussed here are similar to those presented in App. C, where analog simulation is considered.

Example 13.1

Develop a state-equation formulation for the system described by

$$\ddot{y}(t) + a_1\dot{y}(t) + a_0 y(t) = b_0 z(t) \tag{13.3}$$

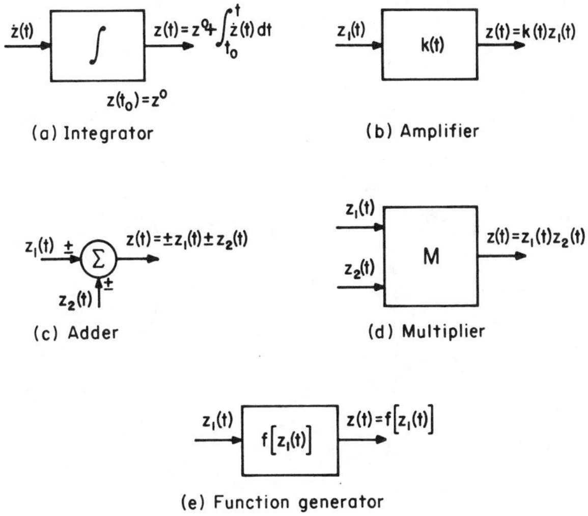

FIG. 13.2. Basic operational blocks for simulation diagrams.

with

$$\dot{y}(t_0) = \dot{y}^\circ \qquad y(t_0) = y^\circ \tag{13.4}$$

Integration of $\ddot{y}(t)$ yields $\dot{y}(t)$, which is integrated to obtain $y(t)$. Thus we solve Eq. (13.3) for $\ddot{y}(t)$ to obtain

$$\ddot{y}(t) = b_0 z(t) - a_1 \dot{y}(t) - a_0 y \tag{13.5}$$

and observe that

$$\dot{y}(t) = \dot{y}^\circ + \int_{t_0}^{t} \ddot{y}(t)\, dt \tag{13.6}$$

and

$$y(t) = y^\circ + \int_{t_0}^{t} \dot{y}(t)\, dt \tag{13.7}$$

The simulation diagrams corresponding to Eqs. (13.5), (13.6) and (13.7) are interconnected as shown in Fig. 13.3. Arbitrarily defining the output of each integrator as a state-vector element, as shown in Fig. 13.3, allows us to describe the system in the desired form by

$$\dot{x}_1(t) = x_2(t) \qquad x_1^\circ = y^\circ \tag{13.8}$$

$$\dot{x}_2(t) = b_0 z(t) - a_0 x_1(t) - a_1 x_2(t) \qquad x_2^\circ = \dot{y}^\circ \tag{13.9}$$

$$y(t) = x_1(t) \tag{13.10}$$

It is clear from Eqs. (13.8), (13.9) and (13.10) that the vectors and matrices

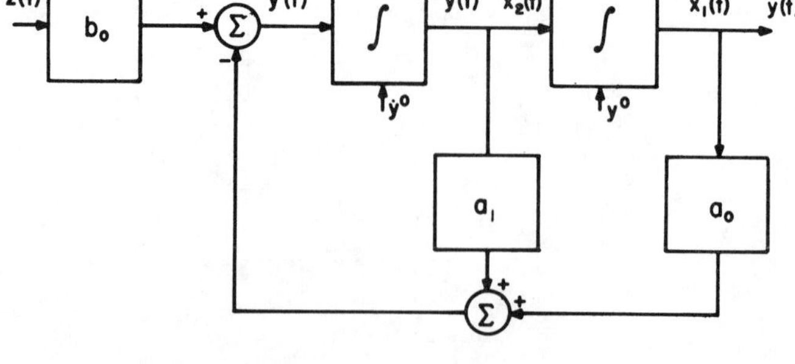

FIG. 13.3. Simulation diagram for Ex. 13.1.

used in Eqs. (13.1) and (13.2) are defined by

$$\mathbf{x}'(t) = [x_1(t), x_2(t)] = [y(t), \dot{y}(t)] \tag{13.11}$$

$$\mathbf{A} = \begin{bmatrix} 0 & 1 \\ -a_0 & -a_1 \end{bmatrix} \qquad m(t) = z(t) \qquad \mathbf{B} = \begin{bmatrix} 0 \\ b_0 \end{bmatrix} \tag{13.12}$$

and

$$\mathbf{C} = [1 \quad 0] \qquad \mathbf{D} = \mathbf{0} \tag{13.13}$$

Note that one or more of a_0, a_1, and b_0 are functions of t if the system is time varying.

Example 13.2

Obtain a state equation formulation for the system described by

$$\ddot{y}(t) + a_1\dot{y}(t) + a_0 y(t) = b_1\dot{z}(t) + b_0 z(t) \tag{13.14}$$

with

$$\dot{y}(t_0) = \dot{y}^{\circ} \qquad y(t_0) = y^{\circ} \tag{13.15}$$

One solution to this problem is obtained directly by using the answer in Ex. 13.1 modified to include an input vector $\mathbf{m}(t)$ with elements $\dot{z}(t)$ and $z(t)$. This results in

$$\begin{bmatrix} \dot{x}_1 \\ \dot{x}_2 \end{bmatrix} = \begin{bmatrix} 0 & 1 \\ -a_0 & -a_1 \end{bmatrix} \begin{bmatrix} x_1 \\ x_2 \end{bmatrix} + \begin{bmatrix} 0 & 0 \\ b_0 & b_1 \end{bmatrix} \begin{bmatrix} z \\ \dot{z} \end{bmatrix} \qquad \begin{bmatrix} x_1^{\circ} \\ x_2^{\circ} \end{bmatrix} = \begin{bmatrix} y^{\circ} \\ \dot{y}^{\circ} \end{bmatrix} \tag{13.16}$$

$$y = [1 \quad 0] \begin{bmatrix} x_1 \\ x_2 \end{bmatrix} \tag{13.17}$$

Note that this formulation depends upon $\dot{z}(t)$, which may not be available. Although it can perhaps be developed by approximate differentiation, this accentuates noise at high frequencies. Dynamic range difficulties are encountered wherever $z(t)$ changes rapidly.

An alternate solution procedure uses the first integrator to integrate $\dot{z}(t)$ along with $\ddot{y}(t)$. Assuming b_1 is a constant, Eq. (13.14) is rearranged as

$$\ddot{y} - b_1\dot{z} = b_0 z - a_1\dot{y} - a_0 y \tag{13.18}$$

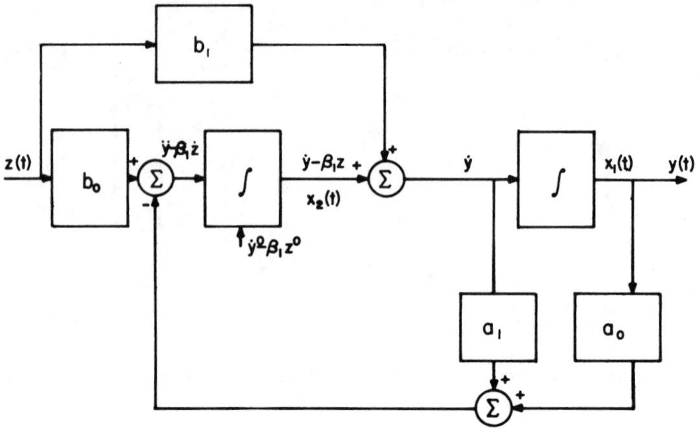

FIG. 13.4. Simulation diagram for Ex. 13.2.

Observe that

$$\int_{t_0}^{t} \dot{z}(t)\,dt = z(t) - z^{\circ} \tag{13.19}$$

Integrating Eq. (13.18) yields

$$\dot{y} = \dot{y}^{\circ} + \int_{t_0}^{t} (\ddot{y} - b_1\dot{z})\,dt + b_1 z - b_1 z^{\circ} \tag{13.20}$$

and

$$y = y^{\circ} + \int_{t_0}^{t} \dot{y}\,dt \tag{13.21}$$

The resulting simulation diagram is shown in Fig. 13.4. Assigning the integrator outputs as state variables provides the state equations as

$$\begin{bmatrix} \dot{x}_1 \\ \dot{x}_2 \end{bmatrix} = \begin{bmatrix} 0 & 1 \\ -a_0 & -a_1 \end{bmatrix} \begin{bmatrix} x_1 \\ x_2 \end{bmatrix} + \begin{bmatrix} b_1 \\ b_0 - a_1 b_1 \end{bmatrix} z \qquad \begin{bmatrix} x_1^{\circ} \\ x_2^{\circ} \end{bmatrix} = \begin{bmatrix} y^{\circ} \\ \dot{y}^{\circ} - b_1 z^{\circ} \end{bmatrix} \tag{13.22}$$

$$y = [1 \quad 0]\begin{bmatrix} x_1 \\ x_2 \end{bmatrix} \tag{13.23}$$

Examples 13.1 and 13.2 illustrate that any linear DE is readily put in state-variable form. Our next example illustrates the procedure for handling simultaneous sets of DE's.

Example 13.3

Obtain a state equation formulation for the system described by

$$\ddot{y}_1 + a_1\dot{y}_1 + a_2 y_2 = b_1\dot{z}_1 + b_2 z_1 + b_3 z_2 \tag{13.24}$$

$$\dot{y}_2 + a_3 y_2 + a_4 y_1 = b_4 z_2 \tag{13.25}$$

with

$$\dot{y}_1(t_0) = \dot{y}_1^\circ \qquad y_1(t_0) = y_1^\circ \qquad y_2(t_0) = y_2^\circ \tag{13.26}$$

The general procedure used in solving problems of this nature is to set up a simulation diagram for each variable, in this case y_1 and y_2, and interconnect these diagrams as required by the equation coefficients. Rearranging Eq. (13.24) yields

$$\ddot{y}_1 - b_1\dot{z}_1 = b_2 z_1 + b_3 z_2 - a_2 y_2 - a_1\dot{y}_1 \tag{13.27}$$

Integrating Eq. (13.27) provides

$$\dot{y}_1 = \dot{y}_1^\circ + \int_{t_0}^{t} (\ddot{y}_1 - b_1\dot{z}_1)\,dt + b_1 z_1 - b_1 z_1^\circ \tag{13.28}$$

and

$$y_1 = y_1^\circ + \int_{t_0}^{t} \dot{y}_1\,dt \tag{13.29}$$

Also, from Eq. (13.25)

$$\dot{y}_2 = b_4 z_2 - a_4 y_1 - a_3 y_2 \tag{13.30}$$

and

$$y_2 = y_2^\circ + \int_{t_0}^{t} \dot{y}_2\,dt \tag{13.31}$$

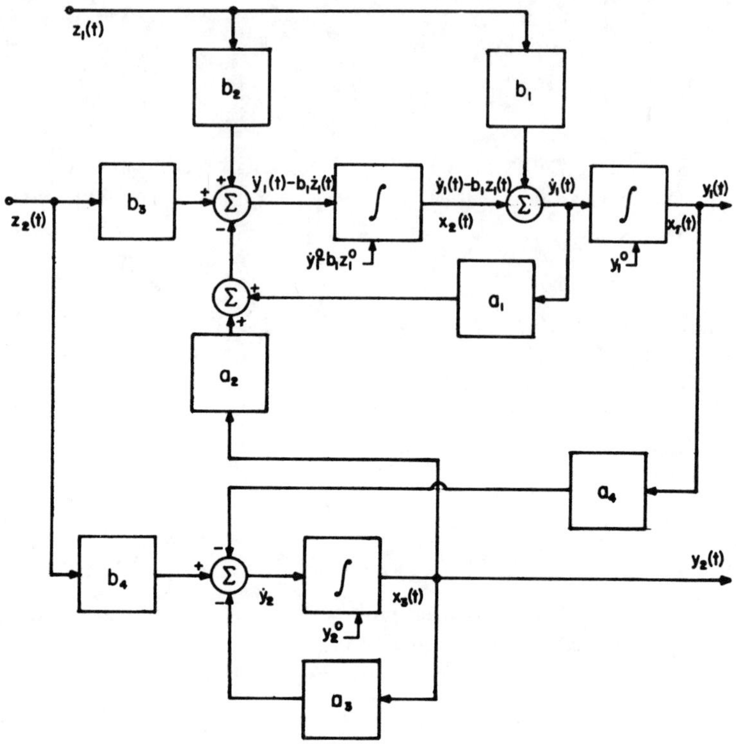

FIG. 13.5. Simulation diagram for Ex. 13.3.

The corresponding simulation diagram is shown in Fig. 13.5. Defining the integrator outputs as state variables in the usual way and writing the corresponding state equations yields

$$\begin{bmatrix} \dot{x}_1 \\ \dot{x}_2 \\ \dot{x}_3 \end{bmatrix} = \begin{bmatrix} 0 & 1 & 0 \\ 0 & -a_1 & -a_2 \\ -a_4 & 0 & -a_3 \end{bmatrix} \begin{bmatrix} x_1 \\ x_2 \\ x_3 \end{bmatrix} + \begin{bmatrix} b_1 & 0 \\ b_2 - a_1 b_1 & b_3 \\ 0 & b_4 \end{bmatrix} \begin{bmatrix} z_1 \\ z_2 \end{bmatrix} \tag{13.32}$$

and

$$\begin{bmatrix} y_1 \\ y_2 \end{bmatrix} = \begin{bmatrix} 1 & 0 & 0 \\ 0 & 0 & 1 \end{bmatrix} \begin{bmatrix} x_1 \\ x_2 \\ x_3 \end{bmatrix} \qquad \begin{bmatrix} x_1^\circ \\ x_2^\circ \\ x_3^\circ \end{bmatrix} = \begin{bmatrix} y_1^\circ \\ \dot{y}_1^\circ - b_1 z_1^\circ \\ y_2^\circ \end{bmatrix} \tag{13.33}$$

It is seen from these examples that the simulation diagram for any system described by one or more linear DE's is easily obtained. Defining the integrator outputs as state-vector elements leads directly to the desired state equations. Equivalent alternate forms of these equations may be obtained, as shown in detail later, by rearranging the simulation diagram in any way which does not alter the input or output variables.

13.2 DEVELOPMENT OF THE STATE EQUATIONS FROM TRANSFER FUNCTIONS

A transfer function is an alternate way of representing a DE relationship between the input and output variables of a linear time-invariant system. Since we can always convert from the transfer function to the equivalent DE and vice versa, it is clear that the state equations can be obtained for any system described in transfer function form. It is not necessary to convert a transfer function to the equivalent DE, however, although this is one possible approach. It is implied when we work with the transfer function representation that all initial conditions are zero. If we wish to take the transfer function as a mathematical model of the system, but desire to introduce non-zero initial conditions as well, we can always go back to the equivalent DE and obtain the desired answer from it in any legitimate way. We assume for the present that all initial conditions are zero. This leads to a simple simulation diagram development.

Consider the general transfer function for an nth-order system

$$\frac{C(s)}{R(s)} = \frac{b_n s^n + b_{n-1} s^{n-1} + \cdots + b_1 s + b_0}{s^n + a_{n-1} s^{n-1} + \cdots + a_1 s + a_0} \tag{13.34}$$

Cross multiplying Eq. (13.34) yields

$$C(s)(s^n + a_{n-1} s^{n-1} + \cdots + a_1 s + a_0) = R(s)(b_n s^n + b_{n-1} s^{n-1} + \cdots + b_1 s + b_0) \tag{13.35}$$

Dividing both sides of Eq. (13.35) by s^n and putting all terms except $C(s)$ on the right-hand side gives

$$C(s) = R(s)\left(b_n + \frac{b_{n-1}}{s} + \cdots + \frac{b_1}{s^{n-1}} + \frac{b_0}{s^n}\right) - C(s)\left(\frac{a_{n-1}}{s} + \frac{a_{n-2}}{s^2} + \cdots + \frac{a_1}{s^{n-1}} + \frac{a_0}{s^n}\right) \tag{13.36}$$

Thus $c(t)$ may be obtained by summing weighted integrations of $r(t)$ and $c(t)$ as shown in Fig. 13.6. Taking the integrator outputs as state variables provides

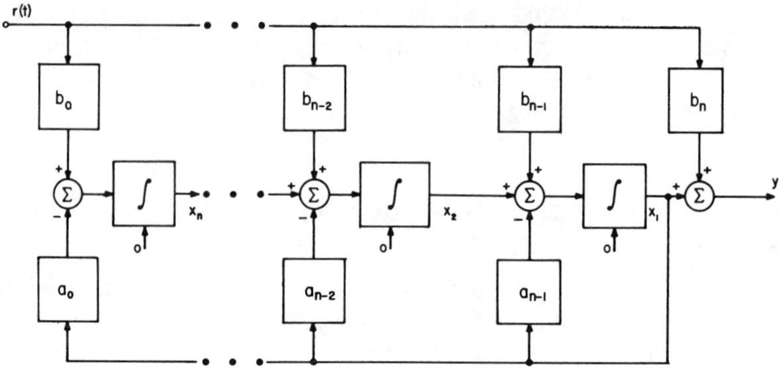

FIG. 13.6. Simulation diagram for a general transfer function.

$$
\begin{bmatrix} \dot{x}_1 \\ \dot{x}_2 \\ \cdot \\ \cdot \\ \cdot \\ \dot{x}_{n-1} \\ \dot{x}_n \end{bmatrix} = \begin{bmatrix} -a_{n-1} & 1 & 0 & \dots & 0 \\ -a_{n-2} & 0 & 1 & \dots & 0 \\ \cdots & \cdots & \cdots & \cdots & \cdots \\ \cdots & \cdots & \cdots & \cdots & \cdots \\ \cdots & \cdots & \cdots & \cdots & \cdots \\ -a_1 & 0 & 0 & \dots & 1 \\ -a_0 & 0 & 0 & \dots & 0 \end{bmatrix} \begin{bmatrix} x_1 \\ x_2 \\ \cdot \\ \cdot \\ \cdot \\ x_{n-1} \\ x_n \end{bmatrix} + \begin{bmatrix} b_{n-1} \\ b_{n-2} \\ \cdot \\ \cdot \\ \cdot \\ b_1 \\ b_0 \end{bmatrix} r \tag{13.37}
$$

with $\mathbf{x}^{\circ} = \mathbf{0}$ and

$$
c = [1 \quad 0 \quad 0 \quad \dots \quad 0 \quad 0] \begin{bmatrix} x_1 \\ x_2 \\ \cdot \\ \vdots \\ \cdot \\ x_{n-1} \\ x_n \end{bmatrix} + b_n r \tag{13.38}
$$

Development of the state equations from transfer functions is thus seen to be straightforward.

Many forms of state-equation representations are possible. Any legitimate rearrangement of gains, summations, and integrations on the simulation diagram can be used as a means for developing equivalent results. It is also possible to go back to the DE representation and define state in a variety of ways to get alternate solutions. If the transfer function is expanded into partial fraction form and a simulation diagram developed for each component, the desired output may be obtained as the sum of the outputs of the individual simulation diagrams. The state-vector elements are defined as the integrator outputs in the usual way. This process is illustrated by the following example.

Example 13.4

Obtain a set of state equations for the system defined by the transfer function expressed in partial fraction form as

$$\frac{C(s)}{R(s)} = \frac{1}{s+2} + \frac{3}{(s+1)^2} - \frac{1}{s+1} + \frac{2s+3}{(s+1)^2+1} \tag{13.39}$$

The simulation diagram for a term of the form

$$\frac{C_1(s)}{R(s)} = \frac{k_i}{s+\lambda_i} \tag{13.40}$$

is readily developed by cross-multiplying, dividing both sides by s, and rearranging, to obtain

$$C_1(s) = \frac{k_i R(s)}{s} - \frac{\lambda_i C_i(s)}{s} \tag{13.41}$$

The corresponding simulation diagram is shown in Fig. 13.7. The first and third terms in Eq. (13.39) are of this form, and the second term may be realized by two such arrangements in cascade, as shown in Fig. 13.9. The last term in Eq. (13.41) corresponds to a complex-conjugate pole pair and may be simulated in several forms. Perhaps the easiest way to expand the term

$$\frac{C_2(s)}{R(s)} = \frac{b_1 s + b_2}{s^2 + a_2 s + a_3} \tag{13.42}$$

is to cross-multiply and simplify in the usual way to obtain

$$C_2(s) = \frac{b_1 R(s)}{s} + \frac{b_2 R(s)}{s^2} - \frac{a_2 C_2(s)}{s} - \frac{a_3 C_2(s)}{s^2} \tag{13.43}$$

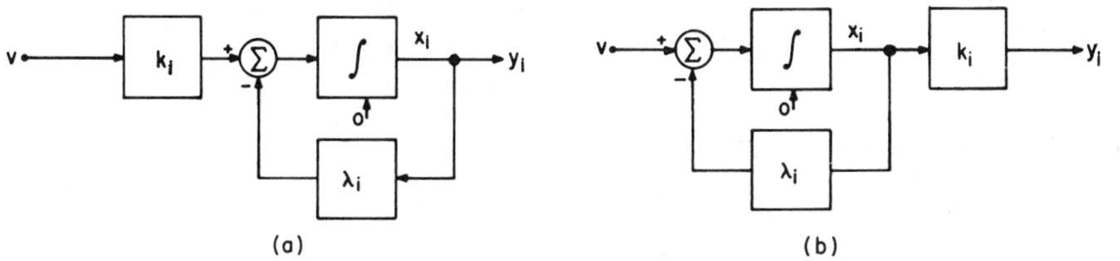

FIG. 13.7. General simulation diagrams for a partial-function pole.

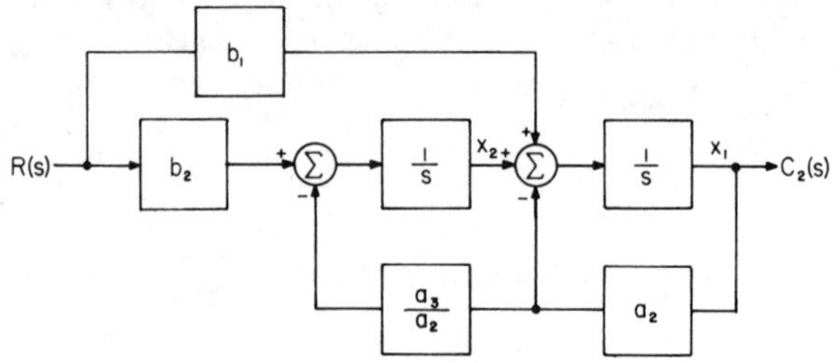

FIG. 13.8. General simulation diagram for a complex pole pair.

The corresponding simulation diagram is shown in Fig. 13.8. Combining these results leads to the total diagram shown in Fig. 13.9. The resulting state equations are

$$\begin{bmatrix} \dot{x}_1 \\ \cdots \\ \dot{x}_2 \\ \dot{x}_3 \\ \cdots \\ \dot{x}_4 \\ \dot{x}_5 \end{bmatrix} = \left[\begin{array}{c:cc:cc} -2 & 0 & 0 & 0 & 0 \\ \hdashline 0 & -1 & 1 & 0 & 0 \\ 0 & 0 & -1 & 0 & 0 \\ \hdashline 0 & 0 & 0 & -1 & 1 \\ 0 & 0 & 0 & -1 & -1 \end{array}\right] \begin{bmatrix} x_1 \\ \cdots \\ x_2 \\ x_3 \\ \cdots \\ x_4 \\ x_5 \end{bmatrix} + \begin{bmatrix} -1 \\ \cdots \\ -1 \\ 3 \\ \cdots \\ 2 \\ 1 \end{bmatrix} r \tag{13.44}$$

with $\mathbf{x}^{\circ} = \mathbf{0}$ and

$$C = \left[\begin{array}{c:cc:cc} 1 & 1 & 0 & 1 & 0 \end{array}\right] \begin{bmatrix} x_1 \\ \cdots \\ x_2 \\ x_3 \\ \cdots \\ x_4 \\ x_5 \end{bmatrix} \tag{13.45}$$

where the vectors and matrices have been partitioned to emphasize the mathematical structure derived from the component partial fractions.

13.3 GENERAL SOLUTION OF THE STATE EQUATIONS

Consider the vector set of first order DE's

$$\dot{x} = \mathbf{A}(t)\,\mathbf{x} + \mathbf{B}(t)\,\mathbf{m} \qquad \mathbf{x}(t_0) = \mathbf{x}^{\circ} \tag{13.46}$$

where $\mathbf{x}$, $\mathbf{m}$, $\mathbf{A}$ and $\mathbf{B}$ are as previously defined. The elements of $\mathbf{A}$ and $\mathbf{B}$ are functions of the independent variable, t. Let $\mathbf{\Phi}(t, t_0)$ denote the solution of the matrix DE developed from the homogeneous part of Eq. (13.46), or

$$\dot{\mathbf{\Phi}}(t, t_0) = \mathbf{A}(t)\mathbf{\Phi}(t, t_0) \qquad \mathbf{\Phi}(t_0, t_0) = \mathbf{I} \tag{13.47}$$

$\mathbf{\Phi}(t, t_0)$ is called the *transition matrix* of the system described by Eq. (13.46). It describes the functional relationship between a complete set of boundary conditions imposed upon $\mathbf{x}(t_0) = \mathbf{x}^\circ$ and state $\mathbf{x}(t)$, assuming $\mathbf{m}(t) = \mathbf{0}$. The general solution of Eq. (13.46) for $\mathbf{x}(t)$ in terms of $\mathbf{\Phi}(t, t_0)$, $\mathbf{x}^\circ$, and $\mathbf{m}(t)$ is

$$\mathbf{x}(t) = \mathbf{\Phi}(t, t_0)\mathbf{x}^\circ + \int_{t_0}^{t} \mathbf{\Phi}(t, \tau)\mathbf{B}(\tau)\mathbf{m}(\tau)d\tau \tag{13.48}$$

This result is a generalization of the superposition principle to the solution of vector sets of DE's. Thus linear system response consists of a component due to the boundary value of state, $\mathbf{x}^\circ$, plus a component due to the control input applied on the interval $[t_0, t]$.

Proof that Eq. (13.48) is a valid solution of Eq. (13.46) is straightforward. We need only to show that $\mathbf{x}(t)$ as given by Eq. (13.48) satisfies Eq. (13.46),

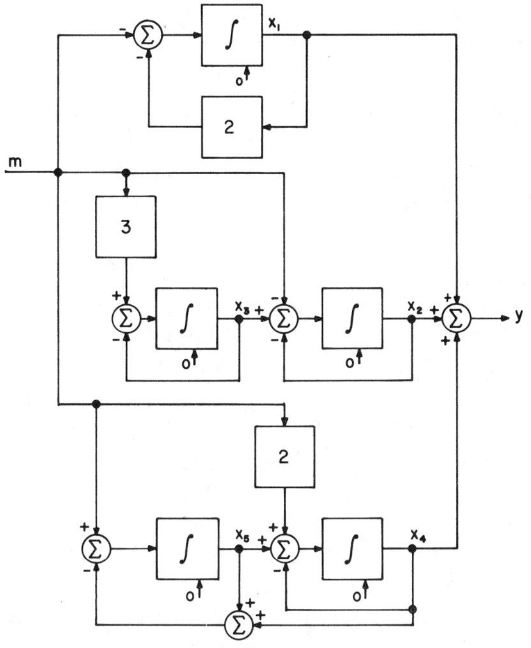

FIG. 13.9. Total simulation diagram for Ex. 13.4.

including the boundary conditions. Differentiating both sides of Eq. (13.48) with respect to t provides

$$\dot{\mathbf{x}}(t) = \dot{\mathbf{\Phi}}(t, t_0)\mathbf{x}^{\circ} + \mathbf{\Phi}(t, t)\,\mathbf{B}(t)\,\mathbf{m}(t) + \int_{t_0}^{t} \dot{\mathbf{\Phi}}(t, \tau)\,\mathbf{B}(\tau)\,\mathbf{m}(\tau)d\tau \qquad (13.49)$$

but $\mathbf{\Phi}(t, t) = \mathbf{I}$ and we may substitute for $\dot{\mathbf{\Phi}}(t, t_0)$ and $\dot{\mathbf{\Phi}}(t, \tau)$ using Eq. (13.47) to obtain

$$\dot{\mathbf{x}}(t) = \mathbf{A}(t)\,\mathbf{\Phi}(t, t_0)\mathbf{x}^{\circ} + \mathbf{B}(t)\,\mathbf{m}(t) + \mathbf{A}(t) \int_{t_0}^{t} \mathbf{\Phi}(t, \tau)\,\mathbf{B}(\tau)\,\mathbf{m}(\tau)d\tau$$

$$= \mathbf{A}(t)\left[\mathbf{\Phi}(t, t_0)\mathbf{x}^{\circ} + \int_{t_0}^{t} \mathbf{\Phi}(t, \tau)\,\mathbf{B}(\tau)\,\mathbf{m}(\tau)d\tau\right] + \mathbf{B}(t)\,\mathbf{m}(t)$$

$$= \mathbf{A}(t)\,\mathbf{x}(t) + \mathbf{B}(t)\,\mathbf{m}(t) \qquad (13.50)$$

Evaluation of Eq. (13.48) at $t = t_0$ provides $\mathbf{x}(t_0) = \mathbf{x}^{\circ}$, as desired, since $\mathbf{\Phi}(t_0, t_0) = \mathbf{I}$ and the integral can contribute nothing over a zero time interval for any bounded input. Thus $\mathbf{x}(t)$ as given by Eq. (13.48) satisfies Eq. (13.46), including the boundary condition, and is a general solution form valid for all linear systems. If $\mathbf{A}(t)$, $\mathbf{B}(t)$, and $\mathbf{m}(t)$ are known, $\mathbf{\Phi}(t, t_0)$ may be obtained by integrating Eq. (13.47), and $\mathbf{x}(t)$ is obtained, in turn, by expanding the integrand in Eq. (13.48) and carrying out the integration term by term. Although this is not a trivial exercise, particularly if n is large and the dependence of $\mathbf{A}$ and $\mathbf{B}$ upon t is complex, the method is clear. Development of computational solutions based upon Eq. (13.48) using an analog or digital computer is relatively straightforward.

It is important to consider the physical significance of the elements making up $\mathbf{\Phi}(t, t_0)$. The general element $\phi_{ij}(t, t_0)$ relates the response $x_i(t)$ to a unit boundary value of $x_j(t_0)$, assuming all other boundary conditions are zero. The elements in the jth column of $\mathbf{\Phi}(t, t_0)$ thus indicate the effect of $x_j(t_0) = 1$ upon each $x_i(t)$ in turn. We may alternately consider $\phi_{ij}(t, t_0)$ as a sensitivity coefficient describing how $x_i(t)$ depends upon the value of boundary condition $x_j(t_0)$. Transition functions are basic tools without which many important developments in the modern control field would be impossible. The following example illustrates how the transition matrix is developed and used in the time varying case.

Example 13.5

Consider the system shown in Fig. 13.10, which is described by

$$\dot{x}_1 = k(t)x_2 \qquad k(t) = 1 - t \qquad (13.51)$$

$$\dot{x}_2 = -x_2 + m \qquad (13.52)$$

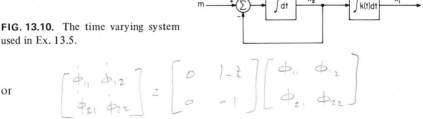

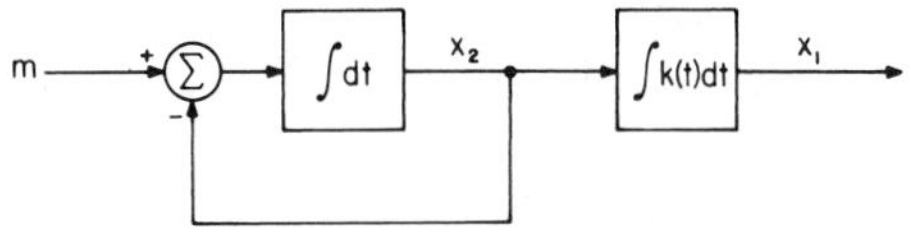

FIG. 13.10. The time varying system used in Ex. 13.5.

or $\begin{bmatrix} \dot\phi_{11} & \dot\phi_{12} \\ \dot\phi_{21} & \dot\phi_{22} \end{bmatrix} = \begin{bmatrix} 0 & 1-t \\ 0 & -1 \end{bmatrix} \begin{bmatrix} \phi_{11} & \phi_{12} \\ \phi_{21} & \phi_{22} \end{bmatrix}$

$$\dot{x} = Ax + Bm \qquad A = \begin{bmatrix} 0 & 1-t \\ 0 & -1 \end{bmatrix} \qquad B = \begin{bmatrix} 0 \\ 1 \end{bmatrix} \qquad (13.53)$$

Find the transition matrix for this time varying system. Assuming $m(t) = u(t)$ and $x'(0) = [1 \quad 2]$, find $x(t)$ for all $t \geq 0$.

The transition matrix is defined by Eq. (13.47), which may be expanded in this case to provide

$$\dot\phi_{11}(t, t_0) = (1 - t)\phi_{21}(t, t_0) \qquad \dot\phi_{12}(t, t_0) = (1 - t)\phi_{22}(t, t_0)$$

$$\dot\phi_{21}(t, t_0) = -\phi_{21}(t, t_0) \qquad \dot\phi_{22}(t, t_0) = -\phi_{22}(t, t_0) \qquad (13.54)$$

Because $\phi(t_0, t_0) = I = \begin{bmatrix} 1 & 0 \\ 0 & 1 \end{bmatrix}$!

with boundary conditions $\phi_{11}(t_0, t_0) = \phi_{22}(t_0, t_0) = 1$ and $\phi_{12}(t_0, t_0) = \phi_{21}(t_0, t_0) = 0$. Two of these equations are trivial. Since ϕ_{21} starts out at zero and its derivative depends only upon itself, $\phi_{21}(t, t_0) = 0$. Thus, $\dot\phi_{11} = 0$ for all t and $\phi_{11}(t, t_0) = 1$. The equation for ϕ_{22} is of the standard exponential form, so $\phi_{22}(t, t_0) = \epsilon^{-(t - t_0)}$. Substituting this ϕ_{22} into the equation for ϕ_{12} and integrating provides

$$\phi_{12}(t, t_0) = \int_{t_0}^{t} (1 - \alpha)\epsilon^{-(\alpha - t_0)} d\alpha \qquad (13.55)$$

where α is used as the dummy variable of integration to avoid notational confusion. Thus

$$\phi_{12}(t, t_0) = \epsilon^{t_0} \int_{t_0}^{t} (\epsilon^{-\alpha} - \alpha\epsilon^{-\alpha}) d\alpha$$

$$= \epsilon^{t_0} \left[\frac{\epsilon^{-\alpha}}{-1} + (1 + \alpha)\epsilon^{-\alpha} \right]\Bigg|_{t_0}^{t} = -t_0 + t\epsilon^{-(t - t_0)} \qquad (13.56)$$

We know that $\mathbf{x}(t)$ is given by Eq. (13.48). The response to the given initial conditions is developed by expanding the first term on the right side of Eq. (13.48) to obtain

$$x_{1i}(t) = \phi_{11}(t, t_0)x_1^\circ + \phi_{12}(t, t_0)x_2^\circ \tag{13.57}$$

$$x_{2i}(t) = \phi_{21}(t, t_0)x_1^\circ + \phi_{22}(t, t_0)x_2^\circ \tag{13.58}$$

Substituting the known $\mathbf{x}^\circ$ and $\mathbf{\Phi}(t, t_0)$ elements into Eq. (13.57) and (13.58) yields $\mathbf{x}_i(t)$, the response component due to initial-state alone. The response component due to the given input is obtained by expanding and integrating the second term on the right side of Eq. (13.48). Note that

$$\mathbf{B}(\tau)\, m(\tau) = \begin{bmatrix} 0 \\ 1 \end{bmatrix} \tag{13.59}$$

for all $\tau \geq 0$ and

$$\mathbf{\Phi}(t, \tau)\, \mathbf{B}(\tau)\, m(\tau) = \begin{bmatrix} \phi_{12}(t, \tau) \\ \phi_{22}(t, \tau) \end{bmatrix} = \begin{bmatrix} -\tau + t\epsilon^{-(t-\tau)} \\ \epsilon^{-(t-\tau)} \end{bmatrix} \tag{13.60}$$

Integrating the terms on the right side of Eq. (13.60) with respect to τ over the range $[0, t]$ yields the response component caused directly by the control input as

$$\mathbf{x}_m(t) = \begin{bmatrix} t - \dfrac{t^2}{2} - t\epsilon^{-t} \\ 1 - \epsilon^{-t} \end{bmatrix} \tag{13.61}$$

The total solution for the given initial conditions and a unit-step input is found by adding Eqs. (13.57) and (13.58) to (13.61), which yields

$$\mathbf{x}(t) = \begin{bmatrix} 1 + t\epsilon^{-t} + t - \dfrac{t^2}{2} \\ 1 + \epsilon^{-t} \end{bmatrix} \tag{13.62}$$

Obviously a solution to Ex. 13.5 can be obtained much more readily by direct integration without finding the transition functions, but our primary concern here is to illustrate how $\mathbf{\Phi}(t, t_0)$ is used. If the initial conditions or control input are changed from the given values, only a small part of the work has to be repeated once the system transition function is known.

Several characteristics of transition functions are of considerable importance. In the homogeneous case $\mathbf{m}(\tau) = \mathbf{0}$ and it is clear from Eq. (13.48) that

$$\mathbf{x}(t_2) = \mathbf{\Phi}(t_2, t_1)\,\mathbf{x}(t_1) \tag{13.63}$$

and

$$\mathbf{x}(t_3) = \mathbf{\Phi}(t_3, t_2)\,\mathbf{x}(t_2) \tag{13.64}$$

Substituting $\mathbf{x}(t_2)$ from Eq. (13.63) into Eq. (13.64) yields

$$\mathbf{x}(t_3) = \mathbf{\Phi}(t_3, t_2)\,\mathbf{\Phi}(t_2, t_1)\,\mathbf{x}(t_1) = \mathbf{\Phi}(t_3, t_1)\,\mathbf{x}(t_1) \tag{13.65}$$

from which we see that

$$\mathbf{\Phi}(t_3, t_2)\,\mathbf{\Phi}(t_2, t_1) = \mathbf{\Phi}(t_3, t_1) \tag{13.66}$$

If we let $t_3 = t_1$ in Eq. (13.66) we obtain

$$\mathbf{\Phi}(t_1, t_2)\,\mathbf{\Phi}(t_2, t_1) = \mathbf{\Phi}(t_1, t_1) = \mathbf{I} \tag{13.67}$$

or

$$\mathbf{\Phi}(t_1, t_2) = \mathbf{\Phi}^{-1}(t_2, t_1) \tag{13.68}$$

Thus inverting a transition matrix is equivalent to interchanging the order of the time arguments. Note that when we solve $\mathbf{\Phi}(t_2, t_1)$ directly by integrating Eq. (13.47), we impose the boundary conditions at $t = t_1$ and integrate to $t = t_2$. The identity presented in Eq. (13.68) allows us, should we choose, to impose boundary conditions at $t = t_2$, integrate to $t = t_1$, and invert the resulting matrix to obtain the same answer.

Another relationship which often proves useful in the development and use of transition functions and general-state-equation solutions concerns the adjoint equations. Consider the linear homogeneous state-equation set

$$\dot{\mathbf{x}} = \mathbf{A}(t)\mathbf{x} \qquad \mathbf{x}(t_0) = \mathbf{x}^\circ \tag{13.69}$$

and the corresponding transition matrix defined by

$$\dot{\mathbf{\Phi}}(t, t_0) = \mathbf{A}(t)\,\mathbf{\Phi}(t, t_0) \qquad \mathbf{\Phi}(t_0, t_0) = \mathbf{I} \tag{13.70}$$

The state variables adjoint to those described by Eq. (13.69) are defined by

$$\dot{\boldsymbol{\lambda}} = -\mathbf{A}'(t)\boldsymbol{\lambda} \qquad \boldsymbol{\lambda}(t_0) = \boldsymbol{\lambda}^\circ \tag{13.71}$$

The adjoint transition matrix is obtained from

$$\dot{\boldsymbol{\theta}}(t, t_0) = -\mathbf{A}'(t)\,\boldsymbol{\theta}(t, t_0) \qquad \boldsymbol{\theta}(t_0, t_0) = \mathbf{I} \tag{13.72}$$

The relationships between $\mathbf{x}$, $\boldsymbol{\lambda}$, $\boldsymbol{\Phi}$, and $\boldsymbol{\theta}$ are important in the general study of linear systems, particularly in the areas of optimal control theory and computational solution methods. Note that

$$\frac{d}{dt}(\boldsymbol{\lambda}'\mathbf{x}) = \dot{\boldsymbol{\lambda}}'\mathbf{x} + \boldsymbol{\lambda}'\dot{\mathbf{x}} = (-\mathbf{A}'\boldsymbol{\lambda})'\mathbf{x} + \boldsymbol{\lambda}'\mathbf{A}\mathbf{x}$$

$$= -\boldsymbol{\lambda}'\mathbf{A}\mathbf{x} + \boldsymbol{\lambda}'\mathbf{A}\mathbf{x} = 0 \tag{13.73}$$

Thus the inner product of $\mathbf{x}$ and $\boldsymbol{\lambda}$ is a constant, or

$$(\boldsymbol{\lambda}, \mathbf{x}) = \boldsymbol{\lambda}'\mathbf{x} = (\mathbf{x}, \boldsymbol{\lambda}) = \mathbf{x}'\boldsymbol{\lambda} = \mathbf{x}^{\circ\prime}\boldsymbol{\lambda}^{\circ} = \text{constant} \tag{13.74}$$

The $\mathbf{x}$ and $\boldsymbol{\lambda}$ variables have an interesting and useful inverse relationship. The same sequence of steps may be used to show that

$$\boldsymbol{\theta}'(t, t_0)\,\boldsymbol{\Phi}(t, t_0) = \text{constant} = \mathbf{I} \tag{13.75}$$

from which we conclude

$$\phi^{-1}(t, t_0) = \phi(t_0, t) = \boldsymbol{\theta}'(t, t_0) \tag{13.76}$$

This relationship is useful in developing computational solutions based upon Eq. (13.48).

It is important to keep several characteristics of transition matrices and the adjoint variables in mind. The $\phi_{ij}(t, t_0)$ element of a transition matrix relates $x_i(t)$ to $x_j(t_0)$ assuming all $x_k(t_0) = 0$ for $k \neq j$. These functions are closely related to the impulse-response functions often used to describe systems, but there is an important distinction as well. The impulse response function is defined only for $t > t_0$, where t_0 is the time when the impulse is applied, but $\phi_{ij}(t, t_0)$ is defined for all t and all t_0, and is thus a more general function. When $t > t_0$, we interpret $\boldsymbol{\Phi}(t, t_0)$ as an operator which converts $\mathbf{x}(t_0)$ to $\mathbf{x}(t)$, indicating how initial state $\mathbf{x}^{\circ}$ influences system response at time t. When $t < t_0$, $\boldsymbol{\Phi}(t, t_0)$ still converts boundary conditions $\mathbf{x}^{\circ}$ to state $\mathbf{x}(t)$. A physical interpretation of this operation is that it yields that $\mathbf{x}(t)$ which, if imposed as an initial condition upon the system at time t, would result in $\mathbf{x}(t_0)$ at $t = t_0$.

In many optimal control theory applications the objective is to use Eq. (13.48) or some relationship of comparable form to develop an iterative computational solution procedure by which system state at terminal time t, which is often fixed, can be influenced using initial conditions at t_0 and control over the interval $[t_0, t]$. Note that $\boldsymbol{\Phi}(t, \tau)$ appears in the integrand, τ is the variable of integration, and τ is also the point where the boundary condition $\mathbf{\Phi}(\tau,\tau) = \mathbf{I}$ is imposed to obtain the function $\mathbf{\Phi}(t,\tau)$. Since t is the fixed parameter in such applications, $\mathbf{\Phi}(t,\tau)$ is most appropriately obtained by integrating the adjoint equation to find $\theta(\tau, t)$, and transposing the result. This avoids the need to integrate repeatedly with initial conditions imposed at time τ, as would be necessary using Eq. (13.70) directly, or the equally difficult process of repeated matrix inversion required if Eqs. (13.70) and (13.68) are used together. Of course we assume a general time-varying system, since $\mathbf{\Phi}(t,\tau)$ reduces to $\mathbf{\Phi}(t - \tau)$ for the time-invariant case, and the result obtained by integrating Eq. (13.70) over the interval $[t_1, t_2]$ is also obtained by integrating over any other interval of equal length, leading to considerable simplification when the system does not vary with time.

An interesting and often useful alternate general solution for the transition matrix may be obtained using the matrix exponential function. The series representation of ϵ^α, the basic component in linear time-invariant system solutions, is

$$\epsilon^\alpha = 1 + \alpha + \frac{\alpha^2}{2!} + \cdots = \sum_{n=0}^{\infty} \frac{\alpha^n}{n!} \tag{13.77}$$

The matrix exponential corresponding to the square matrix $\mathbf{A}$, denoted by $\epsilon^{\mathbf{A}}$, is defined consistent with the scalar form as

$$\epsilon^{\mathbf{A}} = \mathbf{I} + \mathbf{A} + \frac{\mathbf{A}^2}{2!} + \cdots = \sum_{n=0}^{\infty} \frac{\mathbf{A}^n}{n!} \tag{13.78}$$

Just as the nth partial sum of the series for ϵ^α approaches ϵ^α as $N \to \infty$, the elements of the nth partial sum of the series for $\epsilon^{\mathbf{A}}$ uniformly approach limits as $N \to \infty$. Thus the series representation of $\epsilon^{\mathbf{A}}$ may be integrated or differentiated term by term whenever appropriate. Note that

$$\mathbf{\Phi}(t,0) = \epsilon^{\mathbf{A}t} \tag{13.79}$$

is a solution of the time-invariant homogeneous equation

$$\dot{\mathbf{\Phi}}(t,0) = \mathbf{A}\mathbf{\Phi}(t,0) \qquad \mathbf{\Phi}(0,0) = \mathbf{I} \tag{13.80}$$

as shown by direct substitution. This provides an alternate expression for the transition matrix.

The matrix exponential solution is in series form, which is awkward to use directly, although the fact that this simple solution exists is often used in the development of general proofs relating to system performance. The *Sylvester expansion theorem*[1] allows us to convert from the series form to the usual closed-solution form without undue difficulty. The theorem states that any general function of a time invariant matrix expressed in a convergent power series form can be reduced using the identity

$$f(\mathbf{A}) = \sum_{0}^{\infty} c_k \mathbf{A}^k = \sum_{i=1}^{n} f(s_i)\mathbf{F}(s_i) \tag{13.81}$$

where

$$\mathbf{F}(s_i) = \prod_{\substack{j=1 \\ j \neq i}}^{n} \frac{\mathbf{A} - s_j \mathbf{I}}{s_i - s_j} \tag{13.82}$$

The s_i and s_j are the eigenvalues of $\mathbf{A}$, and they are assumed distinct. Note that $\mathbf{F}(s_i)$ does not depend upon the specific $f(\mathbf{A})$ as long as $f(\mathbf{A})$ can be represented by a convergent power series. Use of Eq. (13.81) allows us to obtain the transition matrix from the matrix exponential form. It is the operational equivalent of evaluating the inverse Laplace transform to solve a matrix equation, so the obviously close relationship between Eq. (13.82) and the evaluation of residues is not accidental.

Example 13.6

Consider a system described by the time-invariant homogeneous equation

$$\dot{\mathbf{x}} = \mathbf{A}\mathbf{x} \qquad \mathbf{x}(0) = \mathbf{x}^{\circ} \qquad \mathbf{A} = \begin{bmatrix} 6 & -1 \\ 3 & 2 \end{bmatrix} \tag{13.83}$$

Find $\mathbf{\Phi}(t, 0)$ using the Sylvester expansion theorem.

The eigenvalues of $\mathbf{A}$ are easily found as $s_1 = 3$ and $s_2 = 5$. Direct application of Eqs. (13.81) and (13.82) yields

$$\mathbf{\Phi}(t,0) = \epsilon^{\mathbf{A}t} = \frac{\epsilon^{3t}\begin{bmatrix} 1 & -1 \\ 3 & -3 \end{bmatrix}}{-2} + \frac{\epsilon^{5t}\begin{bmatrix} 3 & -1 \\ 3 & -1 \end{bmatrix}}{2}$$

$$= \frac{1}{2}\begin{bmatrix} 3\epsilon^{5t} - \epsilon^{3t} & \epsilon^{3t} - \epsilon^{5t} \\ 3\epsilon^{5t} - 3\epsilon^{3t} & 3\epsilon^{3t} - \epsilon^{5t} \end{bmatrix} \tag{13.84}$$

This result is readily checked by solving for $\mathbf{\Phi}(t, 0)$ in the usual way.

The *Cayley-Hamilton theorem*[1] which states that "the square matrix **A** satisfies its own characteristic equation," is often useful in transforming functions of a matrix from one form to another. We will illustrate how this works with several examples.

Example 13.7

Show that the matrix

$$\mathbf{A} = \begin{bmatrix} 0 & 1 \\ -6 & -5 \end{bmatrix} \tag{13.85}$$

satisfies its characteristic equation.

The characteristic polynomial of **A** is readily found by setting $|[s\mathbf{I} - \mathbf{A}]| = 0$ to be

$$P(s) = s^2 + 5s + 6 = 0 \tag{13.86}$$

Substituting **A** for s yields

$$P(\mathbf{A}) = \mathbf{A}^2 + 5\mathbf{A} + 6\mathbf{I} = \mathbf{0}$$

$$= \begin{bmatrix} -6 & -5 \\ 30 & 19 \end{bmatrix} + 5\begin{bmatrix} 0 & 1 \\ -6 & -5 \end{bmatrix} + 6\begin{bmatrix} 1 & 0 \\ 0 & 1 \end{bmatrix} = \mathbf{0} \tag{13.87}$$

Example 13.8

Find A^{-1} for the matrix given in Eq. (13.85) using the Cayley-Hamilton theorem.

We know that **A** satisfies its characteristic equation, so

$$\mathbf{A}^2 + 5\mathbf{A} + 6\mathbf{I} = \mathbf{0} \tag{13.88}$$

Pre-multiplying both sides of Eq. (13.88) by $\mathbf{A}^{-1}$ provides

$$\mathbf{A} + 5\mathbf{I} + 6\mathbf{A}^{-1} = \mathbf{0} \tag{13.89}$$

or

$$\mathbf{A}^{-1} = \frac{-\mathbf{A} - 5\mathbf{I}}{6} = \frac{1}{6}\begin{bmatrix} -5 & -1 \\ 6 & 0 \end{bmatrix} \tag{13.90}$$

This often provides a convenient method for instrumenting matrix inversion.

Example 13.9

Use the Cayley-Hamilton theorem to reduce the matrix polynomial

$$f(\mathbf{A}) = 2\mathbf{A}^4 + \mathbf{A}^3 - \mathbf{A}^2 + 3\mathbf{A} + \mathbf{I} \tag{13.91}$$

to an equivalent form not involving powers of $\mathbf{A}$ above the first, where $\mathbf{A}$ is as given in Eq. (13.85). What general conclusions can you draw about reduction of matrix polynomials?

We see from Eq. (13.88) that

$$\mathbf{A}^2 = -5\mathbf{A} - 6\mathbf{I} \tag{13.92}$$

and $\mathbf{A}^4$ is obtained as $(\mathbf{A}^2)^2$ from Eq. (13.92), or

$$\begin{aligned} \mathbf{A}^4 &= 25\mathbf{A}^2 + 60\mathbf{A} + 36\mathbf{I} \\ &= -125\mathbf{A} - 150\mathbf{I} + 60\mathbf{A} + 36\mathbf{I} \\ &= -65\mathbf{A} - 114\mathbf{I} \end{aligned} \tag{13.93}$$

Multiplying both sides of Eq. (13.92) by $\mathbf{A}$ provides $\mathbf{A}^3$ as

$$\mathbf{A}^3 = -5\mathbf{A}^2 - 6\mathbf{A} = -5(-5\mathbf{A} - 6\mathbf{I}) - 6\mathbf{A} = 19\mathbf{A} + 30\mathbf{I} \tag{13.94}$$

Combining the results of Eqs. (13.92), (13.93), and (13.94) allows us to write Eq. (13.91) in the equivalent form

$$f(\mathbf{A}) = -103\mathbf{A} - 191\mathbf{I} \tag{13.95}$$

which is easily reduced to a matrix of coefficients by substituting in $\mathbf{A}$.

It is clear from these calculations that we can always use the matrix characteristic equation to solve for $\mathbf{A}^n$ and all higher powers of $\mathbf{A}$, where $\mathbf{A}$ is assumed $n \times n$, thereby reducing any polynomial in $\mathbf{A}$ to an equivalent form involving only $\mathbf{A}^m$, $0 < m < n$. This observation leads directly to a proof of Sylvester's expansion theorem.[1]

An interesting and useful alternate procedure for evaluating functions of a matrix is possible based upon a different use of the Cayley-Hamilton theorem. Suppose we have a general polynomial $f(\mathbf{A})$ which is of a higher degree than n, where $\mathbf{A}$ is assumed $n \times n$. If $f(s)$ is divided by the characteristic polynomial of $\mathbf{A}$, denoted by $P(s)$, we obtain

$$\frac{f(s)}{P(s)} = Q(s) + \frac{R(s)}{P(s)} \tag{13.96}$$

where $R(s)$ denotes the remainder term. Multiplying both sides of Eq. (13.96) by $P(s)$ provides

$$f(s) = P(s)Q(s) + R(s) \tag{13.97}$$

If we now set $P(s) = 0$, we observe that

$$f(s) = R(s) \tag{13.98}$$

and, since $P(\mathbf{A}) = \mathbf{0}$, $f(\mathbf{A}) = R(\mathbf{A})$. The following example shows how this principle may be put to use.

Example 13.10

Use the division method discussed in the previous paragraph to reduce the matrix polynomial given in Eq. (13.91) to the equivalent lower-order form presented in Eq. (13.95), assuming $\mathbf{A}$ is given by Eq. (13.85).

The polynomial of interest is

$$f(s) = 2s^4 + s^3 - s^2 + 3s + 1 \tag{13.99}$$

and the characteristic polynomial of $\mathbf{A}$ is

$$P(s) = s^2 + 5s + 6 \tag{13.100}$$

Dividing $f(s)$ by $P(s)$ provides

$$\frac{2s^4 + s^3 - s^2 + 3s + 1}{s^2 + 5s + 6} = 2s^2 - 9s + 32 + \frac{-103s - 191}{s^2 + 5s + 6} \tag{13.101}$$

Thus $f(\mathbf{A}) = R(\mathbf{A}) = -103\mathbf{A} - 191\mathbf{I}$, which checks our previous result.

The basic division procedure illustrated by Ex. 13.10 may be applied, with minor modification, to reduce the matrix exponential solution to a closed form, thereby bypassing use of the Sylvester expansion theorem. In the case of a convergent infinite series, as represented by the power series expansion of the matrix exponential in the region for which it converges, it is not feasible to carry out the division, but we can use the fact that this solution structure is available to determine the closed-form solution, which is all that really matters. As before, we know that if $f(s)$ is represented by a convergent infinite series in the region of interest, it could be divided out and rearranged as

$$f(s) = P(s)Q(s) + R(s) \tag{13.102}$$

Since $P(s_i) = 0$ for each of the eigenvalues s_i of the matrix $\mathbf{A}$, substitution of each s_i in turn into Eq. (13.102) provides n simultaneous equations from which the n coefficients in $R(s)$, which is known to be of $n - 1$ order in s_i, can be determined. Of course we have assumed the s_i are distinct, but this is not necessary. When a root $(s - s_i)^\alpha$ occurs, we may differentiate $f(s)$ and $R(s)$ to obtain the relationships needed by setting

$$\left.\frac{d^k f(s)}{ds^k}\right|_{s=s_i} = \left.\frac{d^k R(s)}{ds^k}\right|_{s=s_i} \qquad k = 0, 1, \ldots, \alpha - 1 \tag{13.103}$$

Once the required coefficients in $R(s)$ have been obtained, it follows from the Cayley-Hamilton theorem that $f(\mathbf{A}) = R(\mathbf{A})$, just as in the less general polynomial case.

Example 13.11

Find $\epsilon^{\mathbf{A}t}$, where $\mathbf{A}$ is the matrix given in Eq. (13.85), using the Cayley-Hamilton remainder method.

The characteristic equation of $\mathbf{A}$ is given by Eq. (13.86), and $\mathbf{A}$ has eigenvalues $s_1 = 3, s_2 = 2$. Since $\mathbf{A}$ is 2×2, $R(s)$ is of the first order in s, or

$$R(s) = a_0 + a_1 s \tag{13.104}$$

We know that

$$f(s_1) = \epsilon^{-3t} = a_0 - 3a_1 \tag{13.105}$$

and

$$f(s_2) = \epsilon^{-2t} = a_0 - 2a_1 \tag{13.106}$$

Solving Eqs. (13.105) and (13.106) for a_0 and a_1 provides

$$a_0 = 3\epsilon^{-2t} - 2\epsilon^{-3t} \qquad a_1 = -\epsilon^{-3t} + \epsilon^{-2t} \tag{13.107}$$

Thus we conclude that

$$f(\mathbf{A}) = \epsilon^{\mathbf{A}t} = (3\epsilon^{-2t} - 2\epsilon^{-3t})\mathbf{I} - (\epsilon^{-3t} - \epsilon^{-2t})\mathbf{A}$$

$$= \begin{bmatrix} 3\epsilon^{-2t} - 2\epsilon^{-3t} & \epsilon^{-3t} + \epsilon^{-2t} \\ +6(\epsilon^{-3t} - \epsilon^{-2t}) & -2\epsilon^{-2t} + 3\epsilon^{-3t} \end{bmatrix} \tag{13.108}$$

Example 13.12

Find $\epsilon^{\mathbf{A}t}$ using the remainder method where $\mathbf{A}$ is given by

$$\mathbf{A} = \begin{bmatrix} 0 & 1 & 3 \\ 6 & 0 & 2 \\ -5 & 2 & 4 \end{bmatrix} \tag{13.109}$$

The characteristic equation for this matrix is

$$P(s) = s^3 - 4s^2 + 5s - 2 = (s - 1)^2(s - 2) \tag{13.110}$$

Since $P(s)$ is of the third order, $R(s)$ is of the second order, or

$$R(s) = a_0 + a_1 s + a_2 s^2 \tag{13.111}$$

Setting $f(s_1) = R(s_1)$ and $f(s_2) = R(s_2)$ where $s_1 = 1$ and $s_2 = 2$, yields

$$\epsilon^t = a_0 + a_1 + a_2 \tag{13.112}$$

$$\epsilon^{2t} = a_0 + 2a_1 + 4a_2 \tag{13.113}$$

The third relationship needed to solve for the a_i is obtained by differentiating both sides of

$$f(s) = \epsilon^{st} = a_0 + a_1 s + a_2 s^2 = R(s) \tag{13.114}$$

with respect to s and evaluating the result at $s = s_1$, which yields

$$t\epsilon^t = a_1 + 2a_2 \tag{13.115}$$

We thus must solve

$$\begin{bmatrix} \epsilon^t \\ \epsilon^{2t} \\ t\epsilon^t \end{bmatrix} = \begin{bmatrix} 1 & 1 & 1 \\ 1 & 2 & 4 \\ 0 & 1 & 2 \end{bmatrix} \begin{bmatrix} a_0 \\ a_1 \\ a_2 \end{bmatrix} \tag{13.116}$$

for the a_i. This yields

$$\begin{aligned} \begin{bmatrix} a_0 \\ a_1 \\ a_2 \end{bmatrix} &= \begin{bmatrix} 1 & 1 & 1 \\ 1 & 2 & 4 \\ 0 & 1 & 2 \end{bmatrix}^{-1} \begin{bmatrix} \epsilon^t \\ \epsilon^{2t} \\ t\epsilon^t \end{bmatrix} \\ &= \begin{bmatrix} 0 & 1 & -2 \\ 2 & -2 & 3 \\ -1 & 1 & -1 \end{bmatrix} \begin{bmatrix} \epsilon^t \\ \epsilon^{2t} \\ t\epsilon^t \end{bmatrix} = \begin{bmatrix} \epsilon^{2t} - 2t\epsilon^t \\ 2\epsilon^t - 2\epsilon^{2t} + 3t\epsilon^t \\ -\epsilon^t + \epsilon^{2t} - t\epsilon^t \end{bmatrix} \end{aligned} \tag{13.117}$$

Thus

$$\epsilon^{\mathbf{A}t} = f(\mathbf{A}) = R(\mathbf{A}) = a_0\mathbf{I} + a_1\mathbf{A} + a_2\mathbf{A}^2 \tag{13.118}$$

But

$$\mathbf{A}^2 = \begin{bmatrix} 0 & 1 & 3 \\ 6 & 0 & 2 \\ -5 & 2 & 4 \end{bmatrix} \begin{bmatrix} 0 & 1 & 3 \\ 6 & 0 & 2 \\ -5 & 2 & 4 \end{bmatrix} = \begin{bmatrix} -9 & 6 & 14 \\ -10 & 10 & 26 \\ -8 & 3 & 5 \end{bmatrix} \tag{13.119}$$

Combining Eqs. (13.109), (13.117), (13.118), and (13.119) provides

$$\epsilon^{\mathbf{A}t} = \begin{bmatrix} (9\epsilon^t + 7t\epsilon^t - 8\epsilon^{2t}) & (-4\epsilon^t - 3t\epsilon^t + 4\epsilon^{2t}) & (-8\epsilon^t - 5t\epsilon^t + 8\epsilon^{2t}) \\ (22\epsilon^t + 28t\epsilon^t - 22\epsilon^{2t}) & (-10\epsilon^t - 12t\epsilon^t + 11\epsilon^{2t}) & (-22\epsilon^t - 20t\epsilon^t + 22\epsilon^{2t}) \\ (-2\epsilon^t - 7t\epsilon^t + 2\epsilon^{2t}) & (\epsilon^t + 3t\epsilon^t - \epsilon^{2t}) & (3\epsilon^t + 5t\epsilon^t - 2\epsilon^{2t}) \end{bmatrix} \tag{13.120}$$

This result may be checked using the Sylvester expansion theorem, but solution using the Cayley-Hamilton method is generally much less involved.

When the state equations are time invariant, the Laplace transform may be used to determine the response $\mathbf{x}(t)$ for an arbitrary known transformable input and a general set of initial conditions, or to find $\mathbf{\Phi}(t, t_0)$. Let

$$\dot{\mathbf{x}} = \mathbf{A}\mathbf{x} + \mathbf{B}\mathbf{m} \qquad \mathbf{x}(t_0) = \mathbf{x}^{\circ} \tag{13.121}$$

describe the system of interest, where $\mathbf{A}$ and $\mathbf{B}$ are constant matrices and $\mathbf{m}(t)$ is an arbitrary Laplace transformable input vector. Laplace transforming both sides of Eq. (13.121) and rearranging provides

$$\mathbf{x}(s) = (s\mathbf{I} - \mathbf{A})^{-1}[\mathbf{x}^{\circ} + \mathbf{B}\mathbf{m}(s)] \tag{13.122}$$

When $\mathbf{m}(s)$ is known we may expand the right side of Eq. (13.122) and inverse transform it term by term to obtain $\mathbf{x}(t)$. When $\mathbf{m}(t)$ has not been specified the general solution is given by Eq. (13.48), where $\mathbf{\Phi}(t, t_0)$ reduces to $\mathbf{\Phi}(t - t_0)$, and $\mathbf{\Phi}(t, 0)$ is obtained using the identity

$$\mathbf{\Phi}(t, 0) = \mathcal{L}^{-1}[(s\mathbf{I} - \mathbf{A})^{-1}] \tag{13.123}$$

If solution for $t < 0$ is desired, the two-sided Laplace transform may be used to show that the result obtained from Eq. (13.123) is valid for all t if we simply neglect to associate $u(t)$ terms with each element. The solution for $\mathbf{\Phi}(t, t_0) = \mathbf{\Phi}(t - t_0)$ is obtained from the result developed using Eq. (13.123) by substituting $t - t_0$ for t.

Example 13.13

Consider the system described by Eq. (13.121) with

$$\mathbf{A} = \begin{bmatrix} -1 & 0 \\ 0 & -2 \end{bmatrix} \qquad \mathbf{B} = \begin{bmatrix} 0 \\ 1 \end{bmatrix} \qquad m = u(t) \qquad \mathbf{x}^{\circ} = \begin{bmatrix} 2 \\ -3 \end{bmatrix} \tag{13.124}$$

Find $\mathbf{\Phi}(t, t_0)$ for this system and the time response for $t > 0$ using the Laplace transform method.

We know that $\mathbf{\Phi}(t, 0)$ is defined by Eq. (13.123) and

$$(s\mathbf{I} - \mathbf{A})^{-1} = \begin{bmatrix} s+1 & 0 \\ 0 & s+2 \end{bmatrix}^{-1} = \begin{bmatrix} \dfrac{1}{s+1} & 0 \\ 0 & \dfrac{1}{s+2} \end{bmatrix} \tag{13.125}$$

The corresponding inverse transform is

$$\mathbf{\Phi}(t, 0) = \mathcal{L}^{-1} \begin{bmatrix} \dfrac{1}{s+1} & 0 \\ 0 & \dfrac{1}{s+2} \end{bmatrix} = \begin{bmatrix} \epsilon^{-t} & 0 \\ 0 & \epsilon^{-2t} \end{bmatrix} \tag{13.126}$$

and

$$\mathbf{\Phi}(t, t_0) = \mathbf{\Phi}(t - t_0) = \begin{bmatrix} \epsilon^{-(t-t_0)} & 0 \\ 0 & \epsilon^{-2(t-t_0)} \end{bmatrix} \tag{13.127}$$

Although we could obtain $\mathbf{x}(t)$ using Eq. (13.48), it is more convenient to expand

$$\mathbf{x}(s) = (s\mathbf{I} - \mathbf{A})^{-1}[\mathbf{x}^{\circ} + \mathbf{B}m(s)] \tag{13.128}$$

which yields

$$\mathbf{x}(s) = \begin{bmatrix} \dfrac{2}{s+1} \\ \dfrac{-3s+1}{s(s+2)} \end{bmatrix} \tag{13.129}$$

Inverse transforming Eq. (13.129) provides

$$\mathbf{x}(t) = \begin{bmatrix} 2\epsilon^{-t} \\ \frac{1}{2} - \frac{7}{2}\epsilon^{-2t} \end{bmatrix} \qquad t \geq 0 \tag{13.130}$$

It is a simple matter in the time-invariant case to establish an important analogy between the state and adjoint variables as indicated by their pole locations in the complex plane. Consider a homogeneous nth-order state equation with distinct characteristic equation roots s_i which is expressed in terms of the diagonal matrix

$$\mathbf{A} = \begin{bmatrix} s_1 & 0 & 0 & & \cdots \\ 0 & s_2 & 0 & & \cdots \\ \cdots & \cdots & \cdots & \cdots & \cdots \\ \cdots & \cdots & \cdots & s_{N-1} & 0 \\ \cdots & \cdots & \cdots & 0 & s_N \end{bmatrix} \tag{13.131}$$

We lose no generality in this diagonal representation, since the linear transformations required to produce this form do not change the root locations. Note that $\mathbf{\Phi}(t,0)$ is defined by Eq. (13.123) and $(s\mathbf{I} - \mathbf{A})^{-1}$ is a diagonal matrix with diagonal elements

$$\phi_{ii}(s) = \frac{1}{s - s_i} \tag{13.132}$$

Turning our attention to the transition matrix for the adjoint variables, we see that

$$\mathbf{\theta}(t,0) = \mathcal{L}^{-1}[(s\mathbf{I} + \mathbf{A})^{-1}] \tag{13.133}$$

and $(s\mathbf{I} + \mathbf{A})^{-1}$ is a diagonal matrix with diagonal elements

$$\theta_{ii}(s) = \frac{1}{s + s_i} \tag{13.134}$$

Thus we conclude that the poles of the adjoint-variable system are mirror images across the $j\omega$ axis in the s-plane of those for the state equations, since all complex poles must occur in conjugate pairs. This explains the inverse relationships between $\mathbf{x}(t)$ and $\mathbf{\lambda}(t)$, and between $\mathbf{\Phi}(t, t_0)$ and $\mathbf{\theta}(t, t_0)$, at least for the time-invariant case.

13.4 GENERAL RELATIONSHIPS BETWEEN STATE EQUATIONS AND TRANSFER FUNCTIONS

In the previous sections of this chapter we have considered several basic procedures by which the DE or transfer function representation of a system may be reduced to one or more equivalent state-equation forms. In this section we consider general transformations which may be used to modify the state-equation structure. Since we assume an s-domain transfer function exists, our discussion is limited to linear time-invariant systems.

Consider initially the problem of converting from the state equations to an equivalent transfer-matrix form. The state equations for an nth-order m-input p-output system are

$$\dot{\mathbf{x}} = \mathbf{A}\mathbf{x} + \mathbf{B}\mathbf{m} \qquad \mathbf{x}(t_0) = \mathbf{x}^\circ \tag{13.135}$$

$$\mathbf{y} = \mathbf{C}\mathbf{x} + \mathbf{D}\mathbf{m} \tag{13.136}$$

where the vector and matrix dimensions are as previously defined. Laplace transforming Eqs. (13.135) and (13.136) yields

$$s\mathbf{x}(s) - \mathbf{x}^\circ = \mathbf{A}\mathbf{x}(s) + \mathbf{B}\mathbf{m}(s) \tag{13.137}$$

$$\mathbf{y}(s) = \mathbf{C}\mathbf{x}(s) + \mathbf{D}\mathbf{m}(s) \tag{13.138}$$

Solving Eq. (13.137) for $\mathbf{x}(s)$ provides

$$\mathbf{x}(s) = [s\mathbf{I} - \mathbf{A}]^{-1}[\mathbf{x}^\circ + \mathbf{B}\mathbf{m}(s)] \tag{13.139}$$

Substituting $\mathbf{x}(s)$ from Eq. (13.139) into Eq. (13.138) gives

$$\begin{aligned} \mathbf{y}(s) &= [\mathbf{C}(s\mathbf{I} - \mathbf{A})^{-1}\mathbf{B} + \mathbf{D}]\,\mathbf{m}(s) + \mathbf{C}(s\mathbf{I} - \mathbf{A})^{-1}\mathbf{x}^\circ \\ &= \mathbf{H}(s)\,\mathbf{m}(s) + \mathbf{K}(s)\,\mathbf{x}^\circ \end{aligned} \tag{13.140}$$

where $\mathbf{H}(s)$ and $\mathbf{K}(s)$ are $p \times m$ and $p \times n$ matrices, respectively, defined to simplify the solution form as

$$\mathbf{H}(s) = \mathbf{C}[(s\mathbf{I} - \mathbf{A})^{-1}\mathbf{B} + \mathbf{D}] \qquad \mathbf{K}(s) = \mathbf{C}(s\mathbf{I} - \mathbf{A})^{-1} \tag{13.141}$$

It is clear from Eqs. (13.140) and (13.141) that the transfer matrix $\mathbf{H}(s)$ is completely defined in terms of the state-equation matrices $\mathbf{A}$, $\mathbf{B}$, $\mathbf{C}$ and $\mathbf{D}$. Since we did not assume zero initial conditions, the response depends also upon the initial condition matrix $\mathbf{K}(s)$, which is recognized also as a linear transformation carried out upon the Laplace transform of the usual state-transition matrix. The following example illustrates the application of these results.

Example 13.14

Obtain the transfer matrix representation of the system described by

$$\begin{bmatrix} \dot{x}_1 \\ \dot{x}_2 \\ \dot{x}_3 \end{bmatrix} = \begin{bmatrix} -1 & 1 & -1 \\ 0 & -2 & 1 \\ 0 & 0 & -3 \end{bmatrix} \begin{bmatrix} x_1 \\ x_2 \\ x_3 \end{bmatrix} + \begin{bmatrix} 1 & 0 \\ 0 & 1 \\ 1 & 0 \end{bmatrix} \begin{bmatrix} m_1 \\ m_2 \end{bmatrix} \qquad \mathbf{x}^{\circ} = 0$$

$$\begin{bmatrix} y_1 \\ y_2 \end{bmatrix} = \begin{bmatrix} 1 & 1 & 1 \\ 0 & 1 & 1 \end{bmatrix} \begin{bmatrix} x_1 \\ x_2 \\ x_3 \end{bmatrix} \tag{13.142}$$

Since $\mathbf{x}^{\circ} = \mathbf{0}$, we need not find $\mathbf{K}(s)$, and, because $\mathbf{D} = \mathbf{0}$, development of $\mathbf{H}(s)$ reduces to determining

$$\mathbf{H}(s) = \mathbf{C}(s\mathbf{I} - \mathbf{A})^{-1}\mathbf{B} \tag{13.143}$$

But

$$(s\mathbf{I} - \mathbf{A})^{-1} = \begin{bmatrix} s+1 & -1 & 1 \\ 0 & s+2 & -1 \\ 0 & 0 & s+3 \end{bmatrix}^{-1} \tag{13.144}$$

and

$$(s\mathbf{I} - \mathbf{A})^{-1} = \begin{bmatrix} \dfrac{1}{s+1} & \dfrac{1}{(s+1)(s+2)} & \dfrac{-1}{(s+2)(s+3)} \\ 0 & \dfrac{1}{s+2} & \dfrac{1}{(s+2)(s+3)} \\ 0 & 0 & \dfrac{1}{s+3} \end{bmatrix} \tag{13.145}$$

so

$$\mathbf{H}(s) = \begin{bmatrix} 1 & 1 & 1 \\ 0 & 1 & 1 \end{bmatrix} \begin{bmatrix} \dfrac{1}{s+1} & \dfrac{1}{(s+1)(s+2)} & \dfrac{-1}{(s+2)(s+3)} \\ 0 & \dfrac{1}{s+2} & \dfrac{1}{(s+2)(s+3)} \\ 0 & 0 & \dfrac{1}{s+3} \end{bmatrix} \begin{bmatrix} 1 & 0 \\ 0 & 1 \\ 1 & 0 \end{bmatrix}$$

$$= \begin{bmatrix} \dfrac{2(s+2)}{(s+1)(s+3)} & \dfrac{1}{s+1} \\ \dfrac{1}{s+2} & \dfrac{1}{s+2} \end{bmatrix} \tag{13.146}$$

It is clear from Ex. 13.14 that the transfer-matrix representation is easily obtained from the state equations. The inverse problem, that of developing the state equations from the transfer function representation, although straightforward in concept, is a complex undertaking, and considerable research effort is currently being devoted to this problem. Ho and Kalman, in a recent paper,[3] present a procedure, which they claim to be optimally simple, for obtaining the state equations. Fortunately we may generally obtain the state equations from the system or from its elements, so this problem is not discussed further here.

We have found that there is an infinite set of state-equation representations, and it is obvious that some structures are easier to work with, or provide more insight into the physical or mathematical interpretation of system performance, than others. We are thus motivated to consider the effect of general linear-variable transformations upon the state equations. Emphasis is placed upon particular state-equation forms which are most convenient to use. The usual structure of the state equations is given by Eqs. (13.135) and (13.136). When $\mathbf{A}$ has distinct eigenvalues, it is possible to reduce these equations to the normal form

$$\dot{\mathbf{z}} = \mathbf{\Lambda z} + \mathbf{Bm} \qquad \mathbf{z}(t_0) = \mathbf{z}^{\circ} \tag{13.147}$$

$$\mathbf{y} = \boldsymbol{\gamma}\mathbf{z} + \mathbf{Dm} \tag{13.148}$$

where $\mathbf{\Lambda}$ is the diagonal matrix of eigenvalues derived from $\mathbf{A}$. The normal form is easy to work with and conclusions applicable to all linear systems with distinct eigenvalues can be derived with minimum difficulty by starting from it. The state-equation normal form is readily obtained from the usual DE or transfer function description relating input to output by carrying out the following steps:

1. Reduce the system equation to transfer function form (if necessary) and factor the denominator polynomial in s.
2. Expand the transfer function into partial fraction form.
3. Set up a simulation diagram of the result.
4. Define the integrator outputs as states.
5. Write down the corresponding state-equation description, which will be in normal form.

This procedure is illustrated by the following example.

Example 13.15

Consider the system described by

$$\dddot{\theta} + 3\ddot{\theta} + 2\dot{\theta} = 3m + \dot{m} \tag{13.149}$$

Develop the state equations for this system in normal form.

We initiate the solution by writing down the corresponding transfer function form as

$$\frac{\theta(s)}{m(s)} = \frac{s+3}{s^3 + 3s^2 + 2s} = \frac{s+3}{s(s+1)(s+2)} \tag{13.150}$$

The partial fraction expansion is

$$\theta(s) = m(s)\left[\frac{3/2}{s} - \frac{2}{s+1} + \frac{1/2}{s+2}\right] \tag{13.151}$$

This leads to the simulation diagram shown in Fig. 13.11. The corresponding state equations are

$$\begin{bmatrix} \dot{z}_1 \\ \dot{z}_2 \\ \dot{z}_3 \end{bmatrix} = \begin{bmatrix} 0 & 0 & 0 \\ 0 & -1 & 0 \\ 0 & 0 & -2 \end{bmatrix} \begin{bmatrix} z_1 \\ z_2 \\ z_3 \end{bmatrix} + \begin{bmatrix} 3/2 \\ -2 \\ 1/2 \end{bmatrix} m \tag{13.152}$$

and

$$y = [1 \quad 1 \quad 1] \begin{bmatrix} z_1 \\ z_2 \\ z_3 \end{bmatrix} \tag{13.153}$$

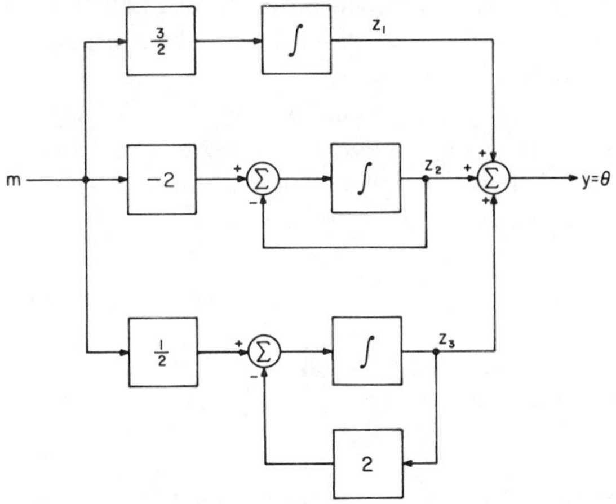

FIG. 13.11. Simulation diagram for the system used in Ex. 13.15.

It is straightforward, as seen from Ex. 13.15, to obtain the state equations in normal form starting from the transfer function description, but a general state-equation description can also be reduced to the normal form by carrying out a linear change of variables. This problem is of sufficient importance that we will consider several aspects of its solution. Suppose we have

$$\dot{\mathbf{x}} = \mathbf{A}\mathbf{x} + \mathbf{B}\mathbf{m} \tag{13.154}$$

where $\mathbf{A}$ is assumed to have nonrepeated eigenvalues. A general linear transformation of variables is described by

$$\mathbf{x} = \mathbf{E}\mathbf{z} \tag{13.155}$$

Substituting this result into Eq. (13.154) yields

$$\mathbf{E}\dot{\mathbf{z}} = \mathbf{A}\mathbf{E}\mathbf{z} + \mathbf{B}\mathbf{m} \tag{13.156}$$

Assume that $\mathbf{E}$ is a nonsingular transformation such that $\mathbf{E}^{-1}$ exists. Premultiplying both sides of Eq. (13.156) by E^{-1} provides

$$\dot{\mathbf{z}} = \mathbf{E}^{-1}\mathbf{A}\mathbf{E}z + \mathbf{E}^{-1}\mathbf{B}\mathbf{m} \tag{13.157}$$

and if Eq. (13.157) is to be of normal form, it is required that

$$\mathbf{E}^{-1}\mathbf{A}\mathbf{E} = \mathbf{\Lambda} \tag{13.158}$$

If $\mathbf{A}$, $\mathbf{E}$ and $\mathbf{\Lambda}$ are $n \times n$, Eq. (13.158) provides n^2 equations in the n^2 unknowns

e_{ij} from which the elements of **E**, denoted by e_{ij}, may be determined. It would be a formidable task to obtain **E** in this way, and, fortunately, it is not necessary to do so.

Assume **A** is symmetric with distinct eigenvalues, $\lambda_1, \lambda_2, \ldots, \lambda_n$. Let $\mathbf{x}_1, \mathbf{x}_2, \ldots, \mathbf{x}_n$ denote the corresponding normalized eigenvectors and let **X** denote the eigenvector matrix obtained by using the normalized eigenvectors as columns, or

$$\mathbf{X} = [\mathbf{x}_1, \mathbf{x}_2, \ldots, \mathbf{x}_n] = \begin{bmatrix} x_{11} & x_{12} & \cdots & x_{1n} \\ x_{21} & x_{22} & \cdots & x_{2n} \\ \cdots & \cdots & \cdots & \cdots \\ x_{n1} & x_{n2} & \cdots & x_{nn} \end{bmatrix} \tag{13.159}$$

Note that **AX** can be written in partitioned form as

$$\mathbf{AX} = [\mathbf{Ax}_1, \mathbf{Ax}_2, \ldots, \mathbf{Ax}_n] \tag{13.160}$$

However,

$$\mathbf{Ax}_i = \lambda_i \mathbf{x}_i \tag{13.161}$$

is the relationship from which the eigenvectors were obtained, so

$$\mathbf{AX} = [\lambda_1 \mathbf{x}_1, \lambda_2 \mathbf{x}_2, \ldots, \lambda_n \mathbf{x}_n] \tag{13.162}$$

and

$$\mathbf{X}'\mathbf{AX} = \begin{bmatrix} \lambda_1 & 0 & \cdots & 0 \\ 0 & \lambda_2 & \cdots & 0 \\ \cdots & \cdots & \cdots & \cdots \\ 0 & 0 & & \lambda_n \end{bmatrix} = \mathbf{\Lambda} \tag{13.163}$$

This provides the desired transformation by which $\mathbf{\Lambda}$ may be obtained for **A** symmetric. It requires only that the normalized eigenvector matrix be determined and used in a straightforward matrix product operation. Since **X** is normalized,

$$\mathbf{X}'\mathbf{X} = \mathbf{XX}' = \mathbf{I} \tag{13.164}$$

so **A** may be obtained from the diagonal eigenvalue matrix if $\mathbf{\Lambda}$ is premultiplied by **X** and postmultiplied by **X**′, as seen from Eq. (13.162).

In most control systems **A** is not symmetric, and the result presented in Eq. (13.163) is not applicable in such cases. It is possible to show, although the development is not as readily carried out as for the symmetric case, that $\mathbf{\Lambda}$ may be obtained for any $\mathbf{A}$ with distinct eigenvalues using

$$\mathbf{T}^{-1}\mathbf{A}\mathbf{T} = \mathbf{\Lambda} \tag{13.165}$$

where $\mathbf{T}$ is a matrix of eigenvectors analogous to $\mathbf{X}$, but these eigenvectors are not orthogonal in the general case and they need not be normalized.

Example 13.16

Consider a linear homogeneous set of state equations with

$$\mathbf{A} = \begin{bmatrix} 0 & 1 \\ -2 & 3 \end{bmatrix} \tag{13.166}$$

Show that $\mathbf{\Lambda}$ can be obtained for this case using Eq. (13.165).

Setting $|(\lambda\mathbf{I} - \mathbf{A})| = 0$ provides the characteristic equation as

$$\lambda^2 - 3\lambda + 2 = (\lambda - 2)(\lambda - 1) = 0 \tag{13.167}$$

Thus the eigenvalues are $\lambda = 2$ and $\lambda = 1$. Setting $(\lambda_i\mathbf{I} - \mathbf{A})\mathbf{x} = \mathbf{0}$ provides the corresponding eigenvectors as $\mathbf{x}_1' = (\alpha, \alpha)$ and $\mathbf{x}_2' = (\beta, 2\beta)$. Thus we will use

$$\mathbf{T} = \begin{bmatrix} 1 & 1 \\ 1 & 2 \end{bmatrix} \tag{13.168}$$

$\mathbf{\Lambda}$ is obtained as

$$\mathbf{\Lambda} = \mathbf{T}^{-1}\mathbf{A}\mathbf{T} = \begin{bmatrix} 2 & -1 \\ -1 & 1 \end{bmatrix}\begin{bmatrix} 0 & 1 \\ -2 & 3 \end{bmatrix}\begin{bmatrix} 1 & 1 \\ 1 & 2 \end{bmatrix} = \begin{bmatrix} 1 & 0 \\ 0 & 2 \end{bmatrix} \tag{13.169}$$

We have assumed in our discussion thus far that $\mathbf{A}$ has distinct eigenvalues. In many problems of interest $\mathbf{A}$ has one or more eigenvalues of multiplicity greater than one and in such cases it is not possible to convert $\mathbf{A}$ to a diagonal form using a linear nonsingular transformation. It is, however, always possible to transform any square matrix **A** into the Jordan canonical form **J** using

$$\mathbf{J} = \mathbf{T}^{-1}\mathbf{A}\mathbf{T} = \begin{bmatrix} L_{k1}(\lambda_1) & 0 & \cdots & 0 \\ 0 & L_{k2}(\lambda_2) & & 0 \\ \cdots & \cdots & \cdots & \cdots \\ 0 & 0 & \cdots & L_{kr}(\lambda_r) \end{bmatrix} \tag{13.170}$$

where k_i denotes the multiplicity of eigenvalue λ_i of the $n \times n$ matrix **A**. It is necessary that

$$\sum_{i=1}^{r} k_i = n \tag{13.171}$$

where r is the number of sub-matrices in **J**. Each $L_{ki}(\lambda_i)$ is a $k_i \times k_i$ matrix with λ_i for each principal diagonal component and 1 for each component on the first super-diagonal, with all other components 0, or

$$L_{ki}(\lambda_i) = \begin{bmatrix} \lambda_i & 1 & 0 & \cdots\cdots & 0 \\ 0 & \lambda_i & 1 & \cdots\cdots & 0 \\ 0 & 0 & \lambda_i & \cdots\cdots & 0 \\ \cdots & \cdots & \cdots & \cdots & \cdots \\ 0 & 0 & 0 & \cdots \;\; \lambda_i & 1 \\ 0 & 0 & 0 & \cdots \;\; 0 & \lambda_i \end{bmatrix} \tag{13.172}$$

Note that $L_1(\lambda_i) = \lambda_i$, or a nonrepeated root gives rise to a single diagonal element in Eq. (13.170). We may thus use Eq. (13.170) to obtain **J** for systems with multiple roots in the same way we would obtain $\mathbf{\Lambda}$ for the general non-symmetric case. Only the form of the result differs.

Transformations of the type

$$\mathbf{J} = \mathbf{T}^{-1}\mathbf{A}\mathbf{T} \tag{13.173}$$

are general linear transformations and they are often called *similarity transformations.* Matrix **A** is said to be *similar* to $\mathbf{\Lambda}$. Similar matrices have the same characteristic equations and thus the same eigenvalues. Any square matrix can be converted to the Jordan canonical form

$$\mathbf{J} = \begin{bmatrix} \lambda_1 & \alpha_1 & 0 & \cdots & 0 \\ 0 & \lambda_2 & \alpha_2 & \cdots & 0 \\ \cdots & \cdots & \cdots & \cdots & \cdots \\ 0 & 0 & 0 & \cdots & \alpha_{n-1} \\ 0 & 0 & 0 & \cdots & \lambda_n \end{bmatrix} \tag{13.174}$$

where the α_i are either 1 or 0. Thus every square matrix is similar to the Jordan form. When the eigenvalues of $\mathbf{A}$ are distinct, $\mathbf{J}$ reduces to $\mathbf{\Lambda}$, which is the diagonal matrix of eigenvalues.

The Jordan canonical form can be developed directly from the system transfer function. Consider a general transfer function of the form

$$\frac{y(s)}{m(s)} = \frac{N(s)}{D(s)} = \frac{a_n s^n + a_{n-1}s^{n-1} + \cdots + a_1 s + a_0}{b_n s^n + b_{n-1}s^{n-1} + \cdots + b_1 s + b_0} \tag{13.175}$$

Assume that $D(s)$ has all simple roots with the exception of one, which has multiplicity α. Then $D(s)$ may be written in factored form as

$$D(s) = (s - \lambda_1)^\alpha (s - \lambda_2) \cdots (s - \lambda_{n-\alpha}) \tag{13.176}$$

The partial fraction expansion of $y(s)/m(s)$ is

$$\frac{y(s)}{m(s)} = k_0 + \frac{k_{1\alpha}}{(s - \lambda_1)^\alpha} + \frac{k_{1\alpha-1}}{(s - \lambda_1)^{\alpha-1}} + \cdots + \frac{k_{11}}{s - \lambda_1} + \frac{k_2}{s - \lambda_2} + \cdots + \frac{k_{n-\alpha}}{s - \lambda_{n-\alpha}} \tag{13.177}$$

where the coefficients are obtained in the usual way. The output $y(s)$ is readily developed using a simulation diagram to represent Eq. (13.177) term by term. Note that

$$\frac{1}{(s - \lambda_1)^\beta} = \frac{1}{(s - \lambda_1)^{\beta-1}} \cdot \frac{1}{s - \lambda_1} \tag{13.178}$$

so the terms associated with the repeated root may be obtained by cascading a sequence of α identical simulation elements. The general simulation diagram for this situation is shown in Fig. 13.12. It is straightforward to extend this general procedure to situations involving two or more repeated roots.

Example 13.17

Consider a system with transfer function

$$\frac{C(s)}{R(s)} = \frac{2}{s^3(s + 1)^2(s + 2)} \tag{13.179}$$

Find the Jordan canonical form for this system using the partial fraction expansion technique.

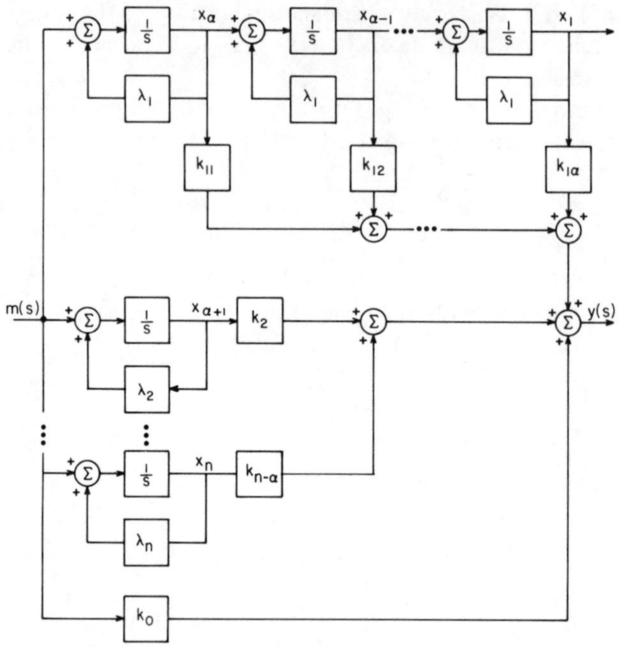

FIG. 13.12. General simulation diagram development of the Jordan canonical form.

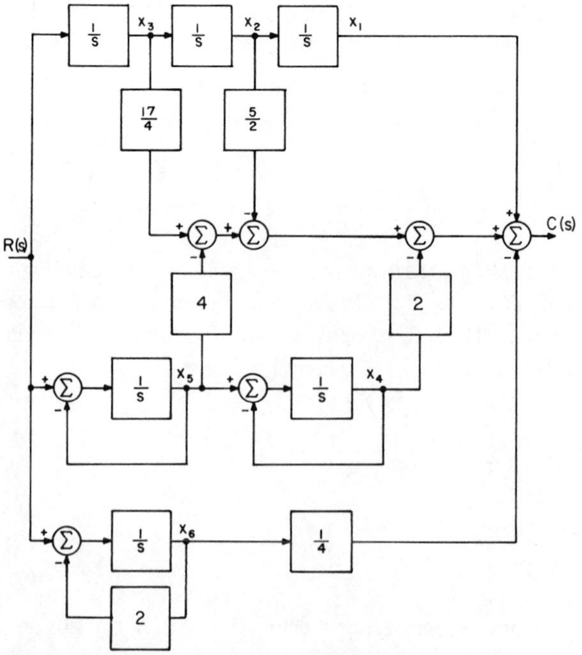

FIG. 13.13. Simulation diagram for the system considered in Ex. 13.17.

Expanding Eq. (13.179) into partial functions provides

$$\frac{2}{s^3(s+1)^2(s+2)} = \frac{k_{11}}{s} + \frac{k_{12}}{s^2} + \frac{k_{13}}{s^3} + \frac{k_{21}}{s+1} + \frac{k_{22}}{(s+1)^2} + \frac{k_3}{s+2}$$

$$= \frac{17/4}{s} - \frac{5/2}{s^2} + \frac{1}{s^3} - \frac{4}{s+1} - \frac{2}{(s+1)^2} - \frac{1/4}{s+2} \qquad (13.180)$$

The corresponding simulation diagram is shown in Fig. 13.13. The state equations are written down directly from the simulation diagram as

$$\begin{bmatrix} \dot{x}_1 \\ \dot{x}_2 \\ \dot{x}_3 \\ \dot{x}_4 \\ \dot{x}_5 \\ \dot{x}_6 \end{bmatrix} = \left[\begin{array}{ccc:cc:c} 0 & 1 & 0 & 0 & 0 & 0 \\ 0 & 0 & 1 & 0 & 0 & 0 \\ 0 & 0 & 0 & 0 & 0 & 0 \\ \hdashline 0 & 0 & 0 & -1 & 1 & 0 \\ 0 & 0 & 0 & 0 & -1 & 0 \\ \hdashline 0 & 0 & 0 & 0 & 0 & 2 \end{array}\right] \begin{bmatrix} x_1 \\ x_2 \\ x_3 \\ x_4 \\ x_5 \\ x_6 \end{bmatrix} + \begin{bmatrix} 0 \\ 0 \\ 1 \\ 0 \\ 1 \\ 1 \end{bmatrix} m \qquad (13.181)$$

and

$$y = \left[1, -\frac{5}{2}, \frac{17}{4}, -2, -4, -\frac{1}{4}\right] \begin{bmatrix} x_1 \\ x_2 \\ x_3 \\ x_4 \\ x_5 \\ x_6 \end{bmatrix} \qquad (13.182)$$

where $y(t) = c(t)$ and $m(t) = r(t)$.

13.5 CONTROLLABILITY AND OBSERVABILITY

When systems modeled in transfer-function or state-variable form are connected in cascade it is always possible that one or more pole-zero cancellations will occur. When this happens the corresponding modes of system operation become directly coupled and cannot be independently influenced by the input. Interest in this problem was stimulated primarily by Kalman, who has published many papers in this area.[4-6] Although several specialized interpretations of controllability and observability have been proposed, we will consider only the basic concepts here.

Controllability is defined as: *a system S is controllable in a region R of state space if, and only if, system state can be driven from* $\mathbf{x}^1$ *to* $\mathbf{x}^2$ *using some allowable control input over the closed interval* $[t_1, t_2]$, *where* $\mathbf{x}^1$ *and* $\mathbf{x}^2$ *are arbitrary points within R.* In many cases the control input and R are not limited and t_2 may approach infinity. A system is *completely controllable* if any $\mathbf{x}^2$ can be reached from $\mathbf{x}^1$ in finite time. It is clear from these definitions that systems in which the state-vector elements cannot be independently influenced by the control signals available are uncontrollable.

Observability is defined as: *a system S is observable if, and only if, all elements of its state can be uniquely determined over a finite time interval by observing the time variations of its available outputs.* In other words unobservable systems have one or more dynamic modes which in no way influence the observed outputs. An arbitrary input may be used during the observation process.

The controllability and observability characteristics of a system S with distinct eigenvalues are readily apparent from the state equations expressed in normal form, or

$$\dot{\mathbf{z}} = \mathbf{\Lambda z} + \boldsymbol{\beta}\mathbf{m} \tag{13.183}$$

$$\mathbf{y} = \boldsymbol{\gamma}\mathbf{z} + \mathbf{Dm} \tag{13.184}$$

S is controllable if, and only if, $\boldsymbol{\beta}$ has no rows with all zero elements, and observable if, and only if, $\boldsymbol{\gamma}$ has no columns with all zero elements. It is thus a simple matter to determine these characteristics from the state equations in normal form since this form eliminates interactions between the state-vector elements.

Virtually all physical systems are controllable and observable within some reasonable performance domain, according to the strict definitions, but perhaps more significantly, the mathematical models we use for systems, which are approximations of the actual situation, may indicate lack of controllability or observability, or both. At the very best this means that the actual system is *almost* uncontrollable, or *almost* unobservable. Either of these conditions lead to difficulties as assuredly as if our mathematical model were exact. It is not any more desirable to design systems which are almost impossible to control than systems which are completely impossible to control. Thus we cannot use the fact that system models are approximations as justification for ignoring the principles of controllability and observability. These characteristics should be considered during the development of any system design.

13.6 DESIGN OF CONTROL SYSTEMS USING STATE-VARIABLE FEEDBACK

Discussion of state-variable feedback (SVF) as a design procedure is introduced at this point, rather than in the latter part of Chap. 11 where the standard minor-loop design procedures are discussed, for two main reasons. First, and most important, we are now familiar with the state representation of systems and the use of SVF to synthesize desired response characteristics is a logical extension of what we have been doing. Second, the material in Chap. 11 emphasizes the use of Bode diagram methods and it is generally assumed there that only one minor loop will be used. In this section we shall emphasize the use of SVF with the gains as design parameters as a means of synthesizing desired transfer functions. Although it is possible to structure this problem in terms of the state equations, it is generally more convenient to work in terms of transfer functions and that notation is used throughout.

The concept of SVF is an improvement on the Guillemin-Truxal (GT) synthesis method[7] which has been available for many years. The use of SVF as a design procedure is considered in detail from several points of view by Schultz and Melsa,[8] and it is not our intention here to go beyond the introductory stage of the problem. Those who wish further detail are directed to Ref. 8.

Since the GT synthesis method is so closely related to the use of SVF, a brief review of the basic concepts involved is appropriate. The first, and generally most difficult, step in the procedure is the reduction of the system specs to a desired closed-loop transfer function. Many factors must be considered here and considerable engineering judgment is required, but this is true in other design procedures as well, so should not be interpreted as a disadvantage of the GT method. The engineer considers the plant or plants available for establishing control and the performance specs together, and chooses a reasonable, desired closed-loop transfer function. Since the equalizer must be realizable, the order of the closed-loop transfer function chosen cannot be less than that of the plant alone. Consider the arrangement shown in Fig. 13.14, which is described by

$$\frac{C(s)}{R(s)} = \frac{G_c(s)\,G_p(s)}{1 + G_c(s)\,G_p(s)} = G(s) \qquad (13.185)$$

where $G(s) = C(s)/R(s)$ is introduced to simplify the subsequent notation. If $G(s)$ has been chosen, we may solve Eq. (13.185) for the required compensation network $G_c(s)$ in terms of $G(s)$ and $G_p(s)$ to obtain

$$G_c(s) = \frac{G(s)}{G_p(s)\,[1 - G(s)]} \qquad (13.186)$$

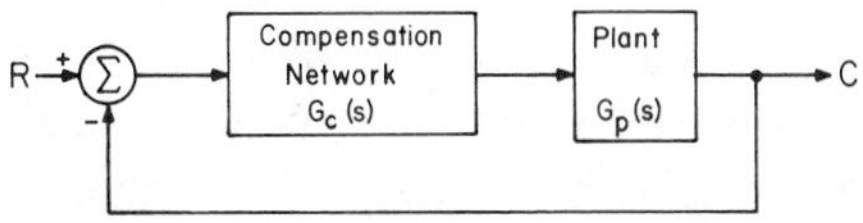

FIG. 13.14. The standard feedback arrangement used in the Guillemin-Truxal design procedure.

Since $G_p(s)$ and $G(s)$ are assumed known, Eq. (13.186) may be used to obtain the required compensation network. One restriction which must be taken into consideration when choosing $G(s)$, to make sure that $G_c(s)$ will be realizable, is readily seen from Eq. (13.185). Note that

$$\lim_{s \to \infty} G(s) = G_c(s)\,G_p(s) = 0 \tag{13.187}$$

is required for realizability of $G_c(s)\,G_p(s)$. Since $G_c(s)$ must also be realizable, it cannot have a pole at $s = \infty$, and $G_c(s)$ must have at least as many finite poles in the 1hp as it has finite plane zeros. Thus $G(s)$ must approach zero as $s \to \infty$ as least as fast as $G_p(s)$, or

$$\left(N_p - N_z\right)_G \geq \left(N_p - N_z\right)_{G_p} \tag{13.188}$$

Equivalently, the excess of poles over zeros for G must equal or exceed that of the plant.

Example 13.18

Consider a standard unity feedback controller as shown in Fig. 13.14 with

$$G_p(s) = \frac{5}{s(s+1)(1+s/10)} \tag{13.189}$$

Find the $G_c(s)$ required to realize the closed-loop transfer function

$$G(s) = \frac{120(s+2)}{[(s+2)^2+16](s+3)(s+5)} \tag{13.190}$$

Applying Eq. (13.186) and simplifying the result yields

$$G_c(s) = \frac{120(s)(s+1)(s+2)(s+10)}{50(s^4+12s^3+67s^2+100s+60)} \tag{13.191}$$

The denominator of Eq. (13.191) may be reduced to the product of two quadratic terms, each of which provides a pair of complex conjugate roots, or

$$G_c(s) = \frac{120(s)(s+1)(s+2)(s+10)}{50(s^2+10.15s+47)(s^2+1.85s+1.28)} \tag{13.192}$$

Realization of a network providing the desired $G_c(s)$ is a relatively straightforward problem in network synthesis.

The solution developed in Ex. 13.18 illustrates several general characteristics of the GT synthesis procedure. Those zeros desired in $G(s)$ which do not appear as zeros in $G_p(s)$ must be included in $G_c(s)$. It is also generally true that $G_c(s)$

will contain as zeros most of the poles of $G_p(s)$, since seldom will those poles be located correctly to provide the desired closed-loop transfer function. Thus $G_c(s)$ provides the desired zeros, cancels any lhp zeros of $G_p(s)$ which are undesired with an overlapping pole, cancels the poles of $G_p(s)$ which are not in desirable locations (this is often all of them) with overlapping zeros, and provides new poles consistent with the desired $G(s)$. The disadvantages of the GT procedure can be summarized as follows:

1. Determining the desired transfer function $G(s)$ consistent with system specs is rather difficult. Arbitrary decisions made during this phase of the design may, perhaps without the designer's knowledge or intention, cause the required equalizing filter to be unduly complex.
2. Since $G_c(s)$ and $G_p(s)$ provide several pole-zero cancellations, the resulting system is not controllable or observable.
3. Pole-zero cancellations are never perfect, so $G(s)$ is actually of a higher order than the mathematical model indicates.
4. The equalizing filters required are typically more difficult to realize than those used for standard Bode diagram methods. Note that $G_c(s)$ as given by Eq. (13.192) has two complex conjugate pole pairs and requires RLC synthesis.
5. It is easy, as in Ex. 13.18, to specify a system in which $G_c(s)$ has a zero at $s = 0$, thereby cancelling the forward path integration which is desirable for control of static error.

In most applications the actual closed-loop transfer function is of little consequence as long as certain desired performance characteristics are provided. Thus specifying $G(s)$ over-defines the problem, placing unnecessary, and often undesirable, restrictions upon the designer. For these reasons we do not emphasize the GT method. In those cases where specifying $G(s)$ is desirable, the GT method applies directly, and the reader is referred to Ref. 7 for further detail.

The general concept of SVF is similar to the GT method, but avoids some of its disadvantages. The intention is to develop a desired closed-loop transfer function by choosing forward and feedback gains rather than by the synthesis of a cascade-equalizing filter. It is assumed that all system states are available, although one or more state signals can be approximately synthesized if necessary. The only limitation on the transfer functions that can be realized this way is that their pole-zero excess cannot be less than that of the plant. Consider the system shown in Fig. 13.15, where the plant has been broken down such that each state signal is available, $k_1 = 1$ is chosen to provide zero steady-state step-response error, and the forward path gain is retained as a solution parameter. Assume that

$$G(s) = \frac{10}{[(s+1)^2+4](s+2)} = \frac{10}{s^3+4s^2+9s+10} \tag{13.193}$$

is desired. This result is realized by reducing the system in Fig. 13.15 to transfer function form and matching the coefficients of the result to those required in

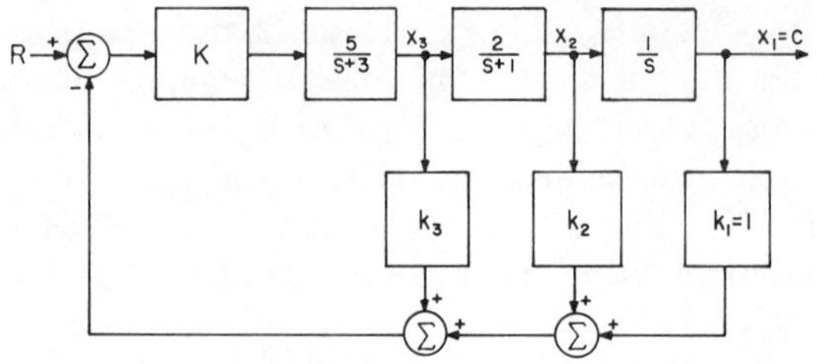

FIG. 13.15. A simple illustration of the state-variable feedback method.

Eq. (13.193). A convenient first step in this procedure is to develop an equivalent single-feedback element, defined as $H_{eq}(s)$, as shown in Fig. 13.16. The closed-loop transfer function becomes

$$\frac{C(s)}{R(s)} = G(s) = \frac{10K}{s^3 + (4 + 5Kk_3)s^2 + (3 + 10Kk_2 + 5Kk_3)s + 10K} \tag{13.194}$$

Equating the coefficients of Eqs. (13.193) and (13.194) provides

$$10K = 10 \qquad 3 + 10Kk_2 + 5Kk_3 = 9 \qquad 4 + 5Kk_3 = 4 \tag{13.195}$$

which are readily solved to obtain

$$K = 1 \qquad k_3 = 0 \qquad k_2 = 0.6 \tag{13.196}$$

Note that it is not always necessary that all states be fed back to obtain the desired $G(s)$.

Design using state-variable feedback for the simple case illustrated in Fig. 13.15 is straightforward. When no zeros other than those already included in $G_p(s)$ are to be provided in $G(s)$, the design procedure may be summarized as follows:

1. Set up the state-variable feedback configuration in standard form assuming all states are available.
2. Leave a variable gain in each feedback path and in the forward transfer function as the parameters to be chosen in realizing $G(s)$. In most cases $G_p(s)$

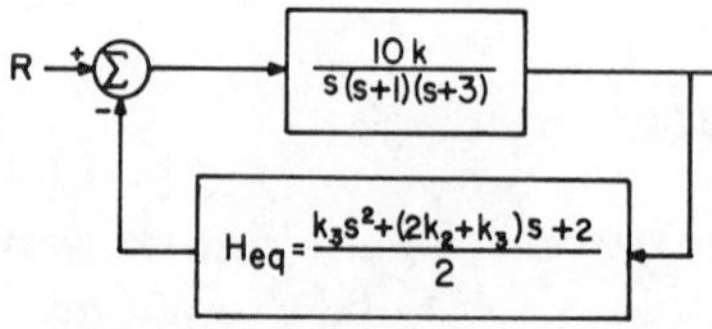

FIG. 13.16. The $H_{eq}(s)$ form of the system shown in Fig. 13.15.

has one integration and $k_1 = 1$ will be used to assure zero steady-state step-input error. When this is not the case it may be desirable to set k_1 at whatever value is required to assure zero steady-state step-response error, which provides a relationship between K and k_1, reducing the number of free parameters to n in this case also.

3. Choose a desired $G(s)$ of the same order as $G_p(s)$, with poles located arbitrarily, and with no zeros other than those of $G_p(s)$.
4. Solve for $G(s)$ in terms of the k_i's and K. The $H_{eq}(s)$ reduction procedure is usually most convenient.
5. Equate the coefficients in the two expressions for $G(s)$ and solve for K and the k_i's.
6. Evaluate the resulting design using Bode diagram, root locus, or simulation methods, as desired.

Now let us consider the case in which $G_p(s)$ has an lhp zero that is undesired in $G(s)$. If the state block in which this zero appears is readily accessible, it may be cancelled by placing an overlapping pole immediately prior to that block and the analysis proceeds just as before. No additional state feedback path is required, since the zero no longer exists as far as observation of the external system is concerned. Of course the resulting system is unobservable but this is not a problem because the unobserved state was not present in the original system, has been introduced only for convenience, and is of smaller magnitude than the signal at the input to the original block containing the undesired zero. Unfortunately in some cases the block containing the undesired zero does not lend itself to cancellation of the zero in an immediately prior element, either because the signals are at an excessive power level, or they are in some form which is awkward to work with.

Cancellation of an undesired plant zero can also be accomplished by adding a pole at a convenient point in the forward path. The pole location can be chosen more or less arbitrarily. Adjustment of this location can later be used, if desired, to arrive at the most acceptable solution. Adding a pole also increases the order of the resulting system and the solution complexity. Consider the case where

$$G_p(s) = \frac{10(s+2)}{s(s+1)(s+3)} \tag{13.197}$$

This is the same as the plant shown in Fig. 13.15 except for the zero at $s = -2$. Assume that this zero is associated with the pole at $s = -1$. We will arbitrarily add a pole at $s = -5$ prior to the plant in an effort to eliminate the undesired zero. The corresponding arrangement is shown in Fig. 13.17. We assume, as before, that $G(s)$ as given by Eq. (13.193) is desired. Since the general transfer function obtained for the configuration shown in Fig. 13.17 will be of fourth-order in the denominator, the zero at $s = -2$ must cancel in the numerator and denominator of $G(s)$. Thus we require that

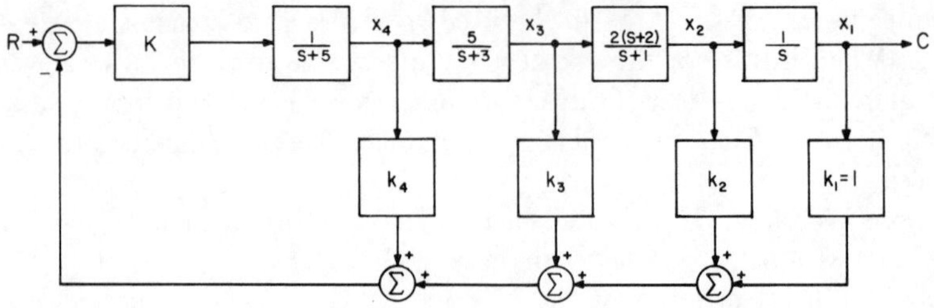

FIG. 13.17. Configuration used to cancel the effect of a plant zero.

$$G(s) = \frac{10(s+2)}{(s+2)(s^3+4s^2+9s+1)} = \frac{10(s+2)}{s^4+6s^3+17s^2+28s+20} \tag{13.198}$$

The $H_{eq}(s)$ for the system shown in Fig. 13.17 is readily found to be

$$H_{eq} = \frac{k_4s^3+(4k_4+5k_3+10k_2)s^2+(3k_4+5k_3+20k_2+10)s+20}{10(s+2)} \tag{13.199}$$

The general form for $G(s)$ in terms of K, k_2, k_3, and k_4 reduces to

$$G(s) = \frac{10K(s+2)}{s^4+(9+Kk_4)s^3+[23+K(4k_4+5k_3+10k_2)]s^2+[15+K(3k_4+5k_3+20k_2+10)]s+20K} \tag{13.200}$$

Equating the coefficients of Eqs. (13.198) and (13.200), observing that $K = 1$ is required, provides

$$9+k_4 = 6 \qquad 23+4k_4+5k_3+10k_2 = 17$$
$$15+3k_4+5k_3+20k_2+10 = 28 \tag{13.201}$$

Solving Eq. (13.201) yields

$$k_4 = -3 \qquad k_3 = 0 \qquad k_2 = 0.6 \tag{13.202}$$

Thus k_2 and k_3 are the same as before and k_4 is that value resulting in cancellation of the zero at $s = -2$. Obviously the k_i may be of either sign.

We have not indicated how the location of the added pole should be chosen. The root locus of the resulting system can be plotted and the location of the added pole adjusted to provide the most desirable root-locus diagram although this involves repeating the solution process several times and the work required may be prohibitive. Development of the root locus is simplified if the added pole is placed at a point where a closed-loop pole is desired, since this forces $H_{eq}(s)$ to have a zero there, simplifying the task of factoring $H_{eq}(s)$. The closed-loop pole chosen may be the one at the same location as the zero to be cancelled or any other real pole which is desired in $G(s)$. The pole may also be located to minimize system sensitivity to parameter changes.[8] Closed-loop performance depends only on the $G(s)$ chosen, and is identical in all cases *as long as the parameters of the system do not change.* The root locus is undoubtedly the most useful way to study and illustrate the effects of parameter changes on performance.

The general procedure used when one or more of the state variables is not accessible involves only a minor modification. The problem is set up in the usual way with feedback of all states assumed possible. The required values of K and the k_i are obtained. The signals from those states which are inaccessible are then synthesized from the signals at the accessible states to provide the desired $H_{eq}(s)$. Nothing more complex than standard block diagram manipulation procedures is required. In general the desired unavailable signals are obtained by processing *prior* state signals to assure that the networks required are realizable. One or more of the feedback paths will now contain a signal processing filter, but that should cause no concern. As an example, suppose we have the arrangement shown in Fig. 13.15 with $k_0 = 1$, $k_2 = 0.6$, and $k_3 = 2$. If x_2 is not available, we can alternately feed back x_3 processed by a network with transfer function

$$G_f(s) = k_3 + \frac{2k_2}{s+1} = \frac{3.2 + 2s}{s+1} \tag{13.203}$$

Of course the resulting system performance is somewhat more subject to parameter variations than would normally be the case, but this should not become a serious consideration unless several states are unavailable.

One final situation which causes some difficulty is the case where the plant (perhaps a hydraulic actuator) has a complex conjugate pole pair. In this case at least one of the states is always inaccessible. The basic procedure is to represent the quadratic transfer function yielding the complex pole pair using whatever state representation seems convenient, assume the states are all available, determine the required feedback gains, and manipulate the block diagram until each signal is obtained from an accessible state using an appropriate processing filter.

The development of systems using state-variable feedback as the fundamental design approach provides an important alternate method of system synthesis. Relatively arbitrary transfer functions can be achieved. Although all states are generally assumed available, it is possible to bypass the need for measuring one or more states which are difficult or impossible to obtain directly.

13.7 CONCLUDING COMMENTS

The analysis and design of systems in state-equation form is a relatively new innovation among control engineers in this country, although Russian scientists have relied heavily upon the state representation, as opposed to our transfer function methods, since interest in control technology was first noted. In the decade from 1950 to 1960 many engineers and applied mathematicians in this country became interested in the stability analysis of NL systems which has led, through Liapunov theory, to the Russian literature dating back prior to 1900. NL systems are conveniently represented in state-equation form and this led naturally to an interest in the general characteristics of state equations for both linear and NL systems. The introduction by Kalman in the late 1950's of the concepts of controllability and observability prompted a considerable effort in these and other closely related areas. Motivation for the use of state equations has also been provided by the rapidly-expanding use of digital computers as a simulation and study tool in the years since 1950, because DE's are first reduced to state form when they are to be integrated numerically on the computer.

The mathematical tools necessary for the analysis of systems described by their state equations are available and they are easily applied to typical situations. Synthesis procedures are also available for both continuous and discrete systems expressed in state form. Since the field is relatively new, potential synthesis methods undoubtedly remain unexplored, and significant developments in this area should appear over the next several years.

Our coverage of state-equation analysis and design is only introductory in nature. Our objective is to indicate an alternate way of looking at systems and to provide a general indication of how system analysis and synthesis is structured using the state-equation model. References have been suggested for further detail. The state equations provide more information than the usual transfer function representation at the expense of somewhat more complex analysis methods. The state representation should be used when the added information is of value and also when there are other factors involved, such as the intention to carry out a simulation study. In any event, the state-equation model is a valuable supplement to the transfer function methods considered in previous chapters.

REFERENCES

1. Hildebrand, F. B., "Methods of Applied Mathematics," Prentice Hall, Inc., Englewood Cliffs, N. J., Sec. 1.22, 1952.
2. Eveleigh, V. W., "Adaptive Control and Optimization Techniques," McGraw-Hill, New York, p. 34, 1967.
3. Ho, B. L., and R. E. Kalman, "Effective Construction of Linear State Variable Models from Input/Output Data," Proceedings of Third Allerton Conference, pp. 449–459, 1965.
4. Kalman, R. E., "On the General Theory of Control Systems," IFAC Proceeding of First International Congress Automation Control, Moscow, 1960, pp. 481–493, Butterworth and Co. (Publishers), Ltd., London, 1961.
5. Kalman, R. E., Y. C. Ho, and K. S. Narendra, "Controllability of Linear Dynamical Systems," in LaSalle et al. (eds.), "Contributions to Differential Equations," Vol. 1, Interscience Publishers, Inc., New York, 1962.

6. Kalman, R. E., "Canonical Structure of Linear Dynamical Systems," Proceedings of National Academy of Science, U.S., vol. 48, No. 4, pp. 596–600, 1962.
7. Truxal, J. G., "Automatic Feedback Control System Synthesis," McGraw-Hill, New York, Chap. 5, 1955.
8. Schultz, D. G., and J. L. Melsa, "State Functions and Linear Control Systems," McGraw-Hill, New York, 1967.

14 NONLINEAR SYSTEMS

The mathematical techniques used up to this point all rely heavily upon the superposition principle, applicable only to linear systems, which allows direct addition of two or more signal components at summing junctions throughout the system. No system is linear for an infinite range of inputs. Our extensive consideration of linear analysis methods can be justified, however, because of the following factors:

1. Many systems are "almost linear" over the signal range of interest and performance predictions based upon a linear model are sufficiently accurate to satisfy design objectives.
2. The use of feedback around nonlinear (NL) elements tends to reduce their effect upon closed-loop system performance and we are concerned primarily with feedback systems.
3. Mathematical procedures for the exact treatment of NL elements are difficult to carry out. The actual performance of "almost linear" systems often does not differ considerably from that predicted using an approximate linear model.

Some systems are highly NL and a linear approximation may yield meaningless results, so it is necessary to consider analysis methods for NL systems. In this chapter we consider several of the fundamental procedures applicable to NL system analysis.

Our discussion is initiated with a careful description of the distinctions between linear and NL devices. Several examples of NL transfer characteristics are presented. Phenomena possible in NL systems but never present in linear systems, such as limit cycles and sub-harmonic oscillations, are discussed and our objectives in developing analytical methods are indicated. A sequence of analysis techniques is considered in the subsequent sections. The *describing function method,* alternately called the *method of harmonic balance* is considered first because it is a direct extension of the linear techniques we have been using. It is one of the simplest tools for determining NL system stability and response characteristics and it is applicable to a wide range of common practical problems. Although describing function accuracy is sometimes rather poor, methods are available for improving precision at the expense of additional calculations. The *phase plane method,* which is most directly applicable to second-order systems, is introduced to illustrate the graphical and computational procedures available for solving more general NL (and linear) problems. Several fundamental *Liapunov stability theorems* are presented and their application is illustrated with examples. The *Popov and critical disc stability criteria* are presented without proof.

The describing function, Popov, and critical disc methods lead directly to analysis and synthesis procedures similar to those we have been using by simply re-interpreting the Nyquist criterion. Thus, these three methods are the ones most directly applicable in practice to the design of controllers in which one or more nonlinear elements are present. Of these methods, the describing function is best known and most widely used, having been available since about 1950, whereas the latter two methods are of relatively recent vintage.

The phase plane analysis of nonlinear systems is directly applicable only to second-order systems with constant inputs and these are severe restrictions which have limited its practical use. It is primarily of academic interest, but is considered here for several reasons. Its application can be extended to higher-order systems, although admittedly only with considerable difficulty, by working with families of phase planes and their solution trajectory patterns. The general characteristics of system equilibrium points are readily illustrated on the phase plane. Graphical procedures for constructing phase plane solutions provide a valuable illustration of the methods by which computational solutions may be obtained using either an analog or a digital computer. This is perhaps the most significant argument in favor of our current interest in phase plane analysis, which was a particularly active research area only a few years ago.

Liapunov theory is also primarily of academic interest. It is not particularly fruitful in the development of straightforward synthesis procedures, although it has been applied to the design of optimum systems and adaptive systems by some researchers. The main justification for its consideration here stems from the fact that it is a powerful analysis tool which often allows us to analytically verify the stability of a final design. Its inclusion is also necessary in the interest of completeness. Those readers who plan to pursue the areas of modern and/or optimal control theory will find that a basic understanding of Liapunov theory provides useful background for that work.

14.1 THE FUNDAMENTAL CHARACTERISTICS OF NONLINEAR SYSTEMS

The analysis of linear time-invariant systems is relatively easily accomplished using a variety of mathematical tools in both the time and complex frequency (transform) domains. Linear systems have long been thoroughly understood, and few significant developments have occurred in this area in recent years. Time-varying linear systems are much more difficult to work with, since transform methods do not generally apply and hand analysis is only feasible in simple cases. Superposition applies for all linear systems, however, so the responses of time-varying systems to individual inputs may be added together or expanded as desired. Since specific situations can be studied using either analog or digital computers activity in this area has been prompted by the rapid growth of computer technology since 1950. NL systems are the most complex general classification, since superposition does not apply. Generalization of results is difficult if not impossible, and each type of NL element has its own peculiar characteristics. Several analytical tools are available for studying NL systems, however, and the most useful of these are covered in subsequent sections. Theoretical and practical consideration of NL systems has been of interest to engineers and mathematicians for over a century. Research activity involving both analytical and computational solution methods has been particularly intense since 1945.

If $y_1(t)$ and $y_2(t)$ are the responses of a linear system to arbitrary input signals $x_1(t)$ and $x_2(t)$, respectively, the response to input $x_1(t) + x_2(t)$ is $y_1(t) + y_2(t)$. This is a mathematical statement of the superposition principle, and it may alternately be taken as the definition of linearity. Only because superposition holds for all linear systems can we arbitrarily add and subtract signals at block diagram summing points and characterize system performance by using standard test inputs such as the unit impulse and unit step functions. Test signal magnitude is irrelevant in linear systems because the input and output magnitudes are directly proportional. The steady-state response pattern of a stable linear system for a sinusoidal input is a sinusoid of the same frequency at a displaced phase angle. This is the basis for our frequency-response characterization of linear systems.

Although frequency-response curves may be determined for NL systems, they do not have the same general significance. The frequency response for NL systems depends upon input amplitude as well as frequency. Thus families of curves are required, the results are less readily obtained, and they are much more cumbersome to use. The output generally contains harmonic components in addition to the fundamental frequency, so it is necessary to define carefully what is meant by output magnitude and phase. The frequency response magnitude curves of NL systems are often discontinuous, exhibiting a *jump resonance phenomenon* as the system goes suddenly from one response mode to another, so frequency response curves may be multivalued. If the harmonic components at the NL element output are fed back to the NL element input, all possible sum and difference frequency components are generated at the output, just as in a communication system modulator. This frequency addition and subtraction phenomenon occasionally yields *sub-harmonic oscillations,* or simultaneous oscillations at two (or more) harmonically related frequencies. More often, NL systems exhibit *limit cycles,* or periodic non-sinusoidal oscillations at a relatively stable amplitude and frequency. All electronic oscillators are NL systems operating in a limit cycle, usually with near sinusoidal output and carefully controlled amplitude and frequency. Exact analytical solutions for the response of NL systems are possible only in a few special cases. In most cases we must resort to a computer study, an approximate analysis of some sort, or one of several available graphical solution procedures when NL system performance is to be evaluated. Whereas linear systems are either unstable or stable with a certain stability margin, there are many types of NL system instability.

The behavior of an electronic oscillator was shown by Van der Pol in 1920 to be described by a normalized NL differential equation (DE) of the form

$$\ddot{q} - \epsilon(1 - q^2)\dot{q} + q = 0 \tag{14.1}$$

where q is the instantaneous capacitor charge in the equivalent circuit shown in Fig. 14.1, and $\epsilon > 0$. The resistance is given by $R = -\epsilon(1 - q^2)$. The coefficients of a linear DE must be independent of the response variable (in this case q) and its derivatives. The factor $1 - q^2$ in the coefficient of $\dot{q}$ makes Eq. (14.1) NL, although performance is nearly linear when $q \ll 1$ over the entire oscillation cycle. We may argue intuitively that this system should have a stable amplitude limit cycle response. The instantaneous rate of energy loss is $\dot{q}^2R$, so as long as $q^2 < 1$ over the entire cycle, energy is gained at each instant except when $\dot{q} = i = 0$. The oscillation amplitude thus increases until eventually $q^2 > 1$ for part of each cycle and R becomes sufficiently positive over enough of the cycle to dissipate energy equal to that gained while R is negative. Large amplitude oscillations decay toward this same equilibrium level because the average value of i^2R is negative for large amplitudes, and a net energy loss occurs each cycle. Of course the oscillations are not sinusoidal, although the harmonic content in the output of most oscillators is small. This problem is considered using the phase plane in Sec. 14.5.

Many NL systems may be represented by a block diagram arrangement in which the linear and NL elements are isolated into separate blocks. This is particularly true when the nonlinearities are piecewise-linear and frequency-independent such that their transfer characteristics are provided by a series of straight line segments. Transfer diagrams representing saturation, a perfect relay, a relay with dead space, and a gain characteristic with dead space are shown in

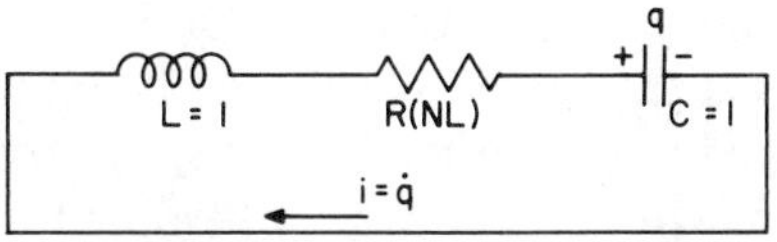

FIG. 14.1. Equivalent circuit of an oscillator using a nonlinear resistance.

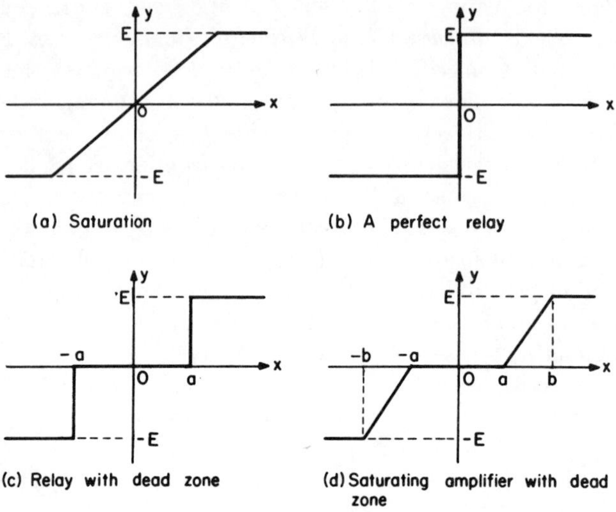

FIG. 14.2. Typical nonlinear transfer characteristics in block diagram form.

Fig. 14.2. The describing function method is particularly appropriate for systems containing a single NL element of this type. Note that NL transfer characteristics may be continuous or discontinuous. Some NL elements, such as the familiar magnetic core hysteresis loop, have double-valued transfer functions. A closed-loop system containing a severe discontinuous NL element such as the perfect relay is called a *soft self-excited* system, since it tends to oscillate in the presence of small disturbance levels. A system containing a relay with a significant dead zone, on the other hand, can not oscillate unless excited by a fairly large disturbance signal, and perhaps not even then. If this type of system can be forced into self-sustained oscillations by a sufficiently large disturbance or initial condition, however, it is called a *hard self-excited* system. The phase plane and describing function analysis procedures each make both soft and hard self-excitation characteristics apparent.

Most nonlinearities arise in components we must use, and are unavoidable. NL elements may also be introduced during design to accomplish specific objectives. Although such nonlinearities could perhaps be avoided, it is not reasonable to oppose their use simply because they complicate the analysis. Procedures for synthesizing NL systems are not well established. Our primary objective here is to introduce basic tools for the analysis of NL systems.

The NL transfer characteristics shown in Fig. 14.2 are all piecewise continuous. Analysis of piecewise linear systems is possible using a linear model for each straight line segment of the NL element and matching boundary conditions at the end points. In many cases symmetry simplifies the problem somewhat, but this is a tedious process at best, and it is not often practically feasible. Analysis procedures applicable to a broad range of NL situations are desirable. They must provide acceptable performance prediction accuracy with a reasonable amount of effort. Instability conditions should be predictable and it is also desirable to predict such phenomena as limit cycles, including their amplitudes and frequencies, jump resonance effects (frequency-response discontinuities), and sub-harmonic generation. Some indication of stability margin should be given and it should be possible to obtain response characteristics for specific inputs. Unfortunately no single analytical procedure yields all of these desired results. All methods provide some measure of stability, since this is the fundamental question, but the particular method chosen in any specific situation should depend upon the results desired and the NL element involved.

14.2 THE DESCRIBING FUNCTION

One approach to NL system analysis which is often used successfully involves assuming an infinite series solution of convenient form and solving for the individual coefficients. The describing function (DF) method is based upon a Fourier series representation of the NL element output in response to a sinusoidal input. The output is generally approximated using only its fundamental frequency component. The DF method is an attempt to extend the standard frequency-response methods of linear analysis to NL systems, and surprisingly accurate results are often obtained. The DF method was apparently developed independently in the late 1940's by researchers in several countries,[1-5] which indicates the widespread interest in NL systems existing at that time. The approximation used in each case is similar to that suggested by Kryloff and Bogoliuboff[6] several years earlier. Applications of the DF method since 1950 have been widespread.

Consider the NL system shown in Fig. 14.3. Application of the DF method to systems of this type is based upon the following assumptions:

1. There is one NL element N in the system. Of course two or more nonlinearities may be considered together as a single element within this general restriction.
2. The characteristics of N are time invariant, so its current output depends only upon present and past input values.
3. $\mathbf{G}(j\omega)$ is low pass and all harmonic components other than the fundamental generated at the output of N in response to a sinusoidal input are severely attenuated as they pass through $\mathbf{G}$ such that only the fundamental frequency output component has significant effect at the input to N.
4. N is symmetrical such that no dc component is generated at its output when a sinusoidal input is applied.

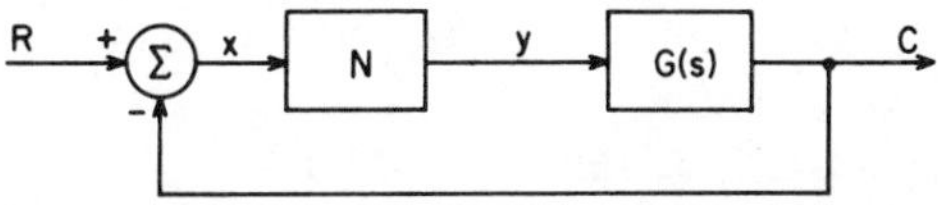

FIG. 14.3. Block diagram of a general NL system.

These assumptions are reasonable in most control system applications. Item 3 appears rather restrictive at first, but most controller open-loop transfer functions are decidedly low pass in nature. Thus the higher harmonic components generated by N in response to a sinusoidal input, which are typically smaller than the fundamental frequency term anyway, are much further reduced in relative magnitude upon passing through $\mathbf{G}$. Assuming all higher harmonics negligible after such processing is generally reasonable. DF analysis errors are usually less than 10%, which is excellent as NL analysis methods go, and this is the true validity measure for the assumptions made. Note that sub-harmonic oscillations are impossible with these assumptions, since frequency differencing cannot take place in N.

Assume the system shown in Fig. 14.3 has no input, is unstable, oscillates in a limit cycle, and satisfies all of our previous assumptions. The output of N, denoted by $y(t)$, is a periodic waveform with fundamental frequency ω_1. Since $\mathbf{G}$ is low pass, only the ω_1 component of $y(t)$ passes through the loop, and $x(t)$, the signal into N, is a sinusoid at ω_1. The amplitude and frequency of oscillation stabilize, assuming an equilibrium condition exists, at those values where the equivalent-loop transmission for signals at the fundamental frequency equals unity. This is the *harmonic balance principle,* and it has been used as a basis for NL system analysis by many researchers using a variety of approaches. It leads us to characterize N by an equivalent fundamental frequency phasor transfer function relating the input sinusoid to the resulting output fundamental frequency component. All higher frequency components of the output are ignored, since they are assumed rejected by $\mathbf{G}$. This assumption is the cornerstone of the DF analysis method.

A typical DF analysis proceeds as follows. Assume $x(t)$ is a sinusoid of amplitude A and frequency ω. Find $y(t)$ in response to $x(t) = A \sin \omega t$ using the known characteristics of N. Decompose $y(t)$ by Fourier series analysis to determine its fundamental frequency component. The equivalent steady-state sinusoidal transfer function for N, called the DF, and denoted by $\mathbf{N}_e(A, j\omega)$ to emphasize its general dependence upon input amplitude and frequency (although this is often shortened to $\mathbf{N}_e(A)$ when N does not depend upon ω, or $\mathbf{N}_e$ for notational simplicity) is defined as the complex number relating the magnitude and phase of $y(t)$ and $x(t)$ in the usual phasor form. In general both the magnitude and phase of $\mathbf{N}_e(A, j\omega)$ depend upon input amplitude as well as frequency, which is the fundamental distinction between linear and NL devices. $\mathbf{N}_e$ and $\mathbf{G}$ may be plotted on a polar diagram and a modified Nyquist criterion applied to see if conditions leading to potential instabilities exist. (Actually $-1/\mathbf{N}_e$ is plotted.) Although dependence of $\mathbf{N}_e$ upon input amplitude leads to some difficulties, the method is straightforward and application does not involve prohibitively extensive calculations.

Our first task is to characterize N by its approximate phasor transfer function $\mathbf{N}_e$. Assume for the present that N is not frequency-dependent. Let

$$x(t) = A \sin \omega t = A \sin \theta \tag{14.2}$$

where $\theta = \omega t$ is used to simplify the subsequent development. Since N is known, the output of N in response to this $x(t)$ may be expanded into a Fourier series as

$$y(\theta) = a_0 + a_1 \sin\theta + a_2 \sin 2\theta + \cdots + \alpha_1 \cos\theta + \alpha_2 \cos 2\theta + \cdots \tag{14.3}$$

Only the $\sin\theta$ and $\cos\theta$ terms are of interest for the assumptions made, so we focus our attention upon a_1 and α_1, which are normalized by defining

$$g(A) = \frac{a_1}{A} \qquad b(A) = \frac{\alpha_1}{A} \tag{14.4}$$

These normalized Fourier coefficients are obtained in the usual way as

$$g(A) = \frac{1}{\pi A}\int_0^{2\pi} f(A \sin\theta) \sin\theta d\theta \tag{14.5}$$

$$b(A) = \frac{1}{\pi A}\int_0^{2\pi} f(A \sin\theta) \cos\theta d\theta \tag{14.6}$$

where the $f(A \sin\theta)$ terms are based upon the functional description of N by

$$y = f(x) \tag{14.7}$$

Thus the DF, or equivalent gain, may be written as

$$N_e(A) = \frac{\sqrt{a_1^2 + \alpha_1^2}}{A} \epsilon^{j\phi} = g(A) + jb(A) \tag{14.8}$$

where

$$\phi = \tan^{-1}\left(\frac{\alpha_1}{a_1}\right) \tag{14.9}$$

In many applications $b(A) = 0$ and $\phi = 0$, which leads to considerable simplification.

Example 14.1

Perhaps the simplest NL element we can choose as an example is the perfect relay shown in Fig. 14.2b, which responds to a sinusoidal input with the square

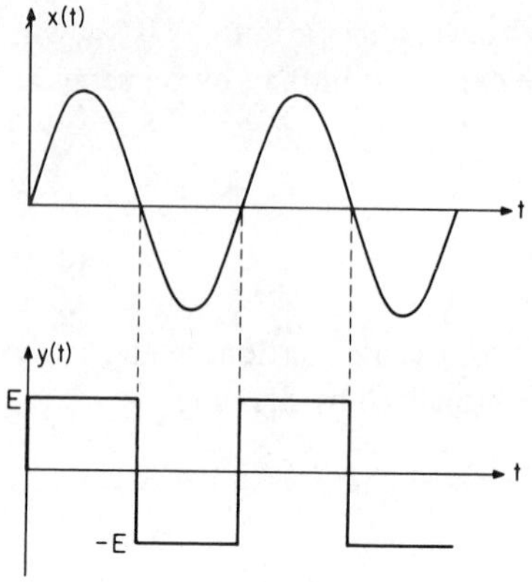

FIG. 14.4. Response of a perfect relay to a sinusoidal input.

wave output waveform shown in Fig. 14.4. Since the axis crossings of the output square wave occur at the same time points as those of the input sinusoid, $b(A)$ is zero. Also, $f(A \sin\theta) = E$ on the interval $[0, \pi]$ and $-E$ on the interval $[\pi, 2\pi]$, so

$$g(A) = \frac{1}{\pi A}\int_0^{\pi} E \sin\theta d\theta - \frac{1}{\pi A}\int_{\pi}^{2\pi} E \sin\theta d\theta$$

$$= \frac{2E}{\pi A}\int_0^{\pi} \sin\theta d\theta = \frac{4E}{\pi A} = \mathbf{N}_e(A) \tag{14.10}$$

Thus $\mathbf{N}_e(A)$ is always real in this case, approaches ∞ as $A \to 0$, and approaches 0 as $A \to \infty$. These limiting gain characteristics are intuitively satisfying, since the relay output power remains constant no matter how the input varies. This also explains why systems containing a near-perfect relay in the signal path almost always exhibit a stable limit cycle.

Example 14.2

As a more general illustration of DF analysis, consider a saturating amplifier with input and output waveforms as shown in Fig. 14.5. Once again due to the symmetry $b(E) = 0$, and

$$g(A) = \frac{1}{\pi A}\left(\int_0^{\theta_1} An_1 \sin^2\theta d\theta + \int_{\theta_1}^{\theta_2} M \sin\theta d\theta + \int_{\theta_2}^{\theta_3} An_1 \sin^2\theta d\theta\right.$$

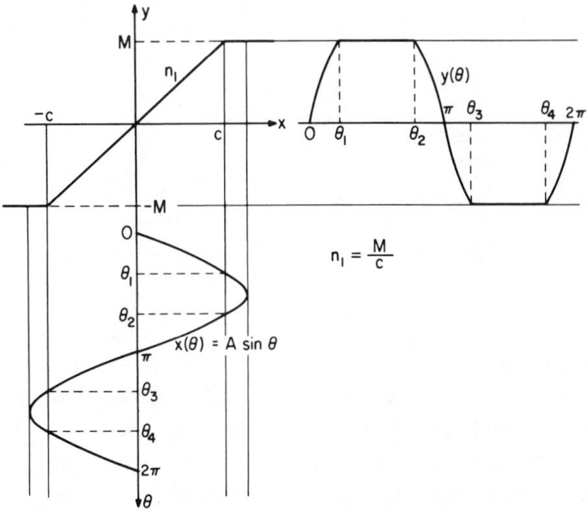

FIG. 14.5. Waveforms for a saturating amplifier with a sinusoidal input.

$$+ \int_{\theta_3}^{\theta_4} M \sin\theta d\theta + \int_{\theta_4}^{2\pi} A n_1 \sin^2\theta d\theta \Bigg)$$

$$= \frac{4}{\pi A}\left(\int_0^{\theta_1} A n_1 \sin^2\theta d\theta + \int_{\theta_1}^{\pi/2} M \sin\theta d\theta\right) \tag{14.11}$$

where the simplified form results from perfect quarter cycle symmetry. Reductions of this type are often possible. Observe that

$$A \sin\theta_1 = c \qquad \theta_1 = \sin^{-1}\frac{c}{A} \tag{14.12}$$

Evaluating the integrals in Eq. (14.11) and simplifying provides

$$g(A) = \mathbf{N}_e(A) = \frac{n_1}{\pi}(2\theta_1 - \sin 2\theta_1) + \frac{4M}{\pi A}\cos\theta_1 \tag{14.13}$$

which applies for all $A \geq c$. Of course $\mathbf{N}_e(A) = n_1$ for $A \leq c$. A sketch of $\mathbf{N}_e$ versus A for this case is shown in Fig. 14.6.

A general DF solution for piecewise linear elements was developed by Sridhar[7], and is summarized by Gibson[8]. Equivalent gains for the characteristics of primary

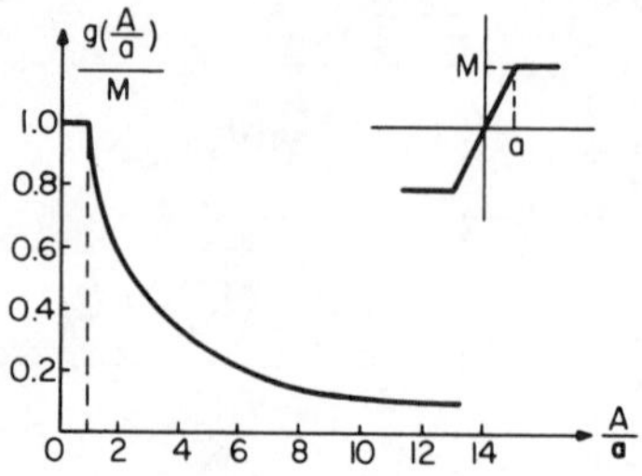

FIG. 14.6. Normalized $N_e(A)$ for a saturating amplifier.

interest are presented in Table 14.1. All of these NL elements are symmetrical, since this is the condition most often encountered. For symmetrical nonlinearities, only those with memory (a multi-valued characteristic like hysteresis, for example), yield $b(A) \neq 0$ and thus result in a phase shift through N. This makes the DF procedure a bit more difficult to apply.

We are now prepared to illustrate DF application. Consider the system shown in Fig. 14.3. Assume $\mathbf{N}_e$ is available. The characteristic equation of this closed-loop system is

$$1 + \mathbf{G}\mathbf{N}_e = 0 \qquad \mathbf{G} = -\frac{1}{\mathbf{N}_e} \tag{14.14}$$

from which the conditions required for sustained oscillations are clearly apparent. If Eq. (14.14) is compared to the equivalent linear system relationship upon which the Nyquist criterion is based, we observe that $-1/\mathbf{N}_e$ replaces the usual -1 point and should be considered a frequency and magnitude-dependent critical point. With this modified interpretation, the Nyquist criterion can be applied directly to DF analysis. We may solve Eq. (14.14) graphically using superposed polar plots of $\mathbf{G}(j\omega)$ and $-1/\mathbf{N}_e(A, j\omega)$. $\mathbf{G}$ depends only on ω, and is easy to plot. Assume $\mathbf{G}$ is open-loop stable, since this is the normal situation encountered. Assume that $\mathbf{N}_e$ depends only upon A, as often true. This condition is relaxed later. For this case the $-1/\mathbf{N}_e(A)$ locus is easily plotted and all intersections between the two loci located. Since ω and A are variable parameters on the $\mathbf{G}$ and $-1/\mathbf{N}_e$ loci, respectively, the amplitude and frequency of the limit cycle defined by each intersection are immediately apparent. Values of A which cause the moving critical point (a point on the $-1/\mathbf{N}_e$ locus) to lie anywhere in the region to the left of an observer proceeding along the $\mathbf{G}$ locus in the direction of increasing ω cannot be sustained by the feedback. Thus any disturbance in this amplitude range decays toward zero. Conversely, values of A for which the critical point lies in the region to the right of the same observer tend to increase due to the feedback, and the system is unstable for these conditions.

We define a *stable limit cycle* as a periodic oscillatory condition which is maintained at or near a given frequency and amplitude in the presence of small disturbances. A stable limit cycle results only when the $-1/\mathbf{N}_e$ locus (the functional critical point) crosses the $\mathbf{G}$ locus moving from the unstable region (right

Table 14.1.

1.

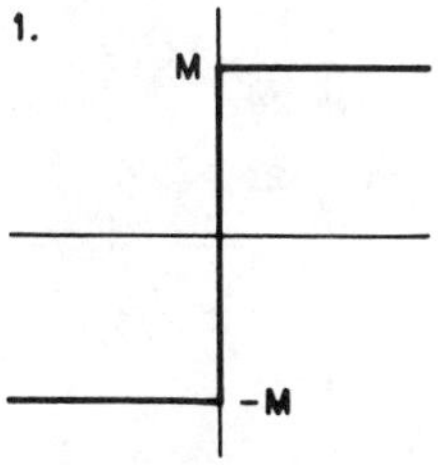

$g(A) = \frac{4M}{\pi A}$

$b(A) = 0$

2.

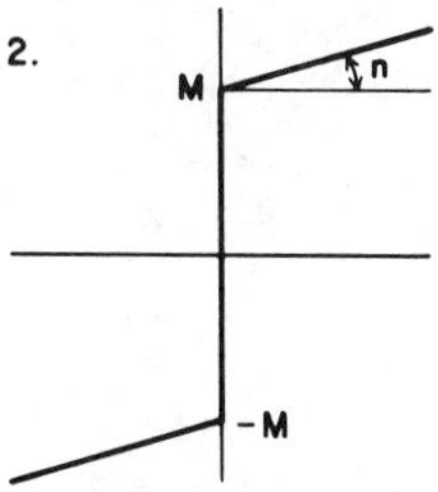

$g(A) = n + \frac{4M}{\pi A}$ n denotes slope of linear characteristic

$b(A) = 0$

3.

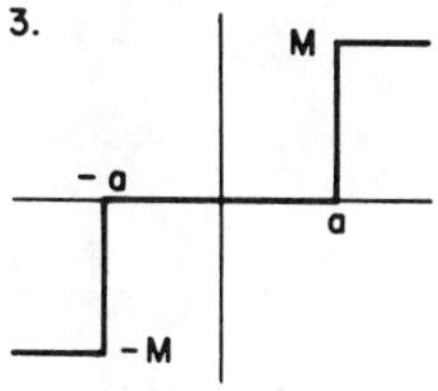

$g(A) = \frac{4M}{\pi A} \cos\theta$

$b(A) = 0$

$A > a \qquad \theta = \sin^{-1}\frac{a}{A}$

4.

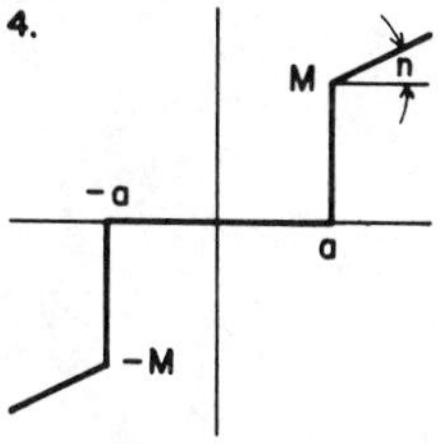

$g(A) = \frac{n}{\pi A}\left[A(\pi - 2\theta) + A\sin 2\theta - 4k\cos\theta\right]$

$b(A) = 0$

$A > a \qquad \theta = \sin^{-1}\frac{a}{A} \qquad k = a - \frac{M}{n}$

5.

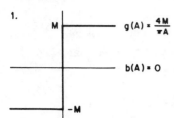

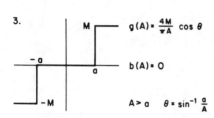

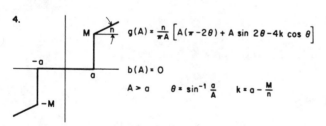

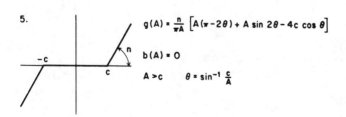

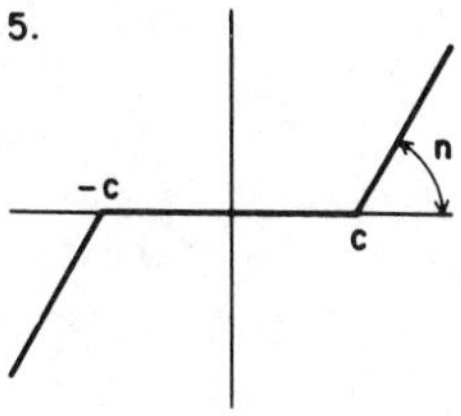

$g(A) = \frac{n}{\pi A}\left[A(\pi - 2\theta) + A\sin 2\theta - 4c\cos\theta\right]$

$b(A) = 0$

$A > c \qquad \theta = \sin^{-1}\frac{c}{A}$

Table 14.1. **Continued.**

6.

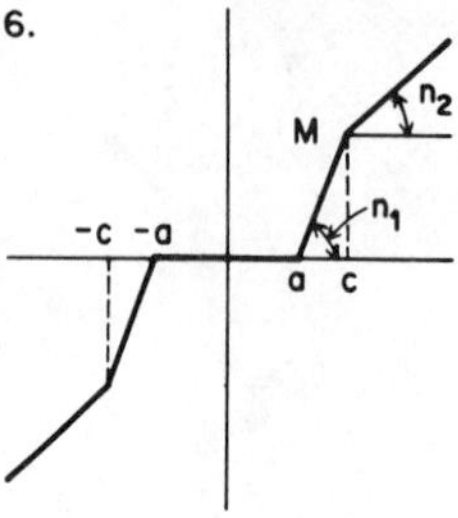

$$g(A) = \frac{n_1}{\pi A}\left[2A(\theta_2 - \theta_1) + A(\sin 2\theta_1 - \sin 2\theta_2) + 4a(\cos\theta_2 - \cos\theta_1)\right] + \frac{n_2}{\pi A}\left[A(\pi - 2\theta_2) + A\sin 2\theta_2 - 4k\cos\theta_2\right]$$

$$b(A) = 0$$

$A > c \quad \theta_1 = \sin^{-1}\frac{a}{A} \quad \theta_2 = \sin^{-1}\frac{c}{A} \quad k = a - \frac{M}{n_2}$

7.

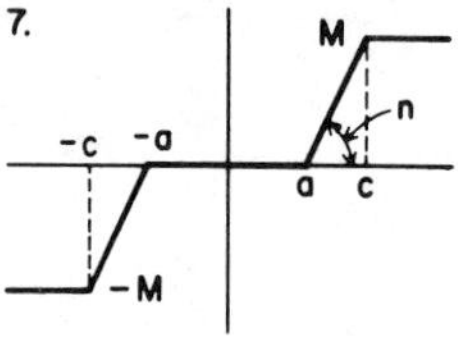

$$g(A) = \frac{n}{\pi A}\left[2A(\theta_2 - \theta_1) + A(\sin 2\theta_1 - \sin 2\theta_2) + 4a(\cos\theta_2 - \cos\theta_1)\right] + \frac{4M}{\pi A}\cos\theta_2$$

$$b(A) = 0$$

$A > c \quad \theta_1 = \sin^{-1}\frac{a}{A} \quad \theta_2 = \sin^{-1}\frac{c}{A}$

8.

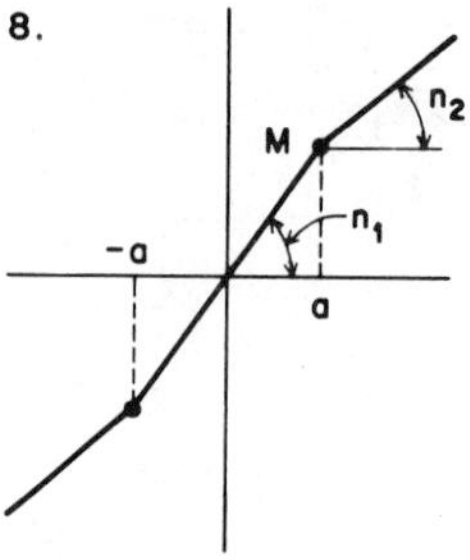

$$g(A) = \frac{n_1}{\pi}(2\theta - \sin 2\theta) + \frac{n_2}{\pi A}\left[A(\pi - 2\theta) + A\sin 2\theta - 4k\cos\theta\right]$$

$$b(A) = 0$$

$A > a \quad \theta = \sin^{-1}\frac{a}{A} \quad k = a - \frac{M}{n_2}$

9

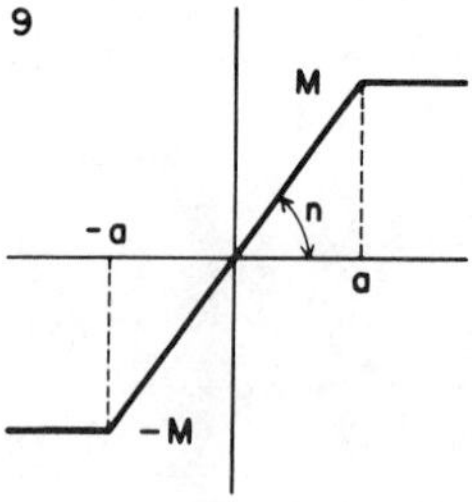

$$g(A) = \frac{n}{\pi}(2\theta - \sin 2\theta) + \frac{4M}{\pi A}\cos\theta$$

$A > a \quad \theta = \sin^{-1}\frac{a}{A}$

10

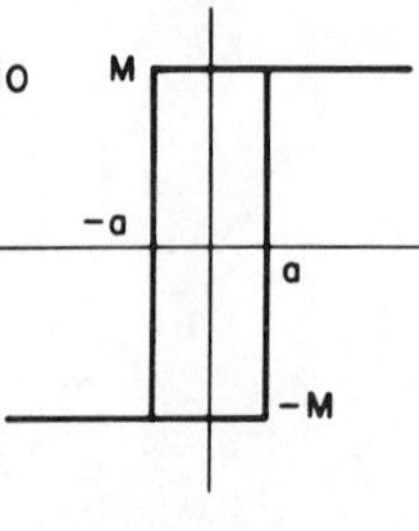

$$g(A) = \frac{4M}{\pi A}\cos\theta$$

$A > a$

$$b(A) = -\frac{4M}{\pi A}\sin\theta$$

$$\theta = \sin^{-1}\frac{a}{A}$$

Table 14.1. Continued.

11	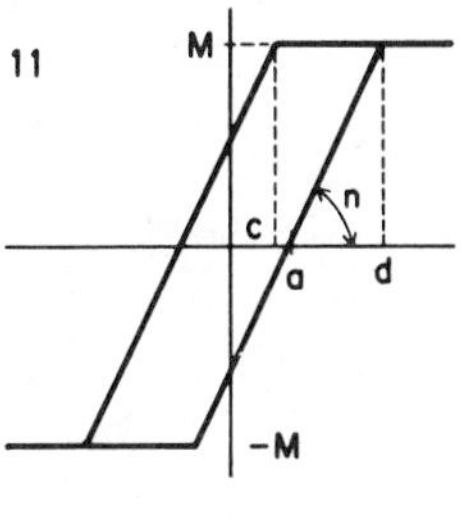	$g(A) = \frac{n}{\pi A}\left[A(\theta_1 + \theta_2) - \frac{A}{2}(\sin 2\theta_1 - \sin 2\theta_2) + 2a(\cos\theta_1 - \cos\theta_2)\right] + \frac{2M}{\pi A}(\cos\theta_1 + \cos\theta_2)$ $b(A) = \frac{n}{\pi A}\left[\frac{A}{2}(\cos 2\theta_2 - \cos 2\theta_1) - 2a(\sin\theta_1 + \sin\theta_2)\right] + \frac{2M}{\pi A}(\sin\theta_2 - \sin\theta_1)$ $A > d \quad \theta_1 = \sin^{-1}\frac{d}{A} \quad \theta_2 = \sin^{-1}\frac{c}{A}$
12	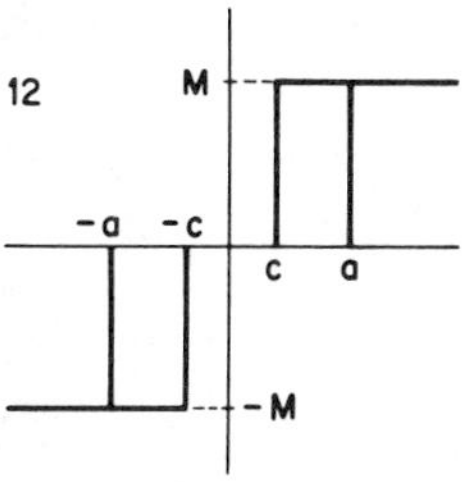	$g(A) = \frac{2M}{\pi A}(\cos\theta_1 + \cos\theta_2)$ $A > c$ $b(A) = -\frac{2M}{\pi A}(\sin\theta_1 - \sin\theta_2)$ $\theta_1 = \sin^{-1}\frac{a}{A} \quad \theta_2 = \sin^{-1}\frac{c}{A}$

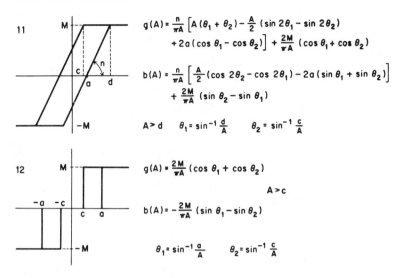

side of **G**) toward the stable region (left side of **G**) as A increases. Limit cycles defined by loci crossings in the opposite direction are unstable, and conditions defined by the corresponding intersection coordinates A and ω cannot be maintained because of disturbances, which are always present. Most of the standard situations are shown in Fig. 14.7. A stable equilibrium point is shown in Fig. 14.7a. The corresponding system will exhibit stable periodic oscillations at A_2 and ω_2, the parameter values taken from the $-1/\mathbf{N}_e$ and **G** loci, respectively, at the intersection point. A system which is stable for small amplitude disturbances, and unstable for disturbances, inputs, or initial conditions above specific critical levels is shown in Fig. 14.7b. The loci intersection in this case is not a stable equilibrium point, however, (since it is a left to right intersection) and oscillations will tend to grow without bound once they are initiated. The corresponding system might perform satisfactorily for small amplitude inputs, but would break into vigorous oscillation if the input exceeds the critical level. The equilibria at points (1) and (2) in Fig. 14.7c are unstable and stable, respectively. This system is stable unless excited by a disturbance at or above the critical level defined by intersection (1), at which time the periodic oscillations would grow to the conditions defined by intersection (2). The system described by the diagrams shown in Fig. 14.7d is unstable for small A, and always breaks into oscillations if left at rest, the amplitude growing to a level defined by the stable limit cycle conditions at intersection (1). If the system is disturbed at a high level, however, amplitude conditions above those defined by the unstable equilibrium at point (2) may be reached and the oscillations will grow without bound.

When $\mathbf{N}_e$ depends upon both frequency and amplitude, for example with a saturating hysteresis type NL element where the hysteresis loop width depends upon frequency, a single $-1/\mathbf{N}_e$ plot is inadequate to define all possible

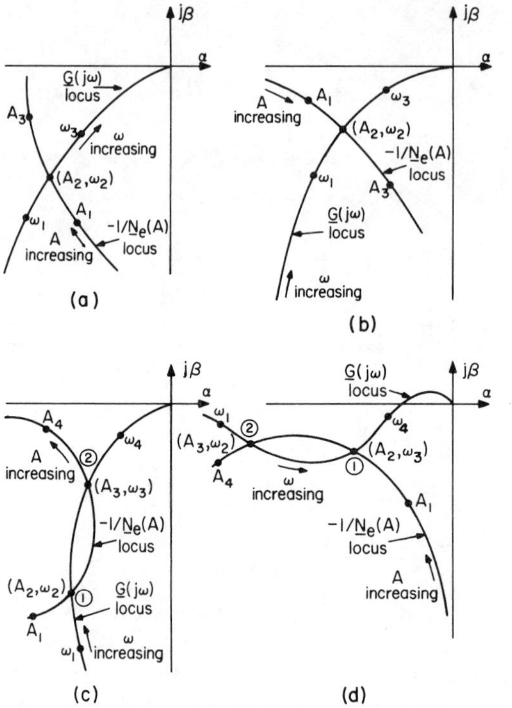

FIG. 14.7. Types of complex-plane loci intersections often encountered.

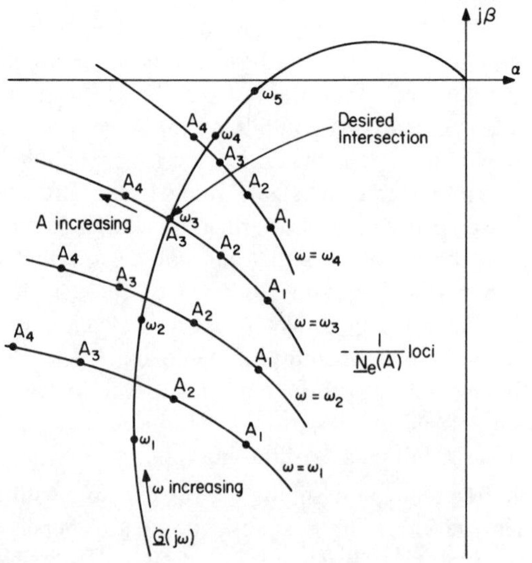

FIG. 14.8. Illustration of the DF procedure when N_e is a function of ω.

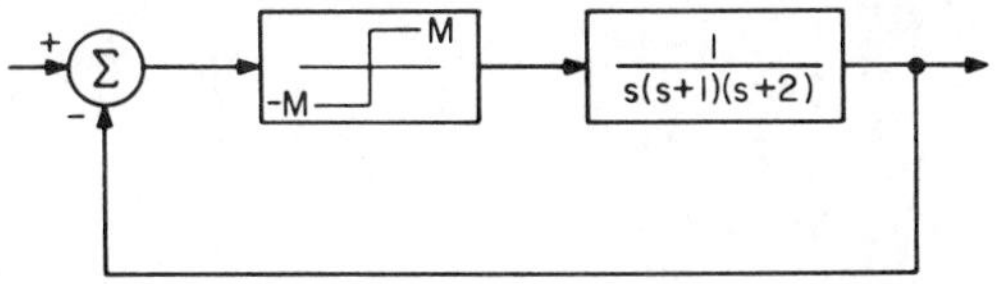

FIG. 14.9. Relay controller discussed in Ex. 14.3.

equilibrium conditions. In this case the standard DF procedure is to plot a family of $-1/\mathbf{N}_e$ loci. On each branch of the $-1/\mathbf{N}_e$ locus, ω is generally fixed and A is the parameter. It is usually possible to estimate with reasonable accuracy the desired intersection frequency, thereby limiting the number of branches which must be plotted. A typical situation is illustrated in Fig. 14.8. Although all branches of the $-1/\mathbf{N}_e$ locus intersect **G** in this case, the ω coordinates on the two curves are identical only at the intersection between **G** and the $-1/\mathbf{N}_e$ locus branch corresponding to $\omega = \omega_3$. This is the only potential equilibrium condition. The system depicted has a stable limit cycle with $A = A_3$ and $\omega = \omega_3$. Stable and unstable limit cycles and multiple equilibria are still possible when $\mathbf{N}_e$ is frequency-dependent. The only real change is in the amount of computational effort involved, since the same basic principles apply in all cases.

Example 14.3[9]

Consider the relay controller shown in Fig. 14.9. The DF for the relay characteristic is shown in Ex. 14.1 to be $\mathbf{N}_e = 4M/\pi A$. Let $M = 1$. Find the amplitude and frequency of the resulting limit cycle.

Polar plots of $\mathbf{G}(j\omega)$ and $-1/\mathbf{N}_e(A) = -\pi A/4$ are shown in Fig. 14.10. Several values of A and ω are noted on the $-1/\mathbf{N}_e$ and **G** loci, respectively. The loci intersection coordinates indicate that the system will oscillate at $\omega \simeq 1.4$ rad/sec and $A \simeq 0.21$ volts peak at the relay input. It may be shown that $\mathbf{G}(j\omega)$ actually crosses the negative real axis at $\omega = \sqrt{2}$ rad/sec. Since **G** is open-loop stable and $-1/\mathbf{N}_e$ crosses **G** from right to left as A increases, the intersection corresponds to a stable limit cycle. Note further that a perfect relay in series with a stable **G** containing at least three more poles than zeros will always yield at least one potential stable limit cycle of this type.

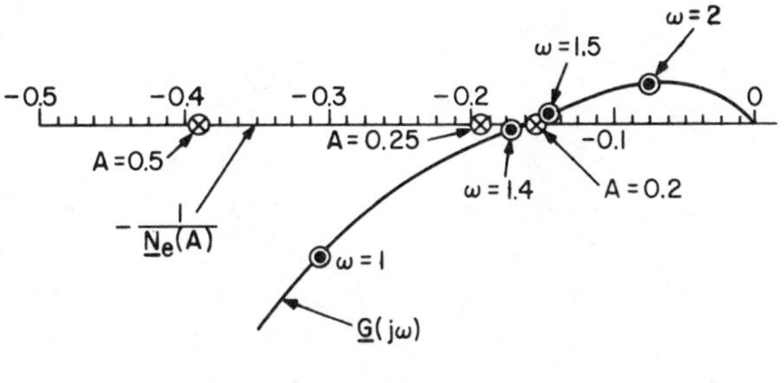

FIG. 14.10. Plots of G and $-1/\mathbf{N}_e$ for the system shown in Fig. 14.9.

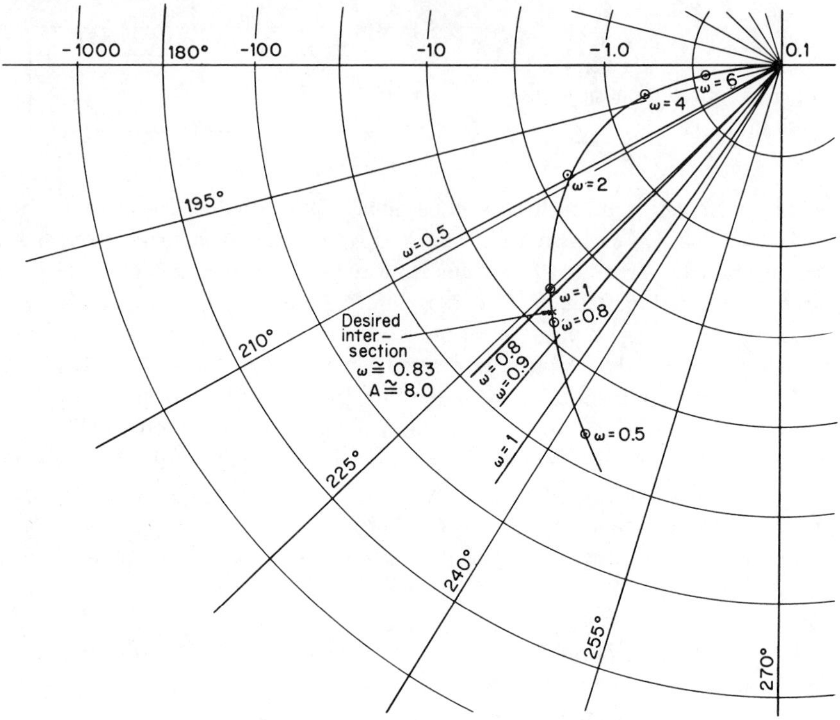

FIG. 14.11. Polar plot for Ex. 14.4.

Example 14.4

Consider a standard unity feedback system with

$$G(s) = \frac{10}{s(s+1)} \qquad N_e(A, s) = \frac{\epsilon^{-s}}{A} \tag{14.15}$$

Find any potential limit cycles for this system.

A polar plot of $\mathbf{G}(j\omega)$ and $-1/\mathbf{N}_e$ is shown in Fig. 14.11. Note that

$$-\frac{1}{\mathbf{N}_e(A, j\omega)} = -A\epsilon^{j\omega} = A\epsilon^{j(\omega+\pi)} \tag{14.16}$$

which shows that the $-1/\mathbf{N}_e$ branches for fixed ω are rays emanating from the origin, and A is the distance from the origin on each ray. The desired solution is defined by that ray for which ω coincides with the value at the intersection with the $\mathbf{G}$ plot, and the corresponding limit cycle parameters are $A \simeq 8.0$ and $\omega \simeq 0.83$. Since the intersection is from right to left as A increases, the limit cycle is stable. Although the $\mathbf{N}_e(A, j\omega)$ characteristic assumed here is extremely simple compared to the typical frequency-dependent DF's encountered in practice, this example illustrates the procedure for solving such problems without involving extensive mathematical manipulations. An interesting application of the frequency-dependent DF technique to find the limit cycle parameters of optimalizing adaptive controllers is provided in Eveleigh.[9]

It should now be clear how to use the DF method as an analysis tool in most situations encountered, but many questions remain unanswered. Our presentation is intended to be introductory in nature, so we will not attempt to provide exhaustive coverage, but several possible extensions of the DF method are mentioned in the next few paragraphs and references providing further detail are mentioned.

What about relative stability? Suppose the $-1/\mathbf{N}_e$ and **G** loci narrowly avoid each other without intersecting, as shown in Fig. 14.12. Although we might conclude that the system is stable, we certainly do not want this situation to exist in a system for which we are primarily responsible. The DF method is approximate. The assumptions we make in applying it are never exactly met, so a reasonable safety margin is desirable. Intersection avoidance distance, just as the margin by which **G** avoids encircling the –1 point in the linear case (G_m and θ_m), provides some measure of relative stability. This measure is empirical in the linear case, and even more so here. A primary design objective is to provide a reasonable avoidance margin, however, and this leads us to the following design procedure based upon the polar plots of **G** and $-1/\mathbf{N}_e$. If the loci intersect, or if the avoidance margin is inadequate, compensation may be added to shape the **G** plot in the region of interest to provide an acceptable avoidance margin. The only distinction between this and the usual linear system design methods arises because we are here concerned with avoiding a locus of points in the complex plane rather than the -1 point. This adds some complexity, of course, but the fundamental principles are unchanged. If an accurate appraisal of the resulting frequency or transient response is desired, a simulation study is the most practical solution.

Application of the DF to typical control problems provides amplitude and frequency prediction accuracy for the resulting limit cycles within about 10% and 5%, respectively. Correction terms for the DF analysis were derived by Johnson[10] and his work has been summarized by many authors.[8,11] A series expansion method is used and it is found that the first frequency correction term is zero, which explains why the DF method typically provides a more accurate prediction of frequency than amplitude. Johnson's method is primarily of theoretical interest because the analysis required is rather extensive, thereby

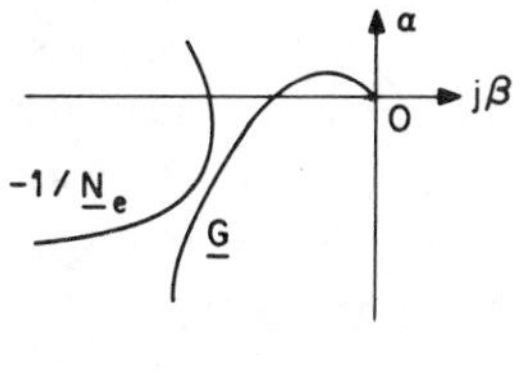

FIG. 14.12. A situation where the DF solution is subject to question.

eliminating a fundamental advantage of the DF method, and his corrections can be developed only for single valued differentiable NL functions.

Some authors[8] suggest alternate DF definitions. In most cases their suggestions are made in an effort to improve the accuracy obtained in solving a specific problem or class of problems. The standard definition used here minimizes the rms error between the actual output and a sinusoidal representation of it, which is a well known characteristic of Fourier series decomposition. We could alternately use a sinusoid with the same peak output as the NL element provides, a sinusoid which minimizes the average error magnitude, etc. The standard DF is generally adequate, is easy to work with, and correction terms may be added to improve accuracy in those few cases where this seems necessary.

The DF analysis procedure is extended to random inputs by Booton[12]. Its application to find the closed-loop frequency response of NL systems has been reported by many authors using a variety of complex analytical procedures. Simulation is suggested as the practical way to obtain frequency response data. The DF has also been used in the analysis of NL elements with two input signals, as opposed to a single sinusoid.[8] Sinusoids at frequencies ω_1 and ω_2, $\omega_2 \gg \omega_1$ have been considered, and an approximate analysis has been made where ω_2 and ω_1 are not significantly different. Random noise plus a sinusoid has also been considered. All of these analyses show that one of the two signals modifies the NL element equivalent gain for the second signal, which has a dominant influence upon system performance. This has led to the use of high frequency "squelching" signals in some NL systems which would otherwise be unstable. All of these topics are highly specialized and will not be pursued here.

The DF method is conceptually straightforward, the analysis required is readily carried out, particularly since tabulated solutions are available, and the results obtained are acceptably accurate in most control system applications. Although many refinements are available, they are primarily of theoretical or highly specialized interest, and we have concentrated upon the basic concepts.

14.3 THE PHASE PLANE

The phase plane is a special type of state-space representation (see Chap. 13) generally used to show the solution trajectories for second-order systems with constant inputs. Even for NL systems it is often possible to describe the response characteristics of the system with acceptable accuracy in the region of interest by showing a few typical phase-plane trajectories. Response trajectories can be sketched on the phase plane using a variety of graphical methods, so the term phase plane is associated with graphical solution methods and with methods for solution presentation as well.

A phase-plane trajectory is the plot of a derivative of a function versus the function. In control applications either the output and its derivative or error and error rate are generally plotted. Time is eliminated from the equations to obtain the phase-plane trajectory, but may be obtained from the phase-plane plot if desired. Since only two variables may be conveniently related on a planar diagram, phase-plane representation and the corresponding graphical analysis procedures are applicable to systems of third order or higher only by using families of curves. Arbitrary initial conditions and/or a constant input are easily handled, but other than piecewise constant inputs must be treated in some other way.

The phase-plane trajectories of typical NL second-order systems may be obtained by the use of an analog or digital computer, analytical solution, and graphical solution. NL DE's are often not significantly more difficult to integrate on an analog computer than linear ones, and the digital computer integrates either with equal facility. Computer study of linear systems is emphasized in previous chapters, and the value of simulation study is dramatically enhanced for NL systems. Analog and digital simulation methods are discussed in Appendices C and E, respectively. Analytical solution of NL problems is usually either impossible or prohibitively difficult. A summary of NL equation forms with known exact solutions is available in Cunningham.[13] Unfortunately nature is cruel, and NL problems with known analytical solutions are seldom encountered in practice.

In many applications an approximate indication of the transient solution characteristics of the system is adequate. Several graphical procedures based upon phase-plane constructions are possible.[8] They are most appropriate when the general solution characteristics are desired and accuracy is not a primary consideration. It is interesting to observe that all graphical solution methods are closely related to the standard digital computer integration algorithms (see App. E). The simplest graphical solution procedure, presented here as an example of what can be done, is called the *isocline method.* The system equations are manipulated to obtain equations for the loci of constant slope, called *isoclines.* These loci are sketched on the phase plane and used as graphical construction aids. Approximate phase trajectories for representative sets of initial conditions can then be sketched without explicitly solving the DE's. This procedure is first illustrated for a linear second-order system so that the known analytical solution provides a comparison basis. Consider a system described by

$$\frac{d^2e}{dt^2} + 2\zeta\omega_0 \frac{de}{dt} + \omega_0^2 e = 0 \tag{14.17}$$

It is easier to work with this equation if ω_0 is first eliminated by defining the change in time scale

$$\tau = \omega_0 t \qquad d\tau = \omega_0 dt \tag{14.18}$$

This is equivalent to setting $\omega_0 = 1$, and yields the normalized form of Eq. (14.17) as

$$\frac{d^2e}{d\tau^2} + 2\zeta \frac{de}{d\tau} + e = 0 \tag{14.19}$$

The two phase variables x_1 and x_2 are defined as

$$x_1 = e \qquad x_2 = \frac{de}{d\tau} \tag{14.20}$$

In this new notation, Eq. (14.19) becomes

$$\frac{dx_2}{d\tau} + 2\zeta x_2 + x_1 = 0 \tag{14.21}$$

Dividing Eq. (14.21) by $x_2 = dx_1/d\tau$ and rearranging provides

$$\frac{dx_2/d\tau}{dx_1/d\tau} = \frac{dx_2}{dx_1} = \frac{-x_1 - 2\zeta x_2}{x_2} \tag{14.22}$$

which may alternately be written as

$$x_2 = \frac{-x_1}{2\zeta + dx_2/dx_1} \tag{14.23}$$

It is apparent from Eq. (14.23) that the isoclines are straight lines through the origin since, for dx_2/dx_1 constant and any specific value of ζ, Eq. (14.23) is of the form $x_2 = ax_1 + b$, with $b = 0$. Several isoclines are shown in Fig. 14.13 for the special case $\zeta = 0.5$. Two phase trajectories are also provided to illustrate how easily the approximate response trajectory for an arbitrary set of initial conditions is obtained once the isoclines are available.

Several factors should be emphasized before this example is abandoned. The transformation $\tau = \omega_0 t$ normalizes the x_2 axis scale in Fig. 14.13 to correspond to that on the x_1 axis, thereby simplifying the plot and the isoclines. Similar

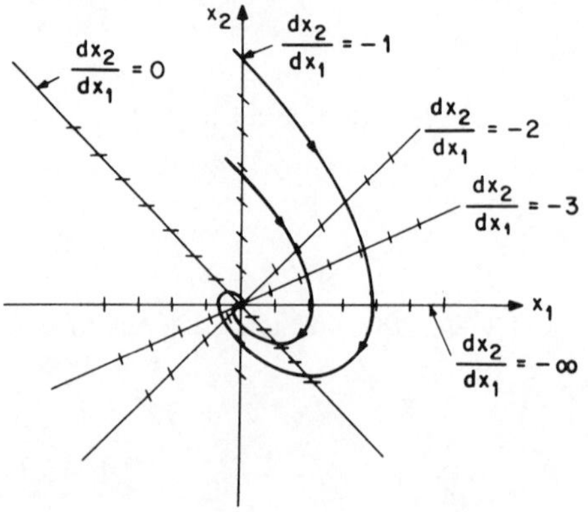

FIG. 14.13. Phase-plane trajectories for an underdamped linear second-order system with $\zeta = 0.5$.

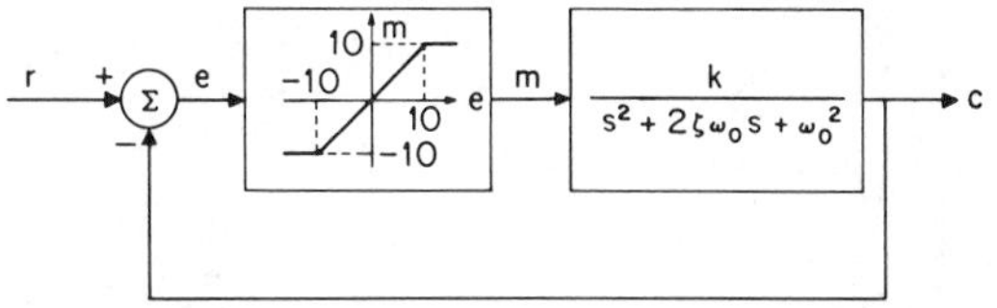

FIG. 14.14. System used to illustrate the isocline method for piecewise linear systems.

scale adjustments are usually appropriate. All solution trajectories proceed from left to right in the upper half plane and from right to left in the lower half plane, since $x_2 = dx_1/dt$. The isoclines depend directly upon ζ. As $\zeta \to 0$, the rate at which the trajectories spiral toward the origin decreases and they become circles for $\zeta = 0$, although graphical analysis is not conclusively accurate in this case. For negative ζ the trajectories spiral away from the origin and the system is obviously unstable. For $\zeta \geq 1$ the system response is overdamped and the solutions approach the origin, but without a characteristic spiral shape. All of these characteristics can be determined from a thorough phase-plane study using isoclines, although in this case they are obvious from our prior understanding of linear second-order system response patterns.

Isoclines can also be used to determine the approximate step response pattern for the piecewise linear system shown in Fig. 14.14. Assume a 15 volt step input is applied to this system at $t = 0$ with all initial conditions zero. Because of the saturation nonlinearity, $m = 10$ is applied to the plant for an unknown length of time until the output reaches 5 volts. Let $K = 10$, $\omega_0 = 1$ and $\zeta = 0.5$. Writing the differential equation in terms of the output variable c, assuming $m = 10$, and carrying out algebraic manipulations analogous to those used previously provides isoclines applicable for the range $c \leq 5$. For $5 \leq c \leq 25, e$ falls in the linear range of the saturating element and the system response is similar to that illustrated in Fig. 14.13 except that the singular point (the point at which the system is in equilibrium, or the point toward which, in this case, the solution converges) is at $c = 150/11$. The isoclines are straight lines in each region, and an approximate solution is readily obtained. Of course this type of problem can also be solved using piecewise linear analysis and matching boundary conditions at the time point (or points) where the breaks in the NL characteristic are encountered. A phase-plane diagram with several isoclines and a few representative trajectories sketched on it contains a great deal of information in readily useful form to supplement the mathematical solution statement. It is suggested that the reader complete the solution of this problem.

A type of NL characteristic often encountered in practice and readily treated using the isocline method is *coulomb* (static) *friction*. The usual viscous (linear) friction force is proportional to velocity. Coulomb friction provides a force independent of velocity magnitude and opposing the velocity in direction. Consider the mechanical system shown in Fig. 14.15. The DE describing this system is

$$M\ddot{x} + Kx \pm f_0 = 0 \tag{14.24}$$

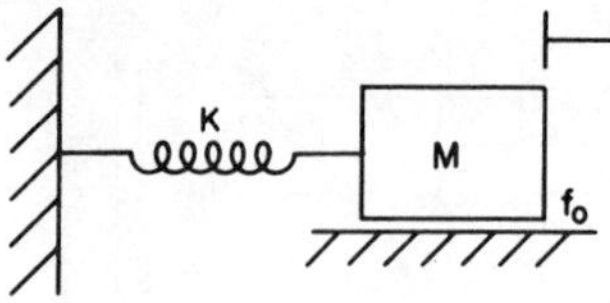

FIG. 14.15. A spring-mass system with coulomb friction f_0.

where f_0 is the static friction level and its sign is taken identical to that of $\dot{x}$. Although this problem is easily treated using a piecewise linear analysis, which shows that the individual phase-plane segments are half circles of consecutively decreasing radius in the upper and lower half planes, the solution is also easily obtained using isoclines. It is convenient to convert Eq. (14.24) to standard form by defining $x_1 = x$ and $x_2 = \dot{x}$, which yields

$$\dot{x}_1 = x_2 \qquad \dot{x}_2 = -\frac{1}{M}(Kx_1 \pm f_0) \tag{14.25}$$

Dividing $\dot{x}_2$ by $\dot{x}_1$ provides

$$\frac{dx_2}{dx_1} = -\frac{1}{Mx_2}(Kx_1 \pm f_0) \tag{14.26}$$

Solving Eq. (14.26) for x_2 yields the isoclines as

$$x_2 = -\frac{Kx_1 \pm f_0}{M\,dx_2/dx_1} \tag{14.27}$$

It is clear from Eq. (14.27) that the isoclines are straight lines. When $\dot{x}$ is positive (in the upper half plane) f_0 is positive and the isoclines pass through $x_1 = -f_0/K$. When $\dot{x}$ is negative (in the lower half plane) f_0 is negative and the isoclines pass through $x_1 = f_0/K$. The general situation is illustrated in Fig. 14.16. Note that as soon as the mass stops within the range where the spring force $|Kx_1| < f_0$, it does not move further. Thus the response trajectories may terminate anywhere in the range $-f_0/K \leq x_1 \leq f_0/K$. An interesting generalization of this problem is obtained by adding a viscous damping term in Eq. (14.24). In this case the trajectories spiral in toward the points $x_1 = \pm f_0/K$ at a rate depending upon the amount of viscous damping in a manner similar to those shown in Fig. 14.13.

The Van der Pol equation is readily treated using isoclines. The steady-state solution is a limit cycle, and all trajectories move toward this limit cycle as $t \to \infty$. One form of the Van der Pol equation is

$$\ddot{q} - \epsilon(1 - q^2)\dot{q} + q = 0 \tag{14.28}$$

Assume, for convenience, that $\epsilon = 1$. Let $x_1 = q$ and $x_2 = \dot{q}$. Making these substitutions and reducing the result to the usual isocline form provides

$$\frac{dx_2}{dx_1} = 1 - x_1^2 - \frac{x_1}{x_2} \tag{14.29}$$

or

$$x_2 = \frac{x_1}{1 - x_1^2 - dx_2/dx_1} \tag{14.30}$$

Isoclines drawn for several values of slope are shown in Fig. 14.17 along with the limit-cycle trajectory and representative solutions approaching the stable limit cycle from both "inside" and "outside". Although the isocline equations are more complex (NL) than any we have considered previously, it is only necessary to plot a few isoclines to obtain a good indication of the solution characteristics, and the limit-cycle trajectory may be sketched without a prohibitive amount of effort. It is recommended that isoclines be plotted only in the region where they are needed to obtain the desired trajectories, thereby saving as much effort as possible.

Time is not directly apparent from the phase-plane plot, and we need a method for obtaining it. Perhaps the most fundamental procedure is based upon the fact that the average velocity for small increments Δx_1 and Δt is given by

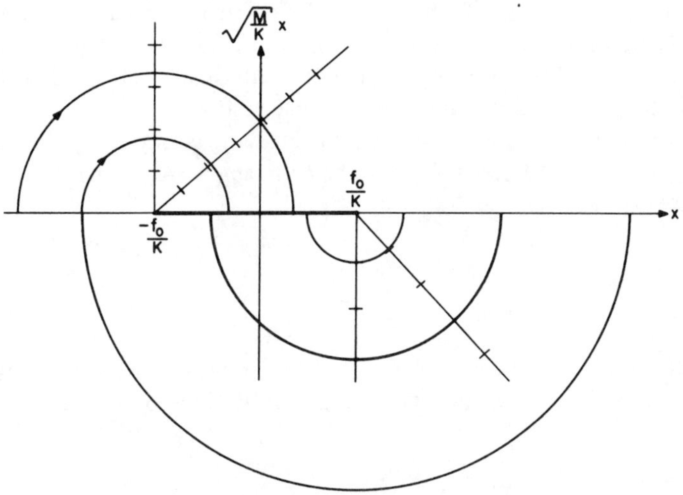

FIG. 14.16. Response trajectories for the system shown in Fig. 14.15.

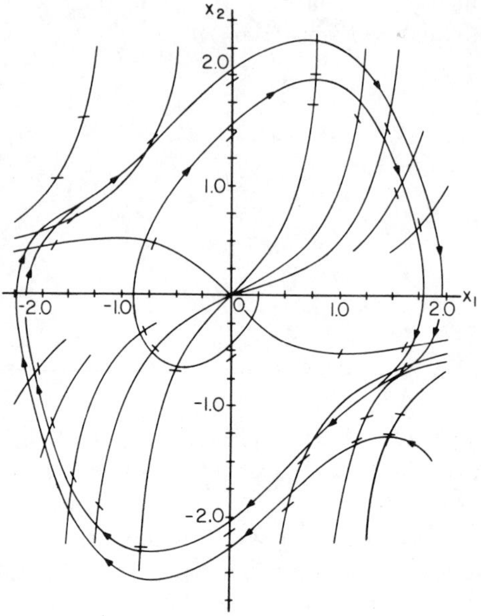

FIG. 14.17. Van der Pol equation phase plane trajectories illustrating a stable limit cycle.

$\bar{x}_2 \simeq \Delta x_1/\Delta t$. Thus we may consider small increments Δx_1 on the phase trajectory, determine $\bar{x}_2$ over that increment, and approximate Δt as

$$\Delta t = \frac{\Delta x_1}{\bar{x}_2} \tag{14.31}$$

This process is repeated as many times as necessary to cover the range of interest and the resulting Δt's are summed to obtain elapsed time. The procedure is illustrated in Fig. 14.18 for the elliptical phase-plane trajectory of a sinusoidal oscillator. The derived $x_1(t)$ curve is clearly sinusoidal. The Δx_1 used are chosen such that $\bar{x}_2$ is obtained with acceptable accuracy.

If fixed time increments along the phase trajectory are acceptable, a triangular template can be used to obtain the data rapidly. Assuming Δt_i, the time change over the ith increment, is small

$$\bar{x}_{2i} \simeq x_{2i} + \frac{\Delta x_{2i}}{2} \tag{14.32}$$

where x_{2i} denotes x_2 at the start of the ith increment, and Δx_{2i} is the change in x_2 over that increment. Substituting Eq. (14.32) into Eq. (14.31) and rearranging provides

$$x_{2i} = \frac{2}{\Delta t_i} \Delta x_{1i} - 2x_{2i} \tag{14.33}$$

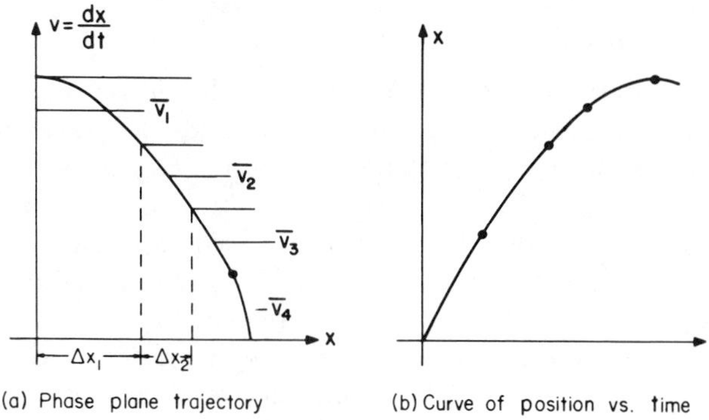

FIG. 14.18. Recovery of time from the phase trajectory.

For a fixed $\Delta t_i = \Delta t$, Eq. (14.33) represents a straight line with slope $2/\Delta t$ and Δx_{2i} intercept of $-2x_{2i}$, where Δx_{2i} and Δx_{1i} are measured from the beginning of the interval. The intersection of this line with the phase trajectory defines the point which satisfies Eq. (14.33) and the original DE, and this point on the trajectory is Δt units from the starting point. If we construct a right triangle template with one acute angle equal to $\tan^{-1}(2/\Delta t)$, where Δt is chosen for acceptable accuracy, this template may be used as illustrated in Fig. 14.19. The points x_{2i}, x_{1i} and $-x_{2i}$ are noted on the phase plane, a pin (or pencil) is placed at the point $(x_{1i} - x_{2i})$, the triangular template is placed against the pin with the right angle aligned along the coordinate axes, and the intersection of the hypotenuse with the phase trajectory is the desired point corresponding to $t_i + \Delta t$. This process may be repeated rapidly to obtain x_1 or $\dot{x}_1 = x_2$ versus t. Typical values of Δt chosen for reasonable accuracy are in the range 0.2 to 0.4. Any time-scaling used in constructing the phase plane plot must be considered in recovering time for the *original* problem.

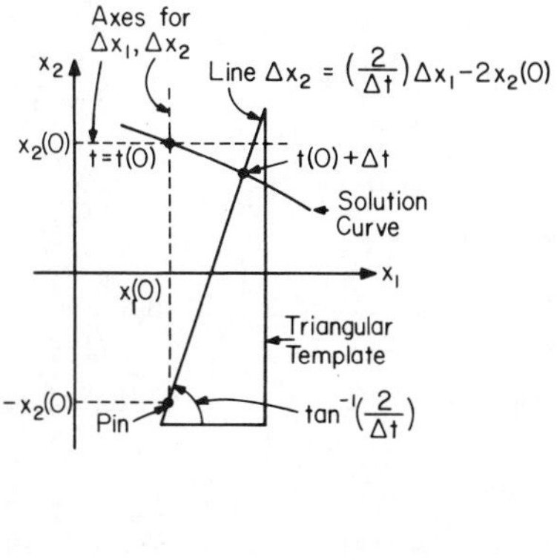

FIG. 14.19. Use of a template to find time points Δt seconds apart on the phase-plane trajectory.

14.4 NL SYSTEM STABILITY AND GENERAL SINGULAR POINT CHARACTERISTICS

Contrary to the usual practice in linear systems, NL system stability is generally defined relative to a region, and NL systems are called stable if, and only if, all trajectories starting in a region R_1 remain within a region R_2, where R_2 is generally larger than R_1. Since this stability interpretation is considerably less restrictive than the one we have been using, it is necessary to introduce precise definitions of several types of NL stability. *Local stability,* alternately referred to as *ϵ neighborhood stability* or *stability in the small,* applies only to the infinitesimal region about a singularity. Linearization of the system equations in the region near a singular point, when valid, provides a direct indication of local stability. *Global stability,* or *stability at large,* refers to stability in the entire state space. This is a broad stability interpretation, it is difficult to prove general theorems related to global performance, and such large dynamic range is not often of practical interest. *Finite stability,* or stability in a limited state-space region, is more useful than either local stability or global stability, but rigorous proofs of finite stability are difficult. It is often possible to imply finite stability for appropriately chosen regions surrounding locally stable singularities. It is not valid, however, to imply finite or global instability for all regions surrounding locally unstable singularities.

The standard linear system stability definition is: *a system is stable if, and only if, its response is bounded for all bounded inputs.* This is called *total stability* in our more general structure. Local, finite, and global stability are not so restrictive, since system state may approach a singularity or limit cycle within the specified region in any of these cases, although system state must remain bounded as $t \to \infty$. A system is *asymptotically stable* in a region if all trajectories starting in that region become arbitrarily close to a singular point in the region as $t \to \infty$. Asymptotic stability is more restrictive than stability and rules out limit cycles. A system is *monotonically stable* if, and only if, some positive definite measure of its state position decreases monotonically as $t \to \infty$. The response of a time-varying system might grow temporarily before converging toward a singularity as $t \to \infty$. Such a system would be globally asymptotically stable, but not monotonically stable.

If a second-order system can be described in terms of the two variables x_1 and x_2 in the form

$$\frac{dx_1}{dt} = P(x_1, x_2) \qquad \frac{dx_2}{dt} = Q(x_1, x_2) \tag{14.34}$$

where P and Q are independent of t, the system is called *autonomous.* The points where $P(x_1, x_2) = Q(x_1, x_2) = 0$ are called *singular points,* and the general characteristics of the phase trajectories near these singular points can often be predicted using an approximate linear representation. A system is in *equilibrium* at each of its singular points, and each equilibrium condition is described as *stable* or *unstable* according to its local stability characteristics. Note that the mechanical example with coulomb friction introduced in Sec. 14.3 is in equilibrium at all points on the real axis in the range $-f_0/K \leq x_1 \leq f_0K$. We will consider only isolated singular points in this section, however. In those cases where an expansion containing linear terms exists, the phase trajectory patterns near isolated singularities may be determined by expanding P and Q in Taylor series form and looking at only the linear terms. A general singular point characterization is often valuable in sketching phase-plane trajectories and in determining limit cycle existence.

Assume that $P(x_1, x_2) = Q(x_1, x_2) = 0$ at the point $\mathbf{x}' = [\alpha, \beta]$. Expanding P and Q about this point provides

$$\frac{dx_1}{dt} = a_1(x_1 - \alpha) + a_2(x_2 - \beta) + a_3(x_1 - \alpha)^2 + a_4(x_1 - \alpha)(x_2 - \beta) + a_5(x_2 - \beta)^2 + \cdots \tag{14.35}$$

and

$$\frac{dx_2}{dt} = b_1(x_1 - \alpha) + b_2(x_2 - \beta) + b_3(x_1 - \alpha)^2 + b_4(x_1 - \alpha)(x_2 - \beta) + b_5(x_2 - \beta)^2 + \cdots \tag{14.36}$$

The derivatives (coefficients) used in this expansion are assumed to exist, and the linear terms are assumed present (a_1, a_2 not both zero; b_1, b_2 not both zero). Liapunov has shown that when the system is structurally stable, only the linear terms of Eqs. (14.35) and (14.36) need be used to determine local stability for each singularity. Structural stability requires that small parameter changes not alter the general solution characteristics. A linear system with a pair of imaginary axis poles is structurally unstable. Most systems are structurally stable and have a linear expansion. Retaining only the first two terms in Eqs. (14.35) and (14.36) linearizes the original equations near the singularity. We may assume $\alpha = \beta = 0$ without loss of generality, since any singularity is readily translated to the origin. Thus we have

$$\frac{dx_1}{dt} = a_1x_1 + a_2x_2 \qquad \frac{dx_2}{dt} = b_1x_1 + b_2x_2 \tag{14.37}$$

Laplace transforming provides

$$\begin{aligned} sX_1(s) - x_1^\circ &= a_1X_1(s) + a_2X_2(s) \\ sX_2(s) - x_2^\circ &= b_1X_1(s) + b_2X_2(s) \end{aligned} \tag{14.38}$$

Eliminating $X_2(s)$ and solving for $X_1(s)$ yields

$$X_1(s) = \frac{(s - b_2)x_1^0 + a_2 x_2^0}{s^2 - (a_1 + b_2)s + (a_1 b_2 - a_2 b_1)} \tag{14.39}$$

Note that we could just as readily solve for $X_2(s)$, since the denominator polynomial (characteristic equation) is the same in both cases, and the roots of that polynomial completely define the general solution form. Six combinations of denominator polynomial zeros are possible.

Case 14.1 Stable Node

The zeros are both real and both located in the lhp. The corresponding solution trajectories are of the general form shown in Fig. 14.20, depending upon $\mathbf{x}^0$. All paths move directly toward the origin after no more than a half cycle of spiral rotation. This relatively direct approach to the origin results because oscillatory energy interchange occurs only for complex conjugate zero pairs.

Case 14.2 Stable Focus

The zeros are a lhp complex pair. The response is oscillatory, with $x_1(t)$ and $x_2(t) = \dot{x}_1(t)$ given by damped sinusoids for any initial condition. Typical phase-plane trajectories are shown in Fig. 14.21. The trajectories approach the origin in a continuing spiral pattern.

Case 14.3 Vortex, or Center

The zeros are on the $j\omega$ axis. This results in an undamped oscillatory solution for $x_1(t)$, and the trajectories form closed elliptical paths in the phase plane as illustrated in Fig. 14.22.

Case 14.4 Unstable Node

The zeros are both real and are both in the rhp. Solutions for $x_1(t)$ and $x_2(t)$ each increase as the sum of two growing exponentials. Typical phase trajectories are shown in Fig. 14.23.

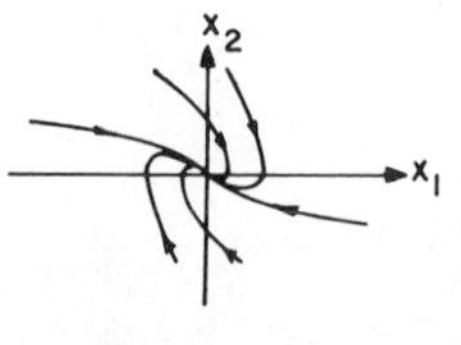

FIG. 14.20. Phase-plane trajectories near a stable node.

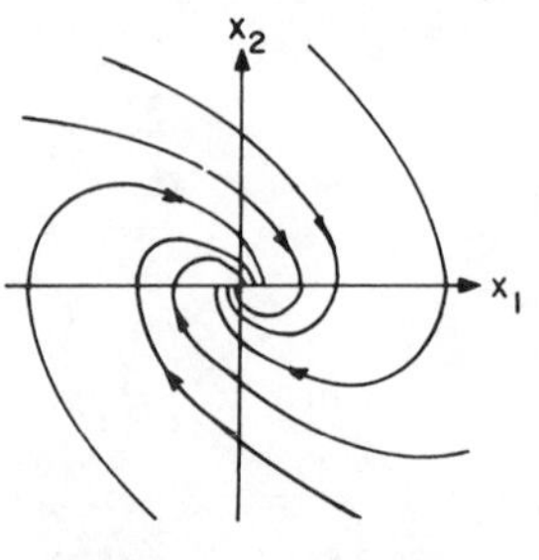

FIG. 14.21. Phase-plane trajectories near a stable focus.

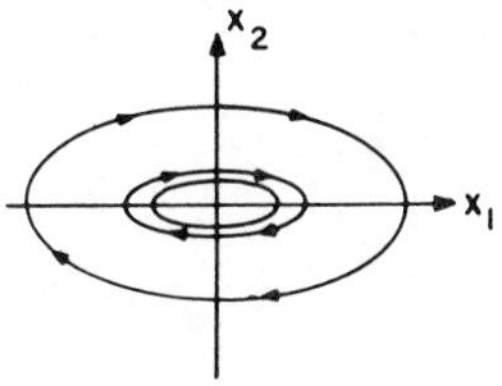

FIG. 14.22. Phase-plane trajectories near a center.

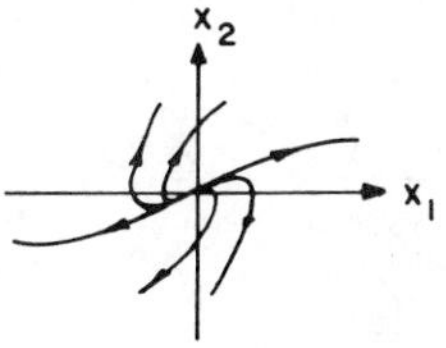

FIG. 14.23. Phase-plane trajectories near an unstable node.

Case 14.5 Unstable Focus

The zeros form a complex conjugate pair in the rhp. The time response is an exponentially increasing sinusoid, and the solution curves in a spiral away from the singularity as shown in Fig. 14.24.

Case 14.6 Saddle Point

The zeros are real, with one in each half plane. For this case Eq. (14.39) becomes

$$X_1(s) = \frac{(s - b_2)x_1^0 + a_2 x_2^0}{(s - s_1)(s + s_2)} \tag{14.40}$$

If the numerator is proportional to $s - s_1$, the growing response mode is unexcited (this is impossible except on paper) and the response is a decaying exponential. Otherwise the response is unbounded. For a numerator term of the form $s - s_1$ to exist in Eq. (14.40) it is necessary that

$$\frac{x_2^0}{x_1^0} = \frac{b_2 - s_1}{a_2} = K_1 \tag{14.41}$$

where K_1 depends upon the coefficients a_1, a_2, b_1, and b_2 in Eq. (14.38). If the numerator of Eq. (14.40) is proportional to $(s + s_2)$, the decaying exponential solution component is never excited, and the response consists of a single term which grows exponentially. For this condition it is necessary that

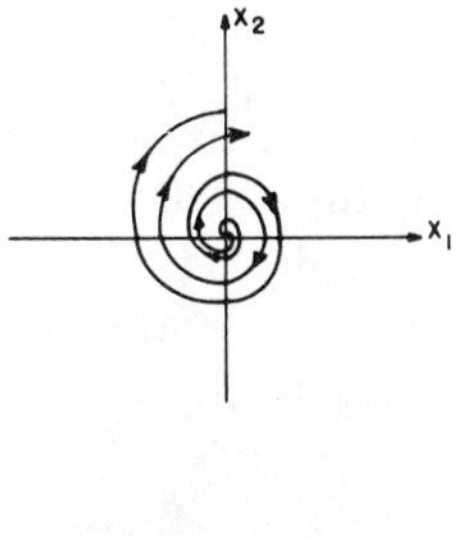

FIG. 14.24. Phase portrait near an unstable focus.

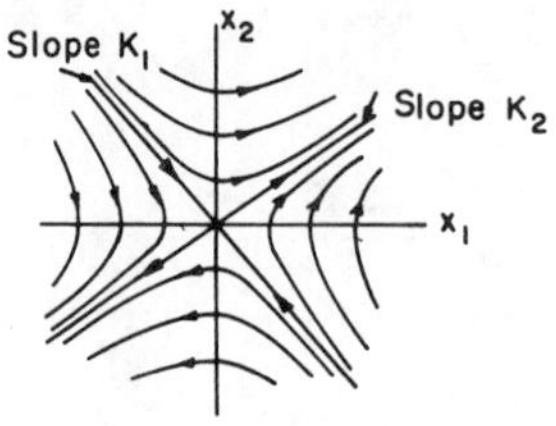

FIG. 14.25. Phase portrait near a saddle point.

$$\frac{x_2^\circ}{x_1^\circ} = \frac{b_2 - s_2}{a_2} = K_2 \tag{14.42}$$

The general form of the phase-plane solutions is shown in Fig. 14.25. Each of the straight lines separating the trajectories into distinct regions is called a *separatrix.*

Knowledge of the general-phase trajectory characteristics near each type of singularity is valuable in solving specific problems. The phase plane may be divided into regions dominated by each singularity and typical trajectories may be constructed using a few isoclines and the known singularity patterns. In complex situations a computer study is appropriate. Trajectories may be obtained for families of initial conditions with little programming effort and a modest expenditure in running time. Singularity classification is also important in establishing the conditions under which limit cycles can (or cannot) exist, as discussed later.

In addition to the usual stable systems in which all solution trajectories approach equilibrium conditions at one or more singularities, and unstable systems with all trajectories approaching infinity as $t \to \infty$, several types of *limit cycles,* characterized by closed-path trajectories, are common in NL systems. Three standard limit cycle forms are defined as follows:

1. *Orbitally stable limit cycle*: All trajectories in the region of interest approach the limit cycle path as $t \to \infty$.
2. *Orbitally unstable limit cycle*: All trajectories move away from the limit cycle as $t \to \infty$.
3. *Orbitally semistable limit cycle*: All trajectories approach the limit cycle from one side and move away from it on the other side as $t \to \infty$.

Orbitally stable and orbitally unstable limit cycles are common. The amplitude control loop of a typical electronic oscillator is NL, and the resulting limit cycle must be orbitally stable for reasonable amplitude stability. Orbitally stable and orbitally unstable limit cycles are common in DF analysis and correspond to the stable and unstable intersections between $\mathbf{G}$ and $-1/\mathbf{N}_e$, respectively. Orbitally unstable limit cycles divide the phase plane into two distinct regions. Orbitally semistable limit cycles cannot exist in structurally stable systems, and are of limited interest.

In some cases the existence of limit cycles can be excluded using the Poincaré index.[14] The Poincaré index of a phase-plane region is obtained as follows. The phase-plane trajectory direction at each point on a closed path around the region of interest is treated as a vector. The Poincaré index is the number of cw (ccw) rotations of the trajectory vector as the closed path is completely traversed in a cw (ccw) direction. It is easily shown that:

1. The index of any closed region containing no singular points is 0.
2. The index of a center, a node, or a focus is 1.
3. The index of a saddle point is -1.
4. The index of any closed region containing two or more singular points is the algebraic sum of the individual singularity indices.

For structually-stable systems, Bendixson[15] provides conditions sufficient to rule out limit-cycle existence in certain closed regions. The sum $\partial P/\partial x_1 + \partial Q/\partial x_2$ must have fixed sign on the region boundary. This is called *Bendixson's first theorem* and the proof is not difficult.[11] In terms of the Poincaré index, no limit cycle is possible in a closed region if any of the following conditions are met:

1. There are no singular points within the region.
2. There is only one singular point; a saddle point with index -1.
3. There are two or more singular points in the region, but no combination of them has a cumulative index of +1.
4. All collections of singular points with total index +1 are limit points of phase trajectories as $t \to \infty$.

Thus all limit cycles must enclose regions with net index of +1. Although this rules out many possibilities, it is unfortunately not easy to prove the existence of a limit cycle when the conditions allow it. *Bendixson's second theorem* relates to this problem, but is intuitively obvious and not very comforting. It is stated as follows. Assume that two finite closed curves C_1 and C_2 may be constructed as shown in Fig. 14.26, with C_1 enclosed by C_2. Denote the region between C_1 and C_2 by D. Assume C_1 has index +1 and D contains no singularity. If a solution trajectory enters D and remains within D as $t \to \infty$, that trajectory approaches a limit cycle in D. Thus if we can find closed curves C_1 and C_2 around a region of index +1 such that D contains no singularities, all trajectories enter D, and no trajectories leave D, the existence of a limit cycle is assured.

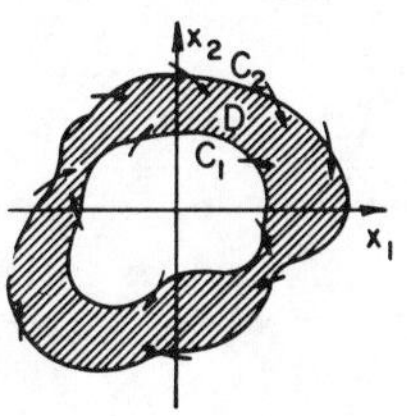

FIG. 14.26. Closed contours used to illustrate Bendixson's second theorem.

Example 14.5[8]

Consider the system described by

$$\ddot{x} - \left(0.1 - \frac{10}{3}\dot{x}^2\right)\dot{x} + x + x^2 = 10 \tag{14.43}$$

a. Define all singularities.
b. Sketch the phase-plane trajectories near each singularity.
c. Apply Bendixson's first theorem to find potential limit cycles.
d. Apply Bendixson's second theorem in an effort to prove the existence of a limit cycle.

Let $x = x_1$ and $\dot{x}_1 = x_2$, and rearrange the result to obtain

$$\frac{dx_2}{dx_1} = \frac{-x_1 + 0.1x_2 - x_1^2 - 10x_2^3/3}{x_2} \tag{14.44}$$

Singularities are located at the points where the numerator and denominator of Eq. (14.44) are both zero, which requires $\mathbf{x}' = [0,0]$ or $\mathbf{x}' = [-1,0]$. Considering the singularity at $\mathbf{x}' = [0,0]$, linearization of Eq. (14.44) yields

$$\frac{dx_2}{dx_1} = \frac{-x_1 + x_2/10}{x_2} \tag{14.45}$$

so $b_1 = -1$, $b_2 = 1/10$, $a_1 = 0$ and $a_2 = 1$, as seen from Eqs. (14.35) and (14.36). Thus the characteristic equation is obtained from Eq. (14.39) as

$$s^2 - \frac{s}{10} + 1 = 0 \quad \text{with roots} \quad s \simeq +\frac{1}{20} \pm j1 \tag{14.46}$$

so the singularity is an unstable focus.

Substitute $x_1 = x_1^* - 1$ in Eq. (14.44) to move the singularity at $\mathbf{x}' = [-1,0]$ to the origin. This results in

$$\frac{dx_2}{dx_1^*} = \frac{-(x_1^*)^2 + x_1^* + 0.1x_2 - 10x_2^3/3}{x_2} \tag{14.47}$$

Retaining only the linear terms yields

$$\frac{dx_2}{dx_1^*} = \frac{x_1^* + 0.1x_2}{x_2} \tag{14.48}$$

so $b_1 = 1$, $b_2 = 0.1$, $a_1 = 0$ and $a_2 = 1$. The corresponding characteristic equation is

$$s^2 - \frac{s}{10} - 1 = 0 \text{ with roots } s \simeq +\frac{1}{20} \pm 1 \tag{14.49}$$

so this singularity is a saddle point.

Although typical-phase trajectories may be obtained using isoclines or alternate graphical techniques, this problem is sufficiently complex that a digital simulation for several starting conditions is more appropriate. The approximate phase-plane portrait is shown in Fig. 14.27. It is apparent that any limit cycle must form a closed path around the unstable focus at the origin. Integration from arbitrary starting conditions near the origin provides the limit cycle path shown. Bendixson's second theorem may be used to prove that a limit cycle exists, since all trajectories near the saddle point separatrix move toward the focus, and all trajectories starting very near the origin move away from it. The stable side of the saddle point "pulls" all trajectories starting far from the origin on the stable side of the separatrix in toward the origin in spite of the unstable focus located there.

The phase-plane method may be extended to systems of third-order or higher, but applications to higher-order systems have not proven particularly fruitful. Cumbersome families of curves must be used to represent the trajectories and evaluation of results is difficult. Much of the analysis applicable to second-order systems has not been extended to higher-order cases, and this is largely due to the relative lack of interest in this area which is at best rather awkward to treat.

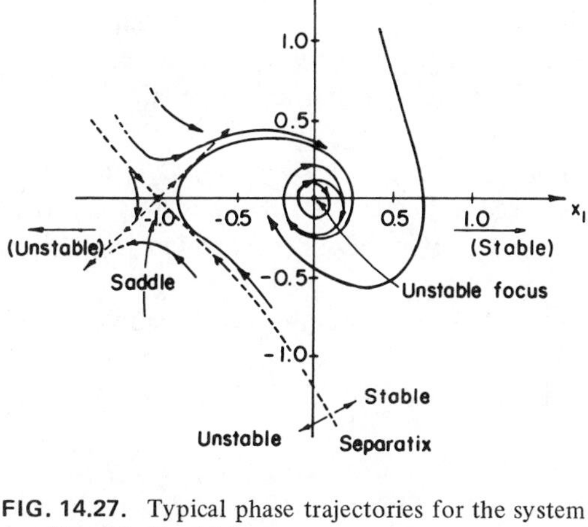

FIG. 14.27. Typical phase trajectories for the system considered in Ex. 14.5.

14.5 LIAPUNOV'S SECOND METHOD

Linear systems can be categorized as stable or unstable in several ways. Most of the standard methods, such as pole locations, are not applicable in the NL case. A noteworthy exception is the interpretation of stability based upon system energy level, or some alternate energy-like characterization, and Liapunov's second method extends the energy related stability concept to NL systems. The stored energy in a homogeneous (undriven) stable linear system is easily shown to be a non-increasing function of time. If we can prove that a positive definite energy-like function of system state, generally referred to as a Liapunov function, or simply a V function, is a monotonically decreasing time function, that system is assuredly stable whether linear or NL. Finding the desired V function is not in general a simple task, and failure to find a V function does not indicate instability, but if at least one V function can be found the system is proved stable.

The original work on the second method was done by Liapunov in about 1890[16], and his work has since been summarized and extended in the literature throughout the world by many authors. Interest in the application of Liapunov theory to NL system analysis and design has been particularly intense since 1950, his work having lain in a relatively dormant state for many years prior to that time. A readable and accurate introduction to Liapunov's work is provided by LaSalle and Lefschetz.[17] We will present several useful theorems and illustrate their application to linear and NL systems. Some basic mathematical review is required first, however.

Liapunov theory is based upon the characteristics of a scalar function $V(\mathbf{x})$, where $\mathbf{x}$ is the system-state vector, and is particularly dependent upon the sign definiteness of scalar functions. We say that $f(x)$, a scalar function of scalar x, is positive (negative) on the closed interval $a \leq x \leq b$ ($[a, b]$) if $f(x) > 0$ (<0) everywhere there. If $f(x) \geq 0$ (≤ 0) everywhere on $[a, b]$, $f(x)$ is said to be non-negative (non-positive) on $[a, b]$. The sign definiteness of $V(\mathbf{x})$, a scalar function of the vector $\mathbf{x}$ is a direct extension of this concept. We revert at this point to the rather terse and concise mathematical format which is common in this area, although an effort is made to clarify the presentation by interpreting each significant result as it is encountered.

Definition 1 Positive (negative) definite

A scalar function $V(\mathbf{x})$ is positive (negative) definite if $V(\mathbf{x}) > 0$ (<0) at all points $\mathbf{x} \neq \mathbf{0}$ in the spherical region $\|\mathbf{x}\| \leq K$ and if, in addition, $V(\mathbf{0}) = 0$.

Note that $\|\mathbf{x}\|$ denotes the usual Euclidian norm of $\mathbf{x}$ given by

$$\|\mathbf{x}\| = (x_1^2 + x_2^2 + \cdots + x_{n-1}^2 + x_n^2)^{1/2} \tag{14.50}$$

Definition 2 Positive (negative) semi-definite

A scalar function $V(\mathbf{x})$ is positive (negative) semi-definite in the region $\|\mathbf{x}\| \leq K$ if $V(\mathbf{x}) \geq 0$ (≤ 0) for all $\mathbf{x}$ in the region and if $V(\mathbf{0}) = 0$.

In the semi-definite case, $V(\mathbf{x})$ may assume the value zero at points other than $\mathbf{x} = \mathbf{0}$, whereas this is disallowed in the definite case. If K is arbitrarily large the definitions apply to the whole space and are said to be global. We will be concerned primarily with global conditions.

Definition 3 Indefinite

A scalar function $V(\mathbf{x})$ is indefinite if it assumes both positive and negative values within the region $\|\mathbf{x}\| \leq K$ for arbitrarily small K.

Thus indefinite functions are neither positive (negative) definite nor positive (negative) semi-deifinte. Note that system energy is a positive definite function of the state displacement from an equilibrium point.

The following functions illustrate these definitions. Obviously

$$V(\mathbf{x}) = \|\mathbf{x}\|^2 = \sum_{i=1}^{n} x_i^2 \tag{14.51}$$

the Euclidean length squared of $\mathbf{x}$, is positive definite, since it is non-negative, and zero only at $\mathbf{x} = \mathbf{0}$. Note, however, that

$$V(\mathbf{x}) = x_1^2 + x_2^2 \tag{14.52}$$

is positive definite only in two-dimensional state space. It is positive semi-definite in spaces of three or more dimensions, since it is then zero at the infinitely many points where $x_1 = x_2 = 0$. The closely related function

$$V(\mathbf{x}) = (x_1 + x_2)^2 \tag{14.53}$$

is positive semi-definite even in two-dimensions, since it is zero everywhere on the line $x_1 = -x_2$. Obviously

$$V(\mathbf{x}) = x_1 + x_2 \tag{14.54}$$

is indefinite.

The so called quadratic form scalar function given by

$$Q(\mathbf{x}) = \mathbf{x}'\mathbf{A}\mathbf{x} \qquad \mathbf{A} = \begin{bmatrix} a_{11} & a_{12} & \cdots & a_{1n} \\ a_{21} & a_{22} & \cdots & a_{2n} \\ \cdots & \cdots & \cdots & \cdots \\ a_{n1} & a_{n2} & \cdots & a_{nn} \end{bmatrix} \tag{14.55}$$

which may alternately be written as

$$Q(\mathbf{x}) = \sum_{i=1}^{n} \sum_{j=1}^{n} a_{ij} x_i x_j \tag{14.56}$$

is often encountered. The sign-definiteness of $Q(\mathbf{x})$ depends only upon $\mathbf{A}$, and this leads to the application of our sign-definite definitions to $\mathbf{A}$. We thus speak, for example, of the positive semi-definite matrix $\mathbf{A}$, thereby implying interest in the quadratic form with which $\mathbf{A}$ is associated.

Note that *any* scalar function of $\mathbf{x}$ which contains *only* terms of second order in $\mathbf{x}$ can be expressed in quadratic form. In fact an infinity of equivalent representations is possible. Consider for example

$$Q(\mathbf{x}) = x_1^2 + 2x_2^2 + 3x_3^2 + 4x_1x_2 + 2x_1x_3 \tag{14.57}$$

Two equivalent choices for $\mathbf{A}$ are

$$\mathbf{A} = \begin{bmatrix} 1 & 2 & 1 \\ 2 & 2 & 0 \\ 1 & 0 & 3 \end{bmatrix} \qquad \mathbf{A} = \begin{bmatrix} 1 & 3 & 0 \\ 1 & 2 & -1 \\ 2 & 1 & 3 \end{bmatrix} \tag{14.58}$$

Only the principle diagonal terms in $\mathbf{A}$ are explicitly determined from Eq. (14.57). Although the sums $a_{ij} + a_{ji}$ are fixed by the coefficients in Eq. (14.57), the off-diagonal terms $(i \neq j)$ can be chosen in infinitely many ways without changing $Q(\mathbf{x})$. It is thus reasonable that we should choose the off-diagonal terms in quadratic form matrices for convenience. We assume, hereafter, that all such matrices are symmetrical unless specified to the contrary. This simplifies the determination of sign definiteness for quadratic forms. *Sylvester's theorem* relates directly to quadratic forms.

Theorem 14.1 Sylvester's Theorem

A necessary and sufficient condition for the positive definiteness of the quadratic form $Q(\mathbf{x}) = \mathbf{x}'\mathbf{A}\mathbf{x}$, $\mathbf{A}$ assumed symmetric, is that each of the following determinants derived from $\mathbf{A}$ be positive:

$$\det[a_{11}]; \quad \det\begin{bmatrix} a_{11} & a_{12} \\ a_{21} & a_{22} \end{bmatrix}; \quad \det\begin{bmatrix} a_{11} & a_{12} & a_{13} \\ a_{21} & a_{22} & a_{23} \\ a_{31} & a_{32} & a_{33} \end{bmatrix}; \ldots; \quad \det[\mathbf{A}] \tag{14.59}$$

If all of these determinants are greater than zero except for one or more which equal zero, $Q(\mathbf{x})$ is positive semi-definite. $\mathbf{A}$ is negative definite or negative semi-definite if $-\mathbf{A}$ is positive definite or positive semi-definite. Obviously the definiteness of $Q(\mathbf{x})$ is global.

Example 14.6

Identify the sign definiteness of the matrix

$$\mathbf{A} = \begin{bmatrix} 2 & 1 & 0 \\ 1 & 3 & 2 \\ 0 & 2 & 3 \end{bmatrix} \tag{14.60}$$

Applying Sylvester's theorem provides

$$\det[2] = 2 \qquad \det\begin{bmatrix} 2 & 1 \\ 1 & 3 \end{bmatrix} = 5 \qquad \det\begin{bmatrix} 2 & 1 & 0 \\ 1 & 3 & 2 \\ 0 & 2 & 3 \end{bmatrix} = 7 \tag{14.61}$$

Thus $\mathbf{A}$ is positive definite.

Example 14.7

Identify the sign definiteness of

$$\mathbf{A} = \begin{bmatrix} 2 & 4 & 0 \\ 0 & 2 & 2 \\ 2 & 0 & 3 \end{bmatrix} \tag{14.62}$$

Before applying Sylvester's theorem we must determine the equivalent symmetrical form of $\mathbf{A}$ as

$$\mathbf{A}_s = \begin{bmatrix} 2 & 2 & 1 \\ 2 & 2 & 1 \\ 1 & 1 & 3 \end{bmatrix} \tag{14.63}$$

The required determinants are

$$\det[2] = 2 \qquad \det\begin{bmatrix} 2 & 2 \\ 2 & 2 \end{bmatrix} = 0 \qquad \det[\mathbf{A}_s] = 0 \tag{14.64}$$

Thus $\mathbf{A}$ as given by Eq. (14.62) is positive semi-definite.

Example 14.8

Identify the sign definiteness of the matrix

$$\mathbf{A} = \begin{bmatrix} -1 & 1 & 0 \\ 1 & -4 & 0 \\ 0 & 0 & -1 \end{bmatrix} \tag{14.65}$$

It is obvious by inspection that this $\mathbf{A}$ cannot be positive definite or positive semi-definite, so we focus our attention upon $-\mathbf{A}$ and obtain the determinants

$$\det[1] = 1 \qquad \det\begin{bmatrix} 1 & -1 \\ -1 & 4 \end{bmatrix} = 3 \qquad \det[-\mathbf{A}] = 3 \tag{14.66}$$

Thus $\mathbf{A}$ as given by Eq. (14.65) is negative definite.

The following definitions will prove useful in considering Liapunov's fundamental theorems.

Definition 4 Stability in the sense of Liapunov

Let $S(\alpha)$ be the open spherical region of radius α about the state-space origin. The origin is *stable in the sense of Liapunov* if for every $S(R)$ there exists an $S(r)$ such that all solutions starting in $S(r)$ remain within $S(R)$ as $t \to \infty$.

Note that Liapunov stability allows limit cycles and solutions which grow temporarily, assuming r and R are appropriately chosen, although it rules out solutions which grow without bound.

Definition 5 Liapunov function

A function $V(\mathbf{x}, t)$, positive definite in a region $S(R)$ about the state-space origin, is called a *Liapunov function* or a *V function* for the system described by the state equations $\dot{\mathbf{x}} = \mathbf{f}(\mathbf{x}, t)$ if $\dot{V} \leq 0$ for all $\mathbf{x}$ contained in $S(R)$ and for all $t \geq a$, where $\dot{V}$ is the total time derivative defined as

$$\dot{V} = \frac{\partial V}{\partial t} + \frac{\partial V}{\partial x_1}\dot{x}_1 + \cdots + \frac{\partial V}{\partial x_n}\dot{x}_n \tag{14.67}$$

Thus a V function is a positive definite scalar function of system state which in some sense measures the distance from the origin. Since its derivative must be negative semi-definite for $t \geq a$, the distance to system state as measured by V cannot increase on the time interval $[a, \infty]$. This leads directly to the following stability theorem.

Theorem 14.2 Stability

If a V function can be defined in a neighborhood $S(R)$ of the state-space origin, the origin is stable in the Liapunov sense.

The validity of Theorem 14.2 is obvious. The following theorem relating to asymptotic stability is somewhat stronger.

Theorem 14.3 Asymptotic Stability

If a Liapunov function $V(\mathbf{x}, t)$ can be defined in some neighborhood $S(R)$ of the state-space origin such that $-\dot{V}$ is positive definite and $V(\mathbf{x}, t) \leq U(\mathbf{x})$ for some positive definite $U(\mathbf{x})$, the origin is asymptotically stable.

Note that the second condition is only necessary for V functions depending explicitly upon t. It rules out all functions which grow without bound as $t \to \infty$ for fixed $\mathbf{x}$, such as $(x_1^2 + x_2^2)\epsilon^t$. A detailed proof is straightforward. Since $-\dot{V}$ is positive definite, $V \to 0$ as $t \to \infty$, but V is only zero at $\mathbf{x} = 0$, so it is necessary that $\mathbf{x} \to 0$ as $t \to \infty$. It is only necessary to structure these statements rigorously.

The conditions of Theorem 14.3 are excessively restrictive. Suppose, for example, we try to apply it to the homogeneous performance of an underdamped series RLC network using stored energy as a V function. Since energy dissipation in R goes to zero each time $i(t)$ passes through zero, this $-\dot{V}$ is not positive definite, and the theorem does not apply directly. Obviously we can neglect the countable number of points at which $-\dot{V}$ is zero, and this observation leads to the following more powerful theorem.

Theorem 14.4 Global Asymptotic Stability

Consider the homogeneous time invariant DE

$$\dot{\mathbf{x}} = \mathbf{f}(\mathbf{x}) \tag{14.68}$$

and the continuously differentiable function $V(\mathbf{x})$. If V has the properties: 1. $V(\mathbf{x}) \geq 0$ for all $\mathbf{x}$; 2. $V(\mathbf{x}) \to \infty$ as $\|\mathbf{x}\| \to \infty$; 3. $-\dot{V}(\mathbf{x}) \geq 0$ for all $\mathbf{x}$, and 4. $\dot{V}(\mathbf{x}) = 0$ only for $\mathbf{x} \epsilon S$, where the set S contains no complete trajectories other than the singular points of Eq. (14.68), then all solutions approach a singular point as $t \to \infty$. Furthermore any singular point which is an isolated local minimum of $V(\mathbf{x})$ is asymptotically stable.

A detailed proof of this theorem is provided by La Salle and Lefschetz,[17] and is somewhat involved. Note that $V(\mathbf{x})$ and $-\dot{V}(\mathbf{x})$ need only be positive semidefinite. Condition 4 allows points where $\dot{V}(\mathbf{x}) = 0$, as encountered in our RLC example, on the solution, but rules out the solution "lingering" at such points unless they are singular points. Thus the solution must approach a point where $V = 0$ as $t \to \infty$, but there may be temporary interruptions in the reduction of V. This rules out limit cycles, for example. Note that condition 2 eliminates all unstable solutions, or solutions which grow without bound. Theorem 14.4 is a powerful one which often proves useful. It is important to note that these stability theorems provide only sufficient stability conditions. Just because we are unable in a particular application to find a V function that works does *not* indicate system instability. It means only that we have failed to prove stability. This leads us to present an instability theorem.

Theorem 14.5 Instability

Consider $V(\mathbf{x})$ defined in a neighborhood $S(r)$ of the origin with $V(\mathbf{0}) = 0$. If V is positive definite in S and if $\dot{V}$ assumes positive values arbitrarily close to $\mathbf{x} = \mathbf{0}$, the state-space origin is unstable.

Many alternate forms of this theorem are possible. Its validity is obvious.

Application of Theorem 14.4 can essentially be reduced to a search for an acceptable V function. Although standard forms have been developed for problems with certain specific structures, and the so-called *variable gradient method* is available[18] for synthesizing V functions, the choice of an appropriate form is primarily a trial and error process in which experience is extremely valuable. For linear systems, $V(\mathbf{x})$ may always be taken as system energy and this choice leads to stability conclusions identical to those provided by the Routh, Hurwitz, or Nyquist criteria.[19]

Example 14.9

$$L\frac{d^2q}{dt^2} + R\frac{dq}{dt} + \frac{q}{C} = 0 \tag{14.69}$$

Choosing stored energy for V yields

$$V = \frac{1}{2}L\left(\frac{dq}{dt}\right)^2 + \frac{1}{2}C\left(\frac{q}{C}\right)^2 \tag{14.70}$$

Reducing Eq. (14.69) to a set of first-order state equations and expressing Eq. (14.70) as a function of state vector $\mathbf{x}$ provides the definitions

$$x_1 = q \qquad x_2 = \dot{q} \tag{14.71}$$

the state equations

$$\dot{x}_1 = x_2 \qquad \dot{x}_2 = -\frac{1}{LC}x_1 - \frac{R}{L}x_2 \tag{14.72}$$

and

$$V(\mathbf{x}) = \frac{1}{2}\left(Lx_2^2 + \frac{x_1^2}{C}\right) \tag{14.73}$$

This $V(\mathbf{x})$ is obviously positive definite and its time derivative is

$$\dot{V}(\mathbf{x}) = Lx_2\dot{x}_2 + \frac{x_1\dot{x}_1}{C} = -Rx_2^2 \tag{14.74}$$

which should be obvious, since Ri^2 is the power dissipation in R. It is clear from Eq. (14.74) that $\dot{V}$ is negative semi-definite as long as $R > 0$, so the origin is globally asymptotically stable by direct application of Theorem 14.4.

Application of Liapunov Theory to linear systems is certainly unnecessary, since there are many alternate ways to test linear system stability, most of which are easier to apply. Linear systems are either unstable, have poles on the $j\omega$ axis, or are globally asymptotically stable, so the number of possible situations in the linear case is severely limited. The following NL example illustrates the more general power of Liapunov theory.

Example 14.10

Consider the NL controller described by

$$\dot{x}_1 = x_2 \qquad \dot{x}_2 = -x_1^3 - x_2 \tag{14.75}$$

and shown in Fig. 14.28. Prove that this system is globally asymptotically stable using a V function of the form

$$V = \alpha x_1^4 + \beta x_1^2 + x_1x_2 + x_2^2 \tag{14.76}$$

What values of α and β are appropriate?

We must first establish limits upon α and β to assure positive definite V. Obviously $\alpha > 0$ and $\beta > 0$ are required. When x_1 and x_2 are very small, αx_1^4 is negligible compared to βx_1^2, and application of Sylvester's theorem to the last three terms of Eq. (14.76) shows that $\beta > 1/4$ is necessary. Evaluating $\dot{V}$ and simplifying provides

$$\dot{V} = (4\alpha - 2)x_1^3x_2 + (2\beta - 1)x_1x_2 - x_1^4 - x_2^2 \tag{14.77}$$

from which it is apparent that $\alpha = \beta = 1/2$ makes $\dot{V}$ negative definite. Since V is positive definite for this choice, the system is globally asymptotically stable.

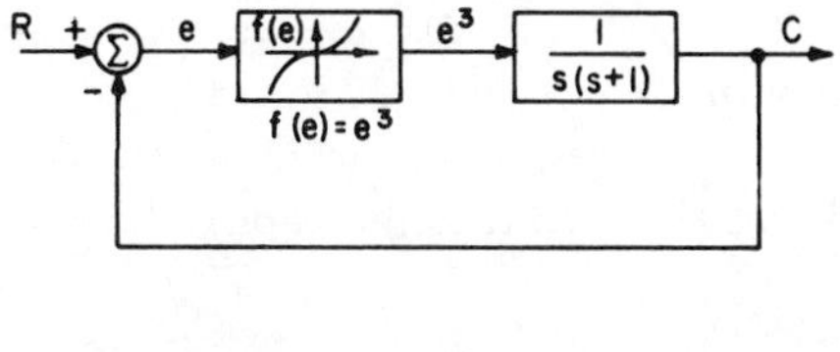

FIG. 14.28. Nonlinear system used in Ex. 14.10.

The reader might certainly ask how we arrived at a V function of the form given in Eq. (14.76). Trial and error with progressively more complex power series representations will soon show that something of this general form is needed. The search for acceptable Liapunov functions is rather disorganized, and trial and error, although not particularly satisfying, is often used. The variable gradient method[18] may sometimes be used to organize the search, but it generally involves computer use, and, if a computer study is to be undertaken, the stability question is easily answered more directly. The reader should observe that global asymptotic stability of the system used in Ex. 14.5 can also be predicted using DF analysis.

14.6 THE CRITICAL DISC AND POPOV STABILITY CRITERIA

The Liapunov stability theorems are fundamentally analysis tools, and are of limited use in developing synthesis procedures. Synthesis based upon a DF analysis of the NL element is relatively straightforward, requiring frequency response shaping of the linear transfer function to avoid complex plane intersections between $\mathbf{G}$ and $-1/\mathbf{N}_e$. Two fundamental theorems leading to similar synthesis methods, each with a simple graphical complex plane interpretation, can be proved, and they will often be found useful in both the analysis and synthesis of NL systems.

We define system stability as assured by the two theorems presented in this section as follows:

Definition 6 Stability

Given an input $r(t)$ such that

$$\int_0^\infty r^2(t)dt < \infty \tag{14.78}$$

a system will be called *stable* if and only if

$$\int_0^\infty c^2(t)dt < \infty \tag{14.79}$$

where $c(t)$ is the system response to $r(t)$.

Although this definition is similar in its implications to previous ones, it is the first general interpretation we have presented in terms of system response to the class of bounded integral square inputs.

Our interest is centered on the class of systems illustrated in Fig. 14.29, where $G(s)$ denotes the Laplace transform of an arbitrary linear subsystem, and

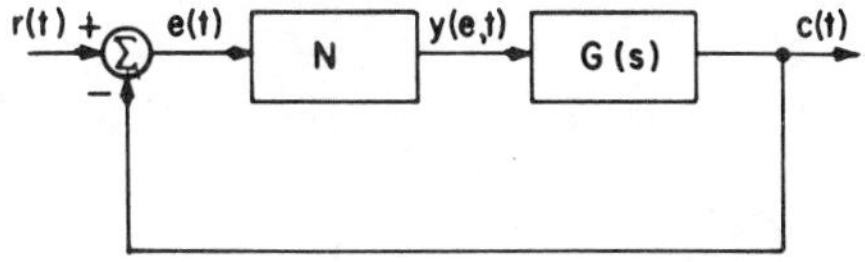

FIG. 14.29. General form of NL system considered in Sec. 14.6.

N is a NL time-varying memoryless element which satisfies the sector criterion illustrated in Fig. 14.30. The plot of $y(e, t)$ versus e must lie for all t between the two straight lines with slopes α and β shown in Fig. 14.30. It must in addition satisfy the Lipschitz condition

$$|y(e_1, t) - y(e_2, t)| \leq \gamma\,|e_1 - e_2| \qquad \gamma < \infty \tag{14.80}$$

which is, for example, automatically satisfied if y is continuous and $|dy/de| < \infty$. Under these conditions, which are not overly restrictive, we may state the following theorem.

Theorem 14.6 The Critical Region Criterion[20,21]

A system of the form shown in Fig. 14.29 with conditions on N and G as stated above is stable if

1. $-\infty < \alpha \leq y(e, t)/e \leq \beta < \infty$.
2. a. For $0 < \alpha \leq \beta < \infty$, the polar plot of $\mathbf{G}(j\omega)$ does not encircle, intersect, or touch the crosshatched disc shown in Fig. 14.31a. b. For $0 = \alpha < \beta < \infty$, the polar plot of $\mathbf{G}(j\omega)$ must lie to the right of the crosshatched area shown in Fig. 14.31b. c. For $-\infty < \alpha \leq \beta < \infty$, the polar plot of $\mathbf{G}(j\omega)$ must lie *within* the crosshatched disc shown in Fig. 14.31c.

This theorem provides a simple stability test for a wide variety of systems. It is often possible to partition our systems into the form required. Application of that part of Theorem 14.6 which applies to the problem of interest is straightforward, and stability assurance is obtained in a manner entirely analogous to the Nyquist procedure. When N is time invariant the following theorem, generally

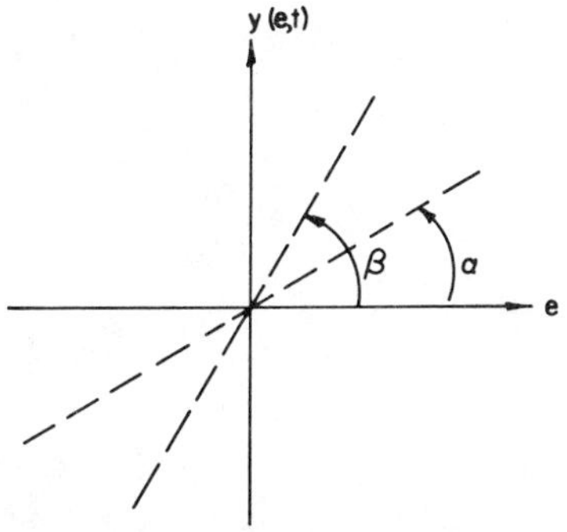

FIG. 14.30. The general sector of interest for the characteristic N.

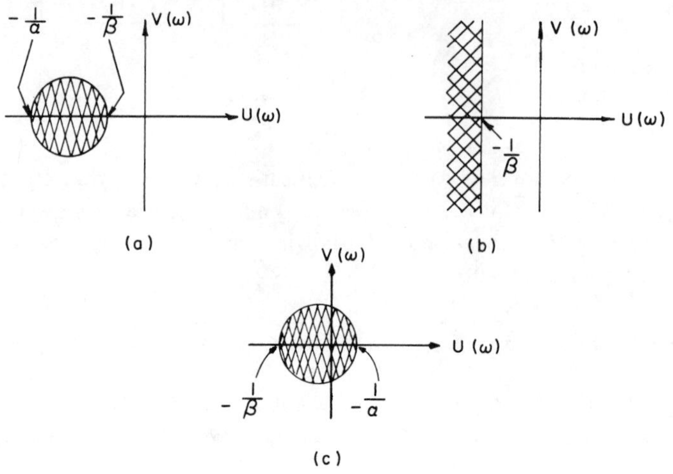

FIG. 14.31. Areas of interest in Theorem 14.6.

called the *Popov criterion* and the first of the many geometrical methods to be developed, also applies.

Theorem 14.7 The Popov Criterion[22]

A system of the form shown in Fig. 14.31 is stable if

1. N is a time invariant NL element satisfying the sector condition of Fig. 14.30.
2. $0 \leq \alpha < N(e)/e \leq \beta < \infty$.
3. a. For $\alpha > 0$, the polar plot of the modified linear system function $U(\omega) + j\omega V(\omega)$, where

$$G(j\omega) = U(\omega) + jV(\omega) \tag{14.81}$$

lies entirely to the right of a straight line with arbitrary positive slope passing through the real axis at point $-1/\beta$, as shown in Fig. 14.32. b. For $\alpha = 0$ condition (a) must be satisfied and $\int_0^\infty |g(\tau)|\, d\tau < \infty$ is also necessary.

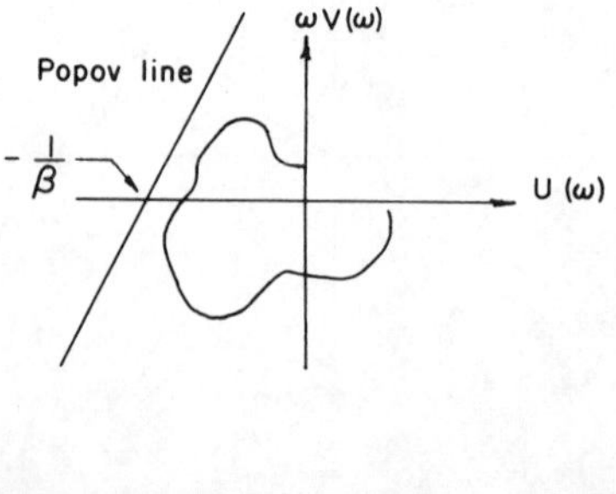

FIG. 14.32. Use of the Popov criterion.

The two theorems presented here are the most useful of a large group developed in recent years and are used to illustrate the rigorous methods available for NL system analysis and synthesis. The proofs are rather involved, and thus not included. Applications are obvious. Both theorems provide sufficient, but not necessary conditions for stability as given by Def. 5. Thus a system may fail all of these tests and still be stable, perhaps even exhibiting a reasonable stability margin. This is frustrating, but must be accepted. Although some instability theorems are available, they are generally far less useful than the stability theorems chosen for discussion.

The results presented here are directly applicable to design. Increasing the margin by which the critical regions are avoided will in general improve the transient performance of NL systems, just as in application of the Nyquist criterion, although the avoidance margin cannot be rigorously related to stability margin in the NL case. Increasing the avoidance margin certainly improves the degree of comfort with which a final design may be submitted, no matter how much the interpretation lacks rigor.

14.7 CONCLUDING COMMENTS

A sampling of material applicable to NL system analysis and design and thought by the author to be representative of the methods available, appropriately straightforward in concept, and of a reasonably useful nature, is presented in this chapter. It is apparent from the many references and the characteristics of some of the results, that we have only scratched the surface of the NL analysis area, intending thereby to provide a useful introduction to the subject, some indication of what has been done, and a feeling for how it relates to the linear methods discussed earlier. The student or practicing engineer interested in NL systems as an area of speciality has much work yet to do,[23] but the area is a challenging one and the potential rewards for further study are great.

It is difficult to choose one of the procedures presented as "most useful." The DF, critical region criterion, and Popov criterion are most directly applicable to synthesis, but the phase plane is important as a data presentation method and it also leads to simple graphical solution techniques. Liapunov theory is a powerful analysis tool which has been applied to design in some cases. All of the methods discussed are of great theoretical interest.

Design of NL systems has historically been accomplished by trial and error, usually on a simulation of some sort, and this method still has great merit despite its lack of rigor. It is impossible to generalize results obtained for NL situations to nearly the extent possible in linear analysis, and NL analysis and synthesis generally require designer ingenuity far beyond that needed in the linear case, which can usually be reduced to a routine procedure. Computer study, together with the various tools presented here, should provide acceptable solutions for most NL problems.

REFERENCES

1. Kochenburger, R. J., A Frequency Response Method for Analyzing and Synthesizing Contractor Servomechanisms, *Transactions AIEE,* vol. 69, Pt. I, 1950.
2. Goldfarb, L. C., On Some Nonlinear Phenomena in Regulatory Systems, Translation of Papers on Stability of Nonlinear Feedback Control Systems, National Bureau of Standards Report, 1691, Washington, D.C., 1952.
3. Oppelt, W., Locus Curve Method for Regulators with Friction, ibid.
4. Tustin, A., The Effects of Backlash and of Speed Development Friction on the Stability of Closed Cycle Control Systems, *Journal of the Institute of Electrical Engineering* (London), vol. 94, p. II A, 1947.
5. Dutilh, J. R., Théorie des Servomécanismes à Relais (in French), *L'Onde Electrique,* vol. 30, 1950.
6. Kryloff, N. and N. Bogoliuboff, "Introduction to Nonlinear Mechanics," translated by S. Lefschetz, Princeton University Press, Princeton, N. J., 1943.
7. Sridhar, R., A General Method for Deriving the Describing Functions for a Certain Class of Nonlinearities, *IRE Transactions on Automatic Control,* vol. AC-5, pp. 135-141, 1960.
8. Gibson, J. E., "Nonlinear Automatic Control." McGraw-Hill, New York, pp. 355-365, 1963.
9. Eveleigh, V. W., "Adaptive Control and Optimization Techniques," McGraw-Hill, New York, p. 387, 1967.
10. Johnson, E. C., Sinusoidal Analysis of Feedback Control Systems Containing Nonlinear Elements, *Transactions AIEE,* vol. 71, part II, 169-181, 1952.
11. Truxal, J. G., "Automatic Feedback Control System Synthesis," McGraw-Hill, New York, pp. 598-612, 1955.
12. Booton, R. C., Jr., The Analysis of Nonlinear Control Systems with Random Inputs, *Proceedings of the Symposium on Nonlinear Circuit Analysis,* vol. II, pp. 369-391, Brooklyn Polytechnic Inst., 1953.
13. Cunningham, W. J., "Introduction to Nonlinear Analysis," McGraw-Hill, New York, Chap. 4, 1958.
14. Andronow, A. A., and C. E. Chaikin, "Theory of Oscillations," Translation edited by S. Lefschetz, Princeton Press, Princeton, N. J., p. 210, 1949.
15. Bendixson, T., "Sur les courbes définies par des equations différentielles," Acta Mathematica, vol. 24, 1901.
16. Liapunov, A. M., "On the General Problem of Stability of Motion," Ph.D. Thesis, Kharkov, 1892.
17. LaSalle, J., and S. Lefschetz, "Stability by Liapunov's Direct Method," Academic Press, New York, 1961.
18. Schultz, D. G., "The Generation of Liapunov Functions," in *Advances in Control Systems,* C. T. Leondes, Editor, vol. 2, Academic Press, New York, 1965.
19. Schultz, D. G. and J. L. Melsa, "State Functions and Linear Control Systems," McGraw-Hill, New York, Chap. 5, 1967.
20. Sandberg, I. W., Some Results on the Theory of Physical Systems Governed by Nonlinear Functional Equations, *Bell System Technical Journal,* vol. 44, p. 871, May-June, 1965.
21. Sandberg, I. W., A Frequency Domain Condition for the Stability of Feedback Systems Containing a Single Time-Varying Nonlinear Element, *Bell System Technical Journal,* vol. 43, p. 1601, July, 1964.
22. Aizerman, M. A., and F. R. Gantmacher, "Absolute Stability of Regulator Systems," Translated by E. Polak, Holden-Day, San Francisco, 1964.
23. Brockett, R. W., The Status of Stability Theory for Deterministic Systems, *IEEE Transactions on Automatic Control,* vol. AC-11, pp. 596-606, July, 1966.

15 SAMPLED DATA SYSTEMS

The systems discussed in the preceding chapters are exclusively continuous, or analog, in nature. In many applications one or more of the signals encountered are in discrete or sampled form. Radar, sonar and time multi-plexed communication systems provide interesting examples. Some highly precise transducers for measuring position, velocity or acceleration provide digital output data. Actuators which change angular or linear position in discrete steps are occasionally used. Undoubtedly the most significant reason for studying sampled data (SD) theory, however, is the modern tendency toward the use of digital signal processing in control systems. Digital equalization allows extensive design flexibility and its use has increased rapidly since 1950.

Our objective in this chapter is to establish the mathematical foundations of SD systems analysis and synthesis. The presentation is introductory in nature. References are provided to assist those readers who wish to emphasize SD systems in their further study. The basic principles of SD system analysis and design are presented and it is shown that a close relationship exists between the methods used in treating continuous and SD systems. Our first step is to establish mathematical models of the sampling and signal reconstruction processes. This allows us to treat the standard analog to digital (A–D) and digital to analog (D–A) conversion processes with some degree of rigor. The Z-transform is

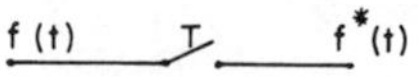

FIG. 15.1. Schematic representation of a sampler.

introduced and its application to SD system analysis is illustrated with examples. Various combinations of sampled and continuous elements in open and closed loop arrangements are considered. The design of SD systems using methods carried over from the continuous system area is considered, thereby establishing a close relationship to analog system design.

15.1 THE SAMPLING PROCESS

A mathematical description of the sampling process is imperative if we are to develop a reasonable foundation for the analysis and design of SD systems. The schematic representation of a sampler is shown in Fig. 15.1. The analog input $f(t)$ is sampled every T seconds by closing the switch for a short interval. The resulting output $f^*(t)$ consists of a sequence of narrow pulses with amplitudes $f(T)$, $f(2T)$, . . . , $f(nT)$ (disregarding the small change in amplitude while the sampling switch is closed), each having duration equal to the interval over which the switch is closed. The * notation used here to denote sampled functions is relatively standard. A typical time function $f(t)$ and the sampled form of that signal, $f^*(t)$, are shown in Fig. 15.2. The *sampling width,* denoted by γ, is the length of time that the sampling switch is closed. The period between samples, denoted by T, is the *sampling interval,* and time points $0, T, 2T, \ldots, nT$ are the *sampling instants.* In most cases the sampling width is sufficiently small that changes in $f(t)$ during the sampling width are negligible, and it is standard practice to assume zero-width, or impulse, sampling in most applications to simplify the mathematical description. It is obvious from Fig. 15.2 that information is lost in the sampling process, although the loss can be kept small by sampling $f(t)$ sufficiently fast. The information loss consideration is a dominant factor in choosing the sampling rate.

The sampling process may be represented as multiplication of $f(t)$ by the pulse sequence sampling function $p(t)$ shown in Fig. 15.3. Thus sampling is a modulation process whereby the "carrier" $p(t)$ is modulated by the signal $f(t)$ to produce an amplitude modulated pulse train of the form shown in Fig. 15.2 and

$$f^*(t) = p(t)f(t) \tag{15.1}$$

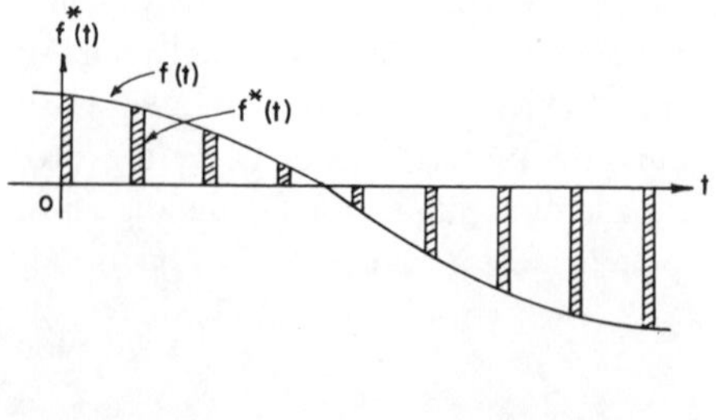

FIG. 15.2. Relationship between $f^*(t)$ and $f(t)$ for a typical case.

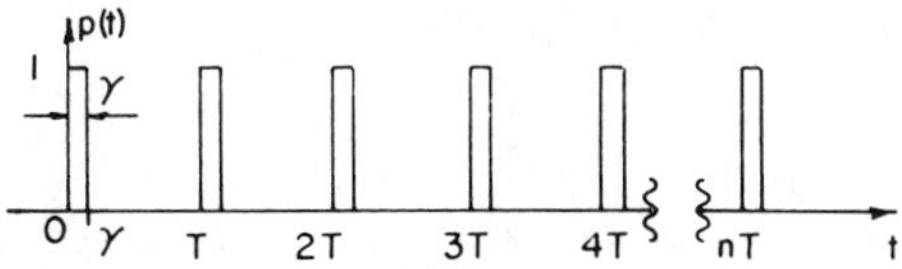

FIG. 15.3. Pulse train representation of the sampling process.

Since $p(t)$ is periodic with period T, it may be expanded in an exponential form of series as

$$p(t) = \sum_{k=-\infty}^{\infty} C_k \epsilon^{j2\pi kt/T} \tag{15.2}$$

where the C_k's are defined by

$$C_k = \frac{1}{T}\int_{-T/2}^{T/2} p(t)\,\epsilon^{-j2\pi kt/T}\,dt \tag{15.3}$$

The C_k's are thus known functions of T and γ for any specific sampling rate and sampling width. Substituting $p(t)$ as given by Eq. (15.2) into Eq. (15.1) provides

$$f^*(t) = \sum_{k=-\infty}^{\infty} C_k f(t)\,\epsilon^{j2\pi kt/T} \tag{15.4}$$

The shifting theorem is

$$\mathfrak{F}[\epsilon^{\lambda t} f(t)] = \mathbf{F}(j\omega - \lambda) \tag{15.5}$$

where $\mathfrak{F}[g(t)]$ denotes the Fourier transform of $g(t)$. Applying Eq. (15.5) to the terms in Eq. (15.4) yields the Fourier transform of $f^*(t)$ as

$$\mathbf{F}^*(j\omega) = \sum_{k=-\infty}^{\infty} C_k \mathbf{F}[j(\omega - k\omega_s)] \tag{15.6}$$

where

$$\omega_s = \frac{2\pi}{T} \tag{15.7}$$

denotes the radian sampling frequency. It is apparent from Eq. (15.6) that the

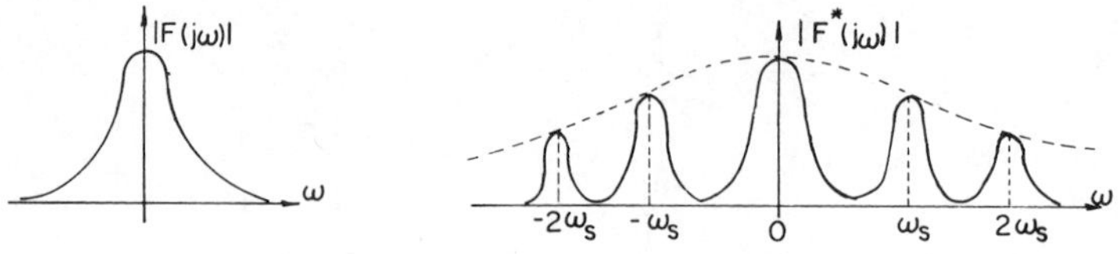

a. Spectrum of f(t). b. Spectrum of f* (t)

FIG. 15.4. Effect of sampling upon the signal spectrum.

sampled output consists of an infinite summation of weighted signal spectra each identical in *shape* to the $f(t)$ spectrum, although multiplied by C_k and shifted $k\omega_s$ rad/sec from the original. This effect is illustrated in Fig. 15.4. The sampling process introduces an infinity of extraneous spectral sidelobes, and some means must be found to extract a reasonable approximation of the $f(t)$ spectrum (the center lobe) from the sampled output if SD systems are to prove useful.

The central signal spectrum shown in Fig. 15.4b can be recovered with little distortion if:

1. The input $f(t)$ has negligible frequency components above some frequency denoted by W_s.
2. The sampling frequency is chosen significantly greater than $2W_s$.
3. A baseband filter which passes all frequencies in the central spectrum without introducing significant amplitude or phase distortion and rejects all spurious components can be realized.

Of course these conditions are never completely met, but they are often approximately satisfied. A frequency can be chosen above which the input signal components are of little interest, the sampling rate set at several times that frequency, and a compromise reconstruction filter designed to pass the information frequencies within acceptable distortion levels and severely attenuate the spurious components.

It is important to interpret *Shannon's sampling theorem*[1] as it relates to SD system performance.

Theorem 15.1

Given a band limited signal $f(t)$ containing no frequency components beyond W rad/sec, $f(t)$ may be recovered completely from the infinite sequence of impulse samples $f^*(t)$ separated by time intervals no greater than $\frac{1}{2}(2\pi/W)$ sec.

Thus any signal may be completely recovered from the sampled form, at least in theory and under ideal conditions, if it is impulse sampled at a frequency at least twice its highest frequency component. It is not feasible in practice to approach the "Shannon limit." Limiting performance could only be achieved by operating upon the *entire* sequence of samples with an *ideal* reconstruction filter. We are usually interested in *real time reconstruction* with little delay (only the past samples are available) using relatively simple filters. Sampling rates ten or more times the theoretical limit are common.

If the sampling frequency is chosen a factor of ten or so above the theoretical limit, the system's transient response period is *much* longer than T. If γ is a small percentage of T, as usually the case, sampling may be assumed to occur instantaneously (impulse sampling) with little error, since then the responses of the system to an impulse and a γ width pulse of equal area are nearly identical under these conditions. The impulse sampling approximation is standard in SD analysis. In the impulse sampled case

$$p(t) = \sum_{n=-\infty}^{\infty} \delta(t - nT) \tag{15.8}$$

where $\delta(t - nT)$ is the unit impulse occurring at $t = nT$. The output of the impulse sampler (impulse modulator) becomes

$$f^*(t) = \sum_{n=-\infty}^{\infty} f(t)\,\delta(t - nT) = \sum_{n=-\infty}^{\infty} f(nT)\,\delta(t - nT) \tag{15.9}$$

Thus $f^*(t)$ is a sequence of impulses, each with area equal to the sampled signal level at the sampling time. Evaluating the Fourier coefficients to obtain the frequency domain representation of $f^*(t)$ provides

$$C_k = \frac{1}{T}\int_{-T/2}^{T/2} \sum_{k=-\infty}^{\infty} \delta(t - nT)\,\epsilon^{-j(2\pi kt/T)}\,dt = \frac{1}{T} \tag{15.10}$$

so

$$\mathbf{F}^*(j\omega) = \frac{1}{T}\sum_{k=-\infty}^{\infty} \mathbf{F}[j(\omega - k\omega_s)] \tag{15.11}$$

All of the spectral lobes are equally weighted by the factor $1/T$, and the sampled signal power is infinite, which we know is impossible. This rather disturbing characteristic results because impulse sampling requires infinitely fast switching plus infinitely large weighting, both of which are impossible. Since we are primarily concerned with the central spectral lobe in most applications, the impulse sampling approximation is reasonable. When the sampling and signal reconstruction operations are treated together, as we shall do subsequently, the impulse sampling approximation is not at all unrealistic.

We are often interested in the response of SD systems to Laplace transformable signals present only for $t \geq 0$. In this case we have

$$f^*(t) = \sum_{n=0}^{\infty} f(nT)\,\delta(t - nT) \tag{15.12}$$

which can be Laplace transformed term by term to obtain

$$F^*(s) = \sum_{n=0}^{\infty} f(nT)\,\epsilon^{-nTs} \tag{15.13}$$

Note that the ϵ^{-nTs} term in Eq. (15.13) may be treated as an ordering variable which indicates where the nth sample appears in time.

Example 15.1

Find a closed form expression for $F^*(s)$ when

$$f(t) = \epsilon^{-\alpha t} \tag{15.14}$$

The corresponding $F^*(s)$ obtained using Eq. (15.13) is

$$F^*(s) = \sum_{n=0}^{\infty} \epsilon^{-\alpha nT}\epsilon^{-nTs} = \sum_{n=0}^{\infty} \epsilon^{-nT(s+\alpha)} \tag{15.15}$$

Observe that

$$\frac{1}{1 - \epsilon^{-T(s+\alpha)}} = 1 + \epsilon^{-T(s+\alpha)} + \epsilon^{-2T(s+\alpha)} + \cdots$$

$$= \sum_{n=0}^{\infty} \epsilon^{-nT(s+\alpha)} \tag{15.16}$$

as easily shown by long division. Thus a closed form expression for $F^*(s)$ is available as

$$F^*(s) = \frac{1}{1 - \epsilon^{-T(s+\alpha)}} \tag{15.17}$$

Similar manipulations will provide a closed form expression for $F^*(s)$ whenever $F(s)$ can be represented by a ratio of finite polynomials in s, as shown in Sec. 15.3.

Note that the development of Eq. (15.17) from Eq. (15.15) is greatly expedited by the impulse sampling approximation. This result would obviously be much more complex if finite duration sampling were used and impossible if the samples were not spaced in time in some orderly fashion. Suppose we alternately start with $f^*(t)$ as given in Eq. (15.4) with the C_k's set equal to $1/T$. Laplace transforming Eq. (15.4) term by term and using the shifting theorem provides

$$F^*(s) = \frac{1}{T} \sum_{n=-\infty}^{\infty} F(s + nj\omega_s) \tag{15.18}$$

where $F(s)$ is the Laplace transform of $f(t)$. Equating $F^*(s)$ as presented in Eqs. (15.13) and (15.18) gives the *Poisson summation rule*

$$\frac{1}{T} \sum_{n=-\infty}^{\infty} F(s + nj\omega_s) = \sum_{n=0}^{\infty} f(nT)\,\epsilon^{-nTs} \tag{15.19}$$

which was discovered by Poisson over 100 years ago. It is not generally possible to obtain a direct closed form equivalent of Eq. (15.18), so $F^*(s)$ as given by Eq. (15.13) is most useful. An SD system analysis and design procedure based upon Eq. (15.18) is discussed, however, in Sec. 15.5.

15.2 SIGNAL RECONSTRUCTION

It is apparent from the sampled signal spectrum shown in Fig. 15.4b that any low pass filter with a sharp cutoff characteristic near the maximum frequency components of $f(t)$ may be used as a smoothing filter to approximately reconstruct $f(t)$ from $f^*(t)$. In some cases the low pass characteristics of the control system plant may be used to provide the primary smoothing effect, although it is far more common to use a reconstruction filter designed specifically for that purpose. The difference between $f(t)$ and the reconstructed signal is called *ripple*, and significant ripple may cause excess wear on system components and reduces operating efficiency, in addition to acting as a source of undesired signal components, or noise. For these reasons the choice of an efficient reconstruction filter is generally assigned a high priority in SD system design. Fortunately several of the standard filter forms are effective and easily realized in hardware. Reconstruction filters are called *data hold* or *data extrapolator* networks, and these terms are used interchangeably throughout our discussion.

The simplest and by far the most common data reconstruction filter is the zero-order hold which converts the sampled waveform into a sequence of steps, as shown in Fig. 15.5. Each step assumes the signal level at the sampling instant and remains at that level for one sampling interval. The zero-order hold unit impulse response, denoted by $g_{0h}(\tau)$, is thus as shown in Fig. 15.6. (A unit impulse results when a unit level signal is sampled.) The corresponding transfer

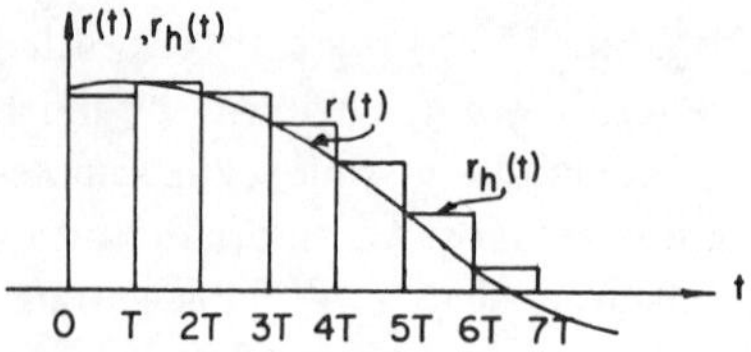

FIG. 15.5. Reconstruction of a waveform using a zero-order hold.

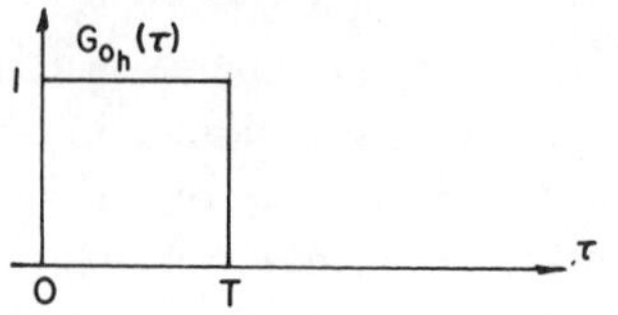

FIG. 15.6. Unit impulse response of a zero-order hold circuit.

function $G_{0b}(s)$ is the Laplace transform of $g_{0b}(\tau)$, or

$$G_{0b}(s) = \frac{1}{s} - \frac{\epsilon^{-Ts}}{s} = \frac{1 - \epsilon^{-Ts}}{s} \tag{15.20}$$

and this representation is generally used in the analysis and synthesis of SD systems. The zero-order hold is easily implemented and the signal reconstruction it provides is often acceptably accurate. It is alternately referred to as a *staircase*, or *boxcar*, filter because of the reconstructed waveform's distinctive shape.

Substituting $j\omega$ for s in Eq. (15.20) and rearranging yields the equivalent phasor form as

$$G_{0b}(j\omega) = T\,\frac{\sin(\omega T/2)}{\omega T/2}\,\epsilon^{-j\omega T/2} \tag{15.21}$$

The corresponding amplitude and phase characteristics are plotted in Fig. 15.7. The amplitude response provides a direct indication of how completely the zero-order hold isolates the main spectral lobe containing the desired signal components. Obviously the frequency response does not compare favorably to that of an ideal sharp cutoff low pass filter, but this is a small price to pay for design simplicity. The sampling frequency is generally chosen sufficiently high that the central lobe containing the desired information falls well within the main signal reconstruction lobe shown in Fig. 15.7, and the side-lobe distortion, although not always negligible, as clear from Fig. 15.5, can usually be held to an acceptable level. The zero-order hold phase characteristics are of particular

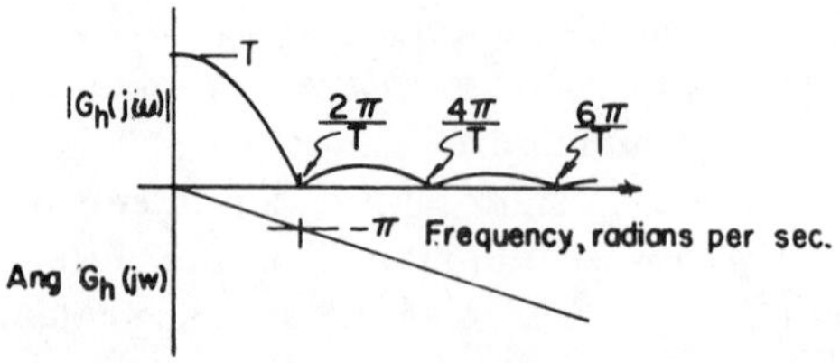

FIG. 15.7. Amplitude and phase versus frequency for a zero-order hold.

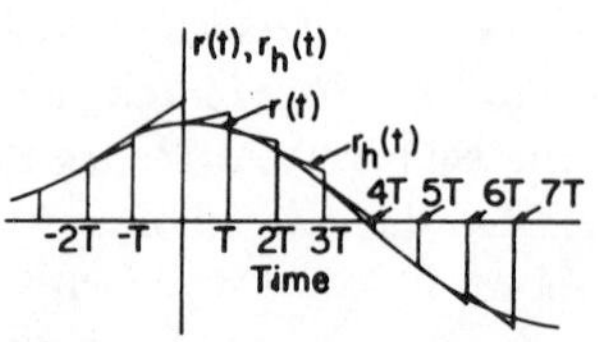

FIG. 15.8. Signal reconstruction using a first-order hold.

interest in determining SD system stability. The network provides a linearly increasing phase lag with frequency corresponding to a perfect time delay of $T/2$ seconds. This $T/2$ second time delay is further illustrated by connecting the center points of the reconstructed staircase waveform shown in Fig. 15.5 to obtain a first order approximation of the smoothed result. Sample and hold phase lag is a major factor in determining SD system stability, and a stable controller may be made unstable when a sampler and zero-order hold are added without changing the gain.

Perhaps the next logical step beyond the zero-order hold is the construction of a hold which extrapolates linearly over each interval using the last two samples to provide

$$r_{1b}(nT + \tau) = r(nT) + \{r(nT) - r[(n - 1)T]\}\frac{\tau}{T} \qquad 0 \leq \tau \leq T \tag{15.22}$$

A network to instrument Eq. (15.22) is called a first-order hold and results in the reconstructed waveform shown in Fig. 15.8. The impulse response of the first-order hold is shown in Fig. 15.9. Note that the output immediately assumes the level of the most recent signal sample (the weight of the most recent impulse) and increases linearly over the interval at a rate equal to $1/T$ times the weight of the most recent input sample minus (due to the negative slope between T and $2T$) the weight of the previous sample. The corresponding $G_{1b}(s)$ is readily evaluated by decomposing $g_{1b}(\tau)$ into its component step and ramp functions to obtain

$$G_{1b}(s) = T(1 + Ts)\left(\frac{1 - \epsilon^{-Ts}}{Ts}\right)^2 \tag{15.23}$$

The first-order hold frequency response is obtained by substituting $s = j\omega$ into Eq. (15.23) and simplifying, which yields

$$G_{1b}(j\omega) = T\sqrt{1 + \omega^2T^2}\left[\frac{\sin(\omega T/2)}{\omega T/2}\right]^2 \epsilon^{-j(\omega T - \tan^{-1}\omega T)} \tag{15.24}$$

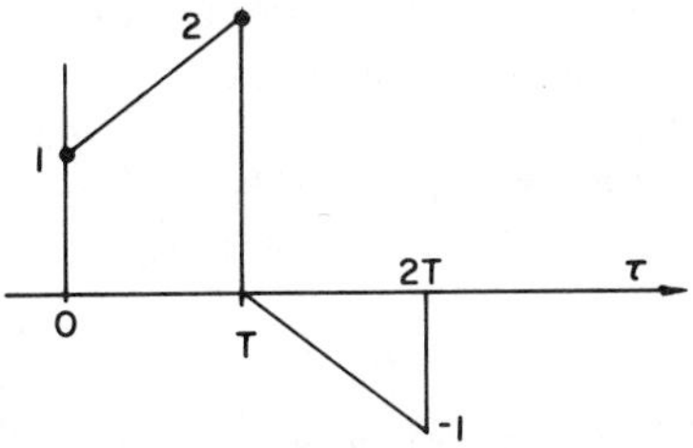

FIG. 15.9. Impulse response of a first-order hold.

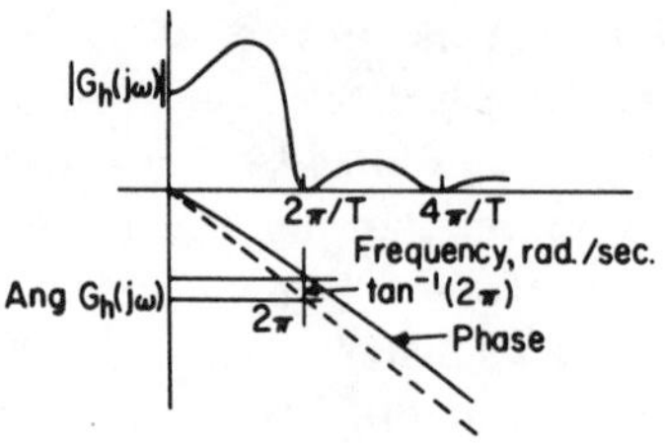

FIG. 15.10. Amplitude and phase versus frequency for a first-order hold.

A plot of magnitude and phase versus ω for the first-order hold is shown in Fig. 15.10. The first-order hold obviously accentuates certain high frequency signal components excessively and introduces significantly more phase lag than the zero-order hold. Since it is also considerably more difficult to realize than a zero-order hold, it is not often used.

Although it is possible to instrument higher order holds, the additional complexity is seldom worth the effort, particularly since older data is used in the reconstruction process, and they introduce progressively larger phase lag than the zero-order hold. The zero-order hold is adequate for most purposes. The first-order hold is introduced primarily for comparison and a bit of added generality.

Realization of the various hold circuits is not particularly difficult. The zero-order hold may be set up as a tracking integrator to which the input is applied for a short time (the observation interval γ) at the start of each sampling interval. The integrator output is driven to the new input level at each sample point, and held there until the next input is applied through appropriate switching circuitry. The higher order holds are realized similarly. Further details are provided in several of the references listed at the end of this chapter.

15.3 *z*-TRANSFORM ANALYSIS

In SD systems we are primarily interested in the system state transition from sample point nT to $(n + 1)T$ as a result of the internal energy conditions present at nT and the known reconstructed signals applied to the various system elements by the one or more hold circuit outputs over the interval $nT \leq t \leq (n + 1)T$. Even a closed loop SD system operates on an *open loop* basis between sampling instants. Equations describing the performance of SD, or discrete, systems are called *difference equations,* and the *z*-transform is applicable to the solution of linear difference equations in much the same way as the Laplace transform is used to solve ordinary differential equations. In fact the *z*-transform is a special interpretation, or a minor modification, of the Laplace transform. Its development and use are discussed in this section.

The Laplace transform of the impulse sampled sequence $f^*(t)$ is given by

$$F^*(s) = \sum_{n=0}^{\infty} f(nT)\,\epsilon^{-nTs} \qquad (15.25)$$

Since this representation of SD signals always contains terms of the form ϵ^{-nTs}, it is convenient to define

$$z = \epsilon^{Ts} \tag{15.26}$$

The sampled sequence may be expressed in terms of z as

$$F^*(s) = F(z) = \sum_{n=0}^{\infty} f(nT)z^{-n} \tag{15.27}$$

where $F(z)$ denotes the *z-transform* of $f^*(t)$. We may think of z as the function of s defined by Eq. (15.26) or as an ordering variable which indicates where the associated signal coefficient $f(nT)$ fits into the impulse sequence.

In the analysis and design of SD systems we will use the z-plane much as the s-plane is used for analog systems, so it is important to indicate the relationships between these two complex planes as defined by the mapping presented in Eq. (15.26). When $s = j\omega$, $z = \epsilon^{j\omega T}$, which is a unit magnitude phasor at angle $\theta = \omega T$. Thus the $j\omega$ axis in the s-plane becomes the unit circle in the z-plane. The mapping is obviously not 1:1, since the unit circle is traversed infinitely many times as $s = j\omega$ progresses from $-j\infty$ to $j\infty$. In fact, as will become clear presently, the entire z-plane is mapped out by the values of s within the horizontal strip bounded by $s = -j\pi$ and $s = j\pi$. Successive strips of width 2π also map into the entire z-plane and this has important implications with respect to information retrieval, as we have already noted. This situation is illustrated by the s-plane and z-plane diagrams shown in Fig. 15.11. The left half of the s-plane maps into the interior of the unit circle in the z-plane and the right half of the s-plane maps into the exterior of the unit circle. The line defined by $\text{Re}(s) = \sigma =$ constant maps into the circle in the z-plane with radius $r = \epsilon^{\sigma T}$. The line $\text{Im}(s) = \omega =$ constant maps into the ray in the z-plane at angle ω radians. These observations should clarify the general mapping characteristics. Stability of SD systems requires that all poles lie in the interior of the z-plane unit circle. This provides a fundamental basis for the z-plane design procedures discussed later.

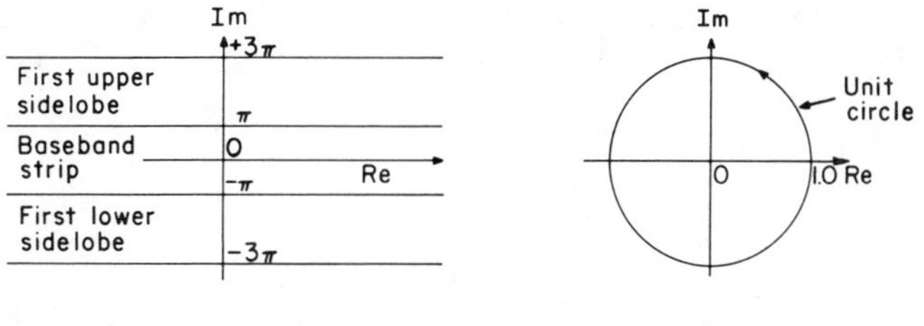

FIG. 15.11. Relationship between the s-plane and the z-plane.

It is interesting to pursue the non-uniqueness of the s-plane to z-plane mapping through the relationship $z = \epsilon^{Ts}$ a bit further. Recall that

$$\sum_{n=0}^{\infty} f(nT)\epsilon^{-nTs} = \frac{1}{T}\sum_{n=-\infty}^{\infty} F(s + jn\omega_s) \tag{15.28}$$

Thus the effect of sampling on the critical frequencies of $F(s)$ causes them to be repeated infinitely many times in the s-plane at intervals of ω_s along the vertical line corresponding to the real part of the original root. Each critical frequency in the baseband strip shown in Fig. 15.11 is mapped into an infinity of critical frequencies by the sampling operation, but, since they all appear on the same vertical line in the s-plane and are separated by $n\omega_s$ units along that line, all of these points map into the same point in the z-plane. Thus the z-plane can be taken as a mapping of the baseband strip in the s-plane with all other strips disregarded for all of our applications, which are low pass in nature. It is important to note also that a carrier signal at frequency 1 mHz, amplitude modulated by a signal with 10 kHz bandwidth, need only be sampled at a 20 kHz rate or higher to preserve the information content as long as the carrier frequency is known, since the strip around $\omega = 2\pi \times 10^6$ can be used as a reference strip just as easily as that around $\omega = 0$. We are not concerned here with such applications of sampling theory, however.

The z-transform of the impulse sampled sequence may be developed mathematically using complex function theory. The impulse sampled sequence is

$$f^*(t) = f(t)\sum_{n=0}^{\infty} \delta(t - nT) = f(t)\,g(t) \tag{15.29}$$

Applying the complex convolution integral to the product in Eq. (15.29) provides

$$\mathcal{L}[f^*(t)] = \frac{1}{2\pi j}\int_{c-j\infty}^{c+j\infty} F(\nu)G(s - \nu)\,d\nu \tag{15.30}$$

where $F(\nu)$ and $G(\nu)$ are the Laplace transforms of $f(t)$ and $g(t)$ as defined in Eq. (15.29). The constant c must be chosen such that all poles of $f(\nu)$ lie to the left of the line $\mathrm{Re}(\nu) = c$. Since $g(t)$ is a sequence of unit impulses, its Laplace transform is

$$G(s) = \sum_{n=0}^{\infty} \epsilon^{-nTs} = \frac{1}{1 - \epsilon^{-Ts}} \tag{15.31}$$

Substituting $G(s - \nu)$ from Eq. (15.31) into Eq. (15.30) provides

$$\mathcal{L}[f^*(t)] = \frac{1}{2\pi j}\int_{c-j\infty}^{c+j\infty} F(\nu)\frac{1}{1 - \epsilon^{-T(s-\nu)}}\,d\nu \tag{15.32}$$

This result may be reduced to a definition of the z-transformation if a method can be provided for evaluating the integral. Contour integration and residue theory may be applied to evaluate the integral in Eq. (15.32) by closing the path at infinity either to the right or the left, as shown in Fig. 15.12, and summing the enclosed residues. Assuming $F(\nu)$ is a ratio of polynomials in ν, it has a finite number of poles and zeros. We must choose c such that the line $\mathrm{Re}(\nu) = c$ is to the right of all poles of $F(\nu)$. Thus if the contour is closed to the left, only N residues need be evaluated, where N is the number of poles of $F(\nu)$. The function

$$G(s - \nu) = \frac{1}{1 - \epsilon^{-T(s-\nu)}} \tag{15.33}$$

has infinitely many poles, since $\epsilon^{-T(s-\nu)} = 1$ at each of the points

$$\nu_m = s - j\frac{2\pi m}{T} \qquad m = 0, \pm 1, \pm 2, \ldots \tag{15.34}$$

Thus if the contour is closed to the right an infinite number of residues must be summed. Fortunately that summation is not difficult, and provides

$$F^*(s) = \frac{1}{T}\sum_{n=-\infty}^{\infty} F\left(s - j\frac{2\pi n}{T}\right) \tag{15.35}$$

This is identical to Eq. (15.18), which was derived by an alternate procedure.

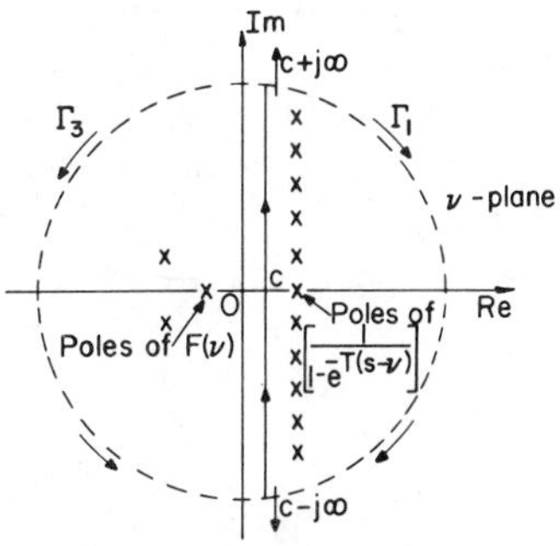

FIG. 15.12. Contours which may be used to evaluate the integral defining the z-transformation of the function $F(\nu)$.

Since Eq. (15.35) cannot be directly reduced to a closed form, it is not a particularly useful solution form.

Closing the contour to the left and summing the residues at the poles of $F(\nu)$ yields

$$F^*(s) = \sum_{\substack{\text{poles of}\\ F(\nu)}} \text{residues of} \left[\frac{F(\nu)}{1 - \epsilon^{-T(s-\nu)}}\right] \tag{15.36}$$

This becomes, upon substitution of $z = \epsilon^{Ts}$,

$$F(z) = \sum_{\substack{\text{poles of}\\ F(\nu)}} \text{residues of} \left[\frac{F(\nu)}{1 - \epsilon^{\nu T} z^{-1}}\right] \tag{15.37}$$

which provides the desired z-transformation definition. It is clear from Eq. (15.37) that $F(z)$ can not have more poles than $F(\nu)$.

Example 15.2

Find the z-transform $F(z)$ corresponding to the impulse sampled sequence derived from the function

$$f(t) = 1 - \epsilon^{-at} \tag{15.38}$$

The corresponding $F(s)$ is

$$F(s) = \frac{1}{s} - \frac{1}{s+a} = \frac{a}{s(s+a)} \tag{15.39}$$

Summing the residues of this function as indicated in Eq. (15.37) yields

$$F(z) = \frac{1}{1 - z^{-1}} - \frac{1}{1 - \epsilon^{-aT} z^{-1}} = \frac{z^{-1}(1 - \epsilon^{-aT})}{(1 - z^{-1})(1 - \epsilon^{-aT} z^{-1})} \tag{15.40}$$

As illustrated by this example, evaluation of z-transforms is straightforward. A number of standard forms are listed in Table 15.1. Extensive z-transform tables are available in several of the references. Complex functions are often most readily treated by breaking them down into simpler components, just as in Laplace transform applications.

Inversion of z-transforms may be carried out by reduction to recognizable terms, as is the standard Laplace transform method, or by direct evaluation of the inversion integral. The z-transform of the impulse sampled sequence $f^*(t)$ is

$$F(z) = \sum_{n=0}^{\infty} f(nT) z^{-n} = f(0) z^0 + f(T) z^{-1} + \cdots + f(nT) z^{-n} + \cdots \tag{15.41}$$

To invert this function requires the development of an explicit expression for $f(nT)$. If both sides of Eq. (15.41) are multiplied by z^{n-1} we obtain

$$z^{n-1}F(z) = f(0)z^{n-1} + f(T)z^{n-2} + \cdots + f[(n-1)T]z^0 + f(nT)z^{-1} + f[(n+1)T]z^{-2} + \cdots \tag{15.42}$$

An applicable theorem developed by Cauchy states that the contour integral

$$I = \frac{1}{2\pi j}\int_\Gamma z^k dz = \begin{cases} 1 & k = -1 \\ 0 & k \neq -1 \end{cases} \tag{15.43}$$

where Γ is any contour enclosing the origin of the z-plane, and k takes on only integer values. Applying this theorem to Eq. (15.42) term by term provides the desired inversion equation as

$$f(nT) = \frac{1}{2\pi j}\int_\Gamma z^{n-1}F(z)\,dz \tag{15.44}$$

We must choose Γ to include all poles of $F(z)$. The integral in Eq. (15.44) may be evaluated by residue methods. In most cases of interest $F(z)$ corresponds to a stable function and all of its poles are within or on the unit circle, so Γ is usually taken as the unit circle.

Example 15.3

Find the sampled sequence corresponding to

$$F(z) = \frac{z-1}{z^2 - z/4 - 1/8} = \frac{z-1}{(z-1/2)(z+1/4)} \tag{15.45}$$

Summing the residues of the integrand

$$z^{n-1}F(z) = \frac{z^{n-1}(z-1)}{(z-1/2)(z+1/4)} \tag{15.46}$$

provides

$$f(nT) = -\frac{4}{3}\left(\frac{1}{2}\right)^n + \frac{5}{3}\left(-\frac{1}{4}\right)^{n-1} \tag{15.47}$$

Table 15.1. Common z-transforms and Advanced z-transforms

No.	$F(s)$	$f(t)$	$F(z)$	$F(z, \Delta) = Z[F(s)\epsilon^{\Delta T s}]$
1.	1	$\delta(t)$	z^{-0}	0
2.	ϵ^{-kTs}	$\delta(t - kT)$	z^{-k}	$z^{(\Delta - 1 - k)}$
3.	$\frac{1}{s}$	1	$\frac{z}{z - 1}$	$\frac{z}{z - 1}$
4.	$\frac{1}{s^2}$	t	$\frac{Tz}{(z - 1)^2}$	$\frac{\Delta T z}{z - 1} + \frac{Tz}{(z - 1)^2}$
5.	$\frac{1}{s^3}$	$\frac{t^2}{2}$	$\frac{T^2 z(z + 1)}{2(z - 1)^3}$	$\frac{T^2 z}{2}\left[\frac{\Delta^2}{z - 1} + \frac{2\Delta + 1}{(z - 1)^2} + \frac{2}{(z - 1)^3}\right]$
6.	$\frac{1}{s + a}$	ϵ^{-at}	$\frac{z}{z - \epsilon^{-aT}}$	$\frac{z\epsilon^{-a\Delta T}}{z - \epsilon^{-aT}}$
7.	$\frac{1}{(s + a)^2}$	$t\epsilon^{-at}$	$\frac{Tz\epsilon^{-aT}}{(z - \epsilon^{-aT})^2}$	$\frac{T\epsilon^{-a\Delta T} z[\epsilon^{-aT} + \Delta(z - \epsilon^{-aT})]}{(z - \epsilon^{-aT})^2}$
8.	$\frac{a}{s(s + a)}$	$1 - \epsilon^{-at}$	$\frac{(1 - \epsilon^{-aT})z}{(z - 1)(z - \epsilon^{-aT})}$	$\frac{z}{z - 1} - \frac{z\epsilon^{-a\Delta T}}{z - \epsilon^{-aT}}$

9.	$\dfrac{a}{s^2(s+a)}$	$\dfrac{1}{a}(at - 1 + \epsilon^{-at})$	$\dfrac{Tz}{(z-1)^2} - \dfrac{(1-\epsilon^{-aT})z}{a(z-1)(z-\epsilon^{-aT})}$	$\dfrac{Tz}{(z-1)^2} + \dfrac{z(\Delta T - 1/a)}{z-1} + \dfrac{\epsilon^{-a\Delta T}z}{a(z-\epsilon^{-aT})}$
10.	$\dfrac{\omega}{s^2+\omega^2}$	$\sin\omega t$	$\dfrac{z \sin\omega t}{z^2 - 2z\cos\omega T + 1}$	$\dfrac{z[z\sin\Delta\omega T + \sin(1-\Delta)\omega T]}{z^2 - 2z\cos\omega T + 1}$
11.	$\dfrac{s}{s^2+\omega^2}$	$\cos\omega t$	$\dfrac{z(z-\cos\omega T)}{z^2 - 2z\cos\omega T + 1}$	$\dfrac{z[z\cos\Delta\omega T - \cos(1-\Delta)\omega T]}{z^2 - 2z\cos\omega T + 1}$
12.	$\dfrac{\omega}{(s+a)^2+\omega^2}$	$\epsilon^{-at}\sin\omega t$	$\dfrac{z\epsilon^{-aT}\sin\omega T}{z^2 - 2z\epsilon^{-aT}\cos\omega T + \epsilon^{-2aT}}$	$\dfrac{z[z\sin\Delta\omega T + \epsilon^{-aT}\sin(1-\Delta)\omega T]\epsilon^{-a\Delta T}}{z^2 - 2z\epsilon^{-aT}\cos\omega T + \epsilon^{-2aT}}$
13.	$\dfrac{s+a}{(s+a)^2+\omega^2}$	$\epsilon^{-at}\cos\omega t$	$\dfrac{z^2 - z\epsilon^{-aT}\cos\omega T}{z^2 - 2z\epsilon^{-aT}\cos\omega T + \epsilon^{-2aT}}$	$\dfrac{z[z\cos\Delta\omega T - \epsilon^{-aT}\cos(1-\Delta)\omega T]\epsilon^{-a\Delta T}}{z^2 - 2z\epsilon^{-aT}\cos\omega T + \epsilon^{-2aT}}$

It is apparent from Eq. (15.47) that the response is oscillatory in nature and tends toward zero as $n \to \infty$. Each of the terms in Eq. (15.47) obviously represent samples taken from an exponentially decaying type of function.

The z-transforms of most functions of interest are ratios of polynomials in z (or z^{-1}). If we are interested only in the first few terms of the sampled sequence, as for example in determining the unit step response of a typical controller, z-transform inversion may be accomplished by long division to obtain an infinite series. This process often proves useful, and is illustrated by the following example.

Example 15.4

Consider the closed-loop system shown in Fig. 15.13. Find the unit step response of this system by obtaining $C(z)$ for an appropriately chosen sampling interval and dividing out the ratio of polynomials to obtain the first few terms in the expansion.

We may readily obtain $C(s)$ as

$$C(s) = \frac{50}{s(s^2 + 5s + 50)} = \frac{50}{s[(s + 2.5)^2 + 43.75]} \tag{15.48}$$

Expanding $C(s)$ in partial fractions provides

$$C(s) = \frac{1}{s} - \frac{s + 5}{(s + 2.5)^2 + 43.75} \tag{15.49}$$

Taking the z-transform using entries 3, 12, and 13 in Table 15.1 yields

$$C(z) = \frac{z}{z - 1} - \frac{z^2 - z\epsilon^{-2.5T}\cos 6.6T + (2.5/6.6)z\epsilon^{-2.5T}\sin 6.6T}{z^2 - 2z\epsilon^{-2.5T}\cos 6.6T + \epsilon^{-5T}} \tag{15.50}$$

Before proceeding further we must choose a value for T. It is apparent that the open loop transfer function for this system crosses unit magnitude at $\omega \simeq 7$. The closed loop BW is approximately 7 rad/sec or $7/2\pi$ Hz. The unit step response may reasonably be expected to have a natural frequency component near this value, and the response pattern should be accurately defined if 10 samples are taken per natural frequency period. Since $7/20\pi \simeq 0.111$, $T = 0.1$ will be

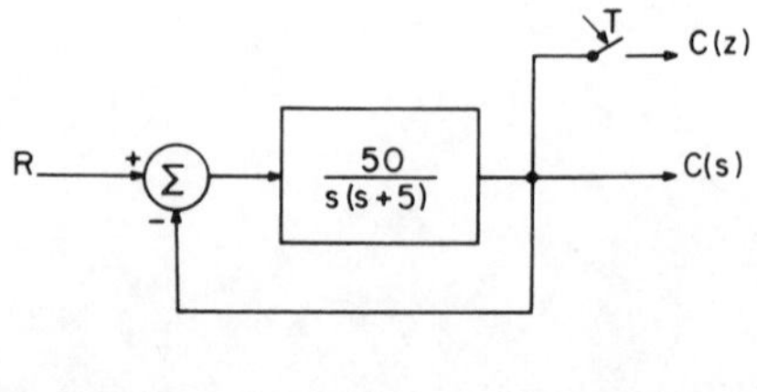

FIG. 15.13. System considered in Ex. 15.4.

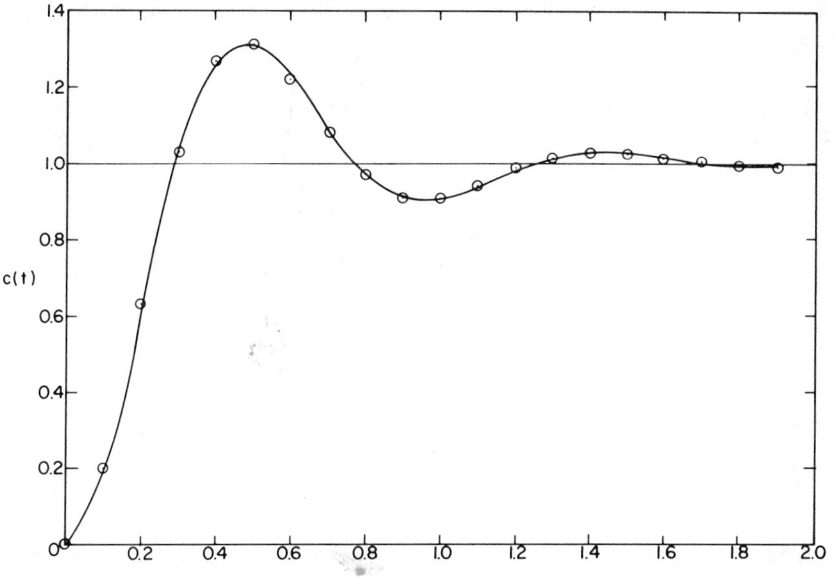

FIG. 15.14. Unit step-response for the system of Ex. 15.4.

used. Substituting this value into Eq. (15.50) and simplifying provides

$$C(z) = \frac{z}{z-1} - \frac{z^2 - 0.434z}{z^2 - 1.23z + 0.606} \tag{15.51}$$

Rather than place $C(z)$ over a common denominator, observe that the first term corresponds to 1 in the time domain, and the second term provides the response error. It is easier and more accurate to divide out the second term in Eq. (15.51) and subtract the result from one to obtain the desired response. The first few terms in the division are obtained as follows:

$$\begin{array}{r|l}
 & 1 + 0.796z^{-1} + 0.374z^{-2} + \cdots \\
\hline
z^2 - 1.23z + 0.606 & z^2 - 0.434z \\
 & \underline{z^2 - 1.23\ z + 0.606} \\
 & \quad + 0.796z - 0.606 \\
 & \quad \underline{\ \ 0.796z - 0.98 \ + 0.48z^{-1}} \\
 & \qquad\qquad + 0.374 - 0.48z^{-1} \\
 & \qquad\qquad\qquad \cdots
\end{array}$$

The corresponding unit step response pattern is shown in Fig. 15.14.

The application of z-transform theory illustrated by Ex. 15.4 is only one of many ways in which it can be used. Note how much more readily the unit step response is obtained this way than by the usual procedure involving the inverse Laplace transform and evaluation of the time response at appropriately chosen points, particularly if a desk calculator or computer can be used. The procedure is a routine numerical operation and involves no transcendental functions once the coefficients have been obtained, so a computer is easily programmed to obtain the solution efficiently, or a secretary may be shown how to do the job on a desk calculator.

It is often convenient to have available a means for obtaining the initial and final values of signals expressed as pulse sequences. Initial and final value theorems serve the same purpose for SD systems with signals and transfer functions in the z-domain as they do in Laplace transform applications. Since

$$F(z) = f(0)z^0 + f(T)z^{-1} + \cdots \tag{15.52}$$

we observe that all terms in the sequence approach zero as $z \to \infty$ except $f(0)$, so

$$f(0) = \lim_{z \to \infty} F(z) \tag{15.53}$$

provides the desired *initial value theorem.* The final value theorem is a bit more difficult to develop. Consider the truncated sequence

$$F_N(z) = \sum_{n=0}^{N} f(nT)z^{-n} \tag{15.54}$$

and a second sequence

$$F'_N(z) = z^{-1}F_{N-1}(z) = \sum_{n=0}^{N-1} f(nT)z^{-n}z^{-1} \tag{15.55}$$

The sequence given in Eq. (15.55) is identical term by term to that given in Eq. (15.54) except that it is delayed T seconds (multiplied by z^{-1}) and truncated one sample sooner. Subtracting $F'_N(z)$ from $F_N(z)$ provides

$$F_N(z) - F'_N(z) = \sum_{n=0}^{N} f(nT)z^{-n} - \sum_{n=0}^{N-1} f(nT)z^{-n}z^{-1} \tag{15.56}$$

In the limit as $z \to 1$ the right side of Eq. (15.56) approaches $f(NT)$, so

$$f(NT) = \lim_{z \to 1} \left[\sum_{n=0}^{N} f(nT)z^{-n} - z^{-1} \sum_{n=0}^{N-1} f(nT)z^{-n} \right] \tag{15.57}$$

As $N \to \infty$, each of the summations in Eq. (15.57) approaches $F(z)$ and $f(NT) \to f(\infty)$, so

$$f(\infty) = \lim_{z \to 1} (1 - z^{-1}) F(z) \tag{15.58}$$

This final value theorem is useful in evaluating SD system steady state performance relative to singularity inputs such as the unit step or unit ramp functions.

Example 15.5

Find the initial and final values of the pulse sequence defined by

$$F(z) = \frac{z}{z^3 - z^2 - 0.25z + 0.25} \tag{15.59}$$

From Eq. (15.53) we see that $f(0) = \lim_{z \to \infty} F(z) = 0$. Applying Eq. (15.58) provides

$$f(\infty) = \lim_{z \to 1} (1 - z^{-1}) F(z) = \lim_{z \to 1} \frac{z - 1}{z} \frac{z}{z^3 - z^2 - 0.25z + 0.25} \tag{15.60}$$

Substituting $z = 1$ yields the indeterminate form 0/0. This difficulty may be overcome either by factoring a $(z - 1)$ term out of the denominator, or by application of L'Hospital's rule. Applying the latter provides

$$f(\infty) = \lim_{z \to 1} \frac{1}{3z^2 - 2z - 0.25} = \frac{1}{0.75} = \frac{4}{3} \tag{15.61}$$

In the study of SD systems it is sometimes desired to investigate the performance at one or more points of interest between samples to provide assurance that the inter-sample response is not oscillatory. The modified z-transform, developed by Barker,[1] is useful for this purpose, as well as in the study of systems which include transport lag. Of course a pure time delay of nT seconds is equivalent to multiplication by z^{-n} in the z-domain, but a delay of λT seconds, where λ is not an integer, must be treated more carefully. Consider the situation illustrated in Figs. 15.15 and 15.16, where λT denotes the time delay.

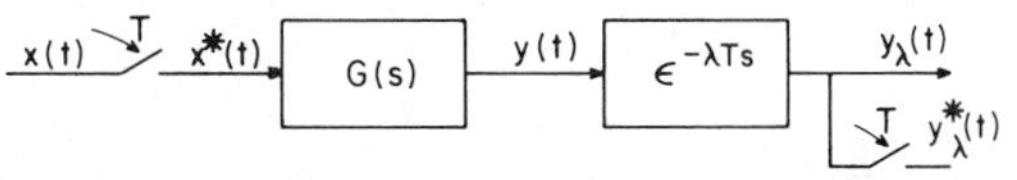

FIG. 15.15. Arrangement used to illustrate the development of the modified z-transform.

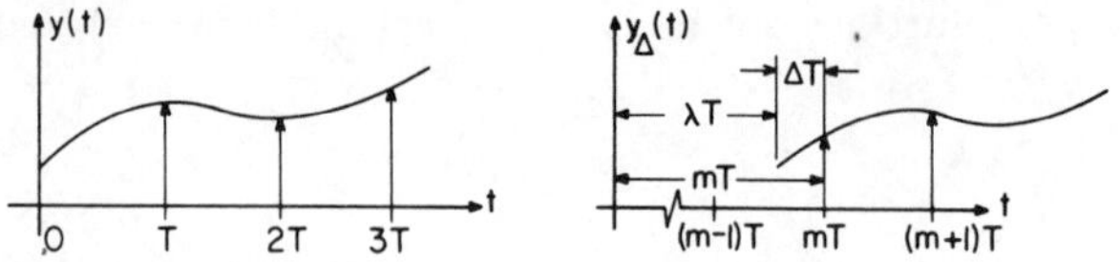

FIG. 15.16. Relationship between $y(t)$, $y_\lambda(t)$, and the sampled forms of these signals.

Choose m, the next higher integer than λ, and define Δ by

$$m = \lambda + \Delta \tag{15.62}$$

Observe that $0 < \Delta < 1$. Thus $y(t)$ may be assumed to have been *delayed* mT seconds and *advanced* ΔT seconds or

$$[y(t - \lambda T)] = Y(s, \lambda) = Y(s)\epsilon^{-\lambda T s} = \epsilon^{-mTs}Y(s)\epsilon^{\Delta Ts} \tag{15.63}$$

The z-transform of ϵ^{-mTs} is simply z^{-m}, so

$$Y(z, \lambda) = z^{-m}Y(z, \Delta) \tag{15.64}$$

where

$$Y(z, \Delta) = Z[Y(s)\epsilon^{\Delta Ts}] \tag{15.65}$$

The $Z[\cdot]$ notation denotes the z-transform of the term in brackets. $Y(z, \Delta)$ may also be expressed directly from the definition in series form as

$$Y(z, \Delta) = \sum_{n=0}^{\infty} y[(n + \Delta)T]z^{-n} \tag{15.66}$$

This relationship could be used to obtain the modified z-transform, although it is generally easier to develop a closed form solution from the z-transform definition by summing residues.

We see from Eqs. (15.37) and (15.65) that

$$Y(z, \Delta) = \sum_{\substack{\text{poles of} \\ Y(\nu)}} \text{residues of } \frac{Y(\nu)\epsilon^{\Delta T\nu}}{1 - \epsilon^{\nu T}z^{-1}} \tag{15.67}$$

The $\epsilon^{\Delta T\nu}$ term causes no difficulty, since Eq. (15.37) was developed by contour integration, closing the contour around the left half plane, and $\epsilon^{\Delta T\nu}$ is zero everywhere on this infinite semi-circle. Application of Eq. (15.67) as the definition of the advanced z-transform is straightforward. The result must be multiplied by z^{-m} for the appropriate m to obtain the desired modified z-transform for a time delay of λT seconds. The advanced z-transforms corresponding to selected functions are included in Table 15.1.

Example 15.6

Find $F(z, \lambda)$ assuming a signal with Laplace transform

$$F(s) = \frac{s+3}{s^2+3s+2} = \frac{s+3}{(s+1)(s+2)} \tag{15.68}$$

is delayed $\lambda T = 0.27$ seconds and sampled with $T = 0.1$ seconds. Choosing $m = 3$, we see that $\Delta = 0.3$ and

$$F(z, \Delta) = \sum_{\substack{\text{poles of} \\ F(\nu)}} \text{residues of } \frac{\epsilon^{0.03\nu}(\nu+3)}{(\nu+1)(\nu+2)} \cdot \frac{1}{1-\epsilon^{\nu T}z^{-1}} \tag{15.69}$$

or

$$F(z, \Delta) = \frac{2\epsilon^{-0.03}}{1-\epsilon^{-0.1}z^{-1}} - \frac{\epsilon^{-0.06}}{1-\epsilon^{-0.2}z^{-1}}$$

$$= \frac{1-0.74z^{-1}}{(1-0.905z^{-1})(1-0.819z^{-1})} \tag{15.70}$$

Thus

$$F(z, \lambda) = z^{-3}F(z, \Delta) = \frac{z^{-3}-0.74z^{-4}}{(1-0.905z^{-1})(1-0.819z^{-1})} \tag{15.71}$$

is the desired modified (delayed) z-transform.

The primary motivation for use of the Laplace transform in solving system problems is the convolution or superposition theorem, which allows the transformed output of a linear system to be expressed as the transformed input times the transform of the impulse response. A similar relationship exists in the z-domain. Consider the system shown in Fig. 15.17. Assume the two samplers

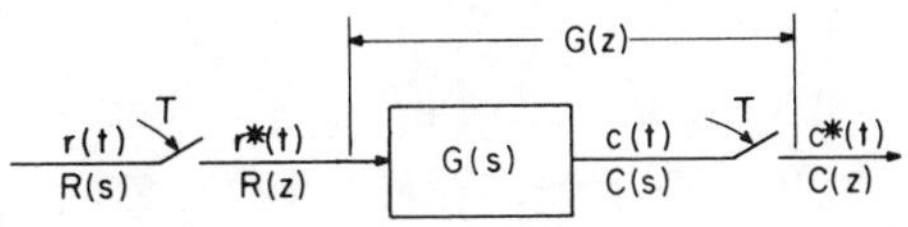

FIG. 15.17. Simple linear system used to derive the pulse transfer function.

provide synchronous impulse sampling. Each sample at the $G(s)$ input results in an impulse response pattern at the $G(s)$ output, and the net output at each sample point is the sum of all such components, or

$$c(mT) = \sum_{n=0}^{\infty} r(nT)\,g[(m-n)T] \tag{15.72}$$

we assume no input exists prior to $t = 0$. ($r(nT) = 0$ for $n < 0$.) Note that $g[(m-n)T] = 0$ for all $m - n < 0$, since the system cannot respond to an input prior to its application. The summation in Eq. (15.72) is analogous to the convolution integral for continuous systems, and is called the *convolution summation.* By definition

$$C(z) = \sum_{m=0}^{\infty} c(mT)z^{-m} \tag{15.73}$$

Substituting $c(mT)$ from Eq. (15.72) into Eq. (15.73) provides

$$C(z) = \sum_{m=0}^{\infty}\sum_{n=0}^{\infty} r(nT)\,g[(m-n)T]z^{-m} \tag{15.74}$$

Letting $k = m - n$, Eq. (15.74) may be rewritten as

$$C(z) = \sum_{k=-n}^{\infty}\sum_{n=0}^{\infty} r(nT)\,g(kT)z^{-n}z^{-k} \tag{15.75}$$

But $g(kT) = 0$ for $k < 0$, so the lower limit on the first summation in Eq. (15.75) may be replaced by 0. Separating the summations provides

$$C(z) = \sum_{k=0}^{\infty} g(kT)z^{-k} \sum_{n=0}^{\infty} r(nT)z^{-n} = G(z)R(z) \tag{15.76}$$

where

$$G(z) = \sum_{k=0}^{\infty} g(kT)z^{-k} \tag{15.77}$$

is the z-transform of the sampled impulse response $g(t)$. *Thus the z-transform of the output pulse sequence is the product of the z-transform of the input pulse sequence and the system impulse response function.* It should be emphasized that this description only relates the output and input at the sample points, and does not define performance at other points over the sampling interval. Although we have assumed a linear continuous network $G(s)$ here, an analogous procedure may be used to define pulse transfer functions for linear digital processors and the results are applied in the same way.

Example 15.7

Find the output pulse sequence $C(z)$ for the network shown in Fig. 15.17 if

$$r(t) = t \qquad G(s) = \frac{s+1}{s^2+5s+6} \tag{15.78}$$

We know that $C(z) = G(z)R(z)$, but $R(z) = Tz/(z-1)^2$ and

$$G(z) = Z[G(s)] = Z\left[\frac{2}{s+3} - \frac{1}{s+2}\right] = \frac{-z}{z-\epsilon^{-2T}} + \frac{2z}{z-\epsilon^{-3T}} \tag{15.79}$$

Thus

$$C(z) = \frac{Tz}{(z-1)^2} \cdot \frac{z^2 + z(\epsilon^{-3T} - 2\epsilon^{-2T})}{(z-\epsilon^{-3T})(z-\epsilon^{-2T})} \tag{15.80}$$

is the desired output pulse sequence.

The z-transform is used in SD systems in a manner closely analogous to the use of the Laplace transform in describing continuous systems. It is easily derived, tables of standard forms are available, and the algebraic manipulations required in system analysis are straightforward. The complex variable z^{-n} may be interpreted as the complex function ϵ^{-sT}, or as is often more useful, as an ordering variable describing the position of an impulse sample in a signal sequence. Modified z-transforms are available to aid in studying system performance between sample points. Many useful theorems may be derived, of which the initial and final value theorems are representative examples. A two-sided z-transform is available for considering systems with random inputs. We will find the z-transform a powerful tool in our subsequent discussions of SD systems.

15.4 SAMPLED DATA SYSTEM CHARACTERIZATION AND ANALYSIS

We are now prepared to apply fundamental z-transform principles to the analysis of SD systems in open or closed loop form. Care must be exercised to distinguish between cascades of networks with and without samplers between them, but no significant difficulty is encountered in either case. Consider the

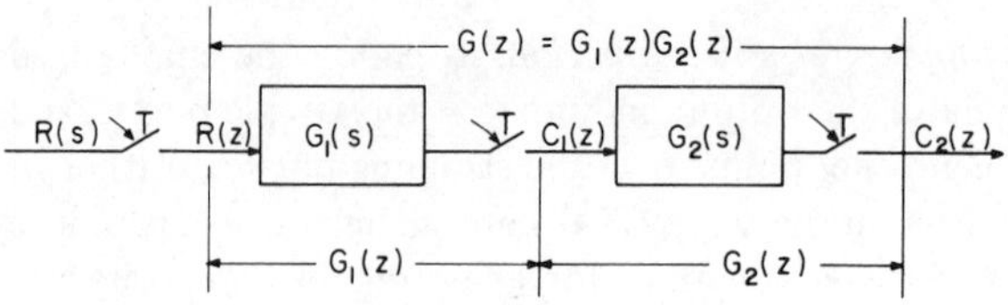

FIG. 15.18. Two sampled elements in cascade.

situation shown in Fig. 15.18 with two linear elements separated by a *synchronous* sampler. (Synchronous sampling means that all samplers close simultaneously.) It is apparent that

$$C_1(z) = R(z)\,G_1(z) \tag{15.81}$$

and

$$C_2(z) = C_1(z)\,G_2(z) = R(z)\,G_1(z)\,G_2(z) \tag{15.82}$$

Thus

$$G(z) = G_1(z)\,G_2(z) \tag{15.83}$$

is the equivalent overall pulse transfer function of two linear elements *separated by a synchronous sampler.* The *non-synchronous* sampled case must be treated using the modified z-transform, resulting in significant added complexity, and is of limited interest.

Consider the cascade of two linear elements shown in Fig. 15.19 without a sampler separating them. In this case

$$C_1(s) = R^*(s)\,G_1(s) \tag{15.84}$$

and

$$C_2(s) = R^*(s)\,G_1(s)\,G_2(s) = R^*(s)\,G_{12}(s) \tag{15.85}$$

so

$$C_2(z) = R(z)\,G_{12}(z) \tag{15.86}$$

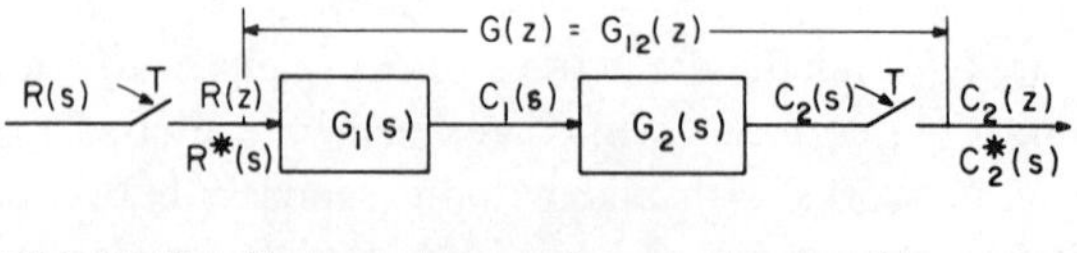

FIG. 15.19. Two continuous elements in cascade.

where

$$G_{12}(z) = Z[G_1(s)\,G_2(s)] \tag{15.87}$$

It is important to note that

$$G_{12}(z) \neq G_1(z)\,G_2(z) \tag{15.88}$$

for elements *not* separated by a synchronous sampler. The truth of Eq. (15.88) is obvious, since the input to $G_2(s)$ is significantly different in the two cases. We will denote the z-transforms of function products by the double subscripted notation defined in Eq. (15.87), or by notation such as $GH(z)$. This notational complexity, although somewhat awkward, is necessary in SD analysis.

Example 15.8

Consider the two systems shown in Figs. 15.18 and 15.19. Let

$$G_1(s) = \frac{1}{s + a} \qquad G_2(s) = \frac{1}{s} \tag{15.89}$$

Find the overall pulse transfer function in each case.

When a sampler separates the two elements

$$G(z) = G_1(z)\,G_2(z) = \frac{1}{1 - \epsilon^{-aT} z^{-1}} \cdot \frac{1}{1 - z^{-1}} \tag{15.90}$$

When there is no sampler between $G_1(s)$ and $G_2(s)$

$$G(z) = G_{12}(z) = z\left[\frac{1}{s(s + a)}\right] = \frac{1}{a}\,\frac{(1 - \epsilon^{-aT})z^{-1}}{(1 - z^{-1})(1 - \epsilon^{-aT} z^{-1})} \tag{15.91}$$

It is clear by comparison of Eqs. (15.90) and (15.91) that the results with and without intermediate sampling differ considerably. It is possible to show that $G_1(z)\,G_2(z)$ and $G_{12}(z)$ approach the same limit as $T \to 0$, which we should intuitively expect. Note that the two functions have the same poles, but different zeros. This should also be expected, since the stability of an *open-loop* SD system depends only upon whether the individual components making up that system are stable. The stability of a linear element *does not* depend upon whether the input signal is a sequence of impulses or a continuous function.

Because SD transfer functions in the z-domain can only be expressed from one sample point to the next, and because there are so many sampler location possibilities, there is no unique closed loop transfer function form for SD systems. The various situations which can arise are perhaps best illustrated by considering a particular situation. An *error sampled* system is illustrated in Fig. 15.20.

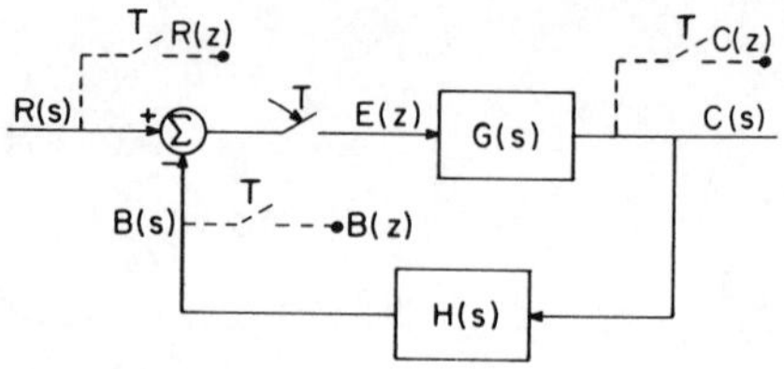

FIG. 15.20. A standard error sampled feedback system.

The dashed samplers shown on the diagram do not exist in the actual system and are introduced only for mathematical convenience. They allow us to consider the z-transform relationship between pulse sequences at significant points in the system. The error pulse sequence for the situation shown in Fig. 15.20 is given by

$$E(z) = R(z) - B(z) \tag{15.92}$$

But

$$B(z) = E(z)\,GH(z) \tag{15.93}$$

where $GH(z)$ is the z-transform of the combination $G(s)\,H(s)$. Using $B(z)$ from Eq. (15.93) in Eq. (15.92) yields

$$E(z) = R(z) - GH(z)\,E(z) \tag{15.94}$$

Solving Eq. (15.94) for $E(z)$ provides

$$E(z) = \frac{R(z)}{1 + GH(z)} \tag{15.95}$$

Since

$$C(z) = E(z)\,G(z) \tag{15.96}$$

we see that

$$C(z) = \frac{R(z)G(z)}{1 + GH(z)} \tag{15.97}$$

In this case we may solve for $C(z)/R(z)$, the overall closed loop pulse transfer function, because the system is sampled at the error point and $R(z)$ appears in Eq. (15.97) in separable form. When neither the input nor the error signal is sampled, this will not be the case.

The fundamental procedures involved in applying z-transform analysis methods to closed loop SD systems are illustrated by the system shown in Fig. 15.20. We start with the signal at a *sample point,* write equations in the z-domain relating that signal to the signals at other sampled points, and solve for the particular signals of interest. Note that all transfer functions in the z-domain relate *sampled* signals at two points in the system, so the equations must be developed using the various sampled points in the system as a frame of reference. The system shown in Fig. 15.21 provides further illustration of this requirement. Starting at the sampled point, we write

$$B(z) = RF(z) - B(z)\,GHF(z) \tag{15.98}$$

or

$$B(z) = \frac{RF(z)}{1 + GHF(z)} \tag{15.99}$$

Also

$$C(z) = B(z)\,G(z) = \frac{RF(z)\,G(z)}{1 + GHF(z)} \tag{15.100}$$

Note that we cannot solve Eq. (15.100) for $C(z)/R(z)$. Pulse transfer functions are readily developed using the same fundamental principles for closed loop SD systems of arbitrary form and complexity.

When the z-transform was developed from the Laplace transform of the impulse sampled sequence, we observed the correspondence between lhp poles in the s-domain and poles within the unit circle in the z-domain. The stability of SD systems may be defined as follows: *A linear SD system is stable if and only if all poles of its transfer function $G(z)$ are located within the unit circle in the z-plane.* Just as in the application of Laplace transform theory, it is desirable to have a procedure for determining the presence of poles outside or on the unit circle without having to factor the denominator polynomial. A modified form of the Routh criterion can be developed for this purpose. Consider the *bilinear transformation* defined by

$$z = \frac{\lambda + 1}{\lambda - 1} \qquad \lambda = \frac{z + 1}{z - 1} \tag{15.101}$$

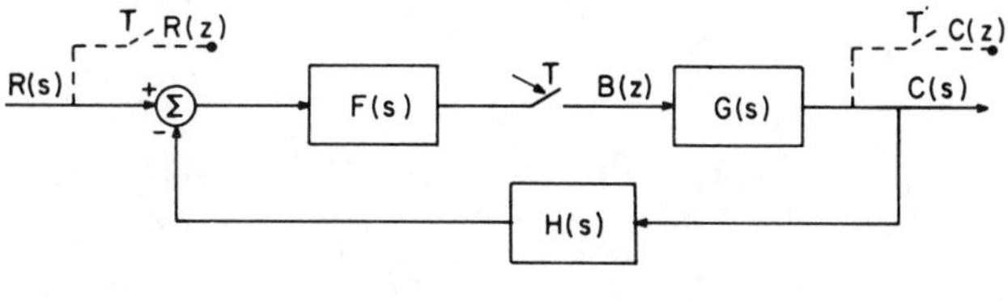

FIG. 15.21. A system without error sampling.

Both z and λ are complex, so let

$$z = x + jy \qquad \lambda = u + jv \tag{15.102}$$

Substituting for z in the λ equation and rationalizing provides

$$\lambda = \frac{x^2 + y^2 - 1 - j2y}{(x - 1)^2 + y^2} \tag{15.103}$$

Observe from the definition of z that $x^2 + y^2 = |z|^2$. Also, the denominator in Eq. (15.103) is non-negative, and zero only at the point $[1, 0]$. Excluding the point $[1, 0]$, for which λ is undefined, the real part of λ assumes the same sign as $x^2 + y^2 - 1$, which is negative for all $|z| > 1$. Thus the bilinear transformation maps points within the unit circle of the z-plane into the left half of the λ-plane, points outside the z-plane unit circle into the right half of the λ-plane, and the z-plane unit circle into the λ-plane imaginary axis. The Routh criterion may be applied to the polynomial in λ to determine stability by carrying out the following steps:

1. Determine the denominator polynomial in z for the transfer function of interest (the system characteristic equation).
2. Substitute for z in terms of λ using the bilinear transformation.
3. Rationalize and simplify the result.
4. Apply the Routh criterion to determine the number of roots of the λ polynomial located in the rhp.
5. These roots are outside the unit circle in the z-plane, and represent unstable poles of the original transfer function.

The following example illustrates this procedure.

Example 15.9

Consider a closed-loop SD system with characteristic equation

$$45z^3 - 117z^2 + 119z - 39 = 0 \tag{15.104}$$

Apply the modified Routh criterion to determine the number of z-plane roots outside the unit circle.

Applying the bilinear transformation to Eq. (15.104) provides

$$45\left(\frac{\lambda + 1}{\lambda - 1}\right)^3 - 117\left(\frac{\lambda + 1}{\lambda - 1}\right)^2 + 119\left(\frac{\lambda + 1}{\lambda - 1}\right) - 39 = 0 \tag{15.105}$$

placing this over the common denominator $(\lambda - 1)^3$, multiplying both sides by

$(\lambda - 1)^3$, and simplifying yields

$$\lambda^3 + 2\lambda^2 + 2\lambda + 40 = 0 \tag{15.106}$$

The corresponding Routh array is

λ^3	1	2	0
λ^2	2	40	0
λ^1	−18	0	
λ^0	40		

Since there are two sign changes in the left column, there are two roots in the right half λ-plane, or outside the z-plane unit circle.

It is obvious from the degree of difficulty encountered in applying the bilinear transformation to the characteristic equation in Ex. 15.9 that this procedure is awkward to use and feasible only for systems of low order. The Schur-Cohn stability criterion,[2] based directly upon the coefficients in the z-domain characteristic equation, eliminates the bilinear transformation step, but a sequence of determinants of dimensions 2×2, 4×4, . . . , $2N \times 2N$ must be evaluated, where N is the highest power of the characteristic polynomial in z. This procedure also involves extensive manipulations for systems of high order. Fortunately, methods analogous to the Nyquist criterion are available for evaluating SD system stability, and the modified Routh or Schur-Cohn criteria are used primarily for simple systems where the answers can be obtained without extensive calculations.

The frequency response of SD systems is easily obtained in the z-domain using the fact that $z = \epsilon^{sT}$. Setting $s = j\omega$ leads to values of z on the unit circle at angle ωT radians. The frequency response is obtained by choosing appropriate values of z on the unit circle, relating each to the corresponding ω coordinate, evaluating $G(z)$, and plotting the results. The frequency response magnitude pattern is symmetrical for complex conjugate values of z, so it is only necessary to evaluate points on half of the z-plane unit circle. The upper half, which corresponds to the ω range $0 \leq \omega \leq \pi/T$, is usually used. The frequency response magnitude and phase patterns repeat infinitely many times as ω covers the range $[0, \infty]$, each complete pattern resulting from an ω change of $2\pi/T$ rad/sec.

Although the frequency response magnitude and phase patterns alone are not indicative of stability of lack thereof (unless there is a pole on the unit circle) the Nyquist criterion may be applied to the polar plot of open loop frequency response data in the usual way to determine stability and relative stability information. If we consider the typical closed loop transfer characteristic

$$\frac{C(z)}{R(z)} = \frac{G(z)}{1 + GH(z)} \tag{15.107}$$

All zeros of the denominator polynomial $1 + GH(z)$ outside the unit circle may be located by making a polar plot of $GH(z)$ as z traverses the contour illustrated in Fig. 15.22 and counting the number of cw rotations about the -1 point. Since the exterior of the unit circle is enclosed in a cw direction, as seen from Fig. 15.22, each zero of $1 + GH(z)$ outside the unit circle results in a cw encirclement of the -1 point by the $GH(z)$ polar plot. Thus

$$N = Z - P \tag{15.108}$$

where

N = the number of cw rotations of the $GH(z)$ plot around the -1 point.
Z = the number of zeros of $1 + GH(z)$ outside the unit circle.
P = the number of poles of $GH(z)$, and thus of $1 + GH(z)$, outside the unit circle.

The poles of $GH(z)$ are known, so Eq. (15.108) may be used to solve for Z as soon as the polar plot is made. If $Z \neq 0$, the system is unstable. (We assume the system is open loop stable.)

Most control systems are low pass in nature, so $GH(z) = 0$ everywhere on the infinite circle contour shown in Fig. 15.22. We thus only need to plot $GH(z)$ on the unit circle and, because

$$GH(z^*) = GH^*(z) \tag{15.109}$$

where * denotes the complex conjugate, only the upper half of the unit circle plot is necessary. Poles on the unit circle are avoided by semicircular detours with infinitesimal radii, just as poles on the $j\omega$ axis are avoided in the continuous case. Application of the Nyquist criterion to SD systems and interpretation of the results is closely analogous to the comparable procedures applied to continuous systems.

Example 15.10

Consider the SD system shown in Fig. 15.23 with $T = 0.03$. Use the Nyquist criterion to determine if this arrangement is stable and the margin by which stability (or instability) is achieved.

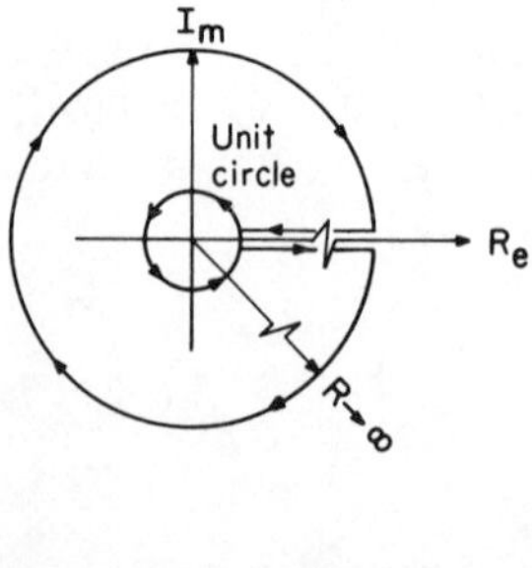

FIG. 15.22. Contour used in applying the Nyquist criterion in the z-domain.

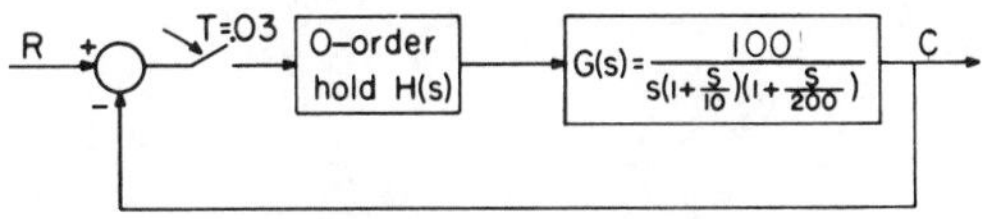

FIG. 15.23. The SD system used in Ex. 15.10.

For this configuration

$$\frac{C(z)}{R(z)} = \frac{GH(z)}{1 + GH(z)} \tag{15.110}$$

But

$$GH(s) = \frac{1 - \epsilon^{-Ts}}{s} \cdot \frac{100}{s(1 + s/10)(1 + s/200)} \tag{15.111}$$

so

$$GH(z) = (1 - z^{-1})\, Z\left[\frac{100}{s^2(1 + s/10)(1 + s/200)}\right] \tag{15.112}$$

Expanding the bracketed term in Eq. (15.112) into partial fractions, evaluating the z-transforms term by term, and simplifying, provides

$$GH(z) = \frac{0.2894z^2 + 0.4702z + 0.01814}{(z - 1)(z - 0.74)(z - 0.0025)} \tag{15.113}$$

Substituting several values of z chosen on the upper half circle in the z-plane, avoiding the pole at $z = 1$ by a small *outward* semi-circle to enclose it within the contour, yields the Nyquist plot shown in Fig. 15.24. A linear scale is used in this case because a wide dynamic range is unnecessary. Only one-half of the total plot is shown, as is the usual practice, and the -1 point is encircled once by the total plot. Our system is thus unstable for $T = 0.03$. It is easily shown that this system is stable in the absence of the sample and hold operation or, what is equivalent, in the limit as $T \to 0$ in the sampled case.

It is clear from Ex. 15.10 that the Nyquist criterion and other standard frequency response methods may be used in SD system analysis and design very much as they are applied to continuous systems. Compensation filters may be added to avoid enclosing the -1 point, and the avoidance margin may be taken as an indication of stability margin with the usual limitations inherent to this interpretation. Magnitudes and angles of the $GH(z)$ components may be obtained graphically from a z-domain plot if desired, and this often provides a rapid

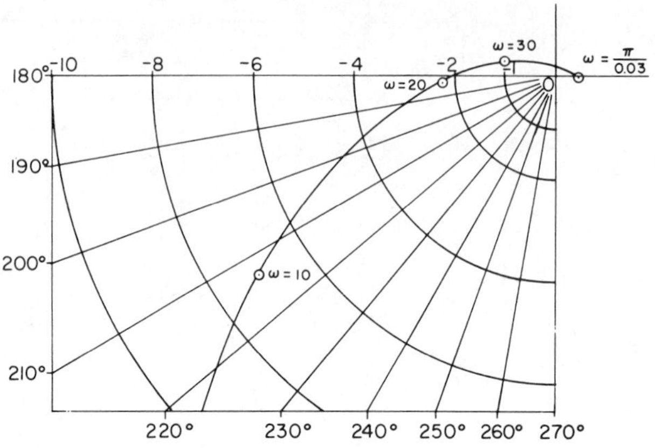

FIG. 15.24. Nyquist plot of $GH(z)$ for the system shown in Fig. 15.23.

approximate solution method. Whereas the Bode diagram often proves useful as an aid in constructing continuous system Nyquist plots, no such helpful tool is available in the z-domain. Points on the Nyquist plots for SD systems are also somewhat more difficult to obtain because $z = \epsilon^{j\omega T}$ is a complex number, whereas $s = j\omega$ is imaginary, but this is a minor complication.

Since the z-plane is a complex plane with the same fundamental characteristics as the s-plane, root loci are constructed in both planes using identical construction rules. The only distinction between the two cases arises because stable poles must lie within the unit circle in the z-plane. The root locus procedure thus applies directly to SD system analysis and design with no significant modifications.

Before considering the design problem, it is appropriate to summarize our observations relating to analysis of SD systems. Since there is no "standard" SD system configuration, the input to output and input to error performance must be evaluated for each specific system form encountered. If all equations are written from sample point to sample point, however, this development is straightforward. The stability of linear SD systems requires that all poles of the z-domain transfer function lie inside the unit circle in the z-plane. The root locus and the Nyquist stability criterion apply directly to SD systems. The Routh stability criterion may be applied to the characteristic equation $1 + A(z) = 0$ by first using the bilinear transformation, although the complexity involved is such that the procedure is only feasible for systems of low order. The Schur-Cohn stability criterion uses the z-domain characteristic equation coefficients directly, but is also complex except for systems of low order. SD system analysis is closely analogous to that of the comparable continuous system. Modification of the standard continuous system design procedures for use in SD system design is relatively straightforward, and is considered in the next section.

15.5 DESIGN OF SD SYSTEMS USING CONVENTIONAL METHODS

Just as in the continuous case, where the Nyquist criterion, Bode diagram, and Evans plot methods originally introduced for analysis purposes lead to synthesis algorithms, the analysis procedures available for SD systems lead directly to several design methods. Our attention is focused upon the single loop error sampled system, because it is the form most often encountered. Although the design of multiple loop systems with several samplers is obviously a more complex problem, the same fundamental concepts apply. For the present we are interested primarily in the design of stable simple systems, with only token consideration given to the more sophisticated aspects of overall performance.

The most significant problem encountered in SD system design using a continuous network $G_e(s)$ for compensation of a system with plant transfer function $G_p(s)$, as shown in Fig. 15.25, results from the fact that

$$Z[G_e(s)G_p(s)] = G_{ep}(z) \neq G_e(z)G_p(z) \tag{15.114}$$

For this reason, it is not clear exactly how changes in $G_e(s)$ influence the roots of the characteristic equation $1 + A(z) = 0$. Several approaches may be taken in an effort to bypass this difficulty, and the degree of approximation required in the resulting mathematical models is typically greater than we are accustomed to from our past experience with continuous systems. It is thus apparent that simulation study to check the performance resulting from a tentative design and serve as an aid in making final adjustments to assure satisfactory response characteristics is often an important step in designing SD systems.

Consider the system shown in Fig. 15.25. The design problem is to choose $G_e(s)$ such that all performance specs are met. By far the simplest design procedure is to ignore the sample and hold operation, designing the system just as though it were continuous. If the sampling frequency is an order of magnitude or so greater than the closed loop system BW and a zero-order hold is used, as is more or less standard practice, this approximation is reasonably valid, and it is often used when these conditions are met. Relatively conservative values of gain margin and phase margin should be established and the final design checked using the z-transformation or a simulation to assure satisfactory performance. A somewhat more exact approximation of the sampling and zero-order hold operations treats the combination as a time delay of $T/2$ seconds, as shown in Sec. 15.2. If an appropriate adjustment is made in the phase margin spec (see Sec. 9.6) the design may be carried out in the usual way.

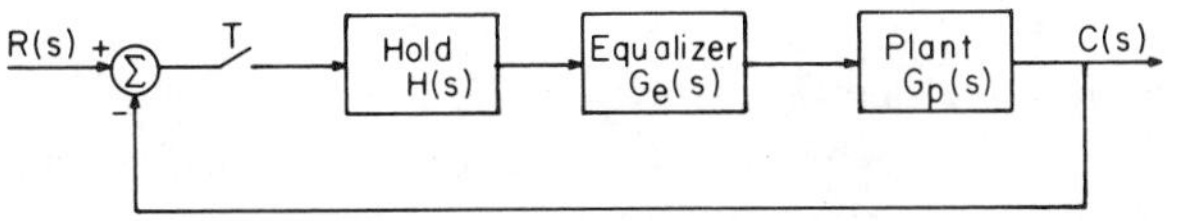

FIG. 15.25. A standard unity feedback single loop error sampled system.

A design method originally proposed by Linvill[3,4] extends the Nyquist criterion in the s-domain to SD system design. We first construct a polar plot of $\mathbf{G}_p(j\omega)$. (Assume the usual unity feedback configuration.) In the absence of equalization, the sampled open loop transfer function is

$$\mathbf{A}(j\omega) = \frac{1}{T} \sum_{n=-\infty}^{\infty} \mathbf{G}_p(j\omega + jn\omega_s) \tag{15.115}$$

where G_p includes any hold circuit used. In most applications ω_s is much larger than the frequencies for which $|G_p| \simeq 1$, so the term in Eq. (15.115) for $n = 0$ is predominant in determining stability, as defined by -1 point encirclement or lack thereof. $\mathbf{A}(j\omega)$ as given by Eq. (15.115) may be approximated near the -1 point using the $N = 0$ term and the lower order frequency sidebands. It is usually adequate to consider three components for $n = -1$, $n = 0$, and $n = 1$. When an equalizer is added, the open loop transfer function becomes

$$\mathbf{A}_e(j\omega) = \frac{1}{T} \sum_{n=-\infty}^{\infty} \mathbf{G}_e(j\omega + jn\omega_s)\,\mathbf{G}_p(j\omega + jn\omega_s) \tag{15.116}$$

$\mathbf{G}_e(j\omega)$ is chosen to provide acceptable closed loop performance. Although this is basically a trial and error design procedure, and may involve considerable algebra, it converges rapidly to the desired solution in those cases where ω_s is much larger than the closed loop BW of the final design. In most cases the total effort required to develop an acceptable design may be minimized by carrying out the following steps:

1. Construct a Bode diagram of G_p with appropriately chosen gain, just as though a continuous design is intended.
2. Approximate the sample and hold operations by a time delay of $T/2$ seconds. Construct a phase versus frequency plot for this time delay on the G_p Bode diagram.
3. Determine G_3 to approximately satisfy all system specs considering the additional phase lag due to the sample and hold.
4. Sketch $\mathbf{A}_e(j\omega)$ as given by Eq. (15.116) and make minor adjustments in G_e as deemed necessary.

Seldom are significant revisions of the first trial design required. The same basic approach may be used to develop a minor loop arrangement. The final design should be checked by z-transform analysis and/or a simulation or experimental study. It should be noted that Linvill's method as outlined here simply adds one step, the corrected Nyquist plot construction, to the first method discussed.

Example 15.11

Consider the design of an error sampled system of the form shown in Fig. 15.25 with a zero-order hold and

$$G_p(s) = \frac{100}{s(1 + s/10)(1 + s/100)} \tag{15.117}$$

The sampling rate is $f_s = 40$. The following specs are to be satisfied:

1. The open loop gain is to remain at 100.
2. The phase margin of the Nyquist plot using $\mathbf{A}_e(j\omega)$ as given by Eq. (15.116) is to be at least 35°.
3. The low frequency steady state error performance is to be degraded no more than necessary.

The Bode plot for $G_p(s)$ is shown in Fig. 15.26. A plot of phase lag versus ω for

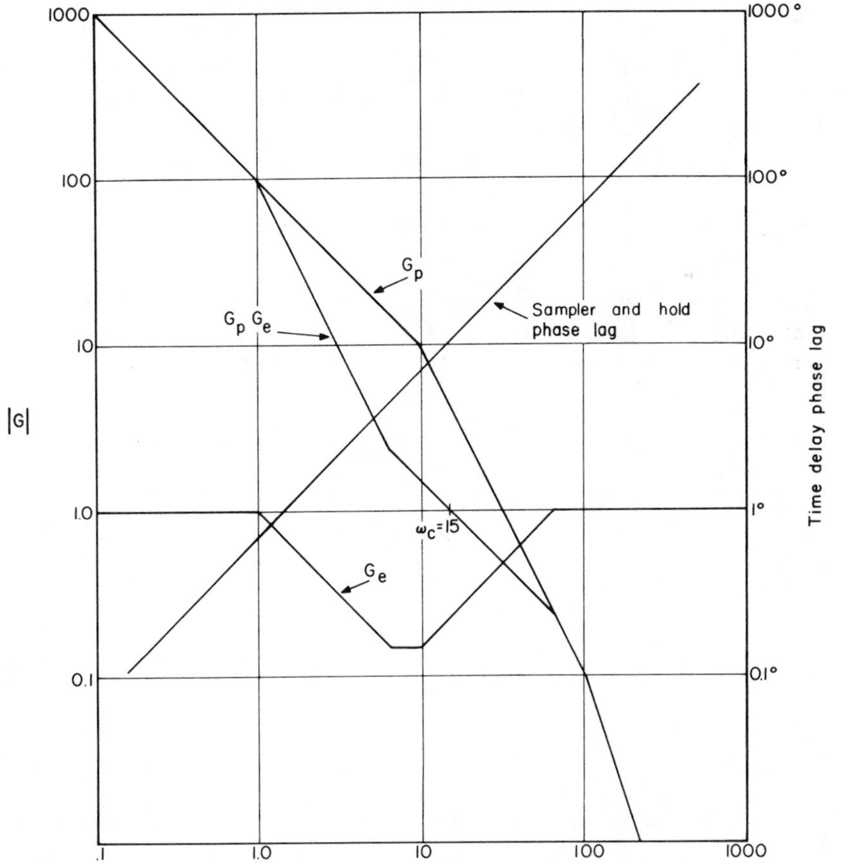

FIG. 15.26. Bode diagram for Ex. 15.11.

a time delay of $T/2 = 0.0125$ seconds is also shown on the same coordinates. It is obvious that any effort to achieve ω_c much above $\omega = 10$ is impractical. Since the phase lag due to the time delay at $\omega = 15$ is 10°, we will disregard the sample and hold operations and choose an equalizer to provide the "equivalent" continuous system with approximately 50° of phase margin, which is about 5° more than necessary. The resulting compensated Bode diagram is also shown in Fig. 15.26. It has a 10:1 span of -1 slope near $\omega_c = 15$. The required equalizer is

$$G_e(s) = \frac{(1 + s/6.5)(1 + s/10)}{(1 + s)(1 + s/65)} \tag{15.118}$$

This system should satisfy the specs nicely. Disregarding the sampler, the open loop transfer characteristic is

$$A(s) = H(s)\,G_e(s)\,G_p(s) = 100\,\frac{(1 - \epsilon^{-0.025\,s})\,(1 + s/6.5)}{s^2(1 + s)(1 + s/65)(1 + s/100)} \tag{15.119}$$

Since $\omega_c = 15$ and $\omega_s = 80\pi \simeq 250$, it is immediately apparent from Fig. 15.26 that near ω_c even the terms in Eq. (15.116) for $n = \pm 1$, which correspond to $\omega_c + \omega_s \simeq 265$ and $\omega_c - \omega_s \simeq -235$ have magnitudes less than 1/100th that of the term for $n = 0$, and may be disregarded with negligible error. A plot

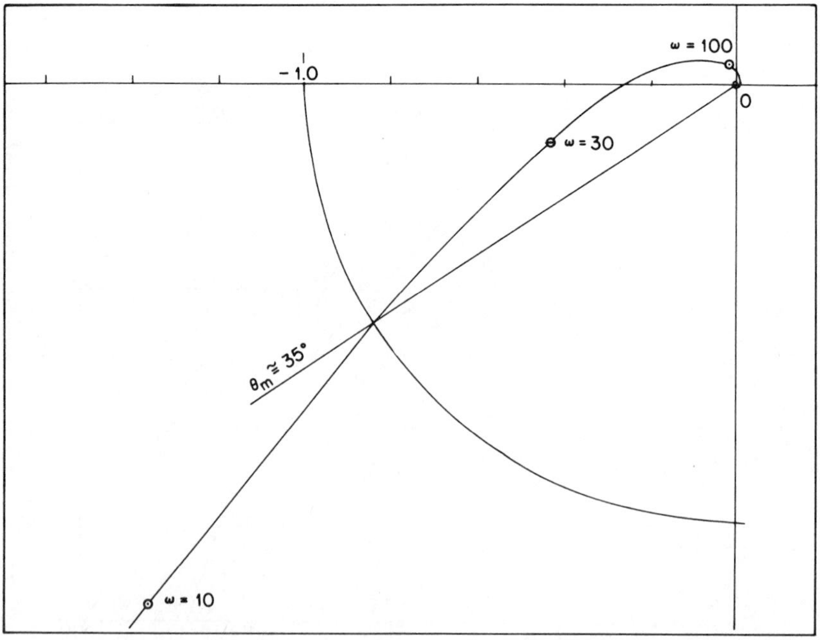

FIG. 15.27. Polar plot for Ex. 15.11.

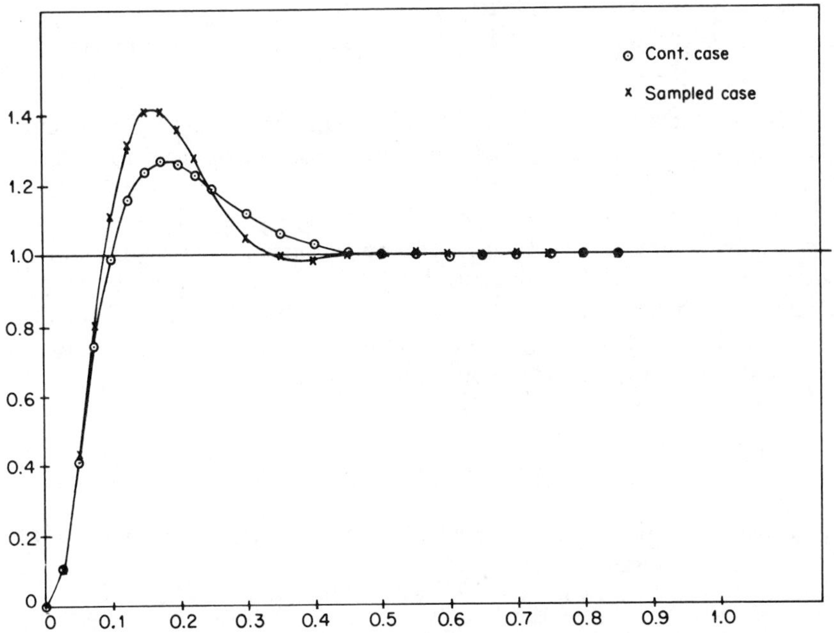

FIG. 15.28. Unit-step response patterns for the system considered in Ex. 15.11.

of $\mathbf{A}(j\omega)/T$ near unit magnitude is shown in Fig. 15.27. The phase margin is found to be approximately 35°, as desired. The unit step response of the equalized system with and without a sample and hold are shown in Fig. 15.28.

Although it is unnecessary to consider any sideband contributions in the system of Ex. 15.11, their effects upon the Nyquist plot are easily shown in the more general case, requiring only a small amount of additional effort. It is not unusual for all sideband contributions to be negligible. This is true if: 1. the system is decidely low pass in nature, and 2. the sampling frequency is significantly larger than the closed loop BW of the final design. Both of these conditions are often satisfied.

If a sampler and hold are used to isolate the equalizer network from the plant, as shown in Fig. 15.29, the open loop transfer characteristic is

$$A(z) = G_e(z)\, HG_p(z) \tag{15.120}$$

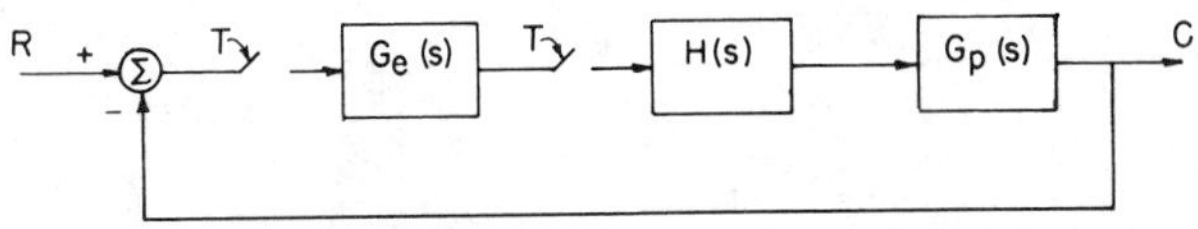

FIG. 15.29. System with the equalizer and plant separated by a sampler and hold.

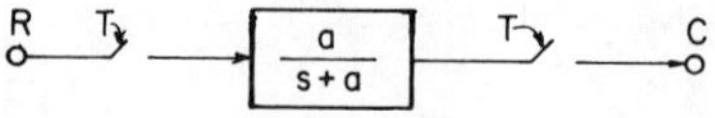

FIG. 15.30. Standard SD lag network.

In this case the equalizer and plant transfer characteristics are separable in the z-domain, and several design possibilities are apparent. We may, for example, make a z-domain Nyquist plot of $HG_p(z)$ by taking values of z on the upper half of the unit circle, and then choose $G_e(z)$ to "shape" that plot as needed near the -1 point to provide the desired closed loop performance characteristics. Lead, lag, and lag-lead networks may be used for this purpose just as in the design of continuous systems. Realization of compensator networks using RC circuits is straightforward, and is discussed in Sec. 15.6. It is sufficient for the present to observe that the standard lag network takes the form

$$G_{\text{lag}}(z) = \frac{z(1 - \epsilon^{-aT})}{z - \epsilon^{-aT}} \tag{15.121}$$

and is realized as shown in Fig. 15.30. A typical lead network is shown in Fig. 15.31, and has transfer function

$$G_{\text{lead}}(z) = \frac{z - \epsilon^{-aT}}{z} \tag{15.122}$$

Combinations of these two networks may also be used. Although it will perhaps seem awkward at first to shape the Nyquist plot directly with the various types of $G_e(z)$ functions available, an acceptable solution is generally possible with little wasted effort. The use of lag for values of z near 1 in the first quadrant of the unit circle to reduce the transfer function magnitude factor, then a lead network later to provide an acceptable margin of -1 point avoidance has become relatively standard in our previous work, and remains so here. Unfortunately a Bode diagram cannot be used directly, but the same combinations of equalizer networks are generally required in SD and continuous systems of similar form, and this helps lead us toward the desired solution.

Standard root locus design methods are also directly applicable to SD systems when the equalizing filter is separated from the plant by a sampler. A pole-zero plot of the plant and hold combination is made, and the resulting root locus diagram constructed. The root locus shape may be altered by assigning equalizer poles and zeros to modify performance as desired. Of course plant poles or zeros

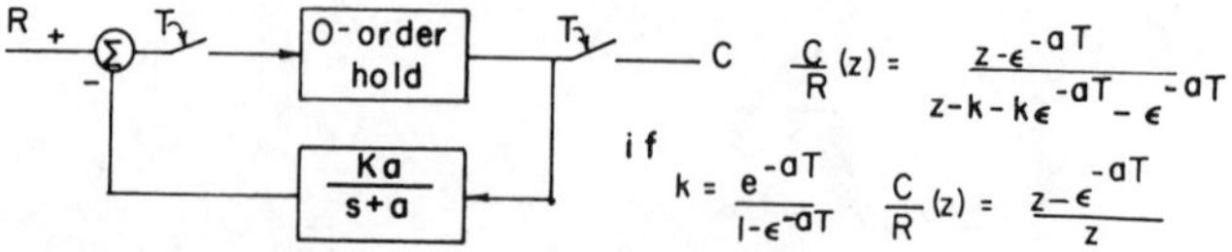

FIG. 15.31. Standard SD lead network.

outside the unit circle cannot be exactly cancelled by compensating equalizer zeros and poles, respectively, and it is not generally desirable to place equalizer poles outside the unit circle, anyway. Realizability requires the equalizer denominator to be of the same power or higher in z than the numerator, since otherwise the output sequence obtained by long division contains terms in positive powers of z which require weighting of input signals that are not yet available to develop the present output. Within these general limitations, however, much freedom of choice remains. It should be recalled that poles near the unit circle in the z-plane result in slowly decaying response patterns, whereas poles near the origin tend to yield response patterns which "settle" rapidly in terms of the number of samples required to approach steady state. The general objective of root locus design is thus to place the equalizer poles and zeros such that the closed loop poles are driven toward the origin and away from the unit circle boundary. One or more configurations which will accomplish this objective are usually apparent.

Example 15.12

Consider a system of the form shown in Fig. 15.29 with

$$G_p(s) = \frac{K}{s(s+1)} \tag{15.123}$$

$H(s)$ is a zero-order hold and $T = 1$ second. The root locus design procedure is initiated by finding the z-transform of $H(s)\,G_p(s)$, which provides

$$HG_p(z) = \frac{K}{100}\left[\frac{z + 0.86}{(z-1)(z-0.606)}\right] \tag{15.124}$$

The corresponding gain root locus is shown in Fig. 15.32. It is clear that the closed loop transfer characteristic can be reduced to a single pole at the origin by introducing an equalizer with transfer function

$$G_e(z) = \frac{z - 0.606}{z + 0.86} \tag{15.125}$$

to cancel the pole at $z = 0.606$ and the zero at $z = -0.86$, and by adjusting the gain to that level which causes the pole at $z = 1$ to move to the origin. The

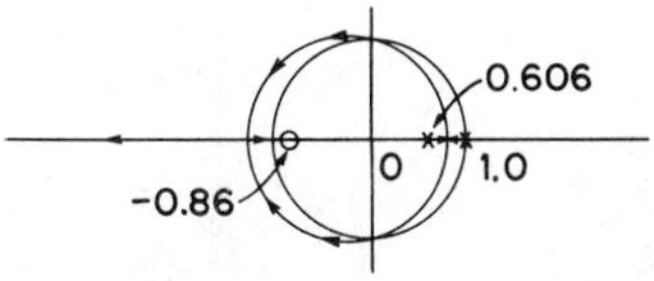

FIG. 15.32. Gain root locus for the system considered in Ex. 15.12.

resulting system has transfer function $C(z)/R(z) = z^{-1}$, and its response follows the input with one sample period delay. Of course this design can never be realized in practice because the pole and zero locations are not precisely known and they cannot be cancelled exactly, but the actual results should not differ very much from the ideal case. Realization of the desired $G_e(z)$ as defined by Eq. (15.125) is straightforward, as shown in Sec. 15.6.

Bode diagram methods cannot be applied directly to the design of systems in the z-domain, because $z = \epsilon^{sT}$ is a transcendental function of s. Use of the bilinear transformation, which maps the z-plane unit circle interior into the left half of an auxilliary plane, usually called the w-plane, converts the open loop transfer function back into a domain in which the standard Bode methods are useful. Although the transformation steps may be somewhat awkward, the advantage of being able to apply the Bode diagram to SD system design makes w-plane synthesis methods attractive. The bilinear transformation is

$$z = \frac{1 + w}{1 - w} \qquad w = \frac{z - 1}{z + 1} = \frac{1 - z^{-1}}{1 + z^{-1}} \tag{15.126}$$

If $z = \epsilon^{j\omega T}$ is substituted into the equation for w we obtain

$$w = \frac{1 - \epsilon^{-j\omega T}}{1 + \epsilon^{-j\omega T}} = j \tan \frac{\omega T}{2} \tag{15.127}$$

from which it is clear that values on the $j\omega$ axis in the s-plane and the unit circle in the z-plane are on the imaginary axis of the w-plane. We also see from Eq. (15.127) that the fictitious frequency v in the $w = u + jv$ plane is given by

$$v = \tan \left(\frac{\omega T}{2} \right) \tag{15.128}$$

Note that v is periodic in ω with period ω_s, the system sampling frequency. The inverse relationship is

$$\omega = \frac{2}{T} \tan^{-1} v \tag{15.129}$$

The relationships given in Eqs. (15.126), (15.128) and (15.129) inter-relate what is being done in one plane and its effect in the other.

The synthesis procedure in the w-plane is carried out as follows:

1. Find $HG_p(z)$, the open loop pulse transfer function of the original hold and plant.
2. Transform $HG_p(z)$ into $HG_p(w)$ through the bilinear transformation given in Eq. (15.126).

3. Plot Bode diagrams of $HG_p(w)$ in the usual way.
4. Using standard techniques, determine $G_e(w)$, the compensation network required to satisfy system specs.
5. Transform $G_e(w)$ into $G_e(z)$, the required equalizer pulse transfer function, and realize this transfer function using the methods presented in Sec. 15.6.

The w-plane synthesis problem is relatively straightforward. Illustration of the various steps involved is provided by the following example.

Example 15.13

Consider the SD system structure shown in Fig. 15.29. Assume $T = 0.1$ second and

$$G_p(s) = \frac{K}{s(1 + s/10)} \tag{15.130}$$

The following specs must be met:

1. $K_v \geq 3$, where K_v is the velocity gain constant.
2. $\theta_m \geq 50°$, where θ_m is the w-plane phase margin.

The z-transform of $HG_p(s)$ is found to be

$$Z\left[\frac{1 - \epsilon^{-Ts}}{s} \cdot \frac{10K}{s(s + 10)}\right] = \frac{0.0368K(z + 0.717)}{(z - 1)(z - 0.368)} = HG_p(z) \tag{15.131}$$

Applying the bilinear transformation to Eq. (15.131) and simplifying yields

$$HG_p(w) = \frac{K}{20} \frac{(1 - w)(1 + w/6.07)}{w(1 + w/0.462)} \tag{15.132}$$

The asymptoic plot of $HG_p(w)$ as given by Eq. (15.132) is taken as the start of the Bode design procedure in the usual way. We must choose K to satisfy the K_v spec. It may be shown[2] that K_v equals twice the gain constant in the w-plane time constant form. Thus

$$K_v = 3 \leq \frac{K}{10} \qquad K \geq 30 \tag{15.133}$$

Let $K = 30$. A Bode plot of Eq. (15.132) for this gain level is shown in Fig. 15.33. It must be kept in mind in choosing the equalizer transfer function that the zero at $w = 1$ is non-minimum phase and contributes additional *phase lag* at all frequencies. Thus, in an equivalent *phase* contribution sense, the change from -2 to -1 slope at $w = 1$ should be considered as adding phase lag at crossover

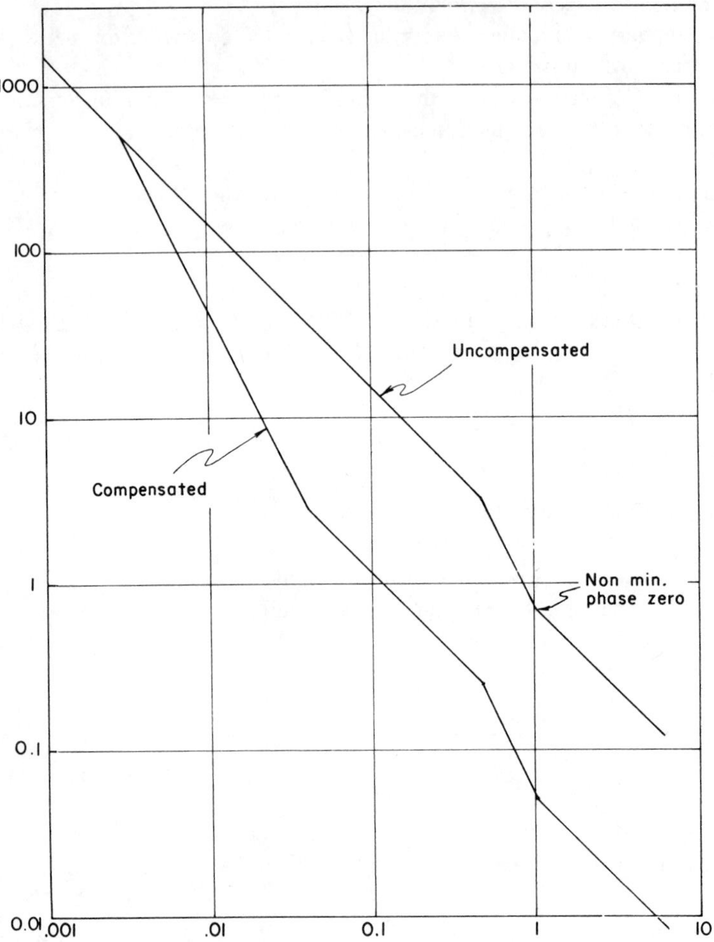

FIG. 15.33. Bode plot of $HG_p(w)$ as given by Eq. (15.131) with $K = 30$.

just as though the slope change were from -2 to -3 and caused by a denominator term. It is obvious from the Bode plot that phase lead compensation will not be particularly effective in this case. Lag compensation using the available -1 slope region is perhaps most appropriate. The lag equalizer providing the compensated shape shown on the diagram results in $\theta_m \simeq 50°$, and has transfer function

$$G_e(w) = \frac{1 + w/0.04}{1 + w/0.003} \tag{15.134}$$

Applying the inverse bilinear transformation to Eq. (15.134) and simplifying yields

$$G_e(z) = 0.078 \frac{z - 0.923}{z - 0.994} \tag{15.135}$$

Application of Bode design methods in the w-plane is illustrated by Ex. 15.13, which is fairly typical of the situations encountered. Specs on gain margin, phase margin, velocity error and other similar characteristics are easily imposed. Since $w = j\infty$ corresponds to $s = j\omega_s/2$ and $z = -1$, $G(w)$ always remains finite as $w \to \infty$. Thus $G(w)$ always has the same number of poles as zeros and one or more of its zeros are often non-minimum phase. Care must therefore be exercised in applying Bode techniques in the w-plane, but this is not a significant handicap. The required transformations are readily carried out, and the resulting equalizers are easily realized.

We have discussed in this section the application of conventional procedures based upon our continuous system experience to the design of SD systems. Little conceptual difficulty is encountered in any of the basic methods considered and the degree of difficulty in their application is not prohibitive. The method based upon a phase lag approximation of the sampler and 0-order hold is without doubt the easiest to apply, but it is also based upon the least accurate system model. Linvill's method, the z-plane root locus method, and synthesis in the w-plane can be carried out to any desired accuracy if sufficient care is taken in modeling and making calculations. Design of SD systems may also be carried out by direct synthesis of pulse transfer functions, thereby taking advantage of the tremendous flexibility available in digital processing. Design from this point of view is considered in Secs. 15.7 and 15.8.

15.6 REALIZATION OF PULSE TRANSFER FUNCTIONS IN RC FORM

It has been shown by Sklansky[5] that any realizable linear pulse transfer function may be realized by a pulsed RC network of the form shown in Fig. 15.34. The general form taken by finite realizable pulse transfer functions is

$$D(z) = \frac{\sum_{i=0}^{p} a_i z^{-i}}{\sum_{j=0}^{q} b_j z^{-j}} \tag{15.136}$$

where the coefficients a_i and b_j are real and $b_0 \neq 0$. If $b_0 = 0$ were allowed, evaluation of $D(z)$ by long division would result in the term $a_0 z/b_1$, which

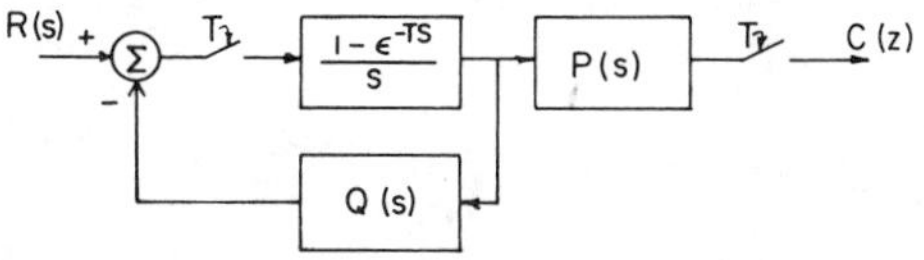

FIG. 15.34. General structure for realizing equalizer filters in RC network form.

requires weighting of the next sample value to obtain the present output. This is obviously impossible. It is necessary that p and q be finite to assure that the system need only store a finite number of past input and output sample values. For typical applications p and q are no larger than 4 or 5. The transfer function of the system shown in Fig. 15.34 is readily obtained as

$$D(z) = \frac{C(z)}{R(z)} = \frac{(1 - z^{-1})Z[P(s)/s]}{1 + (1 - z^{-1})Z[Q(s)/s]} \tag{15.137}$$

We may factor $D(z)$ into the two terms

$$D(z) = D_s(z)D_f(z) \tag{15.138}$$

where

$$D_s(z) = (1 - z^{-1})Z\left[\frac{P(s)}{s}\right] \tag{15.139}$$

and

$$D_f(z) = \frac{1}{1 + (1 - z^{-1})Z[Q(s)/s]} \tag{15.140}$$

Observe that $D_s(z)$ results from the series RC network $P(s)$, and $D_f(z)$ results from the feedback RC network $Q(s)$. Proof that any realizable $D(z)$ can be developed in this form is possible because $D_s(z)$ can be realized with arbitrary zero locations and $D_f(z)$ can be realized to provide arbitrary poles.

All poles of any realizable RC network $P(s)$ are simple and lie on the negative real axis. Thus $P(s)/s$ may be expanded in the form

$$\frac{P(s)}{s} = \sum_i \frac{k_i}{s + s_i} \qquad s_i \text{ real and} \qquad 0 \le s_i < \infty \tag{15.141}$$

One of the s_i is zero, and the k_i are arbitrary. Taking the z-transform of $P(s)/s$ as given by Eq. (15.141) yields

$$Z\left[\frac{P(s)}{s}\right] = \sum_i \frac{k_i}{1 - \epsilon^{-s_iT}z^{-1}} = \sum_i \frac{k_i z}{z - \epsilon^{-s_iT}} \tag{15.142}$$

From Eqs. (15.139) and (15.142) we see that $D_s(z)$ is of the form

$$D_s(z) = \frac{z - 1}{z}\sum_i \frac{k_i z}{z - \epsilon^{-s_iT}} = \sum_i \frac{k_i(z - 1)}{z - \epsilon^{-s_iT}} \tag{15.143}$$

Since one s_i is zero, the corresponding $\epsilon^{-s_i T} = 1$. This pole is cancelled by the $(z - 1)$ term in the numerator of Eq. (15.143). Thus $D_s(z)$ has the following properties:

1. The poles of $D_s(z)$ are in the range $0 < z_i < 1$.
2. The zeros of $D_s(z)$ are arbitrary.

$D_s(z)$ may be chosen to realize *zeros* of $D(z)$ at any desired locations.

Turning our attention to $D_f(z)$ and rearranging Eq. (15.140) provides

$$\frac{z-1}{z} Z\left[\frac{Q(s)}{s}\right] = \frac{1}{D_f(z)} - 1 \tag{15.144}$$

The left hand side of Eq. (15.144) is identical in form to the right hand side of Eq. (15.143), so its poles and zeros are restricted in exactly the same way as those of $D_s(z)$. Since the poles of the left hand side of Eq. (15.144) are zeros of $D_f(z)$, however, we see that

1. The zeros of $D_f(z)$ are in the range $0 < z_i < 1$.
2. The poles of $D_f(z)$ are arbitrary.

Thus $D_f(z)$ may be chosen to realize *poles* of $D(z)$ as desired.

A general algorithm may now be stated for developing finite realizable pulse transfer functions as follows:

1. Factor $D(z)$ into $D_s(z)\,D_f(z)$, where all finite zeros outside the range $0 < z_i < 1$ are associated with $D_s(z)$ and all poles outside the same range are associated with $D_f(z)$.

2. Assign poles and zeros inside the range $0 < z_i < 1$ to either $D_s(z)$ or $D_f(z)$ such that $D_s(z)$ has no poles and $D_f(z)$ has no zeros for infinite values of z. This assures realizability. If there are not enough such factors in $D(z)$, add arbitrary poles in the range $0 < z_i < 1$ to $D_s(z)$ and identical zeros to $D_f(z)$, thereby satisfying this requirement.

3. Find $P(s)$ and $Q(s)$ from the inversion formulas

$$P(s) = sZ^{-1}\left[\frac{zD_s(z)}{z-1}\right] \tag{15.145}$$

and

$$Q(s) = sZ^{-1}\left\{\frac{z[1 - D_f(z)]}{(z-1)D_f(z)}\right\} \tag{15.146}$$

where the notation Z^{-1} denotes the process of obtaining the function of s corresponding to the associated z-transform.

Application of this procedure is straightforward, as illustrated by the following example.

Example 15.14

Realize the pulse transfer function

$$D(z) = \frac{z}{z-1} \tag{15.147}$$

which represents an integrator.

The zero at the origin and the pole at unity must be assigned to $D_s(z)$ and $D_f(z)$, respectively. Thus our first trial is

$$D_s(z) = z \qquad D_f(z) = \frac{1}{z-1} \tag{15.148}$$

Neither $D_s(z)$ nor $D_f(z)$ has acceptable performance at $z = \infty$, so we arbitrarily add a pole at $z = \alpha$, $0 < \alpha < 1$, to $D_s(z)$ and a compensating zero to $D_f(z)$, resulting in

$$D_s(z) = \frac{z}{z-\alpha} \qquad D_f(z) = \frac{z-\alpha}{z-1} \tag{15.149}$$

Application of Eqs. (15.145) and (15.146) provides

$$P(s) = \frac{(1-\alpha)s + a}{(1-\alpha)(s+a)} \tag{15.150}$$

and

$$Q(s) = \frac{-a}{s+a} \tag{15.151}$$

where

$$\alpha = \epsilon^{-aT} \tag{15.152}$$

defines the solution parameter a in terms of α. We may choose α arbitrarily in the range $0 < \alpha < 1$ to facilitate the realization of $P(s)$ and $Q(s)$ with reasonable component sizes and impedance levels.

Note that $P(s)$ and $Q(s)$ as obtained using this synthesis procedure generally require a gain other than unity. When this gain is greater than one, it can often be incorporated in the sample and hold operations. In many applications a realization with $P(s) = 1$ is possible, and in this case only one sample and hold combination is required.

15.7 SAMPLED-DATA SYSTEM DESIGN USING DIGITAL COMPENSATORS

A digital compensator is an arbitrary realizable device which accepts sampled inputs and provides sampled outputs. The mathematical model of a digital compensator thus has a sampler in the input and output signal paths. Because a digital computer may be used to carry out processing in extreme cases, tremendous flexibility is available in designing systems using digital compensation. Digital compensators may also often be realized using simple RC networks as we have already seen. In the design of systems using digital compensation we may concentrate upon obtaining desirable response characteristics within the limits imposed by pulse transfer function realizability. This is a distinctly different approach from that normally used in continuous systems, where closed loop stability is the overwhelming consideration. Stability of the digital system is assured by proper interpretation of the design relationships.

The general principles of digital compensator design will be developed using the configuration shown in Fig. 15.35, from which we see that

$$\frac{C(z)}{R(z)} = K(z) = \frac{D(z)G(z)}{1 + D(z)G(z)} \tag{15.153}$$

If a particular realizable $K(z)$ is desired we may solve Eq. (15.153) for $D(z)$ to obtain

$$D(z) = \frac{K(z)}{G(z)[1 - K(z)]} = \frac{\sum_{i=0}^{p} a_i z^{-i}}{\sum_{j=0}^{q} b_j z^{-j}} \tag{15.154}$$

Recall that realizability requires $b_0 \neq 0$ in this general expression. This, plus the fact that poles and zeros of $F(z)$ outside the unit circle cannot be cancelled

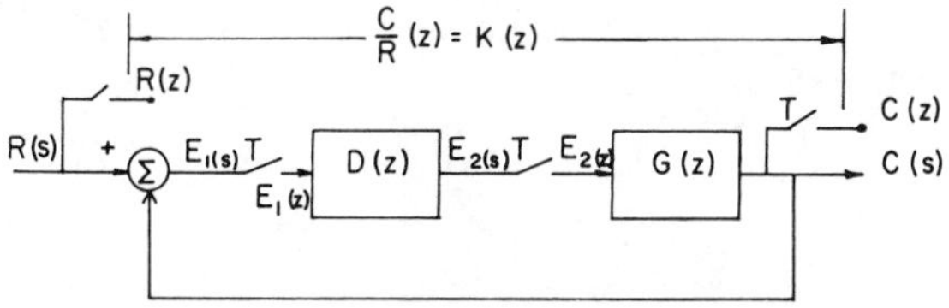

FIG. 15.35. A typical sampled data-system with digital compensation.

using compensating zeros and poles, respectively, in $D(z)$, results in several limitations upon the transfer characteristics which can be realized in particular situations. We will develop general rules relating to what can be done using digital equalization and how it should be accomplished.

A standard for closed loop system performance is provided by the so-called *minimal prototype response functions.* These functions provide zero steady state sample point error and settle to zero error in a minimum number of samples for test inputs such as step, ramp, or acceleration signals. The acceleration prototype also has zero steady state error for a step or ramp input, and the ramp prototype has zero final error for a step input, although the transient responses for singularity inputs of lower order than that for which the system is designed are generally very poor. Obviously $G(z)$, $D(z)$, and $K(z)$ must all be physically realizable and the pole-zero *cancellations* occurring between $G(z)$ and $D(z)$ must all take place within the unit circle. The general form for a realizable plant transfer function is

$$G(z) = \frac{p_m z^{-m} + \cdots + p_n z^{-n}}{q_0 + q_1 z^{-1} + \cdots + q_b z^{-b}}$$

$$= z^{-m} \frac{p_m + \cdots + p_n z^{-n+m}}{q_0 + q_1 z^{-1} + \cdots + q_b z^{-b}} \tag{15.155}$$

where $q_0 \neq 0$ and $m \geq 0$. If $m = 0$, the plant output can be influenced without delay by an impulse input. For $m = m_1$, however, it requires m_1 sample periods of delay before any output effect is observed, as caused by a transport lag, for example. This delay cannot be compensated for with any realizable $D(z)$, so $K(z)$ must have at least as many sample periods of delay as $G(z)$, or $K(z)$ must contain the factor z^{-r}, where $r \geq m$.

Each minimal prototype responds to a specific test input with zero steady state error at all *sample points* following a transient period of minimum duration. Error between sample points may persist indefinitely. It is also possible to design systems with zero error for all time following a minimum duration transient period, and that problem is considered later. Referring to Fig. 15.35, we see that

$$E_1(z) = R(z)[1 - K(z)] \tag{15.156}$$

Step, ramp, and parabolic input functions are all represented in the z-domain by functions of the form

$$R(z) = \frac{R_N(z^{-1})}{(1 - z^{-1})^m} \tag{15.157}$$

where $R_N(z^{-1})$ is a general polynomial in z^{-1} not containing factors of the form $(1 - z^{-1})$. Substituting $R(z)$ from Eq. (15.157) into Eq. (15.156) and setting the final error equal to zero using the final value theorem leads to the requirement

$$1 - K(z) = (1 - z^{-1})^m B(z) \tag{15.158}$$

where $m = 1, 2,$ and 3 for the step, ramp, and parabolic inputs respectively. $B(z)$ is an unspecified ratio of polynomials in z^{-1}. For response in minimum time $B(z) = 1$ is chosen. Assuming a step input, the corresponding minimal prototype transfer function is

$$K_s(z) = z^{-1} \tag{15.159}$$

and the system follows the step input with one sample period delay. The minimal prototype ramp response is similarly found to be

$$K_r(z) = 2z^{-1} - z^{-2} \tag{15.160}$$

and that for a parabolic input is

$$K_p(z) = 3z^{-1} - 3z^{-2} + z^{-3} \tag{15.161}$$

Note that all systems which provide zero error at the sample points after a finite number of steps must have transfer characteristics consisting *only* of a numerator polynomial in z^{-n}.

The minimal prototype responses are perhaps most appropriately used as a frame of reference for actual designs, since minimal prototype performance is often not possible. Additional sample periods of delay [multiplication of the results in Eqs. (15.159), (15.160) and (15.161) by z^{-n}] must be allowed when time delay is present in the plant. Any zero of the plant outside the unit circle must be included in $K(z)$, since zeros of the plant not cancelled by poles of $D(z)$ are also zeros of $K(z)$ [see Eq. (15.153)] and cascade cancellation of poles and zeros outside the unit circle, which may seem possible during analysis, cannot actually be realized. In addition to these limitations, although minimal prototypes respond ideally to the specific inputs for which they are designed, their performance for inputs with changes *more* severe than those for which the response is minimal (a step or ramp applied to the parabolic prototype, for example) is generally very erratic for the first few samples, as easily shown by dividing out the product $K_p(z)/(1 - z^{-1})^m$ for $m = 1$ and $m = 2$.

Before considering closed-loop compensation to obtain a desired transfer characteristic (perhaps, but not necessarily, a minimal prototype form) consider the limits on open-loop system design. Given a plant $G(z)$ with an arbitrary arrangement of poles and zeros, cascade compensation of transfer characteristics is possible only within the following limitations:

1. $K(z)$ *must* contain all poles and zeros of $G(z)$ which are not *within* the unit circle.

2. $K(z)$ *must* contain at least as many units of response time delay as exhibited by the plant itself.

It is clear that prototype performance is only possible in this case when $G(z)$ contains no poles outside or on the unit circle and does not have a time delay in excess of one sampling period. Somewhat more flexibility is available in establishing closed-loop transfer functions, since poles of $G(z)$ outside the unit circle need not appear in $K(z)$ in this case.

Returning to the closed loop system shown in Fig. 15.35, the effects of poles and zeros of $G(z)$ on and outside the unit circle can best be determined by assuming

$$G(z) = \frac{z - z_0}{z - z_p} F(z) \tag{15.162}$$

where all poles and zeros of $F(z)$ are within the unit circle and z_0 and z_p denote the location of a pole and zero, respectively, of $G(z)$ outside the unit circle. Letting $K(z)$ denote the desired closed-loop transfer function, as before, we may substitute $G(z)$ from Eq. (15.162) into Eq. (15.154) to obtain

$$D(z) = \frac{(z - z_p)K(z)}{(z - z_0)F(z)[1 - K(z)]} \tag{15.163}$$

Since the plant pole and zero at z_p and z_0, respectively, are outside the unit circle, they cannot be perfectly cancelled by choice of $D(z)$, so no attempt will be made to realize $D(z)$ as given by Eq. (15.163). We observe, instead, that

1. $K(z)$ must contain among its zeros all zeros of the plant $G(z)$ outside or on the unit circle.
2. $1 - K(z)$ must contain as its zeros all those poles of the plant $G(z)$ outside or on the unit circle.

When these conditions are met the remainder of $D(z)$ as given by Eq. (15.163) may be realized and the desired response $K(z)$ can be closely approximated.

In summary, the following conditions must be satisfied in the compensation of closed loop SD systems:

1. Physical realizability requires that the denominator of $K(z)$ contain a constant term (the denominator may consist only of a constant).
2. If the numerator of the plant transfer function $G(z)$ contains a factor z^{-m} representing a minimum of m sample periods delay, the numerator of $K(z)$ must also contain the factor z^{-m}.
3. For minimal prototype performance $K(z)$ consists only of a numerator polynomial of minimum order consistent with all other requirements.
4. If $G(z)$ has zeros outside or on the unit circle, they must be included in $K(z)$.

5. If $G(z)$ has poles outside or on the unit circle, they must be included as factors of $1 - K(z)$.

6. If the closed loop system is to respond to the singularity inputs (steps, ramps, etc.) with zero steady state error, it is necessary that $1 - K(z)$ contain the factor $(1 - z^{-1})^n$, where n takes on the values one, two, and three for the step, ramp, and parabolic inputs, respectively.

7. The plant must contain a number of integrations consistent with the desired performance.

The following example, in which these conditions are put to use, illustrates the general digital compensator design procedure.

Example 15.15

Consider a system with plant transfer function

$$G(z) = \frac{(1 + z^{-1})(1 + 2z^{-1})z^{-1}}{(1 - z^{-1})(1 - 0.5z^{-1})(1 - 0.3z^{-1})} \tag{15.164}$$

which has two zeros and one pole not in the interior of the unit circle. Determine an equalizer $D(z)$ such that the unity feedback combination has zero sample point error for a unit ramp input after a minimum number of samples.

Since $G(z)$ contains the factor z^{-1} in its numerator, the lowest order term in the $K(z)$ numerator must also be z^{-1}. Minimum sampling point settling time is desired, so $K(z)$ must be of the form

$$K(z) = (1 + z^{-1})(1 + 2z^{-1})(a_1 z^{-1} + a_2 z^{-2} + \cdots) \tag{15.165}$$

which contains the two necessary zeros from $G(z)$, has z^{-1} as the lowest order term, and consists of a numerator polynomial only. For zero steady-state ramp error it is required that

$$1 - K(z) = (1 - z^{-1})^2(1 + b_1 z^{-1} + b_2 z^{-2} + \cdots) \tag{15.166}$$

This also satisfies the requirement that $1 - K(z)$ contain the $G(z)$ pole at $z = 1$.

Since the system is to provide zero ramp response error in minimum time, $K(z)$ should contain no more terms than necessary to satisfy all restrictions. The required a_i and b_i are evaluated by expanding the right sides of Eqs. (15.165) and (15.166), substituting the expanded $K(z)$ from Eq. (15.165) into Eq. (15.166), and equating coefficients. It is clear from Eq. (15.165) and (15.166) that a_1, a_2, b_1, and b_2 are necessary. If only a_1 and b_1 are used, three equations in two unknowns result, and these equations cannot be satisfied. When four terms are carried, we get four equations in four unknowns. If further terms are included we get fewer equations than unknowns, and one or more of the coefficients can be chosen arbitrarily. This provides added flexibility, but does not result in minimum time response. Expanding Eqs. (15.165) and (15.166), substituting $K(z)$ from Eq. (15.165) into Eq. (15.166) and equating coefficients of like power in z^{-1} provides the requirements

$$\begin{aligned} -2b_1 + 1 + b_2 &= -a_2 - 3a_1 & \qquad b_1 - 2 &= -a_1 \\ -2b_2 + b_1 &= -3a_2 - 2a_1 & \qquad b_2 &= -2a_2 \end{aligned} \tag{15.167}$$

Solving for the a_i and b_i yields

$$a_1 = \frac{19}{36} \qquad a_2 = -\frac{13}{36} \qquad b_1 = \frac{53}{56} \qquad b_2 = \frac{13}{18} \tag{15.168}$$

Thus the desired transfer function is

$$K(z) = \frac{1}{36}(19z^{-1} + 44z^{-2} - z^{-3} - 26z^{-4}) \tag{15.169}$$

$K(z)$ as given by Eq. (15.169) provides the unit step and unit ramp responses shown in Figs. 15.36 and 15.37. The unit step response exhibits the large overshoot transient typical of ramp prototypes. The severe transient response associated with minimal response prototypes often causes saturation of the plant

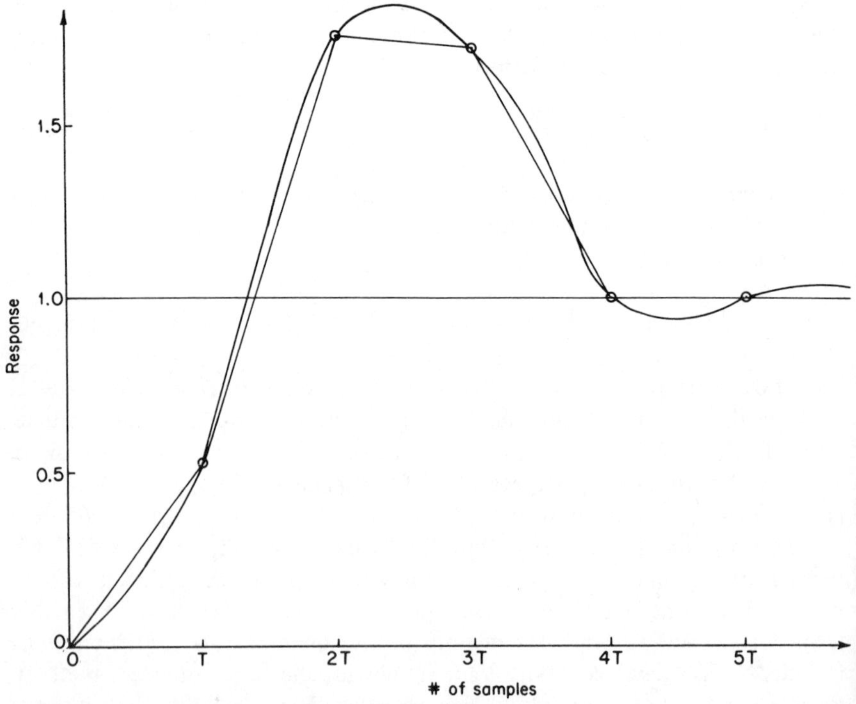

FIG. 15.36. Unit step response of the system considered in Ex. 15.15.

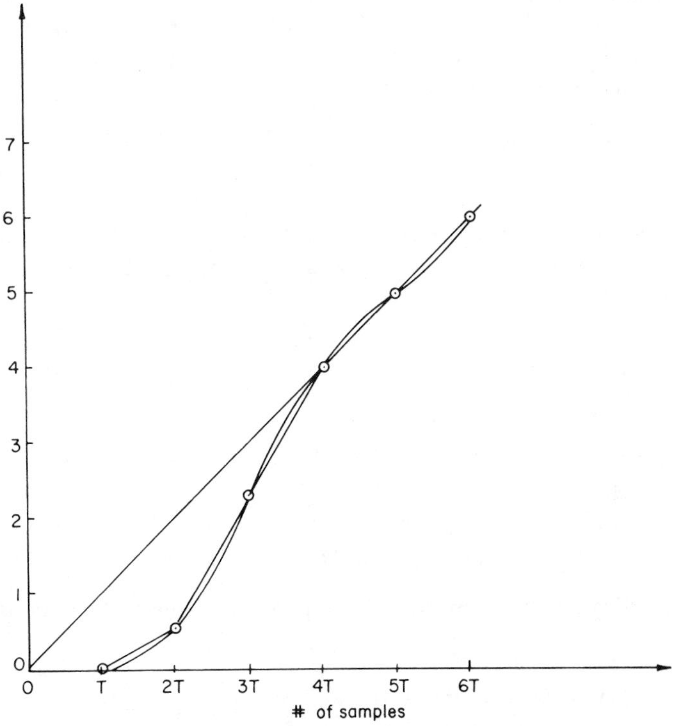

FIG. 15.37. Unit ramp response of the system considered in Ex. 15.15.

elements, and minimal prototypes are not as desirable as it would first appear. A method sometimes used to "tone down" the minimum prototype performance involves the addition of a term, called a *staleness factor,* in the denominator of the desired transfer function. We thus might use

$$K(z) = \frac{K_p(z)}{(1 - az^{-1})^N} \tag{15.170}$$

where $K_p(z)$ is the reference prototype design, a is a constant such that $-1 < a < 1$, and N is a positive integer. It is not common to use values of N other than 1, and a is generally taken as positive. Two or more staleness factors may be used if desired. As a increases from zero, which provides the prototype response, toward unity, the input to output performance becomes increasingly sluggish. As a progresses from zero toward -1, the response becomes more oscillatory.

Example 15.16

Consider the general effect of using a staleness factor upon the unit step response of the minimal prototype for a unit step input.

The desired transfer function becomes

$$K(z) = \frac{kz^{-1}}{1 - az^{-1}} \tag{15.171}$$

where k is included as a parameter to allow setting the steady state response for a unit step input to unity. Application of the final value theorem shows that $k = 1 - a$ is required, so

$$K(z) = \frac{(1 - a)z^{-1}}{1 - az^{-1}} \tag{15.172}$$

is desired. For a unit step input the response $C(z)$ is given by

$$C(z) = \frac{(1 - a)z^{-1}}{(1 - z^{-1})(1 - az^{-1})} \tag{15.173}$$

Dividing out $C(z)$ provides the response as a general function of the parameter a as

$$C(z) = (1 - a)z^{-1} + (1 - a^2)z^{-2} + (1 - a^3)z^{-3} + \cdots \tag{15.174}$$

The corresponding response patterns are shown for several values of a in Fig. 15.38. It is clear that values of a in the range $0.3 \le a \le 0.6$ are most effective in improving the transient response. This is particularly apparent when the unit step response of a ramp prototype is considered.

Since the staleness factor simply adds one more pole in $K(z)$ inside the unit circle and makes a gain change necessary to assure the desired steady-state performance, determination of the $D(z)$ required to provide the modified $K(z)$ is carried out exactly as before. The value of staleness factor used may be chosen to provide a compromise between the response characteristics achieved for the standard types of inputs, with response for those inputs most often encountered given greatest consideration.

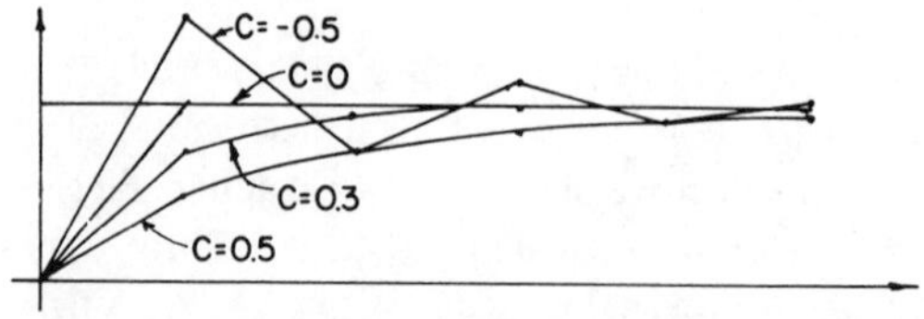

FIG. 15.38. The effects of staleness factors upon the minimal prototype unit step response.

The minimal prototype systems all exhibit, in general, ripple between sample points despite their optimum sampled performance. We may also design ripple free systems by requiring that the transfer function between input and the plant drive signal consist only of a finite number of powers of z^{-1}. If this is done the plant drive signal becomes a constant after a finite number of samples for any of the singularity inputs, and ripple free performance is assured. Obviously this is only possible if $G(z)$ contains the requisite number of integrations to provide the desired output with a constant input. For the system shown in Fig. 15.35,

$$E_2(z) = \frac{C(z)}{G(z)} \tag{15.175}$$

and

$$C(z) = K(z)R(z) \tag{15.176}$$

Combining Eqs. (15.175) and (15.176) provides

$$\frac{E_2(z)}{R(z)} = \frac{K(z)}{G(z)} \tag{15.177}$$

from which it is clear that $K(z)$ must contain *all* of the zeros of $G(z)$ if the transfer function is to consist only of a numerator polynomial in z^{-n}. Since the minimal prototypes are developed requiring that $K(z)$ contain only those $G(z)$ zeros located outside and on the unit circle, this requirement for ripple free design is more severe. Because $K(z)$ must, in general, be more complex than for minimal prototype design, ripple free designs will usually require more sample periods to settle into steady state. The ripple free design procedure may be summarized as follows:

1. All rules for minimal prototype design must be satisfied.
2. $G(z)$ must be capable of generating a continuous output equal to the input for a constant excitation signal.
3. $K(z)$ must contain as its zeros *all* zeros of $G(z)$.

Application of these rules to design involves the same steps required for minimal prototype design. The additional zeros required in $K(z)$ increase the algebraic complexity somewhat. The ripple free design for a parabolic input will follow an input ramp or step with ripple free characteristics and the ramp prototype is ripple free for step inputs, although the transient performance for lower order singularity inputs than that for which the system was designed is very poor. Staleness factors may be used to effect transient response improvements just as in the minimal prototype case.

Several design procedures which have proved fruitful for certain types of continuous systems may be carried over to the SD area. State variable feedback can be used and the feedback coefficients adjusted to achieve the desired performance rather than designing a cascade equalizer. Systems may be designed with finite settling time other than in a minimum number of sample periods, and the added design flexibility used to optimize performance relative to some criterion. For example, we could design a ramp following system with two or three extra samples of settling time and choose the coefficients to minimize the mean square sample point error for a unit step input, thereby assuring reasonable transient performance. Many alternate possibilities are apparent. Feedforward design also applies to digital systems. These topics and many others are considered in the various references listed at the end of the chapter.

15.8 SAMPLED DATA ANALYSIS USING STATE TRANSITION FUNCTIONS

The state transition methods introduced in Chap. 14 are readily applied to the analysis and synthesis of sampled systems. These methods are particularly useful in the efficient development of a program for simulating digital systems (or for studying analog systems using a near-equivalent digital model) on the digital computer. State transition methods also apply directly to the study of systems with multiple samplers, including multi-rate systems, and certain types of nonlinear sampled systems. Our intention in this section is to introduce the fundamental concepts of state transition analysis.

Consider the standard error sampled linear time invariant configuration shown in Fig. 15.39. The performance of this system is described by

$$\dot{\mathbf{x}} = \mathbf{A}\mathbf{x} + \mathbf{B}m \tag{15.178}$$

where m changes only at the sample points. Let m_n denote the value of m on the interval $[nT, (n+1)T]$, and let $\mathbf{x}^n$ denote the state at $t = nT$, with $x_1{}^n = C(nT)$. We may then write

$$\mathbf{x}(nT + t) = \mathbf{\Phi}(t)\mathbf{x}^n + \int_0^t \mathbf{\Phi}(t - \tau)\mathbf{B}m_n \, d\tau \qquad 0 \leq t \leq T \tag{15.179}$$

where $\mathbf{\Phi}(t)$ is the usual transition matrix obtained by solving

$$\dot{\mathbf{\Phi}}(t) = \mathbf{A}\mathbf{\Phi}(t) \qquad \mathbf{\Phi}(0) = \mathbf{I} \tag{15.180}$$

Since m_n is a scalar constant, it may be taken outside the integral sign if desired, and

$$m_n = r(nT) - x_1(nT) = r^n - x_1{}^n \tag{15.181}$$

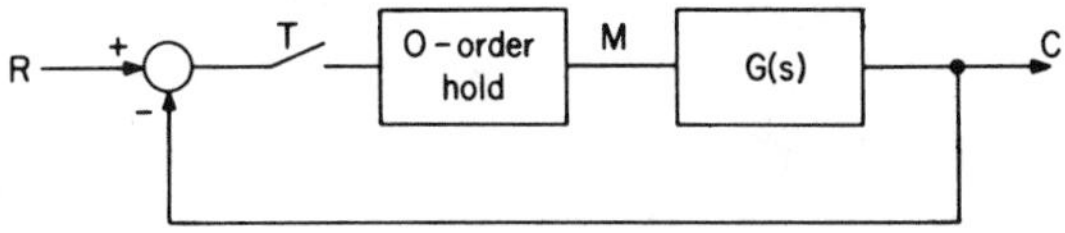

FIG. 15.39. A standard error sampled system.

If the control input is known, Eqs. (15.179) and (15.181) may be used to develop a recursive relationship which provides system state on the time interval between any pair of sample points. Since $\mathbf{\Phi}(t - \tau)$ and $\mathbf{B}$ are characteristics of the system not depending upon state or input, the integral in Eq. (15.179) need only be evaluated once. Define

$$\mathbf{\Phi}_m(t) = \int_0^t \mathbf{\Phi}(t - \tau)\mathbf{B}\,d\tau \tag{15.182}$$

and Eq. (15.179) reduces to

$$\mathbf{x}(nT + t) = \mathbf{\Phi}(t)\mathbf{x}^n + m_n\mathbf{\Phi}_m(t) \tag{15.183}$$

In most applications only system state at the sample points is of interest, and $\mathbf{x}[(n + 1)T] = \mathbf{x}^{n+1}$ is obtained from Eq. (15.183) as

$$\mathbf{x}^{n+1} = \mathbf{\Phi}(T)\mathbf{x}^n + m_n\mathbf{\Phi}_m(T) \tag{15.184}$$

where m_n is given by Eq. (15.181). This result may be used recursively to obtain system state at successive sample points by hand analysis, and it also provides a basis for an efficient digital computer program to automatically provide this data. (See App. E.)

Stability of the closed loop system shown in Fig. 15.39 can also be determined from Eq. (15.184) without difficulty. Since m_n depends upon $x_1(nT)$, Eq. (15.184) may be rewritten as

$$\begin{aligned}\mathbf{x}^{n+1} &= \mathbf{\Phi}(T)\mathbf{x}^n + r^n\mathbf{\Phi}_m(T) - x_1{}^n\mathbf{\Phi}_m(T)\\ &= \mathbf{\Phi}(T)\mathbf{x}^n + r^n\mathbf{\Phi}_m(T) - [\mathbf{\Phi}_m(T), 0, \ldots, 0]\mathbf{x}^n\\ &= \mathbf{\Phi}_c(T)\mathbf{x}^n + r^n\mathbf{\Phi}_m(T)\end{aligned} \tag{15.185}$$

where

$$\mathbf{\Phi}_c(T) = \mathbf{\Phi}(T) - [\mathbf{\Phi}_m(T), 0, \ldots, 0] \tag{15.186}$$

is a modified transition function which takes the feedback into consideration. The partitioned matrix $[\mathbf{\Phi}_m(T), \mathbf{0}, \ldots, \mathbf{0}]$ is obtained by adding enough zero columns to the right of column matrix $\mathbf{\Phi}_m(T)$ to give as many columns as there are in $\mathbf{\Phi}(T)$. The closed loop sampled system is stable if and only if all eigenvalues of $\mathbf{\Phi}_c(T)$ are in the unit circle.

The general transition matrix approach may be applied to multi-rate systems, to variable rate systems, and to certain types of nonlinear systems in which the nonlinearity outputs are sampled. The same basic concepts we have used in our analysis of the simple situation shown in Fig. 15.39 are applicable in all cases. One interesting application of these principles is in the simulation of continuous systems on a digital computer, as discussed in App. E. A sampler and 0-order hold are arbitrarily introduced at one or more points in the system, the sampling rate is chosen sufficiently high to assure that the sampled system response accurately approximates that of the original continuous system, the recursive transition matrix equation is developed, and the response for the desired inputs obtained by iterative application of the result. Several samplers may be introduced to limit the difficulty in obtaining the required transition matrices if desired, and systems with single valued frequency insensitive nonlinearities are similarly treated by placing a sampler and 0-order hold at each nonlinear element output. The reader is referred to App. E for an example illustrating this procedure.

Application of the state transition method for the analysis, simulation, and synthesis of systems involving two or more samplers, including multi-rate systems, follows the same general guidelines introduced here. The transfer function elements between samplers are reduced to transition matrix representations and the results algebraically interrelated to obtain the response patterns for arbitrary initial conditions and the desired inputs. Prototype response patterns can be developed in this way and synthesis procedures analogous to those previously considered in the z-plane are available.[6,7]

15.9 CONCLUDING COMMENTS

It has been shown in this chapter that the analysis and design of SD systems is not radically different than that of the corresponding continuous systems. Many of the standard design procedures for continuous systems can be applied with minor modifications to SD systems as well. The z-transform and the z-plane are analogous in the SD case to the Laplace transform and the s-plane for continuous systems, and comparable stability interpretations have been made. Of the standard design procedures discussed, Linvill's method and the approximation of the sampler and 0-order hold by a time delay are most often used.

Design of SD systems may also be carried out using methods entirely distinct from those used in continuous systems. The minimal prototype and ripple free design procedures are based upon algebraic manipulations of the z-transformed transfer functions to provide specific response patterns. Several additional methods of this nature are possible, although not discussed here. So-called dead-beat systems may be designed, for example, which provide finite time zero error response without overshoot for step or ramp inputs. Equalizer design may be carried out entirely in the time domain[6,7] using state representation and transition matrices to provide minimal prototypes, ripple free response, etc. Since this is essentially an overlap of the z-domain methods considered in Sec. 15.7, it has not been presented here.

We have not considered the response of SD systems between sample points. A straightforward application of advanced z-transforms provides inter-sample information. Although the algebra involved is somewhat tedious, overall performance, including that between samples, should generally be checked before a design is put to use. Of course a simulation study is also appropriate here, as in checking many other aspects of system performance.

REFERENCES

1. Oliver, B. M., J. R. Pierce and C. E. Shannon, The Philosophy of Pulse Code Modulation, *Proceedings IRE,* vol. 36, No. 11, pp. 1324–1331, November, 1948.
2. Tou, J. T., "Digital and Sampled-Data Control Systems," McGraw-Hill Book Co., Inc., New York, p. 238, 1959.
3. Linvill, W. K., Sampled Data Control Systems Studies through Comparison with Amplitude Modulation, *Transactions AIEE,* vol. 70, pt. II, pp. 1779–1788, 1951.
4. Linvill, W. K. nd J. M. Salzer, Analysis of Control Systems Involving a Digital Computer, *Proceedings IRE,* vol. 41, no. 7, pp. 901–906, 1953.
5. Sklansky, J., Pulsed RC Networks for Sampled-data Systems, *IRE Convention Record,* pt. 2, pp. 81–99, March 1956.
6. Tou, J. T., "Modern Control Theory," McGraw-Hill Book Co., Inc., New York, 1964.
7. Dorf, Richard C., "Time Domain Analysis and Design of Control Systems," Addison-Wesley, Inc., Reading, Mass., 1965.

Appendix A
TABLE OF SELECTED LAPLACE TRANSFORM PAIRS

Appendix A. Table of Selected Laplace Transform Pairs

No.	$F(s)$	$f(t)$
1.	1	$\delta(t)$ unit impulse at $t = 0$
2.	$\frac{1}{s}$	$u(t)$ unit step at $t = 0$
3.	$\frac{1}{s^2}$	$t\,u(t)$ unit ramp at $t = 0$
4.	$\frac{1}{s^n}$	$\frac{t^{(n-1)}}{(n-1)!}$ for integer $n > 0$
5.	$\frac{1}{s+\alpha}$	$e^{-\alpha t}$
6.	$\frac{1}{(s+\alpha)(s+\gamma)}$	$\frac{e^{-\alpha t} - e^{-\gamma t}}{\gamma - \alpha}$
7.	$\frac{s + a_0}{(s+\alpha)(s+\gamma)}$	$\frac{(a_0 - \alpha)e^{-\alpha t} - (a_0 - \gamma)e^{-\gamma t}}{\gamma - \alpha}$
8.	$\frac{1}{s(s+\alpha)(s+\gamma)}$	$\frac{1}{\alpha\gamma} + \frac{\gamma e^{-\alpha t} - \alpha e^{-\gamma t}}{\alpha\gamma(\alpha - \gamma)}$

Appendix A. **Table of Selected Laplace Transform Pairs**

No.	$F(s)$	$f(t)$
9.	$\dfrac{s+a_0}{s(s+\alpha)(s+\gamma)}$	$\dfrac{a_0}{\alpha\gamma}+\dfrac{a_0-\alpha}{\alpha(\alpha-\gamma)}e^{-\alpha t}+\dfrac{a_0-\gamma}{\gamma(\gamma-\alpha)}e^{-\gamma t}$
10.	$\dfrac{s^2+a_1s+a_0}{s(s+\alpha)(s+\gamma)}$	$\dfrac{a_0}{\alpha\gamma}+\dfrac{\alpha^2-a_1\alpha+a_0}{\alpha(\alpha-\gamma)}e^{-\alpha t}-\dfrac{\gamma^2-a_1\gamma+a_0}{\gamma(\alpha-\gamma)}e^{-\gamma t}$
11.	$\dfrac{1}{(s+\alpha)(s+\gamma)(s+\delta)}$	$\dfrac{e^{-\alpha t}}{(\gamma-\alpha)(\delta-\alpha)}+\dfrac{e^{-\gamma t}}{(\alpha-\gamma)(\delta-\gamma)}+\dfrac{e^{-\delta t}}{(\alpha-\delta)(\gamma-\delta)}$
12.	$\dfrac{s+a_0}{(s+\alpha)(s+\gamma)(s+\delta)}$	$\dfrac{a_0-\alpha}{(\gamma-\alpha)(\delta-\alpha)}e^{-\alpha t}+\dfrac{a_0-\gamma}{(\alpha-\gamma)(\delta-\gamma)}e^{-\gamma t}+\dfrac{a_0-\delta}{(\alpha-\delta)(\gamma-\delta)}e^{-\delta t}$
13.	$\dfrac{s^2+a_1s+a_0}{(s+\alpha)(s+\gamma)(s+\delta)}$	$\dfrac{\alpha^2-a_1\alpha+a_0}{(\gamma-\alpha)(\delta-\alpha)}e^{-\alpha t}+\dfrac{\gamma^2-a_1\gamma+a_0}{(\alpha-\gamma)(\delta-\gamma)}e^{-\gamma t}+\dfrac{\delta^2-a_1\delta+a_0}{(\alpha-\delta)(\gamma-\delta)}e^{-\delta t}$
14.	$\dfrac{1}{s^2+\beta^2}$	$\dfrac{1}{\beta}\sin\beta t$
15.	$\dfrac{1}{s^2-\beta^2}$	$\dfrac{1}{\beta}\sinh\beta t$
16.	$\dfrac{s}{s^2+\beta^2}$	$\cos\beta t$

17.	$\dfrac{s}{s^2 - \beta^2}$	$\cosh \beta t$
18.	$\dfrac{s + a_0}{s^2 + \beta^2}$	$\dfrac{1}{\beta}(a_0 + \beta^2)^{1/2} \sin(\beta t + \psi)$ $\psi \triangleq \tan^{-1} \dfrac{\beta}{a_0}$
19.	$\dfrac{1}{s(s^2 + \beta^2)}$	$\dfrac{1}{\beta^2}(1 - \cos \beta t)$
20.	$\dfrac{s + a_0}{s(s^2 + \beta^2)}$	$\dfrac{a_0}{\beta^2} - \dfrac{(a_0^2 + \beta^2)^{1/2}}{\beta^2} \cos(\beta t + \psi)$ $\psi \triangleq \tan^{-1} \dfrac{\beta}{a_0}$
21.	$\dfrac{s + a_0}{(s + \alpha)(s^2 + \beta^2)}$	$\dfrac{a_0 - \alpha}{\alpha^2 + \beta^2} e^{-\alpha t} + \dfrac{1}{\beta}\left[\dfrac{a_0^2 + \beta^2}{\alpha^2 + \beta^2}\right]^{1/2} \sin(\beta t + \psi)$ $\psi \triangleq \tan^{-1} \dfrac{\beta}{a_0} - \tan^{-1} \dfrac{\beta}{\alpha}$
22.	$\dfrac{s}{(s^2 + \beta^2)(s^2 + \lambda^2)}$	$\dfrac{\cos \beta t - \cos \lambda t}{\lambda^2 - \beta^2}$

Appendix A. **Table of Selected Laplace Transform Pairs (Continued)**

No.	$F(s)$	$f(t) \qquad 0 \le t$
23.	$\dfrac{1}{(s+\alpha)^2+\beta^2}$	$\dfrac{1}{\beta} e^{-\alpha t} \sin \beta t$
24.	$\dfrac{s+a_0}{(s+\alpha)^2+\beta^2}$	$\dfrac{1}{\beta} [(a_0-\alpha)^2+\beta^2]^{1/2} e^{-\alpha t} \sin(\beta t+\psi)$ $\psi \triangleq \tan^{-1} \dfrac{\beta}{a_0-\alpha}$
25.	$\dfrac{s+\alpha}{(s+\alpha)^2+\beta^2}$	$e^{-\alpha t} \cos \beta t$
26.	$\dfrac{1}{s[(s+\alpha)^2+\beta^2]}$	$\dfrac{1}{\beta_0^2} + \dfrac{1}{\beta_0 \beta} e^{-\alpha t} \sin(\beta t - \psi)$ $\psi \triangleq \tan^{-1} \dfrac{\beta}{-\alpha}$ $\beta_0^2 \triangleq \alpha^2 + \beta^2$
27.	$\dfrac{s+a_0}{s[(s+\alpha)^2+\beta^2]}$	$\dfrac{a_0}{\beta_0^2} + \dfrac{1}{\beta \beta_0} [(a_0-\alpha)^2+\beta^2]^{1/2} e^{-\alpha t} \sin(\beta t + \psi)$ $\psi \triangleq \tan^{-1} \dfrac{\beta}{a_0-\alpha} - \tan^{-1} \dfrac{\beta}{-\alpha}$ $\beta_0^2 \triangleq \alpha^2 + \beta^2$

28.	$\dfrac{s^2 + a_1 s + a_0}{s[(s+\alpha)^2 + \beta^2]}$	$\dfrac{a_0}{\beta_0^2} + \dfrac{1}{\beta\beta_0}[(\alpha^2 - \beta^2 - a_1\alpha + a_0)^2 + \beta^2(a_1 - 2\alpha)^2]^{1/2} e^{-\alpha t} \sin(\beta t + \psi)$ $\psi \triangleq \tan^{-1} \dfrac{\beta(a_1 - 2\alpha)}{\alpha^2 - \beta^2 - a_1\alpha + a_0} - \tan^{-1} \dfrac{\beta}{-\alpha}$ $\beta_0^2 = \beta^2 + \alpha^2$
29.	$\dfrac{1}{(s+\gamma)[(s+\alpha)^2 + \beta^2]}$	$\dfrac{e^{-\gamma t}}{(\gamma - \alpha)^2 + \beta^2} + \dfrac{1}{\beta[(\gamma-\alpha)^2 + \beta^2]^{1/2}} e^{-\alpha t} \sin(\beta t - \psi)$ $\psi \triangleq \tan^{-1} \dfrac{\beta}{\gamma - \alpha}$
30.	$\dfrac{s + a_0}{(s+\gamma)[(s+\alpha)^2 + \beta^2]}$	$\dfrac{a_0 - \gamma}{(\alpha - \gamma)^2 + \beta^2} e^{-\gamma t} + \dfrac{1}{\beta}\left[\dfrac{(a_0 - \alpha)^2 + \beta^2}{(\gamma - \alpha)^2 + \beta^2}\right]^{1/2} e^{-\alpha t} \sin(\beta t + \psi)$ $\psi \triangleq \tan^{-1} \dfrac{\beta}{a_0 - \alpha} - \tan^{-1} \dfrac{\beta}{\gamma - \alpha}$
31.	$\dfrac{s^2 + a_1 s + a_0}{(s+\gamma)[(s+\alpha)^2 + \beta^2]}$	$\dfrac{\gamma^2 - a_1\gamma + a_0}{(\alpha - \gamma)^2 + \beta^2} e^{-\gamma t} + \dfrac{1}{\beta}\left[\dfrac{(\alpha^2 - \beta^2 - a_1\alpha + a_0)^2 + \beta^2(a_1 - 2\alpha)^2}{(\gamma - \alpha)^2 + \beta^2}\right]^{1/2} e^{-\alpha t} \sin(\beta t + \psi)$ $\psi \triangleq \tan^{-1} \dfrac{\beta(a_1 - 2\alpha)}{\alpha^2 - \beta^2 - a_1\alpha + a_0} - \tan^{-1} \dfrac{\beta}{\gamma - \alpha}$
32.	$\dfrac{1}{s(s+\gamma)[(s+\alpha)^2 + \beta^2]}$	$\dfrac{1}{\gamma\beta_0^2} - \dfrac{1}{\gamma[(\alpha - \gamma)^2 + \beta^2]} e^{-\gamma t} + \dfrac{1}{\beta\beta_0[(\gamma - \alpha)^2 + \beta^2]^{1/2}} e^{-\alpha t} \sin(\beta t - \psi)$ *(Continuation on next page)*

Appendix A. **Table of Selected Laplace Transform Pairs (Continued)**

No.	$F(s)$	$f(t) \qquad 0 \le t$
		$\psi \triangleq \tan^{-1} \dfrac{\beta}{-\alpha} + \tan^{-1} \dfrac{\beta}{\gamma - \alpha}$ $\beta_0{}^2 \triangleq \alpha^2 + \beta^2$
33.	$\dfrac{s + a_0}{s(s+\gamma)[(s+\alpha)^2 + \beta^2]}$	$\dfrac{a_0}{\gamma \beta_0{}^2} + \dfrac{\gamma - a_0}{\gamma[(\alpha - \gamma)^2 + \beta^2]} e^{-\gamma t} + \dfrac{1}{\beta \beta_0} \left[\dfrac{(a_0 - \alpha)^2 + \beta^2}{(\gamma - \alpha)^2 + \beta^2}\right]^{1/2} e^{-\alpha t} \sin(\beta t - \psi)$ $\psi \triangleq \tan^{-1} \dfrac{\beta}{a_0 - \alpha} - \tan^{-1} \dfrac{\beta}{\gamma - \alpha} - \tan^{-1} \dfrac{\beta}{-\alpha}$ $\beta_0{}^2 \triangleq \alpha^2 + \beta^2$
34.	$\dfrac{s^2 + a_1 s + a_0}{(s+\gamma)(s+\delta)[(s+\alpha)^2 + \beta^2]}$	$\dfrac{\gamma^2 - a_1\gamma + a_0}{(\delta - \gamma)[(\alpha - \gamma)^2 + \beta^2]} e^{-\gamma t} \quad \dfrac{\delta^2 - a_1 \delta + a_0}{(\gamma - \delta)[(\alpha - \delta)^2 + \beta^2]} e^{-\delta t}$ $+ \dfrac{1}{\beta} \left\{\dfrac{(\alpha^2 - \beta^2 - a_1\alpha + a_0)^2 + \beta^2(a_1 - 2\alpha)^2}{[(\delta - \alpha)^2 + \beta^2][(\gamma - \alpha)^2 + \beta^2]}\right\}^{1/2} e^{-\alpha t} \sin(\beta t + \psi)$ $\psi \triangleq \tan^{-1} \dfrac{\beta(a_1 - 2\alpha)}{\alpha^2 - \beta^2 - a_1\alpha + a_0} - \tan^{-1} \dfrac{\beta}{\gamma - \alpha} - \tan^{-1} \dfrac{\beta}{\delta - \alpha}$
35.	$\dfrac{1}{(s^2 + \lambda^2)[(s+\alpha)^2 + \beta^2]}$	$\dfrac{1}{[\beta_0{}^2 - \lambda^2)^2 + 4\alpha^2\lambda^2]^{1/2}} \left[\dfrac{1}{\lambda} \sin(\lambda t - \psi_1) + \dfrac{1}{\beta} e^{-\alpha t} \sin(\beta t - \psi_2)\right]$

		$\psi_1 \triangleq \tan^{-1} \dfrac{2\alpha\lambda}{\beta_0^2 - \lambda^2}$; $\psi_2 \triangleq \dfrac{-2\alpha\beta}{\alpha^2 - \beta^2 + \lambda^2}$; $\beta_0^2 \triangleq \alpha^2 + \beta^2$
36.	$\dfrac{s + a_0}{(s^2 + \lambda^2)[(s + \alpha)^2 + \beta^2]}$	$\dfrac{1}{\lambda}\left[\dfrac{a_0^2 + \lambda^2}{(\beta_0^2 - \lambda^2)^2 + 4\alpha^2\lambda^2}\right]^{1/2} \sin(\lambda t + \psi_1)$ $+ \dfrac{1}{\beta}\left[\dfrac{(a_0 - \alpha)^2 + \beta^2}{(\beta_0^2 - \lambda^2)^2 + 4\alpha^2\lambda^2}\right]^{1/2} e^{-\alpha t} \sin(\beta t + \psi_2)$ $\psi_1 \triangleq \tan^{-1} \dfrac{\lambda}{a_0} - \tan^{-1} \dfrac{2\alpha\lambda}{\beta_0^2 - \lambda^2}$; $\psi_2 \triangleq \tan^{-1} \dfrac{\beta}{a_0 - \alpha} - \tan^{-1} \dfrac{-2\alpha\beta}{\alpha^2 - \beta^2 + \lambda^2}$ $\beta_0^2 \triangleq \alpha^2 + \beta^2$
37.	$\dfrac{1}{(s + \alpha)s^2}$	$\dfrac{e^{-\alpha t} + \alpha t - 1}{\alpha^2}$
38.	$\dfrac{s + a_0}{(s + \alpha)s^2}$	$\dfrac{a_0 - \alpha}{\alpha^2} e^{-\alpha t} + \dfrac{a_0}{\alpha} t + \dfrac{\alpha - a_0}{\alpha^2}$
39.	$\dfrac{s^2 + a_1 s + a_0}{(s + \alpha)s^2}$	$\dfrac{\alpha^2 - a_1\alpha + a_0}{\alpha^2} e^{-\alpha t} + \dfrac{a_0}{\alpha} t + \dfrac{a_1\alpha - a_0}{\alpha^2}$
40.	$\dfrac{1}{(s + \alpha)^2}$	$te^{-\alpha t}$

Appendix A. **Table of Selected Laplace Transform Pairs (Continued)**

No.	$F(s)$	$f(t) \qquad 0 \le t$
41.	$\dfrac{s + a_0}{(s + \alpha)^2}$	$[(a_0 - \alpha)t + 1]e^{-\alpha t}$
42.	$\dfrac{1}{(s + \alpha)^n}$	$\dfrac{1}{(n-1)!} t^{n-1} e^{-\alpha t}$ $\quad n$ is a positive integer
43.	$\dfrac{1}{s(s + \alpha)^2}$	$\dfrac{1 - (1 + \alpha t)e^{-\alpha t}}{\alpha^2}$
44.	$\dfrac{s + a_0}{s(s + \alpha)^2}$	$\dfrac{a_0}{\alpha^2} + \left(\dfrac{\alpha - a_0}{\alpha} t - \dfrac{a_0}{\alpha^2}\right) e^{-\alpha t}$
45.	$\dfrac{s^2 + a_1 s + a_0}{s(s + \alpha)^2}$	$\dfrac{a_0}{\alpha^2} + \left(\dfrac{a_1\alpha - a_0 - \alpha^2}{\alpha} t + \dfrac{\alpha^2 - a_0}{\alpha^2}\right) e^{-\alpha t}$
46.	$\dfrac{1}{(s + \gamma)(s + \alpha)^2}$	$\dfrac{1}{(\gamma - \alpha)^2} e^{-\gamma t} + \dfrac{(\gamma - \alpha)t - 1}{(\gamma - \alpha)^2} e^{-\alpha t}$
47.	$\dfrac{s + a_0}{(s + \gamma)(s + \alpha)^2}$	$\dfrac{a_0 - \gamma}{(\alpha - \gamma)^2} e^{-\gamma t} + \left[\dfrac{a_0 - \alpha}{\gamma - \alpha} t + \dfrac{\gamma - a_0}{(\gamma - \alpha)^2}\right] e^{-\alpha t}$

48.	$\dfrac{s^2 + a_1 s + a_0}{(s+\gamma)(s+\alpha)^2}$	$\dfrac{\gamma^2 - a_1\gamma + a_0}{(\alpha-\gamma)^2} e^{-\gamma t} + \left[\dfrac{\alpha^2 - a_1\alpha + a_0}{\gamma - \alpha} t + \dfrac{\alpha^2 - 2\alpha\gamma + a_1\gamma - a_0}{(\gamma-\alpha)^2}\right] e^{-\alpha t}$
49.	$\dfrac{s + a_0}{(s+\alpha)(s+\gamma)s^2}$	$\dfrac{a_0 - \alpha}{\alpha^2(\gamma-\alpha)} e^{-\alpha t} + \dfrac{a_0 - \gamma}{\gamma^2(\alpha-\gamma)} e^{-\gamma t} + \dfrac{a_0}{\alpha\gamma} t + \dfrac{\alpha\gamma - a_0(\alpha+\gamma)}{\alpha^2\gamma^2}$
50.	$\dfrac{s^2 + a_1 s + a_0}{(s+\alpha)(s+\gamma)s^2}$	$\dfrac{\alpha^2 - a_1\alpha + a_0}{\alpha^2(\gamma-\alpha)} e^{-\alpha t} + \dfrac{\gamma^2 - a_1\gamma + a_0}{\gamma^2(\alpha-\gamma)} e^{-\gamma t} + \dfrac{a_0}{\alpha\gamma} t + \dfrac{a_1\alpha\gamma - a_0(\alpha+\gamma)}{\alpha^2\gamma^2}$
51.	$\dfrac{s^2 + a_1 s + a_0}{(s+\alpha)^2 s^2}$	$\left[\dfrac{\alpha^2 - a_1\alpha + a_0}{\alpha^2} t + \dfrac{2a_0 - a_1\alpha}{\alpha^3}\right] e^{-\alpha t} + \dfrac{a_0}{\alpha^2} t + \dfrac{a_1\alpha - 2a_0}{\alpha^3}$
52.	$\dfrac{1}{(s^2+\beta^2)s^2}$	$\dfrac{1}{\beta^2} t - \dfrac{1}{\beta^3} \sin\beta t$
53.	$\dfrac{1}{(s^2-\beta^2)s^2}$	$\dfrac{1}{\beta^3} \sinh\beta t - \dfrac{1}{\beta^2} t$
54.	$\dfrac{s + a_0}{(s^2+\beta^2)s^2}$	$\dfrac{a_0}{\beta^2} t + \dfrac{1}{\beta^2} - \dfrac{1}{\beta^3} (a_0^2 + \beta^2)^{1/2} \sin(\beta t + \psi)$ $\psi \triangleq \tan^{-1} \dfrac{\beta}{a_0}$
55.	$\dfrac{s^2 + a_1 s + a_0}{(s^2+\beta^2)s^2}$	$\dfrac{a_0}{\beta^2} t + \dfrac{a_1}{\beta^2} - \dfrac{1}{\beta^3} [(a_0 - \beta^2)^2 + a_1^2\beta^2]^{1/2} \sin(\beta t + \psi)$

(Continuation on next page)

Appendix A. Table of Selected Laplace Transform Pairs (Continued)

No.	$F(s)$	$f(t) \qquad 0 \le t$
		$\psi \triangleq \tan^{-1} \dfrac{a_1\beta}{a_0 - \beta^2}$
56.	$\dfrac{1}{[(s+\alpha)^2+\beta^2]s^2}$	$\dfrac{1}{{\beta_0}^2}\left[t - \dfrac{2\alpha}{{\beta_0}^2} + \dfrac{1}{\beta} e^{-\alpha t} \sin(\beta t - \psi)\right]$ $\psi \triangleq 2 \tan^{-1} \dfrac{\beta}{-\alpha}$ ${\beta_0}^2 \triangleq \alpha^2 + \beta^2$

Appendix B
METHODS FOR FACTORING POLYNOMIALS

In the solution of differential equations using either classical or Laplace transform methods, it is necessary to factor polynomials of the form

$$s^m + a_{m-1}s^{m-1} + \cdots + a_1 s + a_0 = 0 \tag{B.1}$$

We are concerned primarily with values of m from two through five. Difficulty increases rapidly with m. For the quadratic case, $m = 2$, and

$$a_2 s^2 + a_1 s + a_0 = 0 \tag{B.2}$$

The roots are given by

$$s_{1,2} = \frac{-a_1 \pm \sqrt{a_1^2 - 4a_0 a_2}}{2a_2} \tag{B.3}$$

When $m = 3$, there must be at least 1 real root, and it is generally best to find it, divide it out, and factor the remaining quadratic using Eq. (B.3). Trial and error is perhaps the most effective procedure for finding real roots. The following example illustrates the steps involved.

Example B.1

Find the three roots of the equation

$$s^3 + 3.5s^2 + 5s + 3 = 0 \tag{B.4}$$

There must be at least one real root and it should be found first by trial and error. Try $s_1 = -1$. Substituting $s = -1$ into Eq. (B.4) provides, term by term,

$$-1 + 3.5 - 5 + 3 = 0.5 \tag{B.5}$$

Thus $s_1 = -1$ is fairly close, but not the desired value. Try $s_1 = -2$. This yields

$$-8 + 14 - 10 + 3 = -1 \tag{B.6}$$

Since -1 and -2 provide positive and negative results, respectively, the desired s_1, which must yield 0, is bracketed by $-2 < s_1 < -1$. Setting $s_1 = -3/2$ provides

$$\frac{-27}{8} + \frac{7}{2}\left(\frac{9}{4}\right) - \frac{15}{2} + 3 = 0 \tag{B.7}$$

Thus $s_1 = -3/2$ is a root. Factoring this root out of Eq. (B.4) gives

$$\left(s + \frac{3}{2}\right)(s^2 + 2s + 2) = 0 \tag{B.8}$$

The quadratic in Eq. (B.8) has roots at $s_2 = -1 + j1$ and $s_3 = -1 - j1$.

When $m = 4$, the polynomial may have 4 real roots, 2 real roots and a complex conjugate pair, or 2 complex conjugate pairs. When 2 (or 4) of the roots are real, the first 2 may be obtained by trial and error, divided out, and the resulting quadratic factored. The only problem arises when 2 complex conjugate pairs are encountered. In this case it is convenient to use organized trial and error to break down the 4th order equation into the product of 2 quadratic terms, each of which can be factored using Eq. (B.3). Although several relatively elite iterative procedures have been developed to expedite this problem, they are hardly worth committing to memory unless we face many problems of this type, so our discussion is limited to the direct trial and error method. An example illustrates the procedure.

Example B.2

Find the roots of

$$s^4 + 4s^3 + 11s^2 + 14s + 10 = 0 \tag{B.9}$$

This equation may be represented as the product of two general quadratics, or

$$(s^2 + \alpha s + \beta)(s^2 + \gamma s + \delta) = s^4 + 4s^3 + 11s^2 + 14s + 10 = 0 \tag{B.10}$$

Expanding the left side of Eq. (B.10) and equating coefficients of like powers of s on the two sides of the expression provides the relations which must be satisfied by α, β, γ and δ as

$$\begin{aligned} &\text{(a)}\ \beta\delta = b_0 = 10 && \text{(c)}\ \alpha\delta + \beta\gamma = b_1 = 14 \\ &\text{(b)}\ \beta + \delta + \alpha\gamma = b_2 = 11 && \text{(d)}\ \alpha + \gamma = b_3 = 4 \end{aligned} \tag{B.11}$$

These equations are highly nonlinear and trial and error is perhaps the most appropriate solution method. It is suggested that values of β and δ be chosen arbitrarily to satisfy Eq. (B.11a), then Eqs. (B.11b) and (B.11c) may be solved for α and γ and Eq. (B.11d) used as a check on the results. The values of β and δ are then adjusted to improve the results, and the process repeated if necessary,

Suppose we try $\beta = 1$ and $\delta = 10$. Substituting these values into Eqs. (B.11b) and (B.11c) provides

$$\begin{aligned} 10\alpha + \gamma &= 14 \\ 11 + \alpha\gamma &= 11 \end{aligned} \tag{B.12}$$

which requires that $\alpha = 0$ and $\gamma = 14$ or $\alpha = 1.4$ and $\gamma = 0$. Neither result satisfies Eq. (B.11d). It is obvious that β and δ must be chosen more nearly equal. Try $\beta = 2$ and $\delta = 5$. This yields

$$\begin{aligned} 5\alpha + 2\gamma &= 14 \\ 7 + \alpha\gamma &= 11 \end{aligned} \tag{B.13}$$

and the corresponding values of α and γ are either $\alpha = 4/5$ and $\gamma = 5$ or $\alpha = 2$ and $\gamma = 2$. The second pair of values satisfy Eq. (B.11d), so

$$(s^2 + 2s + 2)(s^2 + 2s + 5) = s^4 + 4s^3 + 11s^2 + 14s + 10 = 0 \tag{B.14}$$

The quadratic terms in Eq. (B.14) are readily factored to obtained the roots.

This example illustrates the trial and error process for factoring a 4th order equation into the product of 2 quadratic terms. We simply guess 2 of the coefficients, check the results, adjust the guesses as indicated, and repeat if necessary. The procedure generally converges rapidly to the desired solution, although considerable algebra may be involved. It is suggested, because of this algebraic difficulty, that real roots be extracted first whenever possible.

When $m = 5$, divide out a real root and solve the remaining fourth order equation as suggested in the previous paragraphs. When $m > 5$, try to factor out real roots until an equation of fourth or second order is obtained. If it is impossible to do this (if for example, a sixth order equation having three conjugate root pairs is encountered), a quadratic term may be separated as follows. Divide the polynomial $P(s)$ by the trial divisor

$$Q_1(s) = s^2 + c_1 s + d_1 \tag{B.15}$$

where

$$c_1 = \frac{a_1}{a_2} \qquad d_1 = \frac{a_0}{a_2} \tag{B.16}$$

The a_i are as defined in Eq. (B.1). This yields

$$\frac{P(s)}{Q_1(s)} = s^{m-2} + b_{m-2}s^{m-2} + \cdots + b_2 s^2 + b_1 s + b_0 + (ps + q) \tag{B.17}$$

where $(ps + q)$ is the remainder. Divide $P(s)$ by a second trial divisor

$$Q_2(s) = s^2 + c_2 s + d_2 \tag{B.18}$$

where

$$c_2 = \frac{a_1 - a_0 b_1/b_0}{b_0} \qquad d_2 = \frac{a_0}{b_0} \tag{B.19}$$

This process is continued until the remainder term becomes insignificant. If convergence is slow, guesses in the correct direction may prove quicker. If the remainder increases on any given step, an alternate guess will usually yield an improved answer.

Example B.3

Find the roots of

$$s^4 + 4s^3 + 11s^2 + 14s + 10 = 0 \tag{B.20}$$

which was considered using an alternate approach in Ex. B.2. The trial division process yields

$$
\begin{array}{r|l}
 & s^2 + 2.73s + 7.63 \\
\hline
s^2 + \frac{14}{11}s + \frac{10}{11} & s^4 + 4s^3 + 11s^2 + 14s + 10 \\
 & s^4 + 1.27s^3 + 0.91s^2 \\
\cline{2-2}
 & \quad 2.73s^3 + 11.09s^2 + 14s \\
 & \quad 2.73s^3 + 3.46s^2 + 2.48s \\
\cline{2-2}
 & \qquad 7.63s^2 + 11.56s + 10 \\
 & \qquad 7.63s^2 + 9.69s + 6.94 \\
\cline{2-2}
 & \qquad\qquad 1.87s + 3.06
\end{array}
$$

Thus

$$c_2 = \frac{14 - 10(2.73)/7.63}{7.63} = \frac{14 - 3.58}{7.63} = 1.37 \tag{B.21}$$

and

$$d_2 = \frac{10}{7.63} = 1.31 \tag{B.22}$$

Repeating the division yields

$$
\begin{array}{r|l}
 & s^2 + 2.63s + 6.09 \\
\hline
s^2 + 1.37s + 1.31 & s^4 + 4s^3 + 11s^2 + 14s + 10 \\
 & s^4 + 1.37s^3 + 1.31s^2 \\
\cline{2-2}
 & \quad 2.63s^3 + 9.69s^2 + 14s \\
 & \quad 2.63s^3 + 3.60s^2 + 3.45s \\
\cline{2-2}
 & \qquad 6.09s^2 + 10.55s + 10 \\
 & \qquad 6.09s^2 + 8.35s + 7.98 \\
\cline{2-2}
 & \qquad\qquad 2.20s + 2.02
\end{array}
$$

Although the convergence is not rapid, the results are getting better, and the result obtained in Ex. B.2 is duplicated if the suggested iterative procedure is repeated several more times.

It is a relatively simple matter to program a digital computer to factor polynomials using the procedure illustrated in Ex. B.3, or any of several alternate methods. Many computer libraries include a standard root finding program, which reduces the problem to reduction of the data to an appropriate input format. One interesting computer approach is based upon the standard constrained steepest descent computational algorithm,[1] and involves the following steps:

1. Assume $P(s)$ has a quadratic factor of the form $s^2 + \alpha s + \beta$.

2. Divide $P(s)$ by $s^2 + \alpha s + \beta$, carrying α and β are parameters, to obtain $R(s) = ps + q$, where p and q are general functions of α and β.

3. Determine $\nabla'\mathbf{p} = [\partial p/\partial\alpha, \partial p/\partial\beta]$ and $\nabla'\mathbf{q} = [\partial q/\partial\alpha, \partial q/\partial\beta]$.

4. Determine $\nabla\mathbf{p}_q$ and $\nabla\mathbf{q}_p$, where $\nabla\mathbf{p}_q$ denotes the orthogonal part of $\nabla\mathbf{p}$ with respect to $\nabla\mathbf{q}$ and $\nabla\mathbf{q}_p$ is defined similarly relative to $\nabla\mathbf{p}$. Note that changes in the parameter vector $\mathbf{x}' = [\alpha, \beta]$ in the $\nabla\mathbf{p}_q$ direction will, to a first order approximation, increase p without changing q. Similarly, changes in the $\nabla\mathbf{q}_p$ direction increase q without changing p.

5. Take a corrective step in $\mathbf{x}' = [\alpha, \beta]$ defined by

$$\delta\mathbf{x} = -k_1\nabla\mathbf{p}_q \operatorname{sgn} p - k_2\nabla\mathbf{q}_p \operatorname{sgn} q \tag{B.23}$$

where k_1 and k_2 are positive constants chosen as indicated later. Note that $\operatorname{sgn} p$ is the standard signum or sign function, defined as +1 when p is positive and -1 when p is negative.

6. Since the predicted change in p resulting from a parameter change $\delta\mathbf{x}$ is

$$\delta p_p = \nabla\mathbf{p}'\delta\mathbf{x} = -k_1 \operatorname{sgn} p \nabla\mathbf{p}'\nabla\mathbf{p}_q \tag{B.24}$$

the k_1 needed to give 1 step convergence on a predicted basis is

$$k_1 = \frac{-p}{\nabla\mathbf{p}'\nabla\mathbf{p}_q} \tag{B.25}$$

Similarly,

$$k_2 = \frac{-q}{\nabla\mathbf{q}'\nabla\mathbf{q}_p} \tag{B.26}$$

If these values yield poor convergence, they are scaled down to yield reasonably predictable changes.

7. At each computational step the new values of p and q are compared to the old to make sure an improvement has occurred. If not, the k's are scaled down, the test adjustment $\delta\mathbf{x}$ is disallowed, and the process is repeated. If an improvement is obtained, but the remainder is still significant, the procedure is repeated starting from the new value of $\mathbf{x}$.

8. The computational process is terminated when the remainder becomes insignificant.

Application of this gradient procedure is illustrated by the following example.

Example B.4

Find the roots of

$$P(s) = s^4 + 4s^3 + 11s^2 + 14s + 10 = 0 \tag{B.27}$$

using the gradient procedure. Dividing out a general quadratic yields

$$
\begin{array}{r|llll}
 & s^2 + (4-\alpha)s & + (11-\beta-4\alpha+\alpha^2) & & \\ \hline
s^2 + \alpha s + \beta & s^4 + 4s^3 + & 11s^2 + & 14s + 10 & \\
 & s^4 + \alpha s^3 + & \beta s^2 & & \\ \cline{2-3}
 & (4-\alpha)s^3 + & (11-\beta)s^2 + & 14s & \\
 & (4-\alpha)s^3 + & \alpha(4-\alpha)s^2 + & \beta(4-\alpha)s & \\ \cline{2-4}
 & & (11-\beta-4\alpha+\alpha^2)s^2 + & (14-4\beta+\alpha\beta)s + 10 & \\
 & & (11-\beta-4\alpha+\alpha^2)s^2 + & \alpha(11-\beta-4\alpha+\alpha^2)s + & \beta(11-\beta-4\alpha+\alpha^2) \\ \cline{3-5}
 & & & ps + q &
\end{array}
$$

where

$$p = 14 - 4\beta + 2\alpha\beta - 11\alpha + 4\alpha^2 - \alpha^3 \tag{B.28}$$

and

$$q = 10 - 11\beta + \beta^2 + 4\alpha\beta - \alpha^2\beta \tag{B.29}$$

Thus

$$\nabla \mathbf{p}' = [2\beta - 11 + 8\alpha - 3\alpha^2, -4 + 2\alpha] \tag{B.30}$$

and

$$\nabla \mathbf{q}' = [4\beta - 2\alpha\beta, -11 + 2\beta + 4\alpha - \alpha^2] \tag{B.31}$$

Choose $\alpha = \beta = 1$ as a starting point. Then

$$\nabla \mathbf{p}' = [-4, -2] \tag{B.32}$$

$$\nabla \mathbf{q}' = [2, -6] \tag{B.33}$$

Orthogonalizing $\nabla \mathbf{p}$ relative to $\nabla \mathbf{q}$ and $\nabla \mathbf{q}$ relative to $\nabla \mathbf{p}$ yields (see App. D)

$$\nabla \mathbf{p}_q' = [-4.2, -1.4] \tag{B.34}$$

and

$$\nabla \mathbf{q}_p' = [2.8, -5.6] \tag{B.35}$$

From Eqs. (B.25) and (B.26) we obtain

$$k_1 = 0.204 \qquad k_2 = 0.0765 \tag{B.36}$$

Thus, since p and q are both positive,

$$\delta \mathbf{x} = -0.204\begin{bmatrix}-4.2\\-1.4\end{bmatrix} - 0.0765\begin{bmatrix}2.8\\-5.6\end{bmatrix} = \begin{bmatrix}0.652\\0.714\end{bmatrix} \quad \text{(B.37)}$$

The new values of α and β become

$$\mathbf{x}' = [\alpha, \beta] = [1.652, 1.714] \quad \text{(B.38)}$$

which are closer to the desired values $\mathbf{x}' = [2, 2]$. One more iteration of this procedure yields results near the correct values. Choice of starting conditions near $\mathbf{x}' = [2, 5]$ provides convergence to this second quadratic factor.

Although the algebraic detail involved in Ex. B.4 is extensive, the solution is easily automated on a digital computer. In fact, a general recursive relationship can be established for computing p and q. Suppose we start with a general polynomial of the form given in Eq. (B.1). Dividing out the factor $s^2 + \alpha s + \beta$ yields

$$s^{m-2} + b_1 s^{m-3} + b_2 s^{m-4} + \cdots + b_{m-3}s + b_{m-2} + R(s) \quad \text{(B.39)}$$

where

$$R(s) = ps + q \quad \text{(B.40)}$$

The variables p, q, and the b_i are functions of α, β, and the a_i. It is easily shown that the b_i are given by

$$b_i = a_{m-i} - \beta b_{i-2} - \alpha b_{i-1} \qquad i = 0, 1, \ldots, m \quad \text{(B.41)}$$

Obviously $b_0 = 1$ and $b_j = 0$ for $j < 0$. Also,

$$p = a_1 - \beta b_{m-3} - \alpha b_{m-2} \quad \text{(B.42)}$$

and

$$q = a_0 - \beta b_{m-2} \quad \text{(B.43)}$$

The gradients are

$$\nabla \mathbf{p} = -\begin{bmatrix} b_{m-2} + \alpha \dfrac{db_{m-2}}{d\alpha} + \beta \dfrac{db_{m-3}}{d\alpha} \\[2ex] b_{m-3} + \alpha \dfrac{db_{m-2}}{d\beta} + \beta \dfrac{db_{m-3}}{d\beta} \end{bmatrix} \quad \text{(B.44)}$$

and

$$\nabla \mathbf{q} = -\begin{bmatrix} \beta \dfrac{db_{m-2}}{d\alpha} \\ \\ b_{m-2} + \beta \dfrac{db_{m-2}}{d\beta} \end{bmatrix} \tag{B.45}$$

From Eq. (B.41) we see that

$$\frac{db_i}{d\alpha} = -b_{i-1} - \alpha \frac{db_{i-1}}{d\alpha} - \beta \frac{db_{i-2}}{d\alpha} \tag{B.46}$$

and

$$\frac{db_i}{d\beta} = -b_{i-2} - \alpha \frac{db_{i-1}}{d\beta} - \beta \frac{db_{i-2}}{d\beta} \tag{B.47}$$

This provides a complete set of recursive relationships from which a general polynomial factoring program can be developed. The pertinent data for m from 3 through 8, the cases most often encountered, are summarized in Table B.1. Application is straightforward, and convergence is generally rapid.

Table B.1.

$m = 3$

$$b_0 = 1 \qquad p = a_1 - \beta - \alpha a_2 + \alpha^2$$

$$b_1 = a_2 - \alpha \qquad q = a_0 - \beta a_2 + \alpha\beta$$

$$\nabla \mathbf{p} = \begin{bmatrix} -a_2 + 2\alpha \\ \\ -1 \end{bmatrix} \qquad \nabla \mathbf{q} = \begin{bmatrix} \beta \\ \\ -a_2 + \alpha \end{bmatrix}$$

$m = 4$

$$b_0 = 1 \qquad p = a_1 - \beta a_3 + 2\alpha\beta - \alpha a_2 + \alpha^2 a_3 - \alpha^3$$

$$b_1 = a_3 - \alpha \qquad q = a_0 - \beta a_2 + \beta^2 + \alpha\beta a_3 - \alpha^2\beta$$

$$b_2 = a_2 - \beta - \alpha a_3 + \alpha^2$$

$$\nabla \mathbf{p} = \begin{bmatrix} 2\beta - a_2 + 2\alpha a_3 - 3\alpha^2 \\ \\ -a_3 + 2\alpha \end{bmatrix} \qquad \nabla \mathbf{q} = \begin{bmatrix} \beta a_3 - 2\alpha\beta \\ \\ -a_2 + 2\beta + \alpha a_3 - \alpha^2 \end{bmatrix}$$

Table B.1. Continued.

$m = 5 \quad b_0 = 1 \qquad p = a_1 - \beta b_2 - \alpha b_3$

$b_1 = a_4 - \alpha$

$b_2 = a_3 - \beta - \alpha b_1 \qquad q = a_0 - \beta b_3$

$b_3 = a_2 - \beta b_1 - \alpha b_2$

$$\nabla \mathbf{p} = \begin{bmatrix} 2\beta a_4 - a_2 - 6\alpha\beta + 2\alpha a_3 - 3\alpha^2 a_4 + 4\alpha^3 \\ 2\alpha a_4 - 3\alpha^2 - a_3 + 2\beta \end{bmatrix}$$

$$\nabla \mathbf{q} = \begin{bmatrix} -2\beta^2 + \beta a_3 - 2\alpha\beta a_4 + 3\alpha^2\beta \\ -a_2 + 2\beta a_4 - 4\alpha\beta + \alpha a_3 - \alpha^2 a_4 + \alpha^3 \end{bmatrix}$$

$m = 6 \quad b_0 = 1$

$b_1 = a_5 - \alpha \qquad p = a_1 - \beta b_3 - \alpha b_4$

$b_2 = a_4 - \beta - \alpha b_1$

$b_3 = a_3 - \beta b_1 - \alpha b_2 \qquad q = a_0 - \beta b_4$

$b_4 = a_2 - \beta b_2 - \alpha b_3$

$$\nabla \mathbf{p} = \begin{bmatrix} -3\beta^2 - 6\alpha\beta a_5 + 2\beta a_4 + 12\alpha^2\beta - a_2 - 3\alpha^2 a_4 - 5\alpha^4 + 2\alpha a_3 + 4\alpha^3 a_5 \\ -a_3 - 6\alpha\beta - 3\alpha^2 a_5 + 2\beta a_5 + 2\alpha a_4 + 4\alpha^3 \end{bmatrix}$$

$$\nabla \mathbf{q} = \begin{bmatrix} -2\beta^2 a_5 - 2\alpha\beta a_4 - 4\alpha^3\beta + 6\alpha\beta^2 + \beta a_3 + 3\alpha^2\beta a_5 \\ -a_2 - 3\beta^2 - 4\alpha\beta a_5 - \alpha^2 a_4 - \alpha^4 + 2\beta a_4 + 6\alpha^2\beta + \alpha a_3 + \alpha^3 a_5 \end{bmatrix}$$

$m = 7 \quad b_0 = 1$

$b_1 = a_6 - \alpha \qquad p = a_1 - \beta b_4 - \alpha b_5$

$b_2 = a_5 - \beta - \alpha b_1$

$b_3 = a_4 - \beta b_1 - \alpha b_2$

$b_4 = a_3 - \beta b_2 - \alpha b_3 \qquad q = a_0 - \beta b_5$

$b_5 = a_2 - \beta b_3 - \alpha b_4$

$$\nabla \mathbf{p} = \begin{bmatrix} \{(-3\beta^2 + 12\alpha^2\beta - 5\alpha^4)a_6 + (4\alpha^3 - 6\alpha\beta)a_5 + (2\beta - 3\alpha^2)a_4 + 2\alpha a_3 \\ \qquad - a_2 + 12\alpha\beta^2 - 20\alpha^3\beta + 6\alpha^5\} \\ (4\alpha^3 - 6\alpha\beta)a_6 + (2\beta - 3\alpha^2)a_5 + 2\alpha a_4 - a_3 + 12\alpha^2\beta - 5\alpha^4 - 3\beta^2 \end{bmatrix}$$

Table B.1. Continued.

$$\nabla q = \begin{bmatrix} (6\alpha\beta^2 - 4\alpha^3\beta)a_6 + (3\alpha^2\beta - 2\beta^2)a_5 - 2\alpha\beta a_4 + \beta a_3 + 3\beta^3 + 5\alpha^4\beta \\ - 12\alpha^2\beta^2 \\ (6\alpha^2\beta - \alpha^4 - 3\beta^2)a_6 + (\alpha^3 - 4\alpha\beta)a_5 + (2\beta - \alpha^2)a_4 + \alpha a_3 - a_2 \\ + \alpha^5 + 9\alpha\beta^2 - 8\alpha^3\beta \end{bmatrix}$$

$m = 8$

$$\begin{aligned} b_0 &= 1 \\ b_1 &= a_7 - \alpha \qquad\qquad p = a_1 - \beta b_5 - \alpha b_6 \\ b_2 &= a_6 - \beta - \alpha b_1 \\ b_3 &= a_5 - \beta b_1 - \alpha b_2 \\ b_4 &= a_4 - \beta b_2 - \alpha b_3 \qquad q = a_0 - \beta b_6 \\ b_5 &= a_3 - \beta b_3 - \alpha b_4 \\ b_6 &= a_2 - \beta b_4 - \alpha b_5 \end{aligned}$$

$$\nabla p = \begin{bmatrix} \{(12\alpha\beta^2 - 20\alpha^3\beta + 6\alpha^5)a_7 + (12\alpha^2\beta - 3\beta^2 - 5\alpha^4)a_6 + (4\alpha^3 - 6\alpha\beta)a_5 \\ + (2\beta - 3\alpha^2)a_4 + 2\alpha a_3 - a_2 + 4\beta^3 - 30\alpha^2\beta + 30\alpha^4\beta - 7\alpha^6\} \\ \{(12\alpha^2\beta - 3\beta^2 - 5\alpha^4)a_7 + (4\alpha^3 - 6\alpha\beta)a_6 + (2\beta - 3\alpha^2)a_5 + 2\alpha a_4 - a_3 \\ + 12\alpha\beta^2 - 20\alpha^3\beta + 6\alpha^5\} \end{bmatrix}$$

$$\nabla q = \begin{bmatrix} \{(3\beta^3 - 12\alpha^2\beta^2 + 5\alpha^4\beta)a_7 + (6\alpha\beta^2 - 4\alpha^3\beta)a_6 + (3\alpha^2\beta - 2\beta^2)a_5 \\ - 2\alpha\beta a_4 + \beta a_3 + 20\alpha^3\beta^2 - 12\alpha\beta^3 - 6\alpha^5\beta\} \\ \{(9\alpha\beta^2 - 8\alpha^3\beta + \alpha^5)a_7 + (6\alpha^2\beta - 3\beta^2 - \alpha^4)a_6 + (\alpha^3 - 4\alpha\beta)a_5 \\ + (2\beta - \alpha^2)a_4 + \alpha a_3 - a_2 + 4\beta^3 - 18\alpha^2\beta^2 + 10\alpha^4\beta - \alpha^6\} \end{bmatrix}$$

REFERENCES

1. Eveleigh, V. W., "Adaptive Control and Optimization Techniques," Chap. 5, McGraw-Hill Book Co., Inc., New York, 1967.
2. Eveleigh, V. W., "A Gradient Interpretation of Bairstow's Method for Factoring Polynomial Equations," *Proc. 4th Hawaii Int. Conf. on System Sciences,* Honolulu, January, 1971, p. 149.

Appendix C
ANALOG COMPUTERS

The fundamental purpose of the analog computer is to aid in solving our problems, and this should be kept in mind during our discussion to help us maintain the proper perspective. From early times people have used physical aids in solving mathematical problems. A child learns to count on his fingers, for example, and the use of an abacus is a somewhat more sophisticated application of this same concept. Analog and digital computers are devices, generally electronic, which have been developed to help us perform difficult tasks or tasks which, although routine, must be carried out repetitively. We are concerned in this discussion with analog computers only. Digital simulation is considered in App. E.

The word *analog* is used because, in most applications, the observed or measured quantities are analogous to the physical variables of the problem under study. The fundamental philosophy of the analog computer was developed by Lord Kelvin when he used a mechanical differential analyzer to solve integro-differential equations. Most contemporary analog computers use electronic circuits as their basic building blocks, and the measured quantities are voltages and currents. Electronic analog computers are of two main types:

1. Simulations in which each component of the system under study is directly related to a component or unit of the computer. An example is the standard

network analyzer, in which the individual elements of a power distribution system are approximated by lumped parameters. Measurements are made at critical points of the system, and studies are carried out to determine the optimum power loads for generating stations under specific operating conditions.

2. General purpose analog computers in which each element of the computer is capable of performing a certain mathematical operation, and elements are interconnected to represent the equations of the system under study.

The general purpose computer is more flexible than a specific simulation, but more hardware is required to simulate a general problem this way. Direct simulation devices are feasible only when a given situation is to be studied exhaustively, and they are often used in such cases for economic reasons.

Only the general purpose analog computer will be considered here. The basic building blocks needed for the simulation of linear integro-differential equations are discussed initially. Application of these elements to the solution of simple problems is illustrated by examples. Amplitude and frequency scaling of the problem to remain within hardware capability limits are also discussed.

C.1 BASIC SIMULATION BUILDING BLOCKS

Our first objective is to define the basic building blocks of an analog computer that will allow us to solve linear time invariant integro-differential equations of the form

$$\sum_{i=-l}^{m} \alpha_i D^i x(t) = f(t) \tag{C.1}$$

with initial conditions

$$x^j(0) \triangleq \left.\frac{d^j x(t)}{dt^j}\right|_{t=0} = k_j \qquad j = -l, -l+1, \ldots, m-1 \tag{C.2}$$

We have used D in Eq. (C.1) as the standard differential operator shorthand notation, and

$$D^m x(t) \triangleq \frac{d^m x(t)}{dt^m} \qquad D^0 x(t) = x(t)$$

$$D^{-l} x(t) = \int_0^t dy_1 \int_0^{y_1} dy_2 \ldots \int_0^{y_{l-1}} x(y_l)dy_l \tag{C.3}$$

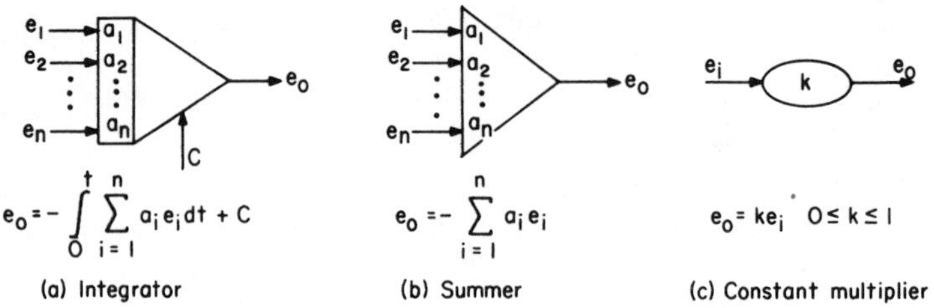

FIG. C.1. Basic elements used in developing analog simulations.

The α_i are assumed constants, and $f(t)$ is a general input forcing function. Four mathematical operations must be performed to solve Eq. (C.1) as given. It is required that we be able to integrate, differentiate, add (subtract), and multiply by a constant. In practice we do not need both integrators and differentiators, since Eq. (C.1) can be reduced to an equivalent integral equation if both sides are integrated m times over the interval $[0, t]$, or to an equivalent differential equation if both sides are differentiated with respect to time l times. Most analog computers use electronic elements as building blocks, and these devices generate "noise," the high frequency elements of which are aggravated by differentiation. For this reason repeated integration, with its associated averaging or noise smoothing effect, is the basic technique used in developing analog solutions. We thus need only three basic building blocks for solving linear equations, and they are: 1. an integrator; 2. a summer, and 3. a constant multiplier. The conventional block diagram representations of these units in analog computer symbols is as shown in Fig. C.1.

A good approximation to each of the three basic analog simulation elements is easily obtained using electronic components. The operational amplifier shown in Fig. C.2 is the fundamental device used in developing integrators and summers. The amplifier has high gain and negligible phase shift over the frequency range of interest. Applying Kirchhoff's current law to node G, assuming the amplifier has zero output impedance and also that Y_g includes the amplifier input admittance component, provides

$$E_g Y_g = (E_i - E_g)Y_i + (E_0 - E_g)Y_f \tag{C.4}$$

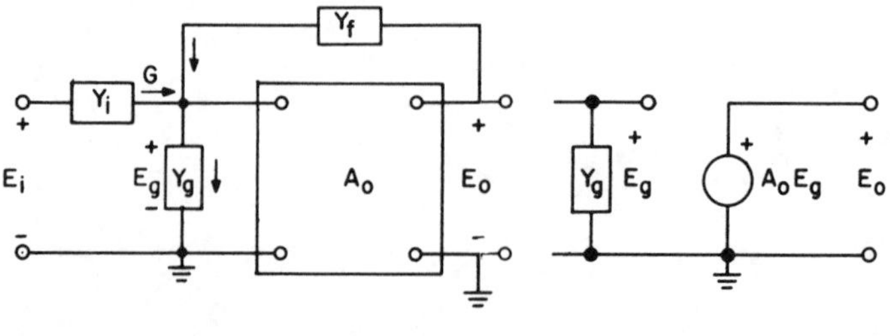

FIG. C.2. Operational amplifier.

or

$$E_g = \frac{E_i Y_i + E_0 Y_f}{Y_i + Y_f + Y_g} \tag{C.5}$$

Also

$$E_0 = A_0 E_g \tag{C.6}$$

Eliminating E_g from Eqs. (C.5) and (C.6) provides the transfer function of the operational amplifier as

$$\frac{E_0}{E_i} = A = \frac{A_0 Y_i}{Y_i + Y_g + Y_f(1 - A_0)} = -\frac{Y_i}{Y_f} \frac{A_0}{A_0 - [1 + (Y_i + Y_g)/Y_f]} \tag{C.7}$$

For a well designed operational amplifier

$$|A_0| \gg 1 + \frac{Y_i + Y_g}{Y_f} = y_\alpha \tag{C.8}$$

and a good approximation of the transfer function is

$$A = -\frac{Y_i}{Y_f} = -\frac{Z_f}{Z_i} \tag{C.9}$$

It is important to stress the two approximations used in deriving A. We have assumed that A_0 is a *real* number such that $|A_0| \gg y_\alpha$. In the frequency range of interest the operational amplifier must have very large gain and the input to

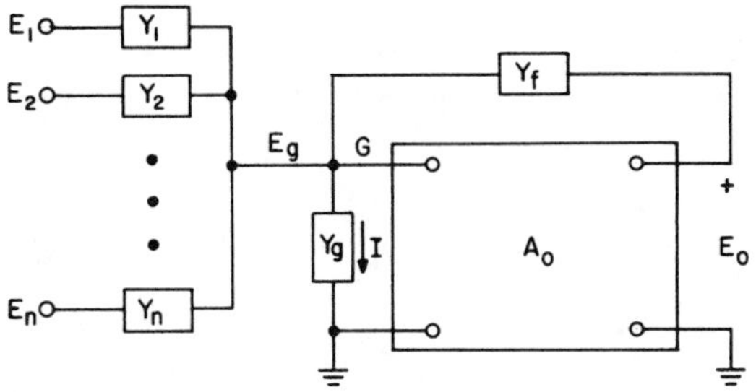

FIG. C.3. Multiple input operational amplifier.

output phase shift, which is nominally 180° because A_0 is negative, must not vary with frequency. If these assumptions are made initially, the operational amplifier transfer function may be derived by an alternate procedure. Consider the multiple input unit shown in Fig. C.3. Since $|A_0|$ is very large, E_g must be near zero to hold E_0 within reasonable bounds. Node G is said to be held at *virtual ground* by the feedback around the amplifier. Thus $E_g \simeq 0$ and $I \simeq 0$, so application of Kirchhoff's current law to node G provides

$$E_0 Y_f + \sum_{i=1}^{n} E_i Y_i = 0 \tag{C.10}$$

or

$$E_0 = -Z_f \sum_{i=1}^{n} \frac{E_i}{Z_i} = \sum_{i=1}^{n} A_i E_i \tag{C.11}$$

where

$$A_i = -\frac{Z_f}{Z_i} \tag{C.12}$$

The result provided by Eq. (C.11) is the multiple input equivalent of Eq. (C.9).

By proper choice of the impedances Z_f and Z_i, integrators and summers can be obtained. In both cases the input element is chosen as a resistor, or

$$Z_i = R_i \qquad i = 1, 2, \ldots, n \tag{C.13}$$

For a summer Z_f is chosen as a resistor and for an integrator Z_f is chosen as a capacitor. The circuit diagram of an integrator is shown in Fig. C.4 and its performance is described by

$$E_0(s) = -\sum_{i=1}^{n} \frac{E_i}{sR_i C_f} + \frac{C}{s} \tag{C.14}$$

where s is the usual Laplace transform operator and $C = v_{Cf}(0)$ is the initial voltage on C_f and represents the problem initial condition.

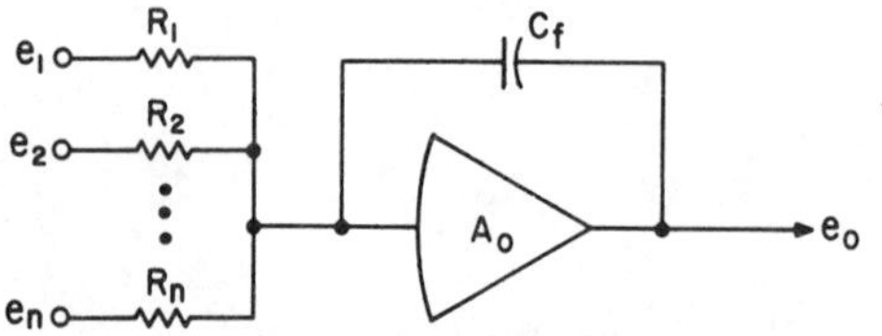

FIG. C.4. Integrator circuit diagram.

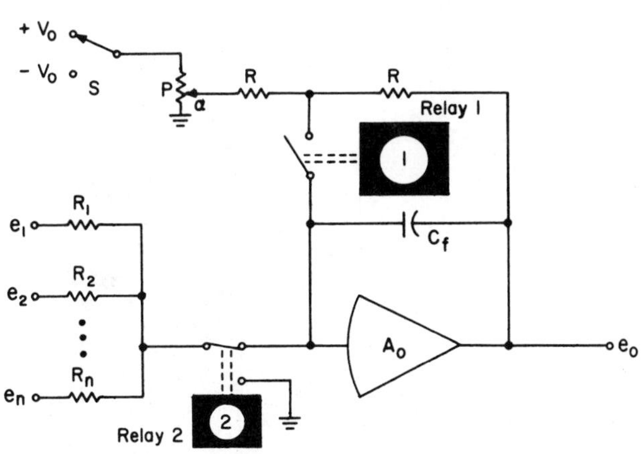

FIG. C.5. Integrator control circuits.

In the time domain

$$e_0(t) = -\frac{1}{C_f}\sum_{i=1}^{n} D^{-1}\frac{e_i}{R_i} + C \tag{C.15}$$

or

$$e_0(t) = -\frac{1}{C_f}\left[\frac{1}{R_1}\int_0^t e_1(\tau)d\tau + \frac{1}{R_2}\int_0^t e_2(\tau)d\tau + \cdots + \frac{1}{R_n}\int_0^t e_n(\tau)d\tau\right] + C \tag{C.16}$$

Control circuitry is included within the computer to set the desired initial voltage on C_f and also to hold the integrator output voltage at constant value when computing is stopped. A simplified diagram of this control mechanism is shown in Fig. C.5. During OPERATE both relays are de-energized and the regular integrating circuit is obtained. If the computing cycle is stopped at some time point, relay 2 is energized removing the integrator inputs. This is called a HOLD condition. To set the initial condition on C_f, switch S and potentiometer α are adjusted to establish potential $-C$ at point P and relays 1 and 2 are both activated. The time constants are chosen such that initial conditions may be established quickly without overloading any of the equipment.

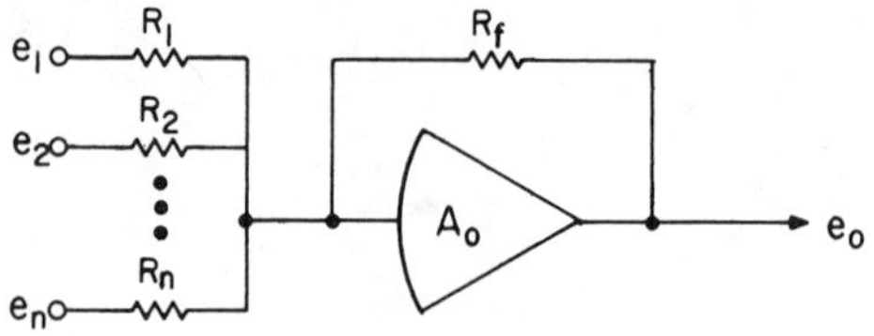

FIG. C.6. Summer circuit diagram.

The circuit diagram of a summer is shown in Fig. C.6, and its performance is described by

$$E_0 = -\sum_{i=1}^{n} \frac{R_f}{R_i} E_i \tag{C.17}$$

Summers are used to obtain the weighted sums of several inputs. They also provide phase inversion.

The last basic building block needed is the constant multiplier. In practice it is only necessary to multiply by constants $0 < k < 1$, since operational amplifier units can be used to provide gain by proper choice of $R_f(C_f)$ and R_i whenever necessary. The standard constant multiplier element chosen is the potentiometer shown in Fig. C.7a. Since accurately calibrated potentiometers are expensive and loading effects vary, gain levels are generally adjusted to the desired values under loaded conditions using a calibrated reference potentiometer and a null detector, as shown in Fig. C.7b. In most analog computers each operational amplifier input resistor must be chosen as one of several standard precise values, and the desired equation coefficients are obtained by combining the most convenient of these values with the appropriate constant multiplier setting.

C.2 SOLUTION OF LINEAR TIME INVARIANT INTEGRO-DIFFERENTIAL EQUATIONS

We are now prepared to apply the basic components introduced in the previous section to solve linear time invariant differential equations. It is

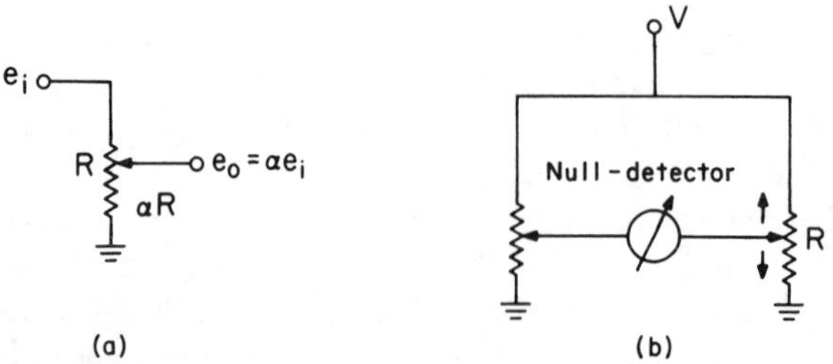

FIG. C.7. Potentiometer used as a constant multiplier.

convenient in the development of a computer simulation diagram to transform integro-differential equations to differential equations by repeated differentiation. We assume this has been done, leaving an equation of the form

$$\sum_{i=0}^{n} \beta_i D^i x(t) = g(t) \tag{C.18}$$

Dividing both sides of Eq. (C.18) by β_n and solving for $D^n x(t)$ gives

$$D^n x(t) = -\sum_{i=0}^{n-1} a_i D^i x(t) + f(t) \tag{C.19}$$

where

$$f(t) = \frac{g(t)}{\beta_n} \qquad a_i = \frac{\beta_i}{\beta_n} \tag{C.20}$$

It is clear from Eq. (C.19) that $D^n x(t)$ can be obtained as the output of a summer with inputs equal to the negative of the terms on the right side of the equation. Although the lower order derivatives of $x(t)$ are not known, they are easily obtained by repeated integration of $D^n x(t)$. Thus $D^n x(t)$ and the lower order derivatives including $x(t)$ itself can be developed using a simulation arrangement which generally involves several feedback paths. The n initial conditions must also be given before the solution is completely specified. The general procedure is illustrated by the following example.

Example C.1

$$\ddot{x} + a_1 \dot{x} + a_0 x = f(t) \tag{C.21}$$

with

$$\dot{x}(0) = C_1 \qquad x(0) = C_0 \tag{C.22}$$

We may obtain $\ddot{x}$ using a summer as shown in Fig. C.8. Although $\dot{x}$ and x are not

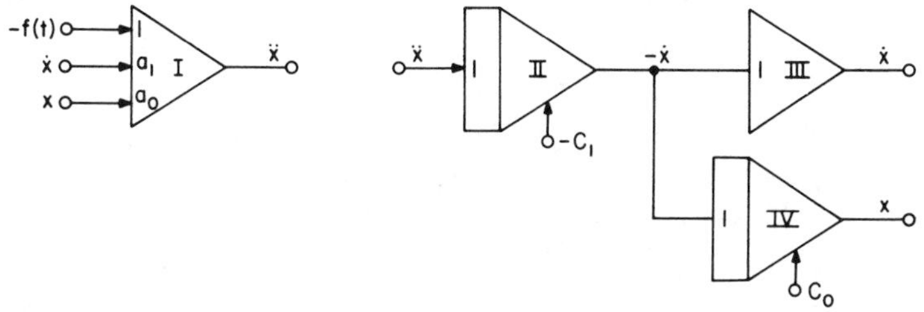

FIG. C.8. The components of $\ddot{x}$ and the development of $\dot{x}$ and x.

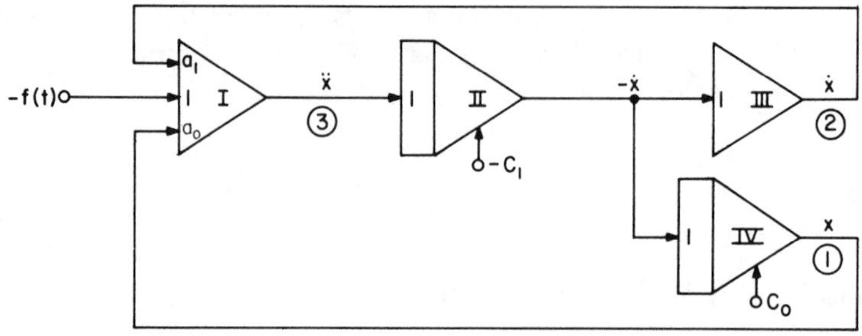

FIG. C.9. Total simulation diagram for Ex. C.1.

known, they are obtained from $\ddot{x}$ by integration and inversion as also shown in Fig. C.8. Note that initial conditions are imposed upon $x(t)$ and $\dot{x}(t)$. Connecting the feedback paths provides a simulation diagram from which the total solution may be obtained as shown in Fig. C.9. This arrangement is obviously not unique. If, for example, $\ddot{x}$ is not required as an output, operational amplifiers 1 and 2 can be combined to provide the equivalent arrangement shown in Fig. C.10. Such simplifications are often possible.

As illustrated by Ex. C.1, the simulation setup is not unique and the choice of a particular arrangement is governed by the following factors:

1. The variables to be observed must be made available.
2. The number of operational amplifiers and other components required should be kept near minimum since only a limited number are available and all devices introduce errors which should be minimized.
3. The simulation should be as closely analogous to the physical system as possible, and it should be possible to make simple adjustments corresponding to changes in system parameters.
4. The open and closed loop characteristic equations of the computer setup and the physical system should have the same stability characteristics.

The requirements are usually contradictory and compromises must be made. The actual setup used is largely governed by the operator's experience. Our primary concern is with simulation simplicity and availability of the desired variables. The following example illustrates several of these principles.

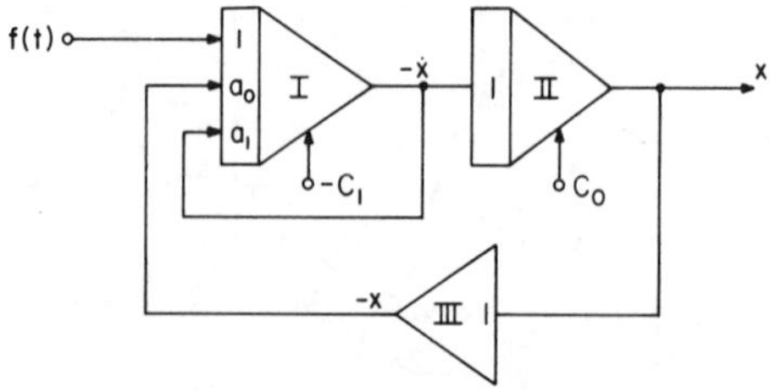

FIG. C.10. Simplified simulation diagram.

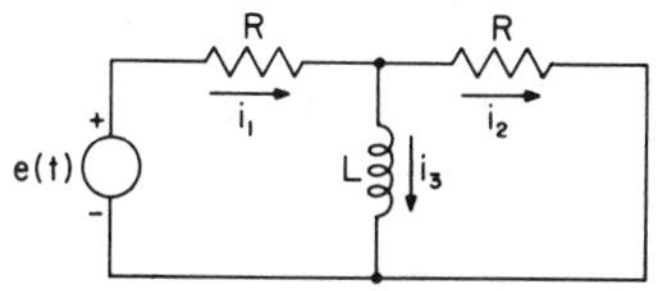

FIG. C.11. The network used in Ex. C.2.

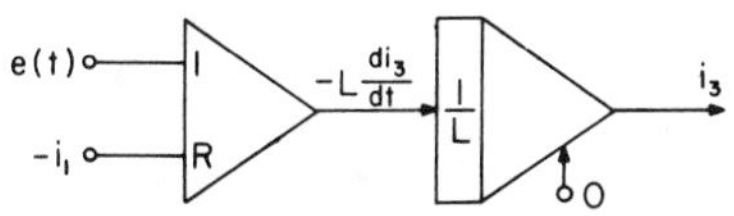

FIG. C.12. Solution for i_3.

Example C.2

Consider the system shown in Fig. C.11, which is described by

$$e(t) = Ri_1 + L\frac{di_3}{dt} \qquad i_3(0) = 0 \tag{C.23}$$

$$0 = Ri_2 - L\frac{di_3}{dt} \tag{C.24}$$

$$i_1 = i_2 + i_3 \tag{C.25}$$

a. Derive a computer setup in which the currents i_1, i_2 and i_3 are all available.

We obtain i_3 directly from Eq. (C.23) as shown in Fig. C.12. Similarly i_1 and i_2 are obtained from Eqs. (C.24) and (C.25) as shown in Fig. C.13, and the combination is shown in Fig. C.14.

b. Obtain the computer setup assuming only i_2 and i_3 are desired.

Two approaches to this problem appear feasible. We may eliminate i_1 from Eqs. (C.23), (C.24) and (C.25) to obtain

$$e(t) = R(i_2 + i_3) + L\frac{di_3}{dt} \qquad i_3(0) = 0 \tag{C.26}$$

$$0 = Ri_2 - L\frac{di_3}{dt} \tag{C.27}$$

Using Eqs. (C.26) and (C.27) as before to solve for i_2 and i_3 yields the diagram

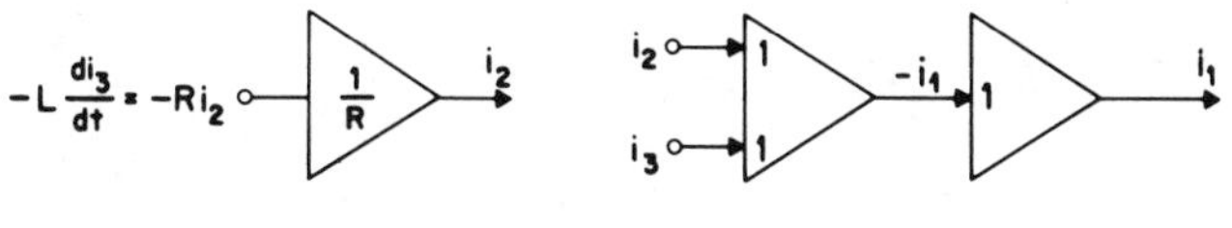

FIG. C.13. Solutions for i_2 and i_1.

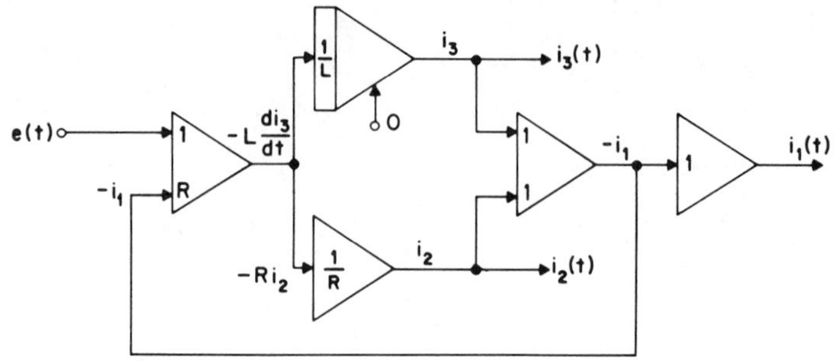

FIG. C.14. The complete simulation diagram.

shown in Fig. C.15. This same result may be obtained by starting from the diagram in Fig. C.14, eliminating all of the hardware associated with the output display of i_1, and simplifying the result as much as possible.

c. Develop a computer stepup assuming i_3 is the only variable to be observed.

Either by eliminating i_2 from Eqs. (C.26) and (C.27) and establishing the corresponding diagram, or by reduction of the diagram shown in Fig. C.15, we obtain the simple simulation arrangement shown in Fig. C.16.

Example C.2 illustrates a spectacular saving in the number of operational amplifiers required by reducing the number of measured variables. This is often, but not always, possible. If, for example, i_1 is the only observed variable, we need four operational amplifiers. In general more operational amplifiers are required as more variables are observed and as more system parameters are made readily adjustable, although the correspondence is not one to one. We also see from Ex. C.2 the general procedure for treating simultaneous linear differential equations, which may be summarized as follows: Each equation is set up as though to solve the problem involving the highest derivative in that equation, and the resulting arrangements are coupled together as required to obtain the overall simulation diagram.

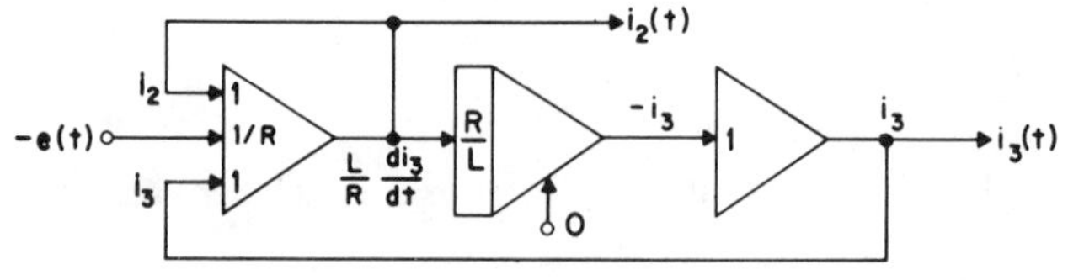

FIG. C.15. Simulation diagram for only i_2 and i_3.

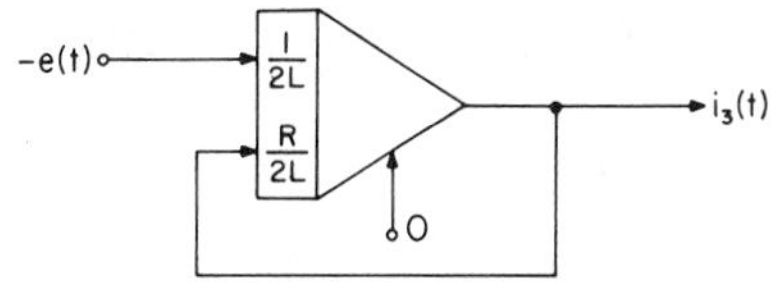

FIG. C.16. Simulation diagram yielding only i_3.

C.3 TIME AND AMPLITUDE SCALING

Although we have thus far ignored the problems of amplitude and time, or frequency, scaling, both are necessary in practice to prevent driving the operational amplifiers and other computer components beyond the range for which they are linear, have negligible drift, and have no significant phase variations with frequency. We will consider time scaling first. Several situations arising which may require time scaling are:

1. When the computer is required to run for long periods of time the errors are integrated, increasing final error. This places a lower bound on the frequencies which may be used and an upper bound on computation time.
2. Operational amplifier phase shift produces an upper bound on frequency, particularly when many operational units are in use.
3. The dynamics of nonlinear elements, such as function generators and multipliers, and of recording devices, such as an x-y plotter, also set an upper bound on the useful frequencies.

Problem frequencies may be obtained by appraisal of the inputs to be considered and by estimating the natural frequencies (pole locations) of the simulated system. It is not inappropriate to simulate the system directly and observe the response to see if time scaling seems desirable.

When time scaling is required we change from *real time t* to *machine time* τ by defining

$$\tau = \gamma t \tag{C.28}$$

where γ is the scale factor between the two time domains. If $\gamma > 1$ the problem solution is slowed down in machine time, whereas the solution runs faster on the machine than it would in real time for $\gamma < 1$. The time scaled equation is obtained from the real time equation by substituting

$$\frac{d^n f(t)}{dt^n} = \gamma^n \frac{d^n f(\tau)}{d\tau^n} \tag{C.29}$$

where $f(t)$ is an arbitrary function appearing in the equation. The initial conditions must also be modified by

$$f^n\,(t = 0) = \gamma^n f^n(\tau = 0) \tag{C.30}$$

The following example illustrates the use of scaling.

Example C.3

Consider

$$\ddot{x} + 4\dot{x} + 400x = 1600 \qquad x(0) = 0 \qquad \dot{x}(0) = 10 \tag{C.31}$$

The natural frequency of this equation is 20 rad/sec, which is rather high. Suppose we wish to reduce the natural frequency of the computer setup to 5 rad/sec. This requires that the ratio of the coefficients associated with the $\ddot{x}$ and x terms in Eq. (C.31) be changed to 25, which is accomplished by choosing $\gamma = 4$ in Eq. (C.28). Converting Eq. (C.31) to the scaled equation in machine time τ using Eqs. (C.29) and (C.30) yields

$$(4)^2 \frac{d^2x(\tau)}{d\tau^2} + (4)(4)^1 \frac{dx(\tau)}{d\tau} + (4)^0 400\, x(\tau) = 1600 \tag{C.32}$$

with

$$x(\tau)\Big|_{\tau=0} = \frac{0}{(4)^0} \qquad \dot{x}(\tau)\Big|_{\tau=0} = \frac{10}{(4)^1} \tag{C.33}$$

which reduces to

$$\frac{d^2x}{d\tau^2} + \frac{dx}{d\tau} + 25x = 100 \qquad x(0) = 0 \qquad \dot{x}(0) = 2.5 \tag{C.34}$$

It is implicitly understood that $x = x(\tau)$ is the machine variable in Eq. (C.34).

Several situations arising in practice which call for amplitude scaling are:

1. The operational amplifiers and recording devices saturate at some signal level, causing the computer to go into an *overload* state or causing errors in the computed and displayed results. This limits maximum signals levels.
2. Noise generated by the electronic devices and indicator sensitivity set a lower limit on the signal amplitudes.

A typical range of signal levels may be ±100V for a large scale computer or ±10V for a transistorized desk top computer. Just as in time scaling, it is necessary to approximate the maximum amplitudes which may occur. The best approximations are obtained by considering the physical system itself. Assume an equation of the form

$$\beta_n D^n x + \beta_{n-1} D^{n-1} x + \cdots + \beta_1 Dx + \beta_0 x = g(t) \tag{C.35}$$

The following rule-of-thumb is often useful. Approximate maximum signal levels are given by

$$x_{max} = \frac{g(0)}{\beta_0} \qquad x_{max}^{(n)} = \omega_n x_{max} \tag{C.36}$$

where $g(0)$ is the level of the input at $t = 0$ if $g(t)$ is a step function or a decaying signal. If $g(t)$ is a sinusoid, $g(0)$ should be its peak value. For this rule to give reasonably accurate results, it is necessary that the response be nearly sinusoidal and the input slowly varying. These conditions are often satisfied. In case of difficulty, signals can always be re-scaled starting from a preliminary simulation diagram. The following example illustrates the scaling procedure.

Example C.4

Consider the time scaled system described by Eq. (C.34). Develop a computer setup in which x and all of its derivatives can be measured assuming a ±10 volt computer is being used.

We will arbitrarily scale the problem to maximum signal levels of 8 volts. The maximum values are approximated using Eq. (C.36) as

$$x_{max} = \frac{g(0)}{\beta_0} = \frac{100}{25} = 4 \text{ volts} \tag{C.37}$$

$$\dot{x}_{max} = \omega x_{max} = (5)(4) = 20 \text{ volts} \tag{C.38}$$

$$\ddot{x}_{max} = \omega^2 x_{max} = (25)(4) = 100 \text{ volts} \tag{C.39}$$

To illustrate the procedure we draw the computer setup corresponding to Eq. (C.34) assuming scaled voltages denoted by $Af(\tau)$, $A_0 x$, $A_1 \dot{x}$ and $A_2 \ddot{x}$, where

$$A = 0.08 \qquad A_0 = 2 \qquad A_1 = 0.4 \qquad A_2 = 0.08 \tag{C.40}$$

This yields the simulation diagram shown in Fig. C.17. The equation corresponding

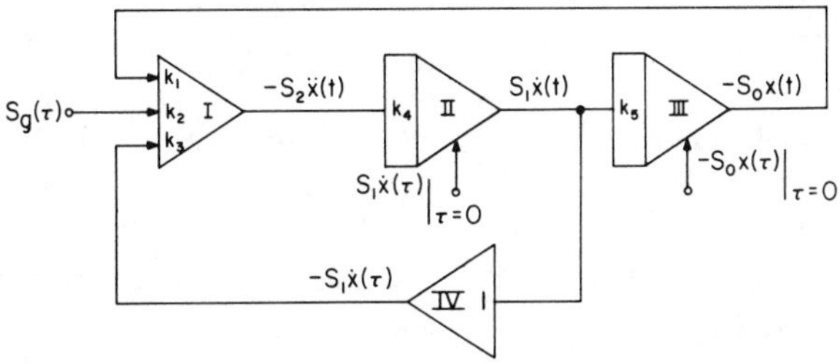

FIG. C.17. Simulation diagram for Ex. C.4.

to this setup is

$$A_2\ddot{x} + k_3 A_1 \dot{x} + k_1 A_0 x = k_2 A g(\tau) \qquad \text{(C.41)}$$

Dividing both sides of Eq. (C.41) by A_2 and equating the resulting coefficients to those of Eq. (C.34) yields the values of k_1, k_2 and k_3 as

$$k_1 = \frac{25\,A_2}{A_0} = 1 \qquad k_2 = \frac{100\,A_2}{A} = 100 \qquad k_3 = \frac{A_2}{A_1} = 0.2 \qquad \text{(C.42)}$$

The required values of k_4 and k_5 are obtained from the relationships between the inputs and outputs of integrators 2 and 3 as

$$k_4 = \frac{A_1}{A_2} = 5 \qquad k_5 = \frac{A_0}{A_1} = 5 \qquad \text{(C.43)}$$

which provides a complete description of the simulation network.

Note that the maximum voltage values were chosen arbitrarily as equal in Ex. C.4 to illustrate how easily scaling is accomplished. Alternate assignments of signal levels are possible, and simplification often results from making level adjustments which eliminate the need for one or more amplifier stages.

Most computer problems must be scaled, and many require both time and amplitude scaling. Once the scale factors are chosen, adjustment of the simulation diagram is straightforward. In most applications a rough preliminary estimate of magnitudes and frequencies is made, the diagram is developed on this basis, the results are checked, and adjustments are made as necessary. This is the only feasible approach when nonlinear systems are simulated, since there is no convenient way to obtain accurate estimates in this case.

C.4 SYSTEM SIMULATION

In many applications we wish to develop a system simulation from a block diagram representation involving several transfer functions. In such cases it is often convenient to simulate the individual transfer function elements independently, thereby retaining a structure corresponding closely to that of the actual system. Signals at the outputs of the individual simulation components are direct analogs of the corresponding system variables. The effects of adjusting critical system parameters such as gain levels and time constants can be readily observed in this way and the simulation study can be used to supplement analytical design procedures with all kinds of valuable data.

It is not necessary to go back to the differential equation description as a first step in developing the computer setup to represent a transfer function of the form

$$\frac{Y(s)}{X(s)} = \frac{a_n s^n + a_{n-1}s^{n-1} + \cdots + a_1 s + a_0}{s^n + b_{n-1}s^{n-1} + \cdots + b_1 s + b_0} \tag{C.44}$$

Observe that this transfer function is not strictly realizable unless $a_n = 0$, since it has finite non-zero transmission as $s \to \infty$, but we often approximate the characteristics of devices by a description of this form which is acceptably accurate over the frequency range of interest. Assume for convenience that all initial conditions are zero. Cross multiplying Eq. (C.44) gives

$$(s^n + b_{n-1}s^{n-1} + \cdots + b_1 s + b_0)Y(s) = (a_n s^n + a_{n-1}s^{n-1} + \cdots + a_1 s + a_0)X(s) \tag{C.45}$$

Dividing both sides of Eq. (C.45) by s^n yields the alternate form

$$[1 + b_{n-1}s^{-1} + \cdots + b_1 s^{-(n-1)} + b_0 s^{-n}]Y(s) = [a_n + a_{n-1}s^{-1} + \cdots + a_1 s^{-(n-1)} + \cdots + a_0 s^{-n}]X(s) \tag{C.46}$$

Thus

$$Y(s) = [a_n + a_{n-1}s^{-1} + \cdots a_1 s^{-(n-1)} + a_0 s^{-n}]X(s) - [b_{n-1}s^{-1} + b_{n-2}s^{-2} + \cdots + b_1 s^{-(n-1)} + b_0 s^{-n}]Y(s) \tag{C.47}$$

It is clear from Eq. (C.47), and the fact that s^{-1} is an integration operator analogous to the D^{-1} used before, that $Y(s)$ may be obtained from $X(s)$ and $Y(s)$ by integration and summation as shown in Fig. C.18. General block diagram

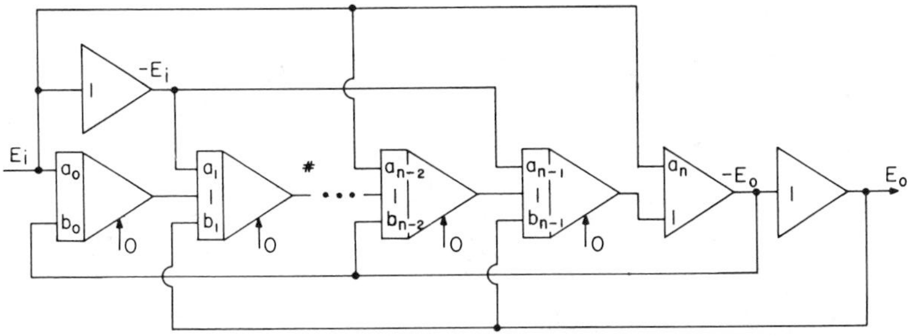

* Note that an even number of integrators is assumed. A minor modification is required if n is odd.

FIG. C.18. Simulation diagram for a general transfer function.

arrangements are simulated by cascading sub-systems of this form together as desired. Considerable simplification is often possible if that seems more appropriate than maintenance of a closely analogous setup.

Certain standard transfer function forms arise so often that special operational amplifier arrangements for realizing them have been developed over the years, and extensive tables of such circuits are available in many references.[1] Several examples are shown in Fig. C.19.

C.5 ADDITIONAL CHARACTERISTICS OF ANALOG COMPUTERS

One of the most significant advantages of analog computers is that they are so readily used to study the response characteristics of a variety of nonlinear

$$\frac{E_o}{E_i} = - \frac{R_2}{R_1} \frac{1}{1 + SC_2R_2}$$

(a)

$$\frac{E_o}{E_i} = - \frac{SC_1R_2}{1 + SC_1R_1}$$

(b)

$$\frac{E_o}{E_i} = - \frac{R_3/R_1}{S^2R_2R_3C_1C_2 + SR_2C_1(1 + R_2/R_1 + R_3/R_2) + 1}$$

(c)

$$\frac{E_o}{E_i} = - \frac{(S + \alpha_0)}{S^2 + \beta_1 S + \beta_0}$$

(d)

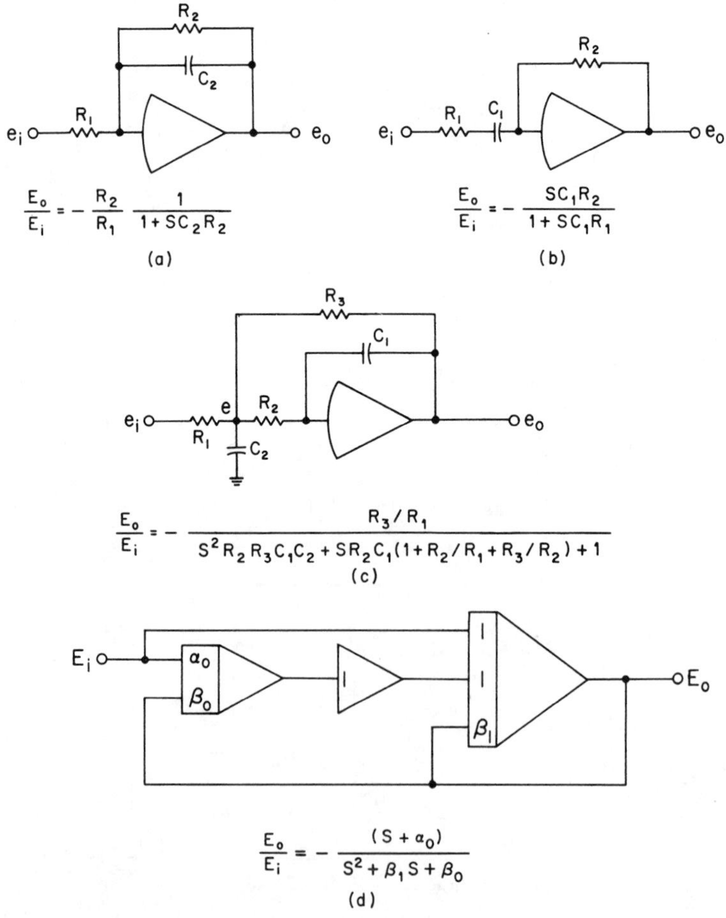

FIG. C.19. Several standard function forms.

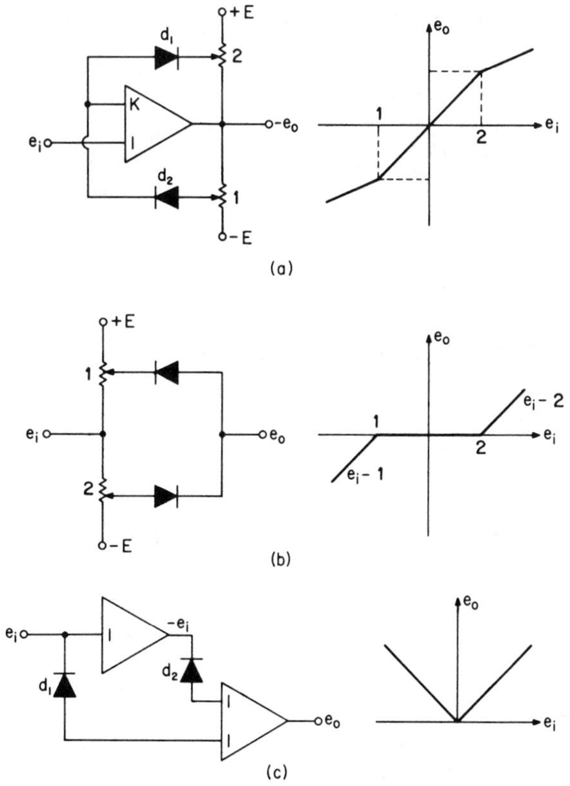

FIG. C.20. Several nonlinear transfer function forms.

systems. Multipliers are an integral part of virtually all analog facilities. Many other nonlinear characteristics, as shown in Fig. C.20, are easily simulated using diodes and other standard devices. It is straightforward to represent saturation, static friction and similar phenomena. Several standard arrangements are given in Ref. 1. Function generators of various types are available which can be used to generate a wide variety of input-output characteristics. It is beyond the scope of our presentation here to consider such highly specialized devices, although it is important to be aware of their existence and have some idea of where additional information may be found if needed.

Most analog computers may be operated in either the COMPUTE or REP-OP mode. When switched from HOLD or STANDBY to COMPUTE, the computer starts from pre-set initial conditions and integrates the given equation continuously until stopped. In REP-OP, the solution is repeated at a pre-set rate, and the response patterns may be recorded or observed on the oscilloscope, as desired. The effects of parameter changes are made apparent quickly in this mode and it is often used for that purpose. The user should always read the instruction manual for the computer he is using to become aware of the options it has to offer, and then use it in that way which best satisfies his needs.

C.6 CONCLUDING COMMENTS

Analog computers solve problems in parallel, so the time required for solution is independent of problem complexity. Only the number of computer components used increases with problem complexity. Solution accuracy is generally acceptable, although the components are non-ideal, noise and nonlinear effects may be present, and loading may alter the outputs somewhat. Since our equations are only an approximation, perhaps describing system performance to within something like 5% accuracy, there is little reason to build a highly precise analog computer. Component accuracies in the range of 0.1 percent are common. The analog computer is most appropriately used to determine approximate performance characteristics. The effects of parameter changes can be made directly available and trial and error design, even for complex systems, is relatively simple.

REFERENCE

1. Korn, G. A. and T. M. Korn, "Electronic Analog Computers," McGraw-Hill Book Co., New York, 1956.

Appendix D
A BRIEF REVIEW OF MATRIX THEORY*

D.1 MATRIX NOTATIONS AND BASIC OPERATIONS

The basic concepts of matrix theory were developed to provide a consistent shorthand notation for treating sets of simultaneous equations. For example the equations

$$\begin{aligned} a_{11}x_1 + a_{12}x_2 &= y_1 \\ a_{21}x_1 + a_{22}x_2 &= y_2 \end{aligned} \tag{D.1}$$

may alternately be written as

$$\mathbf{Ax} = \mathbf{y} \tag{D.2}$$

*The material contained in this appendix has been taken, with minor modification, from "Adaptive Control and Optimization Techniques," by V. W. Eveleigh, McGraw-Hill, 1967.

where

$$\mathbf{A} = \begin{bmatrix} a_{11} & a_{12} \\ a_{21} & a_{22} \end{bmatrix} \qquad \mathbf{x} = \begin{bmatrix} x_1 \\ x_2 \end{bmatrix} \qquad \mathbf{y} = \begin{bmatrix} y_1 \\ y_2 \end{bmatrix} \tag{D.3}$$

and all operations upon **A**, **x**, and **y** are defined consistently with those upon the coefficients in Eq. (D.1). Boldface type is used to denote vectors and matrices. Rules for basic matrix operations are stated briefly without proof, and matrix notation is used hereafter wherever convenient. For a further discussion of matrix theory, refer to Hildebrand.[1]

A matrix is a two-dimensional ordered array of numbers denoted by

$$\mathbf{A} = [a_{ij}] = \begin{bmatrix} a_{11} & a_{12} & \cdots & a_{1n} \\ a_{21} & a_{22} & \cdots & a_{2n} \\ \cdots & \cdots & \cdots & \cdots \\ a_{m1} & a_{m2} & \cdots & a_{mn} \end{bmatrix} \tag{D.4}$$

with elements a_{ij}, i, and j denoting the row and column, respectively, in which the element appears. **A**, as given by Eq. (D.4), is called an *$m \times n$ matrix*. The column matrices **x** and **y** in Eqs. (D.2) and (D.3) are called *vectors*. In this work, vectors are represented by column ($n \times 1$) matrices.

Addition and multiplication of matrices are readily defined. Let **A** and **B** be $m \times n$ and $p \times q$ matrices, respectively. When $m = p$ and $n = q$, their sum, denoted **C** and indicated notationally by

$$\mathbf{C} = \mathbf{A} + \mathbf{B} = \mathbf{B} + \mathbf{A} \tag{D.5}$$

is defined as the matrix with elements

$$c_{ij} = a_{ij} + b_{ij} \tag{D.6}$$

Their product, denoted **D**, and indicated notationally by

$$\mathbf{D} = \mathbf{AB} \tag{D.7}$$

is defined, if and only if $n = p$, as the matrix with elements

$$d_{ij} = \sum_{k=1}^{n} a_{ik} b_{kj} \tag{D.8}$$

D is $m \times q$. **A** and **B** are said to be conformable in the order **AB**. Note that **BA** may not be defined. Only if $m = q$ and $n = p$ are **A** and **B** conformable in either order. In general, even when both are defined,

$$\mathbf{AB} \neq \mathbf{BA} \tag{D.9}$$

Example D.1

Given

$$\mathbf{A} = \begin{bmatrix} 3 & 0 & 1 \\ 2 & 1 & 0 \\ 1 & 3 & 1 \end{bmatrix} \qquad \mathbf{B} = \begin{bmatrix} 2 & 1 & 1 \\ 3 & 0 & -1 \\ -2 & 1 & -1 \end{bmatrix} \tag{D.10}$$

find $\mathbf{D} = \mathbf{AB}$.

We see from Eq. (D.8) that

$$d_{11} = \sum_{k=1}^{3} a_{1k} b_{k1} = 4 \tag{D.11}$$

Evaluating the other d_{ij} similarly provides

$$\mathbf{D} = \begin{bmatrix} 4 & 4 & 2 \\ 7 & 2 & 1 \\ 9 & 2 & -3 \end{bmatrix} \tag{D.12}$$

A close relationship exists between the manipulations used to solve the set of equations

$$\mathbf{Ax} = \mathbf{y} \tag{D.13}$$

using determinants and using matrix methods. The determinant is a scalar and is only defined for a square array of numbers. The most general determinant-evaluation technique uses minors. A number of definitions are in order. The co-factor of element a_{ij} in **A** is defined as A_{ij} and

$$A_{ij} = (-1)^{i+j} M_{ij} \tag{D.14}$$

where M_{ij} is the minor of a_{ij}, obtained as the determinant of the square array reamining when the ith row and jth column of **A** are eliminated. The determinant of **A**, denoted by $|\mathbf{A}|$, is obtained as

$$|\mathbf{A}| = \sum_{k=1}^{n} a_{ik} A_{ik} = \sum_{i=1}^{n} a_{ik} A_{ik} \tag{D.15}$$

Example D.2
Find the determinant of

$$\mathbf{A} = \begin{bmatrix} a_{11} & a_{12} & a_{13} \\ a_{21} & a_{22} & a_{23} \\ a_{31} & a_{32} & a_{33} \end{bmatrix} \tag{D.16}$$

Proceeding step by step using minors provides

$$|\mathbf{A}| = a_{11} \begin{vmatrix} a_{22} & a_{23} \\ a_{32} & a_{33} \end{vmatrix} - a_{21} \begin{vmatrix} a_{12} & a_{13} \\ a_{32} & a_{33} \end{vmatrix} + a_{31} \begin{vmatrix} a_{12} & a_{13} \\ a_{22} & a_{23} \end{vmatrix} \tag{D.17}$$

or

$$|\mathbf{A}| = a_{11}(a_{22}a_{33} - a_{32}a_{23}) - a_{21}(a_{12}a_{33} - a_{32}a_{13}) + a_{31}(a_{12}a_{23} - a_{22}a_{13}) \tag{D.18}$$

Solution of simultaneous sets of equations using determinants is based upon Cramer's rule, which states that the elements of $\mathbf{x}$ in Eq. (D.13) are given by

$$x_i = \frac{\Delta_i}{\Delta} \tag{D.19}$$

with $\Delta = |\mathbf{A}|$ and $\Delta_i = |\mathbf{A}_i|$, where $\mathbf{A}_i$ is the matrix obtained by substituting $\mathbf{y}$ for the ith column of $\mathbf{A}$. Of course we require $\Delta \neq 0$, or $\mathbf{A}$ non-singular. In this context it is important to note that

$$|\mathbf{A}|\,|\mathbf{B}| = |\mathbf{AB}| \tag{D.20}$$

where $\mathbf{A}$ and $\mathbf{B}$ are both $n \times n$ matrices, and the product of two non-singular matrices is also non-singular.

Example D.3
Referring to Eq. (D.13),

$$\mathbf{A} = \begin{bmatrix} 1 & 0 & -2 \\ 0 & 1 & 3 \\ -1 & 2 & -1 \end{bmatrix} \qquad \mathbf{x} = \begin{bmatrix} x_1 \\ x_2 \\ x_3 \end{bmatrix} \qquad \mathbf{y} = \begin{bmatrix} 3 \\ 0 \\ 0 \end{bmatrix} \tag{D.21}$$

Solve for $\mathbf{x}$ using determinants.

For this problem, $\Delta = -9$ and

$$\Delta_1 = \begin{vmatrix} 3 & 0 & -2 \\ 0 & 1 & 3 \\ 0 & 2 & -1 \end{vmatrix} = -21 \qquad \Delta_2 = \begin{vmatrix} 1 & 3 & -2 \\ 0 & 0 & 3 \\ -1 & 0 & -1 \end{vmatrix} = -9$$

$$\Delta_3 = \begin{vmatrix} 1 & 0 & 3 \\ 0 & 1 & 0 \\ -1 & 2 & 0 \end{vmatrix} = 3 \tag{D.22}$$

so $x_1 = 7/3$, $x_2 = 1$, and $x_3 = -1/3$.

A number of special matrices and operations often arise. The zero matrix has all elements equal to zero and is denoted by $\mathbf{0}$. The identity matrix has $a_{ij} = 0$ for $i \neq j$ with $a_{ij} = 1$ for $i = j$ and is denoted by

$$\mathbf{I} = \begin{bmatrix} 1 & 0 & \dots & 0 & 0 \\ 0 & 1 & \dots & 0 & 0 \\ \dots & \dots & \dots & \dots & \dots \\ 0 & 0 & \dots & 1 & 0 \\ 0 & 0 & \dots & 0 & 1 \end{bmatrix} \tag{D.23}$$

$\mathbf{I}$ is always square and has the characteristic that

$$\mathbf{AI} = \mathbf{IA} = \mathbf{A} \tag{D.24}$$

where $\mathbf{A}$ is an arbitrary matrix conformable with $\mathbf{I}$ in the orders indicated. Interchanging rows and columns in a matrix is called taking its transpose. This operation, denoted by a prime sign, arises often. Thus

$$\mathbf{A}' = \begin{bmatrix} a_{11} & a_{21} & \dots & a_{m1} \\ a_{12} & a_{22} & \dots & a_{m2} \\ \dots & \dots & \dots & \dots \\ a_{1n} & a_{2n} & \dots & a_{mn} \end{bmatrix} \tag{D.25}$$

The transpose of a matrix product is readily shown to be

$$(\mathbf{AB})' = \mathbf{B}'\mathbf{A}' \tag{D.26}$$

In general,

$$(\mathbf{AB} \ldots \mathbf{YZ})' = \mathbf{Z}'\mathbf{Y}' \ldots \mathbf{B}'\mathbf{A}' \tag{D.27}$$

In solving matrix equations of the form given by Eq. (D.13) we need an inverse matrix having the characteristic

$$\mathbf{AA}^{-1} = \mathbf{A}^{-1}\mathbf{A} = \mathbf{I} \tag{D.28}$$

To obtain $\mathbf{A}^{-1}$ from $\mathbf{A}$ is a three-step process carried out as follows:

1. Replace a_{ij} by its cofactor A_{ij} for all i and j.
2. transpose the result to obtain the so-called adjoint matrix, and
3. divide by $|\mathbf{A}|$.

Obviously, $\mathbf{A}$ must be square and nonsingular ($|\mathbf{A}| \neq 0$) to have an inverse.

Example D.4

Consider Eq. (D.13) with

$$\mathbf{A} = \begin{bmatrix} 1 & 2 \\ 3 & 4 \end{bmatrix} \qquad \mathbf{x} = \begin{bmatrix} x_1 \\ x_2 \end{bmatrix} \qquad \mathbf{y} = \begin{bmatrix} y_1 \\ y_2 \end{bmatrix} \tag{D.29}$$

Find $\mathbf{x}$ in terms of $\mathbf{y}$ using $\mathbf{A}^{-1}$.

Replacing the elements of $\mathbf{A}$ by their cofactors and transposing provides

$$\mathbf{A}_{\text{adj}} = \begin{bmatrix} 4 & -2 \\ -3 & 1 \end{bmatrix} \tag{D.30}$$

The determinant of $\mathbf{A}$ equals -2. Thus

$$\mathbf{A}^{-1} = \begin{bmatrix} -2 & 1 \\ \frac{3}{2} & -\frac{1}{2} \end{bmatrix} \tag{D.31}$$

and

$$\begin{aligned} x_1 &= -2y_1 + y_2 \\ x_2 &= \frac{3}{2}y_1 - \frac{1}{2}y_2 \end{aligned} \tag{D.32}$$

The rank of **A** is defined as r when $r \times r$ is the largest square array in **A** whose determinant does not vanish. Referring to Eq. (D.13), we define augmented **A**, denoted $\mathbf{A}_a$, as the matrix obtained by adding column vector **y** to **A**. Thus

$$\mathbf{A}_a = \begin{bmatrix} a_{11} & a_{12} & \cdots & a_{1n} & y_1 \\ \cdots & \cdots & \cdots & \cdots & \cdots \\ a_{m1} & a_{m2} & \cdots & a_{mn} & y_m \end{bmatrix} \tag{D.33}$$

It is required, if a solution of Eq. (D.13) or any similar set of equations is possible, that the rank of **A** and $\mathbf{A}_a$ be equal. If **A** and $\mathbf{A}_a$ have rank $r < m$, where m is the number of equations in Eq. (D.13), the complete solution is implied by r of the equations, and any r independent equations may be solved explicity. When $\mathbf{y} = \mathbf{0}$, the trivial solution $\mathbf{x} = \mathbf{0}$ always satisfies Eq. (D.13). An alternate non-trivial solution can only exist in this case if $|\mathbf{A}| = 0$. This gives rise to the characteristic-value problem discussed later. When $r = n$, the number of unknowns, **x** is unique. When $r < n$, an infinity of solutions exists in terms of $n - r$ arbitrary constants.

Example D.5

Which of the following systems of equations have solutions?

$$\text{(a)}\ \begin{aligned} x_1 + x_2 &= 1 \\ x_1 + x_2 &= 2 \end{aligned} \qquad \text{(b)}\ \begin{aligned} x + y &= 1 \\ 2x + y &= 2 \\ 3x + y &= 3 \end{aligned} \qquad \text{(c)}\ \begin{aligned} x + y + z &= 1 \\ 2x + y + z &= 1 \end{aligned} \tag{D.34}$$

Find the solutions where possible.

For part (a),

$$\mathbf{A} = \begin{bmatrix} 1 & 1 \\ 1 & 1 \end{bmatrix} \qquad \mathbf{A}_a = \begin{bmatrix} 1 & 1 & 1 \\ 1 & 1 & 2 \end{bmatrix} \tag{D.35}$$

The rank of $\mathbf{A} = 1$ and the rank of $\mathbf{A}_a = 2$, so no solution exists. For part (b),

$$\mathbf{A} = \begin{bmatrix} 1 & 1 \\ 2 & 1 \\ 3 & 1 \end{bmatrix} \qquad \mathbf{A}_a = \begin{bmatrix} 1 & 1 & 1 \\ 2 & 1 & 2 \\ 3 & 1 & 3 \end{bmatrix} \tag{D.36}$$

and rank $\mathbf{A} = 2 = $ rank $\mathbf{A}_a$. Thus a unique solution exists and can be obtained by solving two of the three equations. The solution is given by $x = 1$, $y = 0$.

For part (c),

$$\mathbf{A} = \begin{bmatrix} 1 & 1 & 1 \\ 2 & 1 & 1 \end{bmatrix} \qquad \mathbf{A}_a = \begin{bmatrix} 1 & 1 & 1 & 1 \\ 2 & 1 & 1 & 1 \end{bmatrix} \tag{D.37}$$

and rank $\mathbf{A}$ = rank $\mathbf{A}_a = 2$. Since the rank is less than the number of unknowns, there is not enough information to find x, y, and z explicitly. However,

$$x = 0 \qquad y = 1 - z \tag{D.38}$$

is the required solution form, and one additional independent relationship between x, y, and z will completely define the solution.

In treating matrix sets of differential equations, it is convenient to have matrix differential operators available. We define the derivative of $\mathbf{A}$ as

$$\frac{d\mathbf{A}}{dt} = \begin{bmatrix} \dfrac{da_{11}}{dt} & \dfrac{da_{12}}{dt} & \cdots & \dfrac{da_{1n}}{dt} \\ \cdots & \cdots & \cdots & \cdots \\ \dfrac{da_{m1}}{dt} & \dfrac{da_{m2}}{dt} & \cdots & \dfrac{da_{mn}}{dt} \end{bmatrix} \tag{D.39}$$

The gradient operator ∇_x is defined as

$$\nabla_x' = \left(\frac{d}{d\mathbf{x}}\right)' = \begin{bmatrix} \dfrac{\partial}{\partial x_1} & \dfrac{\partial}{\partial x_2} & \cdots & \dfrac{\partial}{\partial x_n} \end{bmatrix} \tag{D.40}$$

Thus the gradient of the standard matrix quadratic form is given by

$$\nabla_x(\mathbf{x}'\mathbf{A}\mathbf{x}) = \nabla_x\left(\sum_{i=1}^{n}\sum_{j=1}^{n} a_{ij}x_i x_j\right) = \mathbf{A}\mathbf{x} + \mathbf{A}'\mathbf{x} \tag{D.41}$$

which is readily checked by direct substitution. This operation often proves useful.

Example D.6

Given

$$\mathbf{x} = \begin{bmatrix} x_1 \\ x_2 \\ x_3 \end{bmatrix} \qquad \mathbf{A} = \begin{bmatrix} 1 & 0 & 1 \\ 3 & 2 & 0 \\ 1 & 2 & 1 \end{bmatrix} \tag{D.42}$$

find $\nabla_x(\mathbf{x}'\mathbf{A}\mathbf{x})$.

Using Eq. (D.41),

$$\mathbf{A}\mathbf{x} = \begin{bmatrix} x_1 + x_3 \\ 3x_1 + 2x_2 \\ x_1 + 2x_2 + x_3 \end{bmatrix} \qquad \mathbf{A}'\mathbf{x} = \begin{bmatrix} x_1 + 3x_2 + x_3 \\ 2x_2 + 2x_3 \\ x_1 + x_3 \end{bmatrix} \tag{D.43}$$

and

$$\nabla_x(\mathbf{x}'\mathbf{A}\mathbf{x}) = \begin{bmatrix} 2x_1 + 3x_2 + 2x_3 \\ 3x_1 + 4x_2 + 2x_3 \\ 2x_1 + 2x_2 + 2x_3 \end{bmatrix} \tag{D.44}$$

As a check on Eq. (D.41), suppose no identity were available. Then

$$\mathbf{x}'\mathbf{A}\mathbf{x} = x_1(x_1 + x_3) + x_2(3x_1 + 2x_2) + x_3(x_1 + 2x_2 + x_3) \tag{D.45}$$

Operating upon this result directly using Eq. (D.40) provides the same answer.

In our discussion thus far we have assumed all elements of the matrices considered were scalars. This need not be the case in general. It is equally valid to consider matrices with elements which are matrices and to define all the standard operations upon such matrices. Consider

$$\mathbf{A} = \begin{bmatrix} a_{11} & a_{12} & a_{13} \\ a_{21} & a_{22} & a_{23} \\ a_{31} & a_{32} & a_{33} \end{bmatrix} \qquad \mathbf{B} = \begin{bmatrix} b_{11} & b_{12} & b_{13} \\ b_{21} & b_{22} & b_{23} \\ b_{31} & b_{32} & b_{33} \end{bmatrix} \tag{D.46}$$

$\mathbf{A}$ and $\mathbf{B}$ are divided into submatrices, or partitioned, to provide

$$\mathbf{A} = \begin{bmatrix} \mathbf{a}_1 & \mathbf{a}_2 \\ \mathbf{a}_3 & \mathbf{a}_4 \end{bmatrix} \qquad \mathbf{B} = \begin{bmatrix} \mathbf{b}_1 & \mathbf{b}_2 \\ \mathbf{b}_3 & \mathbf{b}_4 \end{bmatrix} \tag{D.47}$$

where

$$\mathbf{a}_1 = \begin{bmatrix} a_{11} & a_{12} \\ a_{21} & a_{22} \end{bmatrix} \quad \mathbf{a}_2 = \begin{bmatrix} a_{13} \\ a_{23} \end{bmatrix} \quad \mathbf{a}_3 = \begin{bmatrix} a_{31} & a_{32} \end{bmatrix} \quad \mathbf{a}_4 = a_{33}$$

and the $\mathbf{b}$'s are defined similarly. It is a simple matter to show that the sum and product of the matrices in Eq. (D.47) are identical to those obtained using Eq. (D.46). In many problems there are practical reasons for considering certain combinations of matrix elements together as submatrices. Obviously $\mathbf{A}$ and $\mathbf{B}$ can only be added and multiplied in partitioned form if their submatrices are of the same dimensions and conformable in the multiplication order required for normal expansion.

Although this is by no means an exhaustive review of matrix methods, the basic operations discussed should provide a minimum background for our purposes. Many excellent references are available for further study. Certain specific matrix transformations useful in developing standard state vector forms are developed in the text when the need for them becomes apparent.

D.2 LINEAR VECTOR SPACES

As discussed in the previous section, a vector is a column matrix, and all matrix operations applicable to column matrices apply directly to vectors. In addition, however, there are several special characteristics of vectors worth considering, and these are summarized in this section. Vectors prove useful in many engineering problems, and they are often used to denote coordinates in spaces of various kinds.

We define n-dimensional Euclidian space, E^n, as a general n space within which the various matrix and vector rules outlined here are satisfied. Vectors with n elements are used to define points in E^n. As with matrices, two vectors are equal if and only if all of their components are equal. We say that $\mathbf{x} \leq \mathbf{y}$ if $x_i \leq y_i$ for all i. It is often convenient to define unit vectors in E^n for pusposes similar to their use with rectangular coordinates in three space to assign direction to the scalar lengths of vector components. Although any vector of unit length might be called a unit vector, the term is generally reserved for vectors with one element equal to unity and all other elements equal to zero. The inner or scalar product of two n-element vectors $\mathbf{x}$ and $\mathbf{y}$ is denoted by $(\mathbf{x}, \mathbf{y})$ and defined as

$$(\mathbf{x}, \mathbf{y}) = (\mathbf{y}, \mathbf{x}) = \mathbf{x}'\mathbf{y} = \mathbf{y}'\mathbf{x} = \sum_{i=1}^{n} x_k y_k \tag{D.48}$$

Note in particular that

$$(\mathbf{x}, \mathbf{x}) = \sum_{k=1}^{n} x_k^2 = (L_x)^2 = \|\mathbf{x}\|^2 \tag{D.49}$$

where L_x and $\|\mathbf{x}\|$ each denote the length of $\mathbf{x}$, sometimes called the norm of $\mathbf{x}$. This length interpretation in n space is analogous to the usual physical interpretations in two and three dimensions. The distance D_{xy} between the two points in E^n defined by the vectors $\mathbf{x}$ and $\mathbf{y}$ is the norm of $\mathbf{x} - \mathbf{y}$, or

$$D_{xy} = \|\mathbf{x} - \mathbf{y}\| = \|\mathbf{y} - \mathbf{x}\| = \left[\sum_{k=1}^{n} (x_k - y_k)^2\right]^{1/2} \tag{D.50}$$

Two inequalities closely related to these geometric concepts and often useful in practice are the triangle inequality

$$\|\mathbf{x} + \mathbf{y}\| \leq \|\mathbf{x}\| + \|\mathbf{y}\| \tag{D.51}$$

and the Schwarz inequality

$$\|(\mathbf{x}, \mathbf{y})\| \leq \|\mathbf{x}\| \, \|\mathbf{y}\| \tag{D.52}$$

Both Eqs. (D.51) and (D.52) are readily proven from the basic definitions.

Consider the set of m n-element vectors, $\mathbf{x}_1, \mathbf{x}_2, \ldots, \mathbf{x}_m$. A linear combination of these $\mathbf{x}_i$ is defined as

$$\mathbf{x} = \sum_{i=1}^{m} c_i \mathbf{x}_i \tag{D.53}$$

where the c_i are constants not all equal to zero. Vector $\mathbf{x}_k$ from the given set is said to be linearly dependent on the others if

$$\mathbf{x}_k = \sum_{i=1}^{m} c_i \mathbf{x}_i \tag{D.54}$$

for some set of constants c_i. In other words if $\mathbf{x}_k$ can be expressed as a linear combination of the other vectors in the set, it is linearly dependent upon them or, equivalently, not an independent member of the set. If r and only r of the $\mathbf{x}_i$ are independent, the vector set is said to be of rank r. If $r < m$, $m - r$ of the vectors in the set are linearly dependent upon the other r. A vector set of rank

n is said to span E^n, which means that any vector in E^n can be obtained as a linear combination of the vectors in the set.

Directly from Eq. (D.54) it is apparent that an m-element set of vectors in E^n is independent if and only if no set of constants c_i, not all zero, exists such that

$$\sum_{i=1}^{m} c_i \mathbf{x}_i = \mathbf{c}'\mathbf{X}' = 0 \qquad \mathbf{X} = [\mathbf{x}_1, \mathbf{x}_2, \ldots, \mathbf{x}_m] \tag{D.55}$$

For the special case $m = n$, existence of a nontrivial solution for $\mathbf{c}$ requires that

$$\begin{vmatrix} x_{11} & x_{12} & \cdots & x_{1n} \\ x_{21} & x_{22} & \cdots & x_{2n} \\ \cdots & \cdots & \cdots & \cdots \\ x_{n1} & x_{n2} & \cdots & x_{nn} \end{vmatrix} = |\mathbf{X}| = 0 \tag{D.56}$$

A more general test is desired, however, to treat those cases where $m \neq n$, since the determinant is undefined for nonsquare arrays. Assume a solution for the c_i does exist which satisfies Eq. (D.55). Successively forming the scalar product of both sides of Eq. (D.55) with $\mathbf{x}_1, \mathbf{x}_2, \ldots, \mathbf{x}_m$ provides

$$\begin{aligned} c_1(\mathbf{x}_1, \mathbf{x}_1) + c_2(\mathbf{x}_1, \mathbf{x}_2) + \cdots + c_m(\mathbf{x}_1, \mathbf{x}_m) &= 0 \\ c_1(\mathbf{x}_2, \mathbf{x}_1) + c_2(\mathbf{x}_2, \mathbf{x}_2) + \cdots + c_m(\mathbf{x}_2, \mathbf{x}_m) &= 0 \\ \cdots\cdots\cdots\cdots\cdots\cdots\cdots\cdots\cdots\cdots\cdots \\ c_1(\mathbf{x}_m, \mathbf{x}_1) + c_2(\mathbf{x}_m, \mathbf{x}_2) + \cdots + c_m(\mathbf{x}_m, \mathbf{x}_m) &= 0 \end{aligned} \tag{D.57}$$

We see from Eq. (D.57) that a nontrivial solution for the c_i exists only if

$$G = \begin{vmatrix} (\mathbf{x}_1, \mathbf{x}_1) & \cdots & (\mathbf{x}_1, \mathbf{x}_m) \\ \cdots & \cdots & \cdots \\ (\mathbf{x}_m, \mathbf{x}_1) & \cdots & (\mathbf{x}_m, \mathbf{x}_m) \end{vmatrix} = 0 \tag{D.58}$$

G is called the Gram determinant. Thus a set of vectors is linearly dependent if and only if its Gram determinant vanishes, and it is linearly independent if and only if $G \neq 0$.

Example D.7

Are the vectors $\mathbf{x}_1' = (1, 1, 0)$, $\mathbf{x}_2' = (0, 1, 1)$, and $\mathbf{x}_3' = (1, 2, 1)$ linearly independent? If not, express $\mathbf{x}_3$ as a linear function of $\mathbf{x}_1$ and $\mathbf{x}_2$.

This problem can be approached in several ways. Applying Eq. (D.56) yields

$$|\mathbf{x}| = \begin{vmatrix} 1 & 0 & 1 \\ 1 & 1 & 2 \\ 0 & 1 & 1 \end{vmatrix} = 0 \tag{D.59}$$

Thus the vector set is not linearly independent. Using Eq. (D.58) instead of Eq. (D.56) provides

$$G = \begin{vmatrix} 2 & 1 & 3 \\ 1 & 2 & 3 \\ 3 & 3 & 6 \end{vmatrix} = 2(3) - 1(-3) + 3(-3) = 0 \tag{D.60}$$

which yields the same conclusion. Note that $\mathbf{x}_3 = \mathbf{x}_1 + \mathbf{x}_2$, indicating the particular linear dependence present in this case.

The n-element vector set $\mathbf{x}_1, \mathbf{x}_2, \ldots, \mathbf{x}_n$ forms a basis for E^n or spans E^n if any vector in E^n can be written as a linear combination of the $\mathbf{x}_i$. Any linearly independent n-vector set is a basis for E^n and every basis for E^n contains n vectors. The simplest basis for E^n is the set of n unit vectors, although the alternate possibilities are limitless.

Two vectors $\mathbf{x}$ and $\mathbf{y}$ are called orthogonal if and only if $(\mathbf{x}, \mathbf{y}) = 0$. In two or three space, orthogonality implies that $\mathbf{x}$ and $\mathbf{y}$ are perpendicular. If in addition $(\mathbf{x}, \mathbf{x}) = (\mathbf{y}, \mathbf{y}) = 1$, $\mathbf{x}$ and $\mathbf{y}$ are said to be orthonormal. It is often desirable to generate an orthonormal vector set from the independent set $\mathbf{x}_1, \mathbf{x}_2, \ldots, \mathbf{x}_m$. This is readily accomplished as follows. Select one of the vectors, say $\mathbf{x}_1$, and define

$$\mathbf{y}_1 = \frac{\mathbf{x}_1}{\|\mathbf{x}_1\|} \tag{D.61}$$

as the first member of the orthonormal set. Choose a second member, say $\mathbf{x}_2$, and define

$$\mathbf{z}_2 = \mathbf{x}_2 - c_{12}\mathbf{y}_1 \tag{D.62}$$

$\mathbf{z}_2$ must be orthogonal to $\mathbf{y}_1$, so

$$(\mathbf{y}_1, \mathbf{z}_2) = (\mathbf{y}_1, \mathbf{x}_2) - c_{12}(\mathbf{y}_1, \mathbf{y}_1) = 0 \tag{D.63}$$

is required. Thus $c_{12} = (\mathbf{y}_1, \mathbf{x}_2)$, since $(\mathbf{y}_1, \mathbf{y}_1) = 1$. Consequently,

$$\mathbf{z}_2 = \mathbf{x}_2 - (\mathbf{y}_1, \mathbf{x}_2)\mathbf{y}_1 \tag{D.64}$$

Normalizing $\mathbf{z}_2$ provides a second element for the orthonormal set as

$$\mathbf{y}_2 = \frac{\mathbf{z}_2}{\|\mathbf{z}_2\|} \tag{D.65}$$

Next define

$$\mathbf{z}_3 = \mathbf{x}_3 - c_{13}\mathbf{y}_1 - c_{23}\mathbf{y}_2 \tag{D.66}$$

$\mathbf{z}_3$ must be orthogonal to $\mathbf{y}_1$ and $\mathbf{y}_2$, so

$$(\mathbf{y}_1, \mathbf{z}_3) = (\mathbf{y}_1, \mathbf{x}_3) - c_{13} = 0 \tag{D.67}$$

and

$$(\mathbf{y}_2, \mathbf{z}_3) = (\mathbf{y}_2, \mathbf{x}_3) - c_{23} = 0 \tag{D.68}$$

since $\mathbf{y}_1$ and $\mathbf{y}_2$ are orthonormal. Solving Eqs. (D.67) and (D.68) for c_{13} and c_{23} and substituting into Eq. (D.66) provides

$$\mathbf{z}_3 = \mathbf{x}_3 - (\mathbf{y}_1, \mathbf{x}_3)\mathbf{y}_1 - (\mathbf{y}_2, \mathbf{x}_3)\mathbf{y}_2 \tag{D.69}$$

The third orthonormal element is obtained as

$$\mathbf{y}_3 = \frac{\mathbf{z}_3}{\|\mathbf{z}_3\|} \tag{D.70}$$

This procedure is repeated until a complete orthonormal set is obtained. Physically, we are simply subtracting all components from each vector in turn in the direction of vectors previously defined and then normalizing the resulting vector to unit length.

Example D.8

Orthonormalize the three-vector set $\mathbf{x}_1' = (1, 0, 0)$, $\mathbf{x}_2' = (1, 1, 0)$, and $\mathbf{x}_3' = (0, 1, 1)$. Is your answer unique?

Choose $\mathbf{y}_1 = \mathbf{x}_1$ as the first element of the orthonormal set. Define

$$\mathbf{z}_2 = \mathbf{x}_2 - c_{12}\mathbf{x}_1 \tag{D.71}$$

Setting $(\mathbf{y}_1, \mathbf{z}_2) = 0$ provides $c_{12} = 1$, so $\mathbf{z}_2' = (0, 1, 0) = \mathbf{y}_2'$. Define

$$\mathbf{z}_3 = \mathbf{x}_3 - c_{13}\mathbf{y}_1 - c_{23}\mathbf{y}_2 \tag{D.72}$$

Solving for c_{13} and c_{23} as in Eq. (D.69) provides $c_{13} = 0$ and $c_{23} = 1$. Thus $\mathbf{z}_3' = (0, 0, 1) = \mathbf{y}_3'$. This answer is obviously not unique, since $\mathbf{x}_2$ or $\mathbf{x}_3$ could have been normalized to obtain the first element of the orthonormal set, resulting in at least two alternate solutions.

D.3 MATRIX SETS OF DIFFERENTIAL EQUATIONS—THE EIGENVALUE PROBLEM

In physical systems we are often faced with solving for values of λ yielding nontrivial solutions to the linear vector set of equations

$$\mathbf{A}\mathbf{x} = \lambda\mathbf{x} \quad \text{or} \quad (\mathbf{A} - \lambda\mathbf{I})\mathbf{x} = \mathbf{0} \tag{D.73}$$

where $\mathbf{A}$ is an $n \times n$ matrix of constants, $\mathbf{x}$ is an $n \times 1$ vector and λ is a constant. This is the so-called eigenvalue problem encountered in the analysis of linear time invariant homogeneous vector sets of first order differential equations using transform mathematics. Consider how this comes about in a typical case. Let the transfer function of a control system be given by

$$\frac{C(s)}{R(s)} = \frac{1}{s^2 + 3s + 2} \tag{D.74}$$

The characteristic equation of this system defines the general structure of its unforced natural response to initial conditions, and is obtained by setting the denominator of Eq. (D.74) to zero, or

$$s^2 + 3s + 2 = (s + 1)(s + 2) = 0 \tag{D.75}$$

Thus the system described by Eq. (D.74) has poles at $s = -1$ and $s = -2$, and its unforced response is obtained readily as the sum of two decaying exponential terms. Suppose we rewrite Eq. (D.74) in the time domain as

$$\ddot{c} + 3\dot{c} + 2c = r \tag{D.76}$$

Consider for the moment the special case $r = 0$. Define

$$c = x_1 \tag{D.77}$$

and

$$\dot{x}_1 = x_2 \tag{D.78}$$

Rewriting Eq. (D.76) in terms of x_1 and x_2 gives

$$\dot{x}_2 = -3x_2 - 2x_1 \tag{D.79}$$

Taking the Laplace transform of both sides of Eqs. (D.78) and (D.79) yields

$$0x_1 - x_2 = -sx_1 + x_1(0) \tag{D.80}$$

$$2x_1 + 3x_2 = -sx_2 + x_2(0) \tag{D.81}$$

If $x_1(0) = x_2(0) = 0$, we note that the vector form of Eqs. (D.80) and (D.81) is equivalent to Eq. (D.73) with s replacing λ. This same type of problem arises in the analysis of all linear time invariant systems. A nontrivial solution to Eqs. (D.80) and (D.81) exists if and only if

$$\begin{vmatrix} s & -1 \\ 2 & s+3 \end{vmatrix} = s^2 + 3s + 2 = 0 \tag{D.82}$$

which is the characteristic equation defining the pole locations of the original system. We anticipated this result, of course, since the method of system representation can hardly change its general response characteristics.

Certain characteristics of linear-system solutions are of general interest. Values of λ for which nontrivial solutions to Eq. (D.73) exist are called *eigenvalues* of the system and correspond to the usual characteristic equation roots or system pole locations. There are n eigenvalues, not necessarily distinct, for an nth order system. When each eigenvalue is substituted into Eq. (D.73) a solution for $\mathbf{x}$ results. These solutions are called *eigenvectors*. The column of eigenvalues is called a *modal column*, and the matrix of eigenvectors is called a *modal matrix*. An eigenvalue of multiplicity α defines α eigenvectors.

Example D.9

Find the eigenvalues and eigenvectors of

$$\mathbf{A} = \begin{bmatrix} 1 & 0 & -1 \\ 1 & 2 & 1 \\ 2 & 2 & 3 \end{bmatrix} \tag{D.83}$$

For this matrix [see Eq. (D.73)]

$$|s\mathbf{I} - \mathbf{A}| = \begin{vmatrix} s-1 & 0 & 1 \\ -1 & s-2 & -1 \\ -2 & -2 & s-3 \end{vmatrix} = s^3 - 6s^2 + 11s - 6$$

$$= (s-3)(s-2)(s-1) = 0 \tag{D.84}$$

Substituting $s = 1$ in $(s\mathbf{I} - \mathbf{A})\,\mathbf{x} = \mathbf{0}$ provides

$$x_3 = 0 \qquad -x_1 - x_2 = 0 \qquad -2x_1 - 2x_2 = 0 \tag{D.85}$$

or $x_3 = 0$ and $x_1 = -x_2$. Arbitrarily choosing $x_1 = 1$ gives $\mathbf{u}_1' = (1, -1, 0)$. The corresponding second and third eigenvectors are obtained similarly as $\mathbf{u}_2' = (-2, 1, 2)$ and $\mathbf{u}_3' = (1, -1, -2)$.

Example D.10

Find the eigenvalues and eigenvectors for

$$\mathbf{A} = \begin{bmatrix} 2 & 2 & 1 \\ 1 & 3 & 1 \\ 1 & 2 & 2 \end{bmatrix} \tag{D.86}$$

Expanding $s\mathbf{I} - \mathbf{A}$ to get the characteristic equation yields

$$|s\mathbf{I} - \mathbf{A}| = s^3 - 7s^2 + 11s - 5 = (s - 1)^2(s - 5) = 0 \tag{D.87}$$

Substituting $s = 1$ into $(s\mathbf{I} - \mathbf{A})\mathbf{x} = \mathbf{0}$ provides

$$\begin{aligned} -x_1 - 2x_2 - x_3 &= 0 \\ -x_1 - 2x_2 - x_3 &= 0 \qquad \text{or} \qquad x_1 = 2x_2 - x_3 \\ -x_1 - 2x_2 - x_3 &= 0 \end{aligned} \tag{D.88}$$

Two eigenvector elements can be chosen arbitrarily. Choosing $x_2 = 1$, $x_3 = -2$ requires $x_1 = 0$, and choosing $x_3 = 0$, $x_2 = 1$ requires $x_1 = -2$. The corresponding eigenvectors are $\mathbf{u}_1' = (0, 1, -2)$ and $\mathbf{u}_2' = (-2, 1, 0)$. Substituting $s = 5$ into the same equations provides

$$\begin{aligned} 3x_1 - 2x_2 - x_3 &= 0 \\ -x_1 + 2x_2 - x_3 &= 0 \qquad \text{or} \qquad x_1 = x_2 = x_3 \\ -x_1 - 2x_2 + 3x_3 &= 0 \end{aligned} \tag{D.89}$$

and $\mathbf{u}_3 = (1, 1, 1)$ is the corresponding third eigenvector.

Examples D.9 and D.10 illustrate two important points. At least one element of each eigenvector may always be chosen arbitrarily so eigenvectors are readily scaled as desired. When the characteristic equation has a root of multiplicity n, n elements of the corresponding n eigenvectors may be specified arbitrarily.

It is a well-known characteristic of linear systems that the nonhomogeneous equation

$$(\lambda\mathbf{I} - \mathbf{A})\mathbf{x} = \mathbf{c} \tag{D.90}$$

has a general solution, assuming that one exists, consisting of an algebraic sum of the terms appearing in the homogeneous solution, which is equivalent to observing that the numerator terms in a transfer function alter only the relative magnitudes of the response components but not the solution form. Functional solution form is determined exclusively by the input and the denominator terms. This invariance of solution structure explains why the eigenvalue problem is of such general importance.

When $\mathbf{A}$ is a symmetric matrix, a situation occasionally arising naturally and one which can often be developed by a properly chosen transformation, the following general characteristics of the eigenvalues and eigenvectors hold:

1. Any two eigenvectors corresponding to distinct eigenvalues are orthogonal.
2. All eigenvalues of a real symmetric matrix are real.
3. The eigenvectors of any $n \times n$ symmetric matrix $\mathbf{A}$ form a basis for E^n.
4. There exists at least one orthonormal set of eigenvectors of $\mathbf{A}$ which provides a basis for E^n.
5. If an eigenvalue has multiplicity r, the corresponding eigenvectors provide a basis for an r-dimensional subspace.
6. For eigenvalue λ_i of multiplicity r there are r linearly independent eigenvectors with eigenvalue λ_i in any set of n orthonormal eigenvectors of $\mathbf{A}$.
7. If one or more eigenvalues have multiplicity $r \geq 2$, there are an infinite number of distinct sets of orthonormal eigenvectors of $\mathbf{A}$ providing a basis for E^n, corresponding to the infinity of orthonormal selections possible to generate the subspace with dimensions $r \geq 2$.

These characteristics of symmetric matrices often prove useful in developing general theorems relating to linear systems.

In the standard quadratic form

$$Q(\mathbf{x}) = \mathbf{x}'\mathbf{Z}\mathbf{x} = \sum_{i=1}^{n}\sum_{j=1}^{n} a_{ij}x_i x_j \tag{D.91}$$

note that a_{ij} and a_{ji} both multiply $x_i x_j$. Thus $\mathbf{A}$ can always be treated as an equivalent symmetric matrix $\mathbf{B}$ by defining

$$b_{ij} = b_{ji} = \frac{a_{ij} + a_{ji}}{2} \qquad i \neq j \tag{D.92}$$

with $b_{ii} = a_{ii}$. The resulting matrix of terms b_{ij} provides the same result as that given in Eq. (D.91) and is symmetric. $Q(\mathbf{x})$ is defined as positive-definite if $Q(\mathbf{x}) > 0$ for $\mathbf{x} \neq \mathbf{0}$ and positive-semidefinite if $Q(\mathbf{x}) \geq 0$ for $\mathbf{x} \neq \mathbf{0}$ and if there is at least one $\mathbf{x} \neq \mathbf{0}$ for which $Q(\mathbf{x}) = 0$. $Q(\mathbf{x})$ is negative-definite or negative-semidefinite, respectively, if $Q(\mathbf{x}) < 0$ or $Q(\mathbf{x}) \leq 0$ for $\mathbf{x} \neq \mathbf{0}$ and if there is at least one $\mathbf{x} \neq \mathbf{0}$ for which $Q(\mathbf{x}) = 0$. $Q(\mathbf{x})$ is positive-definite, positive-semidefinite, or indefinite if and only if all eigenvalues of **B** are positive, all eigenvalues of **B** are nonnegative and at least one eigenvalue equals zero, or **B** has both positive and negative eigenvalues, respectively, where **B** is the equivalent symmetric matrix with terms defined by Eq. (D.92). $Q(\mathbf{x})$ is also positive-definite if all principal minors of **A** are positive, or if

$$a_{11} > 0 \qquad \begin{vmatrix} a_{11} & a_{12} \\ a_{21} & a_{22} \end{vmatrix} > 0 \ \ldots \ |\mathbf{A}| > 0 \tag{D.93}$$

Consider the homogeneous set of differential equations

$$\dot{\mathbf{x}} = \mathbf{A}\mathbf{x} \tag{D.94}$$

Assume **A** symmetric with distinct eigenvalues $\lambda_1, \lambda_2, \ldots, \lambda_n$. Let $\mathbf{x}_1, \mathbf{x}_2, \ldots, \mathbf{x}_n$ be the corresponding normalized eigenvectors. Let **X** denote the normalized eigenvector matrix, defined as

$$\mathbf{X} = [\mathbf{x}_1, \mathbf{x}_2, \ldots, \mathbf{x}_n] = \begin{bmatrix} x_{11} & x_{12} & \cdots & x_{1n} \\ x_{21} & x_{22} & \cdots & x_{2n} \\ \cdots & \cdots & \cdots & \cdots \\ x_{n1} & x_{n2} & \cdots & x_{nn} \end{bmatrix} \tag{D.95}$$

where $\mathbf{x}'_k = [x_{1k}, x_{2k}, \ldots, x_{nk}]$. Then

$$\mathbf{X}'\mathbf{X} = I \tag{D.96}$$

since the eigenvectors of a symmetric matrix are orthogonal and the $\mathbf{x}_i$ have been assumed normalized. Also,

$$\mathbf{AX} = [\mathbf{Ax}_1, \mathbf{Ax}_2, \ldots, \mathbf{Ax}_n] \tag{D.97}$$

However, from Eq. (D.94) we see that

$$\mathbf{Ax}_i = \lambda_i \mathbf{x}_i \tag{D.98}$$

since the λ_i are obtained by solving

$$(\lambda\mathbf{I} - \mathbf{A})\mathbf{x} = \mathbf{0} \tag{D.99}$$

and Eq. (D.99) is used for each λ_i to define the corresponding $\mathbf{x}_i$. Thus Eq. (D.97) reduces to

$$\mathbf{AX} = [\lambda_1\mathbf{x}_1, \lambda_2\mathbf{x}_2, \ldots, \lambda_n\mathbf{x}_n] \tag{D.100}$$

and

$$\mathbf{X}'\mathbf{A}\mathbf{X} = \begin{bmatrix} \lambda_1 & 0 & \dots & 0 \\ 0 & \lambda_2 & \dots & 0 \\ \dots & \dots & \dots & \dots \\ 0 & 0 & \dots & \lambda_n \end{bmatrix} = \mathbf{\Lambda} \tag{D.101}$$

where $\mathbf{\Lambda}$ denotes the diagonal matrix of eigenvalues. $\mathbf{A}$ may be expressed in terms of $\mathbf{X}$ and $\mathbf{\Lambda}$ by premultiplying both sides of Eq. (D.101) by $\mathbf{X}$ and post-multiplying both sides by $\mathbf{X}'$ to obtain

$$\mathbf{A} = \mathbf{X}\mathbf{\Lambda}\mathbf{X}' \tag{D.102}$$

Thus any symmetric matrix with distinct eigenvalues may be expressed in terms of a diagonal matrix of its eigenvalues and the normalized eigenvector matrix. When $\mathbf{A}$ is not symmetric, the more general form of Eq. (D.101) is obtained as

$$\mathbf{\Lambda} = \mathbf{X}^{-1}\mathbf{A}\mathbf{X} \tag{D.103}$$

and

$$\mathbf{A} = \mathbf{X}\mathbf{\Lambda}\mathbf{X}^{-1} \tag{D.104}$$

results directly from Eq. (D.103).

Example D.11

Consider the equations defined by Eq. (D.94) with

$$\mathbf{A} = \begin{bmatrix} 0 & 1 \\ -2 & 3 \end{bmatrix} \tag{D.105}$$

Find the eigenvalues of $\mathbf{A}$, the normalized eigenvector matrix $\mathbf{X}$, and $\mathbf{\Lambda}$.

Setting $(\lambda\mathbf{I} - \mathbf{A}) = \mathbf{0}$ provides the characteristic equation as

$$\begin{vmatrix} \lambda & -1 \\ 2 & \lambda - 3 \end{vmatrix} = 0 = \lambda^2 - 3\lambda + 2 = 0 = (\lambda - 2)(\lambda - 1) \tag{D.106}$$

Thus the eigenvalues are $\lambda = 2$ and $\lambda = 1$. Substituting these values into the original equations provides the eigenvectors $\mathbf{x}_1' = (\alpha, \alpha)$ and $\mathbf{x}_2' = (\beta, 2\beta)$.

Normalizing them yields

$$\mathbf{X} = \begin{bmatrix} \frac{\sqrt{2}}{2} & \frac{\sqrt{5}}{5} \\ \frac{\sqrt{2}}{2} & \frac{2\sqrt{5}}{5} \end{bmatrix} \tag{D.107}$$

Λ is the diagonal matrix of eigenvalues, or

$$\Lambda = \begin{bmatrix} 1 & 0 \\ 0 & 2 \end{bmatrix} \tag{D.108}$$

Equation (D.103) may be used as a final check on the results.

REFERENCE

1. Hildebrand, F. B., Methods of Applied Mathematics, Prentice Hall, Inc., Englewood Cliffs, N. J., 1952, Chap. 1.

Appendix E
DIGITAL SIMULATION METHODS

A simulation study is often an important step in the evaluation and synthesis of complex control systems. Simulation provides a direct check upon analytical results, and also yields information, such as the unit step response characteristics, which it is not practical to develop by hand. The digital computer has become a standard tool for simulation study of continuous systems in recent years[1] for the following reasons:

1. It is an extremely accurate and flexible tool, providing a straightforward simulation of nonlinear systems, and allowing all sorts of input signals to be chosen.
2. It is easily programmed in a variety of efficient languages.
3. It is highly reliable.
4. It is readily available, often through a time sharing system using remote access user interactive terminals from which the study may be carried out in a manner similar to that used on analog computers.

When all of these factors are considered, the recent trend toward digital computer use in simulation is readily understood.

We will not get involved with programming languages in this appendix. It is assumed that the user has a digital computer available and that he is either familiar with the language it requires, or will study the rudiments of that language on his own. Our primary purpose here is to indicate several general approaches which may be used in developing an adequate digital simulation program. Once these basic approaches are understood, their reduction to a program using any of the many languages available is straightforward. In presenting an example to illustrate the concepts involved, some language must be chosen, and APL is picked for this purpose because of its great power and flexibility for scientific use. Although no detailed discussion of APL will be presented here, its symbolism is almost self explanatory, and its use will cause no problems.

A digital computer simulation can be initiated either from the system differential equations expressed in state vector form, or from a system block diagram reduced to its simplest form involving only integrators, amplifiers, attenuators, summers and, if necessary, single valued nonlinear elements for which the output can be expressed as an algebraic function of the input. Reduction of a general system description to these forms is discussed in Chap. 14, so we will also assume that the reader can get to this point on his own. We will not waste a lot of time attempting to develop an optimally efficient integration algorithm, assuming that a relatively simple, although perhaps somewhat naive, approach which leads to the answer without too much fuss is most desirable.

The digital computer works with discrete data, so it is appropriate to make a discrete approximation of the analog signals for integration purposes. The simplest possible discrete signal representation is to assume that all signals remain constant over successive intervals of time, approximating the signals by sequences of pulses, each T units in duration, as shown in Fig. E.1. It should be noted that this signal approximation is identical to that obtained using a perfect impulse sampler and a zero-order hold, as discussed in Chap. 15. The error introduced by this signal representation is denoted by the cross-hatched region in Fig. E.1. The circumflex is used in the figure to denote the discrete approximation of $f(t)$ obtained by sampling and zero-order holding it. If $f(t)$ and $\hat{f}(t)$ are integrated, the difference between the two results is measured by the *net* area between the two curves. It is obvious that if $f(t)$ is "smooth," this net area is reduced by decreasing T, the time interval between successive observations. Thus the rectangular pulse discrete approximation accuracy can be controlled by adjusting T.

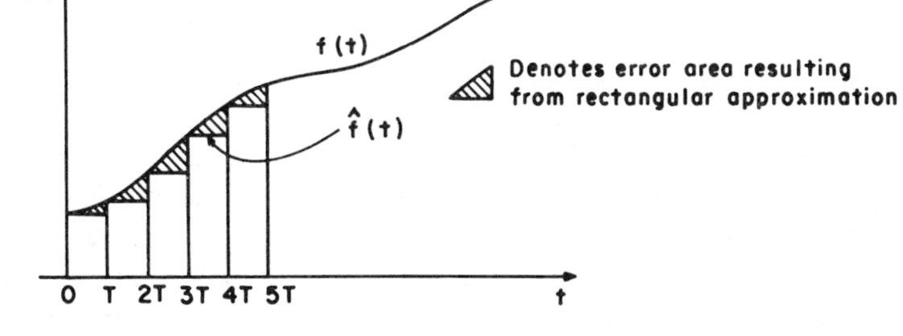

FIG. E.1. Approximation of a continuous signal by a pulse sequence.

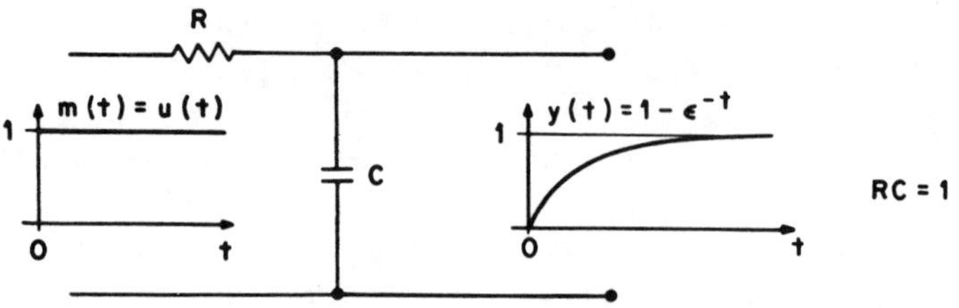

FIG. E.2. RC network used to illustrate the discrete signal approximation integration algorithm.

Suppose we wish to find the unit step response of the series RC network shown in Fig. E.2. The differential equation describing this situation is

$$\dot{y}(t) = m(t) - y(t) \qquad y(t_0) = y^0 \tag{E.1}$$

where $y(t)$ denotes the voltage across C and $m(t)$ is the input voltage. If $m(t) = u(t)$ and $y^0 = 0$, the solution for $y(t)$ is readily found to be $y(t) = 1 - \epsilon^{-t}$. A close approximation to this $y(t)$ may be obtained by integrating Eq. (E.1) on a digital computer. We start by assuming that $\dot{y}(t) = m(t) - y(t)$ is constant over successive intervals of T seconds duration. Since $m(t) = 1$, we have

$$\dot{y}(t) = 1 - y(t) \tag{E.2}$$

The integral of a constant C_0 over the time interval $[t_0, t_0 + T]$ is simply

$$\int_{t_0}^{t_0+T} C_0\,dt = C_0 T \tag{E.3}$$

Thus the discrete signal approximation to Eq. (E.2) becomes

$$y[(n+1)T] = y(nT) + [1 - y(nT)]T \tag{E.4}$$

which will be written as

$$y_{n+1} = y_n + (1 - y_n)T = T + (1 - T)y_n \tag{E.5}$$

This equation, called a *difference equation,* simply says that the new value of y equals the old value plus the change caused over the interval T by the slope term (assumed constant) on the right side of Eq. (E.2). Obviously $y(t)$ changes on the interval and the result is not exact. If T is small, however, the change in $y(t)$ can be neglected. Setting $T = 0.01$, which is only 1 percent of the time constant $RC = 1$, allows $y(t)$ to change only from 0 to 0.01 over the interval $[0, T]$, and the approximation $\dot{y}(t) = 1.0$ (remember $y^0 = 0$ was assumed) over that interval is accurate everywhere to within 1 percent or better. Application of recursive formula Eq. (E.5) for the first several computer iterations yields

$$y_1 = y(0.01) = 0.01$$

$$y_2 = 0.01 + (0.99)(0.01) = 0.0199$$

$$y_3 = 0.01 + (0.99)(0.0199) \quad \text{etc., etc.}$$

The computer is able to carry out these calculations very rapidly, and data points can be printed out for later use as desired. We obviously don't need every data point to provide a general picture of the response pattern, so there is no need to print them all out. After 300 consecutive applications of Eq. (E.5), the computed $y(3) = 0.9509591$, whereas $y(3)$ is actually 0.95021. The difference is certainly insignificant for most practical purposes.

The general procedure for solving a first order differential equation of the form given in Eq. (E.2) using the digital computer is straightforward. The following steps are suggested:

1. Assume $\dot{y}(t)$ is a constant, equal to its value at the start of the time interval, over successive time intervals of duration T seconds.
2. Evaluate the right hand side of the $\dot{y}(t)$ expression using the input (assumed known) and the most recent value of y.
3. Set up a recursion relationship of the form

$$y_{n+1} = y_n + \dot{y}_n T \tag{E.6}$$

4. Program the digital computer to iteratively apply this expression, thereby obtaining an approximation to $y(t)$.
5. Choose T such that $y(t)$ cannot change significantly over the interval for any expected signal levels.
6. Print out data at points which are not so far apart that there will be any doubt about what $y(t)$ is doing, nor so close together as to give excessive redundancy.

This general procedure is readily extended to the case of N simultaneous first order differential equations. Each equation is treated in exactly the same way, and all are solved simultaneously. Although there is much more algebra involved, no new concepts need to be introduced. Thus if we have two equations, for example, we set up the recursion relationships

$$y_{n+1}^1 = y_n^1 + \dot{y}_n^1 T \tag{E.7}$$

$$y_{n+1}^2 = y_n^2 + \dot{y}_n^2 T \tag{E.8}$$

At the nth step we evaluate $\dot{y}_n^1$ and $\dot{y}_n^2$, and then use Eqs. (E.7) and (E.8) to obtain y_{n+1}^1 and y_{n+1}^2, carrying out all calculations in parallel throughout. Extension to more dimensions is straightforward.

Several improvements can be made on the rectangular pulse signal approximation used in developing these simple integration algorithms. We can, for example, develop trapezoidal signal approximations rather easily in a variety of ways, thereby providing improved accuracy for a given value of T at the expense of an increase in the number of calculations required per iteration. Suppose we use Eq. (E.6) to obtain an *estimate* of y_{n+1}, denoted by $\bar{y}_{n+1}$, or

$$\bar{y}_{n+1} = y_n + \dot{y}_n T \tag{E.9}$$

(We can also devise similar algorithms by extrapolating to y_{n+1} from the values of y_n and y_{n-1}, etc.) Our first *estimate* of $\dot{y}_{n+1}$, denoted by $\bar{\dot{y}}_{n+1}$, is determined using m_{n+1} (known) and $\bar{y}_{n+1}$, as obtained using Eq. (E.9). We now use the *average* of $\dot{y}_n$ and $\bar{\dot{y}}_{n+1}$ to obtain y_{n+1}, or

$$y_{n+1} = y_n + \dot{y}_{n\,\mathrm{avg}} T \tag{E.10}$$

where

$$\dot{y}_{n\,\mathrm{avg}} = \frac{\dot{y}_n + \bar{\dot{y}}_{n+1}}{2} \tag{E.11}$$

In most cases this reduces to a simple structure, and accuracy is considerably improved. The concept is easily applied, and extends readily to the integration of n simultaneous first-order equations. Integration algorithms of this sort are classified as *predictor-corrector* methods, for obvious reasons.

It is a simple matter to develop integration algorithms based upon these fundamental principles for solving simultaneous sets of state equations. Accuracy is controlled by choice of T for a given algorithm and application, and accuracy improves as T is decreased until computer round-off error becomes significant. Round-off errors do not often cause troubles in simulation studies. One way to choose T which is often used is the following. Pick a value of T, denoted by T_1, which seems reasonable. Choose a representative signal input. Integrate to find the response over the period of interest. Repeat using $T_2 = T_1/2$. If the two response patterns are essentially the same, T_1 is not too large. It may be determined whether T_1 is smaller than necessary by reversing the process.

Many standard iterative programs are available for integrating sets of state equations. These are often library items available to all users, and it is only necessary that the information (the state equation dimension, coefficients, etc.) be provided to the program in an appropriate format. We see, however, that even when no ready made program is available we can easily generate our own at various levels of sophistication. In fact the methods discussed thus far, although

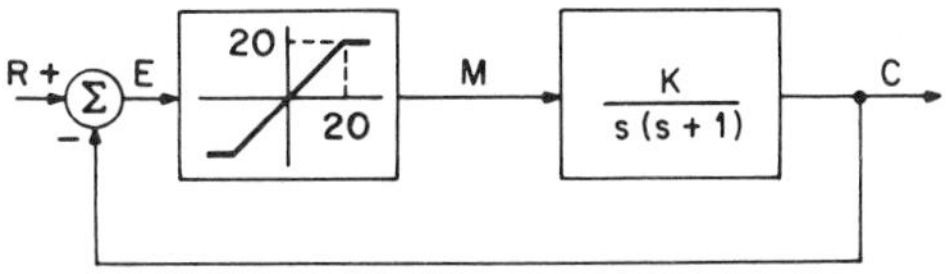

FIG. E.3. Controller with saturation used to illustrate the transition function simulation method.

they provide the foundation for most integration algorithms in common use, are not the most powerful tools we have available for setting up digital simulations.

We will now discuss a more sophisticated approach to the simulation problem based upon the use of state transition functions.[2] The mathematical background necessary for this development is presented in Chap. 14 and App. D, and that material should be reviewed if it is not familiar to the reader. Rather than attempt a completely abstract general discussion, we will carry a particular problem through to illustrate the various steps necessary. Suppose we set out to find the step response of the system shown in Fig. E.3. The first step is to observe that if a T second interval sampler and zero-order hold (the rectangular signal approximation scheme illustrated in Fig. E.1) are added to this system at the point marked E, the signal M at the saturation nonlinearity output will be constant over each time interval. We can develop a general expression relating C to M using the transition matrix corresponding to the transfer function $K/s(s+1)$. This expression can be reduced to a difference equation which, when iteratively applied on the digital computer, will allow us to determine the approximate response $c(t)$ of this system for any given input $r(t)$. Since only one sampling and signal reconstruction process is involved (note that the simulation procedure discussed in the previous paragraphs would require *two* signal sampling and discrete reconstruction approximations for this second-order system), it should be anticipated that this method will provide results which are superior to those obtained before, and indeed it does.

The first step is to assume a sampler and zero-order hold network are added at the error summing junction, as shown in Fig. E.4. Next, we develop the transfer function $K/s(s+1)$ using integrators and summers as shown in Fig. E.5 to facilitate writing down the state equations describing this plant. We define the state variables x_1 and x_2 as shown on the diagram in Fig. E.5. Note that $x_2 = c$. The state equations are seen to be

$$\begin{aligned} \dot{x}_1 &= Km \\ \dot{x}_2 &= x_1 - x_2 \end{aligned} \qquad \text{or} \qquad \dot{\mathbf{x}} = \mathbf{A}\mathbf{x} + \mathbf{B}m \tag{E.12}$$

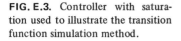

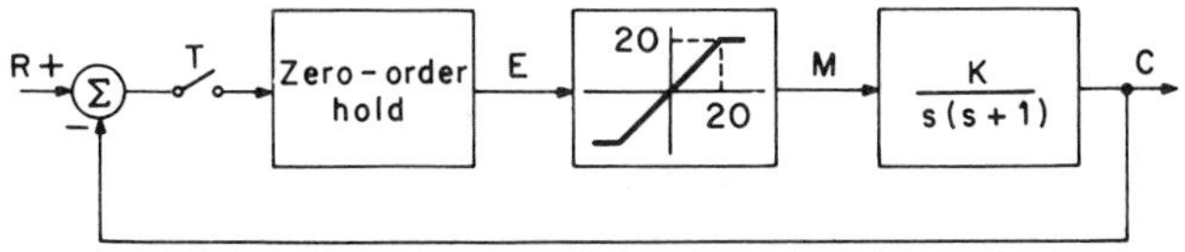

FIG. E.4. The controller with a sampler and zero-order hold added.

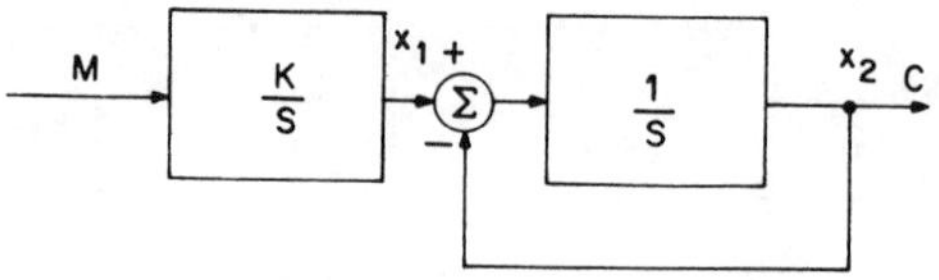

FIG. E.5. Block diagram representation of $K/s\,(s+1)$ using only integrators and summers.

where

$$\mathbf{A} = \begin{bmatrix} 0 & 0 \\ 1 & -1 \end{bmatrix} \qquad \mathbf{B} \begin{bmatrix} 1 \\ 0 \end{bmatrix} \tag{E.13}$$

The general solution to Eq. (E.12) may be written in terms of the initial state and the control input as

$$\mathbf{x}(t) = \mathbf{\Phi}(t, t_0)\,\mathbf{x}(t_0) + \int_{t_0}^{t} \mathbf{\Phi}(t, \tau)\,\mathbf{B}m(\tau)d\tau \tag{E.14}$$

where $\mathbf{\Phi}(t, t_0)$ is the usual state transition matrix obtained by solving

$$\dot{\mathbf{\Phi}}(t, t_0) = \mathbf{A}\mathbf{\Phi}(t, t_0) \qquad \mathbf{\Phi}(t_0, t_0) = \mathbf{I} \tag{E.15}$$

Since $m(t)$ is constant over each interval for the discrete system approximation shown in Fig. E.4, it is more appropriate to reduce Eq. (E.14) to the discrete form which will allow us to iteratively update system state from one sampling point to the next. This yields

$$\mathbf{x}_{n+1} = \mathbf{\Phi}(T)\mathbf{x}_n + m_n \int_0^T \mathbf{\Phi}(T, \tau)\mathbf{B}\,d\tau \tag{E.16}$$

Carrying out the integration in Eq. (E.16) provides the compact algebraic expression

$$\mathbf{x}_{n+1} = \mathbf{\Phi}(T)\mathbf{x}_n + m_n\mathbf{\Phi}_m(T) \tag{E.17}$$

where

$$\mathbf{\Phi}_m(T) = \int_0^T \mathbf{\Phi}(T, \tau)\mathbf{B}\,d\tau \tag{E.18}$$

is introduced to simplify the notation. Values of $\mathbf{\Phi}(T)$ and $\mathbf{\Phi}_m(T)$ are readily obtained in specific applications by solving Eqs. (E.15) and (E.18), respectively. The m_n required over each interval is given by

$$m_n = N(r_n - c_n) \tag{E.19}$$

where $N(\cdot)$ denotes the memoryless nonlinear block input to output relationship given, in this case, by

$$m_n = \begin{cases} 20 & r_n - c_n > 20 \\ r_n - c_n & -20 \leq r_n - c_n \leq 20 \\ -20 & r_n - c_n < -20 \end{cases} \tag{E.20}$$

The value of c_n is obtained from

$$c_n = [0 \quad 1]\mathbf{x}_n = x_n^2 \tag{E.21}$$

The computer is programmed to translate system state from one sample point to the next by iteratively applying Eqs. (E.17), (E.20), and (E.21).

Since Eq. (E.15) is a linear constant coefficient equation for the case under consideration, it may be solved using the Laplace transform. Thus

$$\mathbf{\Phi}(t,0) = \mathcal{L}^{-1}[(s\mathbf{I} - \mathbf{A})^{-1}] \tag{E.22}$$

But

$$s\mathbf{I} - \mathbf{A} = \begin{bmatrix} s & 0 \\ -1 & s+1 \end{bmatrix} \tag{E.23}$$

so

$$(s\mathbf{I} - \mathbf{A})^{-1} = \begin{bmatrix} \dfrac{1}{s} & 0 \\ \dfrac{1}{s(s+1)} & \dfrac{1}{s+1} \end{bmatrix} \tag{E.24}$$

and

$$\mathbf{\Phi}(T) = \mathbf{\Phi}(T,0) = \begin{bmatrix} 1 & 0 \\ 1 - \epsilon^{-T} & \epsilon^{-T} \end{bmatrix} \tag{E.25}$$

Also,

$$\Phi_m(T) = \int_0^T \begin{bmatrix} 1 & 0 \\ 1 - \epsilon^{-(T-\tau)} & \epsilon^{-(T-\tau)} \end{bmatrix} \begin{bmatrix} K \\ 0 \end{bmatrix} d\tau = K \begin{bmatrix} T \\ T - 1 + \epsilon^{-T} \end{bmatrix} \tag{E.26}$$

Suppose we let $K = 1$ to illustrate the basic computational steps. For this choice ω_c, the frequency at which the magnitude of $1/s(s+1)$ is unity, is approximately 0.9 rad/sec. The closed loop system bandwidth is just slightly larger than ω_c, or about 1 rad/sec. As a general "rule of thumb" in simulating closed loop systems, it is reasonable to choose the sampling frequency $\omega_s = 2\pi/T$ about 100 ω_c. This would require that we use $\omega_s = 100$ or $T = 0.0628$. Suppose we choose instead $\omega_s = 14$, or $T = 0.45$, a factor of 7 from the suggested value. This will clearly illustrate the effect of simulation inaccuracy. Evaluating Eqs. (E.25) and (E.26) for $T = 0.45$ yields

$$\Phi(T) = \Phi(0.45) = \begin{bmatrix} 1 & 0 \\ 0.36237 & 0.63763 \end{bmatrix} \tag{E.27}$$

and

$$\Phi_m(T) = \Phi_m(0.45) = \begin{bmatrix} 0.45 \\ 0.08763 \end{bmatrix} \tag{E.28}$$

Note that $\Phi(T)$ and $\Phi_m(T)$ are used many times in the algorithm, small differences between large numbers often arise, and small inaccuracies can cause errors to grow more rapidly than we might expect. The computer should generally be used to evaluate these functions to the maximum number of significant figures available.

Assume a step input of magnitude 25 is applied to the system shown in Fig. E.4 at $t = 0$ with all initial conditions zero. The initial value of E is 25, and the corresponding m obtained using Eqs. (E.20) and (E.21) is 20. Thus

$$\mathbf{x}_1 = \begin{bmatrix} 1 & 0 \\ 0.36237 & 0.63763 \end{bmatrix} \begin{bmatrix} 0 \\ 0 \end{bmatrix} + 20 \begin{bmatrix} 0.45000 \\ 0.08763 \end{bmatrix} = \begin{bmatrix} 9.0 \\ 1.7526 \end{bmatrix} \tag{E.29}$$

The next iteration yields

$$x_2 = \begin{bmatrix} 1 & 0 \\ 0.36237 & 0.63763 \end{bmatrix} \begin{bmatrix} 9.0 \\ 1.7526 \end{bmatrix} + 20 \begin{bmatrix} 0.45000 \\ 0.08763 \end{bmatrix} = \begin{bmatrix} 18 \\ 6.13144 \end{bmatrix} \tag{E.30}$$

Repeating this process over the first 10 seconds of real time response provides the output labeled *a* in Fig. E.6. Also shown for comparison is the response labeled *b*, which was obtained using $T = 0.2$. Since there is a significant difference between the two responses, $\omega_s = 14$ is not large enough, and the errors introduced by the sampling process, introduced for our convenience in the model, cannot be neglected. This is as expected, however. Setting $\omega_s = 100$ results in a small change in the response pattern from that labeled *b*, noticeable only near the peak overshoot point, so ω_s in the range 30–50 is acceptable here. Thus our "rule of thumb" that $\omega_s \simeq 100\omega_c$ for accurate simulation using this algorithm proves somewhat conservative in this case. The results shown were obtained on the IBM System 360 Model 50 computer using the APL language. Solution using $T = 0.45$ requires approximately one second of computer running time (about 7 cents worth) and the smaller T solutions require proportionately longer running times. Although this problem is significantly more complicated than the exponential solution obtained using the rectangular approximation in the earlier paragraphs, the running time here is actually less than was used to obtain that response pattern out to only $t = 3$.

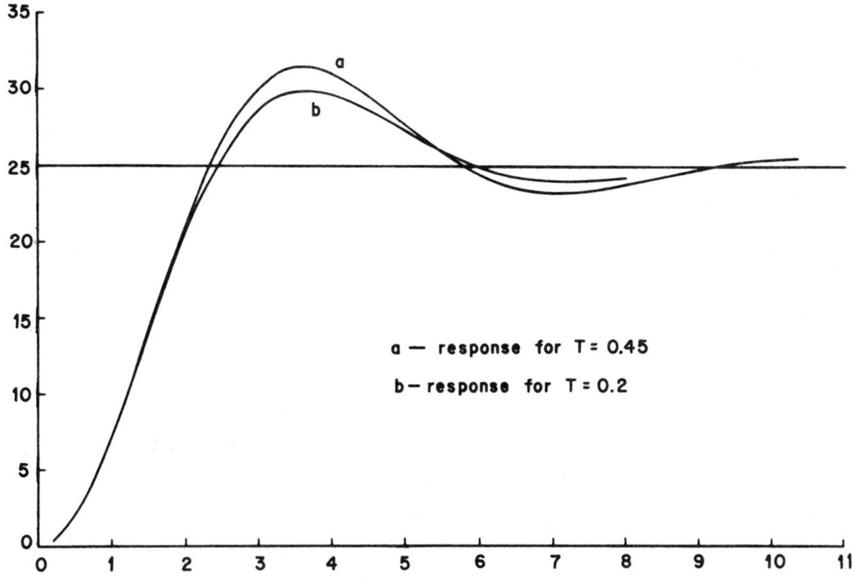

FIG. E.6. Step responses obtained using $T = 0.45$ and $T = 0.2$.

It is interesting to see how easily this problem is set up using the APL language. The program statements, as typed in from the remote console, are:

```
      ∇SIM
[1]   E ← R − X[2]
[2]   → (|E > 20)/5
[3]   M ← E
[4]   → 6
[5]   M ← 20 × E ÷ |E                                   (E.31)
[6]   X ← (Φ +.× X) + M × ΦM
[7]   T ← T + DELT
[8]   'T IS  ' ; T ; ' C IS  ' ; X[2]
[9]   I ← I + 1
[10]  → (I < IMAX)/1 ∇
```

It is also necessary to input the initial conditions

```
X ← 0 0      T ← 0      DELT ← 0.45      I ← 0          (E.32)
```

the values of $\mathbf{\Phi}_m(T)$

```
ΦM ← .45     .08763                                      (E.33)
```

the transition matrix elements

```
Φ ← 2  2 ρ  1  0  .36237  .63763                         (E.34)
```

the stopping condition

```
IMAX ← 22                                                (E.35)
```

and the input

```
R ← 25                                                   (E.36)
```

The computer is commanded to solve the problem by typing in SIM and striking the carriage return. The data is printed out in about a minute or so. The calculations may be terminated at any time from the console if the user wishes, as, for example, if the results are obviously in error, indicating a program "bug."

This simple example illustrates the basic steps involved in the transition function simulation method. The procedure is summarized here for emphasis:

1. Set up a block diagram of the system.
2. Introduce a sampler and zero-order hold at the input to each nonlinear element, if any, and, in general, at those points on the diagram where signals are added together, as at the error summing junction.
3. Develop the state equations for those parts of the system between sampling points.
4. Find the transition functions needed.
5. Set up the iterative algebraic equations relating state at $(n + 1)T$ to state and the input at nT.
6. Program the computer to solve these equations.

Step 2 requires further explanation. Introduction of a sampler is *required* at the input to each nonlinearity to allow a complete mathematical description. It is *convenient* to introduce samplers at each summing point as well, since it may otherwise be necessary to factor polynomials in s to locate the eigenvalues needed in developing the transition matrices. It is desirable to avoid factoring whenever possible. As more samplers are introduced, however, more *inaccuracy* results from signal reconstruction errors, and it may be necessary to decrease T, thereby increasing solution time, to obtain acceptable simulation accuracy.

It should be noted that a correction term can readily be added to the integration algorithm used in this example.[2] Instead of approximating $m(t)$ between samples by a constant value (the usual rectangular approximation) it can be represented by a linear time function (the usual trapezoidal approximation). Several methods are available for determining the slope to be used. One which is simple and works well is as follows. Use Eq. (E.17) as is to obtain a first *estimate* (prediction) of the new state assuming m_n is constant. In other words, use

$$\bar{\mathbf{x}}_{n+1} = \mathbf{\Phi}(T)\mathbf{x}_n + m_n \mathbf{\Phi}_m(T) \tag{E.37}$$

where $\bar{\mathbf{x}}_{n+1}$ denotes the first estimate of $\mathbf{x}_{n+1}$. Assume for the present that the saturation nonlinearity is eliminated. Extension to the nonlinear case is not difficult later if desired. Define

$$\delta m_n = r_{n+1} - \bar{x}_{2,n+1} - (r_n - x_{2,n}) \tag{E.38}$$

which becomes, for a step input

$$\delta m_n = x_{2,n} - \bar{x}_{2,n+1} \tag{E.39}$$

Thus the first estimate of the change in m between the nth and $(n + 1)$th samples is obtained for this situation as the negative of the estimated output change. It is now desired to superpose this input effect upon the previous results. The ramp input signal needed is given over the interval $[nT, (n + 1)T]$ by

$$\bar{m}_n(\tau) = \frac{\delta m_n}{T} \tau \tag{E.40}$$

where $\tau = t - nT$ is zero at the nth sample point. The state response at $\tau = T$ [or $t = (n + 1)T$] due to the influence of this corrective signal is

$$\overline{\boldsymbol{\delta}\mathbf{x}}_{n+1} = \int_0^T \boldsymbol{\Phi}(T,\tau)\mathbf{B}\,\overline{m}_n(\tau)\,d\tau$$

$$= \frac{\delta m_n}{T}\int_0^T \tau\boldsymbol{\Phi}(T,\tau)\,\mathbf{B}\,d\tau$$

$$= \frac{\delta m_n}{T}\hat{\boldsymbol{\Phi}}_m(T) \tag{E.41}$$

where

$$\hat{\boldsymbol{\Phi}}_m(T) = \int_0^T \tau\boldsymbol{\Phi}(T,\tau)\,\mathbf{B}\,d\tau \tag{E.42}$$

may be evaluated ahead of time and stored along with $\boldsymbol{\Phi}_m(T)$. The new value of state becomes

$$\mathbf{x}_{n+1} = \overline{\mathbf{x}}_{n+1} + \overline{\boldsymbol{\delta}\mathbf{x}}_{n+1} \tag{E.43}$$

Thus considerable improvement in accuracy is available by adding two simple calculations to the original results. The ratio of the correction term generated at each step to the signal level at that step can be used as a check upon the integration step size, if desired.

In most applications the rate of change of the input or inputs to each simulation block is known. An algorithm is easily developed to take advantage of this fact, and yields performance improvements similar to the predictor-corrector methods.[2,4] Assume the input signal is $\mathbf{m}(t)$, a vector with M components. Let

$$\mathbf{m}_n(t) = t\,\dot{\mathbf{m}}_n \tag{E.44}$$

Then $\boldsymbol{\delta}\mathbf{m}_n$ as used in Eq. (E.41) becomes

$$\boldsymbol{\delta}\mathbf{m}_n = T\,\dot{\mathbf{m}}_n \tag{E.45}$$

and the resulting slope-projection algorithm is

$$\mathbf{x}_{n+1} = \boldsymbol{\Phi}(T)\mathbf{x}_n + \boldsymbol{\Phi}_m(T)\mathbf{m}_n + \hat{\boldsymbol{\Phi}}_m(T)\dot{\mathbf{m}}_n \tag{E.46}$$

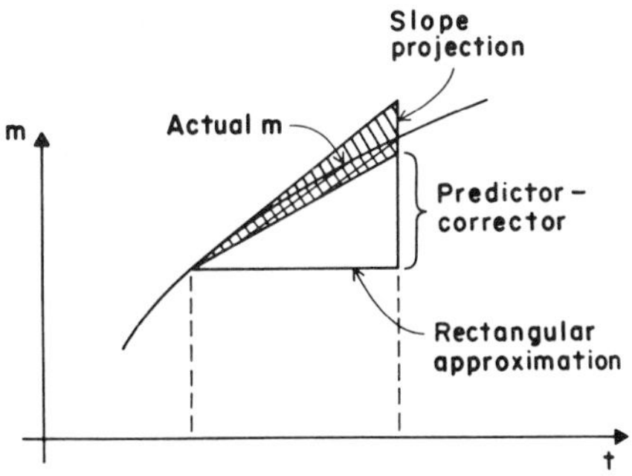

FIG. E.7. Predictor-corrector and slope projection signal reconstruction methods.

This algorithm provides both accuracy and computational efficiency. The predictor-corrector and slope projection signal reconstruction procedures are illustrated in Fig. E.7 on an exaggerated scale. It is typically possible to increase T by a factor of 5–10 from that needed with the zero order hold method when either of these slightly more complex algorithms are used.[2,3,4]

It is possible to develop a variety of more sophisticated integration algorithms based upon similar concepts. The correction process can be repeated two or more times, the slope projection algorithm can be used to develop an original estimate of state and that estimate used to develop a correction, the signal between data points can be represented by a polynomial in time of second order or higher, etc. The results from several examples indicate that such extensions, although perhaps of research interest, are not generally justified in practice. Computational efficiency is not significantly improved, and one main advantage of the approach, its simplicity, suffers appreciably as such changes are made.

Consider the problem of developing the state equations for an arbitrary transfer function $G(s)$. An infinity of possibilities exists. The particular state equations used should thus be chosen for programming convenience.[4] If $G(s)$ is expanded into partial fractions, the result can generally be expressed in terms of the following types of factors, none of which interact with each other in the resulting transition matrix: 1. Integration, K/s; 2. Real root, $K/(s + a)$; 3. Repeated integration, K/s^2; 4. Repeated real root, $K/(s + a)^2$; and 5. A term of the form $(As + B)/[(s + \alpha)^2 + \beta^2]$ resulting from a complex conjugate root pair. Although higher order repeated roots and repeated complex pairs are conceivably possible, they occur so seldom that they are not considered standard elements. A reference library need only include the transition function terms for each of these partial fraction forms, as provided in Table E.1.

An Example

Consider the minor loop configuration shown in Fig. E.8 with

$$G_1(s) = \frac{5}{(1 + s/100)^2} \qquad G_2(s) = \frac{20}{s(1 + s/5)(1 + s/20)}$$

$$H(s) = \frac{0.26s^2}{1 + s/2.6} \tag{E.47}$$

$\frac{C(s)}{R(s)} = G(s)$	Simulation Diagram	State Equations	Transition Functions
1. $\frac{K}{s}$	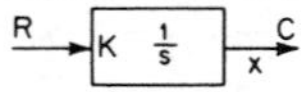	$\dot{x} = Kr \quad B = K$ $C = x \quad m = r$	$\Phi(h) = 1 \qquad \Phi_m(h) = Kh \qquad \hat{\Phi}_m(h) = \frac{Kh^2}{2}$
2. $\frac{K_1}{s^2} + \frac{K_2}{s}$	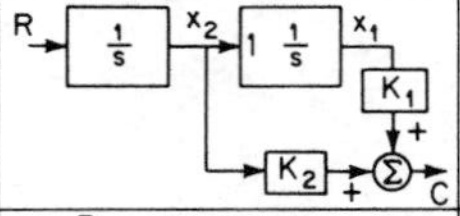	$\dot{x}_1 = x_2$ $\dot{x}_2 = r \quad \underline{B} = \begin{bmatrix} 0 \\ 1 \end{bmatrix}$ $C = K_1x_1 + K_2x_2$	$\underline{\Phi}(h) = \begin{bmatrix} 1 & h \\ 0 & 1 \end{bmatrix} \qquad \underline{\Phi}_m(h) = \begin{bmatrix} h^2/2 \\ h \end{bmatrix} \qquad \hat{\underline{\Phi}}(h) = \begin{bmatrix} h^3/6 \\ h^2/2 \end{bmatrix}$
3. $\frac{K}{s+a}$	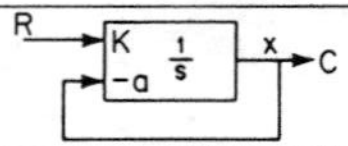	$\dot{x} = -ax + Kr$ $c = x \quad B = K$ $m = r$	$\Phi(h) = \epsilon^{-ah} \qquad \Phi_m(h) = \frac{K}{a}(1-\epsilon^{-ah})$ $\hat{\Phi}_m(h) = -\frac{K}{a^2} + \frac{K}{a^2}\epsilon^{-ah} + \frac{K}{a}h$
4. $\frac{K_1}{(s+a)^2} + \frac{K_2}{(s+a)}$	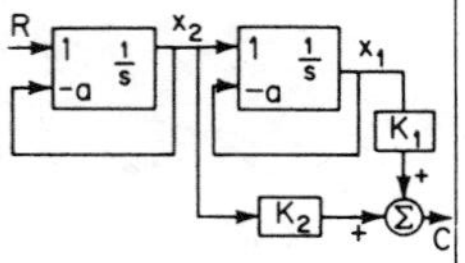	$\dot{x}_1 = -ax_1 + x_2$ $\dot{x}_2 = -ax_2 + r$ $C = K_1x_1 + K_2x_2$ $B = \begin{bmatrix} 0 \\ 1 \end{bmatrix} \quad m = r$	$\underline{\Phi}(h) = \begin{bmatrix} \epsilon^{-ah} & he^{-ah} \\ 0 & \epsilon^{-ah} \end{bmatrix} \qquad \underline{\Phi}_m(h) = \begin{bmatrix} \frac{1}{a^2}(1-\epsilon^{-ah}) - \frac{h}{a}\epsilon^{-ah} \\ \frac{1}{a}(1-\epsilon^{-ah}) \end{bmatrix}$ $\hat{\underline{\Phi}}_m(h) = \begin{bmatrix} \frac{2}{a^3}(\epsilon^{-ah}-1) + \frac{h}{a^2}(1+\epsilon^{-ah}) \\ -\frac{1}{a^2}(1-\epsilon^{-ah}) + \frac{h}{a} \end{bmatrix}$
5. $\frac{K_1s + K_2}{(s+\alpha)^2 + \beta^2}$	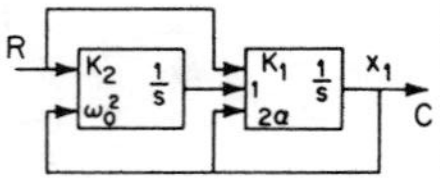 $\omega_0^2 = \alpha^2 + \beta^2$	$\dot{x}_1 = -2\alpha x_1 + x_2 + K_1r$ $\dot{x}_2 = -(\alpha^2+\beta^2)x_1 + K_2r$ $\underline{B} = \begin{bmatrix} K_1 \\ K_2 \end{bmatrix} \quad m = r$	$\underline{\Phi}(h) = \begin{bmatrix} \frac{\omega_0}{\beta}\epsilon^{-ah}\sin(\beta h+\phi_1) & \frac{1}{\beta}\epsilon^{-ah}\sin\beta h \\ -\frac{\omega_0^2}{\beta}\epsilon^{-ah}\sin\beta h & \frac{\omega_0}{\beta}\epsilon^{-ah}\sin(\beta h+\phi_2) \end{bmatrix}$ $\omega_0^2 = \alpha^2 + \beta^2 \qquad \phi_1 = \tan^{-1}(\beta/-\alpha) \qquad \phi_2 = \tan^{-1}(\beta/\alpha)$ $\underline{\Phi}_m(h) = \begin{bmatrix} \frac{K_1}{\beta}\epsilon^{-ah}\sin\beta h + K_2\left[\frac{1}{\omega_0^2} + \frac{1}{\beta\omega_0}\epsilon^{-ah}\sin(\beta h-\phi_1)\right] \\ -K_1\left[1+\frac{\omega_0}{\beta}\epsilon^{-ah}\sin(\beta h-\phi_1)\right] + K_2\left[\frac{2\alpha}{\omega_0^2} + \frac{1}{\beta}\epsilon^{-ah}\sin(\beta h+\phi_2-\phi_1)\right] \end{bmatrix}$ $\hat{\underline{\Phi}}_m(h) = \begin{bmatrix} K_1\left[\frac{1}{\omega_0^2} + \frac{1}{\beta\omega_0}\epsilon^{-ah}\sin(\beta h-\phi_1)\right] + \frac{K_2}{\omega_0^2}\left[h - \frac{2\alpha}{\omega_0^2} + \frac{1}{\beta}\epsilon^{-ah}\sin(\beta h-2\phi_1)\right] \\ K_1\left[\frac{2\alpha}{\omega_0^2} - h - \frac{1}{\beta}\epsilon^{-ah}\sin(\beta h-2\phi_1)\right] + K_2\left[\frac{1}{\omega_0^2}(2\alpha h+1-\frac{4\alpha^2}{\omega_0^2}) + \frac{1}{\beta\omega_0}\epsilon^{-ah}\sin(\beta h+3\phi_2)\right] \end{bmatrix}$

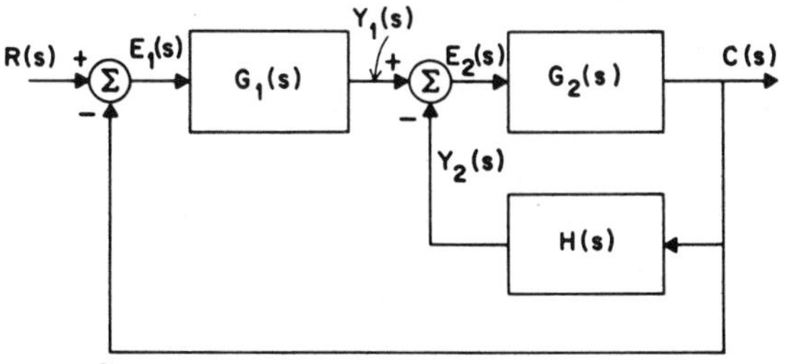

FIG. E.8. A typical situation to which the state transition function method may be applied.

Application of the general simulation procedure to this problem illustrates several of the standard situations encountered in applying the tabulated results. The system is treated in two parts, the first relating $Y_1(s)$ to $E_1(s)$ and the second relating $E_2(s)$ to $C(s)$ and $Y_2(s)$. The first part is very simple, since

$$\frac{5}{(1 + s/100)^2} = \frac{50{,}000}{(s + 100)^2} \tag{E.48}$$

and the corresponding transition functions are obtained directly from entry 4 in Table E.1 as

$$\mathbf{\Phi}(T) = \begin{bmatrix} \epsilon^{-100T} & T\epsilon^{-100T} \\ 0 & \epsilon^{-100T} \end{bmatrix} \tag{E.49}$$

$$\mathbf{\Phi}_m(T) = \begin{bmatrix} (1 - \epsilon^{-100T} - T\epsilon^{-100T}) \\ 1 - \epsilon^{-100T} \end{bmatrix} \tag{E.50}$$

and

$$\hat{\mathbf{\Phi}}_m(T) = \begin{bmatrix} 2 \times 10^{-6}(\epsilon^{-100T} - 1) + 10^{-4}T(1 - \epsilon^{-100T}) \\ -10^{-4}(1 - \epsilon^{-100T}) + \dfrac{T}{100} \end{bmatrix} \tag{E.51}$$

The required algebraic relationships are provided by

$$E_1(s) = R(s) - C(s) \tag{E.52}$$

and

$$Y_1(s) = 50{,}000\,X_1(s) \tag{E.53}$$

The second part of the system must be considered a bit more carefully. Note that

$$\frac{20}{s(1 + s/5)(1 + s/20)} = \frac{2000}{s(s + 5)(s + 20)} = \frac{20}{s} - \frac{80/3}{s + 5} + \frac{20/3}{s + 20} \tag{E.54}$$

and

$$\frac{20}{s(1 + s/5)(1 + s/20)} \cdot \frac{0.26s^2}{(1 + s/2.6)} = \frac{1352s}{(s + 5)(s + 20)(s + 2.6)}$$
$$= \frac{188}{s + 5} + \frac{103.8}{s + 20} - \frac{84.2}{s + 2.6} \tag{E.55}$$

Thus it is necessary to use a 4×1 state vector to completely represent these equations. Choose the expanded form

$$F(s) = \frac{1}{s} + \frac{1}{s + 5} + \frac{1}{s + 20} + \frac{1}{s + 2.6} \tag{E.56}$$

to define the transition functions, with $X_1(s)$ corresponding to the $1/s$ term, $X_2(s)$ to the $1/(s + 5)$ term, etc. Then

$$\mathbf{\Phi}(T) = \begin{bmatrix} 1 & 0 & 0 & 0 \\ 0 & \epsilon^{-5T} & 0 & 0 \\ 0 & 0 & \epsilon^{-20T} & 0 \\ 0 & 0 & 0 & \epsilon^{-2.6T} \end{bmatrix} \tag{E.57}$$

$$\mathbf{\Phi}_m(T) = \begin{bmatrix} T \\ \frac{1}{5}(1 - \epsilon^{-5T}) \\ \frac{1}{20}(1 - \epsilon^{-20T}) \\ \frac{1}{2.6}(1 - \epsilon^{-2.6T}) \end{bmatrix} \tag{E.58}$$

and

$$\hat{\Phi}_m(T) = \begin{bmatrix} \dfrac{T^2}{2} \\ -\dfrac{1}{25} + \dfrac{T}{5} + \dfrac{1}{25}\epsilon^{-5T} \\ -\dfrac{1}{400} + \dfrac{T}{20} + \dfrac{1}{400}\epsilon^{-20T} \\ -\dfrac{1}{6.76} + \dfrac{T}{2.6} + \dfrac{1}{6.76}\epsilon^{-2.6T} \end{bmatrix} \tag{E.59}$$

are the transition functions needed to describe this part of the system, as obtained from entries 1 and 3 of Table E.1. Also

$$C(s) = 20\,X_1(s) - \frac{80}{3}X_2(s) + \frac{20}{3}X_3(s) \tag{E.60}$$

and

$$Y_2(s) = 188\,X_2(s) + 103.8\,X_3(s) - 84.2\,X_4(s) \tag{E.61}$$

define the needed signal inter-relationships. A computer program to obtain the unit step response of this system can be written in a few lines using these functions and any of the standard languages. The unit step response pattern is shown in Fig. E.9. An integration step size $T = 0.001$ provides accuracy within about 1 percent when the rectangular reconstruction algorithm is used, whereas $T = 0.01$ and $T = 0.008$ are reasonable when the predictor-corrector and slope

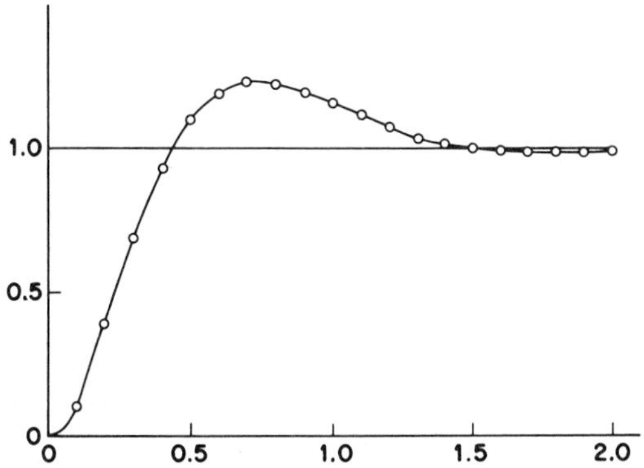

FIG. E.9. Unit step response of the example system.

projection algorithms, respectively, are applied.[4] The slope projection and RK2 algorithms go unstable at $T = 0.01$, as easily shown analytically. A computer running time savings of approximately 6 to 1 is achieved by using the slightly more complicated algorithms.

REFERENCES

1. Benyon, P. R., A Review of Numerical Methods for Digital Simulation, *Simulation,* November, 1968.
2. Eveleigh, V. W., Digital Simulation Based Upon Transition Functions, *Proceedings of 3rd Hawaii International Conference on System Sciences,* Honolulu, Hawaii, January, 1970.
3. Nichol, K. and V. W. Eveleigh, Comparison of Several Digital Simulation Procedures, *Proceedings of 1970 NEC,* Chicago, Illinois, December, 1970.
4. Eveleigh, V. W., Evaluation of a Digital Simulation Procedure Using Transition Functions, *Proceedings of 1970 NEC,* Chicago, Illinois, December, 1970.

Appendix F
PROBLEMS

CHAPTER 2

1. Develop a differential equation description of the RC network shown. Define $v_2(t)$ in terms of the variable used in your equation.

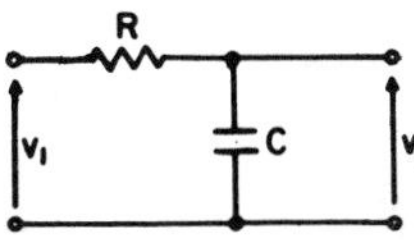

2. Develop an integro-differential equation describing the RLC network shown.
a. Using loop equations. *b.* Using node equations.

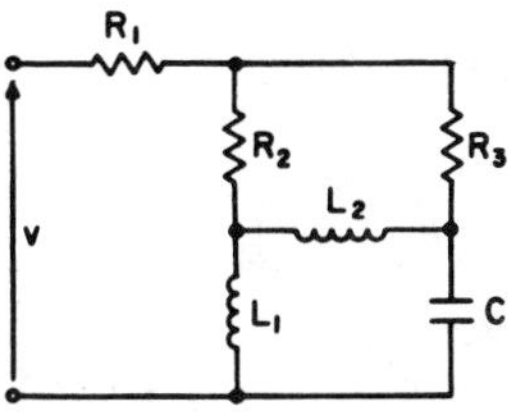

3. Write the differential equation relating motion x of the mass M to the force input $f(t)$. Note that in this and all subsequent translational motion problems it is assumed that each mass moves with only one degree of freedom.

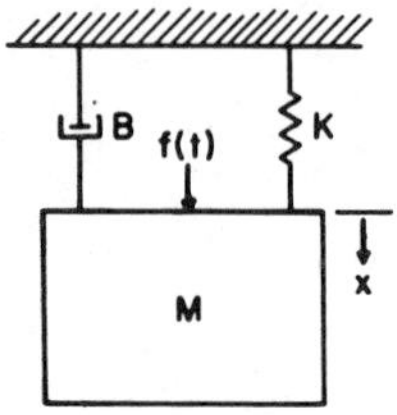

4. Write the differential equations describing how the system shown reacts to the position input $x_0(t)$. The bottom element in the diagram is a viscous damping and spring coupling unit with mass M_2. It may be thought of as an automobile shock absorber and spring combination.

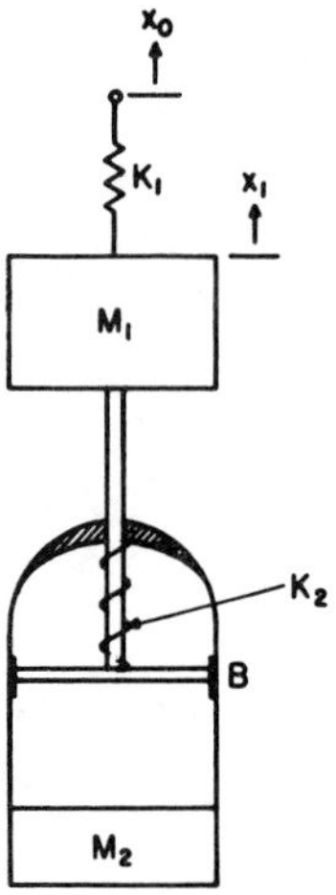

5. Develop the differential equations describing this system.

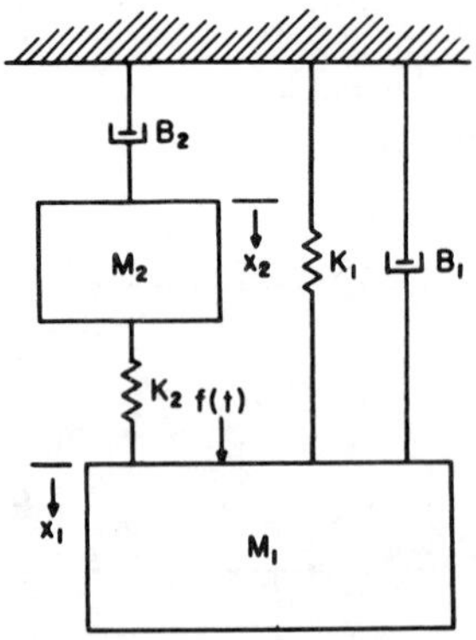

6. In the system introduced in Prob. 2.5, the element values are:

B_1 = 2 lbs./(ft./sec.) K_2 = 0.5 lbs./ft
M_1 = 1 kg B_2 = 50 Newtons/(meter/sec.)
K_1 = 10 Newtons/meter M_2 = 5 lbs.

a. Reduce these to a consistent set of English units.
b. Reduce these to a consistent set of mks units.

7. Write the differential equation describing the system shown. Assume the bar through which force is applied is not flexible, has no mass or moment of inertia, and all displacements are small.

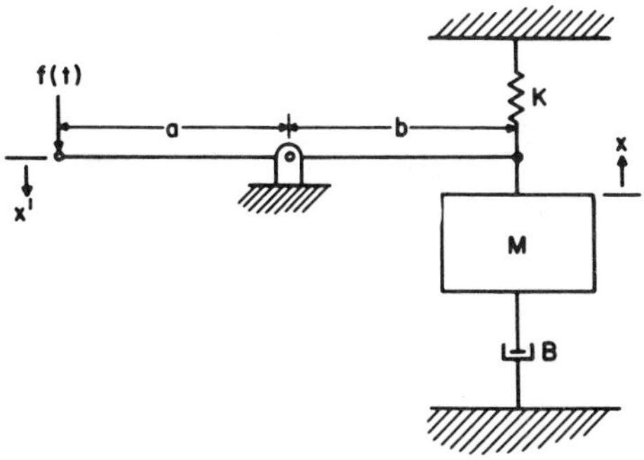

8. Develop electrical analogs for the mechanical systems used in
a. Problem 2.3 *b.* Problem 2.4 *c.* Problem 2.5

9. Develop an electrical analog in which force and voltage are analogous for the mechanical system introduced in Prob. 2.3.

10. Write the differential equations describing the system shown.

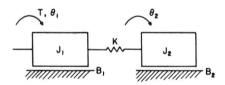

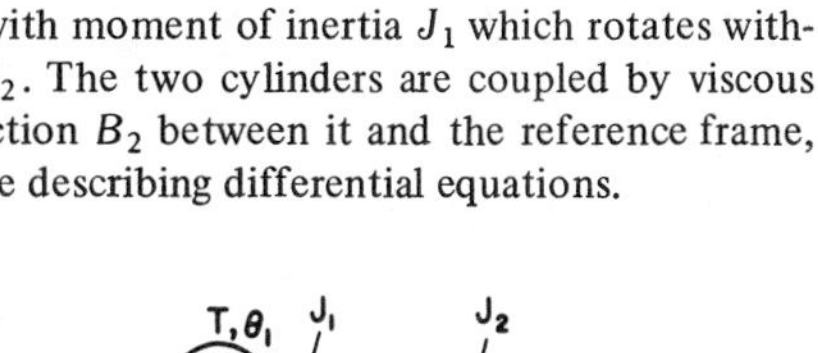

11. Torque $T(t)$ is applied to a small cylinder with moment of inertia J_1 which rotates within a larger cylinder with moment of inertia J_2. The two cylinders are coupled by viscous friction B_1. The outer cylinder has viscous friction B_2 between it and the reference frame, and is restrained by a torsion spring K. Write the describing differential equations.

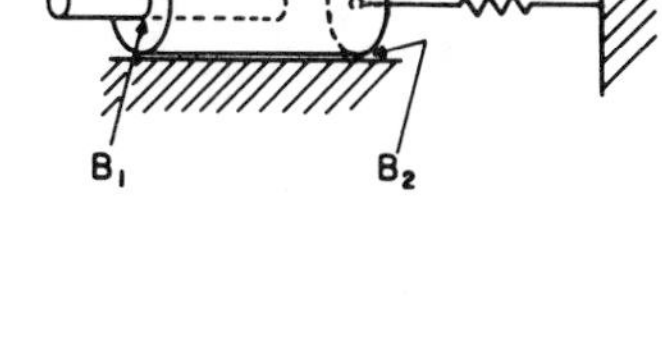

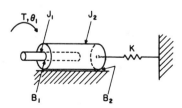

12. Write the differential equations describing the electromechanical system shown. K_e and K_T are the motor back emf and torque constants, respectively.

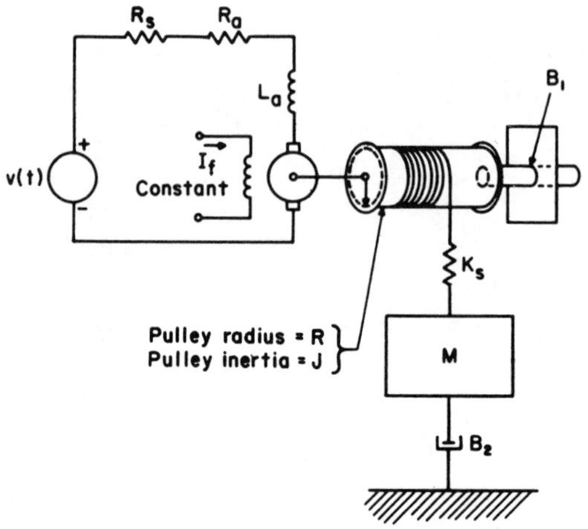

13. *a.* Write the differential equations describing the system shown.
b. What are the equivalent J and B at the input shaft?

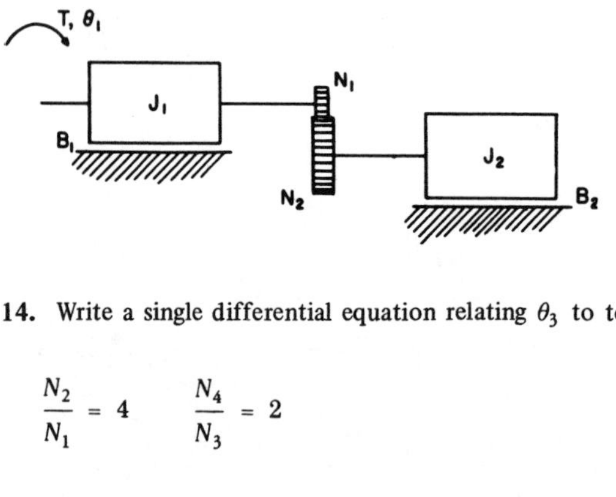

14. Write a single differential equation relating θ_3 to torque input, T_1. The gear ratios are

$$\frac{N_2}{N_1} = 4 \qquad \frac{N_4}{N_3} = 2$$

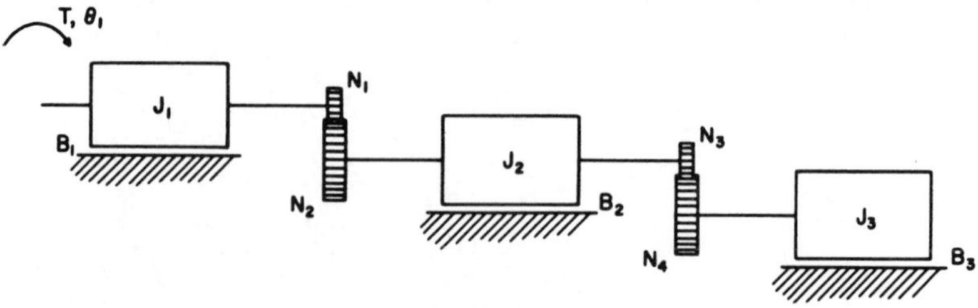

15. Torque $T(t)$ is applied to a shaft as shown. As the shaft turns, linear motion is transferred to a mass M through a cable wrapped on a spool of radius r. Write the describing differential equations.

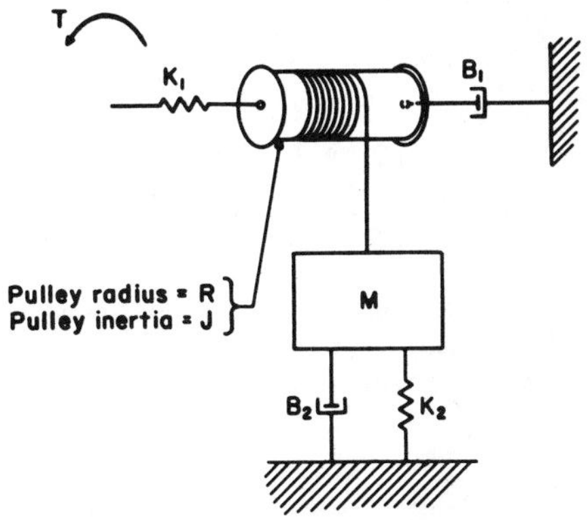

16. The polarized relay shown exerts a force $f(t) = K_i i(t)$ upon the pivoted bar. Assume the relay coil has constant inductance L. The left end of the pivot bar is connected to the reference frame through a viscous damper B_1 to retard rapid motions of the bar. Assume the bar has negligible mass and moment of inertia, and also that all displacements are small. Write the describing differential equations. Note that the relay coil is not free to move.

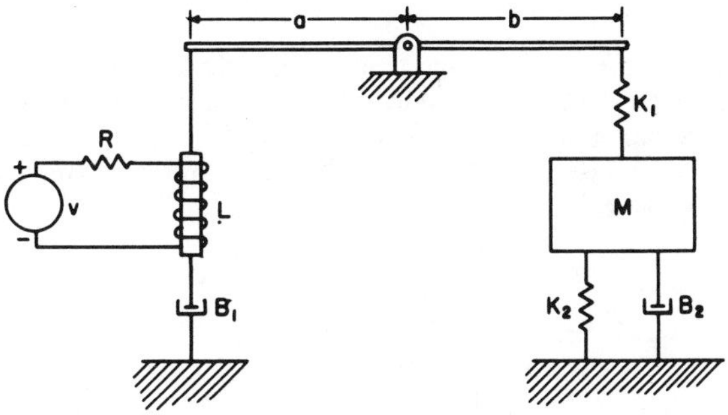

17. Write the differential equations relating motion of the load inertia J to the input voltage $v_f(t)$ for the field controlled motor shown. The armature current is held constant by a regulator, and the torque developed is given by $T = K_T i_f$.

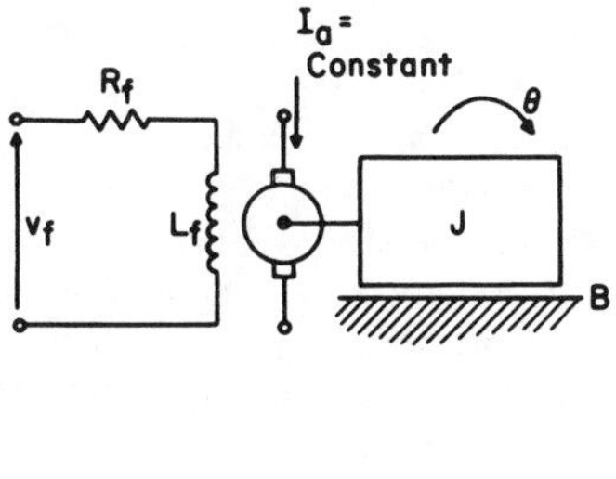

18. The schematic representation of an armature controlled motor is as shown. The field current is held constant. The torque generated is given by $T = K_T i_a$ and the back emf generated in the armature is $v_a = K_e \omega = K_e d\theta/dt$. Write the describing differential equations for this system.

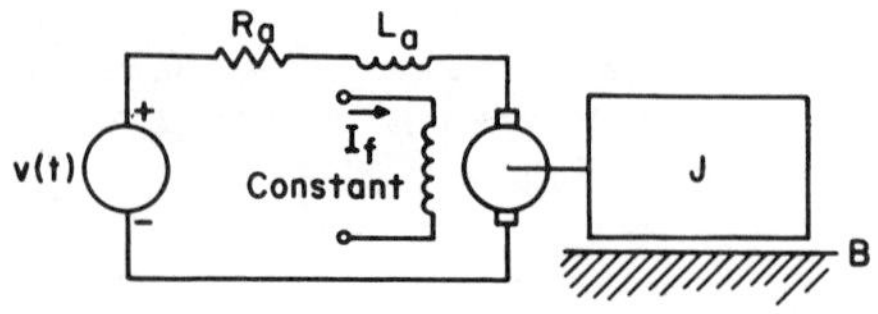

19. Write the differential equations describing the system shown. The unloaded cathode follower gain is K_c. Assume the motor field loading is negligible, or the cathode follower $Z_0 = 0$. The motor torque constant is K_T. Assume μ and r_p as tube parameters.

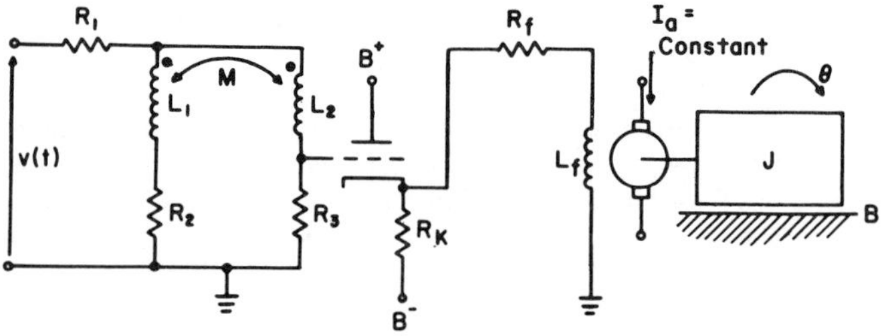

20. It is proposed to use a dc motor driven from solar batteries to supply corrective actions through the action-reaction and momentum conservation principles to control a satallite's orientation in space. (This is a standard approach.) For the motor shown, find the transfer function between voltage input and reaction torque produced. Disregard friction and armature inductance.

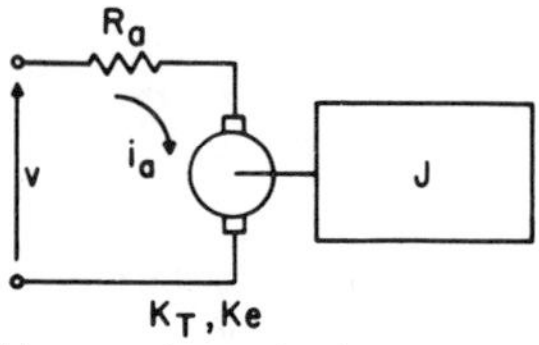

Field current constant

21. Consider the problem of heating a protective tent used for ice-fishing with an oil burning heater. The heater provides $h(t)$ Btu/min, the equipment and people within the tent have thermal capacity C and the tent has thermal resistance R, both in consistent units. Write the differential equation relating temperature within the tent to $h(t)$, assuming the external temperature is constant at T_e.

22. A typical jet airliner must provide the capability to maintain cabin temperature comfortable over a wide range of environmental conditions. Assume the passengers and equipment provide thermal inertia equivalent to 20,000 Kg of water. The heat transfer through the cabin wall is 12 Btu/minute/degree F difference. The electronic equipment on board is assumed operating, and dissipates 500 watts into the cabin. The exterior temperature may range from -50°F to +120°F. It is desired to provide adequate heating and cooling capacity to maintain the cabin at 75°F over the entire range of conditions encountered.

a. What minimum heating and cooling capacity must be provided?

b. Write the differential equations relating the thermal input $q(t)$ to cabin temperature assuming the usual constant-interior-temperature model.

Note: 1 watt = 5.689×10^{-2} Btu/min.

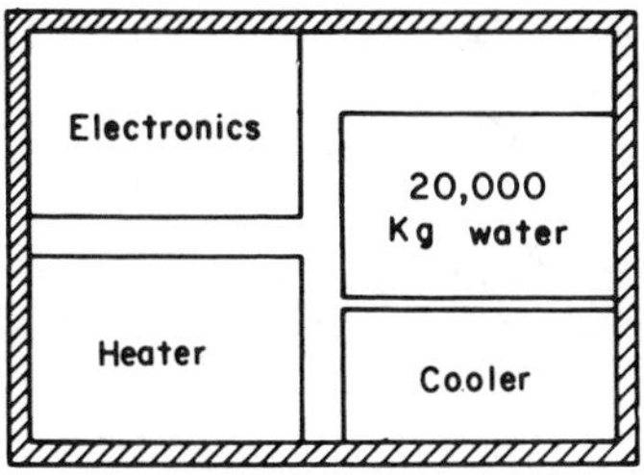

23. A schematic representation of a moving coil microphone is shown. Sound waves striking the diaphragm cause the coil mounted upon it to move within β, the flux field of the permanent magnets. This induces current changes which are observed as the output $v(t)$. The diaphragm is connected to the reference framework through viscous damping element B and a spring with elastance K. Write the differential equations describing the performance of this system around the equilibrium state.

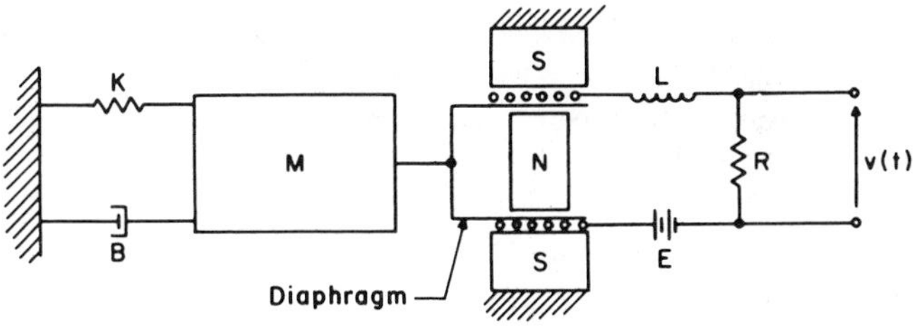

CHAPTER 3

1. Determine which of the following functions are Laplace transformable. Find the abscissa of absolute convergence of the defining integral for each transformable function.

a. t *b.* $t\epsilon^{at}$ *c.* $t \sin \alpha t$ *d.* α^t *e.* t^t

2. Derive Laplace transforms for each of the following functions.

a. $\sin \omega t$ *b.* $\sinh \alpha t$ *c.* $u(t - \alpha)$ *d.* $t\epsilon^{-\alpha t}$

e. $\epsilon^{-\alpha t} \sin \omega t$ *f.* t *g.* t^2

3. Find Laplace transforms for each of the waveforms shown.

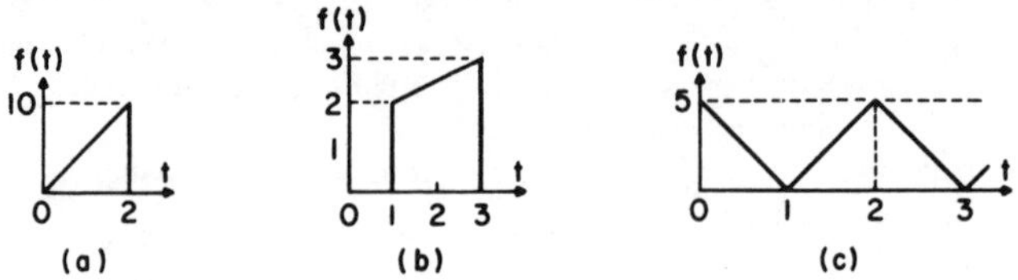

4. Laplace transform each of the following differential equations.

a. $\ddot{x} + 3\dot{x} + 2x = m(t) \qquad x(0^+) = -2 \qquad \dot{x}(0^+) = 4$

b. $4\dddot{x} + 3\ddot{x} + 5\dot{x} + 2x = m(t) \qquad x(0^+) = \ddot{x}(0^+) = 0$

$\dot{x}(0^+) = 3$

5. Consider the multiplier shown. Assume $f_1(t) = 5\epsilon^{-t/2}$. The periodic pulse train shown is $f_2(t)$. Find the Laplace transform of the output $g(t)$.

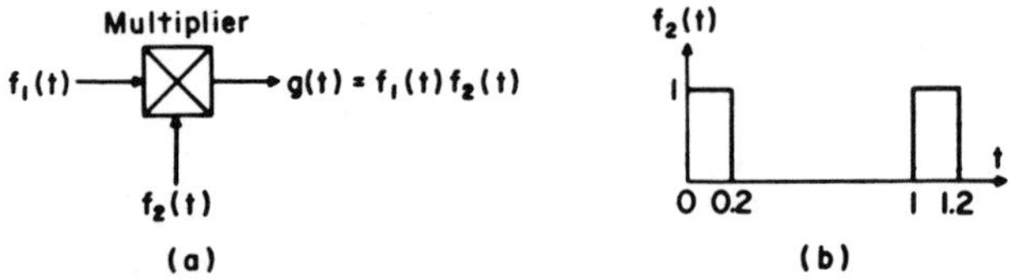

6. Using the facts that

$$\mathcal{L}[t] = \frac{1}{s^2} \qquad \text{and} \qquad \mathcal{L}[\epsilon^{-\alpha t}] = \frac{1}{s + \alpha}$$

develop the Laplace transform of $f(t) = t(\epsilon^{-at} - 2\epsilon^{-bt})$ in two distinct ways (using theorems) without resorting to the defining integral.

7. Factor the following polynomials: (see App. B)

a. $s^3 + 6s^2 + 11s + 6 = 0$

b. $60s^3 + 74s^2 + 30s + 4 = 0$

c. $6s^4 + 19s^3 + 28s^2 + 18s + 4 = 0$

d. $3s^5 + 14s^4 + 41s^3 + 64s^2 + 58s + 20 = 0$

8. Prove that the Laplace transform is a linear transformation, and thus that $\mathcal{L}$ is a linear operator.

9. The Laplace transform of $f(t)$, denoted by $F(s)$, is unique, whenever it exists. Define conditions under which the inverse transform of $F(s)$ is also unique. Give an example in which the inverse transform is not unique.

10. In a given system it is found that the output signal $f(at')$ has Laplace transform

$$F(s) = \mathcal{L}[f(at')] = \frac{s + 3}{s^2 + 7s + 2}$$

If this signal is time scaled by defining $t = at'$, what is the Laplace transform of the scaled signal?

11. Given that

$$\mathcal{L}[\sin \omega_0 t] = \frac{\omega_0}{s^2 + \omega_0^2}$$

Find $\mathcal{L}[t \cos \omega_0 t]$ without using the tables or the defining integral.

12. Prove that

$$\mathcal{L}[tf(t)] = -\frac{d}{ds}[F(s)]$$

13. Expand the following functions into partial fractions:

a. $F(s) = \dfrac{2s + 1}{(s + 1)(s + 2)(s + 3)}$

b. $F(s) = \dfrac{s^2 + 2s + 2}{(s + 1)(s + 2)}$

c. $F(s) = \dfrac{10}{(s + 1)(s + 1 + j1)(s + 1 - j1)}$

d. $F(s) = \dfrac{s^2 + 2s + 2}{(s + 2)(s^2 + 2s + 5)}$

e. $F(s) = \dfrac{1}{s(s + 1)^2(s + 2)}$

f. $F(s) = \dfrac{1}{(s + 1)^2(s^2 + 2s + 5)}$

14. Find the inverse Laplace transforms of the following:

a. $F(s) = \dfrac{5}{(s + 1)(s + 2)}$

b. $F(s) = \dfrac{40}{s^2 + 9s + 20}$

c. $F(s) = \dfrac{s + 1}{(s + 2)(s^2 + 2s + 2)}$

d. $F(s) = \dfrac{1}{(s + 1)^2(s^2 + 4s + 13)}$

e. $F(s) = \dfrac{3s^2 + 2s + 1}{4s^3 + 20s^2 + 27s + 9}$

f. $F(s) = \dfrac{100}{s^5 + 4s^4 + 14s^3 + 62s^2 + 149s + 130}$

15. Sketch the time response patterns for each of the functions given in Prob. 3.14.

16. Program the digital (analog) computer to obtain the time responses for one or more of the functions given in Prob. 3.14 over a time interval approximately equal to 4 times the longest time constant in the functions. Plot your data.

17. Solve the following differential equations using the Laplace transform method:

a. $\dot{x} + 2x = 0$ $\quad x(0^+) = -3$

b. $\ddot{x} + 3\dot{x} + 2x = 10\,u(t)$ $\quad \dot{x}(0^+) = 4 \quad x(0^+) = -2$

c. $\ddot{x} + 5\dot{x} + 4x = (t - 2)\,u(t - 2)$ $\quad \dot{x}(2^+) = 5 \quad x(2^+) = 0$

d. $\dddot{x} + 3\ddot{x} + 4\dot{x} + 2x = \sin 2t$ $\quad \ddot{x}(0^+) = -1 \quad \dot{x}(0^+) = 2 \quad x(0^+) = 4$

18. Use the Laplace transform to solve the following problems:

a. Find $v_2(t)$ in Prob. 2.1 assuming $R = 1M$, $C = 1\,\mu f$, and $v_1(t) = 20\,u(t)$.

b. Find $x(t)$ in Prob. 2.3 assuming $f(t) = 10\,u(t)$, $K = 13$, $B = 4$, and $M = 1$, all in consistent units.

c. Find the transfer function $\theta(s)/V_f(s)$ for the motor and load described in Prob. 2.17. If $L_f = 10$, $R_f = 20$, $K_T = 40$, $J = 2$, $B = 2$, and $V_f(s) = 100/s$, find $\theta(t)$.

19. Apply the initial and final value theorems to each of the following functions. Under what conditions do these theorems apply?

a. $$F(s) = \frac{s^2 + 2s + 3}{s^3 + 2s^2 + 2s + 1}$$

b. $$F(s) = \frac{3s + 1}{s^4 + 5s^3 + 9s^2 + 7s + 2}$$

20. The Laplace transform of the unit impulse may be derived in several ways. Consider the unit impulse as the limit of the triangular waveform shown as $a \to 0$, and derive its Laplace transform by taking the limit of the Laplace transform of the triangular waveform as $a \to 0$.

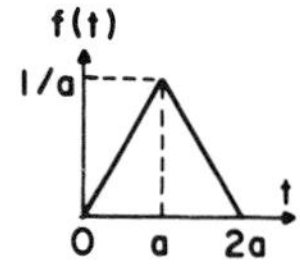

21. It is claimed that the response of an arbitrary network to a unit impulse is adequately approximated by its response to the pulse shown if the pulse duration a is chosen such that $a \ll \tau_{min}$ where τ_{min} denotes the minimum time constant in the network transfer function. Investigate this claim for the network with transfer function $G(s) = 1/(s + 1)$. Sketch the impulse response and compare it to the pulse response for various values of a. What do you conclude? You will no doubt find a digital computer useful in obtaining the data desired.

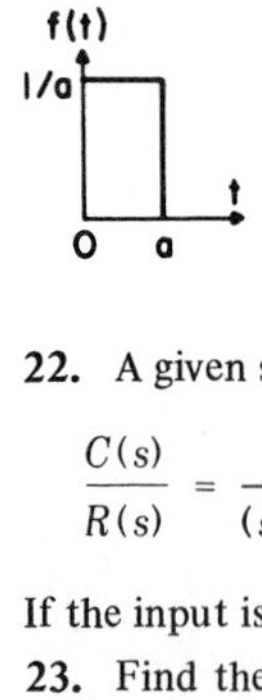

22. A given system has transfer function

$$\frac{C(s)}{R(s)} = \frac{10(s + 1)}{(s + 2)(s^2 + 2s + 5)}$$

If the input is $r(t) = 3 \sin 2t$, what is the *steady-state* output $c(t)$?

23. Find the total (transient and steady-state) response of the network shown to the given sinusoidal input. Assume $v_0(0^+) = 0$. The switch is closed at $t = 0$.

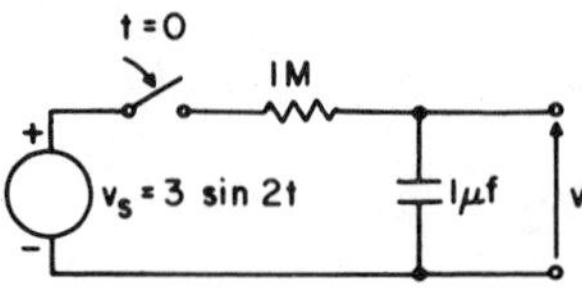

24. Show for the system shown that

$$x_1(s) = \frac{F_0 M_2}{G(s)}$$

where $G(s)$ is a general function of system parameters and s. If, as we can prove, all poles of $x_1(s)$ are in the left half plane for this special situation, what does it imply? Discuss. Note that the ratio of K_2 to M_2 is critical. Why?

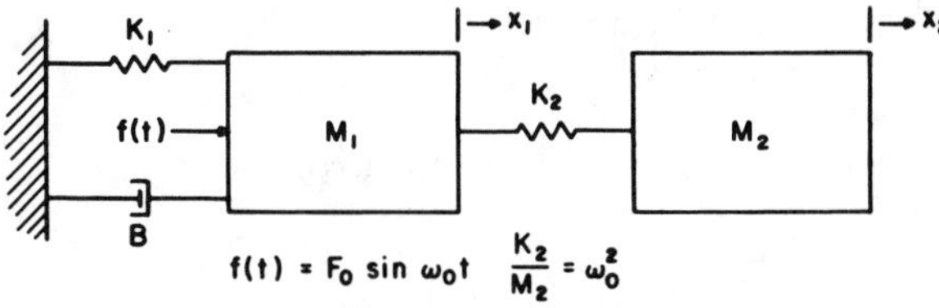

25. A particular controller has transfer function

$$\frac{C(s)}{R(s)} = \frac{3s + 10}{s^2 + 7s + 10}$$

Find the time response $c(t)$ of this system for a unit step input, i.e., $r(t) = u(t)$, if the initial system energy state is zero.

26. The transfer function of a particular system is

$$\frac{C(s)}{R(s)} = \frac{1}{3s^3 + 11s^2 + 12s + 4}$$

Find the unit step response of this system assuming

$\ddot{c}(0^+) = 3 \qquad \dot{c}(0^+) = -1 \qquad c(0^+) = 2$

27. A motor is connected as shown in the diagram. Find the transfer function between V_i and V_0, i.e., solve for $V_0(s)/V_i(s)$. V_0 is the potentiometer output. The potentiometer is driven by the output shaft, providing 0 output at $\theta = 0$ and 100V output at $\theta = 10(2\pi)$. (It is a standard 10 turn helipot). Note that this arrangement may be used as a position controller by subtracting V_0 from a reference voltage representing the desired position and using the difference as an error voltage to drive the amplifier supplying the motor field current.

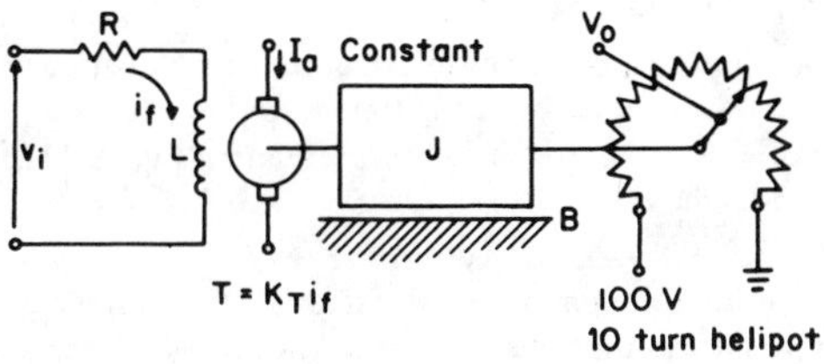

28. A given system has transfer function

$$\frac{C(s)}{R(s)} = G(s) = \frac{s + 2}{s^2 + 4s + 3}$$

Find the output of this system as a function of t in response to the input $r(t) = 1 - \epsilon^{-t}$ assuming zero initial energy state.

29. Consider the system shown. If $g(\tau) = 1 - \epsilon^{-2\tau}$ and $x(t) = \epsilon^{-t}$, find the response $y(t)$ using the superposition integral. Check your result by using $Y(s) = X(s)\,G(s)$.

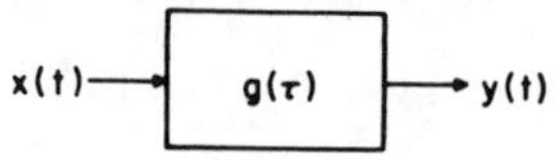

30. A linear system of the type introduced in Prob. 3.29 has transfer function and input as shown.

a. Find $y(t)$ using the convolution integral.

b. Find $y(t)$ using the Laplace transform.

c. Develop a block diagram involving only integration elements with transfer functions $1/s$, delay elements ϵ^{-Ts}, and summers for adding and subtracting signals, as shown in Fig. 4.7, for example, and show that the output of this system is the result you have obtained in parts *a* and *b*. Can the impulse response given here be realized in practice? Explain.

31. A linear system has input as given in Prob. 3.30 and impulse response as shown. Find the output $y(t)$.

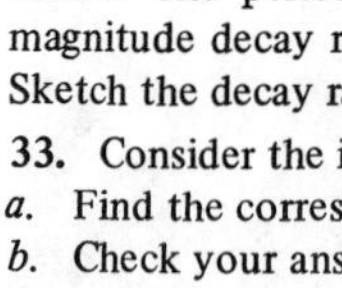

32. Consider the standard second-order system unit step response given in Eq. (3.118). Let T denote one period of the sinusoidal part of this response. Derive an expression for the magnitude decay ratio per period of this sinusoidal solution component as a function of ξ. Sketch the decay ratio versus ξ for $0 \le \xi \le 1$.

33. Consider the impulse response characteristic given in Eq. (3.120).

a. Find the corresponding time response $g(t)$ as a general function of ξ and ω_0.

b. Check your answer by differentiating Eq. (3.118) with respect to time.

c. Sketch $g(t)$ versus t for $\xi = 0.2, 0.4, 0.6$, and 0.8. You may find a digital computer handy here.

34. The unit step response rise time T_r is defined as follows: Locate the point of maximum slope on the unit step response pattern as it progresses from 0 toward its maximum value. Construct a straight line tangent to the curve at this point. Let t_1 and t_2 denote the points where this straight line intersects the time axis and the final value line $g(t) = 1$, respectively. Then $T_r = t_2 - t_1$. In other words, T_r is the time it would take for the unit step response to go from 0 to 1 if it progressed at the maximum rate continuously. Let $\omega_0 = 1$ in Eq. (3.118). Determine T_r for $\xi = 0, 0.2, 0.4, 0.6, 0.8$, and 1.0. Sketch T_r versus ξ over this range.

35. A standard second-order system has unit step response peak overshoot of 90 percent. What is the value of the first local minimum, or undershoot, of $f(t)$?

CHAPTER 4

1. Find the transfer function $V_0(s)/V_i(s)$ for the system shown.

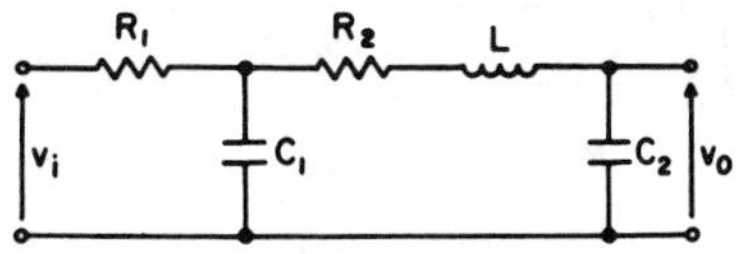

2. *a.* Find the transfer function $V_0(s)/V_i(s)$ for the system shown.
b. If $v_i(t) = u(t)$, and all initial conditions are zero, find $v_0(t)$.

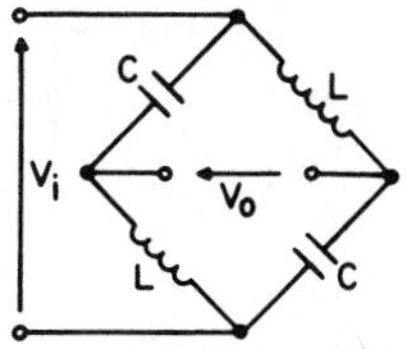

3. Develop the transfer functions relating $F(s)$ to $X_1(s)$ and $X_2(s)$ for the system introduced in Prob. 2.5.

4. Develop a block diagram for the system introduced in Prob. 2.17, reducing the equations to equivalent transfer function elements. One block should relate V_f to I_f, a second should relate I_f to Torque, etc. Keep all parameters in literal form. What are the conditions under which transfer functions can be cascaded in this way?

5. A network with input $v_i(t)$ and unit impulse response $g(\tau) = 1 - \epsilon^{-\tau}$ is connected through an amplifier with gain of 10 to the field circuit of a motor. A unit step input to the field circuit of the motor results in $\theta(t) = \frac{t}{2} - \frac{1}{4}(1 - \epsilon^{-2t})$. None of the elements loads any of the others. Find the overall transfer function $\theta(s)/V_i(s)$.

6. Assuming that M is free to move in only the up-down direction determine the following:
a. The transfer function between f and motion of M. Use x as the distance variable.
b. If M is initially moving downward with velocity 3 units/unit time, what is $x(t)$ for $f(t) = u(t)$?

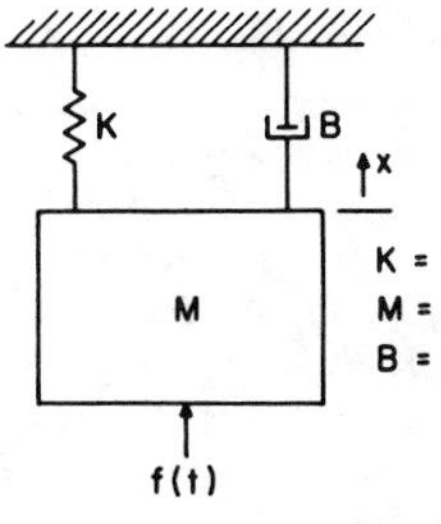

7. Consider the configuration shown. Note that K_s is a shaft torsional constant. Obtain the transfer function $\theta_0(s)/V(s)$.

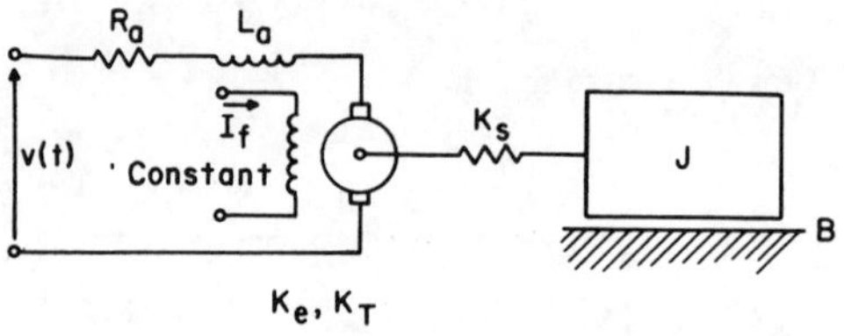

8. *a.* Write the differential equations describing the mechanical network shown.
   b. Find the transfer function between torque input and the position of inertial element J_2.

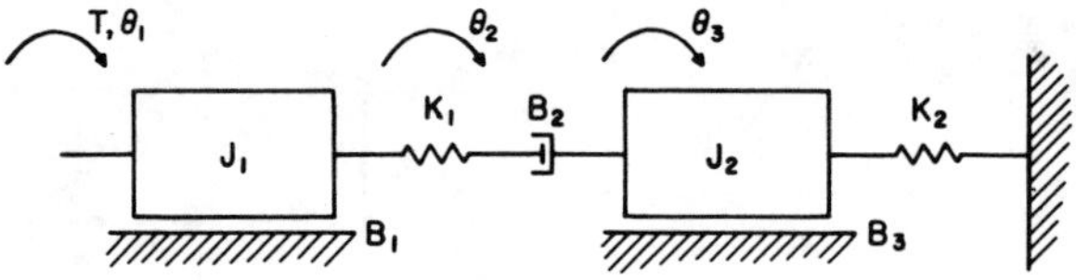

9. *a.* Find the transfer function between V_i and θ_0.
   b. What is the final shaft velocity in terms of system parameters for a unit step input at $t = 0$?

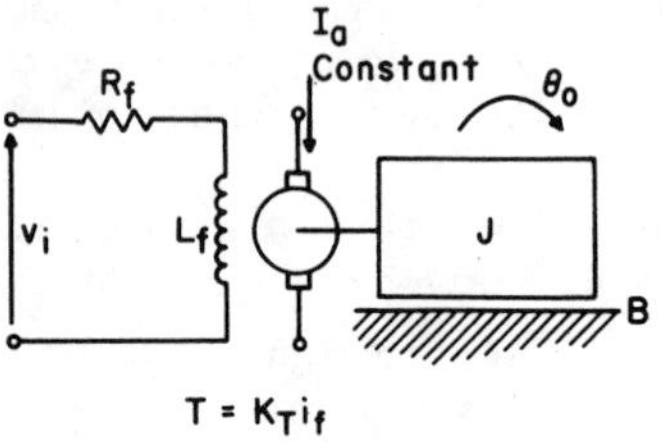

10. Two systems, A and B, have unit step responses

Step Response $_A = 1 - \epsilon^{-3t}$ Step Response $_B = 2(1 - \epsilon^{-t})$

These two units are connected in cascade using a unit gain buffer amplifier such that neither loads the other in any way. What is the unit step response of the series combination?

11. The generator and motor combination shown is used to drive an inertial load from the relatively low power level signal $v_f(t)$.
a. If $R_f = 5$, $L_f = 2$, and $K_{e_1} = 10$, find the open circuit generator output for $v_f(t) = 100\,u(t)$.
b. Derive the transfer function relating $\theta(s)$ to $V_2(s)$ for the motor as a general function of R_{am}, J, K_{e_2}, the back emf constant, and K_T.
c. If $R_{ag} = 0.2$, $R_{am} = 0.3$, $K_{e_2} = 1$, $K_T = 1$ and $J = 10$, find the overall transfer function from $V_f(s)$ to $\theta(s)$ when the motor and generator are connected together. Note that loading can *not* be disregarded here, but only a minor reinterpretation of the result in part *b* is necessary.

d. If $V_f(s) = 20/s$, find $\omega(t)$, the output shaft velocity.

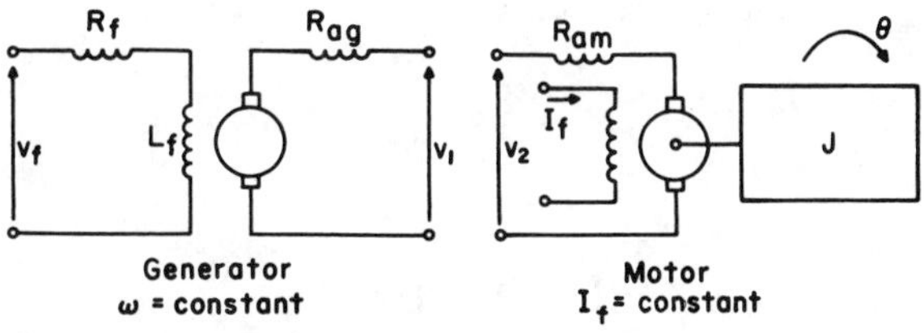

12. Find $C(s)/R(s)$ for the block diagram shown.

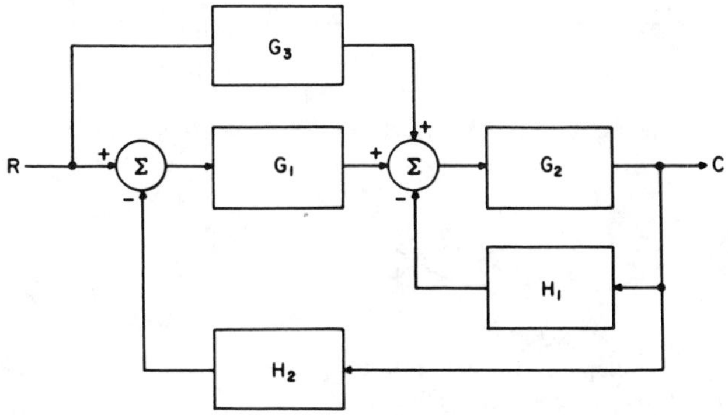

13. Develop a flow graph relating $V_0(s)$ and $V_i(s)$ for the RC network configuration shown. Reduce the flow graph to the transfer function $V_0(s)/V_i(s)$.

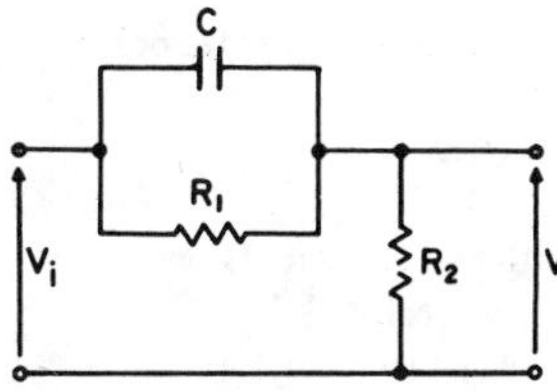

14. Develop a flow graph for the network shown and reduce the result to the transfer function $V_0(s)/V_i(s)$.

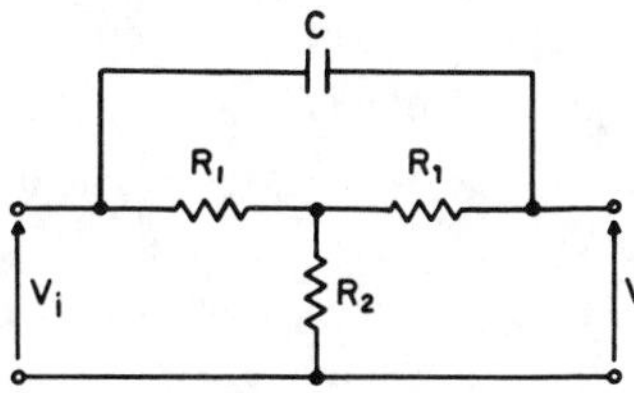

15. Assume the tube in the circuit shown has parameters $\mu = 20$ and $r_p = 10K$. Determine the limit on the value of R for $Y(s)/X(s)$ to be a minimum phase transfer function. (i.e., have poles and zeros only in the 1 hp.)

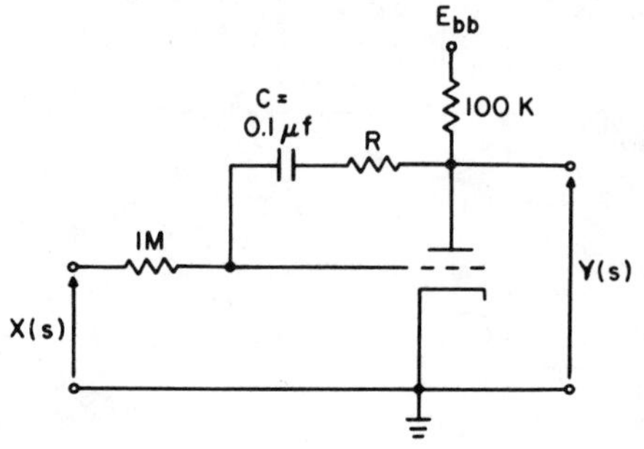

16. Reduce the flow graph given to the transfer function V_0/V_i.

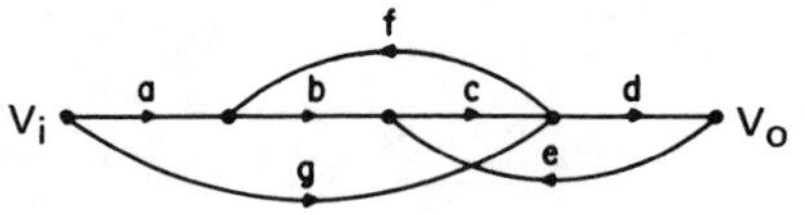

17. Develop a flow graph for the system introduced in Prob. 4.8. Reduce the flow graph to a transfer function relating T and θ_3.

18. Develop a flow graph for the mechanical configuration of Prob. 2.3. Reduce your result to a transfer function relating $X(s)$ and $F(s)$.

CHAPTER 5

1. Consider a general polynomial of the form

$$Q(s) = s^n + a_{n-1}s^{n-1} + \cdots + a_1 s + a_0 = 0$$

With all coefficients real.

a. Prove that $a_i > 0 \quad i = 1, 2, \ldots, n-1$ is required if all roots of $Q(s)$ are in the lhp, irrespective of whether the roots are real or complex. Note that complex roots *must* appear in conjugate pairs.

b. Show with an example that the converse is not true–in other words all a_i may be positive when one or more roots are in the right half plane.

2. Use the Routh criterion to determine the number of roots of the following polynomials in the right and left half planes.

a. $5s^3 + 6s^2 + 6s + 2 = 0$

b. $25s^5 + 105s^4 + 120s^3 + 121s^2 + 21s + 1 = 0$

c. $s^4 + 2s^3 + 2s^2 + 3s + 6 = 0$

d. $s^4 + s^3 + 2s^2 + 9s + 5 = 0$

e. $s^5 + 30s^4 + 2800s^3 + 2.26 \times 10^5 s^2 + 6 \times 10^6 s + 4 \times 10^7 = 0$

f. $s^5 + s^4 + 2s^3 + 2s^2 + 3s + 15 = 0$

g. $s^4 + 2s^3 + 3s^2 + 2s + 2 = 0$

3. Shift the s-plane origin and apply the Routh criterion to determine the root locations for the following polynomials:

a. $s^3 + 5s^2 + 9s + 5 = 0$

b. $s^4 + 4s^3 + 6s^2 + 4s + 2 = 0$

c. $2s^5 + 15s^4 + 48s^3 + 81s^2 + 74s + 30 = 0$

4. How many roots of the following polynomials are there with real parts between 0 and −1?

a. $8s^5 + 44s^4 + 126s^3 + 219s^2 + 258s + 85 = 0$

b. $16s^7 + 60s^5 + 23s^4 + 94s^3 + 194s^2 + 256s + 120 = 0$

5. For what range of K is the system shown stable?

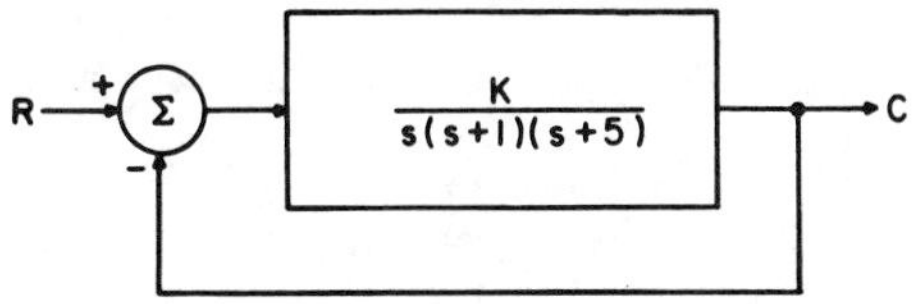

6. Find how the range of K for stable operation of the system shown depends upon τ.

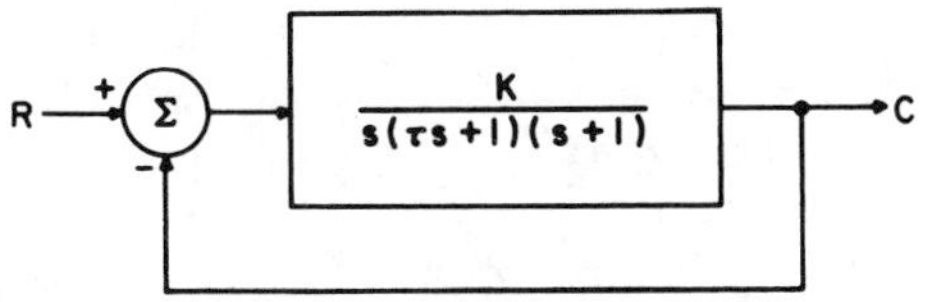

7. Assume

$$G(s) = \frac{K(1 + Ts)^2}{s^3(1 + s)}$$

Where K and T are adjustable system parameters. Sketch the boundary between the stable and unstable regions in the K-T plane.

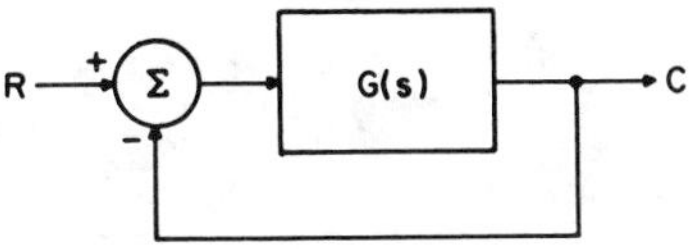

CHAPTER 6

1. Sketch the root locus diagram for each of the following open-loop transfer functions. Although a point by point evaluation using a spirule or angular contribution transparencies is not necessary, you should find OA and AA in each case, and the departure and approach angles for all poles and zeros not on the real axis.

a. $A(s) = \dfrac{K}{s(s + 4)(s + 10)}$

b. $A(s) = \dfrac{K}{s(s^2 + 2s + 2)}$

c. $A(s) = \dfrac{K(s + 2)}{s^2(s + 5)}$

d. $A(s) = \dfrac{K(s + 5)}{s^2(s + 2)}$

e. $A(s) = \dfrac{K(s+6)}{s(s+3)(s^2+2s+26)}$

f. $A(s) = \dfrac{K(s+12)}{s(s^2+16s+100)}$

g. $A(s) = \dfrac{K}{(1+s/2)(1+s/5)(1+s)^2}$

h. $A(s) = \dfrac{K}{(s+5)(s^2+4s+7)}$

i. $A(s) = \dfrac{K}{(s^2+2s+5)(s^2+6s+10)}$

2. *a.* Sketch the root locus for the system with

$$A(s) = \frac{K}{s(s+1)^2}$$

b. Find the departure point of the locus from the real axis, and the corresponding value of K.

c. Find the point where the root locus crosses the $j\omega$ axis and the corresponding value of K.

d. Find the point where the root locus crosses the line $R_e(s) = 1$ and the corresponding value of K.

3. Sketch the root locus for the system shown with

$$G(s) = \frac{K(s+1)}{s^2(s+3)} \qquad H(s) = \frac{s^2+4s+5}{s^3+7s^2+20s+50}$$

In particular, find OA, AA, the departure angle from each pole, the entry angle at each zero, and the approximate point where the locus crosses the $j\omega$ axis.

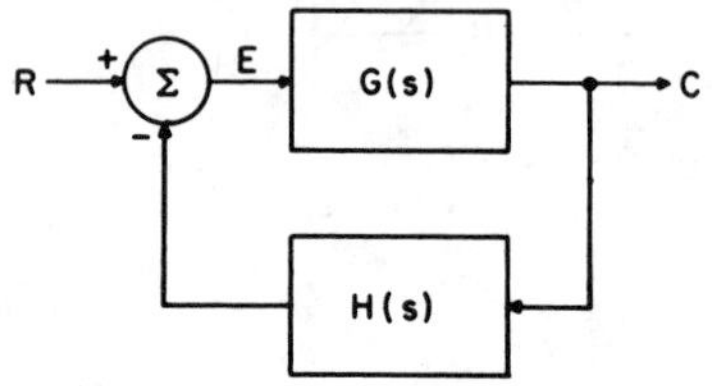

4. Sketch the root locus for a *positive* feedback system with

$$A(s) = \frac{-K}{s(s+1)(1+s/2)}$$

5. A particular system has

$$A(s) = \frac{K}{s(s+1)(s+2)}$$

a. where does the root locus for this system break away from the real axis?

b. where does the root locus cross the $j\omega$ axis?

6. A particular system has

$$A(s) = \frac{K}{s(s+10)(s+100)}$$

a. Sketch the root locus diagram.

b. Determine K such that the closed loop system has a complex conjugate pole pair with damping factor $\xi = 0.5$.
c. Express the closed loop input to output transfer function in factored form for this choice of K. Assume unity feedback.

7. Consider the $A(s)$ given in Prob. 6.6. It is proposed to set $K = 10^4$ and place the equalizing filter $G_e(s)$ in series with $A(s)$, where

$$G_e(s) = \frac{\gamma\tau s + 1}{\tau s + 1} \qquad 0 < \gamma < 1$$

to provide a closed loop complex conjugate pole pair with damping factor $\xi = 0.5$. Determine appropriate values for γ and τ. *Hint:* $\tau = 10$ and $\gamma = 0.1$ are not too far from the desired values. Use trial and error.

8. Program the digital computer to determine the unit step and unit ramp responses for the systems obtained in Probs. 6.6 and 6.7. Discuss the results, attempting to assess the relative merits of these two designs.

CHAPTER 7

1. Find the harmonic (steady-state sinusoidal) transfer functions for the networks shown. In each case sketch the amplitude and phase curves for the network. Perhaps a computer should be used to limit the time spent on this.

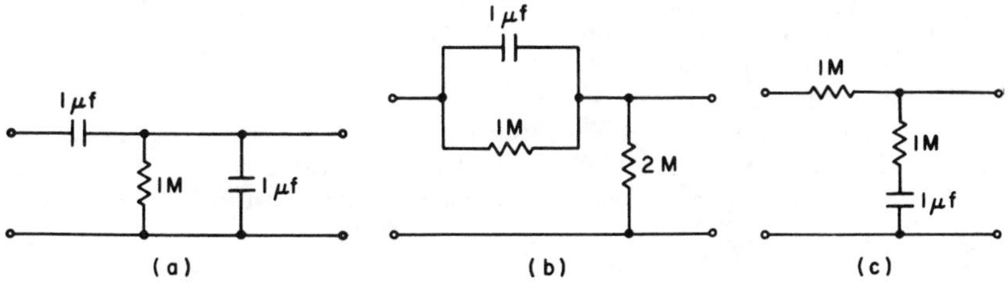

2. Consider the network shown, which consists essentially of two RC lag networks in series.
a. Assume the second network does not load the first, and determine the overall amplitude and phase characteristic as the product and sum, respectively, of those for the individual networks.
b. Find the exact harmonic transfer function and use it to evaluate the error in your approximation in part *a*. You may find a computer of value in obtaining the data used here.

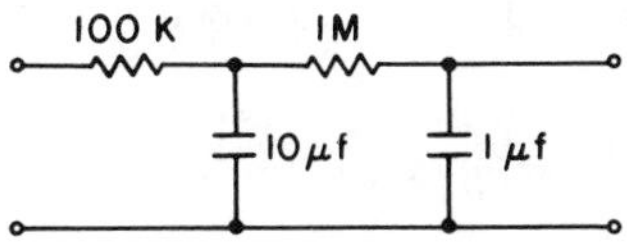

3. A given system has

$$G(s) = \frac{K(1 + s/10)}{s(1 + s/4)(1 + s/80)}$$

The corresponding asymptotic magnitude diagram passes through $|G| = 0.5$ at $\omega = 60$. Find K.

4. Consider the Bode diagram shown. At what frequency does the asymptotic diagram cross unit magnitude?

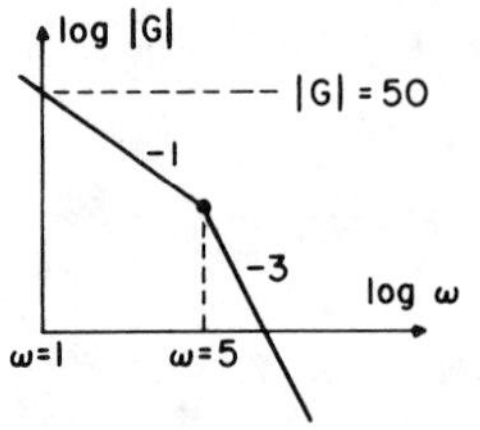

5. Given

$$G(s) = \frac{10{,}000}{s^2 + 4s + 100}$$

Find the actual magnitude and phase shift values at $\omega = 9$.

6. A given system has

$$G(s) = \frac{100(1 + s)}{s^2(1 + s/20)(1 + s/50)}$$

a. Find the corresponding *asymptotic* magnitude at $\omega = 40$.
b. Find the *actual* magnitude at $\omega = 30$.
c. Find the phase shift at $\omega = 30$.

7. The transfer function $G(s)$ from which the Bode diagram shown was developed has no complex poles or zeros. The zero at $\omega = 8$ is in the rhp. What is the phase shift of $G(j\omega)$ at $\omega = 8$, assuming the -1 slope segments go off uninterrupted to $\omega = 0$ and $\omega = \infty$?

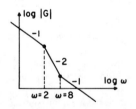

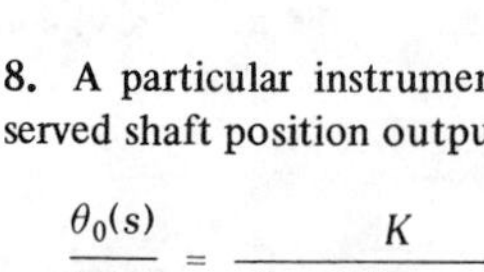

8. A particular instrument servo is known to have a desired shaft position input to observed shaft position output transfer function of the form

$$\frac{\theta_0(s)}{\theta_d(s)} = \frac{K}{s^2 + 2\zeta\omega_0 s + \omega_0^2}$$

The following test data are obtained from a laboratory experimental study:
a. The frequency at which the phase lag is 90° is found to be 10 rad/sec.
b. A 1 radian amplitude sinusoidal input at 10 rad/sec results in a steady-state output sinusoid of .4 radian amplitude.
c. A 1 radian dc input produces a .2 radian steady-state position change.
What are the values of K, ω_0, and ζ? Sketch the magnitude and phase characteristics for this system.

9. Draw asymptotic α diagrams on log-log coordinates for each of the following transfer functions. Sketch in the actual magnitude curves also.

a. $$\frac{Y(s)}{X(s)} = \frac{100}{s(1 + s/10)}$$

b. $$\frac{Y(s)}{X(s)} = \frac{200}{s(1 + s)^2}$$

c. $$\frac{Y(s)}{X(s)} = \frac{10^4 s(1 + s/100)}{(s + 2)(s + 10)^2}$$

d. $$\frac{Y(s)}{X(s)} = \frac{10}{s^2 + 2s + 4}$$

e. $$\frac{Y(s)}{X(s)} = \frac{1000(s + 5)}{s(s^2 + 6s + 25)}$$

10. Program the digital computer to develop data on the magnitude and phase performance of networks with the following transfer functions. Plot your data over a range of plus and minus one decade from the break point in each case on log-log coordinates. Compare your results to those provided in the text. You will use these results often, and may find it convenient to store them inside the back cover of your book for future reference.

a. $$\frac{Y(s)}{X(s)} = \frac{1}{s + 1}$$

b. $$\frac{Y(s)}{X(s)} = \frac{1}{s^2 + 2\zeta s + 1} \quad \text{for } \zeta = 0.1, 0.2, 0.4, 0.7, 1.0$$

11. Sketch phase diagrams for the transfer characteristics given in Prob. 7.9.

12. The approximation $\tan^{-1}\phi \simeq \phi$ is often used in determining the phase shift at critical frequencies in solving control problems. Assume $\phi \geq 0$. For what range of ϕ is the approximation accurate to within 1 percent? To within 10 percent?

13. A given system has transfer function

$$G(s) = \frac{40{,}000(1 + s/10)}{s(1 + s)(s^2 + 8s + 400)}$$

a. Sketch the asymptotic α diagram for this system.
b. Plot enough points on the actual α and β diagrams to give a good representation of what is happening.

14. Using the arc-tan approximation, find the frequency at which the phase shift of the following transfer function is $-135°$.

$$\frac{Y(s)}{X(s)} = \frac{K(s + 1)}{s(1 + s/10)^2(1 + s/50)}$$

15. Find the transfer function $Y(s)/X(s)$ for the network shown. Is it minimum phase? This network is a parallel T circuit, and circuits of this general type are used as equalizing networks in AC servo systems, and as noise filters or "traps" in other applications. Sketch the asymptotic and the actual α diagrams and the β diagram.

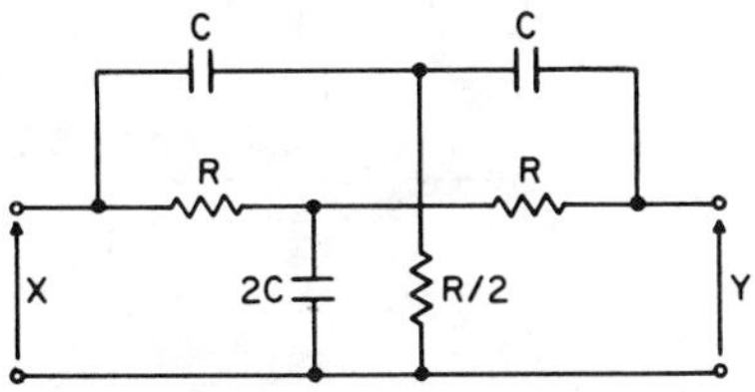

16. Sketch the asymptotic and actual α diagrams and the β diagram for the transfer function

$$\frac{Y(s)}{X(s)} \simeq \frac{100(1 - s/10)}{s(s + 1)(1 + s/5)}$$

How do they compare to those for

$$\frac{Y(s)}{X(s)} = \frac{100(1 + s/10)}{s(s + 1)(1 + s/5)}$$

What do you conclude from this about your ability to determine phase information from the asymptotic α diagram?

17. A particular asymptotic α diagram on log-log coordinates has −1 slope for $\omega < 1$, and −3 slope for all $\omega > 1$. The asymptotic diagram magnitude is 1000 at $\omega = 0.1$. It is known that the double break at $\omega = 1$ is caused by left half plane terms. Can you determine the actual magnitude and phase versus ω values from this information? Why (Why not)?

18. A given transfer function asymptotic α diagram has −1 slope for $\omega < 4$, and −3 slope for $\omega \geq 4$. There are no right half plane factors. There are 3 poles and no zeros. The phase shift at $\omega = 2$ is known to be $-100°$. The asymptotic diagram passes through a magnitude of 20 at $\omega = 2$. What is the transfer function?

19. Consider the asymptotic α diagram shown on log-log coordinates. The asymptotic magnitude is 20 at $\omega = 1$. What is the asymptotic magnitude at $\omega = 4$? At what ω does the asymptotic diagram pass through unit magnitude?

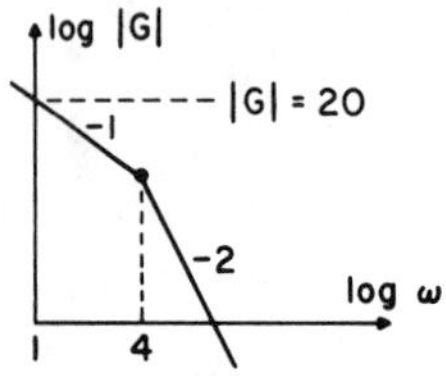

20. The Bode diagram for a particular system is shown. Answer the following:

a. At what value of ω does the break from −1 slope to −2 slope at magnitude 100 occur?
b. At what ω does the diagram cross the magnitude of 0.1?
c. At what magnitude does the diagram cross $\omega = 10$?
d. At what magnitude does the break from −1 slope to −2 slope at $\omega = 1000$ occur?
e. At what ω does the diagram cross the magnitude of 0.01?

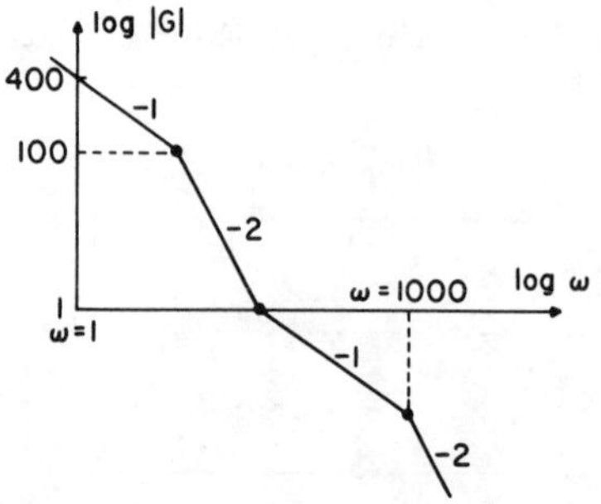

21. Consider an asymptotic α diagram with slope of $-m$, m an integer, $m > 1$, for $0 \leq \omega \leq \omega_1$, slope of -1 for $\omega_1 < \omega < \omega_2$ and slope of $-n$, n an integer, $n > 1$, for $\omega \geq \omega_2$. Find the value of ω for which β is a maximum as a general function of m, n, ω_1 and ω_2.

22. Consider a filter with transfer function

$$G(s) = \frac{1 + s/\omega_1}{1 + \gamma s/\omega_1} \qquad 0 < \gamma < 1$$

Note from the answer to Prob. 7.21 that this filter provides maximum phase lead at $\omega = \sqrt{\omega_1^2/\gamma} = \omega_1/\sqrt{\gamma}$.

a. Use the arc-tan approximation to obtain a general expression for θ_{max} as a function of γ.
b. Sketch θ_{max} versus γ for $0.01 \leq \gamma \leq 1$.

CHAPTER 8

1. Sketch Nyquist diagrams for each of the following open loop transfer functions. In each case indicate whether the system is stable or not.

a. $A = \dfrac{100}{(1 + s)(1 + s/10)}$

b. $A = \dfrac{100}{(1 + s/2)(1 + s/5)(1 + s/20)}$

c. $A = \dfrac{200}{s(1 + s)(1 + s/10)}$

d. $A = \dfrac{100(1 + s/5)}{s^2(1 + s/50)(1 + s/200)}$

e. $A = \dfrac{10{,}000(1 + s/50)^2}{s(1 + s/5)^2(1 + s/400)(1 + s/800)}$

f. $A = \dfrac{2}{s(1 + s/4)^2(s^2/100 + 1)}$

2. Sketch the Nyquist plot for part *c* of the previous problem *enclosing* the pole at the origin in the Bromwich contour.
a. What is your stability conclusion?
b. Does this compare with your previous results? Should it? Discuss.

3. Consider a system with open-loop transfer function

$$A(s) = \frac{K(1 + Ts)^2}{s(1 + 0.1s)^2}$$

a. Using the arc-tan approximation, for what range of values of T can this system be made conditionally stable by appropriate choice of K?
b. If T is set equal to 0.01, sketch the Nyquist plot assuming $K = 100$. For what range of K is the system stable? Note that you only need to know accurately where the Nyquist plot crosses the real axis to obtain this information.
c. Check your results for part *b* using the Routh criterion.

4. Find the gain margin and phase margin for the systems with open loop transfer functions $A(s)$ as given in Prob. 8.1, parts *d* and *e*.

5. Consider a system with

$$A = \frac{100(1 + s/5)}{(s^2 + 100)(1 + s/\gamma)}$$

a. Sketch the Nyquist plot for several values of γ. (*Hint:* Try values on each side of 5).
b. Estimate from your results in part *a* the conditions required on γ for this system to be stable.
c. Use the Routh criterion to check your results for part *b*.

6. Consider a system with

$$A = \frac{500}{s(s^2 + 2s + 100)}$$

a. Sketch the Nyquist diagram.
b. Is this system stable?
c. Can you suggest a simple way in which unstable systems of this type can be modified to assure stability?

7. Consider the system shown with transfer function elements

$$G = \frac{100}{s(s + 1)} \qquad H = 0.1s$$

a. Sketch the Nyquist diagram for this system.
b. Is the system stable? (Careful here)
c. Solve for $C(s)/R(s)$. Where are the closed loop poles located?
d. Discuss the implications of your results in applying the Nyquist criterion.

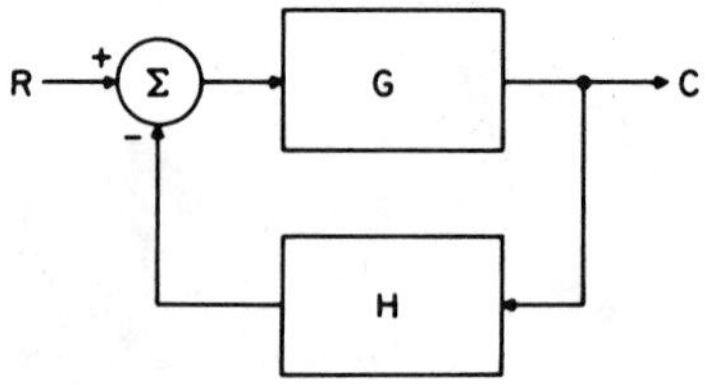

8. The Nyquist diagram for a particular system with open-loop transfer function of the form

$$A(s) = \frac{KP(s)}{Q(s)}$$

where $P(s)$ and $Q(s)$ are general polynomials in s and $Q(s)$ has no rhp factors, is as shown, assuming $K = 100$. For what range of K is this system stable?

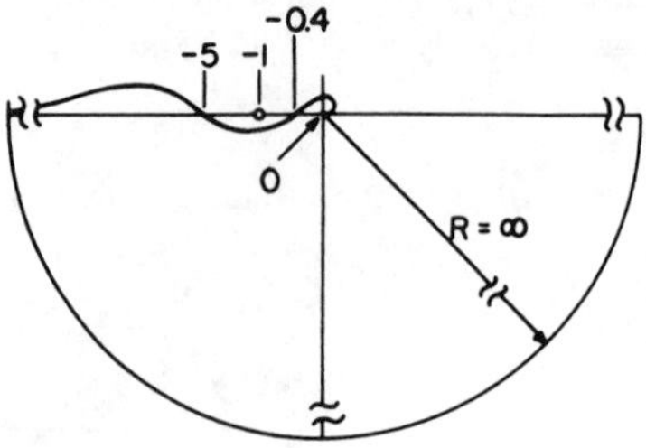

9. Apply the general multiple-loop Nyquist criterion method to determine if the system shown in the given block diagram is stable.

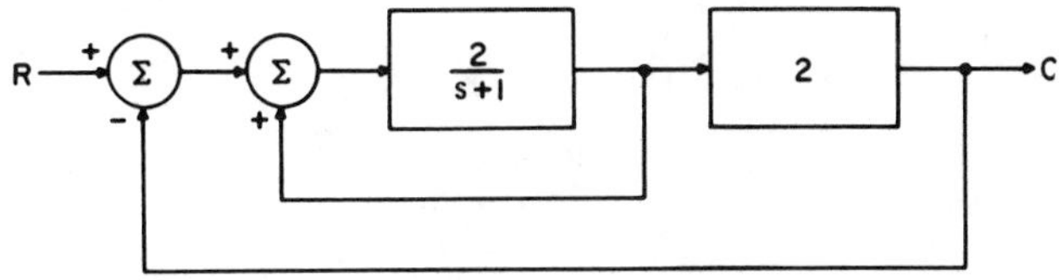

10. Consider the system shown.
a. Apply the multiple loop Nyquist criterion method to check stability.
b. As a check on the results obtained in part *a*, break both feedback loops simultaneously at the point marked *X* on the diagram and apply the Nyquist criterion to the open-loop transfer characteristics from that point back to itself through both feedback paths taken together. Of course your answer must agree with that obtained in part *a*.

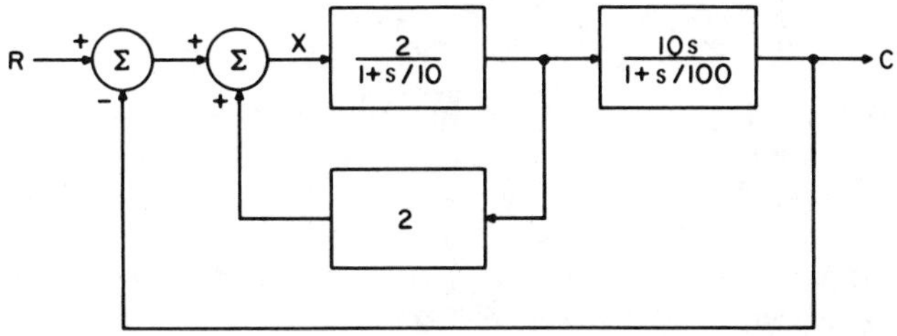

11. Apply the multiple loop Nyquist procedure to determine if the system shown in stable.

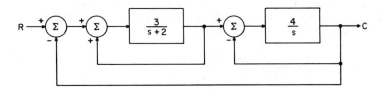

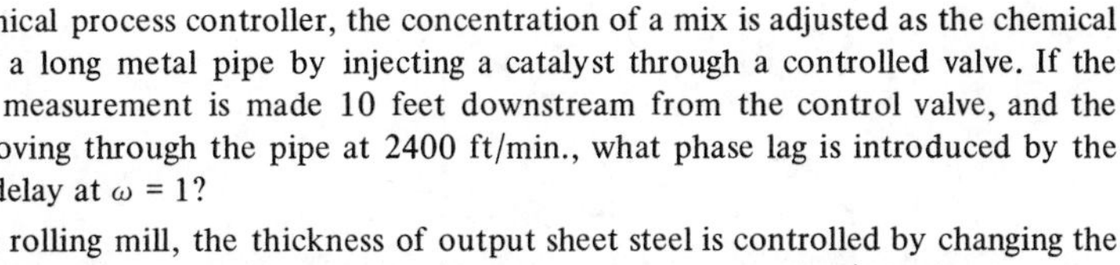

12. In a chemical process controller, the concentration of a mix is adjusted as the chemical flows through a long metal pipe by injecting a catalyst through a controlled valve. If the concentration measurement is made 10 feet downstream from the control valve, and the chemical is moving through the pipe at 2400 ft/min., what phase lag is introduced by the measurement delay at $\omega = 1$?

13. In a steel rolling mill, the thickness of output sheet steel is controlled by changing the distance between rollers through which the steel must pass as it runs through the plant. The control system is shown in schematic form. A field driven dc motor is used to change the roller spacing. The transfer function between field input voltage and roller spacing is measured experimentally and is approximately

$$\frac{T_2(s)}{V_f(s)} = \frac{0.02}{s(1 + 10s)(1 + s)}$$

where T_2 is measured in mils, or thousandths of an inch. The x-ray thickness gauge produces a 1 volt output for a variation of .1 mil in sheet thickness. The amplifier and equalizer have transfer function

$$\frac{V_f(s)}{V_i(s)} = \frac{(1 + s/0.16)(1 + s/0.1)(1 + s)}{(1 + s/0.0065)(1 + s/1.6)}$$

The sheet steel passes through this roller at a velocity of 1000 ft/min.

a. Sketch the asymptotic α diagram for the open-loop transfer function $A(s)$.

b. Sketch the Nyquist diagram for $A(s)$ assuming $d = 0$. Is the system stable if the x-ray thickness gauge is located 10 feet downstream from the rollers?

c. What is the maximum value of d for stability?

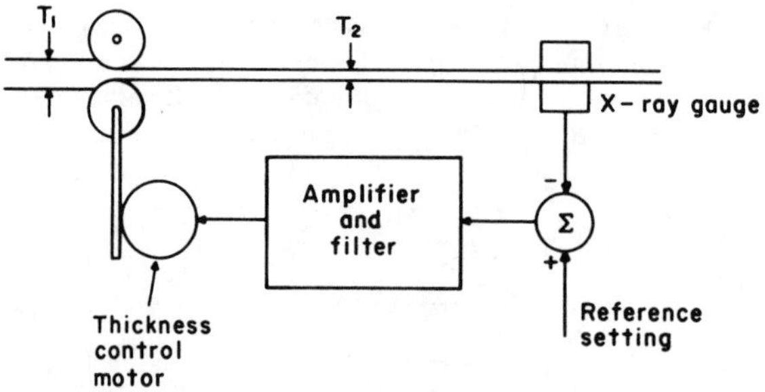

14. Assume that an experiment on the moon is to be controlled from a console located in Houston. A block diagram of the system is as shown, where

$$G(s) = \frac{K}{s(1 + 5s)(1 + s)}$$

and T denotes the time delay associated with radio transmission from earth to moon (or moon to earth). Assume the mean distance from earth to moon is 240,000 mi.

a. Sketch the Nyquist diagram for this system assuming $K = 1$.

b. What is the maximum value of K for stability?

c. Does it seem technically attractive to control systems on the moon from earth? Discuss. What are the general limits upon a remote controller?

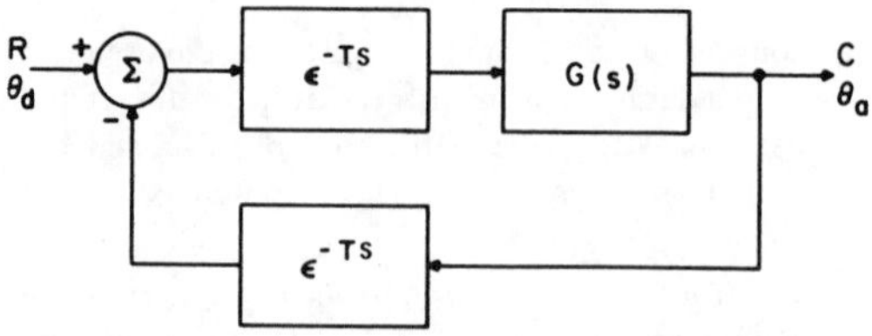

CHAPTER 9

1. Consider the standard unity feedback configuration with forward transfer function $G(s)$. Assume that $G(s)$ is of the form

$$G(s) = K\frac{P(s)}{s^a Q(s)}$$

where $P(s)$ and $Q(s)$ are general polynomials of mth and nth order, respectively, with $n + a \geq m$. The constant term in both $P(s)$ and $Q(s)$ equals one. Assume that a may take on non-negative integer values only, or a values of 0, 1, . . . are allowed. Assume also that $P(s)$ and $Q(s)$ are limited to polynomials such that the closed loop system is stable. The velocity error constant, K_v, is defined as the reciprocal of the steady-state error in response to a unit ramp input. Prove that $K_v = 0$ for the system shown when $a = 0$, $K_v = K$ for $a = 1$, and $K_v = \infty$ for $a \geq 2$.

2. Consider the standard unity feedback configuration introduced in Prob. 9.1. Assume that $G(s)$ is of the general form

$$G(s) = K\frac{P(s)}{sQ(s)}$$

with $P(s)$ and $Q(s)$ general polynomials in s of mth order and nth order, respectively, where $n \geq m$. Assume further that the constant term in both $P(s)$ and $Q(s)$ is unity, and the closed loop system is stable. Consider the asymptotic α diagram for $G(s)$. Prove that the value of K_v for this system is given by the intersection of the −1 slope low frequency segment of the α diagram with the $\omega = 1$ line for the special case where all break frequencies in $G(s)$ are such that $\omega_i \geq 1$. Show in addition that the intersection of the extended low frequency −1 slope segment with the $\omega = 1$ line gives K_v even when the low frequency −1 slope segment does not continue all of the way up to $\omega = 1$.

3. Consider the system shown in which

$$G_e(s) = \frac{100}{s(1 + s/4)(1 + s/50)}$$

A phase margin of 45° is to be provided by the final design.

a. Determine the minimum attenuation which can be used to satisfy the phase margin spec.

b. Will the lag equalizer

$$G_e(s) = \frac{1 + s}{1 + s/0.02}$$

satisfy the phase margin spec? What is the *lag span* of this equalizer? Do you anticipate any problems in realizing this $G_e(s)$?

c. Will the lead equalizer

$$G_e(s) = \frac{1 + s/10}{1 + s/100}$$

satisfy the phase margin spec? What do you conclude about the utility (futility?) of lead equalization to increase the closed loop bandwidth when $G(s)$ has −3 slope or less near the desired crossover frequency?

d. Use an analog or digital computer simulation to compare the unit step responses of the three compensated systems considered here.

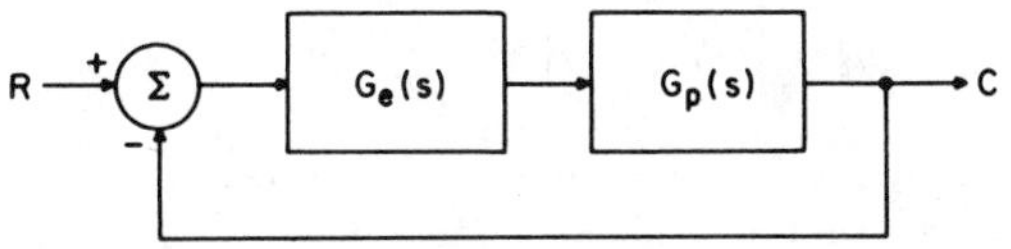

4. Consider a system with $|\mathbf{A}(j\omega)|$ as indicated by the Bode diagram shown.
a. What is ω_c?
b. Assuming that $A(s)$ is *minimum phase*, sketch the Nyquist diagram. What is θ_m?
c. If $A(s)$ is *non-minimum phase* and given by

$$A(s) = \frac{80(1 - s/4)}{s(1 + s/2)}$$

sketch the Nyquist diagram.
d. Discuss the general problem of how a non-minimum phase zero influences the relationships between the Bode and Nyquist diagrams.

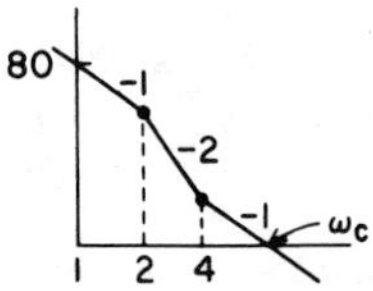

5. A particular unity feedback system has plant transfer function

$$G_P(s) = \frac{100}{s^2(1 + s/10)}$$

a. Will either or both of the equalizers shown stabilize this system? Sketch Bode diagrams of $G_P(s)$ and the $A(s)$ obtained using each of the suggested equalizers.
b. If either (i) or (ii) yields a stable system, what is the resulting phase margin?

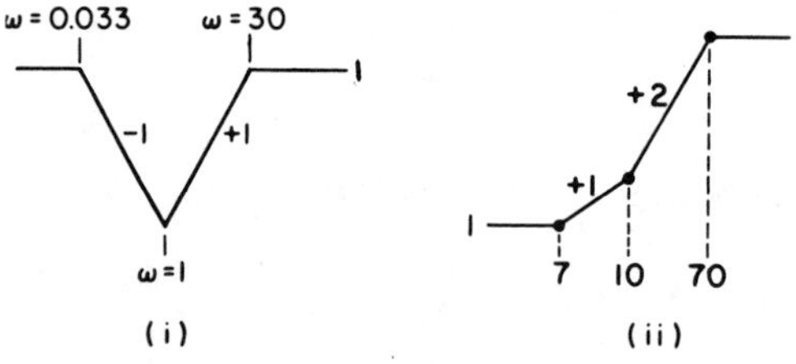

6. Consider the system and the two equalizers suggested in Prob. 9.3.
a. Compare the steady-state errors for a unit magnitude sinusoidal input at $\omega = 0.1$ for the lag and lead compensated systems.
b. Repeat for $\omega = 1$.
c. For standard unity feedback systems, it is claimed that, for all frequencies were $|\mathbf{A}(j\omega)| \gg 1$,

$$\left|\frac{\mathbf{E}(j\omega)}{\mathbf{R}(j\omega)}\right| = \left|\frac{1}{1 + \mathbf{A}(j\omega)}\right| \simeq \frac{1}{|\mathbf{A}(j\omega)|}$$

What is the maximum error in this approximation for $|\mathbf{A}(j\omega)| \geq 10$, and under what conditions does this maximum error occur?
d. It is also observed that in unity feedback systems when $|\mathbf{A}(j\omega)| \gg 1$,

$$\left|\frac{C(j\omega)}{R(j\omega)}\right| = \left|\frac{A(j\omega)}{1 + A(j\omega)}\right| \simeq 1$$

What is the maximum error in this approximation for $|A(j\omega)| \geq 10$, and under what conditions does this maximum error occur?

7. A particular unity feedback system has

$$G_P(s) = \frac{100}{s(1 + s/2)(1 + s/10)}$$

It is desired to realize the open-loop transfer function $A(s)$ for which the Bode diagram is shown.

a. Sketch the asymptotic diagram of the required equalizer.
b. Determine the transfer function $G_e(s)$ in time constant form.

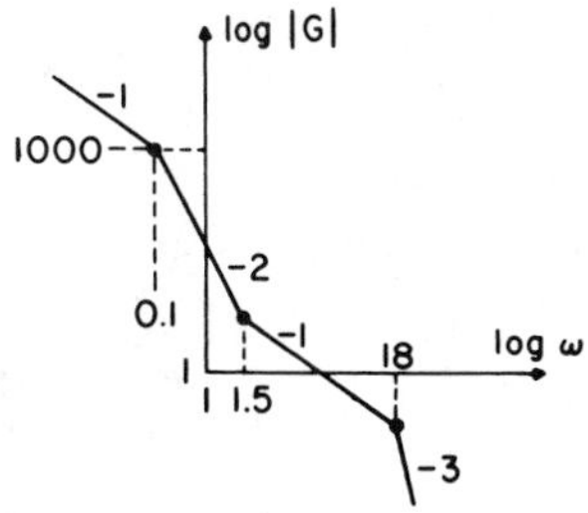

8. Consider the general Bode diagram characteristics shown in the vicinity of crossover. Assume the $-m$ slope region goes back to $\omega = 0$, the $-n$ slope region extends to $\omega = \infty$, and m and n are integers greater than 1. Assuming $A(s)$ is minimum phase, use the arc-tan approximation to find that choice of $\omega_c, \omega_1 < \omega_c < \omega_2$, as a function of ω_1, ω_2, m, and n for which the value of phase lag is maximized.

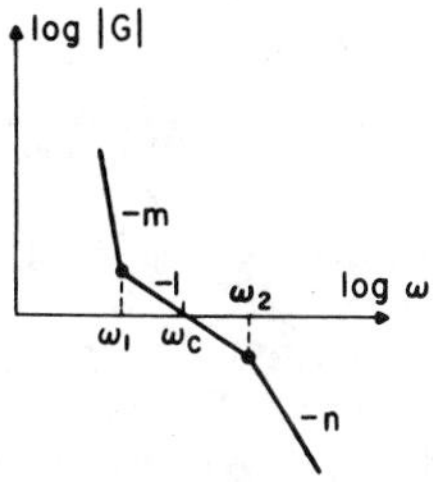

9. We wish to realize the transfer function

$$G_e(s) = \frac{1 + s}{1 + s/0.01}$$

using a lag network of the form shown in Fig. 9.11.

a. Determine appropriate values for R_1, R_2 and C.
b. Discuss conditions under which it may prove difficult to realize a desired lag transfer function.

10. Consider once again the situation introduced in Prob. 9.8. Obtain a general solution for θ_m versus $\gamma = \omega_2/\omega_1$ for the special cases:

a. $m = 2$; $n = 2$

b. $m = 2$; $n = 3$

c. $m = 2$; $n = 4$

11. In a particular unity feedback system

$$G_p(s) = \frac{20}{s(1 + s/5)(1 + s/60)}$$

The following specs must be satisfied:

a. $K_v = 100$

b. $\theta_m \geq 40°$

c. The low frequency error performance is to be degraded no more than necessary by the equalizer chosen.

d. A lag equalizer is to be used.

Determine the required $G_e(s)$. It is not necessary to show how the transfer function $G_e(s)$ is to be realized in network form. In general we will assume, unless indicated to the contrary, that a transfer function description is adequate. Sketch the gain root locus diagram for your final design.

12. Consider a standard unity feedback configuration with

$$G_p(s) = \frac{100}{s(1 + s/50)(1 + s/250)}$$

The following specs must be satisfied:

a. $K_v = 100$

b. A *lag* equalizer is to be used.

c. The steady-state error in response to sinusoidal input signals is not to exceed one part in 70 for all frequencies $\omega \leq 1$.

d. There is to be no region of slope less than −2 for $\omega < \omega_c$.

e. The phase margin is to be maximum consistent with the other specs. What maximum value of phase margin is possible? Specify the equalizer required.

13. A particular unity feedback system has

$$G_p(s) = \frac{100}{s(1 + s/2)(1 + s/100)}$$

The following specs are to be satisfied:

a. $K_v = 100$

b. A *lead* equalizer is to be used.

c. $\theta_m = 40°$ is to be provided.

The lead equalizer gain is to be no larger than required to satisfy all of the above specs. (i.e., the lead span is to be minimized.) Specify the equalizer required.

14. It is desired to design a compensation network for a unity feedback controller with

$$G_p(s) = \frac{10}{s(1 + s/5)(1 + s/20)}$$

The following specs are to be satisfied:

a. $K_v = 100$

b. $\theta_m \geq 45°$

c. Steady-state errors for all sinusoidal inputs with $\omega \leq 0.2$ are not to exceed 1 part in 350.
d. Noise components introduced with the input signal at $\omega = 100$ are to be attenuated at the output by at least a factor of 100.
e. No special noise filter is to be used.
f. Specify the required compensation network.

15. In the steel rolling mill problem first introduced in Chap. 8, assume the configuration is as shown, with

$$G_p(s) = \frac{100}{s(1 + s/5)(1 + s/100)}$$

The value of t_0 is such that the phase lag contributed by the time delay is 1° at $\omega = 1$. The following specs must be met:

$$\lim_{s \to 0} G_e(s) = 1$$

$$\theta_m = 40°$$

The low frequency error performance is to be degraded no more than necessary.
a. Find a $G_e(s)$ which will satisfy these specs.
b. Is it appropriate to try a lead network for $G_e(s)$?

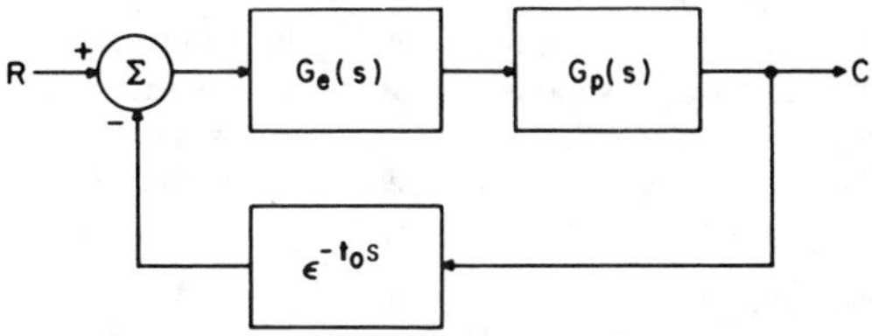

16. Consider a standard unity feedback configuration with

$$G_p(s) = \frac{100}{s(1 + s/10)}$$

It is only required that $\theta_m = 50°$.
a. Determine the attenuation ratio to satisfy this spec.
b. Determine a lag compensator to satisfy this spec without changing the dc gain while degrading low frequency error performance no more than necessary.
c. Determine a lead compensator to satisfy the spec using minimum additional gain (minimum lead span).
d. Use a composite equalizer to satisfy the spec without added gain and with minimum low frequency error performance degradation.
e. Construct root locus diagrams for each of these systems.
f. Use an analog or digital computer simulation to obtain the unit step responses for each of your designs.
g. Discuss the relative merits of these designs, taking all of the data you have gathered into consideration.

17. Consider the configuration shown.
a. Determine $E(s)/R(s)$.
b. How can $F(s)$ be used to improve the low frequency steady-state error performance of this system?

c. Discuss the practical problems encountered in developing this kind of design, including realization of $F(s)$, the effect upon stability, the power levels involved, the cost in terms of redundant circuit components, and performance sensitivity to parameter changes in $F(s)$ or $G_2(s)$.

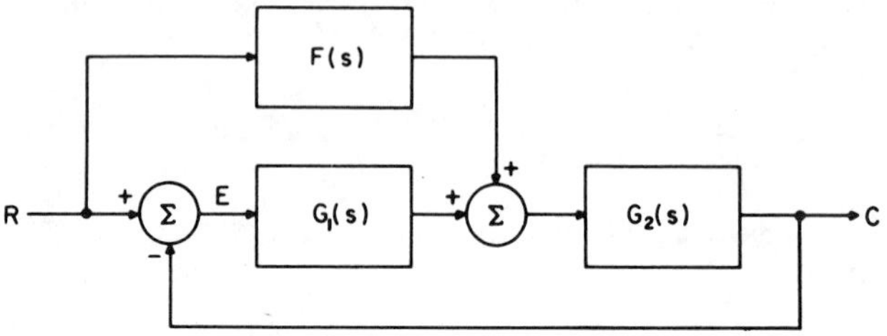

18. Consider a system with

$$G_p(s) = \frac{100}{s(1 + s/5)(1 + s/20)}$$

a. Design a composite equalizer with no net lead to provide $K_v = 100$, $\theta_m = 40°$, and optimum low frequency error performance.
b. Develop an analog or digital simulation of your system.
c. Carry out a parametric study to determine the effects of changes in plant gain, plant time constants, and equalizer time constants upon the unit step response of your system.
d. Discuss your results. Reduce your data to an orderly display.

CHAPTER 10

1. Consider the design of a cascade compensated unity feedback system starting from a plant with transfer function

$$G_p(s) = \frac{10}{s(1 + s/2)(1 + s/10)}$$

The following specs are to be satisfied:
a. $K_v = 100$
b. $M_p \leq 1.4$
c. The equalizer is to contain no gain beyond that necessary to satisfy the K_v spec.
d. ω_p is to be made as large as possible.
e. Low frequency error performance is to be degraded no more than necessary.
Determine the transfer function of a series equalizer to satisfy these specs.
Sketch the magnitude versus ω characteristics [the $M(j\omega)$ plot] for the resulting system.
What is the closed loop BW of your system?
How closely does this closed loop BW compare to the frequency at which $A(j\omega) = \sqrt{2}/2$?

2. The bandwidth (BW) of a standard baseband system is defined as the frequency range from 0 to the value where $|C(j\omega)/R(j\omega)| = M = \sqrt{2}/2$, assuming the zero frequency value of M is unity. Assume

$$\frac{C(s)}{R(s)} = \frac{1}{1 + (2\zeta/\omega_n)s + s^2/\omega_n^2}$$

Let

$$U_b = \frac{BW}{\omega_n}$$

to normalize the BW relative to ω_n. Show that

$$U_B = \frac{BW}{\omega_n} = \sqrt{1 - 2\zeta^2 + \sqrt{2 + 4\zeta^4 - 4\zeta^2}}$$

Plot several points on the curve of U_B versus ζ for $0 \leq \zeta \leq 1$, letting ζ be the abscissa of your plot.

3. A certain closed-loop unity feedback system can be adequately described by the transfer functions

$$\frac{C(s)}{R(s)} = \frac{\omega_n^2}{s^2 + 2\zeta\omega_n s + \omega_n^2}$$

a. Choose values of ζ and ω_n to provide a BW of 100 rad/sec. and an overshoot of 10 percent. (See curve in Chap. 3.)
b. What is the corresponding K_v?
c. What is the approximate unit step response rise time?
Note: From the solution of Prob. 3.34, $T_r \omega_n = 2.045$ when $\zeta = 0.6$.

4. Derive the conditions which must be satisfied by F to make the dynamic error in this system zero. Your result will be analogous to that presented in Eq. (10.8).

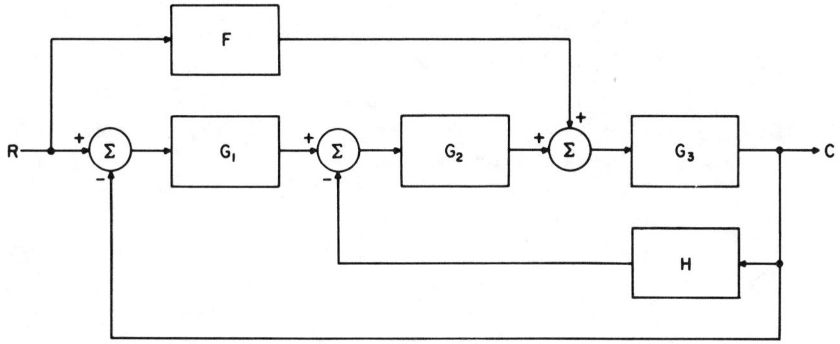

5. Consider a system with plant transfer function

$$G_p(s) = \frac{100}{s(1 + s/5)}$$

The following specs are to be satisfied;
a. The series equalizer is to have unit gain at dc and for all $\omega \geq 50$.
b. $\theta_m = 50°$.
c. The steady-state error in response to sinusoidal input signals for $\omega < 2$ is not to exceed 1 percent.
d. Any feedforward signal is to be coupled to the plant input.
e. What is the steady-state error of your system for a ramp input?

6. Consider the system shown. Find K_v for this system.
a. with s_1 and s_2 open.
b. with s_1 closed and s_2 open.
c. with s_1 and s_2 closed.

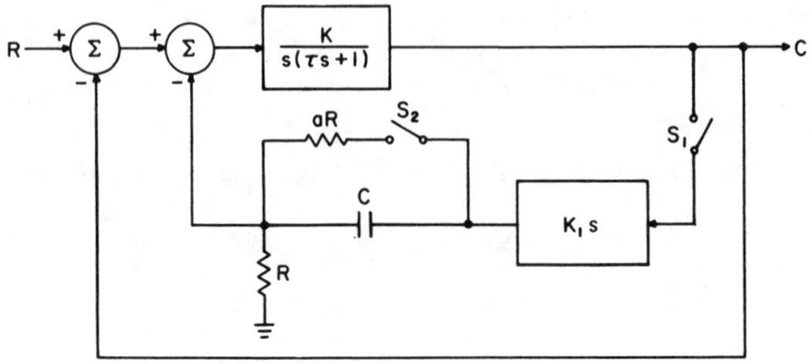

7. A standard unity feedback controller has forward transfer function

$$G(s) = \frac{K}{s(s+10)} \qquad K > 0$$

a. What is the maximum K for which real roots of the characteristic equation are obtained?
b. Assume $K = 9$. A unit step input is applied at $t = 0$. Find values for T_r, T_d, percent overshoot, T_s, and E_s.

8. In the system shown

$$\frac{E(s)}{R(s)} = \frac{s^3 + 11s^2 + 10s}{s^3 + 11s^2 + 10s + 3}$$

a. Find $C(s)$ for $R(s) = 1/s$.
b. Determine values for percent overshoot, T_r, T_d, T_s, and E_s. You will probably want to use a computer here to avoid extensive hand analysis.

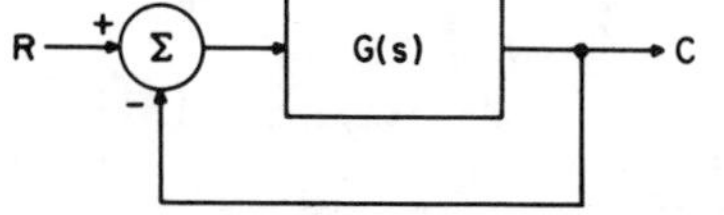

9. A given plant is characterized by

$$G_p(s) = \frac{12}{s(1 + s/4)(1 + s/60)}$$

It is desired to establish a composite cascade equalizer $G_e(s)$ to provide $K_v = 100$, no *net* lead compensation, and maximum closed loop BW consistent with the requirement that the step response percent overshoot is not to exceed 30 percent. Find a satisfactory $G_e(s)$. Trial and error on a computer is the only reasonable solution method.

10. An electronic buffer amplifier has input to output transfer function

$$G(s) = \frac{10}{1 + s/1000}$$

It is desired to place this amplifier in the feedback configuration shown, where $H(s)$ is to be chosen in such a manner that the steady-state error, $R - C$, in response to sinusoidal inputs at $\omega = 1000$ is zero.

a. What are the resulting restrictions upon $H(s)$?
b. Can the desired $H(s)$ be realized in passive form?
c. Is it possible to find an $H(s)$ which makes the error zero at 2 frequencies? Zero over the range $[\omega_1, \omega_2]$?

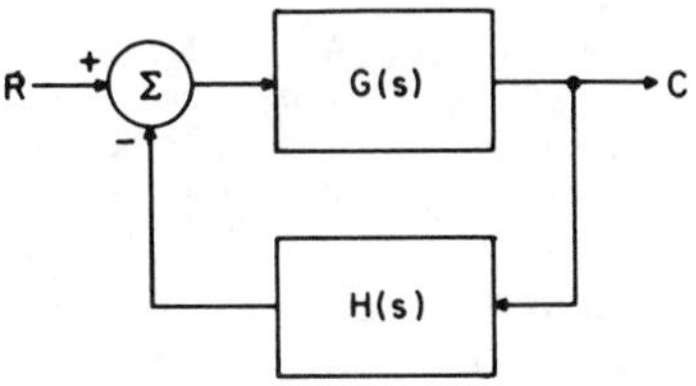

11. The system shown has

$$\frac{C}{R} = \frac{1.25}{s^2 + 5s + 1.25}$$

The gain $K_1 = 0.2$:
a. What are the required values of K_2 and a?
b. What is the damping ratio, ζ? Is this an underdamped system?
c. If K_1 is changed to 2.0, what is the new ξ?
d. Find the pole locations of $C(s)/R(s)$ for the higher gain condition. Place these poles on an s-plane sketch.
e. Estimate the shape of the step response for the high gain condition.
f. What should K_1 be to provide approximately 20 percent peak overshoot, assuming K_2 and a remain fixed?

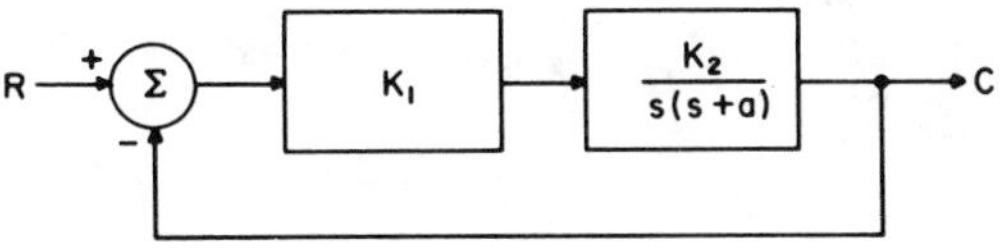

12. Consider the standard second-order underdamped transfer function form

$$\frac{C(s)}{R(s)} = \frac{\omega_0^2}{s^2 + 2\zeta\omega_0 s + \omega_0^2}$$

It is possible to solve directly for the unit step response as

$$1 - \frac{\epsilon^{-\zeta\omega_0 t}}{\sqrt{1-\zeta^2}}\left[\sqrt{1-\zeta^2}\cos\omega_0\sqrt{1-\zeta^2}\,t + \zeta\sin\omega_0\sqrt{1-\zeta^2}\,t\right]$$

Solving the transcendental expressions defining $\omega_0 T_d$ and $\omega_0 T_r$ yields the following data:

ζ	0.0	0.1	0.2	0.3	0.4	0.5	0.6	0.7	0.8	0.9	1.0
$\omega_0 T_d$	1.05	1.089	1.134	1.184	1.240	1.290	1.352	1.433	1.510	1.602	1.68
$\omega_0 T_r$	1.152	1.255	1.370	1.522	1.659	1.860	2.045	2.280	2.55	2.875	3.195

a. Sketch a curve of $\omega_0 T_d$ versus ζ.
b. Sketch a curve of $\omega_0 T_r$ versus ζ.
c. Using the curve from part *b* and the BW/ω_n versus ζ curve developed in Prob. 10.2, plot a curve of $T_r\omega_0$ versus BW/ω_0 for ζ in the range [0, 1]. ζ will be a parameter along your curve. For uniformity of results, plot BW/ω_n on a logarithmic scale on the abscissa using semi-log paper. Show that $T_r BW = 2.2$ is a good approximation to this curve. Thus knowledge of the BW of a *second-order* system is approximately equivalent to knowing the step response rise time. It is often possible to get a close estimate even for higher order systems by using an approximate second-order model.

13. *a.* Write the equations describing the performance of the arrangement shown including the effects of an arbitrary load torque component on the output shaft.
b. Assuming $T_L = 0$, find the transfer function from V_f to θ.
c. Assuming $V_f = 0$, find the transfer function from T_L to θ.
d. The output shaft position is measured by a potentiometer, the corresponding signal compared to the desired input signal in a differential amplifier, and the amplifier output is connected to the field circuit to provide negative feedback. The parameters are chosen such that unity feedback results. The amplifier has zero output impedance and its gain is K_1. There are no significant frequency sensitive characteristics in the feedback or amplifier circuits. Draw a block diagram of this closed loop system Determine the steady-state error resulting from a step input of load torque for this feedback configuration. (Assume the system is stable, of course.)
e. In a second system configuration, a tachometer with gain K_2 is placed on the output shaft, and the tachometer output compared to the reference input in the differential amplifier, with the amplifier output once again connected to the field circuit. The tachometer gain constant and the differential amplifier gain are adjusted to provide the same gain level as before, so we may assume that the tachometer gain is lv/(rad/sec) and the amplifier gain is still K_1, v/v. What is the steady-state error resulting from a step input of load torque for this case? Compare with your answer in part *d*. Note that when tachometer feedback is used, the forward path integration effect is essentially cancelled, and the error performance is significantly altered. Tachometer feedback is often used in designing control systems, but it is important to appreciate the limitations of its use.

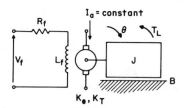

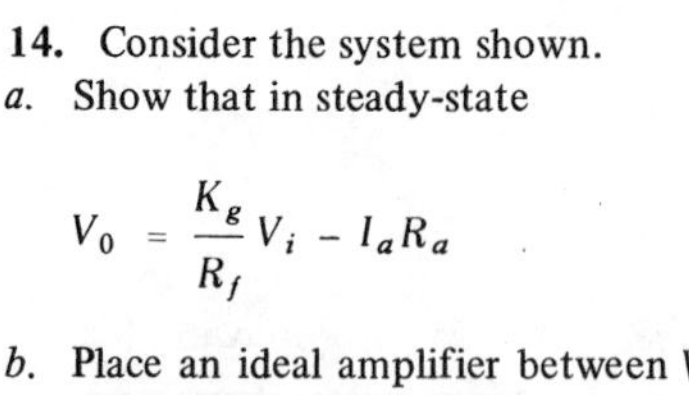

14. Consider the system shown.
a. Show that in steady-state

$$V_0 = \frac{K_g}{R_f} V_i - I_a R_a$$

b. Place an ideal amplifier between V_i and the field circuit with gain $K_1 v/v$. Feed into the amplifier the signal $V_i - V_0$ (i.e., use negative feedback). Develop the equations relating V_0 to V_i and I_a in steady-state.
c. Find the steady-state resistance seen looking back into the generator with feedback present.
d. What price must be paid for this output impedance reduction?

e. Draw the block diagram for this *closed-loop* system with the *two inputs* V_i and I_a. Consider the complete system description not just the steady-state conditions.
f. What adjustment would be necessary to return the steady state gain V_0/V_i to the value it had before feedback? Assume the loop gain is much greater than unity.

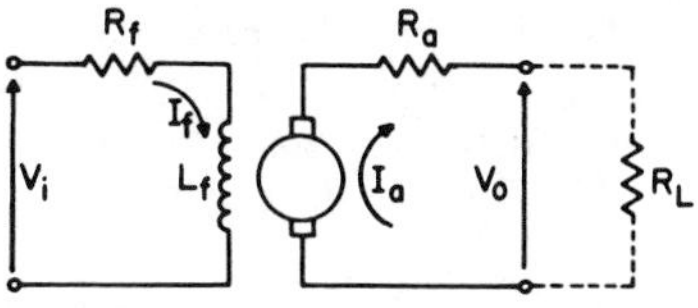

15. Discuss the various factors in closed loop controller design which influence the system's ability to follow the square wave and triangular wave inputs shown. If you were given a set of specs which included one or more restrictions upon your design's tracking capability relative to either or both of these waveforms, how might you proceed in the development of an acceptable solution? What general effects do K, BW, θ_m, M_p, G_m and number of plant integrations have upon tracking capability?

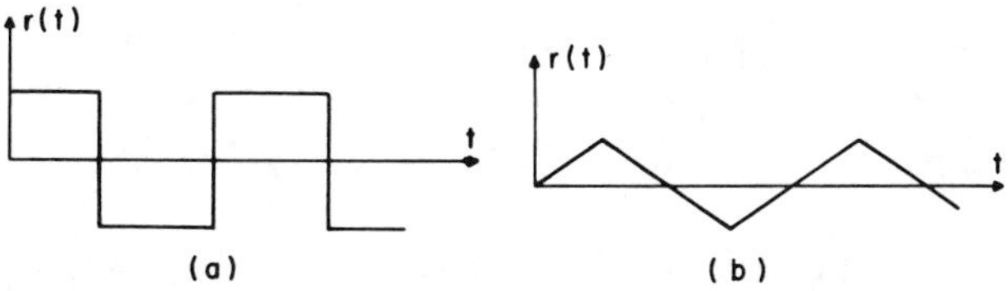

CHAPTER 11

1. Consider the closed loop configuration shown. Assume the signals at the points between all individual transfer functions are available, and the k_i may be set as desired.
a. Sketch the Bode diagram of the forward transfer function A_{Mu} assuming $k_1k_2k_3k_4 = 100$. (i.e., $\prod_i k_i = 100$.)
b. It is desired to realize

$$A_{Mc}(s) = \frac{100(1 + s/5)}{s(1 + s/0.8)(1 + s/62.5)(1 + s/100)}$$

Sketch the A_{Mc} Bode diagram on that for A_{Mu}.
c. Construct the A_m Bode diagram.
d. Assuming $H(s)$ is placed around the block with transfer function $k_3/(1 + s/10)$, what is $H(s)$?
e. If k_3 in the actual system is such that attenuation must be introduced somewhere in the minor loop to realize the desired A_m, should it be placed in the forward path, or in $H(s)$? Why?

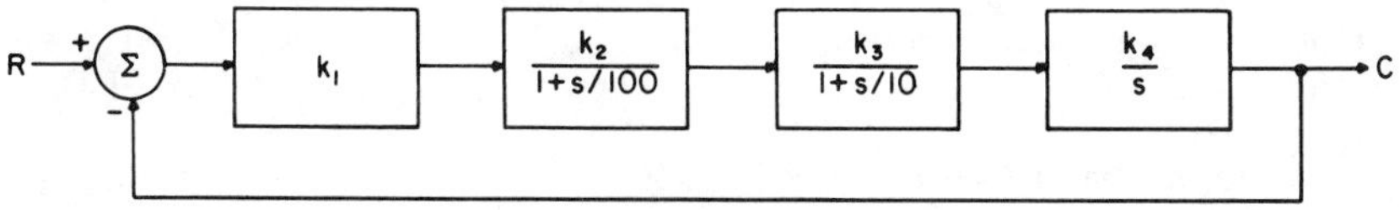

2. Consider the system shown

$$A_m(s) = \frac{3s}{(1 + s/5)^2(1 + s/25)}$$

a. What is $H(s)$?
b. Sketch α diagrams for A_{Mu}, A_m, A_{Mc}.
c. Find θ_m. Apply the usual approximations based upon α diagram shape. The actual θ_m is, in general, larger than this estimate.
d. What is K_v for this system?

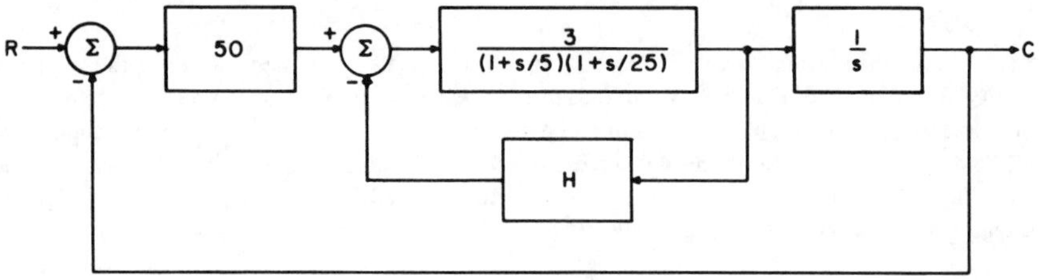

3. A particular plant has transfer function

$$G_p(s) = G_1G_2G_3G_4 = \frac{5}{(1 + s/50)} \cdot \frac{2}{(1 + s/20)} \cdot \frac{2}{(1 + s/5)} \cdot \frac{1}{s}$$

Assume that the signals at all points between these transfer functions are available. The following specs must be satisfied:
α $K_v = 100$
β $\theta_m = 45°$
γ Low frequency error performance is to be degraded no more than necessary.
Develop an equalizer for this system by carrying out the following steps:
a. Choose the additional gain required to satisfy the K_v spec. Disregard any time constant in the amplifier added to provide this gain.
b. Draw Bode diagrams for A_{Mu}, A_{Mc}, and the A_m required.
c. Choose the forward path elements around which $H(s)$ is to be placed.
d. What $H(s)$ is needed?
e. Assume the forward path gain in the minor loop decreases by a factor of 4 *after* your design is complete. Draw a new set of Bode diagrams, and compare the results to the series equalizer case.

4. In a given system the transfer function of the plant and the amplifiers required to drive it is

$$G(s) = \frac{100}{s(1 + s/12)(1 + s/60)(1 + s/200)}$$

Assume the signals are available at any point between the individual terms. Assume the time constants appear in *ascending* order in the forward path.
The following specs must be satisfied:
1. $K_v = 100$
2. $\theta_m = 40°$

Design an appropriate minor loop equalizer.

Discuss the advantages of your minor loop system over a cascade equalized system relative to:
a. Noise introduced in the minor loop forward path.
b. Parameter drift of an element in the minor loop forward path.
c. Load disturbances when the minor loop feedback is *not* derived from the system output.

5. Consider the system shown and the specs:
a. $K_v = 1000$.
b. Steady state errors for $\omega \leq 1$ are not to exceed 0.2 percent.
c. $\theta_m \simeq 40°$.
Specify $H(s)$, k_1 and k_2 to meet the specs.

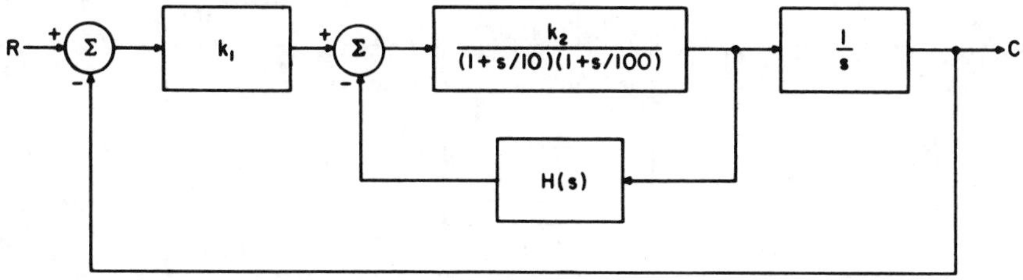

6. Consider the system shown.
a. Construct Bode diagrams for A_{Mu}, A_m and A_{Mc}.
b. A network with transfer function $G_e(s) = (1 + s/100)/(1 + s/500)$ is to be placed in the minor loop to provide additional minor loop phase margin. Should this network be placed in the forward path or the feedback path, assuming it is equally easy to place it in either location? Justify your answer with a complete set of Bode diagrams clearly depicting both cases.
c. A noise component of 0.1 volt magnitude at $\omega = 10$ is introduced at the point shown. What is the magnitude of the resulting output noise effect? Compare this to the equivalent series equalized case.

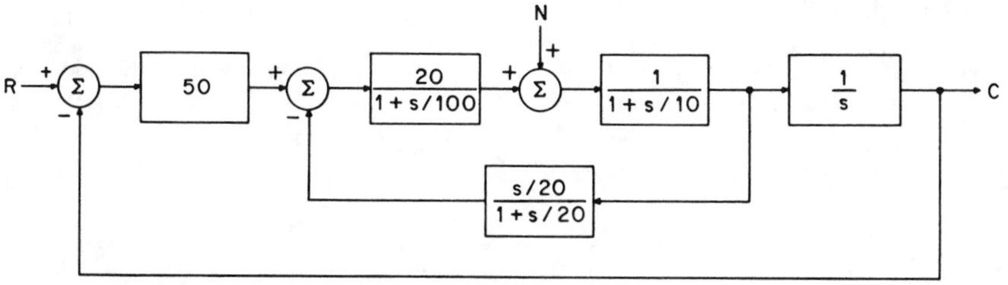

7. The transfer function of a given actuator is

$$G_p(s) = \frac{K}{s(1 + s/2)(1 + s/40)(1 + s/200)}$$

All signals are assumed available for feedback. The following specs are to be satisfied:
a. $K_v = 100$.
b. $\theta_m = 50°$.
c. The minor loop phase margin is to be at least 30°.
The value of K varies, depending upon the system application, over the range

$$2 \leq K \leq 10$$

Design a minor loop equalized system which will satisfy all specs for values of K anywhere within this range. Assume the change in K occurs in the minor loop forward path. Determine a cascade equalizer to do the same job, and compare results.

8. The system shown has a nonlinear element with approximate transfer function $(10/a)/(1 + as)$, where a varies from 0.01 to 0.05 during the course of normal operation.

a. Draw Bode diagrams of A_m, A_{Mu}, and A_{Mc} showing the range of variation caused by the changes in a.

b. Why is minor loop compensation preferable in a case like this?

c. Discuss the effect of finite minor loop gain at $\omega = 0$.

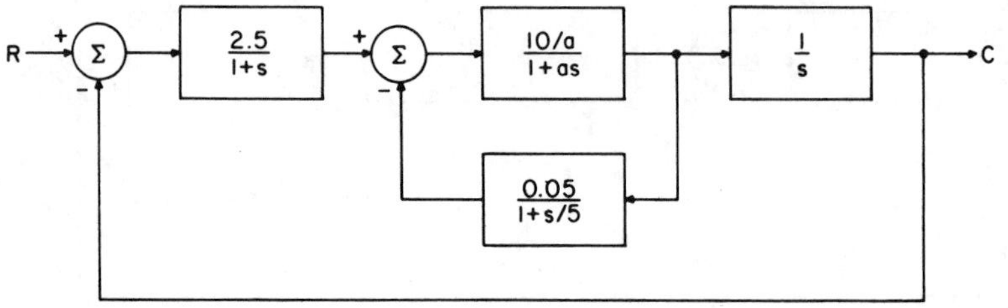

9. Discuss the possibility of using feedforward compensation techniques to improve the low frequency error characteristics of minor loop compensated systems. In particular, is the approach feasible? Does it make any difference whether the feedforward element couples into the minor loop forward path, or outside of it? What about sensitivity to parameter changes?

CHAPTER 12

1. Consider a standard suppressed carrier amplitude modulated signal of the form

$$f(t) = Ag(t) \sin \omega_0 t$$

where $\omega_0 = 2500$ rad/sec. $g(t)$ contains frequency components out to about 250 rad/sec. Demodulation and baseband equalization of this signal are to be accomplished using the arrangement shown. Specify the transfer function of a smoothing filter which you consider appropriate for this task. Justify your choice.

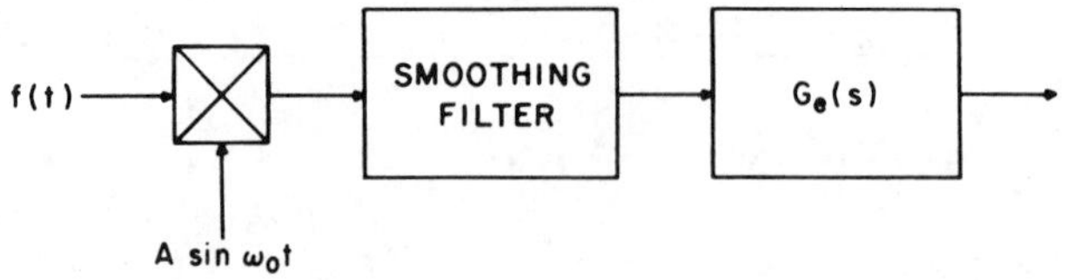

2. In a standard synchro error sensor, the envelope output is given by

$$e(t) = E_{0\,\max} \sin[\theta_e(t)] \simeq E_{0\,\max} \theta_e(t)$$

where $\theta_e(t)$ denotes the instantaneous phase error between input and output. At what value of $\theta_e(t)$ does the approximation error reach 10 percent? Plot the approximation error versus θ for $0 \leq \theta \leq \pi/4$.

3. *a.* What is meant by a unit step input for an ac servomotor?
b. What is meant by a unit ramp input?
c. Sketch unit step and unit ramp signals for the special case $\omega_0 = 2500$.

4. Convert the baseband filter

$$G(s) = \frac{1 + s}{1 + s/10}$$

to an equivalent bandpass filter at a carrier frequency of 400 rad/sec. Do you forsee any difficulty in realizing this filter?

5. A bandpass filter at $\omega_0 = 2500$ is given by

$$G(s) = \frac{2500s}{s^2 + 250s + 6.25 \times 10^6}$$

Find an equivalent baseband representation for this filter.

6. In designing a particular system the following baseband equivalent equalizer is to be realized at $\omega_0 = 2500$.

$$G_e(s_m) = \frac{(1 + s_m/10)(1 + s_m/20)}{(1 + s_m)(1 + s_m/200)}$$

a. Develop the required bandpass filter transfer function $G(s)$.
b. Plot the pole-zero pattern of this filter near $\omega = 2500$.
c. Discuss the problems of realization.

7. An ac system consists of a plant $G_p(s)$ in the forward path of the usual unity feedback configuration. An open loop frequency response test provides the following data:

ω	ω_0	$\omega_0 + 1$	$\omega_0 + 10$	$\omega_0 + 100$	$\omega_0 + 1000$
$\lvert G_p \rvert$	∞	1000	70	0.7	0.001

Assume $G_p(s_m)$ is minimum phase and has no complex conjugate poles or zeros, ω_0 is large, and $|G_p|$ is symmetrical about $\omega = \omega_0$. Determine $G_p(s_m)$.

8. Consider the plant transfer function

$$G_p(s) = \frac{20}{s(1 + s/5)(1 + s/40)}$$

It is desired to develop a cascade equalized unity feedback system to satisfy the following specs:
$K_v = 100$
$\theta_m = 40°$
Low frequency error performance is to be as good as possible. No *net* lead compensation is to be used. Thus once the gain is set to satisfy the K_v spec, $|G_e| \leq 1$ for all ω, where G_e denotes the remainder of the equalizer.
a. Determine $G_e(s)$ to satisfy the specs.
b. Convert your equalizer to an equivalent bandpass transfer function at $\omega_0 = 1000$.
c. Sketch the equalizer pole-zero pattern near $\omega_0 = 1000$.

9. Realize each of the following baseband filters in RLC network form as equivalent bandpass networks assuming $\omega_0 = 377$ rad/sec.

a. $G(s) = \dfrac{1}{1 + s/5}$ *b.* $\dfrac{1 + s/20}{1 + s/2}$ *c.* $G(s) = \dfrac{1 + s/5}{1 + s/50}$

10. Consider the baseband equalizer

$$G(s) = \frac{1 + s/5}{1 + s/50}$$

and the corresponding bandpass RLC network form developed in the previous problem for $\omega_0 = 377$. Carry out each of the following analyses:

a. Plot the actual magnitude and phase curves of $G(j\omega)$ for $-200 \leq \omega \leq 200$.

b. Plot the actual magnitude and phase curves of the equivalent bandpass network over a comparable range. i.e. for $\omega_0 \pm 200$, or for $177 \leq \omega \leq 577$. Let the origin on your previous plot correspond to $\omega = \omega_0 = 377$ on this plot, and place these data on the same diagram.

c. Compare your phase and magnitude curves. What conclusions can you draw? In the absence of any approximation, the results should be identical. Obviously they are not. A detailed study of this situation and the conditions under which various accuracy criteria are satisfied has been carried out by Airato. (See Ref. 2 at the end of Chap. 12.) Some similar work still needs to be done.

11. Consider the design of an instrument servo system using a plant and amplifier with equivalent baseband transfer function

$$G_p(s_m) = \frac{100}{s_m(1 + s_m/10)(1 + s_m/500)}$$

The specs are:

1. $K_v = 100$
2. $\theta_m \geq 40°$

Develop a satisfactory design, specifying all networks used in both transfer function and RLC component topology and value form, in each case assuming $\omega_0 = 2500$.

a. Find a lag equalizer at baseband, and convert to an equivalent RLC bandpass network.

b. Find a composite baseband equalizer, and determine the equivalent RLC bandpass network.

c. Find a baseband lead equalizer, and transform it to equivalent bandpass form.

d. Try the RC network given as entry 1 in Table 12.1.

e. Try the bridged T network given as entry 3 in Table 12.1.

f. Try the twin T network given as entry 4 in Table 12.1.

g. Introduce tachometer feedback around part or all of the plant. Assume the gains may be assigned as desired.

h. Compare your results. Which solutions seem most appropriate? Which of your solutions seem completely useless? What problems are anticipated in realizing the various solutions?

CHAPTER 13

1. Reduce each of the following differential equations to state equation form. Your answers are not to involve $\dot{m}(t)$ or any higher derivatives of $m(t)$.

a. $\dddot{y} + 3\ddot{y} + 2\dot{y} + 4y = 2m$

b. $2\dddot{y} + 4\ddot{y} + 3\dot{y} + 2y = \ddot{m} - 2\dot{m} + m$

The following procedure is suggested:

a. Take the Laplace transform of the equation assuming zero initial conditions.

b. Divide both sides of the equation by the highest power of s (s^n, where n is the order of the original DE).

c. Solve for $y(s)$.

d. Establish a simulation diagram to provide this $y(s)$ with $M(s)$ as the input.

e. Define the integrator outputs as state variables.

f. Write down the state equations by inspection.

2. Develop matrix sets of first order state equations for the systems described by the following transfer functions. Obtain your answers in terms of $m(t)$ alone.

a. $$\frac{C(s)}{R(s)} = \frac{s+1}{s^2+2s+3}$$ *b.* $$\frac{C(s)}{R(s)} = \frac{s^2-3s+1}{s^4+2s^3+3s^2+3s+2}$$

3. *a.* Develop a state equation description in the usual way for the system with transfer function

$$\frac{C(s)}{R(s)} = \frac{s+6}{s^3+6s^2+11s+6}$$

b. Expand $C(s)/R(s)$ into partial fractions, develop a simulation diagram for each partial fraction component, define the integrator outputs as elements of state, and write the state equations.
c. How does $c(t)$ relate to $\mathbf{x}(t)$ for this representation? Discuss the special form of this state equation description.
d. Can you always develop the special form of state equations encountered in part *b*? Suppose you have

$$\frac{C(s)}{R(s)} = \frac{1}{s^2+2s+2}$$

Discuss this situation. If we expand our interpretation of simulation diagrams to allow complex gains, does this help? What about the case

$$\frac{C(s)}{R(s)} = \frac{1}{s^2+2s+1}$$

Discuss this one also. We will present a method for handling these structures later.

4. Develop the partial fraction form of state equation description for the system described by

$$\dddot{y} + 9\ddot{y} + 26\dot{y} + 24y = r(t)$$

5. A given system has transfer function

$$\frac{C(s)}{R(s)} = \frac{3}{s^2+4s+3}$$

a. Develop the state equations in the usual way.
b. Find the transition matrix corresponding to the state equations developed in part *a*.
c. Develop the normal form state equations by using the partial fraction expansion method.
d. Find the transition matrix for the diagonal form developed in part *c*.
e. Discuss the relationship between the normal form transition matrix and that for all other equivalent forms of the state equations.
f. Write down a general expression for $\mathbf{x}(t)$ in terms of the initial conditions, the transition matrix, and the control input. Assuming $\mathbf{x}'(0) = [2, 3]$ and $r(t)$ is a unit step starting at $t = 0$, evaluate your general expression to find $\mathbf{x}(2)$.

6. A given system is described by

$$\dot{\mathbf{x}} = \mathbf{A}\mathbf{x} + \mathbf{B}\mathbf{m}$$

where

$$\mathbf{A} = \begin{bmatrix} -1 & 1 & 0 \\ 0 & 0 & 1 \\ 0 & 0 & 0 \end{bmatrix} \qquad \mathbf{B} = \begin{bmatrix} 0 \\ 1 \\ 2 \end{bmatrix}$$

a. Find the transition matrix $\mathbf{\Phi}(t, t_0)$ for this system.
b. If $\mathbf{x}'(2) = [1, 2, 3]$, find $\mathbf{x}(5)$ and $\mathbf{x}(0)$ assuming $m(t) = 0$.
c. Find $\mathbf{\Phi}(t_0, t) = \mathbf{\Phi}^{-1}(t, t_0)$.
d. Evaluate $\mathbf{\Phi}(2,0)$, $\mathbf{\Phi}(5,2)$, $\mathbf{\Phi}(5,0)$ and $\mathbf{\Phi}(5,2)\mathbf{\Phi}(2,0)$. Of course the last two solutions should be identical.

7. Consider the linear time varying system described by

$$\dot{\mathbf{x}} = \mathbf{A}\mathbf{x} + \mathbf{B}m \qquad \mathbf{A} = \begin{bmatrix} 2 & -\epsilon^{t} \\ \epsilon^{-t} & 1 \end{bmatrix}$$

Show that

$$\mathbf{\Phi}(t, t_0) = \begin{bmatrix} \epsilon^{2(t - t_0)} & -\epsilon^{(2t - t_0)} \sin(t - t_0) \\ \epsilon^{(t - 2t_0)} \sin(t - t_0) & \epsilon^{(t - t_0)} \cos(t - t_0) \end{bmatrix}$$

is the transition matrix for this system.

8. Is it possible for

$$\mathbf{\Psi}(t, t_0) = \begin{bmatrix} \epsilon^{(t - t_0)} & (t - t_0) \\ (t - t_0)^2 & 2\epsilon^{2(t - t_0)} \end{bmatrix}$$

to be a transition matrix? Justify your answer.

9. Prove that $\mathbf{\theta}'(t, t_0)\mathbf{\Phi}(t, t_0) = \mathbf{I}$.

10. Consider the system described by

$$\dot{\mathbf{x}} = \mathbf{A}\mathbf{x} + \mathbf{B}m \qquad \text{with} \qquad \mathbf{A} = \begin{bmatrix} -1 & -3 \\ 1 & -5 \end{bmatrix}$$

a. Find $\mathbf{\Phi}(t, t_0)$.
b. Define the adjoint equations.
c. Find $\mathbf{\theta}(t, t_0)$.
d. Show that, at least for this case, $\mathbf{\Phi}(t_0, t) = \mathbf{\theta}'(t, t_0)$.

11. Prove that $\mathbf{\Phi}(t, t_0) = \epsilon^{\mathbf{A}(t - t_0)}$ is a solution of the linear time invariant homogeneous equation

$$\dot{\mathbf{\Phi}}(t, t_0) = \mathbf{A}\mathbf{\Phi}(t, t_0) \qquad \mathbf{\Phi}(t_0, t_0) = \mathbf{I}$$

12. Show that $\mathbf{A}\epsilon^{\mathbf{A}} = \epsilon^{\mathbf{A}}\mathbf{A}$.

13. Show that $\mathbf{x} = \epsilon^{\mathbf{A}t}\mathbf{C}\epsilon^{\mathbf{B}t}$ is a solution of

$$\dot{\mathbf{x}} = \mathbf{A}\mathbf{x} + \mathbf{x}\mathbf{B} \qquad \mathbf{x}^0 = \mathbf{C}$$

14. Consider the system described by

$$\dot{\mathbf{x}} = \mathbf{A}\mathbf{x} \qquad \mathbf{A} = \begin{bmatrix} 1 & 4 & 0 \\ 0 & -1 & 0 \\ 0 & 3 & 2 \end{bmatrix}$$

Use the matrix exponential and Sylvester's expansion theorem to obtain the transition matrix for this system.

15. Use the Cayley-Hamilton theorem to find $\mathbf{A}^{-1}$ where

$$\mathbf{A} = \begin{bmatrix} 1 & 0 & 2 \\ 3 & 1 & 4 \\ 0 & 2 & 0 \end{bmatrix}$$

16. Reduce the polynomial

$$4\mathbf{A}^5 + 2\mathbf{A}^4 - 3\mathbf{A}^3 + \mathbf{A}^2 - 5\mathbf{A} + 3\mathbf{I}$$

to a second order polynomial in $\mathbf{A}$, where $\mathbf{A}$ is the matrix introduced in the previous problem.

a. Use the Cayley-Hamilton theorem directly.

b. Use the division and remainder method.

17. Use the Cayley-Hamilton remainder method to find the transition function $\mathbf{\Phi}(t,0) = \epsilon^{\mathbf{A}t}$, where

a. $\mathbf{A} = \begin{bmatrix} -1 & 2 & -4 \\ 0 & -2 & 0 \\ 0 & 2 & -3 \end{bmatrix}$ *b.* $\mathbf{A} = \begin{bmatrix} -1 & -3 & 0 \\ 0 & -1 & 0 \\ 2 & 3 & -2 \end{bmatrix}$

18. *a.* Are the vectors $\mathbf{x}' = [1, 2, 3]$, $\mathbf{y}' = [4, 5, 6]$ and $\mathbf{z}' = [5, 4, 3]$ linearly independent?

b. Are the vectors $\mathbf{x}_1' = [1, 3, 0, -2]$ $\mathbf{x}_2' = [0, -2, 1, 1]$ and $\mathbf{x}_3' = [1, 2, -1, -2]$ linearly independent?

c. Find a vector $\mathbf{x}_4$ which is linearly independent from the three vectors given in part *b.*

19. Consider the vectors $\mathbf{x}_1' = [1, k, 0]$, $\mathbf{x}_2' = [1, 0, k]$ and $\mathbf{x}_3' = [k - 1, 0, 1]$. For what values of k are these vectors linearly independent?

20. Find a 3×1 vector orthogonal to both $\mathbf{x}_1' = [k, 1, 0]$ and $\mathbf{x}_2' = [0, 1, k]$.

21. Prove that any non-singular matrix can be used to transform a set of n linearly independent $n \times 1$ vectors into a set of linearly independent vectors.

22. Prove that every subset of a set of linearly independent vectors is linearly independent.

23. Prove that there are n and only n linearly independent n element vectors.

24. *a.* Find $\mathbf{v}_1$ and $\mathbf{v}_2$, the eigenvectors for

$$\mathbf{A} = \begin{bmatrix} 4 & 4 \\ 1 & 4 \end{bmatrix}$$

b. Evaluate $(\mathbf{v}_1, \mathbf{v}_2)$, their inner product.

c. Are the eigenvectors linearly independent?

d. Orthonormalize $\mathbf{v}_1$ and $\mathbf{v}_2$.

25. Find the diagonal matrix of eigenvalues Λ which is similar to

$$A = \begin{bmatrix} -6 & 1 & 0 \\ -11 & 0 & 1 \\ -6 & 0 & 0 \end{bmatrix}$$

26. *a.* Find the eigenvalues for

$$A = \begin{bmatrix} 1 & 1 & 2 \\ 1 & 2 & 3 \\ 2 & 3 & -1 \end{bmatrix}$$

b. Find a set of eigenvectors for this matrix.
c. Are the eigenvectors linearly independent? Orthogonal? Discuss.
d. Orthonormalize the eigenvectors.
e. Show that $X'AX = \Lambda$ by direct evaluation using your results.
f. Starting from your original eigenvector set (before normalizing), show by direct evaluation that $T^{-1}AT = \Lambda$, where T denotes the non-normalized eigenvector matrix.

27. Orthonormalize the vector set $x_1' = [1, 1, 0]$, $x_2' = [1, 0, 1]$ and $x_3' = [1, 1, 1]$. Is your answer unique?

28. Find the eigenvectors and use them to reduce

$$A = \begin{bmatrix} 1 & 0 & -1 \\ 1 & 2 & 1 \\ 2 & 2 & 3 \end{bmatrix}$$

to diagonal form.

29. Prove that any two eigenvectors corresponding to distinct eigenvalues of a real symmetric nonsingular matrix are orthogonal.

30. Prove that if $\lambda_1, \lambda_2, \ldots, \lambda_k$ are distinct eigenvalues of A and if $x_1, x_2, \ldots, x_k$ are the corresponding eigenvectors, these eigenvectors are linearly independent.

31. Show that if all eigenvalues of A are distinct, $T^{-1}AT = \Lambda$, where T is the eigenvector matrix, and Λ is the diagonal matrix of eigenvalues.

32. Consider the system described by

$$\frac{C(s)}{R(s)} = \frac{s^2 + 7s + 12}{s^4 + 5s^3 + 13s^2 + 19s + 10}$$

a. Develop the state equations in the usual way.
b. Find the eigenvalues for the matrix A in your equations.
c. What is the Jordan form matrix for this system?
d. Is the Jordan form convenient to use with complex eigenvalues?

33. Reduce the following equation to normal form, defining all matrices used in the process.

$$\dot{x} = Ax + Bm \qquad A = \begin{bmatrix} 0 & 1 & 0 \\ 0 & 0 & 1 \\ -8 & -14 & -7 \end{bmatrix} \qquad B = \begin{bmatrix} 0 & 1 \\ 2 & 0 \\ 0 & 0 \end{bmatrix}$$

Is this system controllable? Is it necessary to reduce to normal form to tell? Discuss.

34. *a.* Develop an example of a system which is not controllable.
b. Develop an unobservable example.

35. A given system has transfer function

$$\frac{C(s)}{R(s)} = \frac{3}{s^2 + 4s + 3}$$

a. Develop the state equations in the usual way.
b. Develop the state equations in normal form. Define Λ and β.
c. Find $\Phi(t, t_0)$ for the normal form representation.
d. Using the normal form, find the unit step response in the time domain if $z'(0) = [2, -5]$.

36. A given plant has transfer function elements as shown. It is desired to realize

$$\frac{C(s)}{R(s)} = \frac{20}{(s + 1)[(s + 2)^2 + 16]}$$

Set up the state variable feedback design and find all gains required.

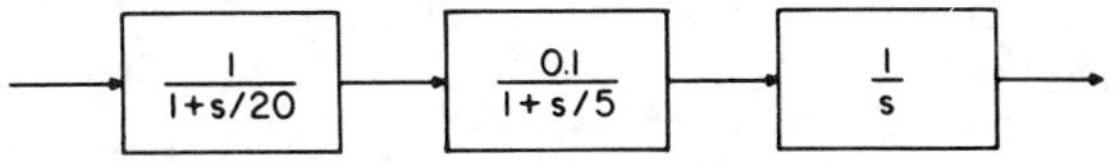

37. It is desired to realize

$$\frac{C(s)}{R(s)} = \frac{100}{(s + 4)[(s + 3)^2 + 16]}$$

in the configuration shown. Find the required K, k_1, k_2, and k_3.

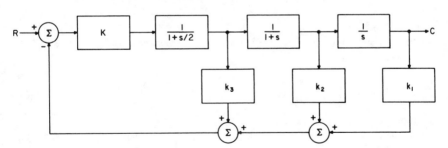

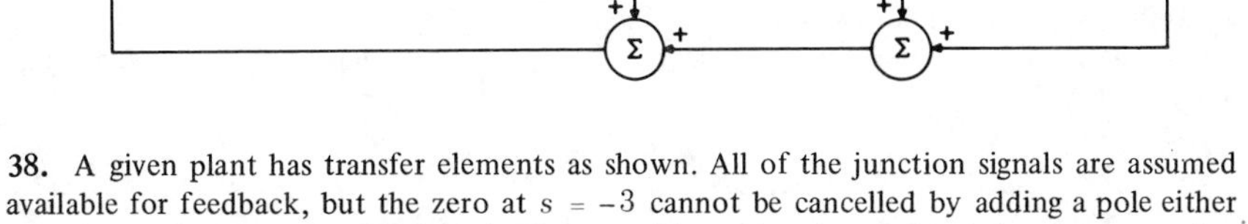

38. A given plant has transfer elements as shown. All of the junction signals are assumed available for feedback, but the zero at $s = -3$ cannot be cancelled by adding a pole either just ahead of or just after it. Cancel this zero with a pole prior to the entire plant to realize the closed loop transfer function.

$$\frac{C(s)}{R(s)} = \frac{13}{(s + 1)[(s + 2)^2 + 9]}$$

Determine all gain constants required.

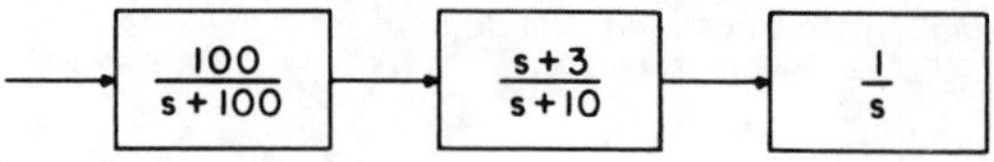

CHAPTER 14

1. *Derive* the *DF* for each of the *NL* elements listed as:

a. Entry 1 in Table 14.1.
b. Entry 3 in Table 14.1.
c. Entry 4 in Table 14.1.
d. Entry 8 in Table 14.1.
e. Entry 10 in Table 14.1.

In each case, reduce your results to the form presented in Table 14.1.

2. Derive the *DF* for the *NL* element defined by

$$y(x) = x \qquad |x| \leq 20$$
$$y(x) = 0 \qquad |x| > 20$$

where y and x denote the output and input, respectively, of N.

3. Consider the system shown.

a. What is the maximum value of K for stability?
b. If $K = 2K_{max}$, find the frequency and amplitude of the resulting limit cycle oscillation.
c. Check your answer to part *c* using either an analog or digital simulation of this system. Discuss the percent errors in your predictions for amplitude and frequency.

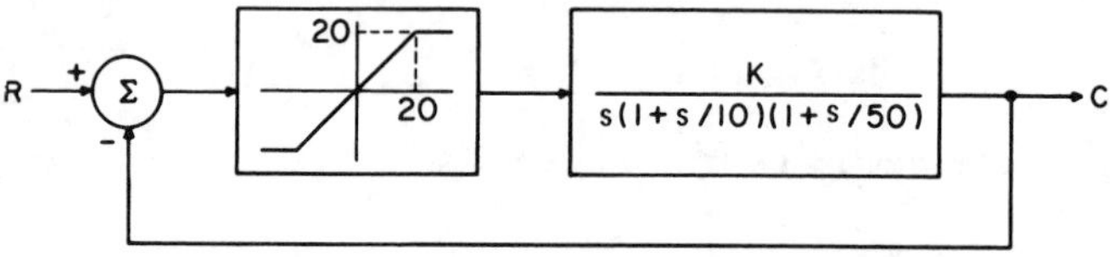

4. In the system shown, the *NL* element has the characteristic shown in entry 8 of Table 14.1, with $n_1 = 1, n_2 = 2, a = 20$. Also

$$G(s) = \frac{20}{s(1 + s/10)(1 + s/20)}$$

a. Is this system unstable? Discuss.
b. If the system is unstable, under what conditions will it oscillate, and at what frequency and amplitude?

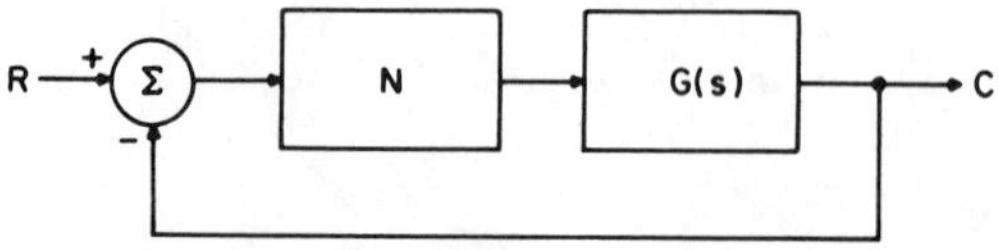

5. A particular unit feedback system has a *NL* element characterized by $N_e(A, \omega) = \epsilon^{-j\pi/4}/A$ in the forward path preceding a plant with transfer function $G_p(s) = 15/s(1 + s/2)$. Find the amplitude and frequency of the closed loop limit cycle. Is this limit cycle stable or unstable?

6. The polar plot for G in the system configuration of Prob. 4 can be constructed from

ω	1	2	3	5	6	10	20	40
$G(j\omega)$	$6.8\angle{-110°}$	$5.3\angle{237°}$	$4\angle{-142°}$	$3.1\angle{-165°}$	$2.5\angle{-186°}$	$1.5\angle{156°}$	$0.7\angle{135°}$	$0.35\angle{-235°}$

The data for N_e in the region of interest were obtained experimentally as (there is no dependence of N_e upon ω)

A	0	1	2	4	10	20	50	100
N_e	$0.1\angle{45°}$	$0.4\angle{30°}$	$0.8\angle{-340°}$	$1\angle{10°}$	$0.7\angle{0°}$	$0.4\angle{-10°}$	$0.2\angle{-20°}$	$0.1\angle{-40°}$

a. Locate all possible equilibria.
b. Classify each equilibrium point as stable or unstable.
c. Determine the frequency and amplitude of oscillation at each point.
d. Discuss the conditions, if at all, under which this system will oscillate.

7. *a.* Discuss the conditions under which limit cycle predictions based upon a *DF* analysis should be reasonable accurate.
b. Under what conditions can the *DF* method be used in the study of systems with two or more nonlinear elements?
c. Postulate and illustrate a situation in which the *DF* analysis predicts stability, but where you would question that answer sufficiently to pursue the matter further.

8. Develop a simple *NL* system example in which the system has two potential stable limit cycles, but is stable for all initial conditions near zero.

9. Consider the Van der Pol equation

$$\ddot{c} - (1 - c^2)\dot{c} + c = 0$$

a. Determine the isoclines in the region around the origin and use them to sketch a few representative trajectories approaching the limit cycle from both sides.
b. Develop an analog or digital simulation of this equation and use it to obtain several representative trajectories.

10. Consider

$$\frac{d^2y}{dt^2} + c_1\frac{dy}{dt} + \omega_0^2(1 - a^2y^2)y + \frac{ky}{c_2} = 0$$

which describes a particular hydraulic controllers response to initial conditions. Let $c_1 = 164$, $c_2 = 0.01$, $k = 1000$, $\omega_0 = 377$ and $a^2 = 0.858$.
a. Use the isocline method to construct several representative phase plane trajectories.
b. Study this equation using a computer.
Note: Some manipulation of this equation and a change in time scale are called for to put the results in a reasonable form.

11. The so-called phase-locked-loop communication system shown may be interpreted as a form of phase tracking controller.
a. Derive the differential equation, with ϕ as the dependent variable, describing this system.
b. Assuming $\tau = 1$ and $K_1K_2 = 2$, sketch several representative solution trajectories on the $\phi, \dot{\phi}$ (phase) plane. You may want to use a computer simulation here.
c. Where are the singularities for this system? What are they?

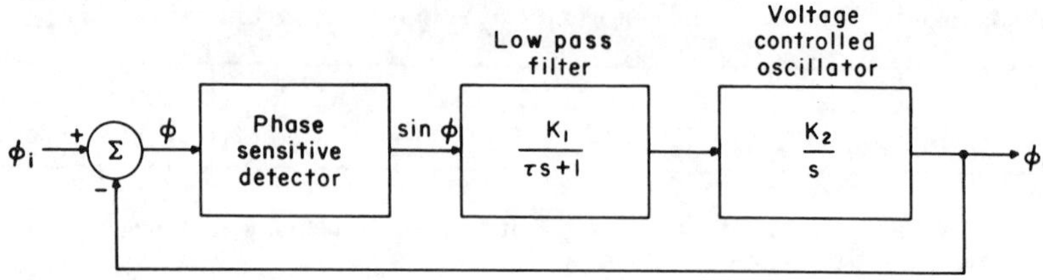

12. Show that the equation describing a pendulum on a weightless shaft with viscous friction is identical to that of the phase-locked-loop introduced in the previous problem.

13. The phase plane plot for a particular system's homogeneous response to the initial conditions $\dot{x}(t_0) = \dot{x}^0 = y^0 = 2$ and $x(t_0) = x^0 = 0$ is as shown. Derive one period of the response pattern for $x(t)$ from this plot. Reasonable care should be taken to provide some degree of accuracy. You may want to make an expanded plot of the trajectory shown to work from.

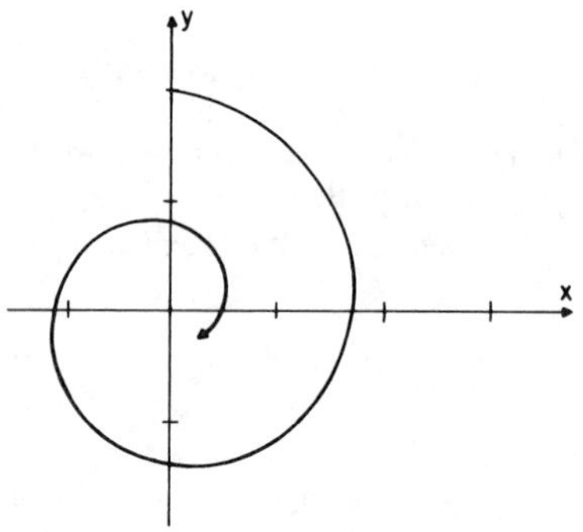

14. A particular system is described by

$$\dot{x}_1 = x_2 - x_1 \qquad \dot{x}_2 = -\frac{x_1^3}{|x_1|}$$

a. Sketch several isoclines for this system.
b. Sketch several solution trajectories.
c. Find and classify all singularities for this system. Discuss.

15. The homogeneous equations describing a particular *NL* system are

$$\dot{x}_1 = 3x_2 - 2x_1 + \frac{x_1^2}{2}$$

$$\dot{x}_2 = -4x_1 - x_2$$

Locate and classify the singularities of this system.

16. Consider the nonlinear equation described in Ex. 14.5. Sketch enough isoclines in the region around the phase plane origin to show, using Bendixson's theorem, that a limit cycle exists in the region around that unstable focus.

17. Consider the Van der Pol equation

$$\ddot{x} - (2 - 3\dot{x}^2)\dot{x} + x = 0$$

a. Find and characterize all singularities of this equation.
b. Apply Bendixson's first theorem to locate all regions within which a limit cycle may exist.
c. Sketch the isoclines. Use them and Bendixson's second theorem to establish the existence of a limit cycle in the region between two bounding curves, C_1 and C_2.

18. One set of Volterra's competition equations is

$$\frac{dx}{dt} = \alpha x - \beta xy$$

$$\frac{dy}{dt} = -\gamma y + \delta xy$$

where x denotes small fish which are preyed upon by large fish y. The small fish increase in number at a rate proportional to their own number, but decreased by the number of predators. The large fish increase in number at a rate proportional to their own number multiplied by the food fish x available, and diminished by their own number. Let $\alpha = 2$, $\beta = 1$, $\gamma = 8$, and $\delta = 2$.
a. Sketch typical solution curves.
b. Find and classify all singular points.

19. A second set of Volterra's competition equations is

$$\frac{dx}{dt} = \alpha x - \beta xy$$

$$\frac{dy}{dt} = \gamma y - \delta xy$$

These two equations describe two fish or animals which prey upon each other. Let $\alpha = 1$, $\beta = 1$, $\gamma = 8$, and $\delta = 2$.
a. Sketch typical solution curves.
b. Find and classify all singular points.

20. Determine the sign definiteness of each of the following quadratic forms:

a. $V(\mathbf{x}) = 3x_1^2 - 2x_1x_2 + 4x_2^2 + 2x_1x_3 - x_2x_3 + 5x_3^2$

b. $V(\mathbf{x}) = x_1^2 - 5x_1x_2 + 2x_2^2 + 4x_1x_3 - 6x_2x_3 + x_3^2$

c. $V(\mathbf{x}) = \mathbf{x}'\mathbf{A}\mathbf{x} \qquad \mathbf{A} = \begin{bmatrix} 2 & 3 & 0 \\ 1 & 2 & 1 \\ 4 & 3 & 5 \end{bmatrix}$

d. $V(\mathbf{x}) = \mathbf{x}'\mathbf{A}\mathbf{x} \qquad \mathbf{A} = \begin{bmatrix} -3 & 1 & 3 \\ -1 & -5 & 0 \\ -1 & 1 & -2 \end{bmatrix}$

21. Consider the system shown.
a. Write the describing equations.
b. Find a V function which can be used to show global asymptotic stability of this system, assuming all parameters are positive.

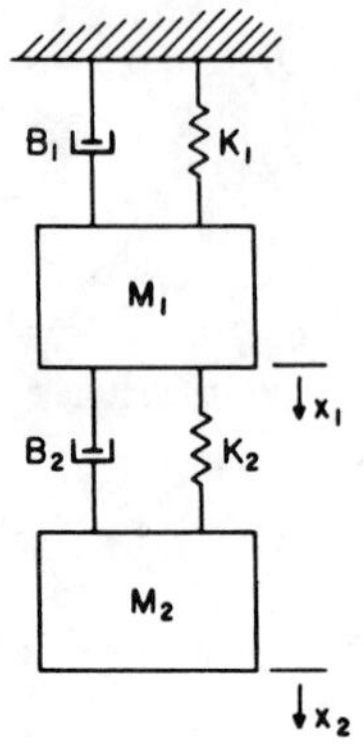

22. Consider the system shown, for which the homogeneous state equations are

$$\dot{x}_1 = x_2 \qquad \dot{x}_2 = -x_2 + x_3$$
$$\dot{x}_3 = -Kx_1 - 4x_3$$

Let

$$V(\mathbf{x}) = 5Kx_1^2 + 2Kx_1x_2 + 20x_2^2 + 8x_2x_3 + x_3^2$$

a. Show that $\dot{V}(\mathbf{x}) = -(40 - 2K)x_2^2$.
b. For what range of K is this system stable?
c. For what range of K is $\dot{V}(\mathbf{x})$ negative definite?
d. Show that $V(\mathbf{x})$ is positive definite for the same range of $K > 0$ for which $\dot{V}(\mathbf{x})$ is negative semi-definite.

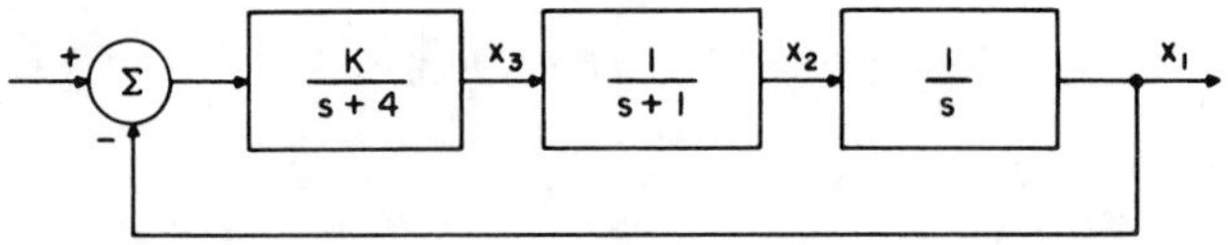

23. It can be shown that a quadratic form Liapunov function can be developed for any stable *linear* system. Suppose that

$$\dot{x}_1 = x_2 \qquad \dot{x}_2 = -Kx_1 - \omega_1 x_2$$

Describe our system. Find $\mathbf{A}$ such that $V(\mathbf{x}) = \mathbf{x}'\mathbf{A}\mathbf{x}$ is positive definite and $\dot{V}(\mathbf{x})$ is negative semidefinite. What are the resulting restrictions upon K and ω_1?

24. A given *NL* system is of the form introduced in Prob. 14.4.
a. Assuming N is of the form shown in Table 14.1, entry 8, with $n_1 = 1$, $n_2 = 1/3$, and

$$G(s) = \frac{50}{s(1 + s/5)(1 + s/20)}$$

use Theorem 14.6 as a basis for the development of a stable design with, in your engineering judgment, a reasonable stability margin. Set $K_v = 50$. Specify the transfer function of your proposed equalizer.

b. Assuming N is of the form shown in Table 14.1, entry 9, with $n = 1, a = 20$ and $M = 20$ and

$$G(s) = \frac{20}{s(1 + s/10)(1 + s/50)}$$

use Theorem 14.6 as a basis for the development of a stable design with, in your engineering judgment, a reasonable stability margin. Set $K_v = 20$. Specify the transfer function of your proposed equalizer. Compare your results to a design based upon a *DF* analysis. What do you conclude?

CHAPTER 15

1. Find closed form solutions for $X^*(s)$ in each of the following cases using the definition

$$X^*(s) = \sum_{n=1}^{\infty} f(nT)\epsilon^{-nTs}$$

a. $X(s) = \dfrac{1}{s + a}$

b. $\dfrac{1}{(s + a)(s + b)}$

c. $X(s) = \dfrac{1}{s^2 + \omega_0^2}$

d. $x(t) = t\epsilon^{-at}$

2. Consider the system shown. Assume impulse sampling at $t = 0$ and at 1 second intervals thereafter. Let

$$f(t) = 2t \qquad 0 \le t \le 10$$
$$f(t) = 0 \qquad \text{elsewhere}$$

a. Write a general expression for $f^*(t)$.
b. Write a general expression for $f_1(t)$.
c. Find f_1 (3.5).

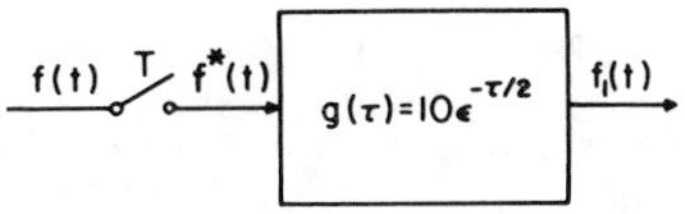

3. Consider the sampling function shown.
a. Find a general expression for the Fourier coefficients C_k in the exponential form as a general function of A, γ, and T.
b. Assume a signal $v(t) = E \cos 10t$ is sampled using this function. Assume that T is such that 10 samples are taken per cycle, and let $\gamma = 0.1T$. Sketch the line spectrum of $f^*(t) = p(t)v(t)$. Assume $\omega_0 \gg \omega_1$.

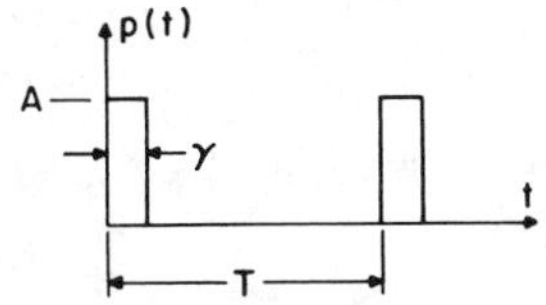

4. Consider the system shown. The output of the sampler is an impulse sequence area modulated with gain of unity. $f(t)$ is as shown.
a. Sketch $|F(j\omega)|$. Estimate a suitable sampling period. Give reasons for your choice.
b. Let $T = 2$ seconds. Determine A and α such that the filter output is "smoothed". Assume various values for α and sketch the actual time response which results. Work in the time domain, using superposition, etc.
c. Sketch $|F^*(j\omega)|$ for $T = 2$. Does this help in choosing α? Discuss.
d. Sketch the output time response if there were no sampling for the filter chosen in part *b*. Are the results similar in each case?
e. If the sampler is a pulse, not an impulse, modulator, how narrow should the pulse be if the impulse approximation is to be valid within 1 percent?

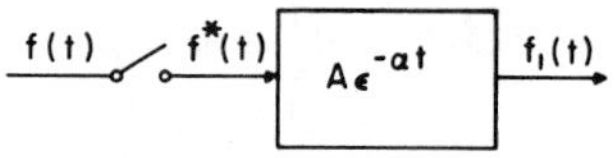

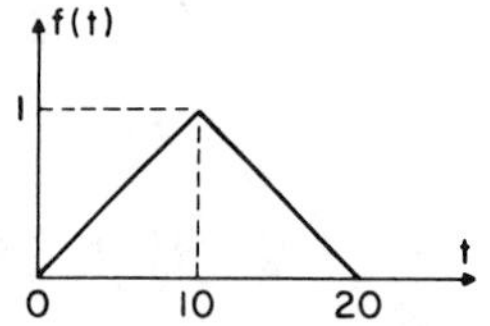

5. Sketch the output obtained when each of the following signals is passed through an impulse sampler followed by a zero-order hold circuit. Let $T = 1$ second.

a. $f(t) = t$ *b.* $f(t) = \sin\left(\frac{t}{10}\right)$

6. Assume a sampling and hold process as follows. The input is sampled instantaneously at $t = 0$ and $f(0) = 3$ is observed. The sampler is followed by a device which puts out a pulse of width $\gamma = 0.1$ starting at $t = 0$, and the pulse has area equal to $f(0) = 3$. Assume that this pulse is applied to a system with $g(\tau) = \epsilon^{-\tau}$. Let $T = 1/4$ second.
a. Find the system output on the interval $0 \leq t \leq 1/4$ for the pulse signal described.
b. Find the system output on the same interval for an input impulse of area 3.
c. Sketch these two responses on the interval $0 \leq t \leq 1/4$. Discuss the sources of inaccuracy introduced by assuming impulse sampling when the actual sampling is of finite duration.

7. The signal $f(t) = A \sin \omega t$ is impulse sampled 4 times per period with the first sample occurring at $t = 0$. A zero-order hold is used. Find the mean square error between the hold output and $g(t) = A \sin(\omega t - \theta)$, where $0 \leq \theta \leq 2\pi$, as a general function of θ. What value of θ results in minimum mean square error? How does this θ relate to T? Discuss this result and its meaning.

8. The frequency response characteristic of an ideal low pass filter, the so-called cardinal data hold, is as shown. Assume the frequency response is

$$G(j\omega) = 1 \qquad -W \leq \omega \leq W$$
$$G(j\omega) = 0 \qquad \text{otherwise}$$

a. Find the impulse response of this network. Put your answer in the standard $(\sin x)/x$ form.

b. Sketch this response.
c. Is the cardinal data hold realizable? Discuss.

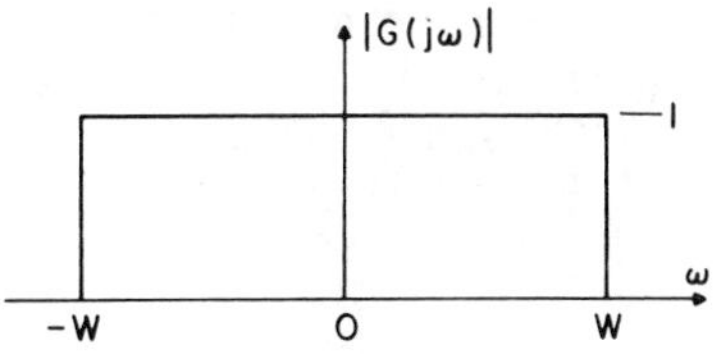

9. The output of the second order hold circuit is given by

$$f(nT + \tau) = r(nT) + \frac{1}{T}\{r(nT) - r[(n-1)T]\}\tau$$
$$+ \frac{1}{T^2}\{r(nT) - 2r[(n-1)T] + r[(n-2)T]\}\frac{(T+\tau)\tau}{2}$$
$$0 \leq \tau \leq T$$

Sketch the impulse response characteristic of this network.

10. You know that

$$F^*(s) = \frac{1}{2\pi j}\int_{c-j\infty}^{c+j\infty} F(p)\frac{1}{1-\epsilon^{-T(s-p)}}\,dp$$

Prove that

$$F^*(s) = \frac{1}{T}\sum_{n=-\infty}^{\infty} F\left(s - j\,\frac{2\pi n}{T}\right)$$

by closing Γ to the right and summing the infinite residues of the function

$$F(p)\,[1 - \epsilon^{-T(s-p)}]^{-1}$$

We thus have 2 alternate forms for $F^*(s)$.

11. *Derive* the z-transforms of the impulse sampled signal sequence obtained from each of the following:

a. $f(t) = t$

b. $f(t) = \cos\omega_1 t$

c. $f(t) = t\epsilon^{-at}$

d. $F(s) = \dfrac{a^2}{s(s+a)^2}$

12. Use the method of residues to find the inverse z-transform for each of the following:

a. $G(z) = \dfrac{z^2(z^2 + z + 1)}{(z - 0.8)(z - 1)(z^2 - z + 0.8)}$

b. $G(z) = \dfrac{z^2}{(z - 0.8)(z - 1)}$

13. The impulse response of a given system is as shown. (Let's not worry about the fact that this is obviously impossible to realize exactly.) Let $r(t) = \epsilon^{-t}$ and denote the sampling period by T seconds. Assume $10/T$ is an integer.

a. Find $G(z)$ in infinite series form and also as a ratio of polynomials in z.

b. Find $R(z)$.

c. Assume $T = 2$. Find the first few terms of $c^*(t)$ by: (i) Multiplying out the first few terms in the series representations of $R(z)$ and $G(z)$. (ii) Expanding $G(z)R(z)$ by long division.

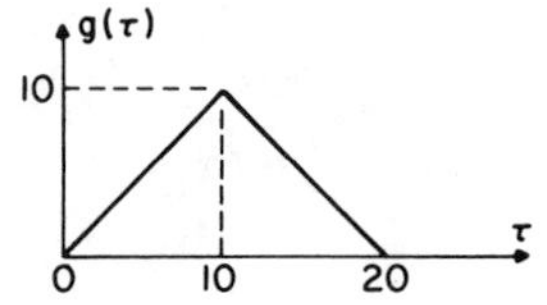

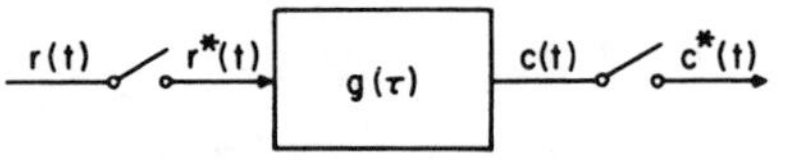

14. Consider the signal and plant

$$R(s) = \frac{1}{s(s+1)} \qquad G(s) = \frac{1}{s+2}$$

a. Assume the input signal is impulse sampled, applied to the plant, and the plant output impulse sampled. Find the output signal $C(z)$.

b. Find $C(z)$ if the input is applied directly to the plant.

c. Compare the two results and discuss. Under what conditions are the two results nearly equivalent?

15. Consider the system shown. Use sampling theory to sketch this systems unit step response function as follows:

a. Choose an output sampling frequency equal to approximately 5 times the system's closed loop crossover frequency.

b. Assume the systems *error* signal is sampled at this rate and observed. Note that the forward signal path is *not* interrupted. We use the error signal, rather than the output, to prevent accumulation of numerical errors caused by the step response output having a pole at $s = 0$, or $z = 1$.

c. Find $E(z)$ for a unit step input.

d. Divide out $E(z)$ by long division to obtain the error response until it approaches steady state.

e. Sketch $c(t)$ for a unit step input.

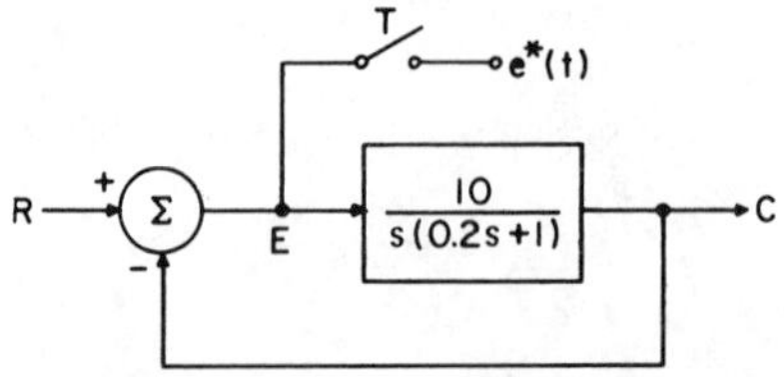

16. Determine the initial and final values of the functions

$$a.\quad G(z) = \frac{z^2(z^2 + z + 1)}{(z - 0.8z + 1)(z^2 + z + 0.8)}$$

$$b.\quad G(z) = \frac{z^2}{(z - 0.8)(z - 1)}$$

$$c.\quad G(z) = \frac{1 + 0.3z^{-1} + 0.1z^{-2}}{1 - 4.2z^{-1} + 5.6z^{-2} - 2.4z^{-3}}$$

17. Given

$$F(z) = \frac{1}{k - \epsilon^{-aT} z^{-1}}$$

Show that $f(t) = \epsilon^{-at} k^{-(1 + t/T)}$.

18. Consider the signal described by

$$F(s) = \frac{1}{s(s + 1)(s + 2)}$$

This signal is passed through a perfect time delay of 0.3 seconds and then sampled at 5 Hz. Find $F(z, \Delta)$ and $F(z, \lambda)$ by using the definition and evaluating residues.

19. Consider the two networks shown.

a. Find $C(z)/R(z)$ for the two cases.

b. Sketch the unit step responses of these two networks for $T = 0.5$, $T = 0.1$, $T = 0.05$ and $T = 0.01$. Use a computer to carry out the numerical details.

c. Compare these two configurations. Generalize relative to sampling rates, time constants, etc.

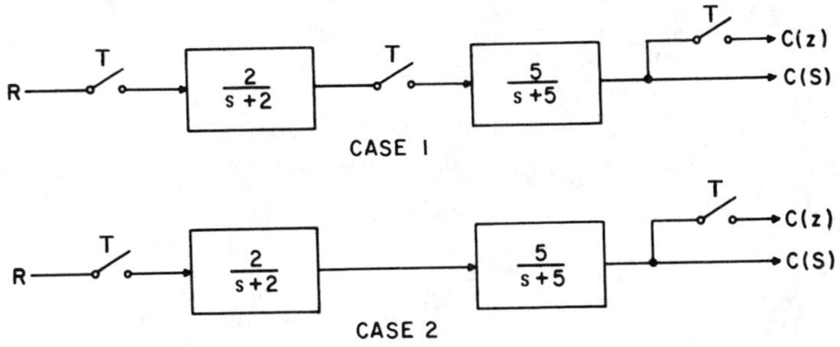

20. Consider the system shown.

a. Find $C(s)$ as a general function of the input and the controller characteristics.

b. Find $C(z)$.

c. Find $C(z)/R(z)$. Does this transfer function always exist? Discuss.

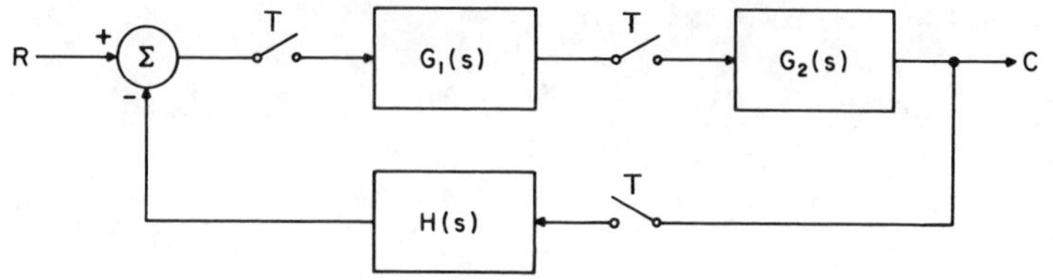

21. For the system shown:

a. Find $C(z)$.

b. Can you obtain a general expression for $C(z)/R(z)$ in this case? Discuss.

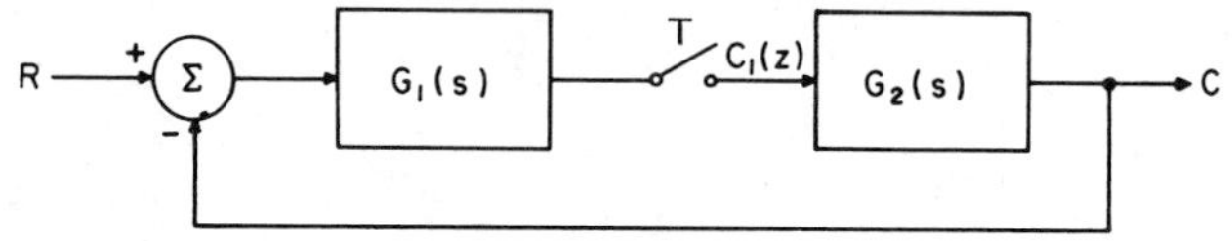

22. Given a closed loop system with open loop transfer function

$$A(z) = \frac{(1 - \epsilon^{-1})z^{-1}}{(1 - z^{-1})(1 - \epsilon^{-1}z^{-1})}$$

a. Find the closed loop characteristic equation.

b. Apply the bilinear transformation to this characteristic equation.

c. Where are the closed loop poles located in the z-plane? In the s-plane? Assume $T = 1$.

23. Consider a unity feedback error sampled system of the type introduced in the previous problem with $T = 0.2$, a zero order hold, and plant transfer function

$$G(s) = \frac{2}{s(1 + 0.1s)(1 + 0.05s)}$$

a. Find the closed loop characteristic equation in the z-domain.

b. Use the bilinear transformation and the Routh criterion to see if the system is stable.

24. Find the phase margin in the s-domain for the system shown. Assume $T = 0.1$, and $K = 2$.

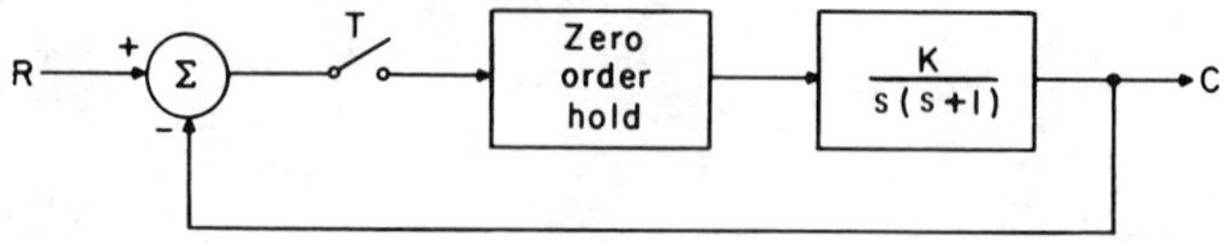

25. Consider a system of the type introduced in Prob. 24. Let $T = 0.693$ and

$$G(s) = \frac{K}{s(s + 1)}$$

a. Sketch the Nyquist diagram for the open loop sampled transfer function with $K = 1$.

b. What is the maximum K for stability?

c. Plot the closed loop frequency response assuming $K = 1$.

d. Assuming $K = 2$, find the closed loop pole locations.
e. Assuming $K = 2$, evaluate $|C(z)/R(z)|$ at $\omega = 1$.
f. Assuming $K = 2$, find the unit step response on the interval $0 \leq t \leq 5$.

26. Consider a system of the type introduced in Prob. 24. Assume $T = 0.51$.
a. Determine the maximum K for stability.
b. Let $K = 2$. Find the general expression for $C(nT)$ assuming that $R(s) = 1/s$.
c. What do you estimate as the lowest positive frequency at which $|C(j\omega)/R(j\omega)|$ possesses a maximum? Justify your answer. Specify the anticipated family of frequencies at which images of this maximum exist. Find the maximum magnitude.
d. Does the presence of the sampler and zero-order hold in this system contribute appreciably to its relative stability and response characteristics? Do you conclude from this that the sampling rate is too high, too low, or about right?

27. Consider the system shown.
a. Disregarding the sampler and zero-order hold, find K for 45° phase margin.
b. Choose the sampling frequency such that the sampler and zero-order hold, treated as a pure time delay, contribute 10° of phase lag at the crossover frequency provided by K as obtained in part *a.*
c. Sketch the Nyquist plot of $G(s)$ and the zero-order hold circuit.
d. Correct this plot for the sampler using Linvill's method. Is the required adjustment significant near the −1 point?

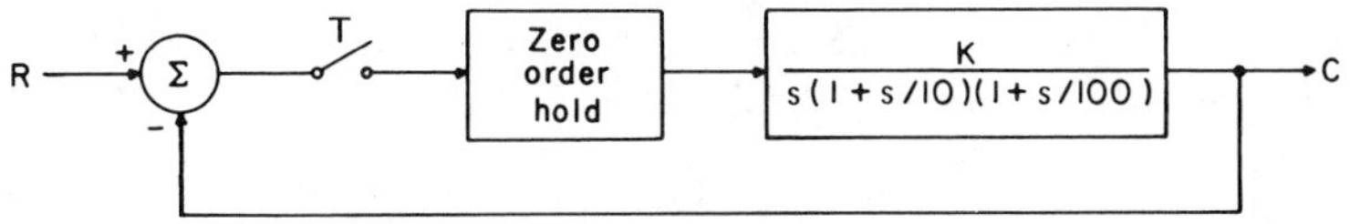

28. In the system shown,

$$G(s) = G_e(s)G_p(s)$$

where

$$G_p(s) = \frac{100}{s(1+s)(1+s/20)}$$

it is desired to choose $G_e(s)$ subject to the following constraints:
a. $G_e(s)$ is to be a lag-lead equalizer with no *net* lead. (i.e., $|G_e| \leq 1$ for all ω.)
b. $G_e(s)$ is to have unit dc gain.
c. Phase margin, $\theta_m \geq 40°$.
d. Low frequency steady state errors are to be minimum.
Assuming $T = 0.01$, treat the sampler and hold as an appropriate time delay, and find $G_e(s)$ to satisfy these specs.

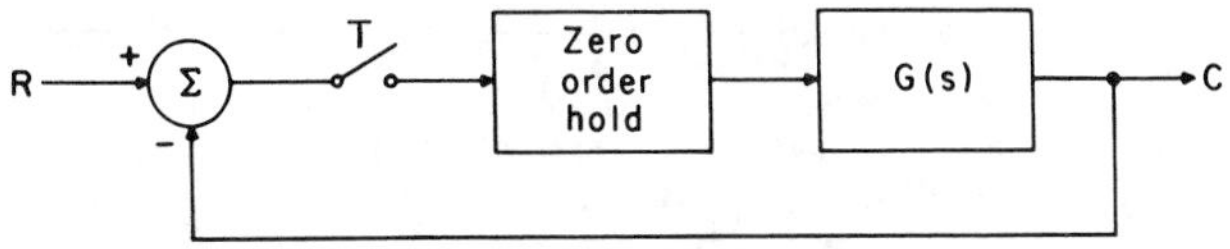

29. Sketch the root locus diagram for a unit feedback sampled system with forward transfer function

$$G(z) = \frac{K(z + 0.12)(z + 0.193)}{(z - 1)(z - 0.368)(z - 0.135)}$$

What is the range of K for stable operation?

30. Consider a standard unity feedback digital system with plant transfer function

$$G_p(s) = \frac{K}{s(1 + s)}$$

The plant is preceded by a zero-order hold circuit. A cascade digital equalizer $D(z)$ is desired to satisfy the following specs:

a. $K_v = 5$ in the *w*-plane.
b. Phase margin $\geqslant 40°$ in the *w*-plane.
c. Gain margin $\geqslant 5$ in the *w*-plane.

Develop an appropriate equalizer $D(z)$.

31. It is desired to obtain

$$D(z) = \frac{z^{-1}(1 + 2z^{-1})(1 + z^{-1})}{(1 - 0.5z^{-1})(1 - 0.25z^{-1})}$$

Indicate how $D(z)$ may be realized using Sklansky's procedure. Determine the $P(s)$ and $Q(s)$ required.

32. Consider the system shown with $T = 0.5$. Find $D(z)$ to provide zero steady state unit step response error and minimum settling time for that input.

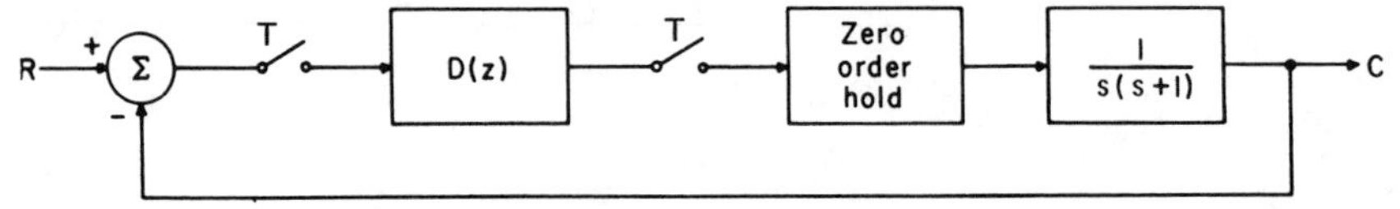

33. Consider the system shown with plant transfer function

$$G(z) = \frac{z^{-2}(1 + 3z^{-1})}{(1 - 2z^{-1})(1 - 0.2z^{-1})(1 - 0.5z^{-1})}$$

a. Determine the transfer function of a digital equalizer $D(z)$ which will provide zero sample point error for a unit step input after a minimum time interval.
b. What is the resulting $K(z) = C(z)/R(z)$?

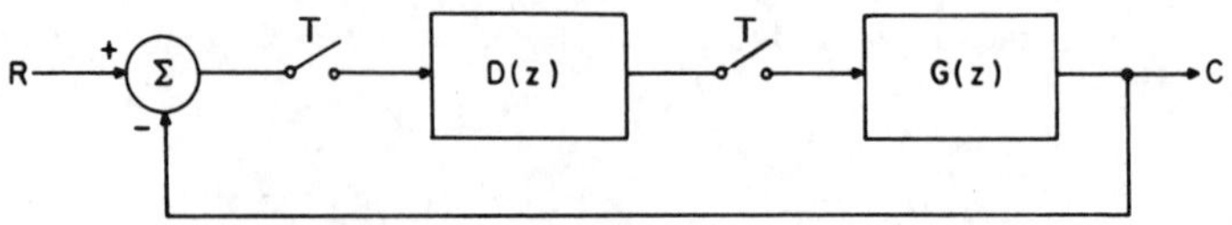

34. Consider a standard unity feedback sampled system with plant transfer function

$$G_p(z) = \frac{z^{-2}(1 - 0.5z^{-1})}{(1 - z^{-1})(1 + 0.2z^{-1})(1 - 0.2z^{-1})^2}$$

a. Find $D(z)$, a cascade digital equalizer, to provide a closed loop unity feedback digital

controller with zero steady state ramp error and minimum settling time for a ramp input.
b. What is the corresponding closed loop transfer function?
c. How does this system relate to the minimal prototype discussed in the text?

35. Consider a standard unity feedback digital controller with plant transfer

$$G_p(z) = \frac{(1 + z^{-1})(1 + 2z^{-1})z^{-1}}{(1 - z^{-1})(1 - 0.5z^{-1})(1 - 0.3z^{-1})}$$

a; Find $D(z)$ such that the closed loop system has zero steady state ramp error and minimum finite settling time for that input.
b. Determine the resulting system's unit step response peak overshoot. Comment.

36. Consider the acceleration (Parabolic signal input) minimal prototype

$$K_p(z) = 3z^{-1} - 3z^{-2} + z^{-3}$$

Suppose we introduce a staleness factor such that

$$K(z) = \frac{AK_p(z)}{1 - az^{-1}}$$

a. Find A such that $K(z)$ yields zero steady state error for a unit step input.
b. Sketch the unit step response patterns for several values of a.

37. Consider a standard unity feedback system with plant transfer function

$$G_p(z) = \frac{(1 + 0.5z^{-1})(z^{-1})}{(1 - z^{-1})(1 - 0.5z^{-1})}$$

Find $D(z)$, a cascade digital equalizer, to provide ripple free steady state response for step and ramp inputs with the shortest possible transient response duration.

38. A standard unity feedback digital controller has plant transfer function

$$G_p(z) = \frac{z^{-1}(1 + 2z^{-1})(1 + 0.5z^{-1})}{(1 - z^{-1})(1 - 2z^{-1})(1 - 0.5z^{-1})}$$

Find $D(z)$ such that this system will be capable of following a unit step input with ripple free response after a minimum time transient period.

39. Consider the system shown. Let (see App. E and Chap. 13)

$$G(s) = \frac{10(1 + s/2)}{s^2(1 + s/20)}$$

Obtain the unit step response of this system as follows:
a. Develop a state equation description for the block labeled $G(s)$ with output $C(s)$ and input $M(s)$.
b. Find the transition matrix for this part of the system.
c. Noting that $m(t)$ = constant from sample point to sample point, develop an iterative relationship for system state of the form

$$\mathbf{x}_{n+1} = \mathbf{\Phi}(T)\mathbf{x}_n + \mathbf{\Phi}_m(T)m_n$$

d. · Use this recursive formula to develop the significant part of $c(t)$ in response to a unit step input. You may want to use a computer here to carry out the numerical details.
e. Discuss the error sources in your solution.

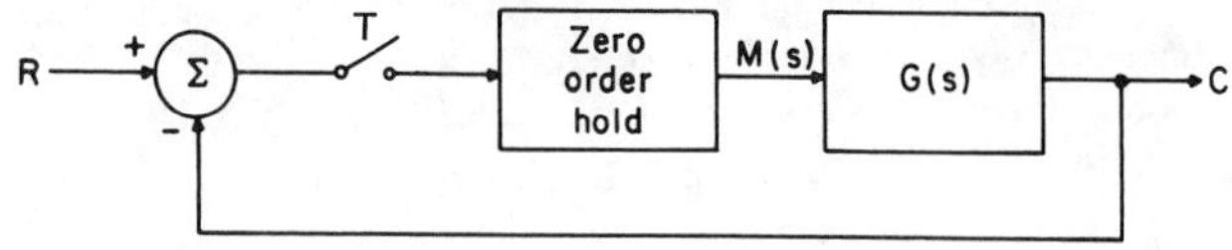

40. Consider the system shown with $T = 1$.

a. Use the modified *z*-transform to determine the unit step response of this system at the sample points and at points half way between the sample points over the interval $0 \leq t \leq 10$.

b. Add a zero order hold prior to the plant. Develop the state equations. Find a general expression for $\mathbf{x}(t)$ over the interval $nT \leq t \leq (n+1)T$ in terms of $e(nT)$ and $\mathbf{x}(nT)$ assuming a unit step input at $t = 0$. Determine the unit step response over the interval where significant changes are taking place. Pay particular attention to the vicinity of the peak overshoot.

c. Discuss the relative merits of the systems with and without a zero-order hold.

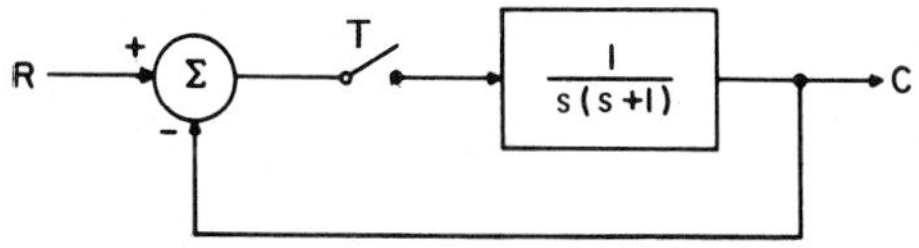

41. Consider the system shown, with $T = 1$ and

$$D(z) = 1 + 0.1z^{-1}$$

$$G(s) = \frac{1 - \epsilon^{-Ts}}{s} \cdot \frac{2}{s(s+2)}$$

$$R(z) = \frac{2}{1 - z^{-1}}$$

a. Using state equations, find the response to the given input assuming all initial conditions are zero. Evaluate far enough in t to indicate the general response pattern.

b. Find $c_{max}(t)$. At what time does it occur?

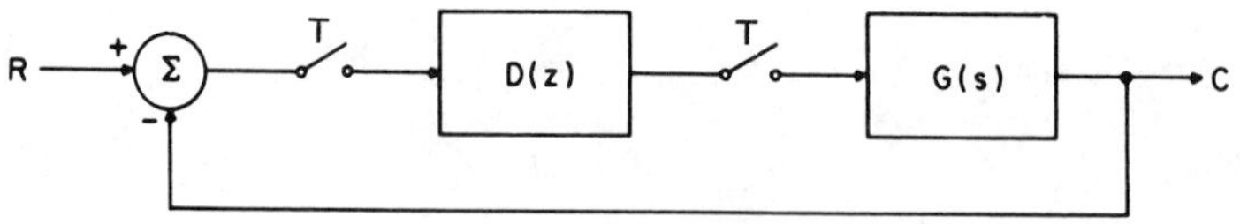

INDEX